# 5th Geoinformation Science Symposium 2017 (GSS 2017)

## Bridging Geospatial Industries and Scientific Research

IOP Conference Series: Earth and Environmental Science
Volume 98



Yogyakarta, Indonesia
27 – 28 September 2017

ISBN: 978-1-5108-5490-1
ISSN: 1755-1307

**Printed from e-media with permission by:**

Curran Associates, Inc.
57 Morehouse Lane
Red Hook, NY 12571

**Some format issues inherent in the e-media version may also appear in this print version.**



Printed by Curran Associates, Inc. (2018)



Institute of Physics
Dirac House, Temple Back
Bristol BS1 6BE UK

Phone: 44 1 17 929 7481
Fax: 44 1 17 920 0979

techtracking@iop.org

**Additional copies of this publication are available from:**

Curran Associates, Inc.
57 Morehouse Lane
Red Hook, NY 12571 USA
Phone: 845-758-0400
Fax: 845-758-2633
Email: curran@proceedings.com
Web: www.proceedings.com

# Table of contents

## Volume 98

**The 5th Geoinformation Science Symposium 2017 (GSS 2017)**

**27–28 September 2017, Yogyakarta, Indonesia**

**Accepted papers received: 5 December 2017**
**Published online: 20 December 2017**

### Preface



### Papers

**Remote Sensing and GIS for Climatology, Hydrology and Coastal Environment**

**Remote Sensing and GIS for Land Use Mapping**



**Remote Sensing and GIS for Disaster Risk Reduction and Management**

## GIS for Precision Agriculture and Regional Development



## Remote Sensing and GIS for Geomorphology, Geology, Soil, and Land Resources

**Image Pre-Processing and Data Acquisition**



**POSTER**

IOP Conf. Series: Earth and Environmental Science **98** (2017) 011001 doi:10.1088/1755-1315/98/1/011001

## Preface

Welcome to The 5th Geoinformation Science Symposium 2017, which takes place from 27 to 28 September 2017 in University Centre, Universitas Gadjah Mada. The Organising Committee has selected "Bridging Geospatial Industries and Scientific Research" as the main theme of this symposium. It is motivated by the fact that, as we all know, Indonesian geospatial industries have been growing fast during the past 10 years. A lot of governmental institutions and private sectors are making use of geospatial data in their business. On the other hand, research in the universities and other scientific institutions is also growing significantly. New study programs focusing on the use and development of geoinformation science and technology have been established, while research funding on the development of this field is also increasing. The problem is that they are not coupling in a synergy, so that the expected outcome of the advances could not be reached optimally.

This meeting aims to embrace researchers, practitioners, lecturers as well and students from various governmental, education and private sectors to share their experiences and also expectations in the engagement in geospatial world. We also invite experts from Australia and Japan university to sit together with the Indonesian Geospatial Agency as well as the Directorate General of Research Reinforcement and Development officials. We hope that this occasion may shape a new perspective on bridging the gap between industries and the scientific research.

During this two-day symposium, more than 70 papers were presented, and about 20 posters were displayed. In addition to these scientific papers, two special sessions focusing on the establishment of Indonesian Association of Remote Sensing Laboratories (InARSeL) and One Geospatial Information Certificate Policy were conducted. I would like to thank Prof. Stuart Phinn from the University of Queensland Australia and to Assoc Prof. Wataru Takeuchi from Tokyo University Japan, who have made time for delivering keynote speeches. I am also grateful to the representative of Geospatial Information Agency (BIG) and the representative of Director General of Reseach Reinforcement and Development, for joining us to represent the Indonesian geospatial industry and scientific research institution respectively. I would also like to mention the Rector of UGM, Dean of the Faculty of Geography, all sponsors, all staff of the Department of Geographical Information Science, organising team, and all participants who have fully supported this event.

Yogyakarta, 27 September 2017

Drs. Projo Danoedoro, M.Sc., Ph.D.

Symposium General Co-Chair
The 5th Geoinformation Science Symposium 2017

IOP Conf. Series: Earth and Environmental Science **98** (2017) 011001 doi:10.1088/1755-1315/98/1/011001

**Conference Photographs and Video**

Photo (see attached files)

Video https://youtu.be/4LR7SkkVZas

**Organized by**

**PUSPICS**
**Department of Geographic Information Science**
**Faculty of Geography**
**Universitas Gadjah Mada**

MAPIN/ISRS

IOP Conf. Series: Earth and Environmental Science **98** (2017) 011001 doi:10.1088/1755-1315/98/1/011001

**Sponsor**

**PT. Bhumi Prasaja**

BNI

Waindo SpecTerra

**PT. Citra Gama Sakti**

IOP Conf. Series: Earth and Environmental Science **98** (2017) 011001 doi:10.1088/1755-1315/98/1/011001

**List of committees**

Steering committee:

- Prof. Dr. Muh Aris Marfai (Dean, Faculty of Geography UGM)
- Prof. Dr. Totok Gunawan (Faculty of Geography UGM)
- Prof. Dr. Hartono (President of Indonesian Geographer Association (IGI) – Faculty of Geography
- UGM)
- Prof. Dewayany (President of Indonesian Society for Remote Sensing (ISRS) – Geospatial Information Agency (BIG))
- Prof. Stuart Phinn (The University of Queensland)
- Dr. R. Suharyadi (Faculty of Geography UGM)
- Dr. Sigit Heru Murti (Faculty of Geography UGM)

General Co-chairs:

- Dr. Projo Danoedoro

Organizing committee:

- Dr. Sudaryatno
- Dr. Retnadi Heru Jatmiko
- Dr. Bowo Susilo
- Dr. Muhammad Kamal
- Dr. Prima Widayani
- Dr. Nur Mohammad Farda
- Dr. Pramaditya Wicaksono
- Ibnu Rosyadi, S.Si.,M.Cs.
- Iswari Nur Hidayati, S.Si., M.Sc.
- Totok Wahyu Wibowo, S.Si., M.Sc.
- Ari Cahyono, S.Si. M.Sc.
- Candra Sari Djati, S.Si., M.Sc.
- Diah Fitria Dewi, SE
- Citra Amalia, S.Pd.
- Novita Padyawati, S.Pt.
- Mini Darini

IOP Conf. Series: Earth and Environmental Science **98** (2017) 011001 doi:10.1088/1755-1315/98/1/011001

## Contact

The 5th Geoinformation Science Symposium Secretariat

PUSPICS, Department of Geographic Information Science

Faculty of Geography

Universitas Gadjah Mada (UGM)

Yogyakarta 55281, Indonesia

Website: http://puspics.ugm.ac.id/snsg2017/

Email: 5thsnsg2017@gmail.com

Person in charge:

Dr. Projo Danoedoro, M.Sc. (projo.danoedoro@geo.ugm.ac.id)

Dr. Pramaditya Wicaksono, M.Sc. (prama.wicaksono@geo.ugm.ac.id)

THE 5TH GEOINFORMATION
SCIENCE SYMPOSIUM 2017

MATION
UM 2017
Scientific Research"
GEOINFORMATION
SCIENCE SYMPOSIUM 2017

ATION
M 2017
ntific Research
Example: Australian Grazing Industry, Landsat/Sentinel for Pasture
AGENDA

THE 5TH GEOINFORMATION
SCIENCE SYMPOSIUM 2017
"Bridging Geospatial Industries and Scientific Research
27 - 28 September 2017
University Club Hotel UGM, Yogyakarta, Indonesia
BNI

THE 5TH GEOINFORMATION
SCIENCE SYMPOSIUM 2017

THE 5TH GEOINFORMATION
SCIENCE SYMP
UM 2017
"Bridging Geospatial Industri
cientific Research"
27 - 28 September 2017
University Club Hotel UGM, Yog
2017

THE 5TH GEOINFORMATION SCIENCE SYMPOSIUM 2017
"Bridging Geospatial Industries and Scientific Research"
27 - 28 September 2017
University Club Hotel UGM, Yogyakarta, Indonesia
2017
CONTACT

THE 5TH GEOINFORMATION
27 - 28 September 2017

THE 5TH GEOINFORMATION
SCIENCE SYMPOSIUM 2017
THE 5TH GEOINF
SCIENCE SYMP
Bridging Geospatial Industri
28 September 2017
sity Club Hotel UGM, Yogyakart

THE 5TH GEOINFORMATION SCIENCE SYMPOSIUM 2017

5TH GEOINFORMATION
NCE SYMPOSIUM 2017
Geospatial Industries and Scientific Research"
tember 2017
lub Hotel UGM, Yogyakarta, Indonesia
BNI
THE
SCIE

THE 5TH GEOINFORMATION SCIENCE SYMPOSIUM
2017
THE 5TH GEOINFORMATION
SCIENCE SYMPOSIUM 2017
"Bridging Geospatial Industries and Scientific Research"
27 - 28 September 2017
University Club Hotel UGM, Yogyakarta, Indonesia
BNI

ATION
M 2017
ientific Research"
Industry a
The 5th Indonesian Geoi
AE⊘CCG

THE 5TH GEOINFORMATION
SCIENCE SYMPOSIUM 2017

Alumni network all over the world
6

THE 5TH GEOINFORMATION SCIENCE SYMPOSIUM
2017
SCIENCE SYMPOSIUM 20
"Bridging Geospatial Industries and Scientific R
28 September 2017
akarta, Indonesia
BNI

SYMPOSIUM 2017
al Industries and Scientific Research"
M, Yogyakarta

THE 5TH GEOINFORMATION SCIENCE SYMPOSIUM
THE 5[TH] GEOINFOR
SCIENCE SYMPOS
"Bridging Geospatial Industries and
University Club Hotel UGM, Yogyakarta, Indon
BNI

THE 5TH GEOINFORMATION SCIENCE SYMPOSIUM
2017
THE 5TH GEOINFORMATION SCIENCE SYMPOSIUM 2017
"Bridging Geospatial Industries and Scientific Research"
27 - 28 September 2017
University Club Hotel UGM, Yogyakarta, Indonesia
BNI

FIRE EXIT

PT. CITRA GAMA SAKTI

The DigitalGlobe
Constellation

IOP Conf. Series: Earth and Environmental Science **98** (2017) 011002
doi:10.1088/1755-1315/98/1/011002

# Peer review statement

All papers published in this volume of *IOP Conference Series: Earth and Environmental Science* have been peer reviewed through processes administered by the proceedings Editors. Reviews were conducted by expert referees to the professional and scientific standards expected of a proceedings journal published by IOP Publishing.

# Automatic Detection of Decadal Shoreline Change on Northern Coastal of Gresik, East Java –Indonesia

**M A Z Fuad[1,2*] and M Fais D A[1]**

[1]Faculty of Fisheries and Marine Science, Brawijaya University, Jl. Veteran- Malang 65145, Indonesia
[2]Marine Resources Exploration and Management Research Group (MEXMA)-Brawijaya University

fuad_maz@ub.ac.id

**Abstract**. The Coastal zone is a dynamic region that has high environmental and economic values. This present research focuses on the analyzing the rate of shoreline change using multi-temporal Landsat Imagery and Digital Shoreline Analysis Systems (DSAS) along the northern part of Gresik coastal area, East Java Indonesia. Five village were selected for analysis; Campurejo, Dalegan, Prupuh, Ngemboh, and Banyuurip. Erosion and Accretion were observed and detected on Multi-temporal satellite Images along the area of interest from 1972 – 2016. Landsat Images were radiometrically and geometrically corrected before using for analysis. Coastline delineation for each Landsat image was performed by MNDWI method before digitized for quantitative shoreline change analysis. DSAS was performed for quantitative analysis of Net Shoreline Movement (NSM) and End Point Rate (EPR). The results indicate that in the study area accretion and abrasion was occurred, but overall abrasion was dominated than accretion. The remarkable shoreline changes were observed in the entire region. The highest abrasion area was occurred in Ngemboh village. From 1972 to 2016, coastline was retreat 242.56 meter to the land and the rate of movement was -5.54m/yr. In contrast, Campurejo area was relatively stable due to the introduction of manmade structure, i.e. Jetty and Groin. The Shoreline movement and the rate of movement in this area were -6.11m and -0.12 m/yr respectively. The research represents an important step in understanding the dynamics of coastal area in this area. By identification and analysis of coastline evolution, the stake holder could perform a scenario for reducing the risk of coastal erosion and minimize the social and economic lost.

**Keywords:** Shoreline Change, DSAS, Landsat Imagery, Gresik Coast

## 1. Introduction

Coastal areas are in one of the most area with highest economical values. In developing countries, 60% of population lives in coastal areas [1]. Coastal erosion in a severe problem, especially for a country facing highly population growth in coastal area [2]. A shoreline is defined as the line of contact between land and a water body [3] and subjected to continually change over time [4].Its considered as one of the most dynamic process in coastal areas [4][5]. Shoreline generally subjected to change due to coastal process, which are controlled by hydrodynamic characteristics, sediment characteristics, beach form, etc. [6]. In addition, the economic and social development also induce the change of coastal area.

Northern coastal area of Gresik, East Java is one of the area that severe from coastal erosion in the past years. Coastal abrasion is a major crisis and it potentially impacts the coastal population and natural environment. The science of shoreline mapping a monitoring has developed significant over the past 70 years [7] and those technique have progressed by automation with technological advance in order to reduce uncertainty. In addition, remotely sensed data combined with GIS technique can provides valuable result of coastline change mapping and monitoring. Monitoring of coastline areas is needed over time and can detect some change of the coastal area includes land-use /cover change and shoreline/coastline change using some statistical approach. Those approach including End point rate, Net Shoreline Movement, and Linear Regression Rate. The variability of such approach depend significantly on the methodology used to derive them [8][9].

The objectives of this study were to detect the abrasion/erosion and accretion areas and determine the rate of shoreline change in the Northern coastal area of Gresik, East Java in the periods of 44 years.

## 2. Study Area

The Study area (Fig 1) is the northern coast of Gresik regency, East Java Province- Indonesia.Spreads between with the leght about 10 km and coscist of 5 coastal village, i.e. Campurrejo, Dalegan, Ngemboh, Prupuh and Banyu Urip Villages. The coast is bordered by Java sea in the north and Bengawan Solo delta in the east. As well as northern coast of Java Island,the area of study is exposed to the seasonally reversing monsoon wind. The wet season occurs during October – February, while dry season is facing during April- August. The dominated wind (46%) blows from west in the wet season with the velocity 1-21 knot, in contrast in the dry season wind blows from Southeast with the velocity 1-11 Knot [10].

Coastal setting determines both sea and swell waves generally approach the coast from North west in the wet season and North East direction in the dry season. Although generally NW waves greater importance than any other aprocing waves. The tidal type of the study area are diurnal and the tidal ranges varies from 1,2 – 1.5 m [11][12].

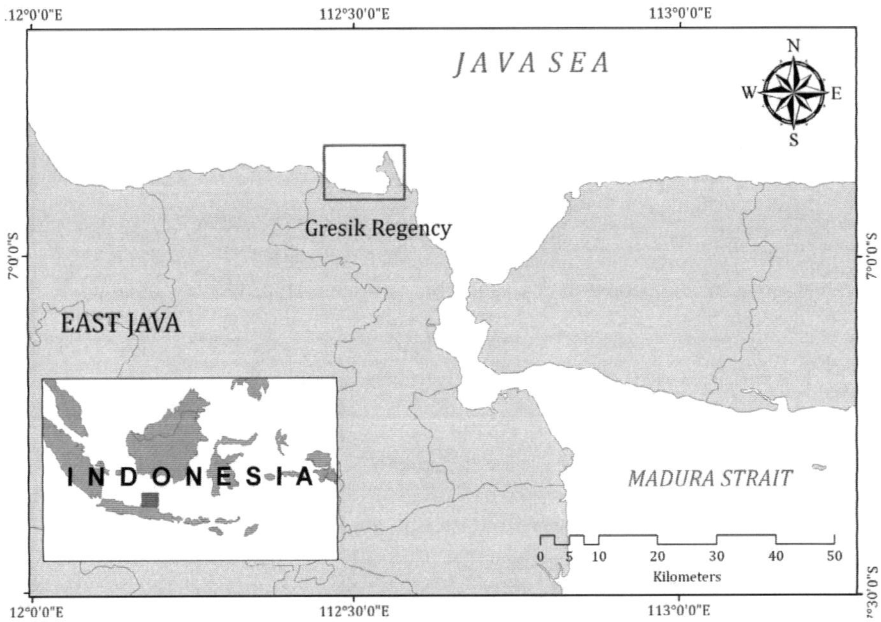


**Figure 1.** Study area

## 3. Methodology

The shoreline change in the study are was analyzed for periods of 44 years (1972-2014), which is classified as long term analysis [13][14] An ortho rectfied Landsat sattelites imagery from first

IOP Conf. Series: Earth and Environmental Science **98** (2017) 012001 doi:10.1088/1755-1315/98/1/012001

generation of landsat (landsat 1) to the latest generatiaon (Landsat 8) were used to delineation of shoreline at disignated time. The Landsat systems has been widely used in remote sensing application for coastal environment changes monitoring and other related studies [15][16]. The landsat images uses for this research were downloaded from USGS Earth Exploree web tool. The details information for each images and coresponding band is given in table 1. The sattelites images were geo-referenced separately using selected control points from topographic map published by Geo-spatial Information Agency, Indonesia. More over some controls points also recognized from fields survey using a Differential –Global Positioning Systems (D-GPS). Image rectification was implemented using WGS 84 datum and its referenced elipsoid. Radiometric calibration was applied to minimized the the possible negative effect of inconsistence of solar irradiance, atmospheric effect, earth-sun distance, and solar zenit angle [17].

**Table 1.** Details of Landsat Images for Shoreline delineation

| Aquisition date | Sattelites | Sensor | *Band* | Resolution (meter) |
|---|---|---|---|---|
| 27/09/1972 | Landsat 1 | MSS | 4 (*Green*)<br>7 (*Near IR*) | 60<br>60 |
| 11/09/1982 | Landsat 3 | MSS | 4 (*Green*)<br>7 (*Near IR*) | 60<br>60 |
| 08/07/1994 | Landsat 5 | TM | 2 (*Green*)<br>5 (*Medium IR*) | 30<br>30 |
| 23/08/2002 | Landsat 7 | ETM+ | 2 (*Green*)<br>5 (*Medium IR*) | 30<br>30 |
| 04/07/2016 | Landsat 8 | OLI | 3 (*Green*)<br>6 (*SWIR 1*) | 30<br>30 |

Automatic shoreline extraction were applied to all multi temporal sattelites images following the Normalized Difference Water Index (MNDWI) method was proposed by [18] which is based on [19] and [20] for Landsat TM and ETM and landsat 8 OLI, respectively. Shoreline produced by this step which is in raster format then converted to vector file using automatic threshold method.

MNDWI transformation for Landsat TM and ETM [18]:

$$\text{MNDWI} = \frac{\text{Green}-\text{NIR}}{\text{Green}+\text{NIR}} \tag{1}$$

MNDWI transformation for Landsat 8 OLI [20]):

$$\text{MNDWI} = \frac{\text{Green}-\text{SWIR 1}}{\text{Green}+\text{SWIR 1}} \tag{2}$$

Shoreline for each images produced by previous step then used as input in Digital Shoreline Analysis System (DSAS) , an ArcGIS extension for quantifing shoreline change in the terms of erosion or accretion [21]. DSAS is an widely used technique for quantitavive methodology for coastal morphology research [16]. In order to extract historical changes of the shoreline, an ortogonal transects with the length of 1.5 km and with interval of 60 m and perpendicular to shoreline were automatically created. In this study, a total of 170 transect were identified along the northern coastal area of Gresik. Each transect then used to measure the position of shoreline overtime from the baseline. The end point rate (EPR) was chosen as the rate change statistic. The EPR is a ratio between the shoreline movement along each transect and time span between them. In order to quantify the length of shoreline movement, the Net Shoreline Movement (NSM) approach was used in this research. This method calculate the distance of coastline from the first year to the last year of observation.

IOP Conf. Series: Earth and Environmental Science **98** (2017) 012001 doi:10.1088/1755-1315/98/1/012001

## 4. Results and Discussion

### *4.1. Decadal Shoreline Evolution*

**1972-1982**

The map of shoreline movement and corresponding table for shoreline dynamics from 1972 to 1982 can be seen in figure 2 and table 2. Negative values in the table indicate the erosion while positive values reperesent accretions. In 1972-1982 most of the area of observations was severe from abrasion. Based on the analysis (table 2), Ngemboh village has the highest average abrasion rate ie EPR -21.18 m / year and NSM -210.81 m. During this first period of detection, there was also an accretion phenomenon. The highest rate of accretion was in Dalegan Village with the average of EPR value 4.84 m / year and NSM 48.14 m.

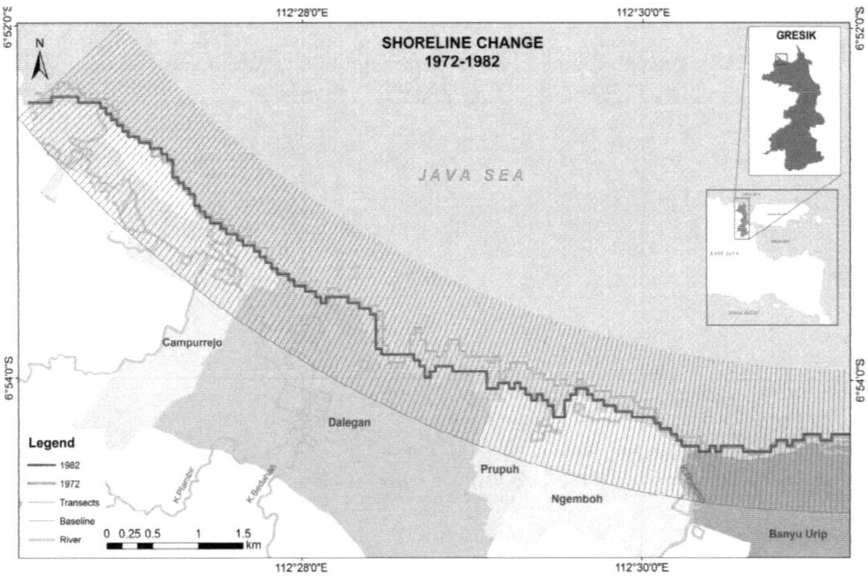


**Figure 2.** Map of Shoreline change 1972-1982

**Table 2.** The Rate and the net movement of shoreline in 1972-1982

| Village | EPR (m/yr) | | NSM (m) | | Remarks |
|---|---|---|---|---|---|
| | Average | | Average | | |
| | + | - | + | - | |
| Campurrejo | 3.12 | -6.58 | 31.02 | -65.47 | Abrasion |
| Dalegan | 4.84 | -13.79 | 48.14 | -137.24 | Abrasion |
| Ngemboh | - | -21.179 | - | -210.81 | Abrasion |
| Prupuh | 2.35 | -12.85 | 23.42 | -127.88 | Abrasion |
| Banyu Urip | 3.26 | -5.27 | 32.43 | -52.49 | Abrasion |

**1982-1994**

In The second period of observation of coastline changes that is 1982-1994, the major phenomenon that occurs was still the same as the previous period of observation. In the research area is still dominated by the event of abrasion (Figure 3 and table 3 ). When compared to the previous periodof observation, the abrasion rate in this period was lower than in 1972-1982. Dalegan Village which in the previous period had the highest level of accretion, but in this period has the highest abrasion rate with average EPR -8.29 m / year and with average NSM -98.01 m. Campurrejo village in this period had the highest level of accretion with the average of EPR is 2.92 m / year and a NSM an average NSM of 34.55 m. Ngemboh village which in the previous period was detected to have the highest abrasion average in this

IOP Conf. Series: Earth and Environmental Science **98** (2017) 012001 doi:10.1088/1755-1315/98/1/012001

period began to decline, as well as the Village Prupuh which in the previous period experienced the highest abrasion rate in this period experienced a similar thing with Ngemboh Village.

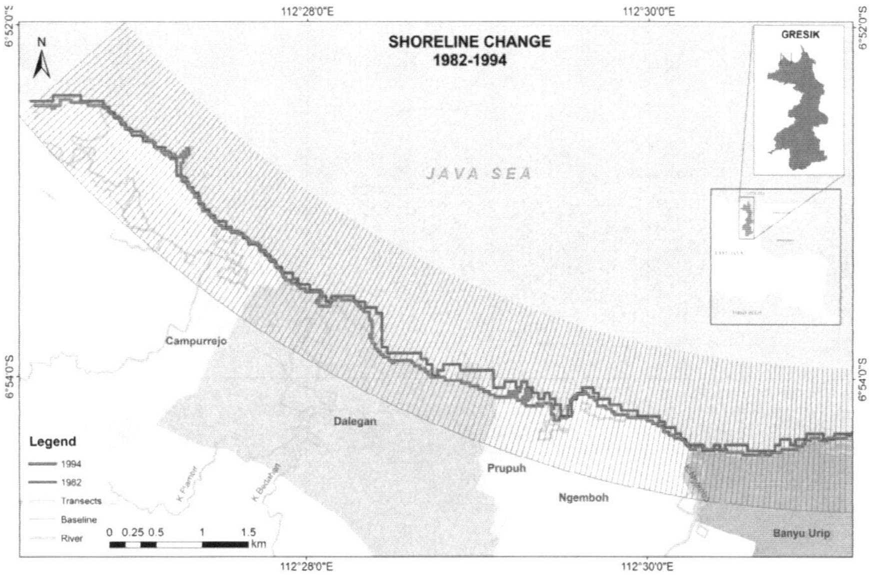


**Figure 3.** Map of Shoreline change 1982-1994

**Table 3.** The rate and the net movement of shoreline in 1982-1994

| Village | EPR (m/yr) Average | | NSM (m) Average | | Remarks |
|---|---|---|---|---|---|
| | + | - | + | - | |
| Campurrejo | 2.92 | -3.22 | 34.55 | -38.01 | Abrasion |
| Dalegan | 0.24 | -8.29 | -8.29 | -98.01 | Abrasion |
| Ngemboh | - | -8.01 | - | -94.73 | Abrasion |
| Prupuh | - | -4.66 | - | -55.06 | Abrasion |
| Banyu Urip | 1.28 | -4.50 | 15.15 | -53.21 | Abrasion |

**1994-2002**

In the periods of 1994-2002, the shoreline at the most of the region began to experience accretion phenomenon while in the previous periods severe from abrassion (fig.4 table 4). The highest average accretion was found in Ngemboh Village with average EPR 12.18 m / year and mean NSM 98.95 m. Ngemboh village which in the previous period tends to experience abrasion, at those periods begins to decrease and reach a peak in 2004. After this year, abrasion was undetected and the shoreline has been stable. As well as Ngemboh village, Dalegan Village also experiencing an accretion process. Moreover, accretion occured in this village was higher compared to previous periods..

The highest rate of abrasion during this period was found in Camprurrejo 1 village, EPR -4.31 m / year with EPR -9.86 m / year and NSM -80.16 m with NSM -35.01 m average. Campurrejo village in the previous period had the highest level of accretion in this period began to experience abrasion, but the abrasion rate occurring in this period was not as high as the abrasion rate detected in 1972-1982.

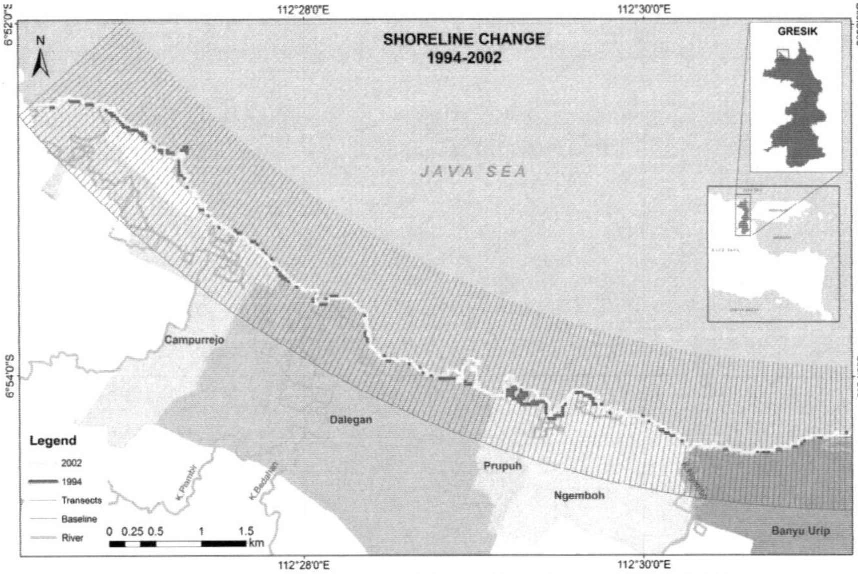


**Figure 4.** Map of Shoreline change 1994-2002

**Table 4.** The rate and the net movement of shoreline in 1994-2002

| Village | EPR (m/yr) | | NSM (m) | | Remarks |
|---|---|---|---|---|---|
| | Average | | Average | | |
| | + | - | + | - | |
| Campurrejo | 4.56 | -4.31 | 37.06 | -35.01 | Accretion |
| Dalegan | 9.31 | -3.90 | 75.65 | -31.65 | Accretion |
| Ngemboh | 12.18 | - | 98.95 | - | Accretion |
| Prupuh | 4.06 | -2.90 | 32.98 | -23.53 | Accretion |
| Banyu Urip | 3.71 | - | 18.54 | - | Accretion |

**2002-2016**

In the last decade of shoreline analysis, all of the study area expect Ngemboh village dominated by acretion process. The highest average of acretion was in the Dalegan Village with the movement of shoreline morethan 107 m in the 14 years. In contrast, ngemboh was facing abrasion for approximately 42 meters in average for this period. The rate of shoreline movement also follows those shoreline change indication, ie. Abrasion. Ngemboh village facing abration for more than 3 m per year. This phenomenon is mainly caused by different landuse that exist in those area. Ngemboh village formerly was dominated by natural coastal type, ie. sandy coast without any protection structure,was vulnerable from coastal erosion. In addition another area of the research study was introduced by traditional man made structure such as traditional fish pond and seawall made from rocks.

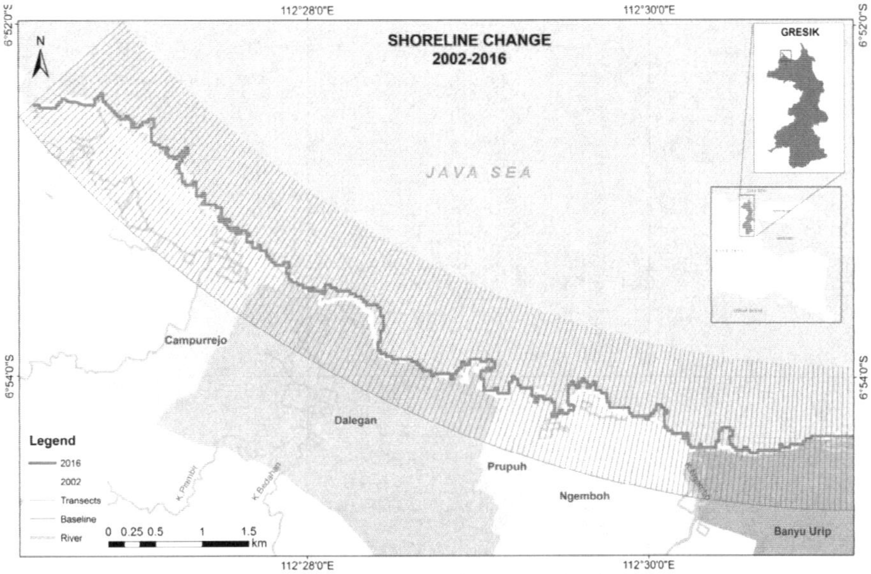


**Figure 5.** Shoreline change 2002-2016

**Table 5.** The rate and the net movement of shoreline in 2002-2016

| Village | EPR (m/yr) Average | | NSM (m) Average | | Remarks |
|---|---|---|---|---|---|
| | + | - | + | - | |
| Campurrejo | 4.11 | -2.74 | 56.95 | -37.94 | Accretion |
| Dalegan | 7.75 | -0.89 | 107.53 | -31.58 | Accretion |
| Ngemboh | 2.29 | -3.04 | 31.76 | -42.13 | Abrasion |
| Prupuh | 6.64 | - | 76.02 | - | Accretion |
| Banyu Urip | 5.92 | -2.17 | 82.05 | -30.04 | Accretion |

*4.2. Long term Shoreline Evolution (1972-2016)*

The longterm shoreline change in the different village in the study area had several characteristics of abrassion. In Dalegan, Prupuh, and ngemboh village shoreline tend to extremely retreart due to abrassion phenomenon. Figure 6 and table 6 revealed that in those area abrassion process was eradicate the coast for more than 240 m for 44 years. In the other hand the abrasion process in Banyuurip and Campurrejo village is not as extremely as other area. This is might be caused by the introduction of man made structure such as Sea wall, groin, and fish pond structure. From the google image (fig. 6), several man made structure also can be identified such as jetty and reclamation area.

**Figure 6.** Long term shoreline change 1972-2016

**Table 6.** The rate and the net movement of shoreline 1972-2016

| Village | EPR (m/yr) | | NSM (m) | | Remarks |
|---|---|---|---|---|---|
| | Average | | Average | | |
| | + | - | + | - | |
| Campurrejo | 1.47 | -3.33 | 64.35 | -103.18 | Abrasion |
| Dalegan | 0.61 | -3.42 | 26.58 | -149.76 | Abrasion |
| Ngemboh | - | -5.54 | - | -242.56 | Abrasion |
| Prupuh | 0.84 | -2.68 | 36.85 | -117.48 | Abrasion |
| Banyu Urip | 0.94 | -1.23 | 59.4 | -53.82 | Abrasion |

## 5. Conclusions

The aims of this study were to identify and characterize shoreline dynamics during longterm periods from 1972-2016 in the region of Northern coastal area of Gresik. The change of shoreline overtime is in response to environmental, economic, and social forces. There were 3 abrasion and acretion regime during this periods. Between 1972-1994 , a clear and notable abrassion trend was detected. The average rate of abrasion for this periods is more than 21m/yr caused the shoreline retreat for almost 210 m for 22 yr. In contrast, in the year of 1994-2002, the research area mostly facing acretion phenomenon. Moroover, in the last periods (2002-2016) of analysis, there were acretion as well as abrasion which identified in the study area.

The present study suggest that multi-temporal sattelites data along with GIS and statistical techniques can be efectively used for detection of shoreline change in the studi area. But the major driving factor for this phenomenon remains unclear. Forecast for the future evolution of coastline is somewhat tricky works because of the economics development in the area.

**References**

[1] Jalila Moussaid, Abderrahmane Ait Fora, Bendahhou Zourarah, Mehdi Maanan and Mohamed Maanan 2015 Using automatic computation to analyze the rate of shoreline change on the Kenitra coast, Morocco *Ocean Engineering* **102** pp 71-77, ISSN 0029-8018, https://doi.org/10.1016/j.oceaneng.2015.04.044.

[2] R S Kankara, S Chenthamil Selvan, Vipin J Markose, B Rajan and S Arockiaraj 2015 Estimation of Long and Short Term Shoreline Changes Along Andhra Pradesh Coast Using Remote Sensing and GIS Techniques *Procedia Engineering* **116** pp 855-862, ISSN 1877-7058, https://doi.org/10.1016/j.proeng.2015.08.374.

[3] Bird E C F1985 *Coastline changes: A Global Review* (Chichester: John Wiley & Sons) ,A Wiley – Inter science publication, Chichesterp.219.

[4] Boak E H and I LTurner 2005 Shoreline definition and detection : a review *J. Coast. Res* **21**(4) pp 688–703.

[5] Mills J P, S J Buckley, H L Mitchell, P J Clarke and J Edwards 2005 A geomatics data integration technique for coastal change monitoring *Earth Surf. Processes Landf* **30**(6) pp 651–664.

[6] Kumar A, A C Narayana and K S Jayappa 2010 Shoreline changes and morphology of spits along southern Karnataka, wes t coast of India :a remote sensing and statistics based approach. *Geomorphology* **120** pp 133–152.

[7] Crowell M 2006 Historical shoreline mapping and analysis : an historica over view. In: Abstract submitted to Shoreline Change Conference II: A Work shop on Managing Shoreline Change. Charleston, USA. ⟨http://www.csc.noaa.gov/shor econf/shorelinechange⟩.

[8] Dolan R, M S Fester and S J Holme 1991 Temporal analysis of shoreline recession and accretion *J.Coast.Res* **7** pp 723–744.

[9] Genz A S, C H Fletcher, R A Dunn, L N Frazer and J J Rooney 2007 The predictive accuracy of shoreline change rate methods and a long shore beach variation on Maui, Hawaii *J. Coast. Res* **23** pp 87–105.

[10] Barcelona A E, D N Sugianto and A Rifai 2015. Wave refraction study on Ujung pangkah waters, Gresik – East Java *J.Oseanografi* **4**(2) pp 434-441.

[11] Fuad M A Z 2015 Aplication of sattelites imagery for Bengawan Solo Delta development monitoring. *Proc. 12th indonesian Marine and Fisheries Science Seminars*, Yogyakarta.

[12] Wyrtky Klaus 1961 Physical Oceanography of the south east asian water *Naga Report* The University of California

[13] Crowell M, S P Leatherman, M Buckley 1993 Shoreline change rate analysis: long term versus short term data *Shore and Beach* **61** pp 13–20.

[14] Anfuso G and J A Martinez Del Pozo 2009 Assessment of coastal vulnerability through the use of GIS tools in South Sicily (Italy) *Environmental Management* **43** pp 533 – 545.

[15] Tucker J C, D M Grant and J D Dykstra 2004 NASA's global orthorectified landsat data set *Photogramm. Eng. Remote Sens* **70** (3) 313e322.

[16] Guneroglu Abdulaziz 2015 Coastal changes and land use alteration on Northeastern part of Turkey. *Ocean & Coastal Management* **118** 225e233. http://dx.doi.org/10.1016/j.ocecoaman.2015.06.019.

[17] Chander G, B L Markham and D L Helder 2009 Summary of current radiometric calibration coefficients for landsat MSS, TM, ETMþ, and EO-1 ALI sensors *Remote Sens. Environ* 113, 893e903.

[18] Xu H 2006 Modification of normalised difference water index (NDWI) to enhance open water features in remotely sensed imagery *International Journal of Remote Sensing* **27**(14) pp 3025–3033. https://doi.org/10.1080/01431160600589179.

[19] McFeeters S K 1996 The use of normalized difference water index (NDWI) in the delineation of open water features *Int. J. Remote Sens* **17** 1425e1432.

[20] Ko B C, H H Kim and J Y Nam 2015 Classification of potential water bodies using Landsat 8 OLI and a combination of two boosted random forest classifiers *Sensors* **15**(6) pp 13763–

13777.
[21] Thieler E R and E A Himmelstoss 2012 *DSAS 4.0: Installation Instructions and User Guide (Updated for Version 4.3)* USGS

# Identification of Mesoscale Convective Complex (MCC) phenomenon with image of Himawari 8 Satellite and WRF ARW Model on Bangka Island (Case Study: 7-8 February 2016)

**Nanda Rinaldy[1*], Immanuel J. A. Saragih[1], Agie Wandala Putra[2], Imma Redha Nugraheni[1], and Banu Wijaya Yonas[1]**

[1] Indonesian School of Meteorology, Climatology, and Geophysics (STMKG)
[2]Agency for Meteorology, Climatology, and Geophysics, Republic of Indonesia (BMKG)

nandarinaldy@gmail.com

**Abstract.** Based on monitoring on 7th and 8th February 2016 there has been a flood that occurred due to heavy rainfall in a long time in some areas of Bangka Island. Mesoscale Convective Complex (MCC) is one type of Mesoscale Convective System (MCS). Previous research on MCC mentions that MCC can cause heavy rain for a long time. This study aims to identify the phenomenon of MCC in Bangka Island both in the satellite imagery and the output of the model. In addition, this study was also conducted to determine the effect of MCC on the weather conditions in Bangka Island. The study area in this research is Bangka Island with Pangkalpinang Meteorological Station as the centre of research. The data used in this research are FNL (Final Analysis) data from http://rda.ucar.edu/, Satellite Image of Himawari-8 IR1 Channel from BMKG, and meteorological observation data (synoptic and radiosonde) from Pangkalpinang Meteorological Station. The FNL data is simulated using the WRF-ARW model, verified using observation data and then visualized using GrADS. The results of the analysis of Himawari-8 satellite image data showed that two MCCs occurred on 7th and 8th February 2016 on Bangka Island and the MCC was nocturnal, which appeared at night which then continued until extinction in the morning the next day. In a peak cloud temperature review with the coordinates of Pangkalpinang Meteorological Station (-2,163 N 106,137 E) when 1st MCC and 2nd MCC events ranged from -60 $^{o}$C to -80 $^{o}$C. The result of WRF-ARW model output analysis shows that MCC area has high humidity value and positive vertical velocity value which indicates the potential of heavy rain for a long time.

**Keywords:** MCC, Himawari 8, Satellite, WRF ARW

## 1. Introduction

Indonesia lies geographically on the equator which certainly has a great potential impact on atmospheric physical conditions during extreme weather events, especially as the Mesoscale Convective Complex (MCC) phenomenon. Extreme weather is a situation in a place at a certain time on a short time scale and is extreme [1]. One of the factors that influence the occurrence of extreme weather is the occurrence of irregularities or the so-called weather anomalies both on a global scale, regional, and local. Mesoscale Convective Complex (MCC) is a combination of Cumulonimbus clouds that are marked by a small shear and a large Convective Available Potential Energy (CAPE), and has a duration of> 6 h (heavy rain ± 3 hours) rainfall and a large area with an index CAPE stability ≥ 1750 J / kg [2]. Weather Research Forecasting (WRF) is a mesoscale weather model created for analysis and forecasting of atmospheric conditions. WRF can model atmospheric conditions in a region so that it can assist in calculating meteorological events better [3]. On 07 and 08 February 2016 there was heavy rain lasting for a long time in Bangka Island and covered a large area causing floods to occur in some areas. In the event of heavy rain is due to the growth of convective clouds that are active over the area of Bangka Island, especially the northern fluctuation that has a very wide range of areas, so that a collection of convective clouds that collide over the area of Pulau Bangka is a Mesoscale Convective Complex (MCC) if analyzed through GMSLPD. The magnitude of the impact caused this incident interesting to be studied using WRF ARW model.

**Figure 1.** Location City PangkalPinang in Bangka Island
Source: Tourism Province of Bangka Belitung Islands

The data used in this research are:

1. Himawari 8 satellite data on 07-08 February 2016 Infrared (IR1) channel with 10 minutes resolution with data extension. Z and NetCDF.
2. WRF-ARW model input data using Final Data (FNL) data on February 07, 2016 and 08 February 2016 obtained from http://rda.ucar.edu website with a spatial resolution of 5 km, 2m grid resolution and resolution temporal 1 hour.
3. 3-hour rainfall data from observation of Pangkal Pinang Meteorological Station on 07-09 February 2016.

In this study Himawari 8 satellite data with the initial extension. Z processed using the application (SATAID). Satellite Interaction and Interactive Diagnosis (SATAID) to display cloud time temperature time series, while NetCDF data will be processed using Grid Analysis and Display System (GrADS) software to display cloud images with IR1 channels that use cloud top temperature to determine cloud type. NC data is extracted from the convert Z data using the sataid2nc.exe application via Command Prompt. After satellite data is used also input data model WRF-ARW form FNL data used to know the values of physical parameters that cannot be known only by using satellite data. In this domain selection is used domain with the location of the center at the Station Meteorology Class I PangkalPinang with coordinates -2.163 N 106,137 E. The results of the WRF-ARW running model are then processed using GrADS data to display streamline, cloud fraction, vertical velocity, based on observation center point at

I PangkalPinang Meteorological Station. In addition to these data are also used additional data supporting observation data of rainfall 3 hours from PangkalPinang Meteorological Station.

## 2. Material and Method

In this study, the authors adopted a research journal conducted by (Putri, R J A, et al) [4]. That is by using parameterization scheme which has been tested in Mesoscale Convective System (MCS) simulation in Nabire region, Papua. As we know that Indonesia has a unique atmosphere in every atmosphere in each region, the authors try to simulate it also with the WRF model on the physical condition of the atmosphere when MCC occurs on the island of Bangka.

This research is focused on location in Bangka Island with coordinate point position at Pangkalpinang Meteorological Station with coordinate -2,163 N 106,137 E. Based on Pangkalpinang Meteorological data in 1998, the climate in Bangka Regency is a tropical climate of type A with rainfall of 107.6 to 343.7 mm per month, and in further study according to Schmidt-Ferguson, in 1999 the variation of rainfall to be between 70 , 10 to 384.50 mm per month, so that the physical condition of the weather in Bangka Island is very much certainly influenced by wind circulation.

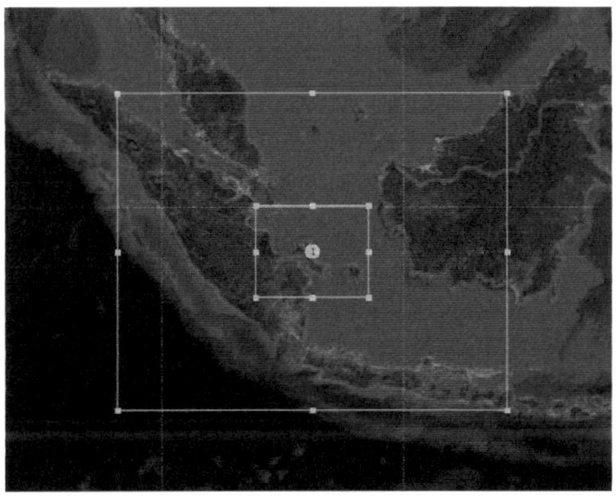

**Figure 2.** Domain research area

Weather Research and Forecasting-Advanced Research (WRF-ARW) version 3.8.1 is used in this study [5, 6, 7]. Used two domains with one-way nesting (Figure 2). Horizontal 18 and 6 grid spacing for both running processes in WRF-ARW. The outermost domain covers most of the islands of Sumatra, Kalimantan, and Java, while the deepest domain (domain 2) is centred on the Pangkalpinang meteorological station. The grid resolution for both domains is 5 m and 2 m respectively. The physical options used in this study are summarized in Table 1.

**Table 1.** Configuration of physical

| Physical Schematic of WRF-ARW Model | Model WRF-ARW Area | |
|---|---|---|
| | Domain 1 (D01) | Domain 2 (D02) |
| Resolution | 18 km | 6 km |
| Mikrofisis | Lin et al. | Lin et al. |
| Short Wave Radiation | Dudhia | Dudhia |
| Long Wave Radiation | RRTM | RRTM |
| Land Surface | Noah LSM | Noah LSM |
| Planetary Boundary Layer | YSU scheme | YSU scheme |
| Cumulus Parameterization | Kain-Fritsch | Kain-Fritsch |

## 3. Result

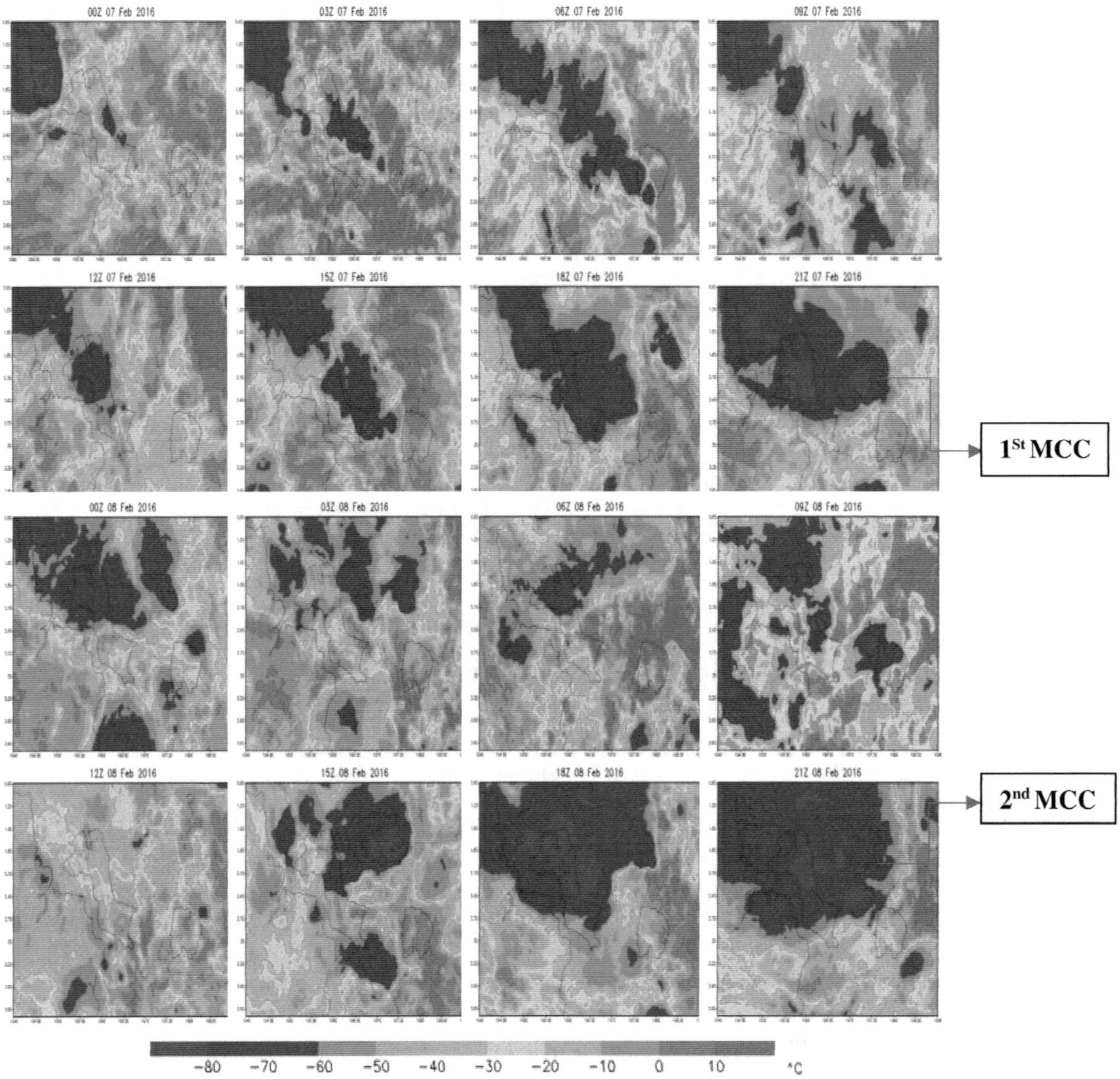


**Figure 3.** Cloud Image with Infrared Channel

The results of cloud classification on Infrared satellite channel images obtained from Himawari-8 satellite show that at the time of heavy rainfall on 07 and 08 February 2016 in Bangka Island area is covered by cloud with very wide coverage with cloud top temperature reaching <-80 °C twice. After analysis using GMSLPD software, it is known that the cloud cover is a Mesoscale Convective Complex (MCC). This is indicated that the cause of heavy rain in the area of Bangka Island is caused by clusters of convective clouds or called 1st MCC and 2nd MCC which each have a lifespan of more than 6 hours.

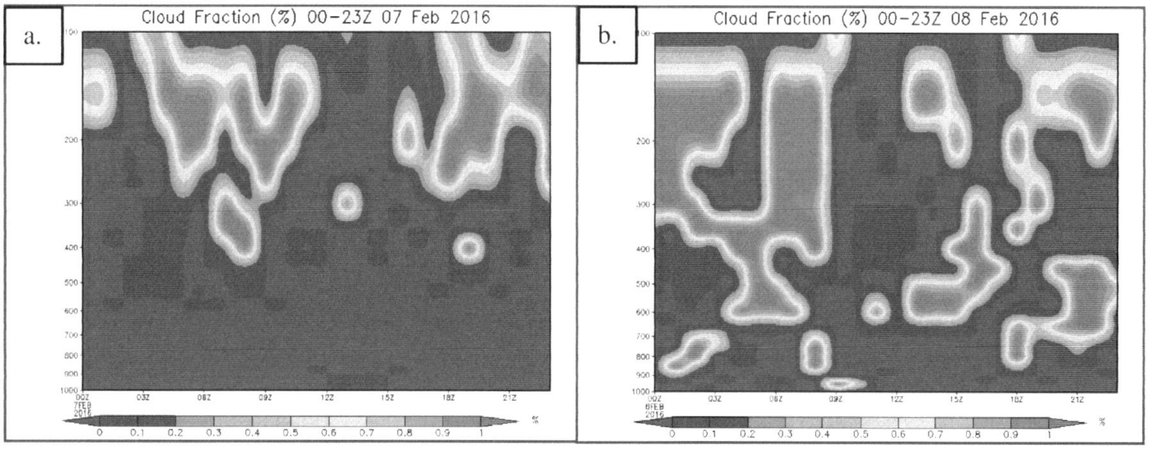


**Figure 4**. **a)** Cloud fraction in vertical profile from 00-23 UTC on February 7, 2016 shows clearly the percentage at layer 1000 mb to 100 mb; **b)** Cloud fraction in vertical profile from 00-23 UTC on February 8, 2016 shows clearly the percentage at layer 1000 mb to 100 mb.

Figure 4 is the result of the Cloud Fraction output (%) of the WRF-ARW centred on the coordinates of the location of the Pangkalpinang meteorological station. In the Cloud Fraction shown in the vertical profile where if we examine at 21 UTC on 07 and 08 February 2016 it clearly shows a pattern that supports MCC formation by marking the percentage value of Cloud Fraction reaching 1% value range meaning to indicate that cloud cover at the layer is thick. Seen to have a significant difference from 07 to 08 February 2016 clearly on 08 February cloud fraction has been formed on the lower layer of 900 mb.

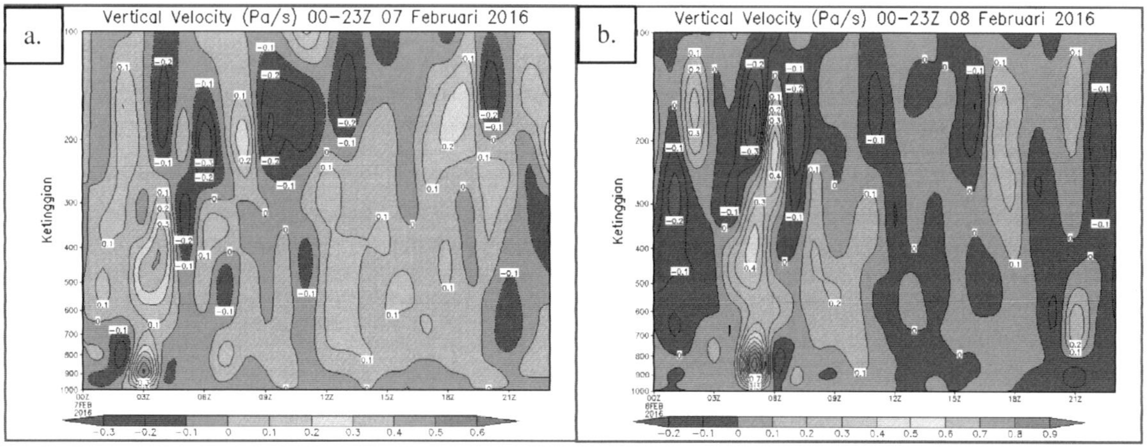


**Figure 5**. **a)** Velocity vertical in vertical profile from 00-23 UTC on February 7, 2016 shows clearly the percentage at layer 1000 mb to 100 mb; **b)** Velocity vertical in vertical profile from 00-23 UTC on February 8, 2016 shows clearly the percentage at layer 1000 mb to 100 mb.

Figure 5 shows velocity vertical during growth, development, and decay of MCC. If the vertical velocity (in units Pa / s) is positive, it indicates there is a movement of the air mass rising vertically to form clouds. Conversely, if a negative vertical velocity indicates there is a decreasing air mass movement. The increasing value indicates the speed of the air mass movement. In the vertical velocity it is seen that if the UTC 21 hour of reference to the WRF results does not show as clear as that produced by the Himawari 8 satellite image of the IR channel which of course has an upwardly uplifting air masses, the results of this WRF model show that Velocity vertical which is shown to indicate the existence of a moving upward air mass on February 07, 2016 at 03 UTC with a value of 0.3 Pa / s, while on February 08, 2016 in the 5th range of UTC with a value reaching 0.9 Pa / s.

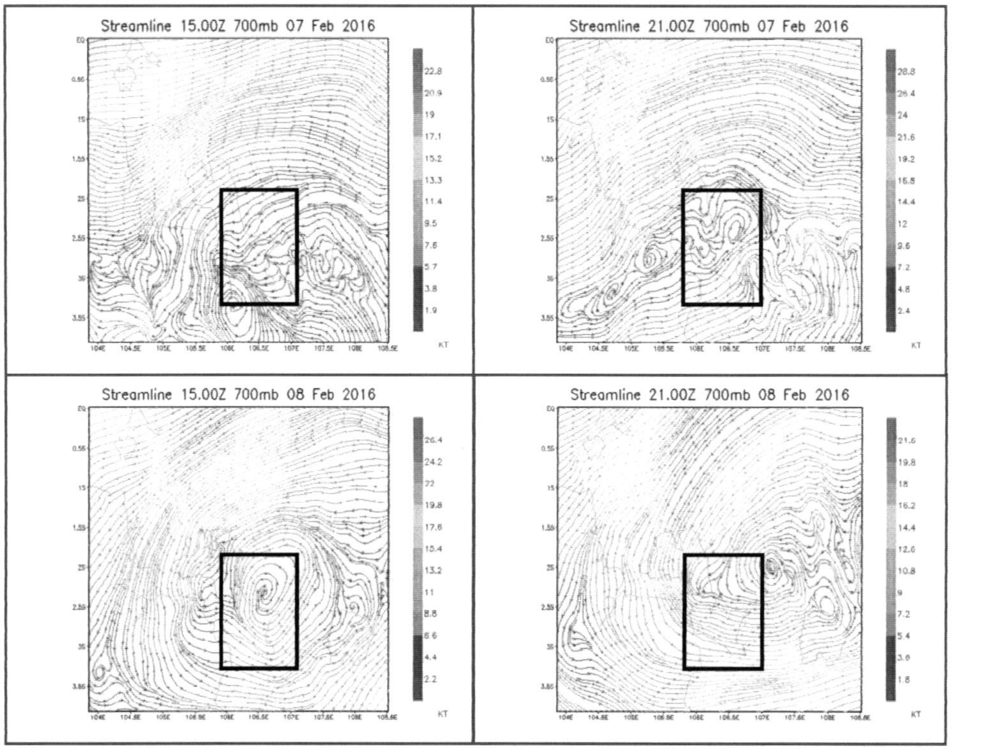


**Figure 6.** Streamline at 15:00 UTC and 21:00 UTC on 07 and 08 February 2016

Streamline results shown from the WRF model were obtained at 6 hours before the peak occurred 1st MCC and 2nd MCC. On 07 February 2016 at 15 UTC and 21 UTC there is a pattern with shear flow followed by a weak wind speed ranging from 1.9 Kt and 2.4 Kt resulting in a buildup of air masses supporting cloud growth, while on February 08, 2016 at 15 UTC and 21 UTC there is a clear pattern over Pangkalpinang area that formed anticyclonic pattern with the lowest wind speed ranged from 2.2 Kt and 1.8 Kt.

**Table 2.** Rainfall Data from February 7 to February 09, 2016 at PangkalPinang Meteorological Station

| Date/Hr (UTC) | 03 | 06 | 09 | 12 | 15 | 18 | 21 | 24 | TOTAL |
|---|---|---|---|---|---|---|---|---|---|
| 7 | 6,8mm | 1,5mm | 5,2mm | 0,1mm | 15,6mm | 13,1mm | 40,9mm | 31,7mm | **114,9mm** |
| 8 | 4,5mm | 53,8mm | 56,5mm | 47,3mm | 1,3mm | 1,1mm | 3,4mm | 16,0mm | **183,9mm** |
| 9 | 22,8mm | 0,8mm | TTU | 0 | 0 | 0 | 0 | 0 | **23,6mm** |

Based on Table 1 it can be seen that the rainfall based on observation observation at the Class I PangkalPinang Meteorological Station shows the amount of rainfall intensity recorded for every three hours according to Universal Time Center (UTC) time system, starting at 00:00 UTC (07.00 WIB) until 00.00 UTC (07.00WIB) the following day. On February 07, 2016 in one day the measured rainfall reached 114.9 mm, then on February 08, 2016 measured rainfall in one day measured to reach 183.9 mm. After February 08, rain falls to a halt on February 09, 2016 which is marked by rainfall at 09.00 UTC (16.00WIB) shows (Unmeasured) TTU. According to BMKG (2010), rainfall intensity ≥ 100 mm per day can be classified as rainy conditions when it is very dense.

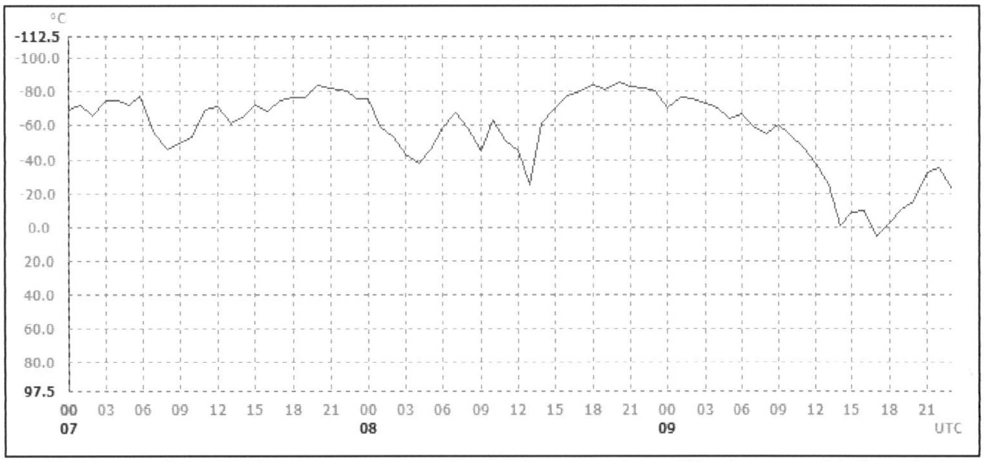


**Figure 7.** Time series cloud top temperature from Himawari 8 satellite images of infrared channels

Can be seen in Figure 2 the cloud top temperature at the point of coordinate location of Meteorological Station PangkalPinang when the MCC saw time series graph is reaching -60 °C to -80 °C. From the temperature has been indicated that the clouds have gone through the tropopause layer indicating in MCC there is Cumulonimbus cloud in the mature phase.

## 4. Conclusions

Based on the result of research, it can be concluded that the existence of MCC can by using the image of Himawari 8 satellite to see the type and pattern of cloud distribution formed, and time series of cloud temperature has shown good approach and according to the pattern, but if viewed from WRF-ARW model result the WRF model is still lacking especially in showing the suitability of the pattern of the Himawari 8 satellite imagery results in helping identify MCCs by looking at the cloud distribution patterns as well as the value of cloud-value of other physical parameters.

## References

[1] Zakir, A et al 2010 Perspektif Operasional Cuaca Tropis, *Puslitbang BMKG*
[2] Maddox, R A 1980 Mesoscale Convective Complexes, *Bull. Amer. Meteor. Soc.* (**61**),**1374-1387**
[3] Hadi, Tri Wahyu et al 2011 Modul pelatihan WRF Fakultas Ilmu dan Teknologi Kebumian. Institut Teknologi Bandung, Bandung
[4] Putri, R J A 2016 Identifikasi Mesoscale Convective System (MCS) Menggunakan Data Satelit Himawari-8 dan Model WRF-ARW ( Studi Kasus Hujan Lebat 21 Maret 2016 di Kabupaten Nabire, Papua), *Proceeding on Seminar Nasional Penginderaan Jauh 2016*
[5] Nurpambudi, Ramadhan et al 2015 Simulasi Mesoscale Convective System menggunakan Model WRF-ARW di Makassar (Studi Kasus Tanggal 6 & 8 Desember 2014) (*Sarjana Terapan Meteorologi Skripsi) STMKG (school of Meteorology Climatology Geophysics), Jakarta*
[6] Kurniawan, Roni et al 2014 Penggunaan Skema Konvektif Model Cuaca WRF (Betts Miller Janjic, Kain Fritsch dan Grell 3D Ensemble) (Studi Kasus: Surabaya dan Jakarta) *Jurnal Meteorologi*

*dan Geofisika* **(15)25-36**
[7] Van Weverberg, et al 2013 The Role of Cloud Microphysics Parameterization in the Simulation of Mesoscale Convective System Clouds and Precipitation in the Tropical Western Pacific *Journal of The Atmospheric Sciences* **(70) 1104-1128**

# Identification of the Sea-Land Breeze Event and Influence to the Convective Activities on the Coast of Deli Serdang

**I J A Saragih[1*], A W Putra[2], I R Nugraheni[3], N Rinaldy[4] and B W Yonas[5]**

[1,3,4,5]School of Meteorology Climatology and Geophysics (STMKG), Jakarta, Indonesia
[2]Meteorological, Climatological, and Geophysical Agency (BMKG), Jakarta, Indonesia

immanuel.saragih@bmkg.go.id

**Abstract.** Located close to the sea indicates that there are influences of the sea-land breeze circulation on the weather condition in Deli Serdang. The purpose of this study is to simulate sea-land breeze occurrence and its influence on the convective activities in Deli Serdang. The research area covers the area of Deli Serdang Regency and the surrounding ocean region in the coordinates 02°57'-03°16'N & 98°33'-99°27'E where Kualanamu Meteorological Station is the centre of the research area at coordinate 03°34'N & 98°44'E and the elevation about 27MAMSL. The research time is a day with the highest rainfall in the highest peak rainy month. The raw data consist of the Himawari-8 satellite image from BMKG, FNL (Final Analysis) data from http://rda.ucar.edu, and meteorological observation data from Kualanamu Meteorology Station. This study indicates that WRF-ARW can simulate the sea-land breeze occurrence on the coast of Deli Serdang well. The existence of the convective index cover in the convergence area proves the sea-land breeze occurred in the coast of Deli Serdang can form the convergence area as the interacted result with the wind from other directions that support convective activities.

**Keywords:** Sea-land breeze, WRF-ARW, Heavy rainfall, Himawari-8

## 1. Introduction

Indonesia is a maritime country located in the tropic area which has a significant temperature difference between land and sea area. The significant temperature difference between land and sea potentially make the wind circulation known as the sea-land breeze [19]. The movement of the sea-land breeze has a correlation with convective activities in a coastal area. The convective activities can be analysed by measuring the convective index which the previous studies measure it by using the calculation of the top cloud temperature from the satellite image [12].

The wind moves due to the difference in the air pressure between two regions. Generally, it moves from the high-pressure air area (High) to the low-pressure air area (Low). The circulation of sea-land breeze can occur in areas adjacent to or near the coast. This circulation is caused by the significant temperature difference between land and sea due to the differences in heat (sunlight) absorption by land and sea [20]. In general, during the daytime, the land absorbs the heat more quickly than the sea. As the result, the air temperature over the land is higher than in the oceans so that the air pressure over the sea will be higher than the land, and it makes the wind circulation from the sea to the land which known as the sea breeze. Meanwhile, at the night when the absence of solar irradiation, the land releases the heat more quickly than the sea. As the result, temperatures over the sea will be higher than the land, so that the air pressure over the land will be higher than the sea, and it makes the wind circulation from the land to the sea which known as the land breeze. There are two conditions expected to occur when the sea-land breeze pass through a coastal area. First, the sea-land breeze can disrupt the growth of the convective clouds. Meanwhile, the second, it can bring in convective air masses and clouds from other regions to accumulate and increase the convective activities in the sea-land breeze destination area [11].

WRF (Weather Research and Forecasting) is a mesoscale numerical weather model which is used to simulate the mesoscale meteorological phenomena for research and weather forecast activities. WRF can be used for weather simulation with a grid width of up to thousands meters [8]. WRF-ARW is a part of WRF. The use of the WRF model for each region requires the determination of parameterization schemes which the accuracy is different.

Deli Serdang is one of the regencies in North Sumatera Province. Deli Serdang is adjacent to the Malacca Strait in the north. Geographically, Deli Serdang is on the East Coast of Northern Sumatra. Located near the sea are the reason of the importance in analysing the sea-land breeze circulation over there and the influence of the weather condition in the coastal of Deli Serdang. Due to the importance of sea-land breeze influence to the weather conditions in the coastal area, further research is needed to improve the weather forecast accuracy. The purpose of this study is to know the capability of WRF-ARW model in simulating the sea-land breeze occurrence as well as to know its influence to the convective activities in Deli Serdang. And finally, the results of this study can be a reference in making the analysis and weather prediction in the coast of Deli Serdang.

Previous studies [12][15] have been done to analyse the convection and convergence patterns in Indonesian Maritime Continent (BMI). Convection and convergence patterns can be analysed by utilizing the results of the infrared channel of Himawari-8 satellite imagery (previously used MTSAT satellites). The cloud top temperature measured using 10.4μm satellite channel data is used to calculate the convection index over the research area. Sakurai has undertaken the formulation of convective index calculations as follows [15].

$$\boldsymbol{CI} = \begin{cases} \boldsymbol{230 - Tb, if\ Tb\ \leq 230K} \\ \boldsymbol{0, if\ Tb \geq 230K} \end{cases} \quad (1)$$

Which CI is the convective index, Tb (Brightness Temperature) is the measured cloud top temperature and K (Kelvin) is the unit value of the cloud top temperature.

## 2. Data and methods

### *2.1. Research area*

The research area covers the area of Deli Serdang Regency and the surrounding ocean region which located at coordinates 02°57'-03°16'N & 98°33'-99°27'E, where Kualanamu Meteorological Station is the centre of research area located at coordinates 03°34'N & 98°44'E and elevation about 27MAMSL (metres above mean sea level).

**Figure 1.** Map of Deli Serdang Regency (Google Maps Views)
[Source: https://www.google.co.id/maps/place/Bandara+Int.+Kualanamu/]

*2.2. Research data*

2.2.1. *Meteorological observation data.* This study uses the meteorological observation data from 2014 to 2016 obtained from the Kualanamu Meteorological Station. These data consist the surface air (synoptic) and the upper air observation data (radiosonde). This study uses the rainfall data (CH) to determine the distribution of the daily and monthly rainfall in Deli Serdang.

*2.2.2. FNL (Final Analysis) data.* This study uses the FNL data accessed from http://rda.ucar.edu which have 1ox1o spatial resolution and 6 hours temporal resolution. This FNL data is the 24-hours data on 1 September 2016 with the spin-up time for 12 hours before and after the main data hours.

*2.2.3. The infrared channel of the Himawari-8 satellite image.* This study uses the infrared (10.4 mm) channel of Himawari-8 satellite image obtained from BMKG. This satellite image data are used to calculate the convective index value over the research area.

*2.2.4. GSMaP data.* GSMaP (Global Satellite Mapping of Precipitation) is consists of the rainfall data measured by the satellite which has the 1-hour temporal resolution and 0.1°x0.1° spatial resolution. The GSMaP data accessed from ftp://hokusai.eorc.jaxa.jp.

*2.3. Data processing*

*2.3.1. Research time identification.* The research time is determined based on rainfall data from 2014 to 2016 obtained from Kualanamu Meteorological Station. In this study, the research time is a day with the highest rainfall in the highest peak rainy month. Based on monthly rainfall data, it is indicated that the highest peak of rainfall occurred in September. Meanwhile, the rainfall data from 2014 to 2016 represent that on 1 September 2016 occurs the highest daily rainfall about 101mm so that 1 September 2016 set as the research time.

*2.3.2. Running process of the WRF-ARW model.* The mesoscale model used in this study is the WRF-ARW Version 3.8.1 model. The WRF-ARW model run by using the Mercator map projection and be divided into two domains. 1st Domain (DO1) located in coordinate -4.82°-11.89°N & 90.32°-107.135°E with a horizontal resolution of 21km. Meanwhile, the 2nd Domain (DO2) located in coordinate 1.15°-5.98°N & 96.31°-101.15°E with 7km horizontal resolution. Here are the domains and parameterization settings used in this study.

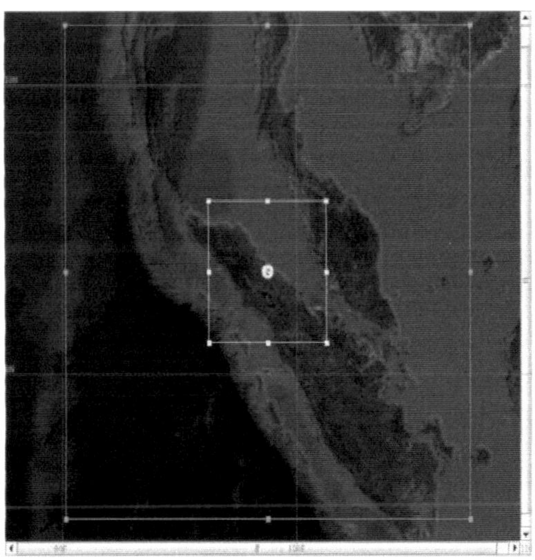

**Figure 2.** Display of domains used in WRF-ARW model

**Table 1.** WRF-ARW configuration used in the research.

| | Domain 1 | Domain 2 |
|---|---|---|
| Type Projection | Mercator | |
| Centre Point Lon | 98.73° | |
| Centre Point Lat | 03.57° | |
| Horizontal Dimension X | 91 | |
| Horizontal Dimension Y | 91 | |
| Grid-Point Distance | 9km | 3km |
| Graphics Data Resolution | 10m | 5m |
| Time Step | 60 | |
| Microphysics Option | WRF Single-Moment 3-class scheme | |
| Cumulus Parameterization Option | Kain-Fritsch scheme | |
| Shortwave Radiation Option | Dudhia scheme | |
| Longwave Radiation Option | RRTM scheme | |
| Boundary-Layer Option | YSU scheme | |

2.3.3. *Verification.* This study compares the selected meteorological parameters of the output of WRF-ARW model and meteorological observation data from Kualanamu Meteorological Station to verify the accuracy of the output of WRF-ARW model.

*2.3.4. Research flowchart*

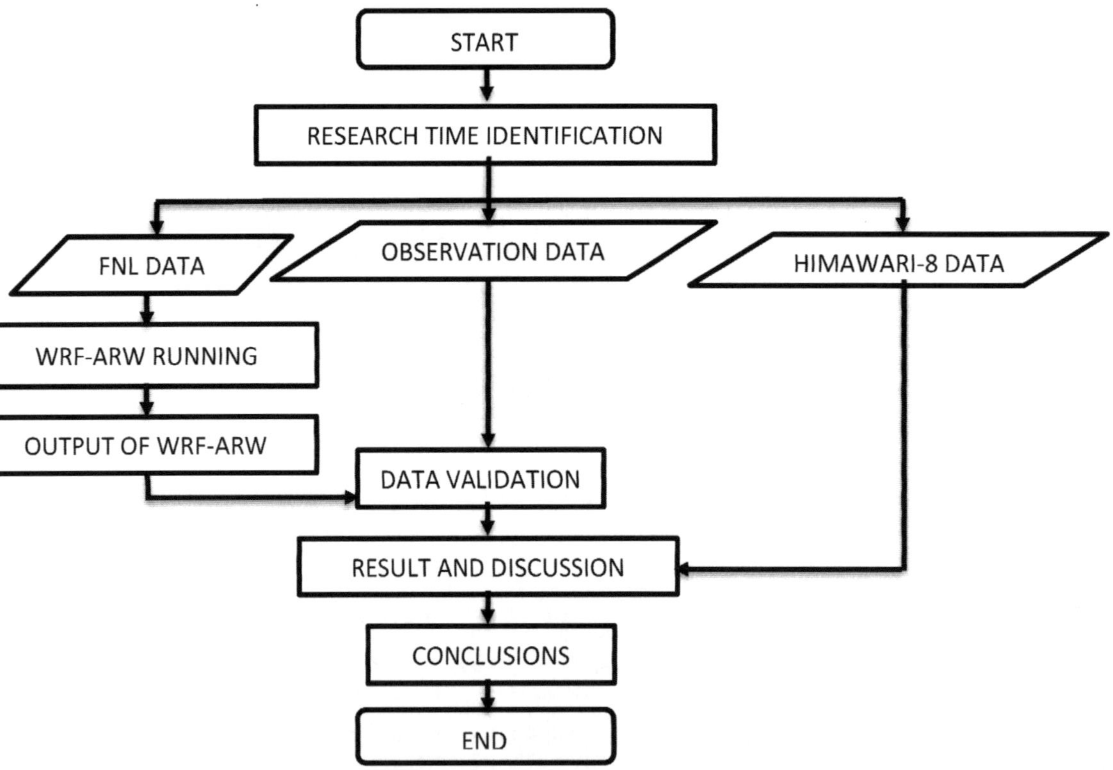


**Figure 3.** Research flowchart

## 3. Result and discussion

### *3.1. Research time identification*

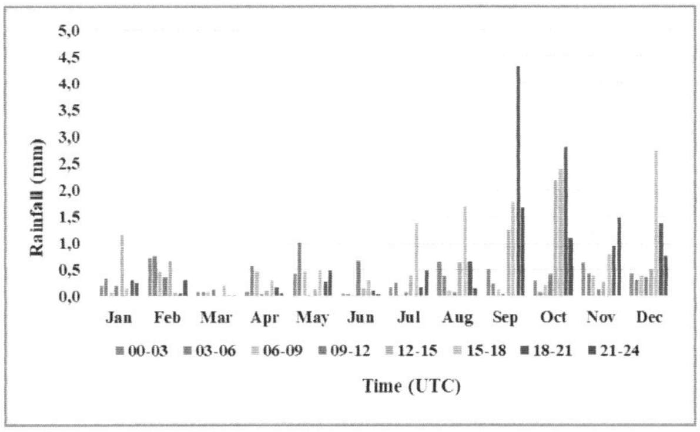


**Figure 4.** Graphic of 3-hour rainfall average in Kualanamu Meteorological Station from 2014 to 2016

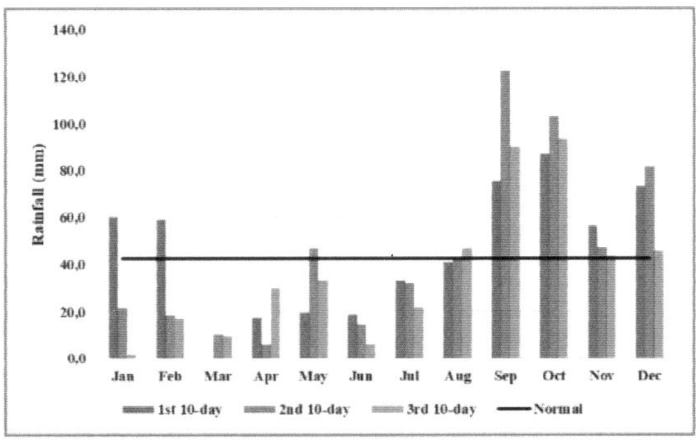


**Figure 5.** Graphic of 10-day rainfall average in Kualanamu Meteorological Station from 2014 to 2016

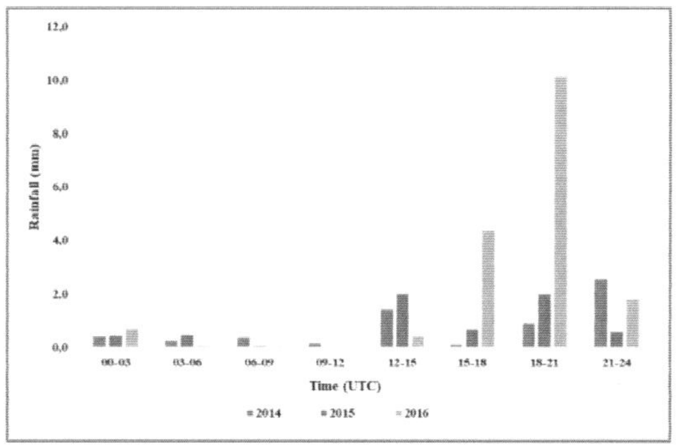


**Figure 6.** Graphic of 3-hour rainfall average in Kualanamu Meteorological Station in September 2016

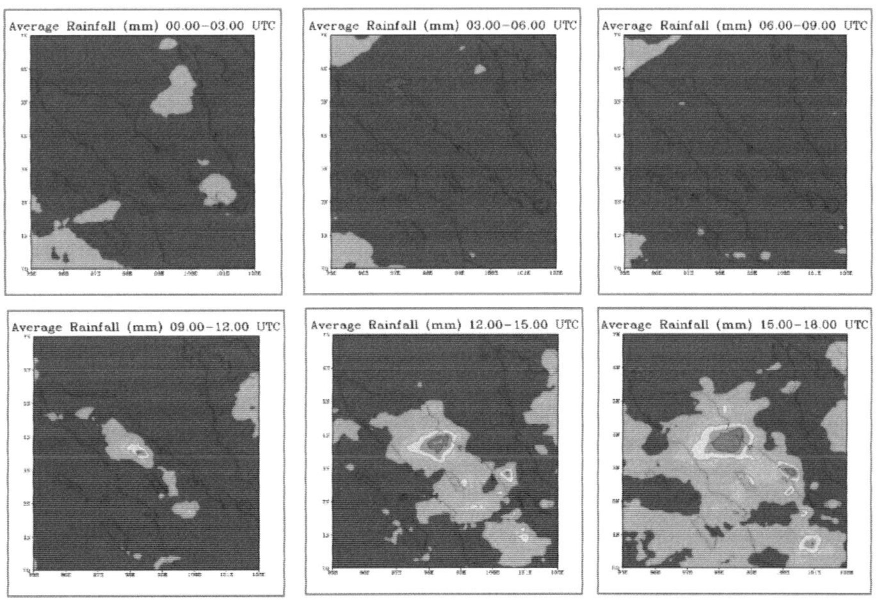

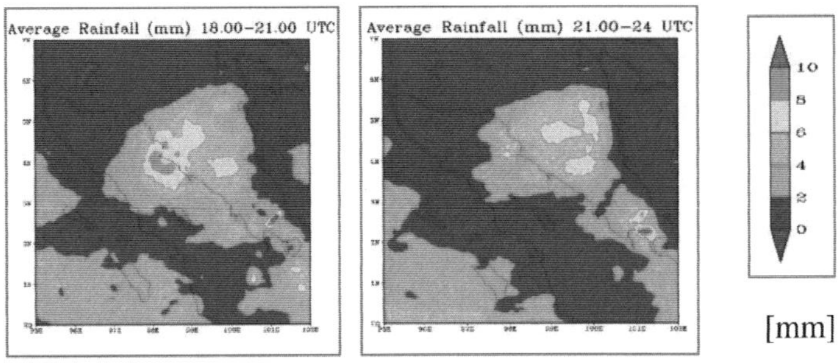


**Figure 7.** Spatial average of GSMaP rainfall in the research area in September 2016

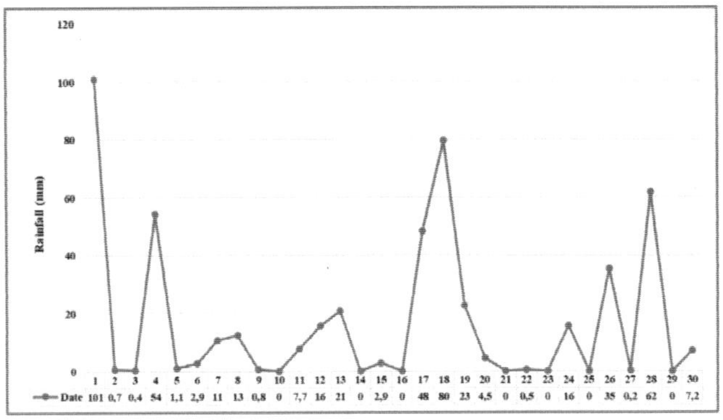


**Figure 8.** Graphic of the daily-rainfall total in Kualanamu Meteorological Station in September 2016

Based on rainfall data from 2014 to 2016 obtained from Kualanamu Meteorological Station can be produced the graphic of 3-hour [Figure 4] and 10-day [Figure 5] rainfall average in Kualanamu Meteorological Station from 2014 to 2016 and also the graphic of 3-hour rainfall average [Figure 6] and the daily-rainfall total [Figure 8] in Kualanamu Meteorological Station in September 2016. Meanwhile, based on GSMaP data can be produced the spatial average of rainfall in the research area (Deli Serdang) during September 2016 [Figure 7]. Figure 4 and 5 indicate that September is the month which has the highest rainfall of the rainfall peak in Kualanamu. Figure 8 represents that 1 September 2016 is the day which has the highest rainfall, about 101mm/day. Meanwhile, Figure 6 and 7 indicate that from 12:00UTC to 24:00UTC there is the significant variation of in the amount of the rainfall in September 2016, while the highest rainfall occurs in the period of 18:00UTC to 21:00UTC. Thus, based on analysis of rainfall data from 2014 to 2016, 1 September 2016 is determined as research time.

### *3.2. Data validation*

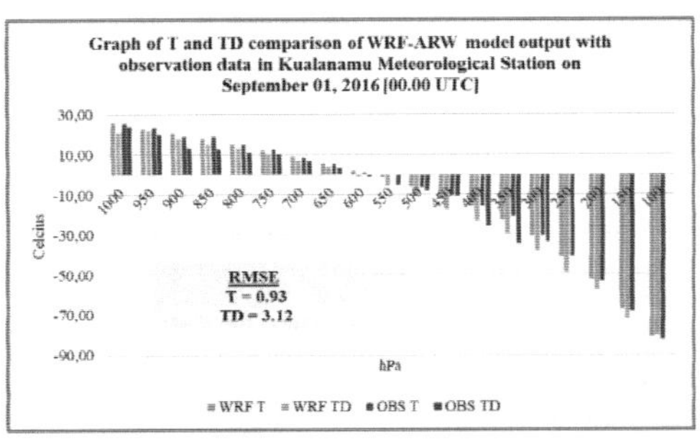


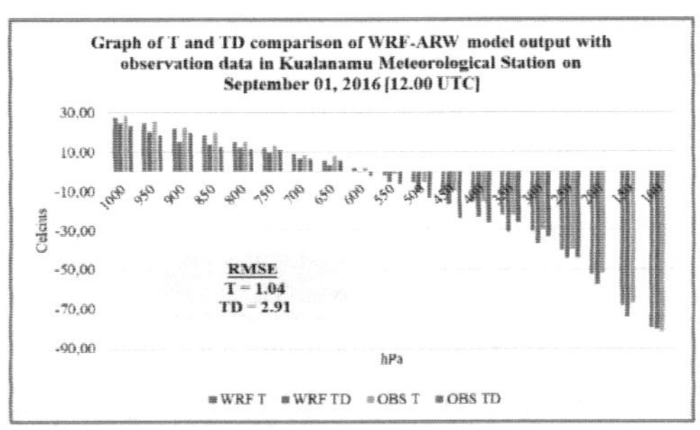

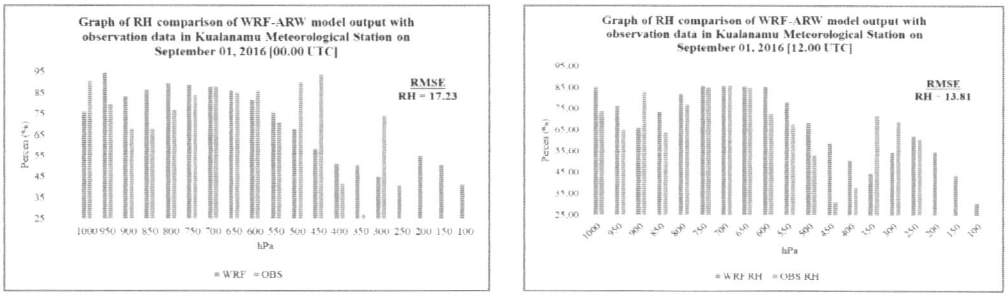


**Figure 9.** Graph of upper air temperature (T), dew point temperature (TD), and relative humidity (RH) comparison of WRF-ARW model output with observation data in Kualanamu Meteorological Station on 1 September 2016 {00:00UTC [Left] and 12:00UTC [Right]}.

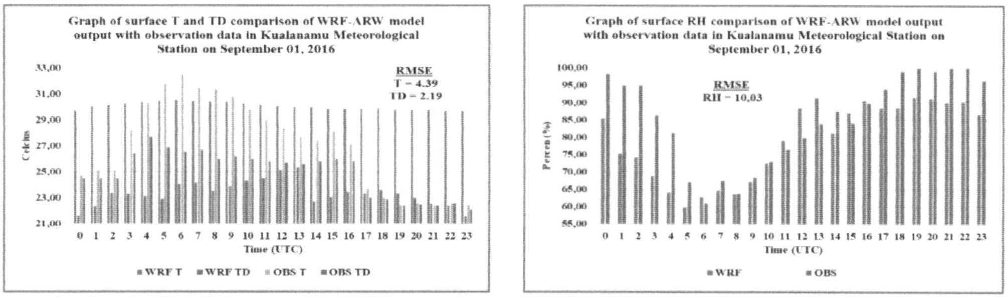


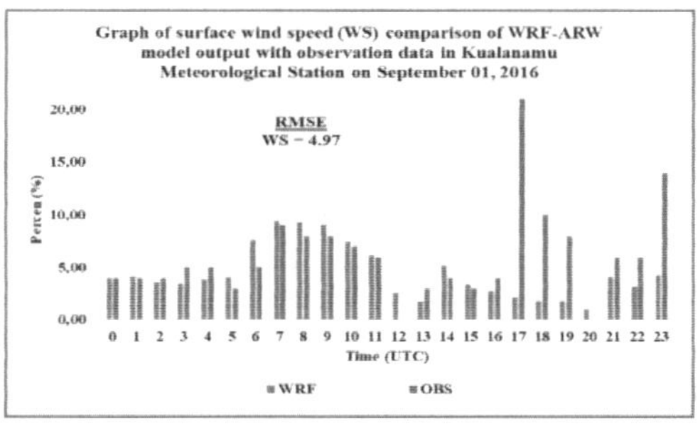


**Figure 10.** Graph of surface air temperature (T), dew point temperature (TD), relative humidity (RH), and wind speed (WS) comparison of WRF-ARW model output with observation data in Kualanamu Meteorological Station on 1 September 2016

In this study, the validation of WRF output data is done by comparing it with the meteorological observation data obtained from Kualanamu Meteorological Station. Based on Figure 9, the RMSE (Root Mean Square Error) value of upper air parameters such as temperature (T) and dew-point temperature (TD) at 00:00UTC and 12:00UTC are 0.93 & 1.04 and 3.12 & 2.91 respectively. For the temperature comparison, these RMSE values categorized well. Meanwhile, the comparison of relative humidity (RH) at 00:00UTC and 12:00UTC are 17.23 and 13.81 respectively, which can be categorized as good enough. Based on Figure 10, the graphs indicate that the RMSE value of surface air parameters such as T, TD, RH, and wind speed (WS) from 00:00UTC to 23:00UTC are 4.39, 2.19, 10.03, and 4.97, which all of these values statistically can be categorized better. Thus, the comparison value of the surface air parameters is better than the upper air parameters. And as the result, it can conclude that the WRF-ARW model used in this study can describe the weather parameters condition in Deli Serdang well.

### *3.3. Analysis of Atmosphere Condition*

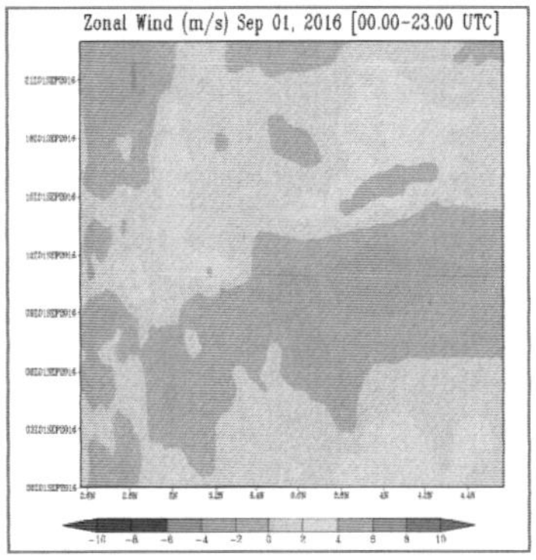


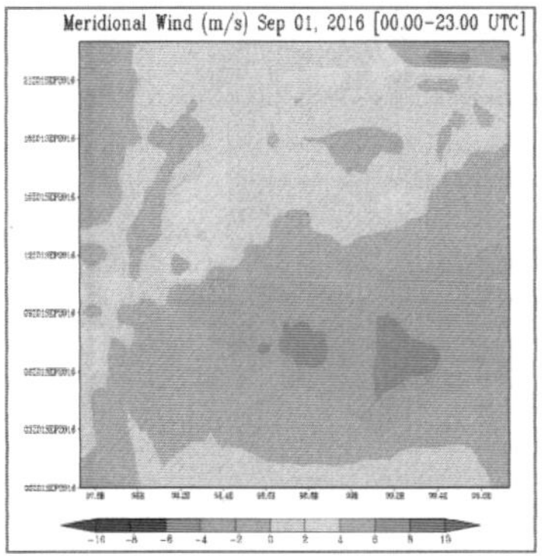


**Figure 11.** Zonal (02.57°N to 04.57°N 98.73°E) [Left] and meridional (03.57°N 97.73°E to 99.73°E) [Right] wind in the research area on 1 September 2016.

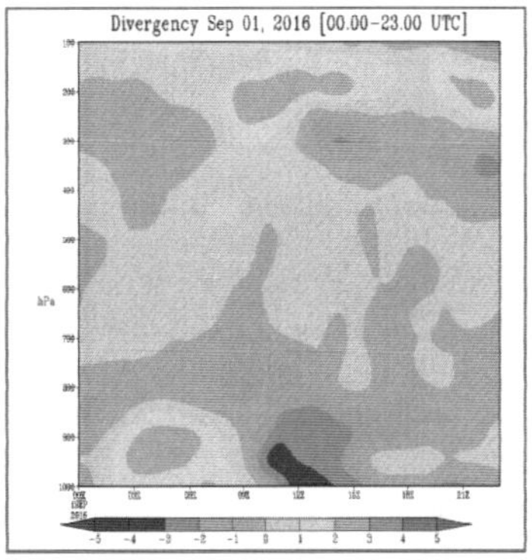


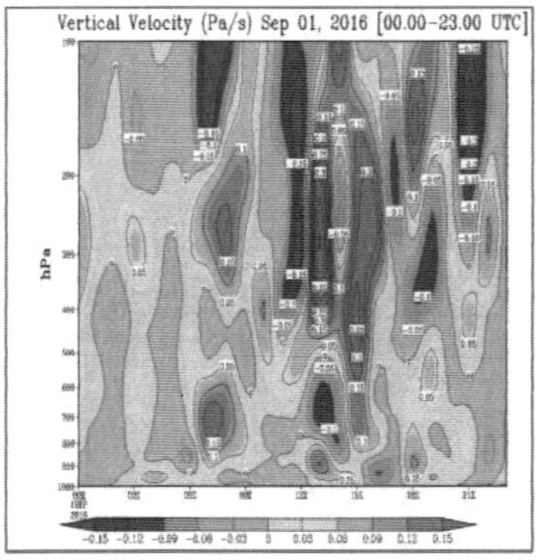


**Figure 12.** The divergence [Left] and the vertical velocity [Right] values in the research area on 1 September 2016.

Figure 12 shows the divergence value [Left] and vertical velocity value [Right] on 1 September 2016 from 00:00UTC to 23:00UTC at 1000hPa to 100hPa. The positive divergence-value indicate the spread of air masses, whereas the negative divergence-value indicate the buildup of air mass or commonly known as convergence. In the left part of Figure 11, the negative and positive divergence patterns appear less regular. Strong negative divergence occurs in the lower layers at the level of 1000hPa to 900hPa while strong positive divergence occurs at 200hPa. The presence of convergence in the lower layers and divergences in the upper layer indicates the occurrence of a strong convective event that is always identical with the less friendly weather condition. The vertical velocity value is related to the direction of vertical wind movement. In the right part of Figure 11 presents the vertical velocity pattern that varies every hour on 1 September 2016. At 00:00UTC to 12:00UTC is not seen the significant vertical velocity value. At 12:00UTC, the positive vertical velocity, about 0.05 to 0.35, indicate the upward vertical air movement (updraft). This condition occurrence until 15:00UTC which indicating any convective activities or the formation of convective clouds (rain clouds) over the area. At 17:00UTC until 21:00UTC, the negative vertical velocity value looks to be dominant, in the range of values from -0.05 to -0.25. It shows the occurrence of vertical downward movement (downdraft) which indicates the occurrence of rain over the area in this time span.

IOP Conf. Series: Earth and Environmental Science **98** (2017) 012003 doi:10.1088/1755-1315/98/1/012003

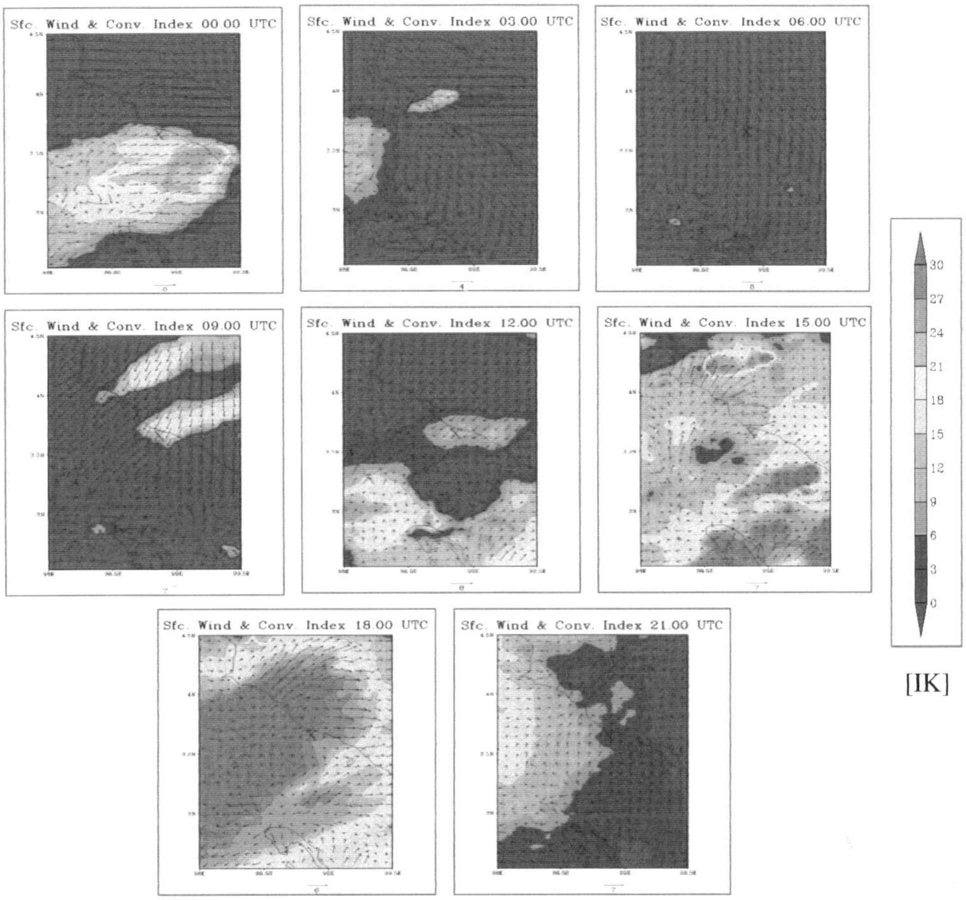


**Figure 13.** Spatial average of GSMaP rainfall in the research area on 1 September 2016

Figure 11 indicates the wind pattern during the rain -peak on 1 September 2016. The condition at 03:00UTC shows that the transition of land breeze currents into the sea breeze has begun. The condition at 06:00UTC shows that the moving of sea breeze is quite fast, about 10-14kt, which entered the West Coast of Sumatra including the region of Deli Serdang. The condition at 09:00UTC shows that the sea breeze is getting louder and finally began to weaken at 12:00UTC. The condition at 12:00UTC shows that there is an encounter-wind in the opposite direction but its speed is relatively low. The condition at 15:00UTC shows the beginning of the rising of the land breeze speed and the weakening of the sea breeze speed. Visible land breeze speeds are increasing and occur until at 24:00UTC. If analysed from its convective activities, from 00:00UTC until before 09:00UTC, there is no significant convective activities even though the sea breeze flows at a relatively strong speed. This has occurred because of the absence of the wind from the opposite direction that blocks the sea breeze to form the convergence area. It is because in September the Eastern wind (Easterly) was occurred in Deli Serdang, so the sea breeze is reinforced by the movement of the wind from the north (Asia). At 12:00UTC, the convective activities began to appear. The land breeze that flows from the south meets the Eastern winds moving from the north to form a convergence area scattered in the area around Deli Serdang. This is increasingly evident at 18:00UTC where in addition to meeting the wind from the opposite direction, the land breeze is influenced by the surface topography and make the wind shear around the Deli Serdang area. At 21:00UTC, it is shown that the Eastern wind is weakening and the land breeze is clearly visible. In general, it is seen that the wind that blows from land to sea on the coast of Deli Serdang began clearly visible from 18:00UTC to 03:00UTC which the speed is about 2-12kt. The wind patterns shown are in accordance with the theory of sea-land breeze circulation where the land breezes usually occur about three hours after sunset and increase in speed until sunrise and still continue to occur in a few hours after sunrise. Meanwhile, the sea breeze appears near the shore a few hours after sunrise and reach the maximum condition when the difference between land and sea temperature reaches the maximum value. However, this is not entirely consistent with Riehl's theory [14] where ideally the direction of the wind circulation of the sea breeze is perpendicular to the shoreline. Besides that, observed the existence of convergence that occur between two of sea breeze and between sea breeze and the wind from other direction blowing from the north (Eastern wind), this is in accordance with the theory of Wirjohamidjojo and Sugarin [21].

## 4. Conclusions

The conclusions of this study are as follows. In general, the occurrence of sea-land breeze over the coast of Deli Serdang can be simulated well by WRF-ARW model with the configuration used in this research. The sea breeze tends to be stronger than the land breeze. The sea breeze speed can reach about 10-16kt while the speed of the land breeze only about 6-10kt. The sea-land breeze that occurs on the coast of Deli Serdang can be the convergence form in the transition time or when the encountering-wind from another direction and it will support the convection activity at Deli Serdang. This is evidenced by the high value of the convective index in the convergence area. The recommendation of this study is the research time needs to be done in other months and not only in the heavy rain occurrence, so it will be known the variation of the influence of the sea-land breeze circulation on the convective activities on the coast of Deli Serdang.

## References

[1] Antonelli M and Rotunno R 2007 Large-eddy simulation of the onset of the Sea breeze *Journal of the Atmospheric Sciences* **64**(12) pp 4445-4457.

[2] Azorin-Molina C, Tijm S, E E Ebert, S M Vicente-Serrano and M J Estrela 2014 Sea breeze thunderstorms in the eastern Iberian Peninsula. Neighborhood verification of HIRLAM and HARMONIE precipitation forecasts *Atmospheric research* **139** pp 101-115.

[3] Challa V S, J Indracanti, M K Rabarison, C Patrick, J M Baham, J Young ... and A Yerramilli 2009 A simulation study of mesoscale coastal circulations in Mississippi Gulf coast *Atmospheric Research* **91**(1) pp 9-25.

[4] Chen F, S Miao, M Tewari, J W Bao and H Kusaka 2011 A numerical study of interactions between surface forcing and Sea breeze circulations and their effects on stagnation in the greater Houston area *Journal of Geophysical Research: Atmospheres* **116**(D12).

[5] Dewita A, A S A Bakar and K Dwicahyo 2015 Pemanfaatan WRF-ARW untuk simulasi potensi angin sebagai sumber energi di Teluk Bone *Jurnal Material dan Energi Indonesia* **5**(02).

[6] Firth R, J Kala, T J Lyons and J Andrys 2017 An analysis of regional climate simulations for Western Australia's wine regions-Model evaluation and future climate projections *Journal of Applied Meteorology and Climatology*.

[7] Fitriyawita M and M Saepudin 2016 *Kajian interaksi aktivitas konvektif angin darat dan angin laut menggunakan WRF-ARW di Kalimantan Barat.*

[8] Hadi Tri Wahyu 2011 *Pelatihan Model WRF* (Bandung: Institut Teknologi Bandung).

[9] Hahmann A N, C LVincent, A Peña, J Lange and C B Hasager 2015 Wind climate estimation using WRF model output: Method and model sensitivities over the sea *International Journal of Climatology* **35**(12) pp 3422-3439.

[10] Mori S, H Jun-Ichi, Y I Tauhid, M D Yamanaka, N Okamoto, F Murata ... and T Sribimawati T 2004 Diurnal land–sea rainfall peak migration over Sumatera Island, Indonesian Maritime Continent, observed by TRMM satellite and intensive rawinsonde soundings *Monthly Weather Review* **132**(8) pp 2021-2039.

[11] Molina Cesar Azorin 2014 Sea breeze Thunderstorm in The Eastern Iberian Peninsula, Neighborhood Verification of HIRLAM and HARMONIE Precipitation Forecasts *Atmospheric Research* **139** pp 101-115.
[12] Nuryanto D E 2011 Analisis pseudo-vektor pada aktivitas konvektif benua maritim Indonesia *Jurnal Meteorologi dan Geofisika* **12**(2).
[13] Papanastasiou D K, D Melas and I Lissaridis 2010 Study of wind field under Sea breeze conditions; an application of WRF model *Atmospheric Research* **98**(1) pp 102-117.
[14] Riehl Herbert 1954 *Tropical Meteorology* (New York: Mc. Graw-Hill Book Company Inc).
[15] Sakurai N, F Murata, M D Yamanaka, S Mori, J I Hamada, H Hashiguchi ... and B Suhardi 2005 Diurnal cycle of cloud system migration over Sumatera Island *Journal of the Meteorological Society of Japan. Ser. II* **83**(5) pp 835-850.
[16] Steele C J, S R Dorling, R Von Glasow and J Bacon 2013 Idealized WRF model sensitivity simulations of Sea breeze types and their effects on offshore windfields *Atmospheric Chemistry and Physics* **13**(1) p 443.
[17] Subarna D 2010 Simulasi cuaca daerah Padang *Berita Dirgantara* **9**(3).
[18] Trilaksono N J 2012 *Studi pola konveksi diurnal di Pulau Sumatra menggunakan model Weather Research and Forecasting-Advanced Research WRF (WRF-ARW).*
[19] Tjasyono H K, Bayong and Woro B Sri 2014 *Atmosfer Ekuatorial* (Jakarta: Badan Meteorologi Klimaatologi dan Geofisika).
[20] Winarso Paulus A 2011 *Analisa Cuaca I* (Jakarta: Akademi Meteorologi dan Geofisika).
[21] Wirjohamidjojo, Soeryadi and Sugarin 2008 *Praktek Meteorologi Kelutan* (Jakarta: Badan Meteorologi dan Geofisika).
[22] Yulihastin E 2012 Mekanisme hujan harian di Sumatera *Berita Dirgantara* **13**(3).
[23] https://www.google.co.id/maps/place/Bandara+Int.+Kualanamu/ consulted on 1 September 2017

# Re-assessing Rainwater Harvesting Volume by CHIRPS Satellite in Semarang Settlement Area

**Yosef Prihanto[1*], Raldi H Koestoer[2], Dwita Sutjiningsih[3]**

[1] School of Environmental Science, Universitas Indonesia, Jl. Salemba Raya 4 Jakarta Pusat 10430, putranusa212@yahoo.com
[2] School of Environmental Science, Universitas Indonesia, Jl. Salemba Raya 4 Jakarta Pusat 10430, ralkoest@yahoo.co.uk
[3] Department of Civil Engineering, Faculty of Engineering, Universitas Indonesia, Depok 16424, dwita@eng.ui.ac.id

putranusa212@yahoo.com

**Abstract.** Semarang City is one of the most influential coastal cities in Java Island. The city is facing increasingly-high water demand due to its development and water problems due to climate change. The spatial physiography and landscape of Semarang City are also exposed the city to water security problem. Hence, rainwater harvesting treatment is an urgent effort to meet the city's water needs. However, planning, implementation and management of rainwater harvesting are highly depended on multitemporal rainfall data. It has not yet been fully compiled due to limited rain stations. This study aims to examine the extent to which CHIRPS satellite data can be utilized in estimating volume of rainwater harvesting 16 sub-districts in Semarang and determine the water security status. This study uses descriptive statistical method based on spatial analyses. Such method was developed through spatial modeling for rainfall using isohyetal model. The parameters used are rainfall, residential rooftop area, administrative area, population, physiographic and altitude units. Validation is carried out by using monthly 10 rain stations data. The results show level of validity by utilizing CHIRPS Satellite data and mapping rainfall distribution. This study also produces a potential map of distribution rainfall volume that can be harvested in 16 sub-districts of Semarang.

**Keywords:** CHIRPS, Rainwater Harvesting, Isohyetal, Urban, Spatial

## 1. Introduction

City of Semarang is one of the most influential coastal cities in Java Island. The city is facing increasingly high-water demand due to its vast development and also water problems due to climate change. The spatial physiography and landscape of Semarang City shown in this paper shows that the city have water security problem. Therefore, rainwater harvesting treatment is suggested as an urgent effort to meet the city's water needs.

UN-Water has formulated an understanding in water security. Water security is defined as capacity of the population to safeguard sustainable access to adequate quantities of acceptable quality of water that can be used to sustain livelihoods, human well-being, and socio-economic development, as well as for ensuring protection against water-borne pollution and water-related disasters, and for preserving ecosystems in a climate of peace and political stability [1]. Basically, water security problem is very

IOP Conf. Series: Earth and Environmental Science **98** (2017) 012004 doi:10.1088/1755-1315/98/1/012004

complex. It not only deals with physical and technological aspects in meeting the water needs, but it also concerns the impact on the environment and social, political, and economic conditions on a group of people, often associated with other community groups. Conflict of interest between groups of people also affects social and political stability that determine the water security itself. Semarang now faces water security problem. One of the answers for this problem is to apply rainwater harvesting.

Many researches and studies on rainwater harvesting have been done in various places in the world [2] [3] [4] [5] [6] [7] [8] [9]. Each of them concluded in benefits and advantages of this method. The studies undertaken include developing methods to choose suitable sites for application of rainwater harvesting [10], calculating potential volumes [2] [3] [4] [6], planning water catchment sizes [11], assessing feasibility and economic efficiency [9] [12], management aspects [7] [9] [13] even aspects of the sustainability of this method [14]. Utilization of remote sensing and geographic information system (GIS) technologies is also widely used in the development of method for identification location and the potential measurement of rainwater harvesting applications [8] [10] [15].

The use of remote sensing in various rainwater-harvesting studies is commonly used for the identification of land use, roof area, flow density and other physical aspects [8] [10] [15]. The use of GIS plays a larger role in modeling and calculating the potential volume of rainwater harvesting. Generally modeling and volume calculations are performed on relatively limited study areas and use only average rainfall values in micro coverage [8] [10] [15]. Whereas planning, implementation and management of rainwater harvesting highly depend on multitemporal rainfall data. In the case of Semarang City, this data has not been fully compiled due to limited rain stations. Based on that condition, this study utilizes CHIRPS satellite data to estimate the volume of rainwater harvesting in Semarang City's 16 sub-districts and then determine water security status for the city.

## 2. Method

### *2.1. Location of the study*

The study was conducted in the City of Semarang. It covers 16 sub-districts and 177 urban villages (*kelurahan*). Semarang is a city on the north coast of Java Island that has unique spatial physiography and landscape. The description of the location and administrative boundaries of Semarang city is presented in Figure 1. The city can be use to represent the condition in many big cities in Indonesia whereas there are continues growth and followed by challenges in water provision.

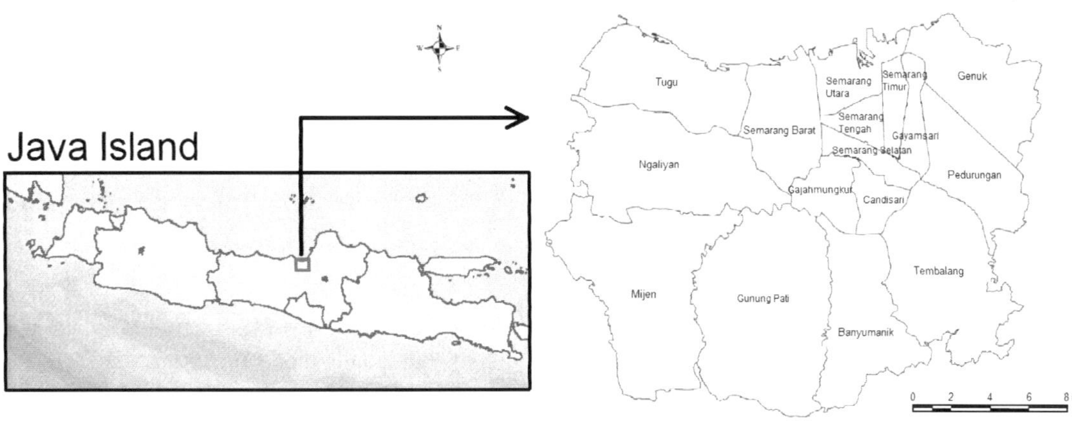


**Figure 1.** Location of study.

### *2.2. Secondary data*

This research uses land-use data obtained from BIG (Geospatial Information Agency). The data is the result of visual interpretation of high-resolution satellite image of recording in 2012. Data of interpretation result have been validated through field test survey in 2013 and 2016. Land-use unit as observation priority is settlement and non-settlement. Population data uses data published by BPS (Central Bureau of Statistics) Semarang City in 2012 and 2015. Population data used are the population and the number of head of household. Average monthly rainfall data from 10 weather stations in Semarang uses data from 2012. CHIRPS dataset of Semarang City and surroundings using recording data for 10 years (2007-2016).

Climate Hazards Group InfraRed Precipitation with Stations (CHIRPS) is the name of a rain satellite developed for global interests. CHIRPS developed by U.S. Geological Survey (USGS) and Climate Hazards Group at University of California, Santa Barbara. Quasi-global gridded products are available from 1981 to near present at 0.05° spatial resolution (~5.3 km). New products are released approximately mid-month of the month following the observation [16].

### *2.3. Method of determining the virtual weather station point*

The isohyetal method in this study was established by determining virtual weather station points. These points are considered to be the location of weather stations whose rainfall data values will be taken into account in model calculations. The nets of those points are set uniformly distributed and covering up to areas outside administrative boundaries of City of Semarang. This study calculates that 35 spatial spots that have a distance of 3 minutes north south and 3 minutes east west can represent study area. The upper left point starts at the coordinates $110^0$13'30'E and $6^0$55'30'S while the bottom right point at the coordinates $110^0$31'30'E and $7^0$7'30'S. Establishing nets with wider coverage than study area has aim to obtain a better spatial distribution modeling result.

### *2.4. Calculation of the average monthly rainfall value*

Monthly rainfall calculation is obtained monthly based on monthly CHIRPS recording data of 10 years, 2007-2016. The monthly rainfall rate in one year at each virtual weather station is tabulated for 10 years. Monthly rainfall figures in the 10-year range tabulated based on the virtual weather station points are then sequenced on the basis of the value, from the lowest to the highest value. The value of the $8^{th}$ order or the value that is considered to have an 80% probability, is the value chosen as average value of rainfall for corresponding virtual weather station point. By the result of this processing then can be found the average monthly rainfall value for each virtual weather station point.

### *2.5. Preparation of monthly rainfall maps by isohyetal method using GIS software*

The result of calculation of the average monthly rainfall value then becomes input data to make isohyetal map using GIS software. Monthly average rainfall data processing from 35 virtual weather station points using the ArcGIS 10.3 application. The module facility used is Spatial Analysis Tools that is Interpolation with Spline method option. The Spline tool uses an interpolation method that estimates the value of a mathematical function that minimizes overall survival curvature, resulting in a smooth surface that passes exactly through the input points. The classification of interpolation results by using 7 rainfall classes each then has a difference of 100 mm wide range. This isohyetal modeling is simulated for every month of the year.

### *2.6. Model validation*

Model validation of the utilization of CHIRPS recording data uses monthly average rainfall data from 10 stations in Semarang city in 2012 record. This validation is only monthly recaptured from one period January to December on the basis of 10 years record. This validation is trying to see and to know how far value deviations occur when this modeling method is used to utilize CHIRPS data.

### *2.7. Estimation the addition area rooftop of the settlement*

Estimated addition of roof area of settlement is done by approach of population, average number of family member, number of head household and average of roof area every head count for family. This estimate is conducted due to unavailability of spatial data of the latest 2017 residential roof area. The calculation of the 2017 roof area is based on data of the roof data in 2012, the population in 2012 data, and the number of head household in 2012 and 2015. The calculation of estimated population of Semarang City in 2017 is based on population data of each urban village. Growth rates are calculated based on the average population growth of each urban village in the last 10 years of observation (2006-2015). Calculation of 2017 population in 2017 can be estimated by using equation (1):

$$P_n = P_o\,(1+r)^n \quad (1)$$

Remarks:
$P_n$ = number of population in year n
$P_o$ = number of population in year 0 or base year
n = number of years between 0 and n
r = population growth rate per year (in %)

*2.8. Estimation of rainwater that can be harvested*
Estimation of rainwater volume that can be harvested is done by multiplying monthly average rainfall of the modeling results (P, in mm/month), the width of the roof or the catchment area (A, in $m^2$), and the runoff coefficient (RC, nondimensional) [17] as shown in equation (2). Roof area is the result of the estimation of the roof area in each urban village area in 2017. Rainfall value is the value of the average rainfall of the proportion of isohyetal calculations and the distribution based on the width of the urban village. The runoff coefficient is set to 0.7 with the consideration that there is no detailed information about the type of roof and its slope. Another consideration in determining the value of runoff coefficient is the average type of residential roof in the study area is still dominated by roof tile [17]. The calculation of the population's water needs, using the reference set by the Ministry of Public Works Indonesian, 120 liters/day/person.

$$\text{RWH potential} = P.A.RC \quad (2)$$

## 3. Result and Discussion
This section presents results of studies and discussions on results obtained. It consists of 6 sections, and each step of study activity will be described.

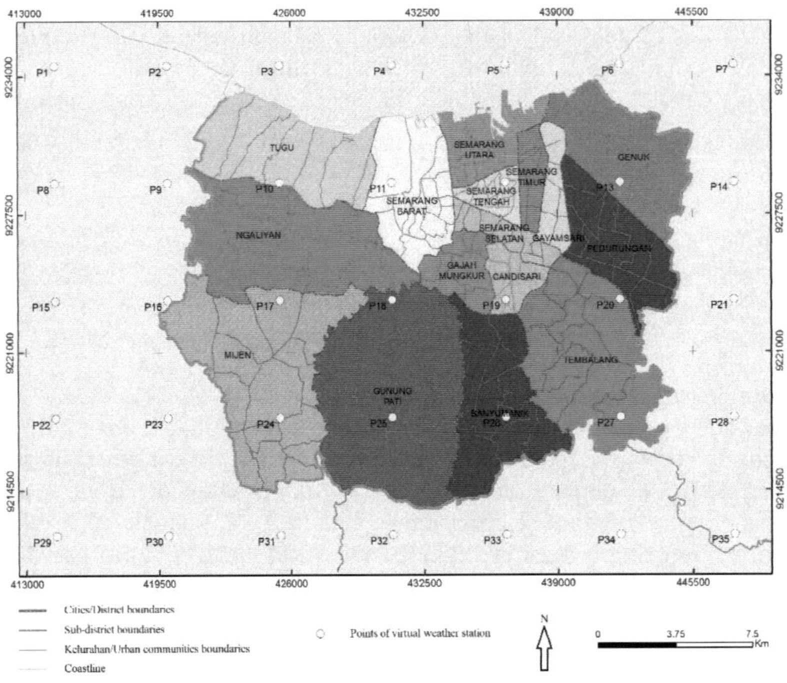


**Figure 2.** Virtual weather station points.

### *3.1. Determining virtual weather station point*

The result of the determination of 35 virtual weather station points (P1-P35) visually distributed is shown in Figure 2. The virtual weather station points form a 7x5 matrix and spread evenly across the entire study area. The coordinates of each virtual weather station point of reference are to record the monthly precipitation value of CHIRPS data within the 10-year recording range. The rainfall number is used as input to the next processing.

### *3.2. Calculation monthly rainfall and monthly rainfall maps by isohyetal method*

The result of rainfall data processing obtained from CHIRPS dataset is further processed using Spatial Analysis Tools facility module that is Interpolation with Spline method option. This processing produces an isohyetal map as illustrated in Figure 3. It only presents visually the results of isohyetal modeling for 6 months only. In this study isohyetal modeling is still done for 12 months in one year. Distribution and differences in rainfall over time can be seen in Figure 3. In July-September it can be observed that the study area's rainfall tends to be at a low value. In December-January actually shows the opposite condition, namely the value of rainfall area tend to be high.

The rainfall distribution sample illustrated in Figure 3 shows the potential volume of rainwater that can be harvested. The quantity is varied in volume each month. Differences of volume each month must be taken into consideration in rain harvest management in Semarang City.

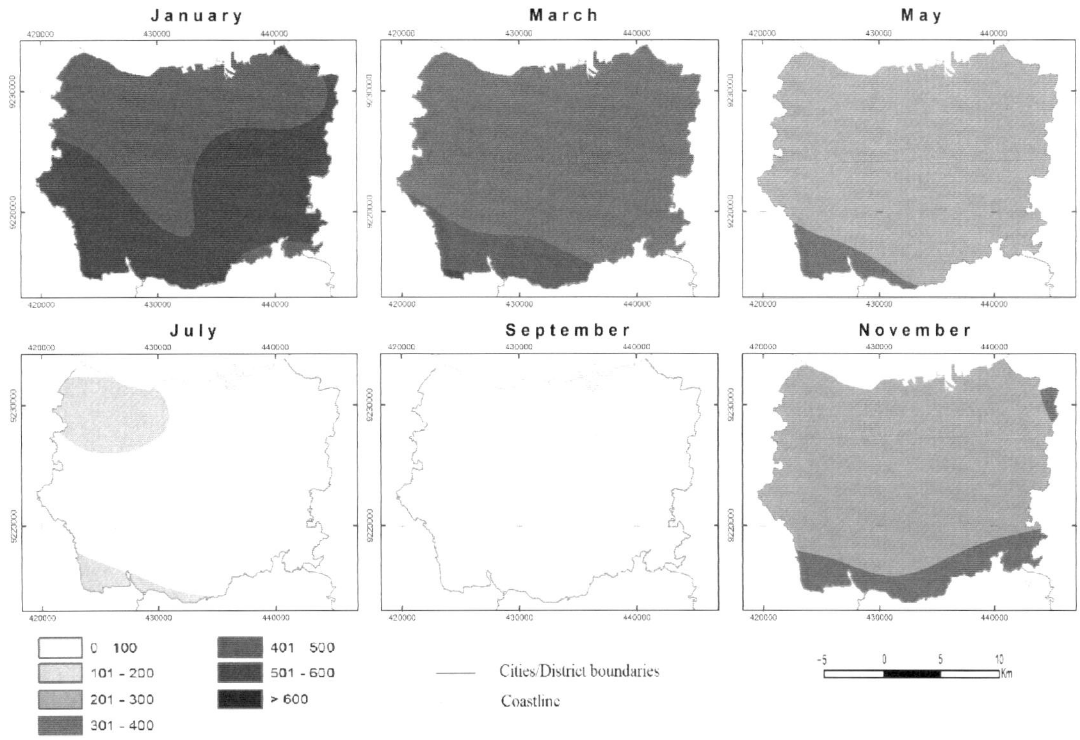


**Figure 3.** Monthly rainfall maps by isohyetal method.

### *3.3. Validation model*

The result of modeling that has been done on CHIRPS data is then validated. Validation is done by comparing precipitation values obtained from the modeling at the existing weather station points. The spatial distribution of the weather station location referred to in Figure 4. In terms of distribution and quantity, these weather stations can be rated very limited.

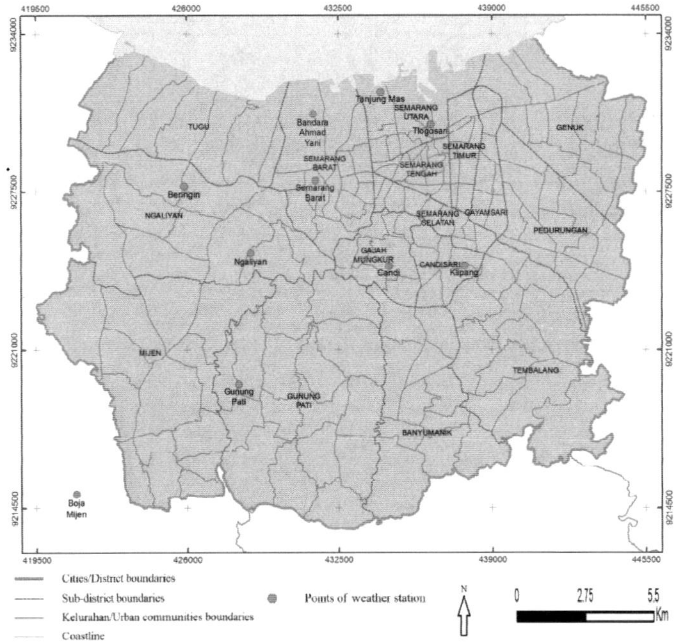


**Figure 4.** Distribution of 10 weather stations in City of Semarang.

Monthly average rainfall data from each weather station is shown in Table 1. Based on this table, it can be seen that wet months occur in January, February, March, October, November, and December. While dry months occur in April, May, June, July, August, and September. Very dry months occur in July, August, and September. Very wet months occur in January and December. Based on the table observation can also be seen some stations have record values of very high rainfall in wet season are station of Gunung Pati and Boja Mijen.

**Table 1.** Average monthly rainfall value weather stations in Semarang area.

| No. | Station | January | February | March | April | May | June | July | August | September | October | November | December | Average |
|---|---|---|---|---|---|---|---|---|---|---|---|---|---|---|
| | | | | | | | (unit in mm) | | | | | | | |
| 1 | Ahmad Yani | 490 | 340 | 190 | 176 | 68 | 78 | 1 | 0 | 4 | 214 | 255 | 318 | 178 |
| 2 | Tanjung Mas | 440 | 308 | 266 | 109 | 173 | 63 | 1 | 0 | 29 | 187 | 256 | 283 | 176 |
| 3 | Tlogosari | 406 | 401 | 318 | 221 | 85 | 116 | 2 | 0 | 13 | 258 | 444 | 191 | 205 |
| 4 | Semarang Barat | 498 | 270 | 220 | 170 | 90 | 110 | 2 | 0 | 3 | 271 | 256 | 327 | 185 |
| 5 | Beringin | 595 | 321 | 245 | 214 | 133 | 98 | 0 | 0 | 0 | 268 | 161 | 334 | 197 |
| 6 | Ngaliyan | 583 | 383 | 256 | 104 | 116 | 95 | 0 | 0 | 0 | 226 | 198 | 458 | 202 |
| 7 | Candi | 460 | 292 | 221 | 193 | 112 | 79 | 2 | 0 | 0 | 202 | 295 | 342 | 183 |
| 8 | Klipang | 313 | 232 | 293 | 187 | 129 | 112 | 20 | 0 | 1 | 75 | 266 | 370 | 167 |
| 9 | Gunung Pati | 682 | 479 | 334 | 285 | 221 | 161 | 0 | 0 | 0 | 218 | 447 | 589 | 285 |
| 10 | Boja Mijen | 1.698 | 1.110 | 691 | 616 | 154 | 343 | 0 | 0 | 31 | 89 | 321 | 555 | 467 |

Source: weather station data.

The mean rainfall value of CHIRPS data modeling in reference to station reference coordinate point is presented in Table 2. In general, distribution of the pattern is similar to the average monthly

IOP Conf. Series: Earth and Environmental Science **98** (2017) 012004 doi:10.1088/1755-1315/98/1/012004

rainfall value distribution pattern with the reference of weather stations. What is apparent in Table 2 is the average value of monthly rainfall in the dry month that tends to have a higher value than the average monthly reference rainfall value. While in the group of average values of monthly rainfall in wet months, generally the model results show lower value than the average monthly reference rainfall value.

**Table 2.** Average monthly rainfall value from isohyetal model of weather stations in Semarang area.

| No. | Station | January | February | March | April | May | June | July | August | September | October | November | December | Average |
|---|---|---|---|---|---|---|---|---|---|---|---|---|---|---|
| | | (unit in mm) | | | | | | | | | | | | |
| 1 | Ahmad Yani | 442 | 283 | 278 | 160 | 85 | 74 | 14 | 23 | 32 | 132 | 272 | 352 | 179 |
| 2 | Tanjung Mas | 462 | 288 | 286 | 158 | 93 | 72 | 13 | 21 | 30 | 138 | 276 | 347 | 182 |
| 3 | Tlogosari | 462 | 288 | 286 | 158 | 93 | 72 | 13 | 21 | 30 | 138 | 276 | 347 | 182 |
| 4 | Semarang Barat | 442 | 283 | 278 | 160 | 85 | 74 | 14 | 23 | 32 | 132 | 272 | 352 | 179 |
| 5 | Beringin | 452 | 284 | 287 | 170 | 91 | 75 | 15 | 24 | 30 | 143 | 278 | 353 | 184 |
| 6 | Ngaliyan | 407 | 291 | 258 | 161 | 95 | 77 | 12 | 22 | 31 | 129 | 230 | 314 | 169 |
| 7 | Candi | 443 | 292 | 280 | 167 | 94 | 80 | 13 | 21 | 31 | 143 | 242 | 338 | 179 |
| 8 | Klipang | 443 | 292 | 280 | 167 | 94 | 80 | 13 | 21 | 31 | 143 | 242 | 338 | 179 |
| 9 | Gunung Pati | 438 | 328 | 294 | 198 | 107 | 84 | 12 | 24 | 30 | 141 | 276 | 372 | 192 |
| 10 | Boja Mijen | 546 | 414 | 412 | 269 | 158 | 104 | 21 | 28 | 31 | 178 | 385 | 509 | 255 |

Source: CHIRPS data processing.

**Table 3.** Value of difference between reference rainfalls compared to model.

| No | Station | January | February | March | April | May | June | July | August | September | October | November | December | Average |
|---|---|---|---|---|---|---|---|---|---|---|---|---|---|---|
| | | (unit in mm) | | | | | | | | | | | | |
| 1 | Ahmad Yani | -48 | -57 | 88 | -16 | 17 | -4 | 13 | 23 | 28 | -82 | 17 | 34 | 1 |
| 2 | Tanjung Mas | 22 | -20 | 20 | 49 | -80 | 9 | 12 | 21 | 1 | -49 | 20 | 64 | 6 |
| 3 | Tlogosari | 56 | -113 | -32 | -63 | 8 | -44 | 11 | 21 | 17 | -120 | -168 | 156 | -23 |
| 4 | Semarang Barat | -56 | 13 | 58 | -10 | -5 | -36 | 12 | 23 | 29 | -139 | 16 | 25 | -6 |
| 5 | Beringin | -143 | -37 | 42 | -44 | -42 | -23 | 15 | 24 | 30 | -125 | 117 | 19 | -13 |
| 6 | Ngaliyan | -176 | -92 | 2 | 57 | -21 | -18 | 12 | 22 | 31 | -97 | 32 | -144 | -33 |
| 7 | Candi | -17 | 0 | 59 | -26 | -18 | 1 | 11 | 21 | 31 | -59 | -53 | -4 | -4 |
| 8 | Klipang | 130 | 60 | -13 | -20 | -35 | -32 | -7 | 21 | 30 | 68 | -24 | -32 | 12 |
| 9 | Gunung Pati | -244 | -151 | -40 | -87 | -114 | -77 | 12 | 24 | 30 | -77 | -171 | -217 | -93 |
| 10 | Boja Mijen | -1152 | -696 | -279 | -347 | 4 | -239 | 21 | 28 | 0 | 89 | 64 | -46 | -212 |

Source: calculation processing of weather station data and CHIRPS data.

The results in Tables 1 and 2 subsequently become the basis to calculate the value of difference between the average monthly rainfall values of the model-making product with the average monthly reference rainfall value. Based on the calculation data presented in Table 3, there is a difference in yield for wet and dry months. For wet months, the precipitation value of the modeling results has a lower value trend than the reference data. For dry months the value of the rainfall model tends to be higher than the reference value. In Table 3, a positive number indicates the higher excess value while the lower one is denoted by a negative value.

The result from calculating the difference in value of precipitation further becomes a reference to calculate value of error or deviation. The calculation result of deviation can be seen in Figure 5 and Table 4. The calculation result shows that the validation value of the model for each weather station varies in value. In general, the number of deviations on 8 stations is below 30%, but 2 stations have deviations of more than 30%, namely the station of Mount Pati and Boja Mijen. These two stations are located at an altitude above 200 m above sea level. Based on the direction of deviation, generally the modeling data is below the reference value; this is indicated by a negative error value.

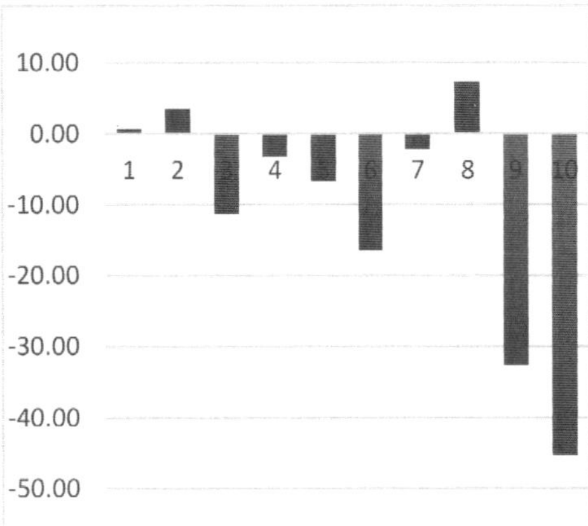


**Figure 5.** Percentage distribution of monthly rainfall error in weather station.

**Table 4.** Percentage monthly rainfall error in weather station.

| No | Weather station | Error (%) |
|---|---|---|
| 1 | Ahmad Yani | 0.56 |
| 2 | Tanjung Mas | 3.41 |
| 3 | Tlogosari | -11.22 |
| 4 | Semarang Barat | -3.24 |
| 5 | Beringin | -6.60 |
| 6 | Ngaliyan | -16.34 |
| 7 | Candi | -2.19 |
| 8 | Klipang | 7.19 |
| 9 | Gunung Pati | -32.63 |
| 10 | Boja Mijen | -45.40 |

*3.4. Land-use of Semarang City and distribution roof top area of the settlement*

The land-use classification used in this study is reclassified and simplified into 9 classes of land-use. The land-use unit used is; regular settlements, irregular settlements, public and social facilities, industrial estates, trade and service areas, agricultural cultivation areas, green areas, water bodies, others/unidentified. The results of land use GIS processing are shown in Figure 7. Figure 7 shows the general distribution of settlements that follow pattern of landscape transportation and physiography of City of Semarang. Figure 6 presents height surface and physiographic of Semarang City.

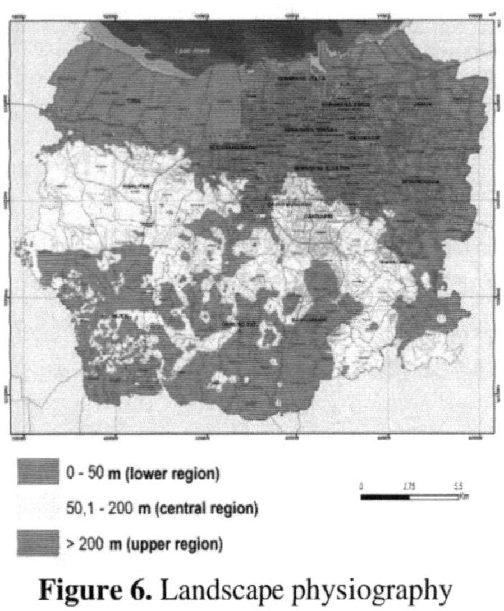


**Figure 6.** Landscape physiography of Semarang City.

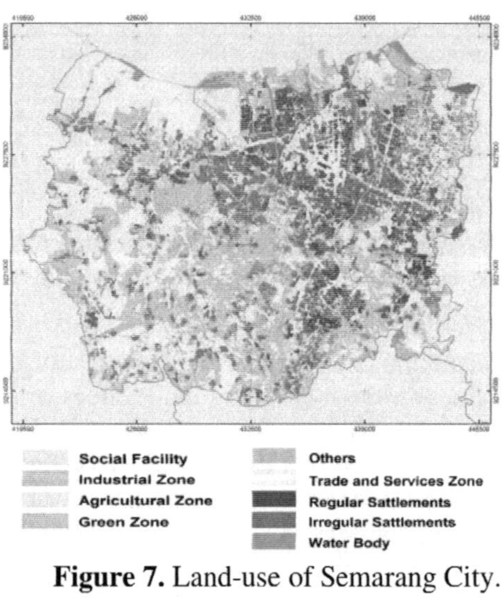


**Figure 7.** Land-use of Semarang City.

### *3.5. Addition area rooftop of the settlement*

The results of population data processing are presented in Table 5 and Table 6. Those tables show estimated population in 2017. Based on that estimation, it can be calculated the number of heads of households by 2017. The number of heads of households is used to estimate the roof area in 2017. The estimation of existing residential roof area in City of Semarang up to 2007 is estimated to reach 3,543.37 ha.

**Table 5.** Distribution of population, number of head household and the projected.

| Sub-district (*Kecamatan*) | Population | | Projected Population | Number of Head household (Hh) | | Average Family members | | Projected number of Head household |
|---|---|---|---|---|---|---|---|---|
| | 2012 (person) | 2015 (person) | 2017 (t=2) | 2012 (Hh) | 2015 (Hh) | 2012 (person) | 2015 (person) | 2017 (Hh) |
| Mijen | 56,570 | 61,405 | 65,715 | 16,669 | 18,916 | 3 | 3 | 20,244 |
| Gunung Pati | 75,027 | 78,641 | 82,461 | 21,501 | 22,760 | 3 | 3 | 23,866 |
| Banyumanik | 128,225 | 132,508 | 137,132 | 35,529 | 44,632 | 4 | 3 | 46,190 |
| Gajahmungkur | 63,430 | 63,707 | 64,384 | 14,885 | 14,941 | 4 | 4 | 15,100 |
| Semarang Selatan | 82,931 | 79,620 | 78,462 | 24,746 | 22,542 | 3 | 4 | 22,214 |
| Candisari | 79,902 | 79,258 | 78,957 | 19,909 | 21,316 | 4 | 4 | 21,235 |
| Tembalang | 142,941 | 156,868 | 166,680 | 43,197 | 45,898 | 3 | 3 | 48,769 |
| Pedurungan | 175,770 | 180,282 | 185,951 | 45,512 | 46,101 | 4 | 4 | 47,551 |
| Genuk | 91,527 | 97,545 | 103,606 | 24,719 | 29,952 | 4 | 3 | 31,813 |
| Gayamsari | 73,584 | 74,178 | 75,789 | 27,173 | 19,790 | 3 | 4 | 20,220 |
| Semarang Timur | 78,889 | 77,331 | 76,129 | 21,842 | 21,709 | 4 | 4 | 21,372 |
| Semarang Utara | 127,921 | 127,752 | 128,366 | 32,108 | 32,865 | 4 | 4 | 33,023 |
| Semarang Tengah | 71,674 | 70,259 | 68,944 | 20,844 | 20,796 | 3 | 3 | 20,407 |
| Semarang Barat | 158,981 | 158,131 | 158,701 | 44,302 | 57,649 | 4 | 3 | 57,857 |
| Tugu | 30,904 | 31,954 | 33,428 | 8,603 | 9,038 | 4 | 4 | 9,455 |
| Ngaliyan | 120,922 | 125,828 | 131,889 | 33,645 | 42,422 | 4 | 3 | 44,465 |
| Semarang City | 1,559,198 | 1.595,267 | 1,636,594 | 435,184 | 471,327 | 4 | 3 | 483,779 |

**Table 6.** Average roof every head household and wide roof of the settlement.

| No | Sub-district (*Kecamatan*) | Average roof every head household | Wide roof of the settlement | |
|---|---|---|---|---|
| | | 2012 ($m^2$) | 2012 (ha) | 2017 (ha) |
| 1 | Mijen | 83.48 | 139.15 | 168.99 |
| 2 | Gunung Pati | 86.16 | 185.25 | 205.62 |
| 3 | Banyumanik | 79.74 | 283.32 | 368.33 |
| 4 | Gajahmungkur | 85.76 | 127.65 | 129.49 |
| 5 | Semarang Selatan | 43.65 | 108.02 | 96.97 |
| 6 | Candisari | 73.43 | 146.19 | 155.93 |
| 7 | Tembalang | 76.51 | 330.50 | 373.13 |
| 8 | Pedurungan | 61.73 | 280.94 | 293.52 |
| 9 | Genuk | 59.19 | 146.31 | 188.30 |
| 10 | Gayamsari | 32.49 | 88.28 | 65.69 |
| 11 | Semarang Timur | 51.52 | 112.54 | 110.11 |
| 12 | Semarang Utara | 55.42 | 177.93 | 183.00 |
| 13 | Semarang Tengah | 39.72 | 82.80 | 81.06 |
| 14 | Semarang Barat | 75.34 | 333.78 | 435.90 |
| 15 | Tugu | 58.07 | 49.96 | 54.91 |
| 16 | Ngaliyan | 142.22 | 478.52 | 632.41 |
| | in Semarang City | 70.57 | 3,071.13 | 3,543.37 |

*3.6. Potential volume of rainwater that can be harvested*

Based on average monthly rainfall data, estimation data of roof area and runoff coefficient can be calculated the estimation of rainfall volume that can be harvested in residential area. The calculation of potential volume of rainwater that can be harvested every month in residential area of Semarang can be seen in Table 7. The visualization of volume distribution per month can be seen in Figure 8. Based on those tables and figures, it appears that there are fluctuations in potential volume of rainfall affected by season. That condition has been identified at the time of making the isohyetal model. Hence, that condition requires the management of potential rainwater that can be harvested during wet months to cover shortfall in dry months.

**Table 7.** Monthly availability of rainwater harvesting in one year.

| | January | February | March | April | May | June | July | August | September | October | November | December |
|---|---|---|---|---|---|---|---|---|---|---|---|---|
| Volume RWH x $10^6$ liter | 12,693 | 12,126 | 9,294 | 7,103 | 6,289 | 3,792 | 2,191 | 1,114 | 1,076 | 3,556 | 6,854 | 9,795 |

The calculation of value of urban water needs in urban settlements refers to value of water demand set 120 liters/day/person. The results show volume of water demand in residential area of each sub-district within a year. A comparison of water demand scores with potential value of rainwater harvesting volume in one year will result in water resilience status of the area.

**Table 8.** Water needs and availability of rainwater harvesting in one year.

| No | Sub-district | One year water needs x $10^6$ liter | One year rainwater harvest x $10^6$ liter | Difference x $10^6$ liter | Status |
|---|---|---|---|---|---|
| 1 | Mijen | 2,878 | 3,837 | 958 | Surplus |
| 2 | Gunung Pati | 3,612 | 4,512 | 900 | Surplus |
| 3 | Banyumanik | 6,006 | 8,299 | 2,293 | Surplus |
| 4 | Gajahmungkur | 2,820 | 2,632 | -188 | Deficit |
| 5 | Semarang Selatan | 3,437 | 2,103 | -1,333 | Deficit |
| 6 | Candisari | 3,458 | 3,207 | -251 | Deficit |
| 7 | Tembalang | 7,301 | 7,982 | 682 | Surplus |
| 8 | Pedurungan | 8,145 | 6,087 | -2,058 | Deficit |
| 9 | Genuk | 4,538 | 4,009 | -529 | Deficit |
| 10 | Gayamsari | 3,320 | 1,659 | -1,661 | Deficit |
| 11 | Semarang Timur | 3,334 | 2,266 | -1,069 | Deficit |
| 12 | Semarang Utara | 5,622 | 3,757 | -1,866 | Deficit |
| 13 | Semarang Tengah | 3,020 | 1,681 | -1,338 | Deficit |
| 14 | Semarang Barat | 6,951 | 9,445 | 2,494 | Surplus |
| 15 | Tugu | 1,464 | 1,160 | -304 | Deficit |
| 16 | Ngaliyan | 5,777 | 13,247 | 7,471 | Surplus |
| | Total Semarang City | 71,683 | 75,883 | 4,201 | Surplus |

The water security status of each sub-district within the administrative area of Semarang City is presented in Table 8. Comparison between amount of community water demand and potential volume of water that can be harvested in settlement areas of each urban village is presented in Figure 9. The results of analysis and calculation of water security status presented in Table 8 indicate that 9 sub-districts are in deficit status if their water sources are only using rainwater harvesting. On the other hand, the other 7 districts are in surplus status only if they utilize rain water as the sole source of water disregarding other water sources. Mostly, districts with deficit status are located in the urban areas with higher population density. Areas with surpluses are generally on the edge with lower population densities and average per-head household roof area is also wider. Figure 9 also illustrates importance of inter-regional water management in addition to time-based rainwater management, as has been mentioned in relation to the potential for monthly rainwater availability in one year (Figure 8).

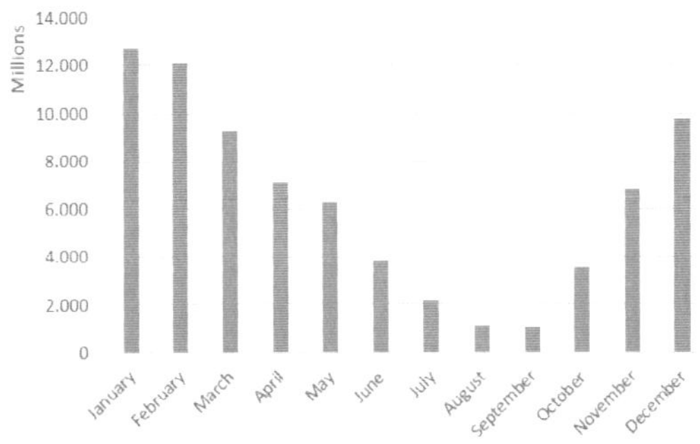


**Figure 8.** Total potential volume of rainwater harvesting monthly a year.

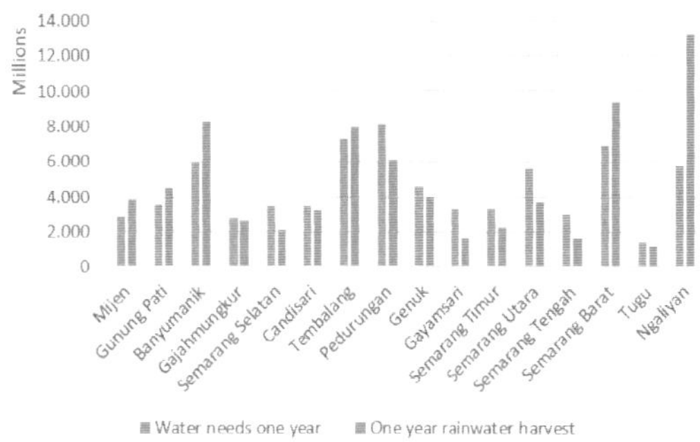


**Figure 9.** Total potential volume of rainwater harvesting and water needs per sub-district for one year.

## 4. Conclusions

Generally, it can be concluded that CHIRPS data can be used to construct rainfall isohyetal model for rainwater harvesting planning. This consideration was obtained based on validation results indicating that average error value of rainfall generated by the model is below 30% (8 weather stations). Only 2 stations have a deviation value exceeding 30%. This needs to be studied further related to the physiographic condition of the area where the weather station is located, which is above 200m height and is an area with hilly physiography. Another consideration that supports the CHIRPS data can be used for rainwater harvesting planning is the estimated value of rainfall generally lower than the reference value. This condition is considered much better as it avoids overly optimistic volume planning.

Water security status of Semarang City assessed based on the potential volume of rainwater harvesting generally shows a surplus. If Semarang City only relies on rainwater as the only source of water, the city is actually able to meet its water needs. However, the surplus condition should consider water management factor in terms of time and distribution of the region. The average monthly rainfall that changes within season and the condition of residential area as well as the extent of roof available in each region makes the potential distribution of rainwater harvesting volume becomes uneven. Based on the results of this study, it can be stated that rainwater harvesting in City of Semarang based on its potential volume can be used as an alternative source of water to meet water needs of the city's population.

## Acknowledgments

Acknowledgments to the Geospatial Information Agency (BIG) for the support of data and software used in this study. Acknowledgments are also conveyed to the Universitas Indonesia (UI), especially the School of Environmental Sciences for support in publication activities and Universitas Diponegoro, especially Urban Planning Studies Program for all supports during fieldwork process. Authors thanks to Sylvia Prisca Delima for her helpful comment on the first draft of this paper.

**References**

[1] UN University 2013 *Water Security & the A Global Water Agenda* UN-Water Analytical Brief Canada.

[2] Zaizen M, Urakawa T, Matsumoto Y, Takai H 2000 The collection of rainwater from dome stadiums in Japan *Urban Water Journal* **1** (4) 355–359

[3] Villarreal E L and Dixon A 2005 Analysis of a rainwater collection system for domestic water supply in Ringdansen Norrköping Sweden *Building and Environment Journal* **40** (9) 1174–1184

[4] Ghisi E 2006 Potential for potable water savings by using rainwater in the residential sector of Brazil *Building and Environment Journal* **41** (11) 1544–1550

[5] Nolde E 2007 Possibilities of rainwater utilisation in densely populated areas including precipitation runoffs from traffic surfaces *Desalination Journal* **215** (1–3) 1–11

[6] Abdulla F A and Al-Shareef A W 2009 Roof rainwater harvesting systems for household water supply in Jordan *Journal of Desalination* **243** (1-3) 195-207

[7] Zhang D, Gersberg R M, Wilhelm C, Voigt M 2009 Decentralized water management rainwater harvesting and greywater reuse in an urban area of Beijing China *Urban Water Journal* **6** (5) 375–385

[8] Akter A and Shoukat A 2015 Potentiality of rainwater harvesting for an urban community in Bangladesh *Journal of Hydrology* **528** 84-93

[9] Jeremy G, Sam T and Diveshkumar M 2012 Mandatory urban rainwater harvesting: learning from experience *Water Science & Technology* **65** (7) 1200-1207

[10] Rida A, Abdullah D, Ghada S 2010 Combining GIS with multicriteria decision making for siting water harvesting ponds in Northern Jordan *Journal of Arid Environments* **74** 1471-1477

[11] Guebaili A, Djebbar Y, Guedri A, Boukhari S 2011 Rainwater harvesting in North Africa: A novel method for reservoir sizing *Journal of Materials and Environmental Science* **2** *(S1) 469-472*

[12] Farrenya R, Gabarrella X, Rieradevall J 2011 Cost-efficiency of rainwater harvesting strategies in dense Mediterranean neighbourhoods *Resources, Conservation and Recycling* **55** 686–694

[13] Zeng B, Tan H, Wu L 2007 A New Approach to Urban Rainwater Management *Journal of China University of Mining & Technology* **17** (1) 82-84.

[14] Jaemin S, Mooyoung H, Tschung-il K, Jee-eun S 2009 Rainwater harvesting as a sustainable water supply option in Banda Aceh *Desalination* **248** 233–240

[15] Winnaar G, Jewitt G P W, Horan M 2007 A GIS-based approach for identifying potential runoff harvesting sites in the Thukela River basin, South Africa *Physics and Chemistry of the Earth* **32** 1058–1067

[16] Funk C, Peterson P, Landsfeld M, Pedreros D, Verdin J, Rowland J, Romero B, Husak G, Michaelsen J, Verdin A, Pedreros P 2014 A Quasi-Global Precipitation Time Series for Drought Monitoring *U.S. Geological Survey Data Series 832* Available online: pubs.usgs.gov/ds/832/ (accessed on 8 August 2014).

[17] Ramon F, Tito M P, Albert G, Carlota T, Joan R, Xavier G 2011 Roof selection for rainwater harvesting: Quantity and quality assessments in Spain *Water Research* **45** 3245-3254

IOP Conf. Series: Earth and Environmental Science **98** (2017) 012005 doi:10.1088/1755-1315/98/1/012005

# Land Cover Analysis by Using Pixel-Based and Object-Based Image Classification Method in Bogor

**Birohmatin Amalisana[1*], Rokhmatullah[2], and Revi Hernina[3]**

[1]Bachelor Program of Geography, Universitas Indonesia, Depok – Indonesia
[2]Department of Geography, Faculty of Mathematic and Natural Science, Universitas Indonesia, Depok – Indonesia

amalisanaa@gmail.com

**Abstract**. The advantage of image classification is to provide earth's surface information like landcover and time-series changes. Nowadays, pixel-based image classification technique is commonly performed with variety of algorithm such as *minimum distance, parallelepiped, maximum likelihood, mahalanobis distance*. On the other hand, landcover classification can also be acquired by using object-based image classification technique. In addition, object-based classification uses image segmentation from parameter such as scale, form, colour, smoothness and compactness. This research is aimed to compare the result of landcover classification and its change detection between parallelepiped pixel-based and object-based classification method. Location of this research is Bogor with 20 years range of observation from 1996 until 2016. This region is famous as urban areas which continuously change due to its rapid development, so that time-series landcover information of this region will be interesting.

**Keywords**: land cover, pixel-based classification, object-based classification

## 1. Introduction

Remote Sensing has an advantage to apply in many sectors. Through the remote sensing data, it can be present representation of earth surface. Remote sensing data is raster image that obtained from sensor of satellite. One of the satellite that provide image data is Landsat which released by USGS (United States Geoligical Survey). Landsat image consist of Landsat 4, 5, 7 and 8. Each Landsat has different sensor component, for example the Landsat 5 has MSS and Thermal Sensor then Landsat 8 has more sensor like OLI, Panchromatic, Cirrus and Thermal. The used of this sensor is relatively similar especially for classification methods or thermal methods. Landsat image has 30 m resolution for each pixel at band multi spectral that include visual band and infrared band. Landsat image was include in optical image type that easily obtained and available in various of time continuously so giving the information about change on the surface easily.

The most application of remote sensing widely used is classification techniques to determine the distribution of earth surface cover. This landcover classification methods using multispectral sensor that contain visual band and infrared so the image can represent the visualization of earth surface in real sight. The other advantage of remote sensing in addition to reaching the area that is difficult to achieve, it also can record the data in various time so this method really better for monitoring function. The classification method which provide by geographic information system include supervised classification and unsupervised classification. Supervised classification method was the most used to interpret landcover from image but this method still has many shortages especially if there a lot of changes on high spatial resolution. supervised and unsupervised classification works based on the value per pixel. In addition of that classification method, there was a method which based on object of the image. This classification still new and hasn't been widely used to remote sensing application. The basic concept of object-based classification is to generalize the shape and texture of a pixel in the image into an object with the meaning area. That generalization was done by segmentation method based on predetermined parameters.

Application of pixel-based and object based classification was located at Bogor City, West Java. Bogor City included in Jabodetabek region because the location was near from the capital city, DKI Jakarta. So, Bogor City has changed and developed quickly. The Landcover change will be observed during 20 years since 1996 until 2016. This research aims to compare two methods of image classification for landcover. The classification is based on pixel and based on object. Pixel based classification had been used for many studies to get the information of landcover, meanwhile the object based classification is being developed to add the variety of image classification methods. Therefore, this compare of two classifications at this research to be used to know the different between these two classifications to representation landcover. The result of this landcover identification can be used to earlier studies for planning, problem solving about environmental and etc.

## 2. Methodology

This research is located in Bogor City, Jawa Barat Province that knows as buffer city or hinterland because the location is near from the capital city of Indonesia, DKI Jakarta. Bogor City is one of the target for resident to build a settlement. This research located at Bogor City, West Java. Bogor City become the main destination of urban who want to live peacefully at the sub-urban area but did not has the long distance with the central of urban area. That phenomenon caused Bogor City has been changed in many ways as a consequence of settlers from rural area. The migrant is increasing population growth which is caused urban probelms and development.

Image classification is defined as the pixel assigning process of raster data specified in classes [1]. In general mean, landcover classification using remote sensing image is to group the pixels into land cover classes based on certain parameters. Landcover identification in this study using remote sensing data such Landsat 5 and Landsat 8 OLI with 30 meters resolution which observed for 20 years from 1996 until 2016. Landsat image classification method is divided into two steps method:

a. Pixel-Based Classification

This method used digital number identification for each pixel that had been classified into landcover type as expected. This research used parallelepiped method to determine the landcover through software ENVI 5.1 and ArcGIS 10.2.1. Pixel-based classification generally done by grouping pixels into predefined classes based on the mean and variance of spectral value each pixels [2]. This classification method is the most widely used and easiest method to apply pixels to the landcover. Pixel-based classification has the disadvantage at extracting information content in high-resolution data and the result of classification is inconsistent.

b. Object-Based Classification

This method is done by using the identification of each object's appearance on the image then grouped into certain types of landcover classes. Object-based classification in this study uses segmentation method to determine land cover with eCognition software application. OBIA approach is carried out through two steps; segmentation and classification. Object was formed through the process of grouping pixels segmented adjacent in the same quality. Identification of landcover in the process of segmentation and classification is derived from the extraction of shapes, textures and spectral information. The process of segmentation is the most important step in object-based classification because in this process the image data will be interpreted into landcover class based on geometry shape and etc. In addition, the segmentation also pay attention to account the color and hue of the object to generalize the landcover class. So there was no pixels are classified singly because the object created as a group of several pixels neighbour that have the same information [2]. Segmentation phase uses parameters that should be as accurate as possible in reality.

Landsat image data processing is done using *software* ENVI 5.1, eCognition dan ArcGIS 10.3. The study workflow used is shown as Figure 1.

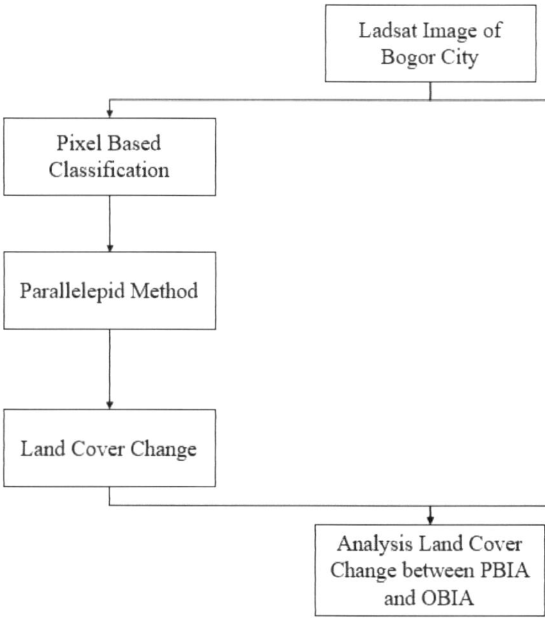


**Figure 1.** Research Workflow

Landsat Image in 1996 and 2016 each is carried out the method of classification based on pixel using parallelepiped and based on object using segmentation method. The class division of landcover is based on four class that is Non Built Area, Vegetation Area, Builtup Area and Water Body Area. After obtaining the result of classification, then carried out the samples for validation using Google Earth Image and field observation in 2016. This validation aims to see the accuracy of Landsat image classification and then will be compared the value of accuracy with detailed description and spatial analysis. Assessment of accuracy of landcover classification in both methods using overall accuracy with the formula below [3]

$$OA = \frac{\sum_{i=1}^{c} Eii}{N} \quad (1)$$

Where e is the number of classes, N is the number of the certain classes and Eii is the error Error matrix diagonal cell. Error matrix is like a confusion matrix which is a square array of number set out in rows and columns and expresses the relationship between the samples in the reference and classified image [3].

## 3. Result and Discussion

### *3.1. Pixel Based Image Classification 1996 and 2016*

Pixel-based classification of Landsat 5 for 1996 and Landsat 8 OLI for 2016 was conducted in the study area using parallelepepid method which available in ENVI 5.1. That four classes become the training area on each image to produce pixel value extraction. Figure 2 shows the maps of landcover in Bogor city in 1996 form Landsat 5 image classification whereas Figure 3 represents the maps of landcover in Bogor City in 2016. The classification has heterogeneous pixel values but still looks fine in class vegetation area and builtup area. While the non built area and water body area has so many mixed pixels who shows a noise on the landcover classification results. This was because the pixel-based classification method could classify the pixels as a one meaning class. The single classified pixel that scattering was not generalizable so it remains identified by 30 meters area according to the given training sample. This classification is deemed to have the possibility of improper interpretation or provide less general landcover information.

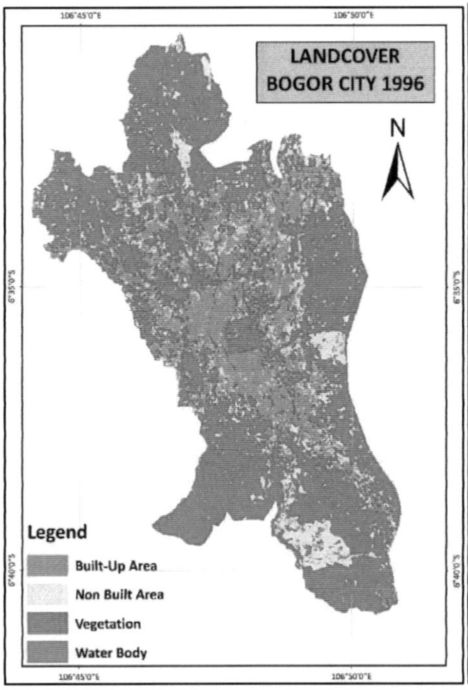


**Figure 2.** Landcover in Bogor City 1996 using Pixel Based Classification

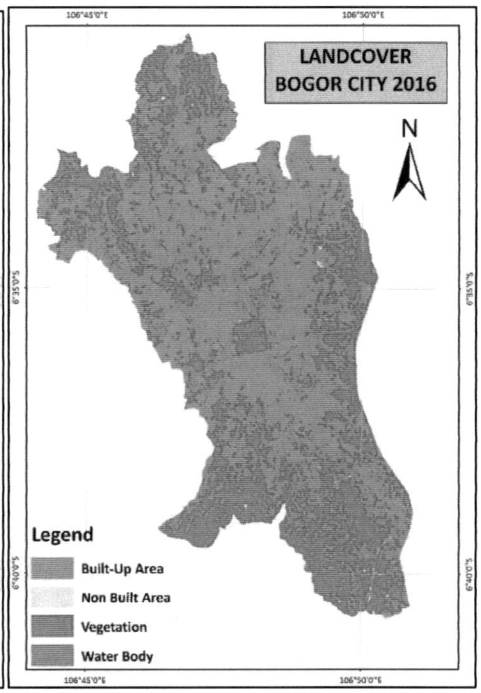


**Figure 3.** Landcover in Bogor City 2016 using Pixel Based Classification

### *3.2. Object Based Image Classification 1996 and 2016*

Object-based classification was through the process of image segmentation into a homogeneous region meaning based on the spectral value of neighbouring pixels and the object's spatial shape shown in Figure 4 and 5.

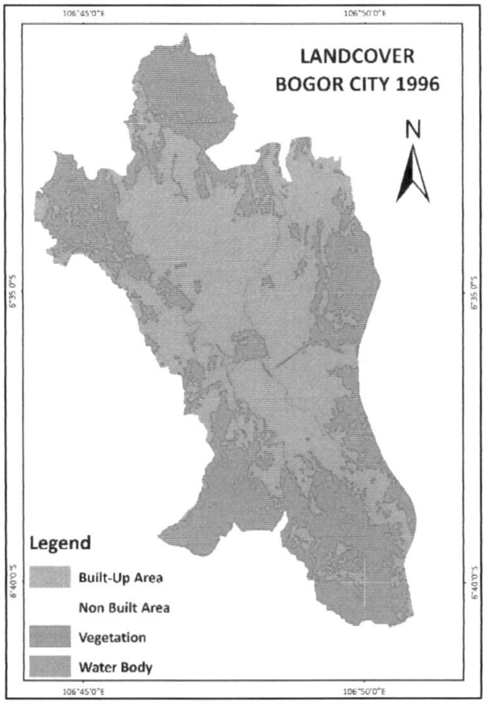


**Figure 4.** Landcover Bogor City 1996 using Object Based Classification

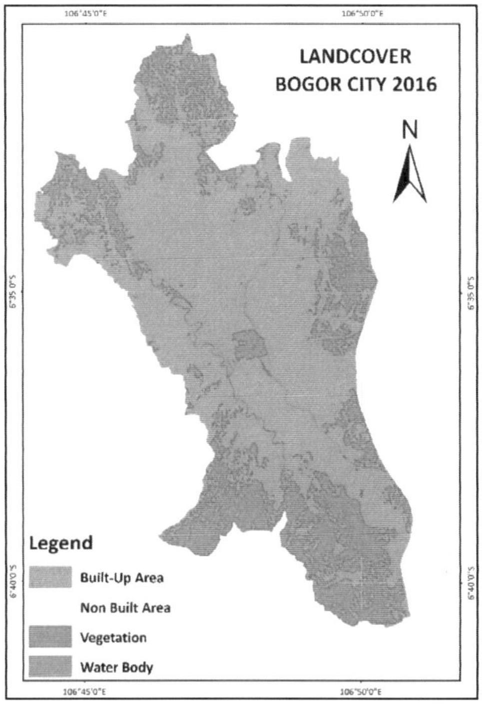


**Figure 5.** Landcover Bogor City 2016 using Object Based Classification

Figure 4 shows the result of landcover classification in 1996, while Figure 5 shows the result of landcover classificationbin 2016 with all of them using object-based segmentation methods. The result of image interpretation of landcover classiffcation is more general and cleaner than landcover classification using pixel-based. There was no pixel noise that scattered as a classified area. As seen on figure 4 and 5 there is no single class pixel classified so the segmentation process helps in generalizing the pixel value into the spectral and the shape correspoding to the neighbour pixel. The shape of the object is also becoming increasingly more clear than the pixel-based classification as an example is water body area that more visible in object-based classification.

### *3.3. Accuracy Landcover Comparation Between Pixel Based Image Analysis (PBIA) and Object Based Image Anaysis (OBIA)*

Classification accuracy is used to determine the precision between training sample in reality and classified data. Training sample is carried out by taking sample on high-resolution image such as Google Earth Image as well as field observation to add validation steps. The accuracy table of these two classifiers can be found in table 1 below.

**Table 1.** Overall Accuracy Presentation for Pixel Based Classification and Object Based Classification

| Accuracy Statistics | Parallelepiped Classification (%) | Segmentation Object Classification (%) |
|---|---|---|
| **Overall Accuracy** | 61.481 | 82.15 |

The comparison between landcover classification based on pixel and object based is shown with the percentage of accuracy value for each other 61.481% and 82.15%. The highest value of accuracy was an object-based classification using segmentation methods. Object-based classification has the advantage of interpreting image with high homogeneity and could identified no single classed pixel distribution. While the pixel-based classification has an accuracy percentage is 61.841% which means there are still many errors that are encountered due to the distribution of single classified pixels. This presentation of accuracy value chows that object-based classification using segmentation more accurate than pixel-based classification.

*3.4. Land Cover Change during 20 years Since 1996 until 2016 at Bogor City*

The changes in landcover that occured in Bogor City from 1996 to 2016 is show as Figure 1 and 2 above. The most striking changes is found on built up area which more wide than before. The area of landcover in 1996 adn 2016 is listed in table 2 below.

**Table 2.** The change of area landcover classes for 20 years in Bogor city

| Landcover Classes | Year ($km^2$) | |
|---|---|---|
| | 1996 | 2016 |
| **Built Up Area** | 21,15 | 68,15 |
| **Vegetation** | 70,548 | 40,058 |
| **Non Built Area** | 0,2341 | 7,815 |
| **Water Body** | 5,1334 | 3,9929 |

The builtup area shows changes to the area more than 100% from 21.15 $km^2$ increased to 68.15 $km^2$. in addition to the built up area, the non built up area also increased more than 100% from 0.2341 $km^2$ to 7.815 $km^2$. While for the vegetation area decreased of almost 50% than before from 70.548 $km^2$ to 40.058 $km^2$. The changes that occur in Bogor City leads to the growth following the needs of living place for human. They shows that urban expansion is leading the changes in Bogor City as a sub urban area.

## 4. Conclusions

Based on the results of the study showed that the classification based on pixels and objects can be used in Landsat images with a resolution of 30 meters to identify land cover. But, object-based classification using segmentation method more accurate than pixel-based classification. Object-based classification results the landcover interpreations with more generous and homogeneous coverage and clear boundaries between classes whereas the pixel-based classification has an errors in the distribution of a single calssified pixels that make the results more heterogeneoous and have a lot of noise.

**Reference**

[1.] Y. H. Araya and C. Hertagen. 2008. *A Comparison of Pixel and Object-based Land Cover Classification : A Case Study of The Asmara Region, Eritrea.* WIT Transaction on Built Environment Vol 100, ISSN 1743-3509 (Geo-Environment Landscape Evolution III)

[2.] Manalu, Johanes R., Sutanto, Ahmad., and Trisakti, Bambang. 2016. *Comparison Between Pixel-based and Object-based Methods For Landcover Classification using PiSAR-L2 Data.* LAPAN : Jurnal Penginderaan Jauh Vol. 13 No 1 Juni 2016 : 49 – 60.

[3.] Roostaei Sh., Alavi S.A., Nikjoo, M.R., and Kamran Valizadeh Kh., 2012. *Evaluation of Object Oriented and Pixel Based Classification Methods for Extracting Changes in Urban Area.* Iran : International Journal of Geomatics and Geoscience Vol 2. No. 3 ISSN 0976 – 4380.

# Spatiotemporal Built-up Land Density Mapping Using Various Spectral Indices in Landsat-7 ETM+ and Landsat-8 OLI/TIRS (Case Study: Surakarta City)

**Risky Yanuar S*, Aulia Yogi H, Prima Widayani**

Department of Geography Information Science, Universitas Gadjah Mada, Sleman, Special Region of Yogyakarta, 55281, Indonesia

risky.yanuar@gmail.com

**Abstract**. Spectral indices variations support for rapid and accurate extracting information such as built-up density. However, the exact determination of spectral waves for built-up density extraction is lacking. This study explains and compares the capabilities of 5 variations of spectral indices in spatiotemporal built-up density mapping using Landsat-7 ETM+ and Landsat-8 OLI/TIRS in Surakarta City on 2002 and 2015. The spectral indices variations used are 3 mid-infrared (MIR) based indices such as the Normalized Difference Built-up Index (NDBI), Urban Index (UI) and Built-up and 2 visible based indices such as VrNIR-BI (visible red) and VgNIR-BI (visible green). Linear regression statistics between ground value samples from Google Earth image in 2002 and 2015 and spectral indices for determining built-up land density. Ground value used amounted to 27 samples for model and 7 samples for accuracy test. The classification of built-up density mapping is divided into 9 classes: unclassified, 0-12.5%, 12.5-25%, 25-37.5%, 37.5-50%, 50-62.5%, 62.5-75%, 75-87.5% and 87.5-100 %. Accuracy of built-up land density mapping in 2002 and 2015 using VrNIR-BI (81,823% and 73.235%), VgNIR-BI (78.934% and 69.028%), NDBI (34.870% and 74.365%), UI (43.273% and 64.398%) and Built-up (59.755% and 72.664%). Based all spectral indices, Surakarta City on 2000-2015 has increased of built-up land density. VgNIR-BI has better capabilities for built-up land density mapping on Landsat-7 ETM + and Landsat-8 OLI/TIRS.

**Keywords**: mapping, Landsat, built-up, Surakarta

## 1. Introduction

The current trend in present time is the acceleration of urbanization and the expansion of urban areas that make the earth's surface turn into an impermeable layer, which is built-up land [22]. Built-up land is an area that is filled and surrounded by building[19]. The growth of built-up land in urban areas has resulted in increased built-up land density [17]. Increased built-up land density, so it is necessary for the supervision of urban development [23]. Supervision on urban development, especially built-up land density can be done through remote sensing [2]. Remote sensing has advantages such as faster, accurate, and has direct access to the earth's surface [26]. Remote sensing can also be used spatiotemporal to observe changes in built-up land density in urban areas [7]. The study of urban areas in remote sensing can be used as a means of decision-making such as urban development planning, disaster management, and urban environmental management [9][21].

Remote sensing is generally used to overcome and utilize the land. There is a way that can be used in remote sensing to produce more accurate and detailed data, ie by using various indices [9]. The built-up land indices are an indices used to extract information on rapidly constructed land through remote sensing imagery [27]. The built-up land indice generates data from the land area built on clearly observed sites. One application of land use indice is to map land constructed by [25]. Other objects that exist in urban phenomenon such as asphalt, vegetation, air, open land can be compared with other objects on curve of spectral reflection [6].

Built-up indice such as Urban Index (UI) [15], Normalized Different Built-up Index (NDBI) [28], VgNIR-BI, VrNIR-BI , and Index Based Built-up Index (IBI) [27]. The difference in each of the built-up land indice located in the band and the formula used. All land indices are built using infrared waves. However, there are several bands that are used on an indice too, but are not used on other built-up land indices. The built-up land indice, such as VgNIR-BI uses green waves, and VrNIR-BI uses red waves. This is different from the UI built-up land indice that using SWIR-2 and NDBI using SWIR-1, while the IBI uses the five waves.

A mid developments on the current assortment of built-up land indices, there is little research that examines the accuracy ratios of each of the built-up land indice. Research on comparison between indices has been done before by [12], on the indice of NDBI, NDISI, and NDII. The purpose of this research is to analyze and to compare the accuracy of built-up indices such as UI, NDBI, VgNIR-BI, VrNIR-BI, and Built-up with spatiotemporal data. Landsat-7 ETM+ representing the state of 2002 and Landsat-8 OLI/TIRS representing the state of 2015.

## 2. Material and Method

### *2.1 Study site and data specifications*

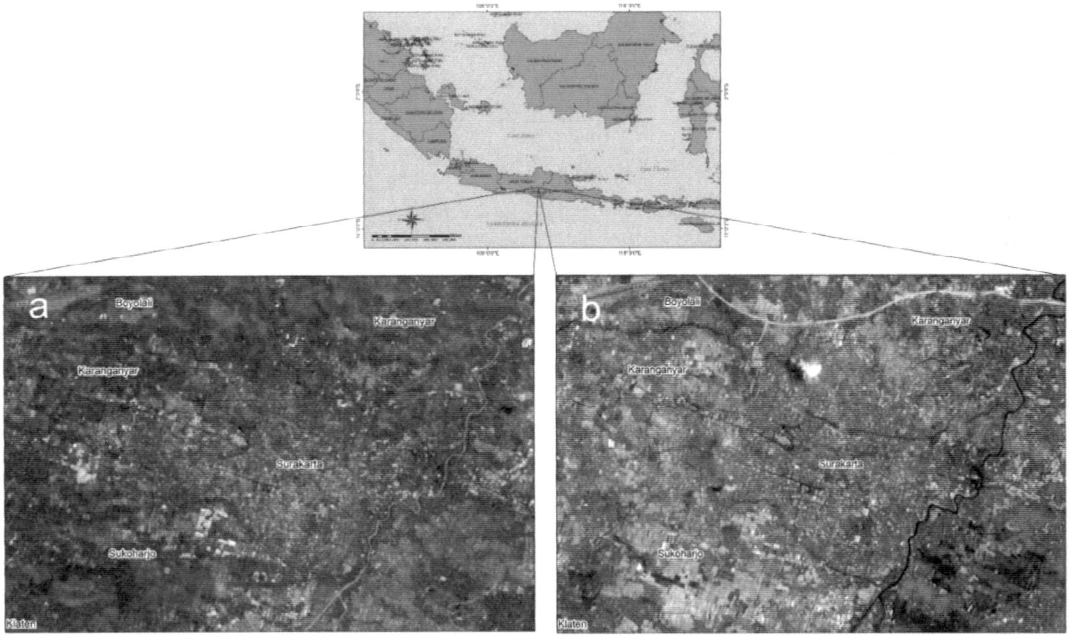


**Figure 1.** (a)Landsat-7 ETM+ 432 and (b)Landsat-8 OLI/TIRS 543 image of the study area: Surakarta City and its surrounding areas. Built-up land appear in dark blue or green-blue, while vegetation appears in red. Dry paddyfield appear white-red, while bodies of water appear in black. Bare land appear light blue. The black line mark the regency boundaries.

For this research, we use Surakarta City and surrounding area, Central Java Province, Indonesia. According to [13], the population density in Surakarta is 11,675 people/km$^2$, which is the highest density in Central Java (the density of Central Java is only 1,038 people/km$^2$), so that the development of the built-up land is quickly developed. Surakarta City with an area of 44 km$^2$, is bordered by Karanganyar and Boyolali regencies in the north, Karanganyar and Sukoharjo regencies in the east and west, and Sukoharjo regency in the south. According to population, Surakarta City is also the third largest city in the southern part of Java after Bandung and Malang [14]. The east side of the city is passed by the Bengawan Solo river.

In this research, the image used is Landsat-7 ETM+ and Landsat-8 OLI L1T (Standard Terrain Correction) image. This image can be found on path 119 and row 65 of the World Reference System (WRS). The Landsat-7 ETM+ image was acquired on May 13, 2002 and contains 8 bands, while Landsat-8 OLI is acquired on October 13, 2015 and contains 11 bands. The bands of Landsat-7 ETM+ image are stored as 8-bit digital numbers, while Landsat-8 OLI images are stored as 16-bit digital numbers [20].

**Table 1.** Feature of Landsat-7 ETM+ and Landsat-8 OLI/TIRS

| Electromagnetic Region | Landsat-7 ETM+ | | Landsat-8 OLI/TIRS | | Spatial Resolution |
|---|---|---|---|---|---|
| | Band | Wavelength (μm) | Band | Wavelength (μm) | |
| Coastal Aerosol | NA | NA | 1 | 0.43 – 0.451 | 30 |
| Blue | 1 | 0.441 – 0.514 | 2 | 0.452 – 0.512 | 30 |
| Green | 2 | 0.519 – 0.601 | 3 | 0.533 – 0.590 | 30 |
| Red | 3 | 0.631 – 0.692 | 4 | 0.636 – 0.673 | 30 |
| Near Infrared | 4 | 0.772 – 0.898 | 5 | 0.851 – 0.879 | 30 |
| Short Wave Infrared (SWIR1) | 5 | 1.547 – 1.749 | 6 | 1.566 – 1.651 | 30 |
| Short Wave Infrared (SWIR2) | 7 | 2.064 – 2.345 | 7 | 2.107 – 2.294 | 30 |
| Panchromatic | 8 | 0.515 – 0.896 | 8 | 0.503 – 0.676 | 15 |
| Cirrus | NA | NA | 9 | 1.363 – 1.384 | 30 |
| Thermal Infrared (TIR) 1 | 6 | 10.31 – 12.36 | 10 | 10.60 – 11.19 | TIRS (100); ETM+ (60) |
| Thermal Infrared (TIR) 2 | NA | NA | 11 | 11.50 – 12.51 | |

Source: Landsat-7 Data User Handbook, 2003; Landsat-8 Data User Handbook, 2016

*2.2 Landsat-7 ETM+ and Landsat-8 OLI/TIRS data pre-processing*

Satellite imagery data should be preprocessing first before used to find the indices value. The aim of preprocessing data to make pixel value in ideal condition so that it can be used for both visual and mathematical analysis [4]. Before performing image processing into spectral indices, Landsat-7 ETM+ and Landsat-8 OLI/TIRS images should be performed radiometric calibration and atmospheric correction. The raw image value calibrated pixel value or digital number (DN) of the multispectral sensor must be converted into Top of Atmosphere (TOA) reflectance.

The TOA spectral radiance of multispectral bands of Landsat-7 ETM+ and Landsat-8 OLI imageries calculated using the formula. In this study, we used the radiometric calibration module available in the Environment for Visualizing Images (ENVI Ver. 5.2, Exelis Visual Information Solutions, Boulder, CO, USA) software to derive the TOA spectral radiance. The pixel value is processed into a TOA spectral radiance, that is:

$$L\lambda\ (Landsat-7) = \left(\frac{(LMAX\lambda - LMIN\lambda)}{(QCALMAX - QCAMIN)}\right) * (QCAL - QCALMIN) + LMIN\lambda \qquad (1)$$

Where $L\lambda$ is the spectral radiance at the sensor's aperture in watts/(meter$^2$*srad*μm); the Greyscale and Brescale are respectively, the rescaled gain and rescaled bias contained in the level 1 product header or ancillary data record both expressed in watts/(meter$^2$*srad*μm); the QCAL is the quantized calibrated pixel value in DN; the LMIN is the spectral radiance that is scale to QCALMAX; the QCALMIN is the minimum quantized calibrated pixel value in DN. The LMIN, LMAX, QCALMAX and QCALMIN values can also be found in the metadata file.

$$L\lambda\ (Landsat-8) = (ML * Qcal) + AL \tag{2}$$

Where $L\lambda$ is the TOA spectral radiance (Watts/(m$^2$*srad*µm)), $ML$ is the band-specific multiplication factor rescaling from metadata, $AL$ is the band-specific additive rescaling factor from metadata, and $Qcal$ is the value of standard pixel products quantized and calibrated (DN) [24]. From TOA spectral radiance, then converted to TOA reflected value.

$$\rho\lambda\ (Landsat-8) = \frac{\rho\lambda'}{\cos(\theta sz)} = \frac{\rho\lambda'}{\sin(\theta se)} \tag{3}$$

Where $\rho\lambda$ = TOA reflectance, $\theta sz$ = angle of sun elevation, and $\theta se$ = angle of sun zenith. TOA reflectance value converted to surface reflectance value.

The image processed into these spectral indices required atmospheric correction to the level of at-sensor reflectance. Atmospheric correction of Landsat-8 OLI/TIRS and Landsat-7 ETM+ image using histogram adjustment. The histogram evaluated is a histogram of reflected values on the sensor in the form of fractional numbers [6]. The method used in histogram adjustment is dark subtraction [28].

Radiometric corrected satellite imagery is done geometric correction process to improve geometric position accuracy and reduce distortion of earth curvature effect [6]. The first thing to do is to determine the GCP (Ground Control Points) and then done the process of geometric correction. This control point is an object visible to the image as well as visible on the reference map used in geometric correction [11]. This control point can be an object visible to the image as well as seen on the reference map used in geometric correction, such as crossing between rivers and road or crossroads and some other objects clearly visible in the image and reference map [18]. The reference map in this study is a map of the Geospatial Information Agency with a projection of the WGS 1984 Transverse Mercator 49M with a UTM grid. The points taken as much as five points for each image with RMS less than 0.5.

*2.3 Spectral indices*

According to [15] UI can provide actual information related to urban conditions from satellite imagery. The spectral UI indice uses near infrared and middle infrared channels such as SWIR2. The UI spectral indice is used to describe the inverse relationship between the brightness levels of urban areas in near infrared and middle infrared [15]. NDBI has an advantage over a UI that has a unique spectral response from built-up land and other types of land cover [28]. UI and NDBI are the spectral indices developed and used to explore in observation the difference between the spectral response of built-up land in near infrared and middle infrared. The difference between NDBI and UI is NDBI using SWIR1, while UI uses SWIR2. Spectral indice based on visible band to see the condition of built-up land are VrNIR-BI and VgNIR-BI. The visible band spectral indice looks to use a combination of red and green channels with near infrared. The atnospherically corrected reflectance values and brightness temperatures are derived from the built-up indices.

$$VrNIR-BI = \frac{Red-NIR}{Red+NIR} \tag{4}$$

$$VgNIR-BI = \frac{Green-NIR}{Green+NIR} \tag{5}$$

$$NDBI = \frac{SWIR1-NIR}{SWIR2+NIR} \tag{6}$$

$$UI = \frac{SWIR2 - NIR}{SWIR2 + NIR} \quad (7)$$

$$Built - up = \left(\frac{SWIR1 - NIR}{SWIR2 + NIR}\right) - \left(\frac{NIR - Red}{NIR + Red}\right) \quad (8)$$

Where $Green, Red, NIR, SWIR1,$ and $SWIR2$ refer to the atmospherically corrected surface reflectance values of band 2, band 3, band 4, band 5 and band 7 in Landsat-7 ETM+ image, while band 3, band 4, band 5, band 6 and band 7 in Landsat-8 OLI/TIRS image, respectively.

*2.4 Extract ground value and linear regression*

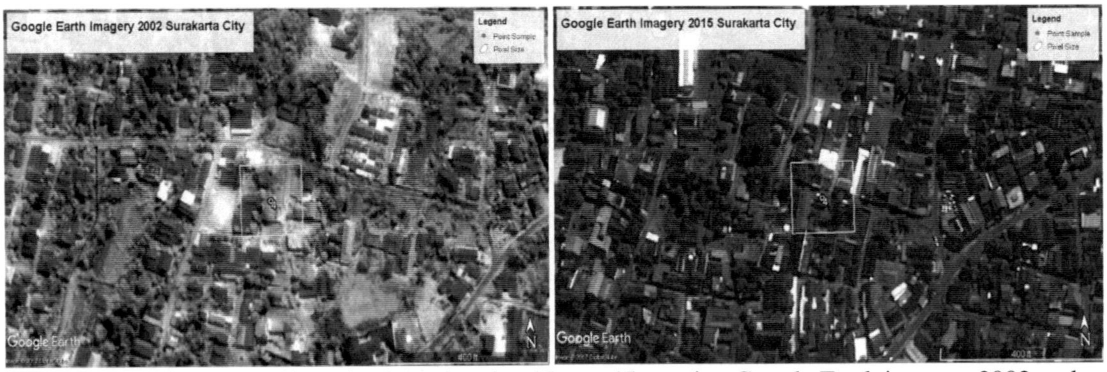


**Figure 2.** Extracting built-up land density size 45 m x 45 m using Google Earth imagery 2002 and 2015 in Surakarta City

The value of the spectral indices (VrNIR-BI, VgNIR-BI, NDBI, UI and Built-up) can not be represented as a built-up land density value, but only limited to the existence of built-up land in the image so it needs a technique to obtain built-up land density information [1]. The linear regression statistic method can predict a value using dependent and linearly independent variables [3][16]. Predicted value of built-up land density through this spectral indice value using a ground value sample in the form of built-up land density value in the field. Ground values extraction using high resolution Google Earth images of 2002 and 2015. The size of ground value extraction measures 45 m x 45 m adjusted to the spatial resolution of Landsat-8 OLI/TIRS and Landsat-7 ETM+ images. Linear regression for built-up land density using 27 ground value samples.

*2.5 Accuracy assessment*

To measure the accuracy and capability of the five spectral indices (VrNIR-BI, VgNIR-BI, NDBI, UI and Built-up) for mapping the developed built-up land density using the standard error of estimate. Standard error of estimate is a measure of the number of regression model errors in predicting the value of Y [10]. Standard error of estimate has a mapping accuracy (percent) using some calculation of bottom, upper, minimum error (%) and maximum error (%) based on confidence level of 95%. Detailed calculation of accuracy test methods can be found in literature [5][16].

Accuracy test of ground built-up land density mapping uses 7 ground value samples extracted from the high resolution Google Earth imagery reference of 2002 and 2015. After all ground values are verified, two ground truth images (Landsat-7 ETM+ and Landsat-8 OLI/TIRS) are generated. Finally, density of the built-up land map was built in 2002 and 2015 using various spectral indices performed comparations.

IOP Conf. Series: Earth and Environmental Science **98** (2017) 012006 doi:10.1088/1755-1315/98/1/012006

## 3. Result and Discussion

### *3.1 Spectral Indices*

Appearance of built-up land on each Landsat-7 ETM+ and Landsat-8 OLI/TIRS indices is shown in figure 3. The bright color on the image shows the existence of the built-up land. Surakarta city has the appearance of higher built-up land compared with the surrounding area. This is seen from the bright color that dominates the Surakarta city compared with the surrounding area.

Visually, Landsat-7 ETM+ has a higher sensitivity to the existence of built-up land compared to Landsat-8 OLI/TIRS. In Landsat-7 ETM+ it is clear that the difference between the built-up land that looks lighter and not built-up land that looks darker. Meanwhile, Landsat-8 OLI/TIRS has no sensitivity to the built-up land such as Landsat-7 ETM+. The existence of built-up land and not built-up land on Landsat-8 OLI/TIRS looks more vague.

The five indices that are processed in Landsat-7 ETM+ and Landsat-8 OLI/TIRS are VgNIR-BI, VrNIR-BI, NDBI, UI, and Built-up have different visual appearance. The NDBI, UI, and Built-up in Landsat-7 ETM+ and VrNIR-BI and UI in Landsat-8 OLI/TIRS show a clear distinction between built-up land and not built-up land. Not all built-up land indices are capable of representing built-up land and not built-up land in various types of imagery.

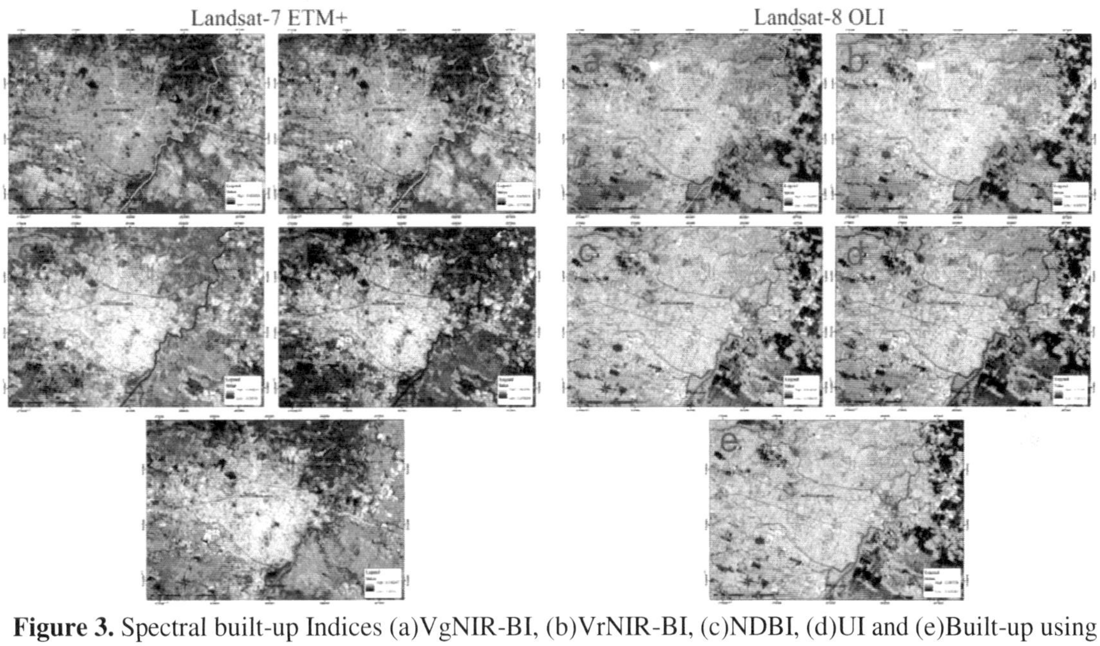


**Figure 3.** Spectral built-up Indices (a)VgNIR-BI, (b)VrNIR-BI, (c)NDBI, (d)UI and (e)Built-up using Landsat-7 ETM+ and Landsat-8 OLI imagery in Surakarta City and its surrounding areas. Built-up land indicated in light color.

The statistical value of built-up land indice image presented in table 2 is able to explain the appearance of the built-up land based on the image pixel value. There are four components in the image statistics that can explain the appearance of objects in the image, namely the minimum value, maximum, mean, and standard deviation. Table 2 shows the statistic values fromm built-up land indice imagery of the Landsat-7 ETM+ and Landsat-8 OLI/TIRS. The value in the table comes from the statistical calculation of all pixel values contained in the image.

**Table 2.** Spectral Indices Statistic Value

| Spectral Indices | Landsat-7 ETM+ (2002) | | | | Landsat-8 OLI/TIRS (2015) | | | |
|---|---|---|---|---|---|---|---|---|
| | Min | Max | Mean | Stdev | Min | Max | Mean | Stdev |
| VgNIR-BI | -0.551 | 0.666 | -0.166 | 0.169 | -0.489 | 0.134 | -0.229 | 0.076 |
| VrNIR-BI | -0.713 | 0.575 | -0.286 | 0.206 | -0.545 | 0.092 | -0.243 | 0.093 |
| NDBI | -0.781 | 0.556 | -0.030 | 0.137 | -0.389 | 0.619 | -0.072 | 0.098 |
| UI | 0.250 | 142.646 | 52.899 | 17.571 | -0.801 | 0.757 | -0.314 | 0.189 |
| Built-up | -1.031 | 0.740 | -0.317 | 0.274 | -0.333 | 0.388 | -0.060 | 0.082 |

Source: Processing data, 2017

The minimum value shows the lowest pixel value in the image. The result of the minimum values of all built-up land indices imagery in Landsat-7 ETM+ and Landsat-8 OLI/TIRS, except UI indice shows a minus value. This minus value represented that there is no built-up land in the pixel. The UI indice in Landsat-7 ETM+ has a minimum value 0.250, it is different value compared to the minimum value in the other indice. The difference value with the other indices is due to the range of pixel values in the larger UI indice or the maximum value that is much greater than the value of other indicees.

The maximum value indicates the highest pixel value in the image that states the existence of built-up land. The high pixel value in the image will make the color on the image brighter. The UI indice in Landsat-7 ETM+ and Landsat-8 OLI/TIRS shows the highest maximum pixel value of 142.646 in Landsat-7 ETM+ and 0.757 in Landsat-8 OLI/TIRS. The maximum value of the UI indice in Landsat-7 ETM+ compared to other indices is due to a large range of values compared to other indices.

Mean is the average pixel value in the image obtained from the total pixel value divided by the number of pixels in the image. When the pixel value in the image has a relatively large value, the resulting value will be greater. The indice with the greatest mean is the UI in Landsat-7 ETM+ that is 52,899, while other indices of Landsat-7 ETM+ and Landsat-8 OLI/TIRS have very small and negative values. The mean value can also explain the description of built-up land in the field. The greater mean value shows that the area of built-up land recorded in the image is getting bigger.

The standard deviation (Stdev) contained in the image statistics shows the pixel value spreads in the image. The results showed that the standard deviation of all indices in Landsat-7 ETM+ and Landsat-8 OLI/TIRS, except that the UI indice in Landsat-7 ETM+ has a greater value than the mean or mean. It shows that the distribution of values within the image has a very wide spread, meaning that the pixel value in the image is very diverse and indicates that built-up land is not only concentrated in a single place, but spread in study area.

### *3.2 Regression between ground value and spectral indices*

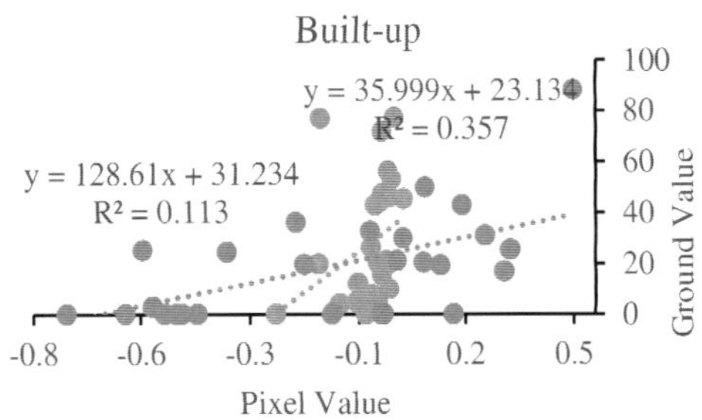


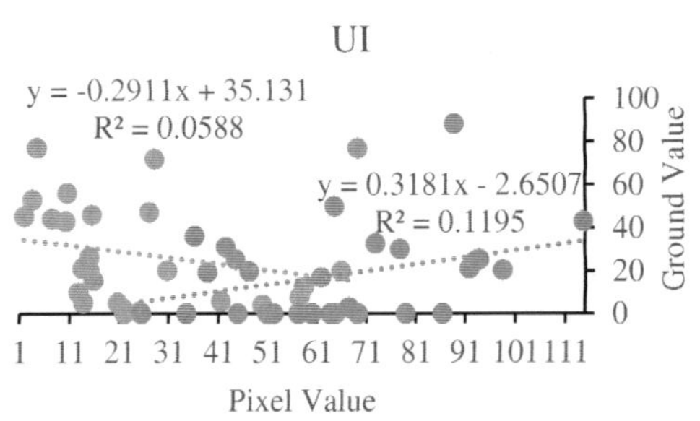

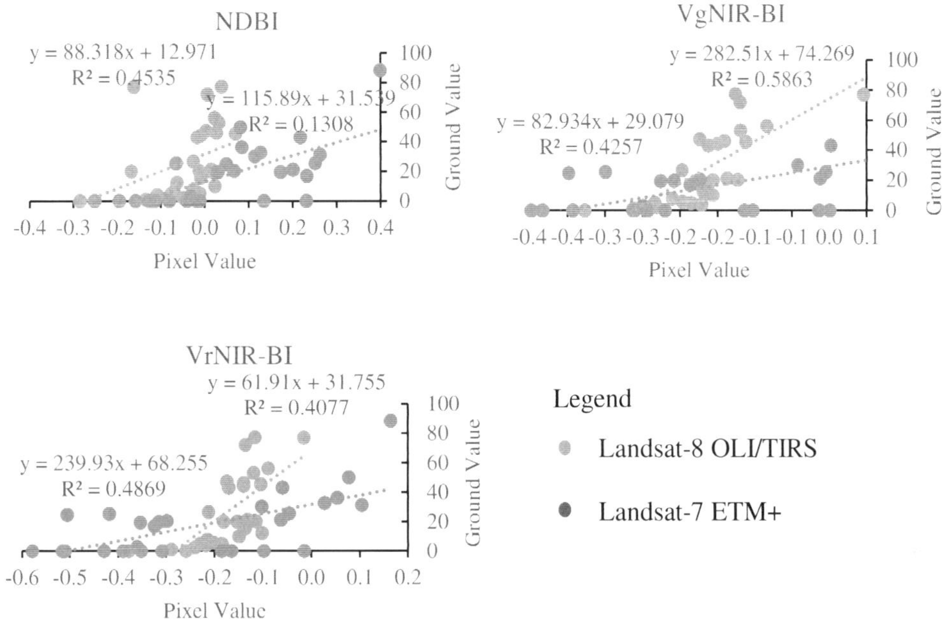


**Figure 4.** Regression spectral built-up indices using Landsat-7 ETM+ and Landsat-8 OLI/TIRS with built-up land density ground value

Figure 4 shows the regression graph between pixel value and field value of each indice in Landsat-7 ETM+ and Landsat-8 OLI/TIRS. The relationship between pixel value representing the built-up land density in the image with its presence in the field can be explained using the regression graph based on $R^2$ value. The smaller value of $R^2$ indicates that the relationship between image and field values is getting smaller.

VgNIR-BI has the greatest $R^2$ value compared to other indice which is 0.5863 in Landsat 8 OLI/TIRS. Similarly, the VrNIR-BI indice has the greatest value in Landsat-7 ETM+ that is 0.4869. This suggests that the VgNIR-BI and VrNIR-BI indices have the strongest correlation between image and field values. Therefore, the two indices are able to explain the existence of built-up land density.

The value of $R^2$ on the UI and Built-up indices has the lowest value compared to the values in the VrNIR-BI, VgNIR-BI, and NDBI indices. The UI indice has the lowest $R^2$ value in Landsat 8 OLI/TIRS of 0.0588 and the Built-up indice has the lowest $R^2$ value at Landsat 7 ETM+ of 0.113. Low regression value states that the existence of objects in the field can not be explained by image value well. Therefore, the UI indice is not good to explain the existence of built-up land density.

### *3.3 Spatiotemporal built-up land density mapping*

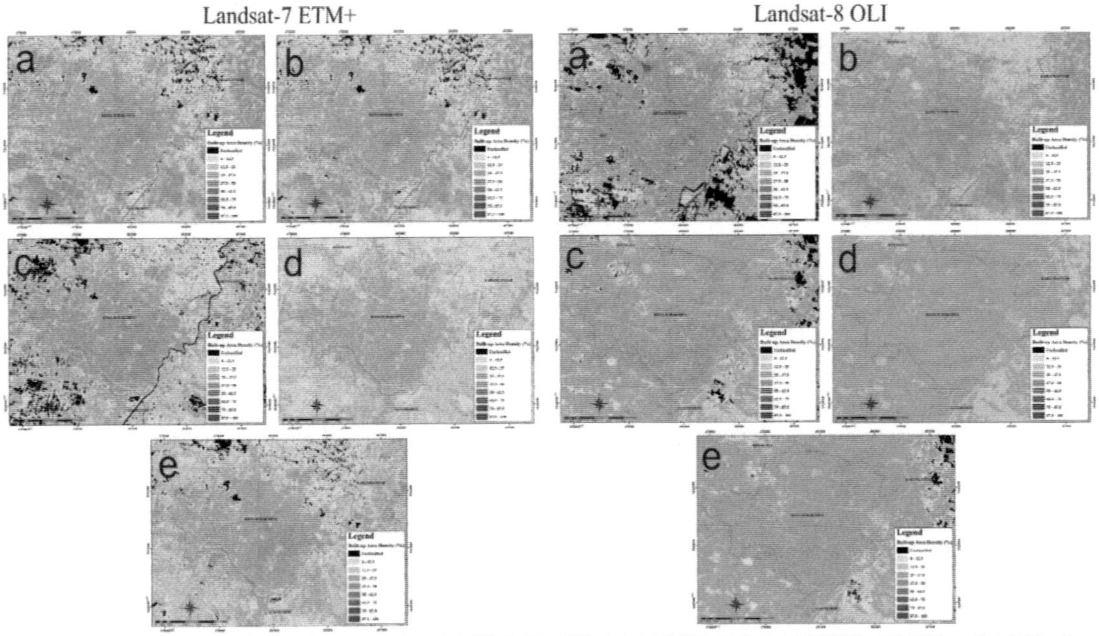


**Figure 5.** Built-up land density mapping (a) VgNIR-BI, (b) VrNIR-BI, (c) NDBI, (d) UI and (e) Built-up using Landsat-7 ETM+ and Landsat-8 OLI/TIRS imagery in Surakarta City and its surrounding area. The imagery are nine classes: unclassified in black, while 0-12.5%, 12.5-25%, 25-37.5%, 37.5-50%, 50-62.5%, 62.5-75%, 75-87.5% and 87.5-100 % increasingly in red.

Figure 5 is a built-up land density mapping based on each indice on Landsat-7 ETM+ and Landsat-8 OLI/TIRS. The classification of built-up land density mapping was built using VgNIR-BI, VrNIR-BI, NDBI, UI, and Built-up in Landsat-7 ETM+ and Landsat-8 OLI/TIRS, this image is divided into 9 classes that is unclassified, 0-12.5%, 12.5-25% , 25-37.5%, 37.5-50%, 50-62.5%, 62.5-75%, 75-87.5% and 87.5-100%. The increasingly red color indicates that the density of built-up land in the area has denser density. The black color on the map indicates that the area has an unclassified density value. The black color appears because the area is not built-up land, but such as surface flows, lakes, or other types of land use.

The visualization of built-up land density on Landsat-7 ETM+ varies greatly on VgNIR-BI, VrNIR-BI, NDBI, UI and Built-up. The NDBI of Landsat-7 ETM+ has the most contrasting built-up land density and shows a very striking difference between built-up and not built-up land, in contrast to the UI indice on Landsat-7 ETM+ which has a very slight difference between built-up and not built-up land. The built-up land density on VgNIR-BI and VrNIR-BI on Landsat-7 ETM+ has almost the same visual appearance.

The capabilities of VgNIR-BI, VrNIR-BI, NDBI, UI and Built-up using Landsat-8 OLI/TIRS for built-up land density mapping do not have significant visual differences such as Landsat-7 ETM+. The visual appearance showing the density of the built-up land using VrNIR-BI in Landsat-8 OLI/TIRS provides a very clear visualization of the difference between built-up land density of Surakarta City. Visually, the combination of green and red band indices in Landsat-8 OLI/TIRS has a good ability to mapping built-up land density when compared to other band.

Based on figure 5, spatiotemporal built-up land density mapping Surakarta City on 2002 and 2015 using VgNIR-BI, VrNIR-BI, NDBI, UI and Built-up in Landsat-7 ETM+ and Landsat-8 OLI/TIRS has visual visibility that explains built-up land density in Surakarta City increased on 2002-2015. The VgNIR-BI indice provide high mapping ability of spatiotemporal developed spatial density that clearly shows the development of built-up land density in Surakarta City on 2002-2015. It is seen that the pattern of dispersion of increased built-up land density in Surakarta City leads to the north, west and south while the eastern direction does not increase. This is because a river bengawan solo in the eastern Surakarta City that inhibits the spread of increased built-up land density.

*3.4 Area percentage of built-up land density in Surakarta City on 2002 and 2015*

**Table 3.** Area Percentage of Built-up Land Density in Surakarta City 2002 and 2015

| Class | Landsat-7 ETM+ 2002 ($km^2$) | | | | | Landsat-8 OLI 2015 ($km^2$) | | | | |
|---|---|---|---|---|---|---|---|---|---|---|
| | Built-up | NDBI | UI | VgNIR-BI | VrNIR-BI | Built-up | NDBI | UI | VgNIR-BI | VrNIR-BI |
| Unclassified | 100.2 | 163.7 | 0.0 | 111.3 | 111.3 | 30.1 | 45.6 | 0.0 | 265.0 | 519.6 |
| 0-12.5 | 389.0 | 390.9 | 444.8 | 332.1 | 345.5 | 126.2 | 127.9 | 0.0 | 209.5 | 66.5 |
| 12.5-25 | 317.8 | 234.1 | 401.8 | 230.0 | 240.7 | 235.7 | 232.1 | 345.3 | 218.3 | 62.7 |
| 25-37.5 | 87.9 | 95.1 | 54.8 | 154.4 | 167.3 | 492.4 | 465.9 | 552.1 | 134.0 | 53.0 |
| 37.5-50 | 6.9 | 17.6 | 0.4 | 68.1 | 34.7 | 15.2 | 27.4 | 4.4 | 58.6 | 48.3 |
| 50-62.5 | 0.0 | 0.4 | 0.0 | 4.3 | 2.3 | 0.4 | 1.1 | 0.0 | 9.9 | 45.1 |
| 62.5-75 | 0.0 | 0.0 | 0.0 | 1.0 | 0.0 | 0.0 | 0.0 | 0.0 | 2.5 | 43.4 |
| 75-87.5 | 0.0 | 0.0 | 0.0 | 0.6 | 0.0 | 0.0 | 0.0 | 0.0 | 0.7 | 38.4 |
| 87.5-100 | 0.0 | 0.0 | 0.0 | 0.0 | 0.0 | 0.0 | 0.0 | 0.0 | 1.4 | 24.6 |

Source: Processing data, 2017

Table 3 explains the value of area built-up land density Surakarta City on 2002 and 2015 varies on different spectral indices. This is due to VgNIR-BI, VrNIR-BI, NDBI, UI and Built-up indice in Landsat-7 ETM+ and Landsat-8 OLI/TIRS have different capabilities to estimate built-up land density. The classification of built-up land density mapping is divided into 9 classes, ie, non-classified, 0-12.5%, 12.5-25%, 25-37.5%, 37.5-50%, 50-62.5%, 62.5-75%, 75-87.5% and 87.5 -100%. The division into 9 classes on the built-up land density mapping in order to make the difference more clear in built-up land density in Surakarta City.

The built-up land density classes with not classified classes has the largest area in VgNIR-BI and VrNIR-BI indices, that are 111.3 $km^2$ in Landsat-7 ETM+ and VrNIR-BI of 519.6 $km^2$ in Landsat-8 OLI/TIRS. The built-up land density classes that is not classified with the lowest area is UI that has value 0 $km^2$ in Landsat-7 ETM+ and Landsat-8 OLI/TIRS. Spectral indice with built-up land density class of 87.5-100% is found only in VgNIR-BI area of 1.4 $km^2$ and VrNIR-BI area of 24.6 $km^2$ in Landsat-8 OLI/TIRS image. According to table five, spectral indices that have a contrasting development of built-up land density are VgNIR-BI and VrNIR-BI, while the non-contrast development of built-up land density is UI.

*3.5 Accuracy assesment of built-up land density mapping*

Based on table 4, the built-up land indice accuracy test values using VgNIR-BI, VrNIR-BI, NDBI, UI and Built-up in Landsat-7 ETM+ and Landsat-8 OLI/TIRS for built-up land density mapping in Surakarta City on 2002 and 2015 have different value. The different value of accuracy test value because each band in Landsat-7 ETM+ and Landsat-8 OLI/TIRS has different spectral reflectance characteristics for built-up land. To determine the capability of each spectral indice, it must analyze the statistical value of maximum error, minimum error, maximum accuracy and minimum accuracy.

Maximum error is the maximum error level on the regression result in predicting the Y value, while the minimum error is the minimum level of error in the regression result in predicting Y. The highest error is 187.84% for NDBI in Landsat-7 ETM+ and 74.517% for UI in Landsat-8 OLI/TIRS. The lowest maximum error value is 52.424% for VgNIR-BI in Landsat-7 ETM+ and 53.655% for NDBI in Landsat-

8 OLI/TIRS. The highest minimum error value is 65.129% for NDBI in Landsat-7 ETM+ and 35.602% for UI in Landsat-8 OLI/TIRS. The lowest minimum error value is 18.176% for VgNIR-BI in Landsat-7 ETM+ and 25.635% for NDBI in Landsat-8 OLI/TIRS.

Maximum accuracy is the maximum level of proximity measurement of quantity to the actual value or 100% difference with minimum error value, while minimum accuracy is the proximity level of quantity measurement to the actual value or 100% difference with maximum error value. The highest maximum accuracy score is 81.823% for VgNIR-BI in Landsat-7 ETM+ and 74.365% for NDBI in Landsat-8 OLI/TIRS. Low maximum accuracy value of 34.87% for NDBI in Landsat-7 ETM+ and 64.398% for UI in Landsat-8 OLI/TIRS. The highest minimum accuracy value is 47.575% for VgNIR-BI in Landsat-7 ETM+ and 46.345% for NDBI in Landsat-8 OLI/TIRS. The lowest minimum accuracy score is -87.84% for NDBI in Landsat-7 ETM+ and 25.483% for UI in Landsat-8 OLI/TIRS.

**Table 4.** Accuracy Assessment of Built-up Land Density Mapping in Surakarta City on 2002 and 2015

| Spectral Indices | Landsat-7 ETM+ 2002 (%) | | | | Landsat-8 OLI/TIRS 2015 (%) | | | |
|---|---|---|---|---|---|---|---|---|
| | Max Error | Min Error | Max Accuracy | Min Accuracy | Max Error | Min Error | Max Accuracy | Min Accuracy |
| VgNIR-BI | 52.424 | 18.176 | 81.823 | 47.575 | 56.020 | 26.765 | 73.235 | 43.980 |
| VrNIR-BI | 60.754 | 21.065 | 78.934 | 39.245 | 64.825 | 30.972 | 69.028 | 35.175 |
| NDBI | 187.840 | 65.129 | 34.870 | -87.840 | 53.655 | 25.635 | 74.365 | 46.345 |
| UI | 163.605 | 56.726 | 43.273 | -63.605 | 74.517 | 35.602 | 64.398 | 25.483 |
| Built-up | 116.070 | 40.244 | 59.755 | -16.070 | 57.216 | 27.336 | 72.664 | 42.784 |

Source: Processing data, 2017

Based on the overall statistical data in table 4, the VgNIR-BI spectral indice has the best ability for built-up land density mapping in Surakarta City on 2002 and 2015 using Landsat-7 ETM+ and Landsat-8 OLI/TIRS, while NDBI has the worst ability for Landsat-7 ETM+ and UI for Landsat-8 OLI/TIRS. VgNIR-BI has a maximum error, minimum error, maximum accuracy and minimum accuracy respectively of 52.424%, 18.176%, 81.823% and 47.545% using Landsat-7 ETM+ and 56.020%, 26.765%, 73.235% and 43.980% using Landsat -8 OLI/TIRS. The combined ability of green and red band in Landsat-7 ETM+ and Landsat-8 OLI/TIRS is the best combination of band for spatiotemporal built-up land density mapping in Surakarta City.

## 4. Conclusions

The observation of the built-up land in Surakarta City is done spatiotemporal using landsat-7 ETM+ representing the condition of 2002 and Landsat-8 OLI representing the state of the year 2015. The development of built-up land in the image can be observed in various ways, such as through image pixel value, visual observation, and measurement of built-up land density. Built-up land indices were attempted to find out the best indices in representing built-up land in the field through regression charts and accuracy tests with field data.

Observation of built-up land through image statistics observed with component image that is the minimum value, maximum, mean, and standard deviation. Individual built-up land indices on Landsat-7 ETM+ and Landsat-8 OLI have varying statistical values in representing built-up land. The variation in the value of the image statistics is due to different channel calculations on various indices. Values in the image statistics on the VgNIR-BI, VrNIR-BI, NDBI, and Built-up indices in Landsat-7 ETM+ and Landsat-8 OLI were normal, but found different values in the UI indice in Landsat-7 ETM+. This shows that statistically the whole indice is able to represent well-constructed land.

Visually, built-up land density changes from 2002 to 2015 are increasing. Built-up land density is higher visible from the gradation of the color of the map is getting red in 2015 in the city of Surakarta. The NDBI indice on Landsat-7 ETM+ has a clear visualization between the built-up land class, as well as the Vr-NIR-BI indice on Landsat-8 OLI. The VgNIR-BI and VrNIR-BI indices on Landsat-7 ETM+ and Landsat-8 OLI imagery are able to represent a development of built-up land is clearly. This is evident from the value of the development of the area of built-up land.

The built-up land indice applied to Landsat-7 ETM+ and Landsat-8 OLI is not fully able to represent well on built-up land. The regression graph and accuracy test were performed to find out the best indice in mapping the density of the built-up land in Surakarta City. The VgNIR-BI indice on Landsat-8 OLI and the VrNIR-BI indice on Landsat-7 ETM+ has the greatest $R^2$ value, so both indices are able to explain the existence of built-up land is great. Based on the results of the accuracy test, the value of VgNIR-BI on Landsat-7 ETM+ and NDBI on Landsat-8 OLI has a high accuracy value. Therefore, the VgNIR-BI has the best ability to built-up land density mapping in Surakarta City in 2002 and 2015 using Landsat-7 ETM+ and Landsat-8 OLI from visual and statistic perspective.

**Acknowledgment**
We are thankful to Mrs. Prima Widayani and Departement of Geography Information Science, Faculty of Geography, Universitas Gadjah Mada whose support to this research. Many thanks are also goes to US Geological Survey (USGS) to give opportunity for the authors to access remote sensing data, such as Landsat-8 OLI/TIRS and Landsat-7 ETM+ and also to Badan Informasi Geospasial (BIG) Indonesia for map of Rupabumi Indonesia.

**References**
[1] Amir Kumar, Arvind Chandra Pandey, and A T Jeyaseelan 2012 Built-up and vegetation extraction and density mapping using WorldView-II *Geocarto International* vol 27(5) pp 557-568.
[2] Bhatta Basudeb 2010 *Advances in Geographic Information Science: Analysis of Urban Growth and Sprawl from Remote Sensing Data* (London: Springer) p 61.
[3] Bintarto R and Hadisumarno 1991 *Geography Analysis Method 4th Ed.* (Jakarta: LP3ES) p 64.
[4] Campbell James B and Wynne Randolph H 2011 *Introduction to Remote Sensing* (New York: The Guilford Press) p 305.
[5] Congalton Russel G and Green Kass 2009 *Assessing the Accuracy of Remotely Sensed Data: Principles and Practices* (London: CRC Press) p 119.
[6] Danoedoro Projo 2012 *Introduction to Digital Image Processing* (Yogyakarta: Andi Press) p 168 and 186.
[7] Dradas Mohsen, Helmi Z M Shafri, Noordin Ahmad, Biswajeet Pradhan, and Sahabeh Safarpour 2015 Spatio-Temporal Analysis of Urban Growth from Remote Sensing Data in Bandar Abbas City, Iran *The Egyptian Journal of Remote Sensing and Space Sciences* vol 18 pp 35-52.
[8] ENVI 2017 *Atmospheric correction: Dark subtraction* Available online: https://www.harrisgeospatial.com/docs/AtmosphericCorrection.html (accessed: 6 September 2017).
[9] Estoque, Ronald C, and Yuji Muraya 2015 Classification and Change Detection of Built-up Lands from Landsat-7ETM+ and Landsat-8 OLI/TIRS Imageries: A Comparative Assessment of Various Spectral Indices *Ecological Indicators* vol 56 pp 205-217.
[10] Freedman D A 2005 *Statistical Models: Theory and Practice* (Cambridge: Cambridge University Press) p 18.
[11] Gao Jay 2012 *Digital Analysis of Remotely Sensed Imagery 4th ed.* (New York: McGraw-Hill) p 158.
[12] Garg Abhisha, Divyansu Pal, Hukum Singh, and Deepak Chander Pandey 2016 A Comparative Study of NDBI, NDISI, and NDII for Extracrion of Urban Impervious Surface of Dehradun (Uttarakhand, India) Using Landsat 8 Imagery *Proc. Int. Conf. on Emerging Trends in Communication Technologies (ETCT)* pp 1-5.

[13] Indonesian Central Bureau of Statistics (BPS) 2016 *Provinsi Jawa Tengah dalam angka 2016* Available online: https://jateng.bps.go.id/website/pdf_publikasi/Provinsi-Jawa-Tengah-Dalam-Angka-2016.pdf (accessed: 6 September 2017).

[14] Indonesian Central Bureau of Statistics (BPS) 2017 *Kota Surakarta dalam angka 2017* Available online: https://surakartakota.bps.go.id/website/pdf_publikasi/Kota-Surakarta-Dalam-Angka-2017.pdf (accessed: 6 September 2017).

[15] Kawamura Makoto, Sanath Jayamanna, and Yuji Tsujiko 1996 Relation Between Social and Environmental Conditions in Colombo, Sri Langka and The Urban Indice Estimated by Satellite Remote Sensing Data *International Archives of Photogrammetry and Remote Sensing* vol 31 pp 321-326.

[16] Kenney J and Keeping E S 1963 *Mathematics of Statistics* (van Nostrand) p 187.

[17] Kneebone Elizabeth 2009 *Job Sprawl Revisited: The Changing Geography of Metropolitan Employment* (Washington DC: Brookings Institution) p. 15

[18] Lillesand Thomas M, Kiefer Ralph W, and Chipman Jonathan W 2003 *Remote sensing and Image Interpretation 5$^{th}$ ed.* (New York: John Wiley & Sons, Inc) p 498.

[19] Pesaresi Martino, Daniele Ehrlich, Ivano Caravaggi, Mayeul Kauffmann, and Christophe Louvrier 2011 Toward Global Automatic Built-up Area Recognition Using Optical VHR Imagery *IEE Journal of Selected Topics in Applied Earth Observations and Remote Sensing* vol 4(4) pp 923-934.

[20] Rashed Tarek and Jürgens Carsten 2010 *Remote Sensing and Digital Image Processing: Remote Sensing of Urban and Suburban Areas* (London: Springer) p 33.

[21] Roy D P, Wulder M A, Loveland T R, Woodcock C E, Allen R G, Anderson, M C, *et al* 2014 Landsat-8 Science and Product Vision for Terrestrial Global Change Research *Remote Sensing Environment* vol 145 pp 154-172.

[22] Sun Genyun *et al* 2016 Combinational Build-up Indice (CBI) for Effective Impervious Surface Mapping in Urban Areas *IEE Journal of Selected Topics in Applied Earth Observations and Remote Sensing* vol 9(5) pp 2081-2092.

[23] USGS 2006 *Multi-resolution Land Characteristic 2001 (MRLC2001) Image Processing Procedure* Available online: (http://landcover.usgs.gov/pdf/image_preprocessing.pdf) (accessed: 5 September 2017).

[24] USGS 2013 *Using the USGS Landsat 8 Product* Available online: (http://landsat.usgs.gov/Landsat8_using_Product.php) (accessed: 5 September 2017).

[25] Wang Zhaoqi, Chencheng Gang, Xueling Li, Yizhao Chen, and Jialong Li 2015 Application of a Normalized Difference Impervious Indice (NDII) to Extract Urban Impervious Surface Feature Based on Landsat TM Images *International Journal of Remote Sensing* vol 36(4) pp 1055-1069.

[26] Wu Mengquan, Jingpu Wang, Xia Wang, and Xibing Sun 2011 Spatio-temporal Analysis of Urban Built-up Area Extension in Yantai City Based on Multi-source Remote Sensing Images *19th International Conference on Geomatics* vol pp. 1-6.

[27] Xu H 2008 A New Indice for Delineating Built-up Land Features in Satellite Imagery *Int. Journal of Remote Sensing* vol 29(14) pp 4269-4276.

[28] Zha Y, J Gao and S Ni 2003 Use of Normalized Difference Built-up Indice in Automatically Mapping Urban Areas from TM imagery *International Journal of Remote Sensing* vol 24(3) pp 583-594.

# Neighbourhood Socio Economic Disadvantage Index's Analysis of the Flood Disasters Area at East Jakarta in 1996 and 2016

**Christina Ranti Ristiani*, Rokhmatuloh, Revi Hernina**

Department of Geography, University of Indonesia, Kampus UI Depok 16424

christina.ranti@gmail.com

**Abstract**. Flood is one of natural disasters that have often happened in East Jakarta. Flood can give several negative impacts and it can affect all aspects of society lives such as economics, political, cultural, socials and others. East Jakarta is an urban area which continuously grows and establishes to become a rapid area. It can be seen from the highest population density in East Jakarta (BPS, 2016) and categorized into a region prone to flooding based on data Prone Flood Map in 1996 and 2016. The higher population exists in East Jakarta, the bigger possibility of the negative effects of disaster it gets. The negative impacts of flood disaster can affect societies especially with socio-economic disadvantage. One of the index to measure socio-economic disadvantage is NSDI (Neighbourhood socio-economic disadvantage index). However, to adjust indicators used in NSDI with Indonesia statistical data compatibility, it needs further assessment and evaluation. Therefore, this paper evaluates previous main indicators used in previous NSDI studies and improves with indicators which more suitable with statistical records in Indonesia. As a result, there will be improved 19 indicators to be used in NSDI, but the groups of indicators remain the same as previous namely; income, education, occupation, housing, and population.

**Keywords**: NSDI, flood, Jakarta

## 1. Introduction

Recently, NSDI has been used to analyse the relationship between the public urban green open space availability and neighbourhood socio-economic disadvantage [1].

When being applied in spatial context, the relationship between NSDI and other phenomenon can be assessed by using Geographically Weighted Regression (GWR). Furthermore, by using Principal Component Analysis (PCA) socio-economic indicators within initial loading can be removed if not correlated with dependent variables [1,2,3].

Therefore, this study will discuss indicators used in previous NSDI application, and evaluate similar indicators which are more suitable with statistical records in Indonesia. The analysis within this paper is purposed to adapt NSDI indicators as possible approach to measure social deprivation in communities located in areas where flood occurred in East Jakarta from 1996-2016.

## 2. Methods

Firstly, all possible socio-economic indicators exist in East Jakarta statistical records will be explored especially those reflect disadvantage such as number of employment and education level.

The data exploration mainly focus on East Jakarta statistical records published in East Jakarta Statistical Bureau website [4]. The next step is to list all possible indicators within the website and compare it with previously used indicators in NSDI (Table 1). If there are many indicators adaptable with previous studies, it will select only the same number of indicators within one group.

## 3. Systematic review and discussion

At first, Australia Bureau of Statistics (ABS) in 2011 developed an index that summarizes variables to illustrate relative disadvantage, this index is called Index of Relative Socio-Economic Disadvantage (IRSD) [3]. In form of vector maps, IRSD can be displayed to rank the regions from the most disadvantaged socio-economic condition to least disadvantaged regions.

If an area has a low score of IRSD, it means that this area has high proportion of relatively disadvantaged people, while high score of IRSD reflects that the area has low incident of disadvantage [3].

The dimension used by ABS to compute IRSD:

1. Income variables,
2. Education variables,
3. Employment variables,
4. Occupation variables,
5. Housing variables, and
6. Other miscellaneous indicators of relative advantage or disadvantage.

However, for some Asian countries the indicators used for every variable by ABS seems out of context such as % people who do not speak English well. These differences can be found when Li and Liu (2016) conducted research to observe the relationship between Urban Public Green Spaces (UPGSs) and Neighborhood Socioeconomic Disadvantage Index (NSDI) [1]. Li and Liu (2016) instead used variables from ABS, they used 14 socioeconomic indicators which was grouped into 5 groups to compute NSDI.

Therefore, this paper develops new indicators for socioeconomic disadvantages based on data availability and compatibility with East Jakarta Municipality condition. The comparison of socioeconomic disadvantage indicators can be observed from Table 1.

**Table 1.** comparison of socio-economic disadvantage indicators [1,4]

| Groups | Indicators, Li and Liu (2016), 14 indicators in 5 groups | Data Compatibility and availability Indonesian Bureau of Statistics/ Government of Jakarta Province<br>Proposed 19 indicators |
|---|---|---|
| 1. **Income** | | |
| | Percentage of household with income below the average (%) | Percentage of poverty (%) |
| | Percentage of household receiving subsistence allowance (%) | Percentage of household receiving subsidized rice (%) |
| | | Percentage of people receiving subsidized health-care (%) |
| 2. **Occupation** | | |
| | Percentage of people living unemployment (%) | Percentage of people living unemployment (%) |
| | Percentage of people in temporary work (%) | Percentage of job seeker (%) |

| | | |
|---|---|---|
| | Percentage of blue collar worker (%) | - |
| | - | Percentage of people working in agriculture (%) |
| 3. **Education** | | |
| | Illiteracy rate (%) | Percentage of people never go to school (%) |
| | Percentage of people receiving no fundamental education (%) | Percentage of people with discontinued school (%) |
| | - | Percentage of people with no certificate (%) |
| | Percentage of people with degree below high school (%) | Percentage of people with primary school degree (%) |
| | - | - |
| 4. **Housing** | | |
| | Percentage of household without kitchens or toilettes (%) | Percentage of household without kitchens or toilettes (%) |
| | Percentage of household without clean energy (%) | Percentage of household without access to clean water (%) |
| | Percentage of low-rent housing households (%) | Percentage of low-rent housing households (%) |
| 5. **Population** | | |
| | Proportion of floating population (%) | Percentage of homeless people (%) |
| | Percentage of adult female living alone (%) | - |
| | Percentage of older people living alone (%) | - |
| | - | Percentage of birth assisted by non-medical (%) |
| | - | Percentage of divorced people (%) |

In their research, Li and Liu (2016) has demonstrated that there is negative correlation among NSDI and UPGSs' abundance, quality, and accessibility in most districts of Shanghai City, China.

Interestingly, Li and Liu (2016) used Principal Component Analysis (PCA) method to analyze the most significant indicators should be involved in their study. PCA had already been used by ABS (2011) to remove non-significant indicators. The formula of NSDI used [1]:

$$NSDI = \sum_{i=1}^{n} C_i \left(\sum_{j=1}^{k} L_j \, . \, x_j\right) \quad (1)$$

where $C_i$ expressed the component I's eigen value; $L_j$ is indicator of $j$'s loading score;$x_j$ is indicator $j$'s standardized value.

In this case to measure the area with disadvantage it will use the percentage of people living with poverty. The consideration behind this, because Indonesian Bureau of Statistics has measured several indicators to be used in poverty computation, hence no need to compute low dimension per capita income.

People living in poverty are the people whom its daily calorie consumption around 2100 kilo calorie or below, plus can only buy certain commodities in housing, cloth, education and health [4].

Since the growth of non-agricultural population proportion reflects rapid economic development [2], therefore in this study, we include the number of population which works in agriculture sector as indicator of disadvantage.

Since the Indonesian Bureau of Statistics has also counted the employed person whom in sick-leave as temporary unemployment, so in this study we do not include temporary unemployment as indicator of socio-economic deprivation, as suggested by previous research [1,2].

### *3.1. Principal Component Analysis*

Principal component analysis (PCA) is a multivariate technique that analyses a data table in which observations are described by several inter-correlated quantitative dependent variables.

The goals of PCA are to (a) extract the most important information from the data table, (b) compress the size of the data set by keeping only this important information, (c) simplify the description of the data set, and (d) analyze the structure of the observations and the variables [5].

PCA works by ranking the component in the variance-covariance which has the highest percentage of variance (%p) among other components. In the multivariate statistics, variance of each variable/factor can be computed by using:

$$\mathrm{var}_j = \frac{\sum_{i=1}^{n}(x_{ij}-\mu_j)^2}{n-1} \tag{2}$$

In PCA, variance and covariance can be formed to be a single matrix consists of eigenvalue ($\lambda_{pn}$) and eigenfactors. Basically, eigenvalue in j column is the variance, therefore the percentage variance of every column can be computed by using;

$$\%p = \lambda_{pn} \cdot (100)/\Sigma\lambda_{pn} \tag{3}$$

where %p is the percentage variance in component j, $\lambda_{pn}$ is variance in component j, and $\sum\lambda_{pn}$ is total variance [6].

To maintain the factor/variable, it can be observed from the factor loading value, which expresses the correlation ($R_{kp}$) between variable *k* with component *p*. The formula to calculate correlation between variable *k* with component *p* is:

$$Rkp = \frac{a_{kp}\sqrt{\lambda_{pn}}}{\sqrt{var_k}} \tag{4}$$

where $\boldsymbol{a_{kp}}$ =eigenvector (factor loading) for variable **k,** principal component ***p*; $\lambda_{pn}$** = *p-th* eigenvalue (component); ***var k*** = variance of variable *k* (diagonal values in the covariance matrix) (Jensen,1996).

In previous study, the socioeconomic indicators with factor loading more than 0.75was selected, and removed the factor with value less than 0.75 [1].

### *3.2. Proposes framework*

Therefore, based on aforementioned systematic review and discussion, it would be best for implementation of NSDI to replace earlier indicators used by Li and Liu (2016) with newly 19 indicators (Table1).

IOP Conf. Series: Earth and Environmental Science **98** (2017) 012007 doi:10.1088/1755-1315/98/1/012007

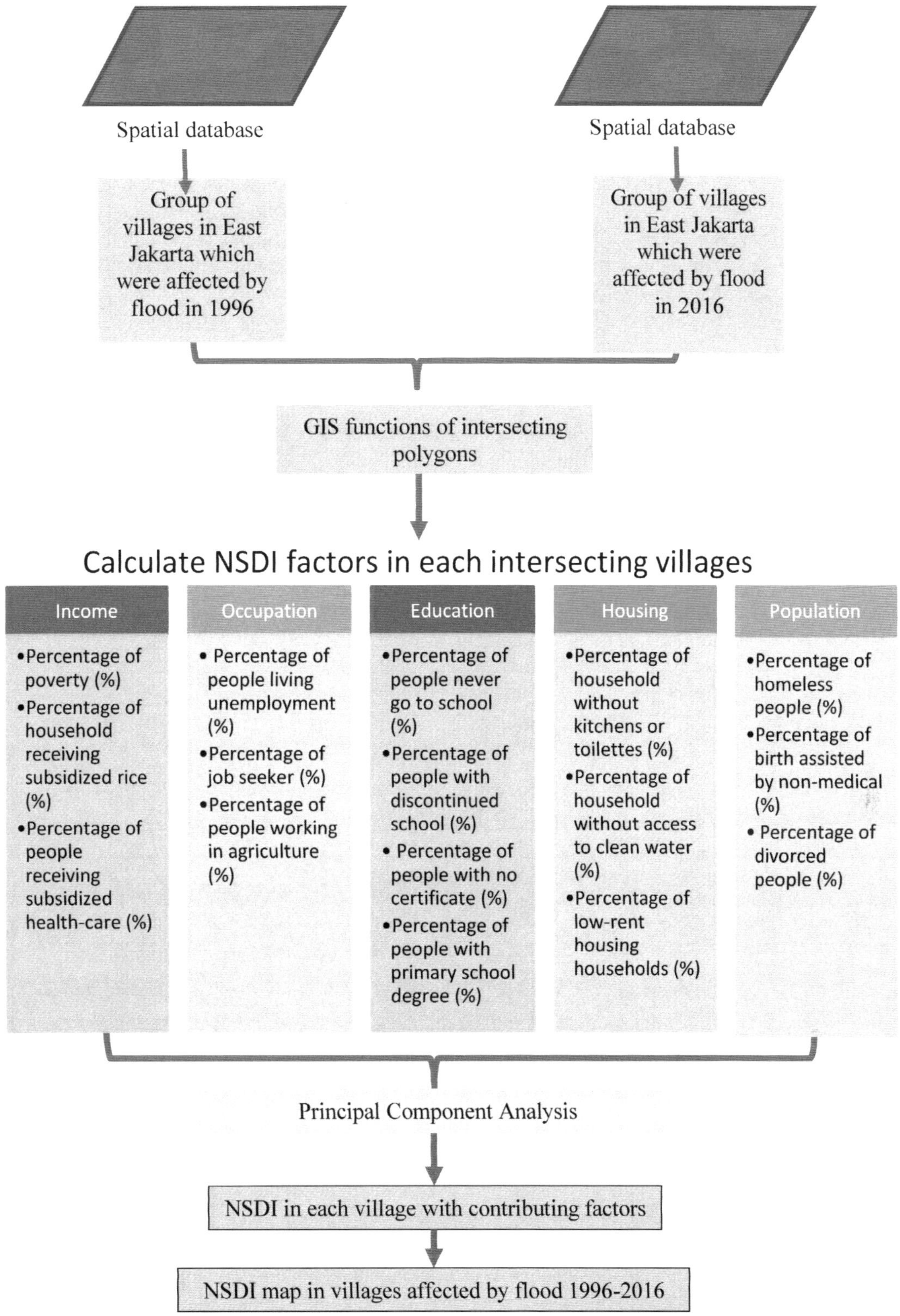


**Figure 1**. Proposed framework for NSDI computation in flood affected villages

## 4. Conclusion

Based on systematic review and discussion, proposed framework (Figure 1) to map villages which was affected by flood events from 1996-2016 in East Jakarta Municipality possible to apply with the condition that all proposed data are available.

## References

[1] Li H and Liu Y 2016 Neighborhood socioeconomic disadvantage and urban public green spaces availability: A localized modeling approach to inform land use policy, *Land Use Policy*, 57, p 470

[2] Su S, Gong Y, Tan B, Pi J, Weng M, and Cai Z 2017 Area Social Deprivation and Public Health: Analyzing the Spatial Non-stationary Associations Using Geographically Weighed Regression, *Social Indicators Research* , 133, p 819

[3] Australian Bureau of statistics 2011 Technical Paper Socio-Economic Indexes for Areas (SEIFA) http://www.ausstats.abs.gov.au/ausstats/subscriber.nsf/0/22CEDA8038AF7A0DCA257B3B00116E34/$File/2033.0.55.001%20seifa%202011%20technical%20paper.pdf

[4] Indonesia Satistical Bureau 2017 https://www.bps.go.id/Subjek/view/id/23

[5] AbdiH and Williams LJ 2010 Principal component analysis, *WIREs Comp Stat,* 2, p 433

[6] Jensen J. R. 2005 *Introductory Digital Image Processing: A Remote Sensing Perspective* (Prentice Hall)

# Geographic Information System and Remote Sensing Approach with Hydrologic Rational Model for Flood Event Analysis in Jakarta

**M R Aditya, R Hernina*[*] and Rokhmatuloh**

Department of Geography University of Indonesia, Kampus UI Depok 16424

revie12@yahoo.com

**Abstract**. Rapid development in Jakarta which generates more impervious surface has reduced the amount of rainfall infiltration into soil layer and increases run-off. In some events, continuous high rainfall intensity could create sudden flood in Jakarta City. This article used rainfall data of Jakarta during 10 February 2015 to compute rainfall intensity and then interpolate it with ordinary kriging technique. Spatial distribution of rainfall intensity then overlaid with run-off coefficient based on certain land use type of the study area. Peak run-off within each cell resulted from hydrologic rational model then summed for the whole study area to generate total peak run-off. For this study area, land use types consisted of 51.9 % industrial, 37.57% parks, and 10.54% residential with estimated total peak run-off 6.04 $m^3$/sec, 0.39 $m^3$/sec, and 0.31 $m^3$/sec, respectively.

**Keywords:** Run-off, Impervious surface, Flooding, Jakarta

## 1. Introduction

Nowadays, there are changing views about Geographic Information System (GIS) definition which not only seen as a computerized system but also as integrated software and hardware used to develop and analyze data for problem solving [1].

Recently, GIS platform has been used to integrate spatial database (Digital Elevation Mode (DEM), land use type, soil map, rainfall, slope, and drainage area) to calculate peak run-off with hydrological rational model [2]. Moreover, locations with high risk flood can be assessed by analyzing areas which have total peak run-off estimation based on rational model which was combined with arithmetic overlay [3]. On 9-10 February 2015 there was massive flood event in Jakarta Province, where two days of heavy precipitation from 9 – 10 February 2015 was classified as very exceptional [4].

Therefore, to give more detail about flood event analysis on 9-10 February 2015, this article uses both GIS & remote sensing approach with hydrological rational model to observe peak run-off spatial distribution on specific location in Jakarta Province, Indonesia.

### *1.1. Study Area*

Study area is located in Grogol Petamburan sub-district as a part of West Jakarta Municipality, with population of 223,604 people in 2016 [5].

IOP Conf. Series: Earth and Environmental Science **98** (2017) 012008 doi:10.1088/1755-1315/98/1/012008

This unit of analysis is also positioned between two major rivers flow in Jakarta Province, Pesanggrahan and Ciliwung River. Based on GIS assisted geometry calculation, total area of this spatial unit is 1.53 sq km (Figure 1).

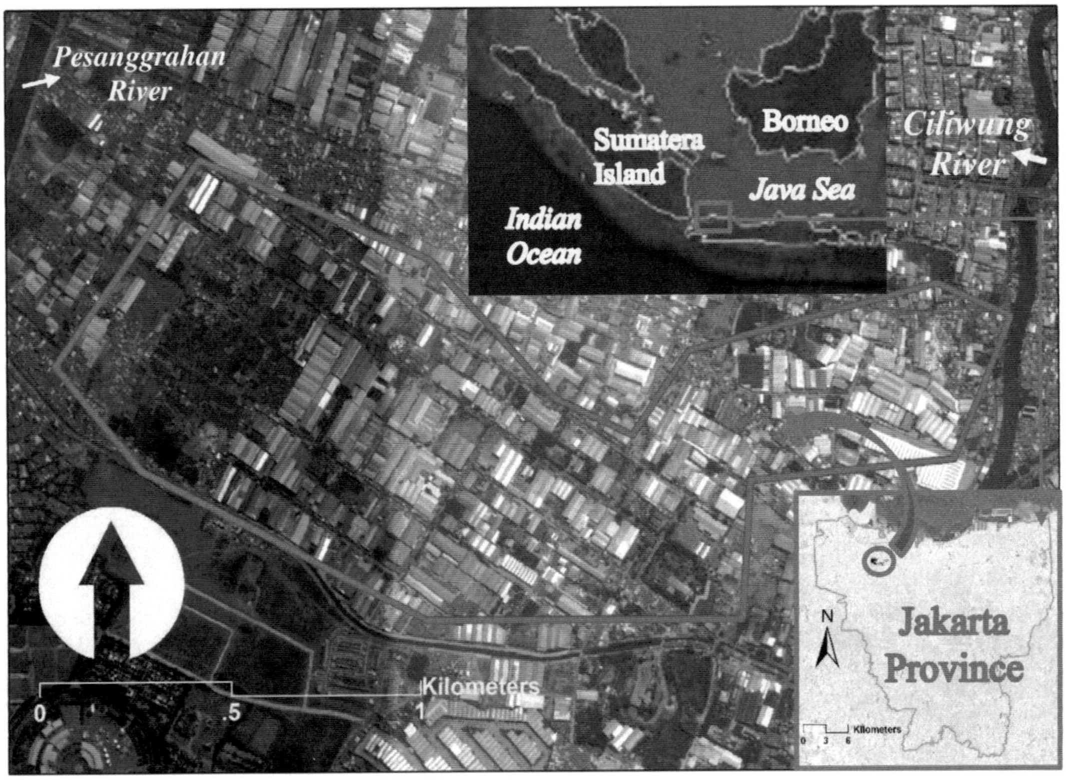


**Figure 1**. Study area

## 2. Methods

To describe each step to gain result of peak run-off spatial analysis from GIS and hydrologic rational integration, a framework has been made in Figure 2.

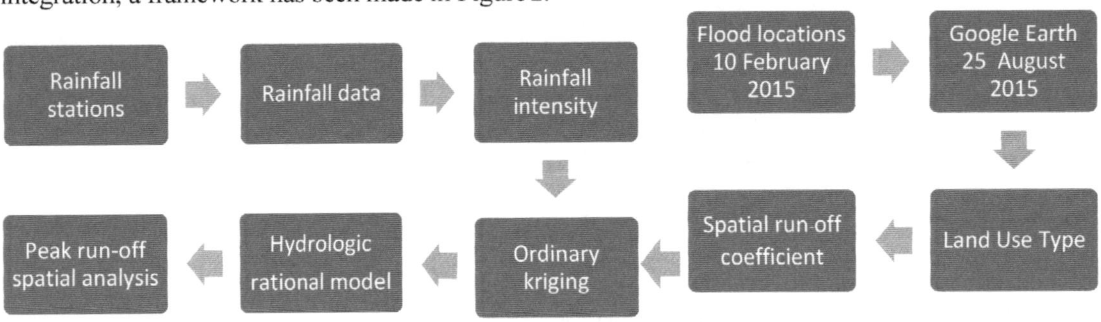


**Figure 2**. Framework for integrated GIS & remote sensing approach with hydrologic rational model

### 2.1. Data

In order to compute rainfall intensity, rainfall data were downloaded from Indonesian Meteorology and Geo-physics Agency website [6] which provided report about the flood event of Jakarta Province within 8-10 February 2015.

In addition, flood locations were downloaded from website of Indonesian Natural Disaster and Mitigation National Agency [7]. This map then digitized to become spatial database of flood locations. Furthermore, to analyze the land use type within flood location, Google Earth images dated 25 August 2015 were downloaded.

**Table 1.** Data used

| Name | Data type | Data processing with GIS or remote sensing technique |
|---|---|---|
| Rainfall data 8-10 February 2015 | spreadsheet | Query selection by attributes |
| Flood locations | map | Digitize |
| Google Earth images 25 August 2015 | RGB images | Unsupervised image classification |

### 2.2. Geographic Information System

In this article, kriging plays an important role, where it interpolates rainfall intensity value based on several rainfall stations.

Once it gets the cell with specific information of rainfall intensity for every pixel, it can be multiplied with run-off coefficient. Moreover, with the help of raster calculator function of ArcGIS software, the value of in every cell can generate the peak run-off information. Another important function of GIS used in this study is masking capability where it could detach selected raster cell of study area from the classified raster cells.

2.2.1. *Ordinary kriging (OK).* Ordinary kriging was used to interpolate rainfall intensity data of 10 February 2015. OK works by using constant mean (μ), while the objective of using ordinary kriging is to make interpolation smoother based on data trend [8]. OK has been applied successfully to interpolate rainfall data both yearly and monthly in Sri Lanka though minimum values were overestimated and maximum values underestimated [9].

### 2.3. Rainfall intensity

To predict rainfall intensity, this article uses Mononobe equation [10]:

$$I_t = \frac{R_{24}}{24}\left(\frac{24}{t}\right)^m \tag{1}$$

where

$I_t$ = Rainfall intensity (mm/hour)
$R_{24}$ = 24-hour rainfall duration t
$m$ = 2/3

### 2.4. Hydrologic rational model

To compute hydrological rational model with per-pixel analysis, Equation (2) [2,3] was used:

$$Qp = C \times I \times A \tag{2}$$

where

$Q_p$ = Peak run-off (m3/sec)
$C$ = Run-off coefficient

IOP Conf. Series: Earth and Environmental Science **98** (2017) 012008 doi:10.1088/1755-1315/98/1/012008

$I$ = Rainfall intensity (mm)
$A$ = Drainage area (in this case pixel unit at spatial resolution 1.6 m x 1.6 m)

*2.5. Image processing*
To acquire land use type in the study area, image processing were performed by employing these steps;
- Download high resolution image from Google Earth dated 25 August 2015.
- Image geometric correction by using orde-3 polynomial method.
- Unsupevised classification using ISODATA clustering to get 3 land use type namely; industrial, residential, and parks.

*2.6. Run-off coefficient*
Run-off coefficient is needed for every pixel in land use type which was generated from unsupervised image classification from Google Earth image. In addition, this article uses run-off coefficient in Table 2.

**Table 2.** Run-off coefficient [11]

| Land use type | C (median) |
|---|---|
| Industrial heavy areas | 0.75 |
| Residential, multi units-detached | 0.50 |
| Parks | 0.125 |

## 3. Result and discussion
Using Equation (1), rainfall intensity from 12 stations can be predicted (Table 3).

Furthermore, it can be observed from Table 3 that the highest rainfall intensity occurred in Tanjung Priok rainfall station while the lowest was in Soekarno-Hatta Airport.

**Table 3.** Rainfall intensity 10 February 2015 (data processing from [6])

| Number | Rainfall stations | Rainfall intensity (mm/hour) |
|---|---|---|
| 1 | Pintu Air Pulogadung | 22.07 |
| 2 | Jembatan Panus, Depok Jawa Barat | 8.44 |
| 3 | Stasiun Meteorologi Kelas I Soekarno-Hatta Cengkareng | **0.73** |
| 4 | BMKG Citeko | 3.16 |
| 5 | BMKG Jl. Angkasa No.2 | 16.07 |
| 6 | Stasiun Klimatologi Bogor | 7.22 |
| 7 | Stasiun Meteorologi Maritim Tanjung Priok | **29.29** |
| 8 | Pintu air Kali Pesanggrahan Kedoya | 17.28 |
| 9 | Pintu air manggarai | 17.36 |
| 10 | Pintu air cideng | 17.04 |
| 11 | Pintu air Sunter | 9.41 |
| 12 | Pintu air karet | 14.44 |
| 13 | Angke hulu | 5.11 |
| 14 | Krukut hulu | 7.30 |
| 15 | Pintu pesanggrahan depok | 17.36 |
| 16 | Pasar Teluk Gong | 16.39 |
| 17 | Waduk Kebon Melati | 14.68 |
| 18 | Rorotan | 21.18 |
| 19 | Ragunan Zoo | 6.41 |
| 20 | Presidential Palace | 7.87 |

### *3.1. Rainfall intensity interpolation with ordinary kriging*

Rainfall intensity interpolation using OK is displayed in Figure 3 (A), where higher rainfall intensity can be observed close to the northeast of Jakarta Province.

In addition, it can be visually observed from Figure 3 (B) that most flooded areas were located in the north-eastern part of Jakarta Province.

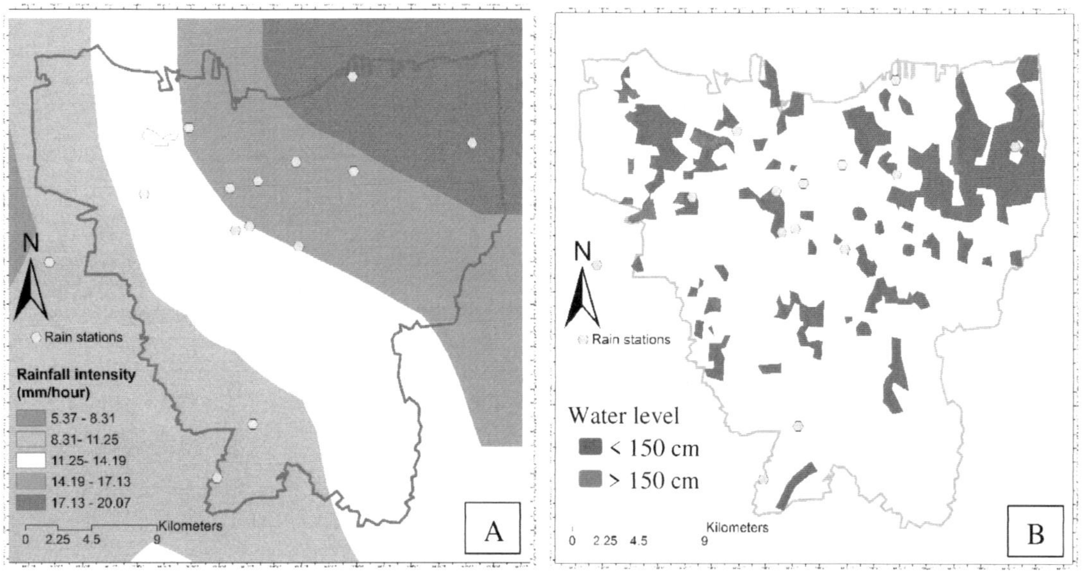


**Figure 3**. (A) Rainfall intensity spatial distribution using OK, (B) spatial distribution of flood event February 2015 (data processing from [6] [7])

### *3.2. Peak run-off spatial distribution*

In terms of spatial analysis, spatial distribution of peak run-off is displayed in Figure 4. Most of the study area was dominated by industrial land use type with approximately an area extent of 0,79 sq km (51.91%), which resulted in estimated peak run-off 0.028 $m^3$/hour/pixel.

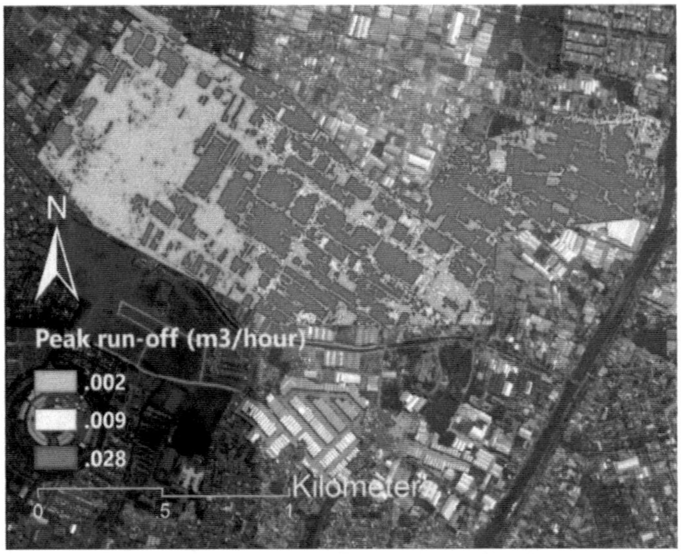


**Figure 4**. Peak run-off spatial distribution

Moreover, the smallest land use type was residential with an area extent of 0.16 sq km, and predicted peak run-off 0.09 $m^3$/hour/pixel. Surprisingly, this area also had parks around 0.58 sq km (37,57%) which was located in the eastern part. For further analysis, the study area which is surrounded by two major rivers, Pesanggrahan and Ciliwung River, make it unfavorable since the water level from those rivers might also exceeded riverbanks. Therefore, for further study the distance from river could be included in the analysis not only limited to spatial hydrology analysis.

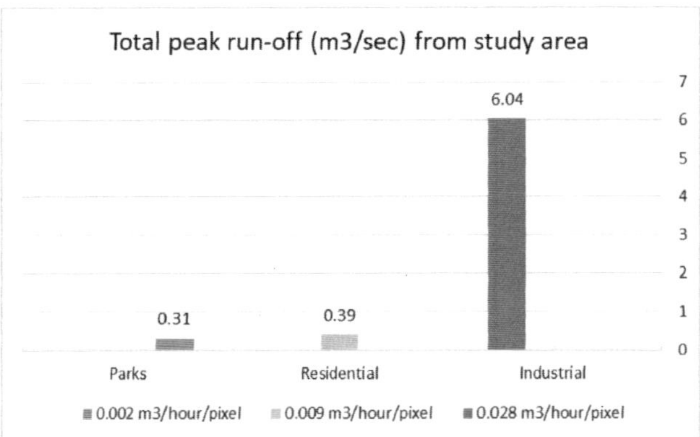


**Figure 5**. Total Peak run-off

In terms of estimated total peak run-off from the study area, the biggest came from industrial land use type with approximately 6.04 $m^3$/sec, followed by residential (0.39 $m^3$/sec) and parks (0.31 $m^3$/sec), respectively (Figure 5).

## 4. Conclusion

By combining both GIS and remote sensing with rational hydrological model, this study has managed to reach its objectives that is to display information about peak run-off spatial analysis. The biggest total peak run-off came from the eastern part of the study area where it was dominated by industrial land use, while the smallest total peak run-off mostly resulted from the western part.

**References**

[1] Wieczorek W F, Delmeric A M 2009 Geographic Information Systems *Comput Stat* **1**(2) p 167
[2] Kyei D A, Forkuor G, Venus V 2015 Modeling Flood Hazard Zones at the Sub-District Level with the Rational Model Integrated with GIS and Remote Sensing Approaches *Water* **7** p 3531
[3] Nyarko BK 2002 Application of Rational Model in GIS for Flood Risk Assessment in Accra *Journal of Spatial Hydrology* **2** p 1
[4] Siswanto et al 2017 A very unusual precipitation event associated with the 2015 floods in Jakarta: an analysis of the meteorological factors, *Weather and climate extremes* **16** p 23
[5] Indonesian Statistical Bureau 2017 West Jakarta Municipality in Figures (Kota Administrasi Jakarta Barat Dalam Angka- Bahasa Indonesia) https://jakbarkota.bps.go.id/backend/pdf_publikasi/Kota-Administrasi-Jakarta-Barat-Dalam-Angka-2017.pdf
[6] Indonesian Agency of Meteorology Climatology and Geophysics 2015 Jakarta Flood Event Analysis Report (Analisa Kejadian Banjir Jakarta- Bahasa Indonesia) http://data.bmkg.go.id/share/Gambar_Foto/Artikel/ANALISIS_KEJADIAN_BANJIR_PROPINSI_DKI_JAKARTA_TANGGAL_09_PEBRUARI_2015.pdf
[7] Jakarta Province Natural Disaster Mitigation Agency 2016 Flooded Area Report Map of February 2015 http://gis.bpbd.jakarta.go.id/documents/58
[8] ESRI 2017 http://desktop.arcgis.com/en/
[9] Wijemannage A L K, Ranagalage M, Perera E N C 2016 Asian Conference of Remote Sensing
[10] Gautama RS 1997 International Mine Water Association Proceedings p 235
[11] California Environmental Protection Agency 2011 Runoff Coefficient (C) Fact Sheet https://www.waterboards.ca.gov/water_issues/programs/swamp/docs/cwt/guidance/513.pdf

# Identification of Fault Poso Earthquake Causes 2017 Mw 6.6 with Gradient Vertical Gravity Satellite Imagery

**B M M Simamora[1*], I Fajerianti[1,2], B Utaminingtyas[1], T M Apriliaji[1], A Fatimah[1], K Nafi[1]**

[1]State College of Meteorology, Climatology and Geophysics
[2]Meteorology, Climatology and Geophysics Agency

buhasimamora089@gmail.com

**Abstract**. May 29, 2017 a devastating earthquake occurred in the district of Poso, Central Sulawesi. This earthquake can be affected significantly by the people. Based on the analysis of the source mechanism, an earthquake caused by a normal fault around the graben zone of Palopo. The purpose of this research is to identify Earthquake-causing fault of Poso 2017 Mw 6,6 by using Second Vertical Derivative (SVD) analysis of gravity. The result of profiles line that bypass the alleged fault orientation indicates the existence of normal fault structure indicating the absolute value of the minimum SVD anomaly is relatively smaller than the maximum SVD anomaly value in the epicenter zone of the earthquake.

**Keywords**: earthquake, SCD, Poso

## 1. Introduction

On May 29, 2017 at 21:35 pm there was a tectonic earthquake arround of Central Poso District with the power of Mw 6.6. The earthquake occurred at the coordinates S 1.28°, E 120.48° at a depth of 10 km. The epicenter on 38 km Northwest Poso, Central Sulawesi. Based on the results of the shock level shakemap analysis, BMKG indicates that this earthquake is potentially felt on the scale of II SIG-BMKG or III-V MMI in Poso, Loeo, Palu, Kasongan, and Toli-Toli. This is in corresponding with the people report that this earthquake is felt on the intensity scale II SIG-BMKG or V MMI in Poso and Torue, Palu and Sigi (III-IV MMI). People in Toli-toli, Pasang Kayu and Tana Toraja also felt on the III MMI scale. Likewise, people in Gorontalo, Boalemo, Pohuwato and Bone Bolango feel the earthquake on the scale of II SIG-BMKG (III MMI). People in Palopo, Masamba, and Balikpapan also feel this earthquake on the scale I SIG-BMKG (II MMI).

The purpose of this research is to identify Earthquake-causing fault of Poso 2017 M 6,6 by using Second Vertical Derivative (SVD) analysis of gravity. Previously, there has been no identification of the fault structure in the Graben Palopo region using gravity anomaly data. Hartati (2012) [1]has previously conducted cesarean identification research using gravity method but did not focus in Central Sulawesi. Hartati (2012)[1] had previously conducted a fault identification study using gravity methods but did not focus on Central Sulawesi.

## 2. Data and method

The data used in this research are satellite gravity anomaly data obtained from CryoSat-2 and Jason-1 satellite observations [2]. This data is a global data with grid every 1 minute degree. Downloaded data in the form of free air anomaly data (FAA). Data downloaded for coordinate constraints S 0.28o - 2.28o, and E 119.48o - 121.48o. Furthermore, bouguer correction was done to obtain a simple bouguer anomaly of the study area. This study is limited by not doing terrain correction. After obtaining the bouguer anomaly value, separation of regional anomalies and residual anomalies was performed. The method of separating anomalies in this study is to use a digital screening approach to separate anomalies that have long wavelengths associated with regional anomalies (Deep effect) and anomalies with shorter wavelengths associated with target structural causes of shallow earthquakes. In this research, we selected separation method using second vertical derivative filter (SVD). SVD is done by performing a convolution between bouguer anomalies and the SVD filter matrix. In this research the SVD filter matrix used is the matrix of Elkins (1951)[3]. The vertical derivative profile of these two gravity anomalies can be used to aid the interpretation of the types of earthquake-causing structures.

## 3. Result

After correcting the bouguer, a simple bouguer anomaly was obtained. Map of this bouguer anomaly can be seen in Figure 1. In the study area of two-degree times two degrees or about 12,392.14 km2 (assuming one degree in the equator = 111.32 km) there are 21,901 observation points. It means there is one observation point in every 1.77 $km^2$. The Poso earthquake with the magnitude of Mw 6.6 empirically calculated using the Wells and Coppersmith equations (1994)[4] is generated by a normal fault (down) with dimensions about 29.34 x 16.34 km. By the length and width of the fault so that the dimensions of the source of the earthquake will be identified by the gravity anomaly analysis performed in this research.

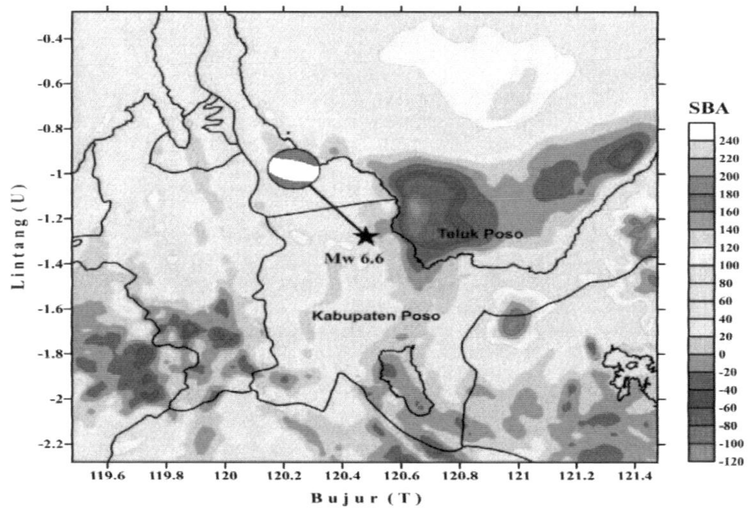


**Figure 1.** Map of the simple bouguer anomaly of Poso District and surrounding areas.

The purpose of this research is to identify Earthquake-causing fault of Poso 2017 M 6,6 by using Second Vertical Derivative (SVD) analysis of gravity. Previously, there has been no identification of the fault structure in the Graben Palopo region using gravity anomaly data. Hartati (2012) [1]has previously conducted cesarean identification research using gravity method but did not focus in Central Sulawesi. Hartati (2012) [1]had previously conducted a fault identification study using gravity methods but did not focus on Central Sulawesi.

The smallest bouguer anomaly in the study area (Figure 1) has a value of -108.19 mGal located in Poso Bay or southeast of the epicenter of the earthquake. The largest bouguer anomaly in the study area has a value of 243.12 mGal on land in eastern bay of Poso. While in the southwest of the epicenter zone there is also an earthquake bouguer anomaly is quite high which is about 80-120 mGal. From the mapping of the bouguer anomaly, it can be seen that the epicenter location of Poso Mw 6.6 earthquake is in the southwest of the low (negative) anomaly in Poso Bay and northeast of the positive anomaly contour. Judging from the geological map, the epicenter area of Poso Mw 6.6 earthquake is located in Graben Palolo zone [5]. The fault orientation of the Poso Mw 6.6 earthquake can be seen through the source mechanism parameters of GCMT which is a fault with northwest-southeast straightness.

If using Poso earthquake epicenter data from BMKG analysis (E 120,48, N -1,28), the position of this earthquake is in zone with anomaly value close to zero (about 5 mGal). The Poso earthquake epicenter zone is flanked by two opposite anomalies. The northeast of the epicenter location tends to have a negative anomaly, while the southwest tend to have a positive anomaly pattern. Earthquakes tend to occur at the border between two different anomalies. A low (negative) anomaly value indicates a region with a low subsurface density while a high (positive) anomaly value indicates otherwise.

Result of mapping of residual anomaly from screening result using SVD method can be seen in Figure 2. There was a significant difference between the contours of bouguer anomaly and its residual anomalies. This suggests that the contour of bouguer anomaly is dominated by inner effects (regional anomalies). From the results of the residual anomaly mapping (shallow effect), it can be seen clearly that the location of the epicenter is in a zone with residual anomalies of about 5-10 mGal. While in the southwest near this epicentral zone, there is a high residual anomaly with a value of about 90-120 mGal. To further deepen the interpretation of this residual anomaly, a four-lane profile is made of paths A-A ', B-B', C-C 'and D-D' as shown in Fig. 2.

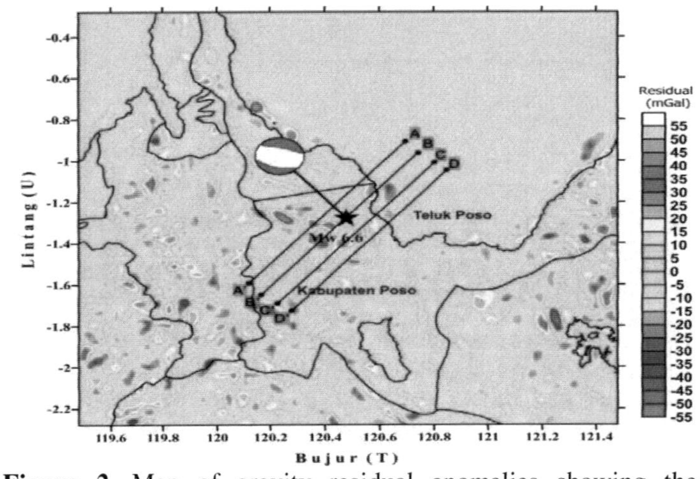


**Figure 2.** Map of gravity residual anomalies showing the shallow effect of filtration by a second vertical derivative method.

The slicing created in the northeast-southwest orientation crosses the orientation direction of the caustic fault of GCMT modeling results. The purpose of making this slicing is to see the SVD profile of each track in order to help interpretation the structure in accordance with the theory obtained from previous studies.

The profile of the four trajectories can be seen in Figure 3. Figures are presented in sequential SVD profiles from A-A' to D-D' paths. Broadly speaking, if we assume the use of epicenter data from BMKG, then the location of the epicenter is at a distance of about 40-60 km from the starting point of the track. The location of the epicenter looks after a positive anomaly and before the anomaly is negative when viewed from the starting point of the track (points A, B, C and D). The B-B' slicing cuts right at the epicenter of the BMKG earthquake and the red star in Figure 3 illustrates the position of the epicenter of Poso Mw 6.6 earthquake on the profile.

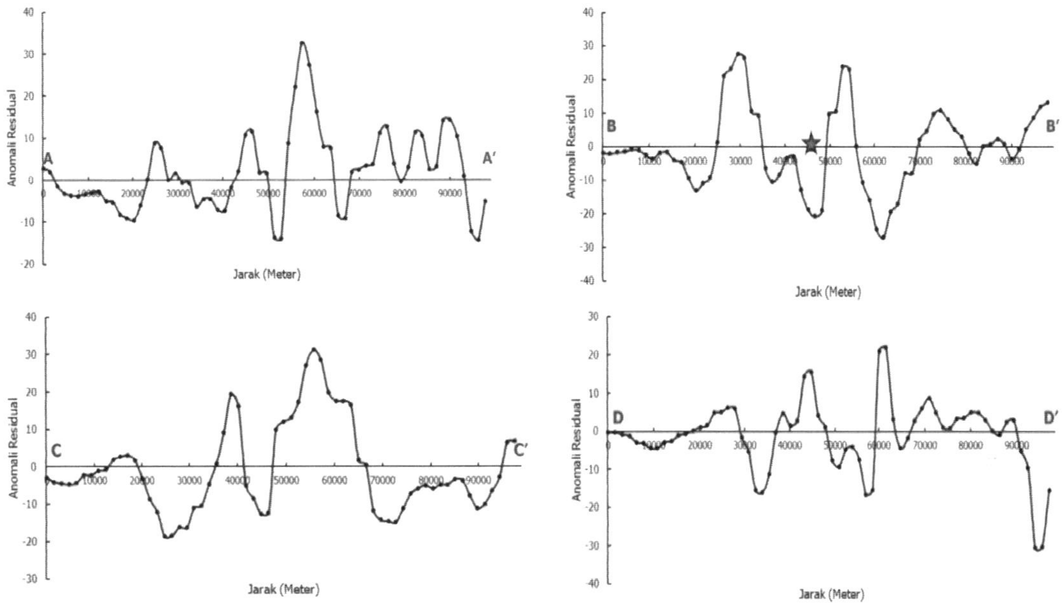


**Figure 3.** Profile residual anomaly (mGal) on the A-A ', B-B', C-C 'and D-D' track respectively

From the residual anomaly profile of the B-B path and supported by the third profile of the other paths, it can be seen that the absolute value of the SVD price is at least less than the absolute value of the maximum SVD. The absolute minimum SVD value in the northeast of the epicenter is 20.8 mGal while the absolute maximum SVD value to the northeast is 23.8 mGal. This reinforces the notion that the Erthquke Poso May 29 2017 Mw 6.6 is caused by the fault structure that is experiencing a downward movement (normal fault). Reynolds (1997) [6]states that the criteria for determining the type of structure of the fault are as follows: normal fault (decreasing) indicates the absolute value of the minimum SVD anomaly is relatively smaller than the maximum SVD anomaly value while for the reverse upward fault.In gravity data, the anomalous values will change vertically as a result of the uneven distribution of mass uniformly vertically, then the derivatives of both will show the magnitude of the gravity effect of the wider structures and lie deeper. Therefore small / local and vague structures can be clarified or more sharpened by curve form rather than regional structures that are more widespread in shape.

## 4. Conclusion

Based on the analysis on the vertical derived profiles of the two vertical gravity images of the satellite in the epicentral zone, it can be concluded that the Poso earthquake of 29 May 2017 with the power of Mw 6.6 is caused by normal fault (decreasing) in the grid zones of Palolo. This norml fault can be shown with the absolute value of the minimum SVD anomaly relatively smaller than the maximum SVD anomaly value in the epicenter zone of the earthquake.

## References

[1] Hartati, Anita. 2012 Identifikasi Struktur Patahan Berdasarkan Analisa Derivative Metode Gayaberat di Pulau Sulawesi. *Skripsi*. FMIPA Universitas Indonesia, Jakarta.
[2] Sandwell D T, Müller R D, Smith W H, Garcia E, & Francis R 2014 New global marine gravity model from CryoSat-2 and Jason-1 reveals buried tectonic structure. Science, 346(6205), 65-

67.

[3] Elkins, T. A. 1951 *The second derivative method of gravity interpretation*. Geophysics, 16(1), 29-50.

[4] Wells D L, & Coppersmith K J 1994 New empirical relationships among magnitude, rupture length, rupture width, rupture area, and surface displacement. *Bulletin of the seismological Society of America*, 84(4), 974-1002.

[5] Simanjuntak T O, Surono dan Supandjono J B 1997 Peta Geologi Lembar Poso, Sulawesi, Skala 1:250.000. Pusat Penelitian dan Pengembangan Geologi, Bandung.

[6] Reynolds J M 1997 *An Introduction to Applied and Environmental Geophysics*. West Sussex, England, John Wiley & Sons Ltd, 796p.

# Land Subsidence Monitoring by InSAR Time Series Technique Derived From ALOS-2 PALSAR-2 over Surabaya City, Indonesia

**A Aditiya[1,3*], W Takeuchi[1] and Y Aoki[2]**

[1] Civil Engineering Department, The University of Tokyo, Tokyo 151-0063, Japan
[2] Earthquake Research Institute, The University of Tokyo, Tokyo 151-0063, Japan
[3] Geospatial Information Authority of Indonesia, Jl. Raya Jakarta-Bogor Km. 46, Cibinong, Indonesia

arif@iis.u-tokyo.ac.jp

**Abstract** Surabaya is the second largest city in Indonesia and the capital of East Java Province with rapid population and industrialization. The impact of urbanization in the big city can suffer potential disasters either nature or anthropogenic such as land subsidence and flood. The pattern of land subsidence need to be mapped for the purposes of planning and structuring the city as well as taking appropriate policy in anticipating and mitigating the impact. This research has used interferometric Synthetic Aperture Radar (InSAR) Small Baseline Subset (SBAS) technique and applied time series analysis to investigate land subsidence occured. The technique includes the process of focusing the SAR data, incorporating the precise orbit, generating interferogram and phase unwrapping using SNAPHU algorithms. The results showed land subsidence has been detected during 2014-2017 over Surabaya city area using ALOS-2/ PALSAR-2 images data. These results reveal the subsidence has observed in several area in Surabaya in particular northern part reach up to ~2 cm/ year. The fastest subsidence occurs in highly populated areas suffer vulnerable to flooding and sea level rise impact. In urban areas we found a correlation between land subsidence with residential or industrial land use. It concludes that land subsidence is mainly caused by ground water consumption for industrial and residential use respectively.

**Keywords:** Land Subsidence, Time Series, InSAR, PALSAR-2, Surabaya

## 1. Introduction

Surabaya which the second largest city in Indonesia with population more than 2.7 million people as well as the capital of East Java province is coastal city that has a delta system and located on low elevation (less than 10 meters above sea level). As the urban area Surabaya has potential antrophogenic hazard such as land subsidence, flood, inundation, drought, crack in the road and building damage. Land subsidence its self is the combination of the natural compaction of sediments, ground water extraction, geothermal fluids, coal, and other solids through mining; and underground construction. Therefore long-term land surface monitoring is needed for urban area to reduce economic loss and sustainable improvement. Most of the major land subsidence areas around the world due to the increasing use of ground water [1]. Even if the hazards associated with subsidence are different from those caused by sudden and catastrophic natural events like floods and earthquakes, because surface sinking is a slow event, expansive damage can occur. The pattern of land subsidence need to be mapped for the purposes of planning and structuring the city as well as taking appropriate actions in anticipating and mitigating the impacts. These characteristics can be determined by several measurements such as direct measurement, levelling, Global Positioning System (GPS), InSAR Interferometric Synthetic Aperture Radar (InSAR) and Gravimetry [2]. The application of satellite radar data have provided the ability to detect and monitor ground deformation with centimeter to milimeter precision at greater spatial detail and ability to cover remote area.

The objective of this study is to investigate surface elevation change of environmental in particular land subsidence episode between 2014 and 2017 in Surabaya city, Indonesia. To recognize the pattern characteristics of land subsidence, active remote sensing technique such as Synthetic Aperture Radar (SAR) has been widely implemented to characterize of urban area. The basic principle of SAR is Sensor emitted the pulse to the object and getting back reflection of electromagnetic waves from the object and calculated as phase differential.

**Figure 1.** Condition of Surabaya City over last 5 years.

## 2. Methodology

### *2.1. Study Area*

Surabaya is located in the northern coast of Java island and mostof the area locates on low elevation between 3 up to 6 m above sea level. Surabaya also places around hill called Bukit Lindah and Gayungan on southwest direction with high elevation 25-50 meters above sea level. Total area cover of Surabaya is 33,306 Ha consists of 31 sub-districts and 160 urban villages. Alike a tropical city Surabaya has two seasons, dry and wet seasons. Between November up to April are months for dry season, July up to October for wet season, while month remaining for transition season. Temperature in Surabaya was 23.6 up to 33.8°C in daily average, maximum air moisture 92% and air pressure 1012.5 hPa. Looking into Surabaya topography, there are large area will be directly affected to impact of sea level rise. The following description of field survey was focused on the areas where those locate nearest distance from coastline. As defined by Local Government of Surabaya that the coastal area is the areas from shoreline continue to inland nearest coastal area. It may be consist of wet and dry land that still have affected by sea behavior.

Soil type in Surabaya was composed by alluvial deposit, which develop from river and coastal features. Mostly Brantas River and Rowo River affected the characteristic for river feature and some part area in east and north of Surabaya continue to along Madura Strait has coastal characteristic. It encompasses clay, silt and clay silty. A thin sheelfish layer was founded in some places. Geologic of

IOP Conf. Series: Earth and Environmental Science **98** (2017) 012010 doi:10.1088/1755-1315/98/1/012010

Municipality Surabaya consists of clay and sandy layer. Alluvial soil that susceptible for cropland dominated condition of soil. The soil was founded in west of Surabaya contain a highly lime soil that less open for plantation.

Based on geological study conducted by Geology Agency, the soil capacity in carring a building construction indicated that soil in Surabaya varies types and uneven spreading areas that differ capacity in place to place in old city part such as in the following districts of Wonokromo, Sawahan, Genteng, Tegalsari, Gubeng, tambaksari, Simikerto, Semampir, Paben Cantikan, Krembangan and Bubutan were classified a clay soil with 10-18 m depth [3]. The building foundation that would be constructed in this type soil has to be grounded on 25-30 m depth. The hilly area in west city mostly contain clay soil so the building foundation has to be grounded on 4 up to 10 mdepth.

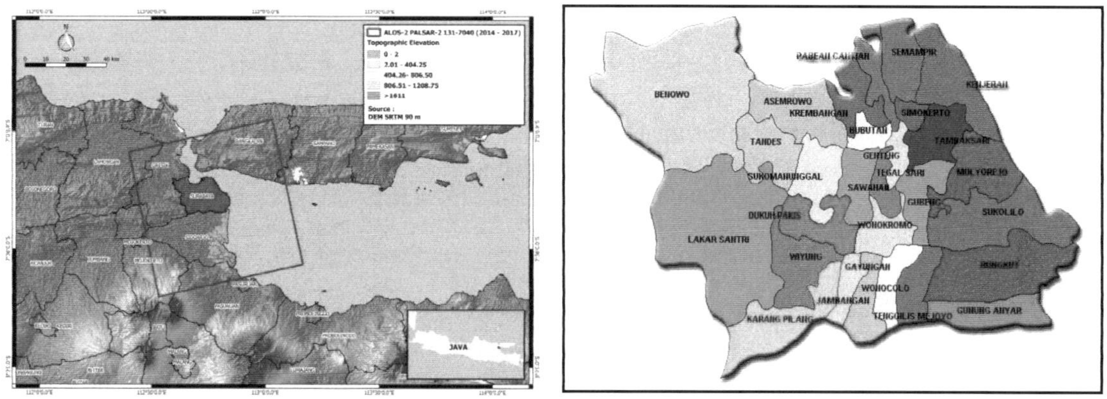


(a) (b)

**Figure 2.** (a) Geographic location of Surabaya, Indonesia and red rectangle of PALSAR-2 image; (b) Administrative sub-district boundary of Surabaya.

### *2.2. Image Data Sets*

The data used in this study is Phased Array Type L-band Synthetic Aperture Radar (PALSAR-2) sensor onboard of the Advanced Land Observing Satellite (ALOS-2) images. The ALOS-2 has worldwide observation scenarios for various missions, e.g., detection of deformation, monitoring forest change, and monitoring sea ice. Its precise orbit control and the gallium nitride (GaN) high- power transmitter/receiver enable accurate analyses in high spatial resolution. We used acquired data period from September 2014 to July 2017. These data sets were acquired in the ascending orbit with an off-nadir angle of 28.2°. Observation parameters for all the images were as follows observation mode Strip Map ; track 131 and right looking direction. These data were used to generate interferogram and to obtain line of sight (LOS) value. Table 1 shows the cover ranges of the ALOS-2/ PALSAR-2 data.

IOP Conf. Series: Earth and Environmental Science **98** (2017) 012010 doi:10.1088/1755-1315/98/1/012010

**Table 1.** List of ALOS2/PALSAR2 data

| Satellite | Orbit | $B_{PERP\,(m)}$ | Track | Off Nadir Angle | Acquisition Date | Look Direction |
|---|---|---|---|---|---|---|
| ALOS-2 | Asc. | 0 (M) | 131 | 28.2° | 20140916 | Right |
| ALOS-2 | Asc. | 127.0 | 131 | 28.2° | 20150707 | Right |
| ALOS-2 | Asc. | 28.8 | 131 | 28.2° | 20150915 | Right |
| ALOS-2 | Asc. | 40.5 | 131 | 28.2° | 20160202 | Right |
| ALOS-2 | Asc. | -37.1 | 131 | 28.2° | 20160705 | Right |
| ALOS-2 | Asc. | -8.5 | 131 | 28.2° | 20160913 | Right |
| ALOS-2 | Asc. | -121.2 | 131 | 28.2° | 20170131 | Right |
| ALOS-2 | Asc. | -173.6 | 131 | 28.2° | 20170704 | Right |

We used level 1.1 format data with spatial resolution reach up to 10 meters for strip map mode [4]. A Shuttle Radar Topography Mission-3 (SRTM-3) version-4.1 (90 m resolution) was used to eliminate the topographic phase, which was downloaded from public consortium for spatial information (CGIAR-CSI).

### *2.3. InSAR Technique*

Interferometric synthetic aperture radar (InSAR) is a reliable microwave remote sensing technique using at least two or more SAR images acquired at different times to generate displacement maps to detect surface changes over a specific area. Most 2 last decades, InSAR has wide used to estimate surface deformation and has a remarkable achievement in geodynamics studies. In the SAR data processing, the interferograms are generated by combining two complex SAR images, the interferometric phase observation per resolution cell is composed by a number of contributors [5].

$$\phi_{int} = \phi_{topo} + \phi_{defo} + \phi_{orb} + \phi_{atm} + \phi_{scat} + \phi_{noise} \quad (1)$$

where $\phi_{int}$ is interferometric phase, $\phi_{topo}$ is topographic phase, $\phi_{defo}$ is deformation phase due to the deformation in the radar line of slight, $\phi_{orb}$ is deterministic flat earth phase and the residual phase signal due to orbit in determination, $\phi_{atm}$ is atmospheric phase, $\phi_{scat}$ is phase due to a temporal and spatial change in the scatter characteristics of the earth surface between the two observation times, and $\phi_{noise}$ is phase degradation factors, caused by e.g., thermal noise, coregistration noise and interpolation noise. Furthermore InSAR approach, two SAR images from slightly different orbit configurations and at different times are combined to exploit the phase difference of the signals LOS (line of sight). InSAR is based on the phase difference which acquired from two different SAR acquisitions. By assuming that the scattering phase is the same in both images, the interferometric phase $\phi$ is a very sensitive measure of the range difference R2 - R1 i.e :

$$\phi = \phi_1 - \phi_2 = \frac{4\pi}{\lambda}(R2 - R1) \quad (2)$$

Here, $\phi_1$ and $\phi_2$ are the phases of the first and second SAR images, respectively. R1 is the distance from the SAR to the scatterer by the first acquisition; R2 is the distance by the second acquisition; and, $\lambda$ is the wavelength, as for L-band ALOS-2 data $\lambda$ is 23.6 cm. The Small Baseline Subset (SBAS) method relies on an appropriate combination of differential interferograms produced by data pairs characterized by a small orbital separation (baseline) in order to reduce the spatial decorrelation phenomena [6]. Cumulative phase can be obtained by solving a linear least squares problem. Solves only for pixels with complete dataset, i.e. all interferograms and acquisitions are available. In a given pixel, the observation equation as showed :

$$\Phi ij = \sum_{n=i}^{j-1} \delta\varphi n \tag{3}$$

where $\Phi_{ij}$ is phase of the interferogram combining acquisitions $i$ and j, $\delta_{\varphi n}$ is phase increment between acquisitions $n$ and $n+1$. Interferograms having mutual small baselines combinations are created based on the available of image. However, this can produce different subsets of InSAR pairs connected in time and separated by large baselines. SBAS method allow to obtain surface deformation and to analyze their space time characteristics. The generation of a linear model will increase the sampling rate and allow the use of all acquisitions included in the different small baseline subsets [7]. While the usual approach to analyze phase differences in classical InSAR processing is to set a coherence threshold to reduce phase noise and preserve spatial resolution. The small baselines method searches to ease phase unwrapping by means of selecting small baselines interferograms and filtering the phases. It creates a network of interferograms to estimate heights and deformation with respect to one single master image [8]. We use multiple SAR acquisitions of the same area to perform time-series analysis based on SBAS method. Interferograms with a maximum spatial baseline of 2600 meter with respect to the first image are phase-unwrapped and inverted for the phase [9].

The first stage was focus to convert the raw data to SLC format and calibration. Furthermore the data was to form interferograms from single-look complex (SLC) images was applied for interferogram calculation. Then the Statistical-cost, Network-flow Algorithm for Phase Unwrapping (SNAPHU) package was used to unwrap phase [10]. At the last stage, the GIAnT was implemented to perform time series InSAR SBAS analysis. The ground displacement was also solved at this stage.

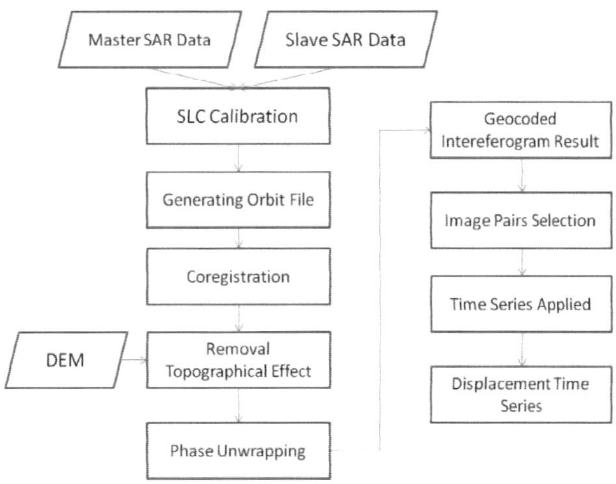


**Figure 3.** Flowchart of InSAR Time Series

IOP Conf. Series: Earth and Environmental Science **98** (2017) 012010 doi:10.1088/1755-1315/98/1/012010

## 3. Result and Discussion

We generated 28 interferogram results from all possible pairs of PALSAR-2 images. Some of the interfrogram images is shown by figure 4. In the interferogram processing not all produce good results, it can be due to unwrapping error and atmospheric effects on the image itself. Furthermore we continue to process time series data for Surabaya region by using GIAnT program [11]. In the SBAS method, we computed all interferograms and the wrapped phase was corrected for spatially- uncorrelated look angle error and noise associated with the master image. After unwrapping stage and filtering spatially correlated noise then it can be calculated mean of velocity line-of-sight (LOS).

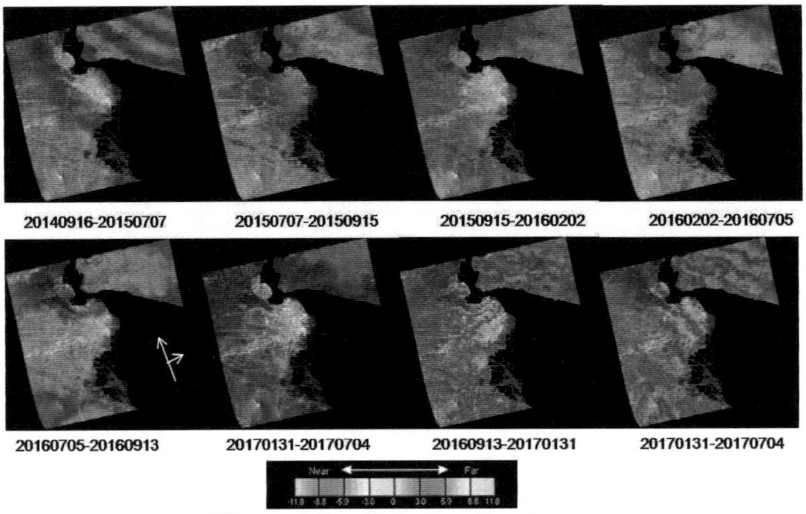


**Figure 4.** Interferogram Results

Figure 5 shows the surface deformation significant occured in northern part of Surabaya which is located in high density populated area. Value for each pixels from 2014 to 2017 with the deformation rates obtained in the interval 10 mm during 3 years. Finally, the results are mapped to geodetic coordinate system as shown figure below.

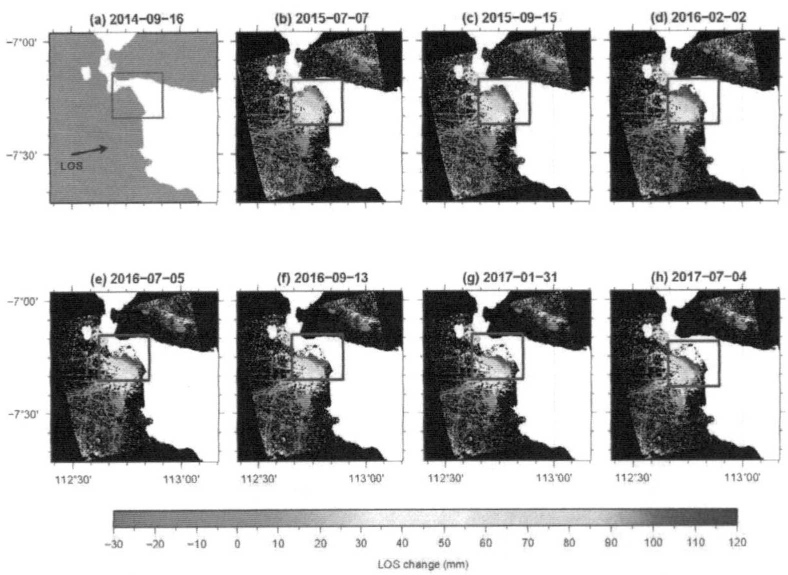


**Figure 5.** Surface Deformation Time Series Result

In order to make better interpretation, the processing was considered SBAS technique which is stable and calculate the differences respect to it. Therefore, blue areas mean deformation occured and uplifting respect to the mean deformation value. Taking into account that no uplifting is expected, we can consider that the light green areas are sinking slower than deep blue which is located in northern part of Surabaya. In other hand, we obtain the center of Surabaya to be subsiding with a rate of ~17 mm/year. We can also see the estimated deformation field which is not as smooth as expected probably due to unwrapping errors when the interferogram lacks of correlation. It can be seen at figure 5 that the oddity occurred in the 2014 to 2015 with the deformation value reach up to 10 cm on points A and 6 cm on point D. However, in the northern part (Kenjeran sub-district) subsidence reach up to 2.2 cm/ year and the southern part is more lower reach up to 0.6 cm/ year. Generally the coherence value of ALOS-2 images is better than ALOS but the temporal resolution is opposite for the area in Indonesia.

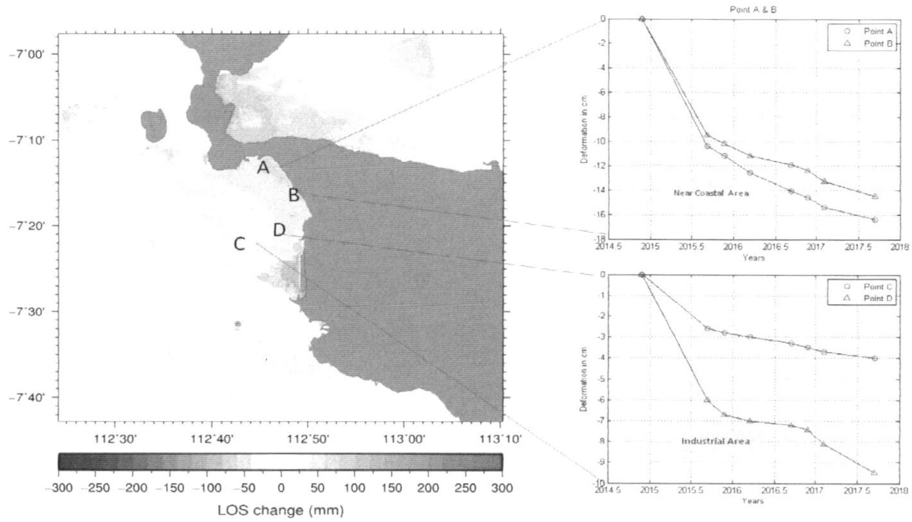


**Figure 6.** Plot of Specified Point in Surabaya

## 4. Conclusions

This work has presented an analysis of the ground subsidence phenomena in Surabaya City. The advanced multi-temporal InSAR technique is applied to this site using 8 ALOS-2 PALSAR-2 images acquired from 2014 to 2017. We identified a few locations undergoing subsidence at rates up to 2,2 cm/year. The average subsidence velocity map has been retrieved by Small Baseline technique processing to reduce spatial decorrelation characteristics among image itself. The urbanization and urban growth which have resulted in more groundwater extraction are mainly responsible for the subsidence in Surabaya. Subsidence in turn leads to flooding and water nuisance. From 2014 to 2017, the estimation of the average subsidence rate is ~1.7 cm/year with the maximum value up to ~2.2 cm/year, potentially suffering damages in the future.

After 3 years, in the regions along Kenjeran, Semampir, and Pabean sub-district the land had sunk up to ~6 cm. If not addressed, subsidence leads to an increase of inundation, both in frequency and spatial extent in particularly western part of Surabaya which is many inhabitant live. Regarding climate change, it can be confronted with flooding more often as a result of sea level rise mean rate up to 5.47 mm/year calculated in the period of 64 years (1925-1989) [12]. However, surface deformation will be correlated greater locally. Groundwater extraction for industrial use and residential consume are responsible for rapid subsidence. Therefore inundation will lead to an increase and will put coastal cities below relative sea level within decades. It is essential to consider human factors where the city is inhabited by more than 2.5 million people and subsidence directly impacts on urban structures and infrastructure. Furthermore, there could also be possible existence of other causes due to anthropogenic factors in almost all subsidence areas and natural factors such as tectonic processes in Southern of Surabaya. Future investigation can be improved by utilizing GPS technique in the significant deformation areas and monitoring long-time subsidence to achieve accurate result.

**Acknowledgement**
We thank to *Indonesia Endowment Fund for Education (LPDP)* for financial support during study period. The ownership of PALSAR-2 data belongs to METI (Ministry of Economy, Trade and Industry) and JAXA. These data were made available by the PALSAR Interferometry Consortium to Study our Evolving Land surface (PIXEL) under a cooperative research contract with the Earthquake Research Institute (ERI) of the University of Tokyo.

**References**
[1] Jesus Pacheco Martinez et al 2015 Application of InSAR and Gravimetry for Land Subsidence Hazard Zoning in Aguascalientes, Mexico *Remote Sensing* pp 17035-17050.
[2] Jiming Guo, Lv Zhou, Chaolong Yao and Jiyuan Hu 2016 Surface Subsidence Analysis by Multi-Temporal InSAR and GRACE: A Case Study in Beijing *Sensor* **16(**1495)
[3] Surabaya Government 2008 *Laporan Akhir Evaluasi Pelaksanaan Pembangunan Surabaya Drainage Master Plan (SDMP) 2018 Kota Surabaya* (Surabaya: Municipality Development Planning Board of Local Goverment Surabaya).
[4] Ryo Natsuaki, Hiroto Nagai, Takeshi Motohka, Masato Ohki, Manabu Watanabe, Rajesh Thapa, Rajesh B., Takeo Tadono, Masanobu Shimada and Shinichi Suzuki 2016 SAR interferometry using ALOS-2 PALSAR-2 data for the Mw 7.8 Gorkha, Nepal earthquake *Earth, Planets and Space* **68(**15).
[5] R F Hanssen 2001 *Radar Interferometry: Data Interpretation and Error Analysis* (Dordrecht: Kluwer Academic).
[6] Paolo Berardino, Gianfranco Fornaro, Riccardo Lanari and Eugenio Sansosti 2002 A New Algorithm for Surface Deformation Monitoring Based on Small Baseline Differential SAR Interferograms *IEEE Transactions on Geoscience and Remote Sensing* pp 2375-2382.
[7] Riccardo Lanari, Casu Riccardo, Francesco Manzo, Mariarosaria; Giovanni Zeni, Paolo Berardino, Michelle Manunta and Antonio Pepe 2007 An Overview of the Small Baseline Subset Algorithm : A DInSAR Technique for Surface Deformation Analysis *Pure and Applied Geophysics* **164** pp 637-661.
[8] Anuphao Aobpaet, Miguel Caro Cuenca, Andy Hooper and Itthi Trisirisatayawong 2009 Land Subsidence Evaluation Using InSAR Time Series Analysis in Bangkok Metropolitan Area in *Fringe 2009 Workshop*, Frascati.
[9] Estelle Chaussard, Falk Amelung, Hasanudin Abidin and Sang-Hoon Hong 2013 Sinking cities in Indonesia: ALOS PALSAR detects rapid subsidence due to groundwater and gas extraction *Remote Sensing of Environment* **128** pp 150-161.
[10] C W Chen and H A Zebker 2000 Network approaches to two-dimensional phase unwrapping: intractability and two new algorithms *Journal of the Optical Society of America* **17** pp 401-414.
[11] Agram P S, R Jolivet, B Riel, Y N Lin, M Simons, E Hetland, M P Doin and C Lassere 2013 New Radar Interferometric Time Series Analysis Toolbox Released *Eos Trans AGU* **97(**7) pp 69-70.
[12] Annisaa Hamidah Imaduddina and Widiyanto Hari Subagyo 2014 Sea Level Rise Flood Zones: Mitigating Floods in Surabaya Coastal Area in *Cities International Seminar: Resilient cities - Beyond Mitigation, Preparedness, Responses and Recovery, Surabaya* pp 123-129.

# Participatory Mapping for Flood Disaster Zoning based on World View-2 Data in Long Beluah, North Kalimantan Province

**Sudaryatno[1], Disyacitta Awanda[2*] and Sufiyana Eka Pratiwi[2]**

[1]Lecturer of Universitas Gadjah Mada, Bulaksumur, Yogyakarta, Indonesia 55281,
[2]Undergraduate Student of Universitas Gadjah Mada, Bulaksumur, Yogyakarta, Indonesia 55281

disyaawanda@hotmail.com

**Abstract**. Flood is one of the most frequent disasters in Indonesia. These conditions cause the necessary efforts to reduce the impact of these hazards. To reduce the impact of these hazards is to understand spatially the impact of previous disasters. Participatory mapping is one of the solutions to be able to assist in reducing the impact of flood disaster by conducting flood zoning so it can be known the range of the flood. The community plays an important role in participatory mapping because the experiences and mental maps of the community are the main sources of information used. North Kalimantan Province has a very large watershed area that is in Kayan watershed, there are several villages, one of them is Long Beluah Village. Kayan watershed has a flood problem annually that affects most of the areas including the Long Beluah Village. This study aims to map the zoning of floods in the village of Long Beluah in a participatory manner using remote sensing World View-2 data within community, so that people also understand the conditions they face. The method for achieving that goal is participatory mapping which means community involvement as well as the ability of community mental maps that will make an important contribution in this research. The results of this study show that flood zoning can be mapped based on experience and community mental maps that the greatest floods in February 2015 inundated most of the community settlements in Long Beluah Village. There are few places from the uninhabited areas of settlements and serve as refugee camps. The participatory zonation map of the participatory floods is quite appropriate with the situation at the time of the greatest flood that hit the village of Long Beluah, so that through the map can be drawn up plans to reduce the impact of such disasters such as evacuation routes and a more strategic refuge point.

**Keywords**: participatory mapping, flood, World View-2, North Kalimantan, disaster

## 1. Introduction

Flood is one of the most frequent disasters in Indonesia. These conditions cause the necessary efforts to reduce the impact of these hazards. Bulungan District, North Kalimantan Province is one of the areas with significant flood hazard. Judging from its territory, Bulungan District has two major watersheds, namely Kayan watershed and Sekatak watershed. Based on data from the Central Bureau of Statistics Bulungan District, Kayan watershed is the largest river basin system with 1,129,717 ha and 1.937,57 m3 / sec flow rate (Table 1).

**Table 1.** Bulungan District's Watersheds

| Watershed | Areas (Ha) | Debit($m^3$/sec) |
|---|---|---|
| Kayan | 1.129.717 | 1.937,57 |
| Sekatak | 198.212 | 371,0 |

Source: Badan Pusat Statistik Kabupaten Bulungan

Based on data from the National Disaster Management Agency, major floods have occurred in Bulungan district in February 2015, where initial information was obtained from the community of Long Beluah Village that the floods at the time were the greatest floods ever (Figure 1). The flood incident in Long Beluah Village in February 2015 was then used as the basis for collecting information on making the Flood Zoning Map of Long Beluah Village. For people with most of their property at home (no banks), flooding is a worrying thing. This shows that people need something that can help in preventing and even avoiding greater damage from floods [1].

**Figure 1.** Latest flood after the biggest one (Ferbuary 2015) on 2016.

Other studies have shown there is a way to estimate the damage from floods called stage-damage curves, which are seen by each type of land cover and have a scale of 0 (no damage) to 1 (complete destruction) which is also the influence of inundation depth [2]. One of the efforts in reducing the damage caused by flood, especially in keeping the property that is in the house is to keep the path or space that let the water flood into the house [1]. In fact, some indigenous houses in Indonesia have done that like traditional houses in Kalimantan. Another effort to be able to effectively stop flooding is a long-term solution. The deeper case of understanding how to cope with floods is not only through flood disaster experts but rather looking at the views of directly affected communities, so that decisions can be directed by understanding the real issues even though the technical measures have not been determined yet [1].

Overcoming flood problems cannot be done instantly so that in managing the flood problem it needs understanding and the steps of decision making and management both short and long term. Implementation of various policies will certainly have consequences on culture or habits that have lasted so long on the community, so this needs to be a particular concern in determining the strategy of flood risk reduction. Handling such vulnerabilities requires not only one concept to be applied in every place, but it is necessary to prepare concept variations so as to fit issues in different places. The hereditary habits of the community can make the plan in dealing with the flood disaster to be ineffective so that the need for an approach that can provide different views and a more dynamic approach [3].

The spatial understanding of the historical impact of the recent floods is one way to reduce future disaster risks. The participation of flood-affected communities is indispensable in gathering spatial information on the distribution of puddles in the village of Long Beluah. The role of society is done through participatory mapping of flood zoning. Knowledge and experience of the community in dealing with flood disaster can provide accurate information in mapping distribution of puddle flood. The community has an important role in providing information related to the disaster, it shows people are able to survive and understand the potential of disaster with their understanding, and the better if able to master the existing technology and resources [4]. There have been several studies that utilize participatory mapping approaches and have proven effective in mapping floods especially in areas where data availability is incomplete [5].

Society participation has a great role for risk management as a complex network. Society participating has a great point and movement in resolving the problem of social science, communities, and decision makers. Information from the community on the extent of damage and perceived impact can be used as a source of information for governments and experts to analyse and measure disaster-related matters. In urban areas this can be assessed from the resilience of buildings that are tailored to the community's involvement with the disaster-related disaster they face [6]. Developed countries like Europe have even gone a long way in reducing flood disaster risk by making prototypes or equipment that will be useful in the future through the OSIRIS Project program [7].

The remote sensing image of Worldview-2 Panchromatic channel has a spatial resolution of up to 50 cm that is capable of displaying a detailed picture of the Earth's surface. The image of Worldview-2 was used in a participatory mapping process with the community so that the community also understood the condition of their area and the dangers they faced. In addition, the use of remote sensing imagery can help people to more easily perform zoning inundation based on mental maps that have been formed.
The result of the flood zoning map is expected to form a better understanding for the people of Long Beluah Village in facing the threat of flood disaster that can occur in the future. The participatory map of the flood zoning is expected to provide more information and understanding for the community about spatial flooding spores so that it is expected to assist in further flood mitigation and disaster preparedness.

## 2. Method

Participatory mapping requires some data to provide an overview of the problem and also the conditions related to the location of the study area to be performed. Required data such as remote sensing imagery, maps that already exist and physical data. This activity also involves some people who better understand the conditions and understanding related to the area of study and flood events at that time. The communities involved provide information on the greatest floods that have occurred and assist in determining the boundaries or extent of the floods in the study area. Most of the communities involved were young villagers who participated in the evacuation process as well as some of the affected residents affected by the floods. Participatory mapping activities with discussions to determine the boundaries of puddles caused by floods, the extent of the affected area and the point of safe location as a place of evacuation.

The image data used is the 2012 View-2 Panchromatic imagery of 2012, so it is necessary to make adjustments related to the changes that occurred until 2015 where in that year the greatest flood occurred. The direct impact of the affected communities is a major source of data in creating a 2015 zoning map in the village of Long Beluah. Other information, such as the inundation depth, becomes additional data for subsequent analysis. Complete data limitations such as altitude data and rainfall data result in limited information that can complement the results of flood zoning maps, so that the public interest becomes the main data used for the preparation of flood zoning maps.

The result of participatory mapping is done by cartographic visualization so that it is more interesting and easy to understand. In addition, the zoning map of the flood becomes the main base in determining the point of evacuation location so that it can be prepared the route or the evacuation hammer to prepare for the possibility of disaster in the future. Simple participatory mapping methods can be a solution when the availability of limited data and limitations in data collection, so that the results of participatory mapping, especially for flood problems can be a reference flood mitigation policy (Figure 2).

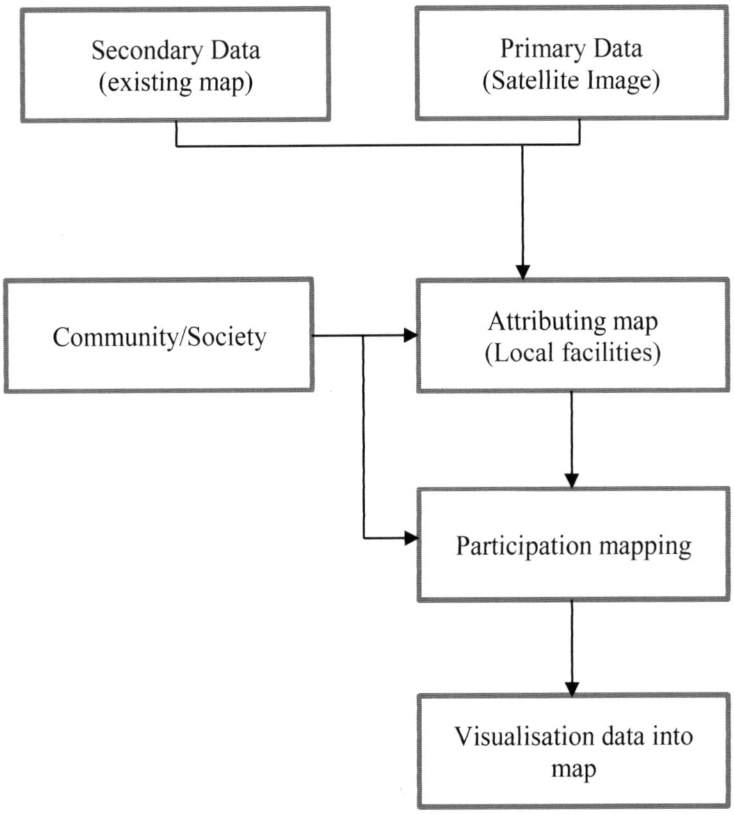


**Figure 2.** Simple workflow of participation mapping.

## 3. Results and Disscusion

Floods that occur in the village of Long Beluah almost every year hit. This is due to the morphology and morphometry of the Kayan watershed which covers most of North Kalimantan. The location of Long Beluah Village is located at the neck of the Kayan watershed, but floods in the upstream also affect the neck of the Kayan watershed. Accumulated rainwater received upstream can not be accommodated by the main river channel so that water overflows (Figure 3). This resulted in overflowing water, especially in the river channel that turned sharply due to geological fault. Soil conditions, which are mostly soil-type clays, make the water less receding quickly. Another factor that can flood the worse is the high rainfall intensity makes the water puddle more widespread. Climatic conditions in the North Kalimantan region, especially in the upstream is very uncertain so that floods can occur at any time.

**Figure 3.** Illustration of inundation flood in Kayan Watershed.

The largest floods hit the village of Long Beluah occurred in 2015 in February. Almost the entire settlement area is submerged until the water level reaches 7 meters. Damage caused by the flood is large enough, especially on goods that have not been saved, so submerged and damaged. In addition, people become very susceptible to disease because the flood water also brings bacteria from unhealthy environments. All community activities are paralyzed so people can only rely on food aid from the government. Evacuation site points that have been built are very limited to meet the needs of refugees, especially clean water. Accessibility is also limited due to the high water inundation making land vehicles unable to pass the usual route and need to use small boats to reach the submerged areas.

### *3.1. Disaster Mitigation*

Understanding of society in the face of flood is very less. The community anticipates after signs of flooding will occur, so it is quite late when the floods hit. Flooding symptoms or signs will most easily be observed by monitoring the intensity of the rain that occurred. If the intensity of rain is so high then the public is aware that it will happen that the flood will occur. In fact, the flood is caused by the upstream rain. The fact that data and also communications between regions resulted in very limited information for early warning. At the upstream almost no complete data available to be able to analyze the intensity of rainfall that occurred. The new Automatic Rainfall Logger tool is installed in Long Beluah Village one year after the great flood that occurred in February 2015. The tool is only one in one Bulungan District, so the data can only present the rain intensity locally only. Very limited inter-regional communication system due to inadequate telecommunication signal towers so that to be able to coordinate and early warning is also very difficult to do.

Prevention action by the community is not much done, but the culture of the community house in the village of Long Beluah has characteristic of Dayak house culture that is House Stage. Stage House has a shape like an ordinary house made of ulin wood (the best wood that is waterproof) and has a buffer or poles under it so that the average stage house is high so it is not easily flooded by water. The culture has existed long ago that indicates that people are already aware how to cope with floods by making their homes higher. In the event of regular flooding (not extreme) the house of the stage is very effective to protect the residence from the puddle, although the flood can close the access of transportation. In the occurrence of extreme floods as happened in February 2015, the stage house is still not enough to prevent water puddles that reaches 7 meters high, so at that time almost the entire settlement area was inundated. Preparation in building the point of evacuation location is done suddenly so that the evacuation process is less coordinated and spontaneous.

The unique morphological and morphometric conditions of the Kayan River make people unable to do much to prevent them. The need for building engineering such as dams to prevent flood disasters occur very effectively. Viewed from the side of society, what they can do is learn the characteristics of the flood disasters they face and prepare solutions to reduce vulnerability to flood hazards. Understanding spatial flood characteristics is important for the community, because by understanding spatially people can determine the right decision to take evacuation measures and disaster risk reduction efforts (Figure 4).

**Figure 4**. Activity of participation village boundary mapping to understand spatial thinking.

### *3.2. Community Participation Flood Zoning*

The flood zoning mapping aims to identify the extent of the impact of the greatest flood that once hit the village of Long Beluah. This activity is one solution of data limitations to do mapping with complete data. Limitations of supporting data to create a complete flooding model are the main reasons for participatory mapping involving communities to be very important. The community has experience and understanding related to the condition and environment of their residence. In addition, mental maps of society are sufficient to describe the situation at the time of the flood. Community involvement can also generate spatial thinking or understanding in tackling disasters.

The communities involved in participatory mapping of the flood zone are the people involved in the evacuation process and also the communities directly affected by the floods. Most of those involved were village youths who were still very well aware of the flood conditions at the time. Any place in Long Beluah Village can be explained in great detail through high-resolution black and white imagery. Simply put, the community can explain through the image to where the water inundated the area of the village of Long Beluah. The image provided has been added to the description of public facilities which are also through participatory village office management. Participants are very helpful in recognizing the area of Desa Long Beluah with the attributes of public facilities. Technically, the delineation is done simultaneously with the explanation of the participants directly directing the flooded points. The results of the flood zoning showed almost all the settlements were submerged in floods.

The shortcomings of the resulting results are the lack of supporting data such as contours, making it difficult to validate by height or topography. Supporting data such as village maps are also not available so that in comparison to village size it is not known what percentage of the submerged villages. It can be calculated only the area of submerged and visually submerged residential areas. The results of the initial delineation still need to be improved so that visualization can be done well.

The visualization of participatory mapping data is made as interesting and clear as it is easily understood by the community (Figure 5). Visualization is more emphasized on the selection of symbol warrants that are shuffled with the existing image so that still looks submerged settlements. The results of the flood zone mapping can also be analyzed as appropriate places for evacuation when floods occur.

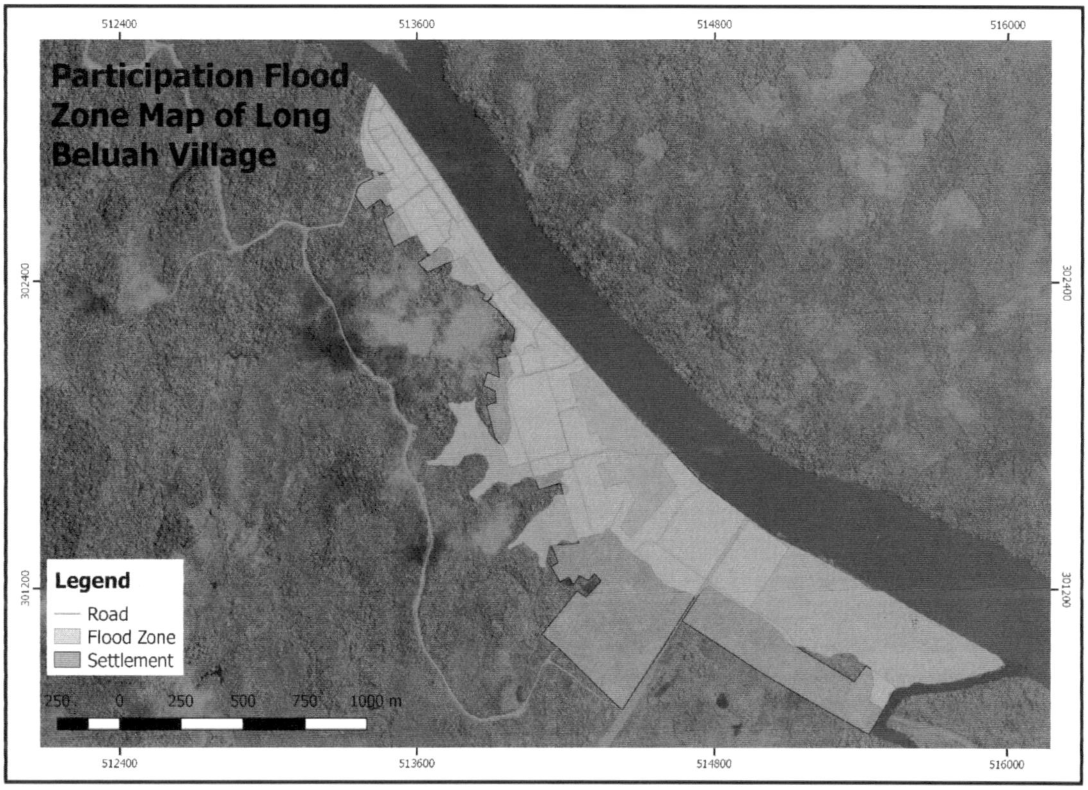


**Figure 5.** Participation Flood Zone Map of Long Beluah Village.

### *3.3. Making Evacuation Route*

Based on the disaster management system, determining the evacuation route is very important to facilitate the process of disaster evacuation more systematically. The determination of the evacuation route is based on the result of the flood zoning map and the local wisdom of the community in facing the previous flood. The location of evacuation sites as evacuation sites is determined based on location height data, evacuation location information from people experiencing previous flood events, consideration of availability of water supply and electricity, consideration of easy access of aid delivery, and weather security.

Directions for evacuation routes have an important role in the evacuation process. This is to facilitate the community through the nearest route to the safe location of the pool, of course in addition to local wisdom of people who are accustomed to face the flood disaster. The installation of evacuation route directions is placed at strategic points that are easily visible.

## 4. Conclusions

Floods that occurred in the Long Beluah Village is a flood that routinely occurs due to morphology and morphometry Kayan watershed. Total precautions require an infratucture engineering capable of withstanding large water accumulations. What can be done now is to reduce the risk of flood hazard and reduce the level of flood disaster. Cultural understanding of society has existed long ago with the existence of traditional houses that have the function of avoiding the house from standing water. The extreme floods that hit indigenous houses were not enough to mitigate the effects of floods, so another precaution was needed. The greatest flood disaster in February 2015 became the basis of reference in preparation for the subsequent floods through spatial understanding of the flood zone maps.

Incomplete supporting data for more detailed mapping makes participatory mapping a prime solution. Community understanding is important information as the main data in the preparation of flood zone maps. Community involvement will fill the space between experts and non-experts so the results will complement each other. The resulting flood zone maps can be used as a possible historical data in the preparation of plans and policies in addressing future flooding problems. Further analysis by using flood zoning can yield some other supporting information in relation to disaster mitigation such as making evacuation routes.

## References

[1] I Kelman 2007 Decision-Making For Flood-Threatened Properties *Flood Risk Management in Europe* ed Selina Begum, Macel J F Stive and Jim W Hall (Netherland: Springer) **Chapter 1** pp 3–19.

[2] D Alkema and H Middelkoop 2007 The Influence of Floodplain Compartmentalization on Flood Risk Within The Rhine-Meuse Delta *Flood Risk Management in Europe* ed Selina Begum, Macel J F Stive and Jim W Hall (Netherland: Springer) **Chapter 1** pp 21–42.

[3] K M De Bruijn, C Green, C Johnson and I McFadden 2007 Evolving Concepts in Flood Risk Management: Searching For A Common Language *Flood Risk Management in Europe* ed Selina Begum, Macel J F Stive and Jim W Hall (Netherland: Springer) **Chapter 1** pp 61–75.

[4] Dagmar Haase 2013 Participatory Modelling of Vulnerability and Adaptive Capacity in Flood Risk Management *Nat Hazards* **67** pp 77–97

[5] W Chingombe, E Pedzisai, D Manatsa, G Mukwada and P Taru 2015 A Participatory Approach in GIS Data Collection for Flood Risk Management, Muzarabani District, Zimbabwe *Arab Jurnal Geoscience* ***8*** pp 1029–1040

[6] Milena Mari ′Lia Nogueira De Andrade and Claudio Fabian Szlafsztein 2015 Community Participation in Flood Mapping in The Amazon Through Interdisciplinary Methods *Natural Hazards* **78** pp 1491–1500

[7] M Eelrich 2007 OSIRIS – An Example of Citizen-Oriented Technology Development in The Area of Dissemination of Information on Flood Risk Management *Flood Risk Management in Europe* ed Selina Begum, Macel J F Stive and Jim W Hall (Netherland: Springer) **Chapter 1** pp 43–60.

# Mapping the Distribution of Potential Land Drought in Batam Island Using the Integration of Remote Sensing and Geographic Information Systems (GIS)

**M Z Lubis[1*], H M Taki[2,3], W Anurogo[1], D S Pamungkas[4], P Wicaksono[5], T Aprilliyanti[1]**

[1] Geomatics Engineering, Politeknik Negeri Batam, Batam 29461, Indonesia
[2] King Abdulaziz University, Department of Urban and Regional Planning, Jedah, Saudi Arabia
[3]University of Indonesia, Department of Geography, Depok, Indonesia
[4] Electrical Engineering, Politeknik Negeri Batam, Batam 29461, Indonesia
[5]Faculty of Geography, Universitas Gadjah Mada, Sekip Utara, 55281 Yogyakarta, Indonesia

zainuddinlubis@polibatam.ac.id

**Abstract.** Potential land drought mapping on Batam is needed to determine the distribution of areas that are very potential to the physical drought of the land. It is because the drought is always threatening on the long dry season. This research integrates remote sensing science with Geographic Information System (GIS). This research aims to map the distribution of land drought potential in Batam Island. The parameters used in this research are land use, Land Surface Temperature (LST), Potential dryness of land on the Batam island. The resulting map indicates the existence of five potential drought classes on the island of Batam. The area of very low drought potential is 2629.45 ha, mostly located in the Sungai Beduk sub-district. High drought potential with an area of 7081.39 ha is located in Sekupang sub-district. The distribution of very high land drought potential is in Batam city and Nongsa sub-district with area of 15600.12 ha. The coefficient of determination ($R^2$) is 0.6279. This indicates a strong positive relationship between field LST and modelled LST.

**Keywords:** Land Drought, Land Surface Temperature, Remote Sensing, GIS

## 1. Introduction

Geographically, Batam island has a very strategic location, namely in international shipping line. Batam City based is located between: - 0° 25' 29 "N - 1° 15' 00" N and 103° 34' 35" E - 104 ° 26' 04"E. Today, drought is increasingly common annually under normal climate conditions. Drought is the availability of water far below the water requirement for the necessities of life, agriculture, economic and environmental activities (Law number 24 year 2007 in disaster management). The problem of water existence in the earth is divided into two main categories, namely the advantages and lack of water [1]. Previous research related to drought potential has been done by [2], by utilizing remote sensing and Geographic Information Systems (GIS) technique. The variables used were vegetation index, brightness index, wetness index, rainfall, hydrogeology and land use map. Next research was done by [3] on integrating remote sensing technique and GIS for potential identification of drought, and revealed that clay rock is one example of rock that has high porosity, but have low permeability, so that the clay rocks have high humidity, but the wells in the rock has 4 low potentials.

Mapping the distribution of land drought potential using Landsat-8 OLI remote sensing imagery with Temperature Vegetation Dryness Index (TVDI) method. The main parameter used to find the potential of land drought is NDVI (Normalized Difference Vegetation Index) which is used to get the level of vegetation density in the research area and LST (Land Surface Temperature) to know the amount of temperature contained in the land is very good to determine the distribution of drought in a region [4,5]. The key elements of drought monitoring and forecasting system using GIS and remote sensing techniques can also be made by providing real-time water cycling of near-terrestrial water and assessment of drought conditions in the field [6,7]. The combination of multi-temporal remote sensing data for drought monitoring for different months in a year or a month in different years, shows significantly varying locations of drought areas, indicating seasonal and seasonal fluctuations, and the use of remote sensing and GIS techniques [8-10] continues to be sustainable, and further enhances the level of accuracy of multispectral data [11]. This research aims to map the distribution of land drought potential in Batam Island, and assess the relationship between land surface temperature of Land Surface Temperature (LST) measured in the field and modelled LST value from Landsat-8 OLI. The parameters used in this research are land use, and LST.

## 2. Methods

### *2.1. Research location*

This research location was on Batam Island (1°05' N 104°02' E) using Landsat-8 OLI recorded on April 2016, the image was obtained by downloading through the www.earthexplorer.usgs.gov. Stages of digital image processing were image cropping, radiometric correction, geometric correction, image composite and image transformation. While the parameter processing was done by scoring, and layout was done using ArcGIS 10.2 software, that produced a map of potential drought in Batam Island. The research location map, produced by Development Planning Agency at Sub-National Level (BAPPEDA), can be seen in Figure 1, and the research flow chart can be seen in Figure 2.

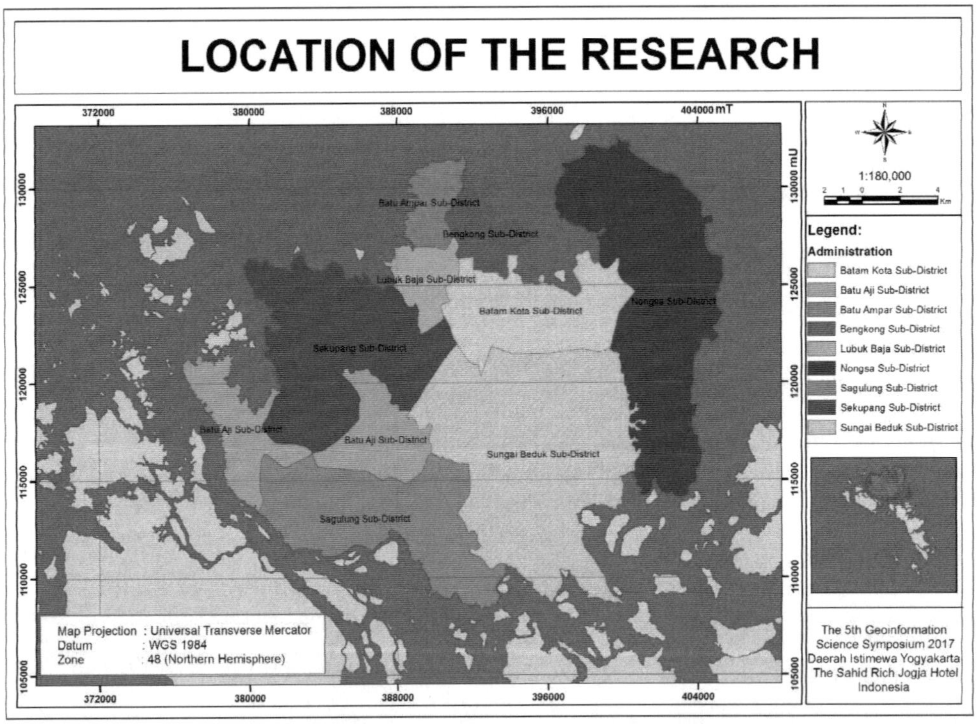


**Figure 1**. Research location in Batam island, Indonesia

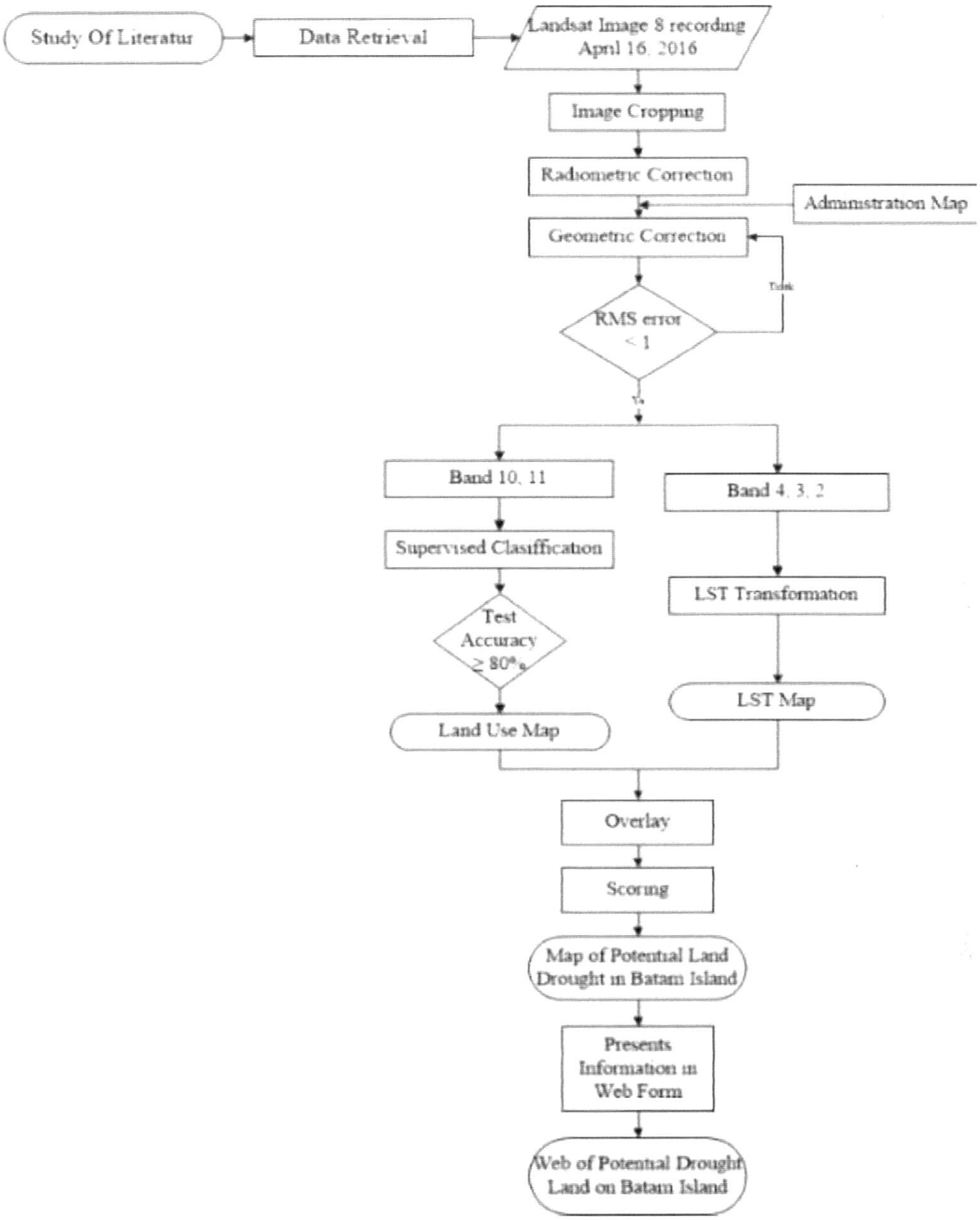


**Figure 2**. Flow Diagram of Research

*2.2. Land Surface Temperature (LST)*

Image processing of Landsat-8 OLI to obtain LST value was done using the following steps (Figure 3):

1. Converts the Digital Number (DN) of bands 10 and 11 into the spectral radiance via radiometric calibration.
2. Converts spectral radiance to brightness temperature in Kelvin unit.
3. Change the brightness temperature unit from K to Celcius.

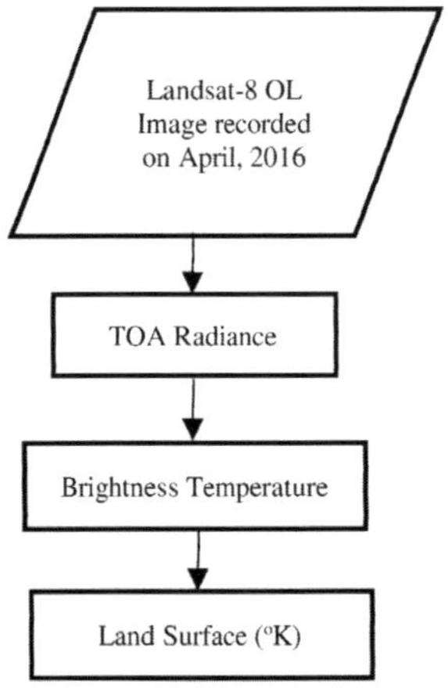


**Figure 3**. Flowchart of obtaining LST using Landsat-8 OLI

### *2.3. Land Use Map*

The processing of land use map started by collecting data from the related institutions. Then, the data or map was digitized to produce the shapefile (vector data), performed attributes editing, and performed the scoring process. The map that has been scored on the basis of the impact of land use on land drought was selected as final map. Flowchart of land used data processing can be seen in Figure 4.

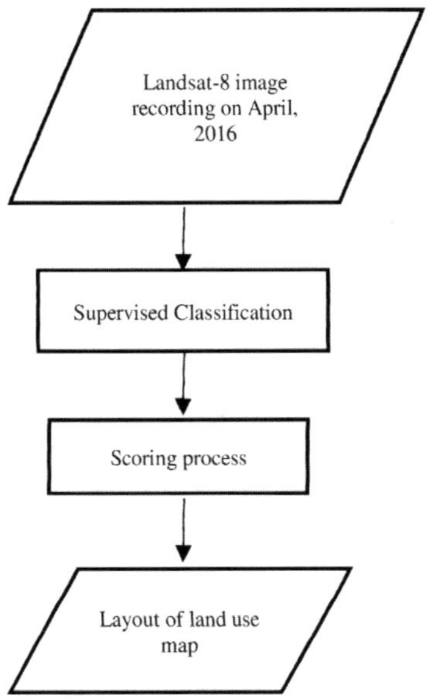


**Figure 4**. Flowchart of land use processing

## 3. Results and Discussion

Map of potential land drought is a map derived from the processing of data/parameters that affect the drought of the land. Parameters used in processing drought potential maps are land use map, and surface temperature map/LST [12-14]. The weight and scoring process of each parameter is different.

IOP Conf. Series: Earth and Environmental Science **98** (2017) 012012 doi:10.1088/1755-1315/98/1/012012

Based on the accumulated score of the parameters that affecting the drought, the categories for potential land drought class were obtained as shown in Table 1.

**Table 1**. Category of potential land drought

| Map | Category | Score | Area (Ha) | Percentage (%) |
|---|---|---|---|---|
| Land Drought Potential | Very Low Drought Potential | 30 – 40 | 2629.45 | 5.92 |
| | Low Drought Potential | 41 – 51 | 9585.52 | 21.59 |
| | Medium Drought Potential | 52 – 62 | 9507.12 | 21.41 |
| | High Drought Potential | 63 – 73 | 7081.39 | 15.95 |
| | Very High Drought Potential | 74 – 84 | 15600.12 | 35.13 |

Based on Table 1 the percentage of each class of potential land drought can be seen. The area of very low drought potential class 2629.45 ha, distributed mainly in Sungai Beduk, Sekupang, and Batu Aji district. Low drought potential class has an area of 9585.52 ha, mostly distributed in Sekupang sub-district. Medium drought potential class has an area of 9507.12 ha, with the dominant location in Sekupang District. High drought potential class have the area of 7081.39 ha, distributed in Sekupang, Sagulung, and Nongsa Sub-district, and. very high drought potential class has an area of 15600.12 ha, with located mostly in Batam City and Nongsa sub-district. Based on the analysis of land use scoring and LST, a map consisting of five drought potential classes was generated. Map of Potential Land Drought in Batam Island 2016 can be seen in Figure 5.

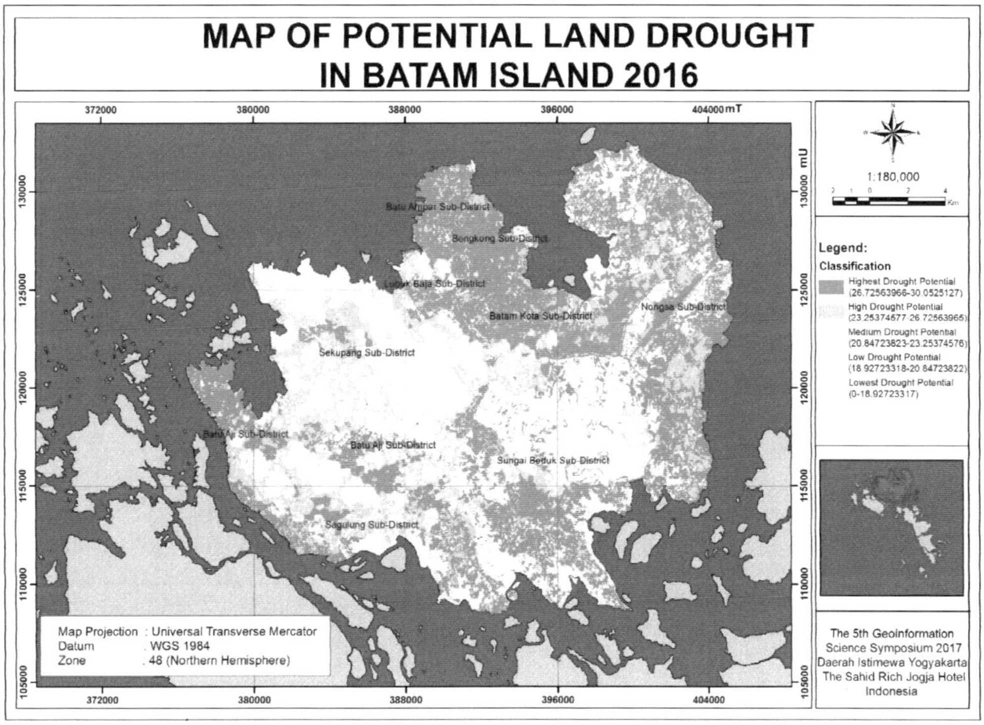


**Figure 5**. Map of Potential Land Drought in Batam Island 2016

Figure 5 gives information on the distribution of the potential land drought in Batam Island 2016. The map has five classes with different areas. In general, the Batam Island has a very high drought potential, as characterized by the red colour. The potential of land drought is similar to previous research conducted in Natuna, Riau Islands (Put the source here). In this research, field validation was also done to find out the relationship of LST from the Landsat-8 OLI with field LST. Sampling points were randomly taken, which represented each of the existing drought potential class. The sample collected in the field is 30, and it can already represent the variation of the entire Batam Island. The distribution of sample point for validation of field temperature can be seen in Figure 6, and scatter plot between field LST and LST value from Landsat-8 OLI image can be seen in Figure 7. Coefficient of determination ($R^2$) is 0.6279 and indicates a positive strong relationship between field and modelled LST. If indicates a positive relationship between LST image with temperature in the field, it will affect the value of surface temperature is very perpendicular and real to the research location using remote sensing techniques, as a parameter to determine the distribution of drought distribution, settlement and vegetation of land in a region [15-19].

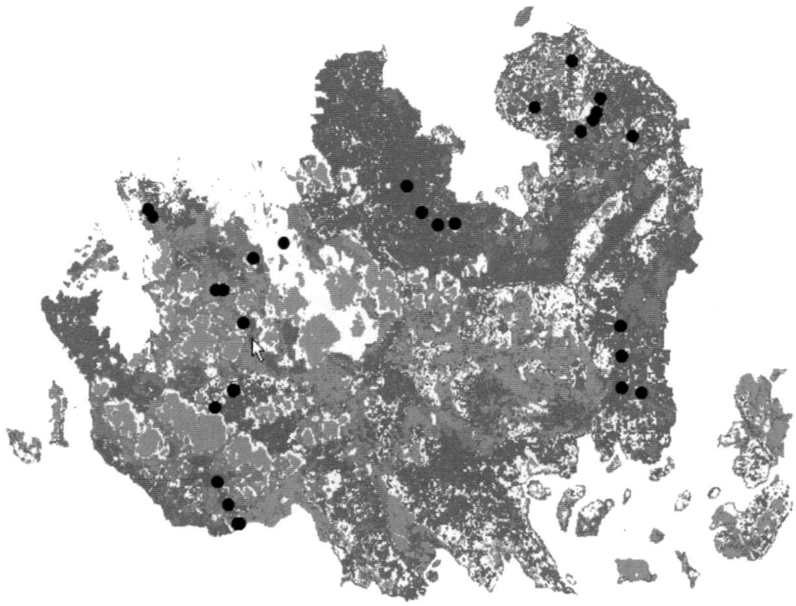

**Figure 6**. The distribution of sample points for validation of field temperature in Batam Island 2016

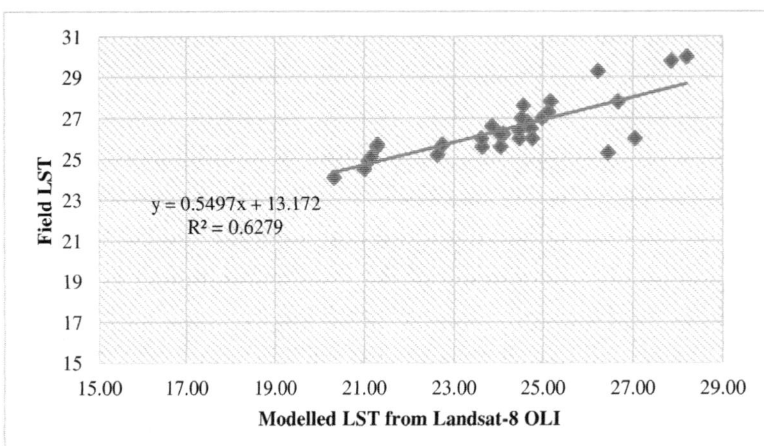


**Figure 7**. Scatter plot between field LST and LST values from Landsat-8 OLI image

## 4. Conclusions

Research on mapping the potential of drought on the island of batam resulted in five classes of potential drought lands from highest drought potential, high drought potential, medium drought potential, low drought potential and lowest drought potential. The highest drought potential was located on Batam City and Nongsa Sub-District.

## Acknowledgments

We would like thank to Development Planning Agency at Sub-National Level (BAPPEDA) who has been willing to provide shapefile data and administration map of Batam City, Indonesia.

## References

[1] Reichle, R. H., Draper, C. S., Liu, Q., Girotto, M., Mahanama, S. P., Koster, R. D., & De Lannoy, G. J 2017 Assessment of MERRA-2 land surface hydrology estimates. *Journal of Climate*, 30(8), 2937-2960

[2] Ahmadalipour, A., Moradkhani, H., Yan, H., & Zarekarizi, M 2017 Remote Sensing of Drought: Vegetation, Soil Moisture, and Data Assimilation. In *Remote Sensing of Hydrological Extremes* (pp. 121-149) Springer International Publishing

[3] Rebello, V. P. A., Getirana, A., Lakshmi, V., & Rotunno Filho, O. C 2017 Monitoring Drought in Brazil by Remote Sensing. In *Remote Sensing of Hydrological Extremes* (pp. 197-218). ` Springer International Publishing

[4] Satgé, F., Espinoza, R., Zolá, R. P., Roig, H., Timouk, F., Molina, J., ... & Bonnet, M. P 2017 Role of Climate Variability and Human Activity on Poopó Lake Droughts between 1990 and 2015 Assessed Using Remote Sensing Data. *Remote Sensing*, 9(3), 218

[5] Liu, M., Xu, X., Xu, C., Sun, A. Y., Wang, K., Scanlon, B. R., & Zhang, L 2017 A new drought index that considers the joint effects of climate and land surface change. *Water Resources Research*, 53(4), 3262-3278

[6] Sheffield, J., Wood, E. F., Chaney, N., Guan, K., Sadri, S., Yuan, X., & Ogallo, L 2014 A drought monitoring and forecasting system for sub-Sahara African water resources and food security. *Bulletin of the American Meteorological Society*, 95(6), 861-882

[7] Belal, A. A., El-Ramady, H. R., Mohamed, E. S., & Saleh, A. M 2014 Drought risk assessment using remote sensing and GIS techniques. *Arabian Journal of Geosciences*, 7(1), 35-53

[8] Hao, C., Zhang, J., & Yao, F Combination of multi-sensor remote sensing data for drought monitoring over Southwest China. *International Journal of Applied Earth Observation and Geoinformation*, 35, 270-283

[9] Zhang, A., & Jia, G Monitoring meteorological drought in semiarid regions using multi-sensor microwave remote sensing data. *Remote Sensing of Environment*, 134, 12-23

[10] Wanders, N., Pan, M., & Wood, E. F Correction of real-time satellite precipitation with multi-sensor satellite observations of land surface variables. *Remote Sensing of Environment*, 160, 206-221

[11] Wicaksono, P 2016 Improving the accuracy of Multispectral-based benthic habitats mapping using image rotations: the application of Principle Component Analysis and Independent Component Analysis. *European Journal of Remote Sensing*, 49, 433-463

[12] Jiménez-Muñoz, J. C., Sobrino, J. A., Skoković, D., Mattar, C., & Cristóbal, J 2014 Land surface temperature retrieval methods from Landsat-8 thermal infrared sensor data. *IEEE Geoscience and Remote Sensing Letters*, 11(10), 1840-1843

[13] Chen, J., Ding, F., Li, Q., Wu, W., Fan, P., & Zhang, X. 2016 Using a modified HUTS algorithm to downscale Land Surface Temperature retrieved from Landsat-8 imagery: A case study of Xiamen City, China. In *Earth Observation and Remote Sensing Applications* (EORSA), *2016 4th International Workshop on* (pp. 38-42). IEEE

[14] Kilic, A., Allen, R., Trezza, R., Ratcliffe, I., Kamble, B., Robison, C., & Ozturk, D 2016 Sensitivity of evapotranspiration retrievals from the METRIC processing algorithm to improved

radiometric resolution of Landsat 8 thermal data and to calibration bias in Landsat 7 and 8 surface temperature. *Remote Sensing of Environment*, 185, 198-209

[15] Du, C., Ren, H., Qin, Q., Meng, J., & Zhao, S 2015 A practical split-window algorithm for estimating land surface temperature from Landsat 8 data. *Remote Sensing,* 7(1), 647-665

[16] Du, C., Ren, H., Qin, Q., Meng, J., & Zhao, S 2015 A practical split-window algorithm for estimating land surface temperature from Landsat 8 data. *Remote Sensing,* 7(1), 647-665

[17] Jia, K., Wei, X., Gu, X., Yao, Y., Xie, X., & Li, B 2014 Land cover classification using Landsat 8 operational land imager data in Beijing, China. *Geocarto International*, 29(8), 941-951

[18] Wicaksono, P., & Hafizt, M 2013 Mapping seagrass from space: Addressing the complexity of seagrass LAI mapping. *European Journal of Remote Sensing*, 46(1), 18-39

[19] Nikam, B. R., Ibragimov, F., Chouksey, A., Garg, V., & Aggarwal, S. P 2016 Retrieval of land surface temperature from Landsat 8 TIRS for the command area of Mula irrigation project. *Environmental Earth Sciences*

IOP Conf. Series: Earth and Environmental Science **98** (2017) 012013 doi:10.1088/1755-1315/98/1/012013

# Analysis of Environmental Vulnerability in The Landslide Areas (Case Study: Semarang Regency)

**Hani'ah[1*], H S Firdaus[1*], and A L Nugraha[1]**

[1] Geodesy Engineering Dept, Faculty of Engineering, Diponegoro University, Tembalang-Semarang, 50277, Indonesia

hanasugiastufirdaus@gmail.com

**Abstract**. The Land conversion can increase the risk of landslide disaster in Semarang Regency caused by human activity. Remote sensing and geographic information system to be used in this study to mapping the landslide areas because satellite image data can represent the object on the earth surface in wide area coverage. Satellite image Landsat 8 is used to mapping land cover that processed by supervised classification method. The parameters to mapping landslide areas are based on land cover, rainfall, slope, geological factors and soil types. Semarang Regency have the minimum value of landslide is 1.6 and the maximum value is 4.3, which is dominated by landslide prone areas about 791.27 $km^2$. The calculation of the environmental vulnerability index in the study area is based on Perka BNPB No. 2/2012. Accumulation score of environmental vulnerability index is moderate value, that means environment condition must be considered, such as vegetation as ground cover and many others aspects. The range of NDVI value shows that density level in conservation areas (0.030 - 0.844) and conservation forest (0.045 - 0.849), which rarely until high density level. The results of this study furthermore can be assessed to reduce disaster risks from landslide as an effort of disaster preventive.

**Key words :** Environmental Vulnerability, Landslide, Landsat 8, NDVI

## 1. Introduction

Landslide is one of the disasters that often occurred in several locations in Indonesia. Semarang Regency is one of the potential locations of landslide disaster, because that regency is dominated by hills and mountains landform. Landslide almost as erosion with a large mass of soil and rocks down from mountain or cliff. Based on data of Regional Disaster Management Agency (BPBD), in Semarang Regency recorded 38 landslide disasters of 113 total accumulation disasters in 2014. Meanwhile, in 2015 recorded 132 landslides of 305 total accumulation disasters, and in 2016 recorded 87 incident of landslides from 199 disasters. The recorded data indicate that landslides disaster has occurred frequently in Semarang Regency, this is because of high rainfall intensity and also the topographic condition of Semarang Regency is dominated by hills and mountains landform with a degree of the slope about (2-40%). Landslide is affected by two factors that are nature aspects and human aspects. Basically, a landslide can be prevented by cultivating plants which have strong roots due to the function of the plants as a ground cover and aggregate soil binder if soil received pressure by high rainfall. The development of human activities caused some landuse changed that function from conservation area to cultivation area or residential area that can increase the risk of landslide disaster. Plants in the cultivation area not forming a dense network of roots so can not hold the soil when heavy rain occurred with high rainfall intensity. In addition, the development of settlements that not follow the rules of conservation can increase the risk of landslide disaster.

Assessment of environmental vulnerability in landslide prone areas is necessary for the identification and assessment of hazards, so that in the research, environmental susceptibility analysis in landslide areas is conducted using remote sensing and geographic information system method. In this study, mapping of potential landslide areas is done previously by scoring and weighting of the parameters landslide. The parameters to mapping landslide areas in Semarang Regency are based on land cover, rainfall, slope, geological factors and soil types. The high human activity caused some changes in land cover, so to obtain actual land cover in study areas using remote sensing is based on Landsat 8 imagery processing. The use of satellite imagery to mapping land cover is very effective due to the wide study area is 95020.67 Ha or about 2.92 % of the total area in Central Java. Scoring and weighting method is spatial analysis which utilizes geographic information system. Values of scoring and weighting are based on the relevant literature review and regulations such as the regulations from Directorate of Volcanology and Disaster Mitigation to mapping accurate landslide areas in Semarang Regency.

The result of landslide prone areas that be analyzed to know about the environmental susceptibility in landslide areas. The vulnerability can be defined as exposure multiplied by sensitivity. Exposed assets such as physical structures and ecological or environmental areas. Each of assets has its own sensitivity, which varies for each type of disasters. Land cover and vegetation density will be analyzed to obtain the environmental vulnerability index in landslide areas. The calculation of the environmental vulnerability index in the study area is based on Perka BNPB No. 2/2012. The results in this study, furthermore can be assessed to reduce disaster risks from landslide as an effort of disaster preventive.

## 2. Data and Study Areas

Data in this study uses Landsat 8 Imagery to get land cover by supervised classification processing and vegetation density by NDVI algorithm. In addition, to mapping landslide area in Semarang Regency also needs some others data such as rainfall data from Indonesia Agency for Meteorology, Climatology and Geophysics in 2016, geological factors, soil types and slope which is processed from DEM SRTM.

### *2.1. Landsat 8 Imagery*

Landsat 8 imagery that used in this study is L1T type products which available to users is a radiometrically and geometrically corrected image. The result is a geometrically rectified product free from distortions related to the sensor (e.g., view angle effects), satellite (e.g., attitude deviations from nominal), and Earth (e.g., rotation, curvature, relief). The image is also radiometrically corrected to remove relative detector differences, dark current bias, and some artifacts [1]. The Level 1 image is presented in units of DNs, which can be easily rescaled to TOA reflectance. Values of TOA reflectance in the image are used as input values to obtain land cover by supervised classification processing and NDVI algorithm to get vegetation density. Landsat 8 have the OLI (Operational Land Imager) sensor that collects image data for 9 shortwave spectral bands over a 190 km swath, with a 30 m spatial resolution for all bands except the 15 m panchromatic band and TIRS (Thermal Infrared Sensor) that 2 spectral bands with a 100 m spatial resolution for all band. Figure 1 shows Landsat 8 image, which is used in this study with path/row 120/65 and date acquired in 19 August 2016.

### *2.2. Study Areas*

In this study, evironmental vulnerability analysis be done in Semarang Regency. Ungaran is regency capital in study area. Semarang Regency located at the geographic positions of $110^0$ 14' 54.74" - $110^0$ 39' 3" East Longitude and $7^0$ 3' 57" - $7^0$ 30' 0" South Latitude. Administratively, Semarang Regency is divided into 19 districts, 27 sub-districs and 208 villages [7]. Semarang Regency is directly adjacent to Semarang City and Demak Regency in north; Boyolali Regency and Magelang Regency in south; Grobogan Regency and Boyolali Regency in east; and Temanggung Regency and Kendal Regency in west. Location of Semarang Regency which is study area in this research that shown in Figure 2. The

IOP Conf. Series: Earth and Environmental Science **98** (2017) 012013 doi:10.1088/1755-1315/98/1/012013

study area is located at an altitude about 500-2000 m above sea level. The lowest altitude is in the Candirejo Village in Priangapus District and while the highest in Bantur Village in Getasan District.

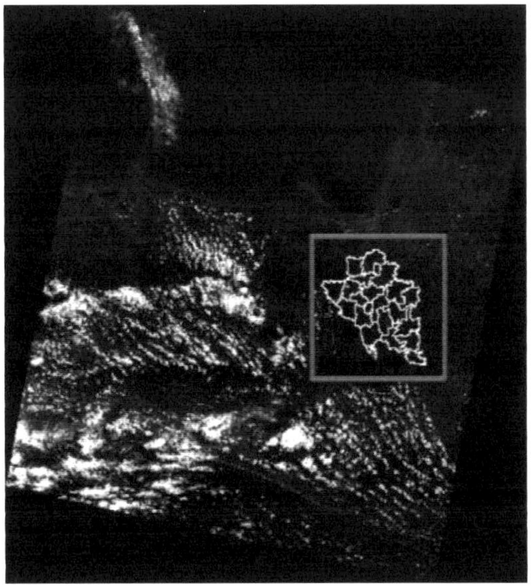

**Figure 1.** Landsat 8 Image with path/row 120/65 and date acquired in 19 August 2016. The image uses composite band 432, which true colour composite. The red square in this image shown study areas in Semarang Regency.

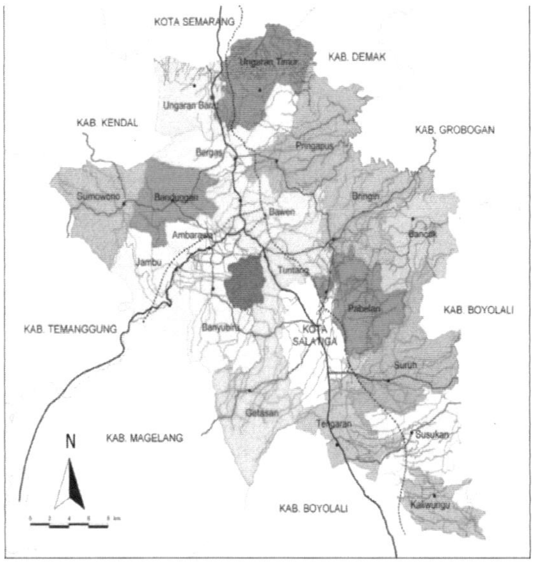


**Figure 2.** Location of Semarang Regency with the boundary areas and 19 district in study area.

## 3. Methods

Basically, the methods that is used to obtain the environmental vulnerability index in the study area, which based on Perka BNPB No. 2/2012. Potential landslide areas in Semarang Regency must be mapped previously with scoring and weighting of the parameters, which included in the spatial analysis. The results of potential landslide areas in Semarang Regency be classified by the level of prone landslide. To calculation value of index environmental vulnerablity in landslide areas, which based on forests, shrubs and swamps land cover in study area. In addition, vegetation density in that locations can be known by NDVI algorithm processing. Overall step or methods that used in this study can be seen in Figure 3.

### *3.1. Pre-processing*

DEM SRTM must be pre-processing to obtain degree of slope from altitude values from raster data. Images pre-processing must be done for Landsat 8 to get values of TOA reflectance in Semarang Regency, before processed to the next stage. Units of DNs can be converted to TOA reflectance using the reflectance scaling factors provided in the metadata file. The following equation is used to convert DN values to TOA reflectance :

$$\rho\lambda' = M\rho * Qcal + A\rho \quad (1)$$

where:
ρλ' = TOA Planetary Spectral Reflectance, without correction for solar angle.
Mρ = Reflectance multiplicative scaling factor for the band.
Aρ = Reflectance additive scaling factor for the band.
Qcal = L1 pixel value in DN

$$\rho\lambda = \frac{\rho\lambda'}{\sin(\theta)} \quad (2)$$

where:
ρλ = TOA Planetary Reflectance
θ = Solar Elevation Angle (from the metadata, or calculated)

### *3.2. Supervised Classification*

Image classification is a process of compiling, sorting or grouping the pixels into the particular criteria or object categories. Image classification is divided by two types, Supervised Classification and Unsupervised Classification [2]. In this type of classification the image analyst "supervises" the pixel categorization process by specifying, to the computer algorithm, numerical descriptors of the various land cover types present in a scene. To do this, representative sample sites of known cover type, called training areas, are used to compile a numerical interpretation key that describes the spectral attributes for each feature type of interest. The resulting classification is necessary to test the accuracy of object classification to the real objects on earth surface. Determination the accuracy of classification can be done by looking at the evaluation values, which calculated by the confusion matrix.

### *3.3. NDVI (Normalized Difference Vegetation Index) algorithm*

Vegetation density in this study has been done using NDVI algorithm to get values of vegetation index. NDVI (Normalized Difference Vegetation Index) is a sensitive vegetation measurement using spectral energy disparity is reflected by vegetation canopy in red electromagnetic spectrum wave and near-infrared spectrum [5]. The NDVI transformation follows the equation :

$$NDVI = \frac{\rho NIR - \rho RED}{\rho NIR + \rho RED} \quad (3)$$

where:
ρNIR = Near-Infrared Canal Reflectance Value
ρRED = Red Canal Reflectance Value

### *3.4. Environmental Vulnerability Index*

Based on Perka BNPB No. 2/2012, indicators of environmental vulnerability consist of forests, shrubs and swamps parameters in study area. Each parameters of environmental vulnerability in this study as a results of supervised classifiaction that shown by land cover in landslide areas. The calculation of the environmental vulnerability index follws the equation [3] :

Environmental Vulnerability = (0.4*Scoring of Conservation Forest)+(0.4*Scoring of Natural Forest) + (0.1*Scoring of Mangrove Forest)+(0.1*Scoring of Shrubs) (4)

where:
values of scoring are determined by wide areas for each parameters in environmental based on Perka BNPB No. 2/2012.

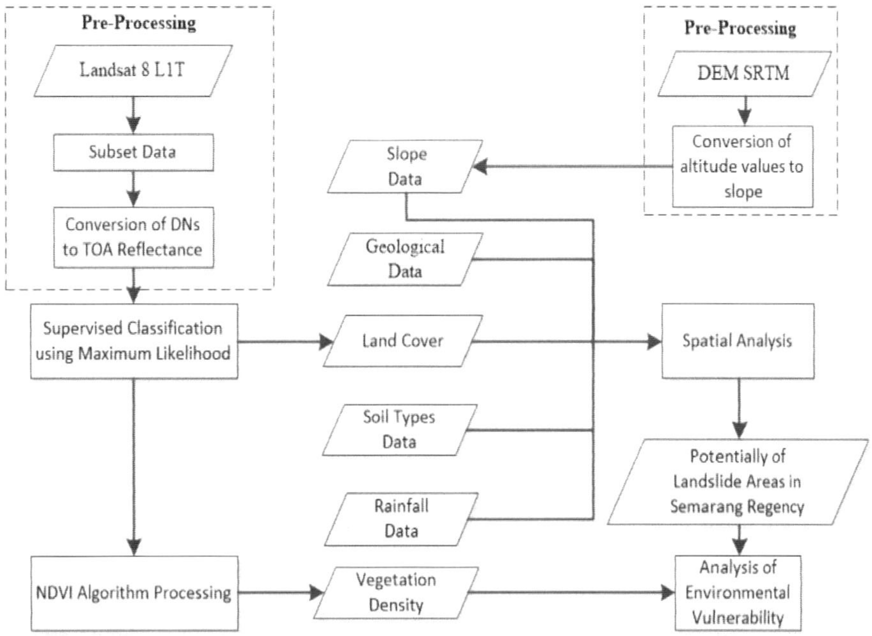


**Figure 3.** Flowchart showing the overall methods used in this study

## 4. Results and Discussion

### *4.1. Slope*

Slope become an important factor that causing the landslide though it should be concerned other factors, which affect each other. Slopes that have ≥ 40 % degree of slope , generally potential of landslide but not necessarily slopes with these conditions causing a landslide. The stability of slope influenced by motion force and restrain force in the slopes. Motion forces are forces that attempt to make the slope slide, while restrain forces are forces that maintain the stability of the slope. If the restrain force greater than motion force, the slope will not be disturbed. Scoring of slope to mapping potential landslide area in this study based on Ministry of Public Works Decree No. 22/PRT/M/2007 about Guidelines for Spatial Planning of Landslide Prone Areas. The slope in Semarang Regency from the results of DEM SRTM processing have (0 - >40 %) degree of slope, which is dominated by (0 – 8 %) degree of slope (845.04 km$^2$) and (9 – 15 %) degree of slope (127.63 km$^2$). In Figure 4 shows that slope in this study area.

### *4.2. Geological Factors*

Geological factor is one of the important parameters that causing a landslide, this is because the mass of rock and soil that do the movement. Geological structure, characteristic of rocks, loss of soil adhesives due to natural processes, and earthquakes are the geological factors that influence the landslides. Geological structure that influences a landslide is the contact of bedrock and weathering rocks, cracks or fracture, layers rock and fractures. In this study, to mapping potential landslide areas using type of rock and characteristic of rock to get values of geological factors. Geological data is obtained from Systematic Geological Map. Type of rocks in this study consist of surficial deposit, volcanic rocks, sedimentary rocks, and intrusive rock. Semarang Regency is dominated by 12 type of volcanic rocks, 6 type of sedimentary rocks, 1 type of surficial deposit, and 1 type of intrusive rock. Surficial deposit in study area is an aluvium (Qa) ; intrusive rock is an andesit (Tma) ; volcanic rocks are Gajahmungkur

Volcanics (Qhg), Kaligesik Volcanics (Qpk), Jongkong Formation (Qpj), The Kemalon and Sangku Volcanics (Qks), Volcanic Breccla (Qvb), Undifferentiated Volcanic Rocks (Qvu), Sumbing Lava (Qls), Gilipetung Volcanics (Qg), Andong & Kendil Volcanics (Qak), Deposit of Cinder Cones (Qcc), Telomoyo Volcanics (Qte) and Qba ; Sedimentary rocks are Kaligetas Formation (Qpkg), Kalibeng Formation (Tmpk), Kerek Formation (Tmk), Many Member of Kalibeng Formation (Tmkb), Penyatan Formation (QTp), Payung Formation (Qp). In Figure 5 shows that distribution type of rocks in the study area.

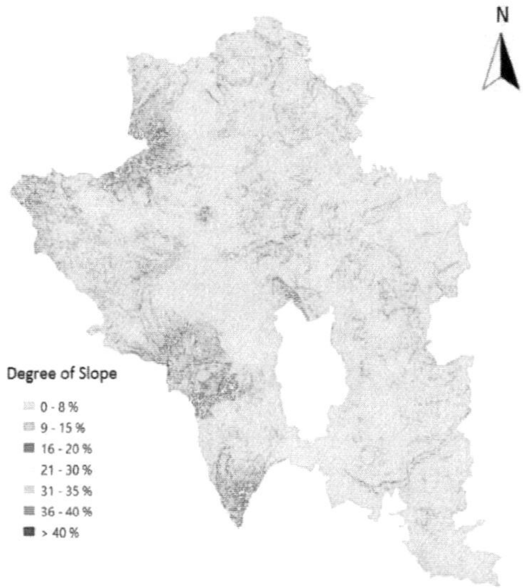


**Figure 4.** Slope in this study area by DEM SRTM processing.

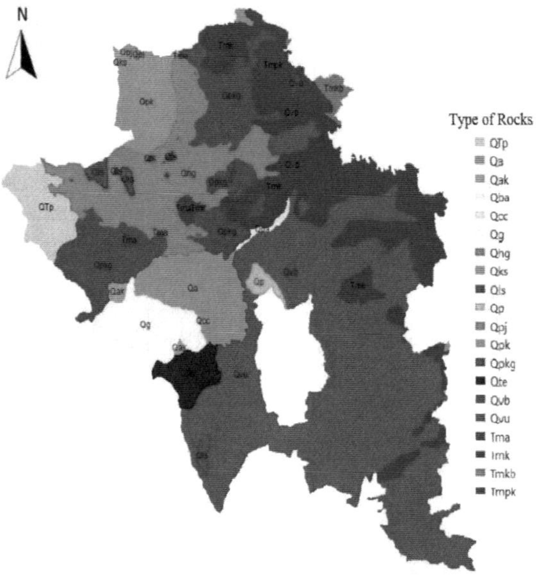


**Figure 5.** Type of rocks in this study as geological factor parameter.

*4.3. Soil Types*

Friable soil conditions will more easily to erosion and landslide because of water circulation into the cross-section soil more easily than massive soil. Land which has potential landslide can be assessed by soil erosion sensitivity (K). K factor is soil erodibility factor which represents both susceptibility of soil to erosion and the rate of runoff, as measured under the standard unit plot condition. This value indicates that soil eroded easily or not based on the physical and chemical characteristic soils. Smaller the values as a smaller sensitivity of the erosion. The values of soil erodibility in this study can be seen in Table 1, K factor values are obtained by study literature that based on Puslitbang Pengairan Bogor, 1985. Soil types in this study have been shown in Figure 6.

**Table 1.** K factors in each soil type in study area based on Puslitbang Pengairan Bogor, 1985

| Soil Types | K factor |
|---|---|
| Dark Brown Alluvial Soil | 0.193 |
| Dark Brown Andosols Soil | 0.278 |
| Grey Grumusols | 0.176 |
| Red Latosols | 0.075 |
| Complex of Dark Grey Andosols and Litosols | 0.271 |
| Complex of Grey Regosols and Dark Grey Grumusols | 0.302 |
| Association of Brown Latosols | 0.175 |
| Association of Yellowish Red Latosols, Dark Brown Latosols, and Litosols | 0.062 |
| Reddish Brown Litosols | 0.075 |
| Dark Brown Mediterranean Soils | 0.323 |

IOP Conf. Series: Earth and Environmental Science **98** (2017) 012013 doi:10.1088/1755-1315/98/1/012013

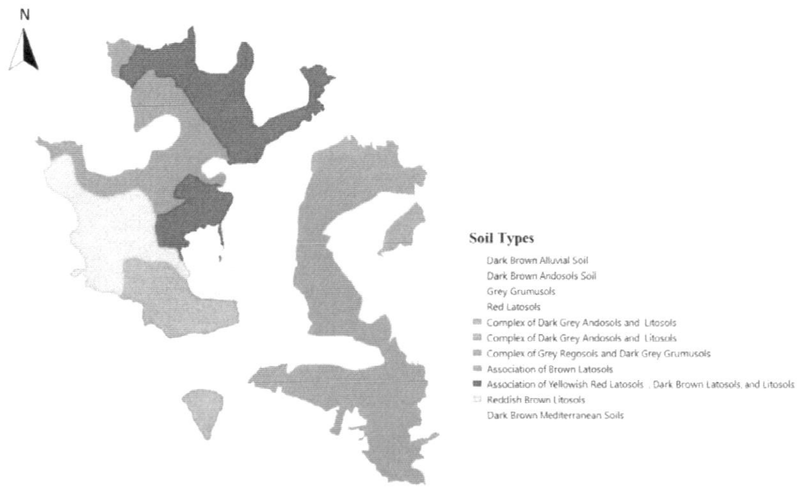


**Figure 6.** Soil Types in Semarang Regency based on Bappeda Semarang Regency data

*4.4. Land Cover*

Human activities cause the change functions of landuse, that changes can be a control factor of ground movement or increase the risk of landslide. This is based on landuse that will be affect land cover conditions. Plants in forest land cover will be able to reduce erosion. When less vegetation covered some areas, the function of roots to hold soil or aggregate soil binder decreases. So, land cover is one of the important factors that causing a landslide. Land cover in study area is processed by supervised classification in Landsat 8 image. The overall accuracy of classification is 88.41 % from the results of the confusion matrix. Percentage of overall accuracy has shown that classification results can represent true land cover in earth surface. Figure 7 shows that land cover in study areas.

*4.5. Rainfall*

Rainfall is the amount of rain that fell to earth by one unit surface area at a certain time period. The volume of rainfall that falls in particular area over long period relatively, indicates that the amount of rainfall intensity. So, the unit of rainfall used $m^3$/unit area and expressed in water level (mm) generally. The rainfall that sinks into the clay soil on the slopes will be restained by more compact and impermeable rocks. The amount of rainfall intensity causes that volume of retained water become increase, so that waters is pressing the soil aggregate and leading sandy clay soil to move. The rainfall in this study was obtained by 8 rainfall observer station that is spread evenly in Semarang Regency. Because of rainfall data is a point, so to mapping rainfall areas in this study uses interpolation method by Inverse Distance Weighted (IDW). The interpolation is a simple deterministic method by considering the surrounding points. The assumption of this method is the value of interpolation will be more similar to the near sample data, besides that farther one. The weight will change linearly according to the distance of sample data [4]. The rainfall intensity in study area about 2591 – 4276 mm/year that can be seen in Figure 8.

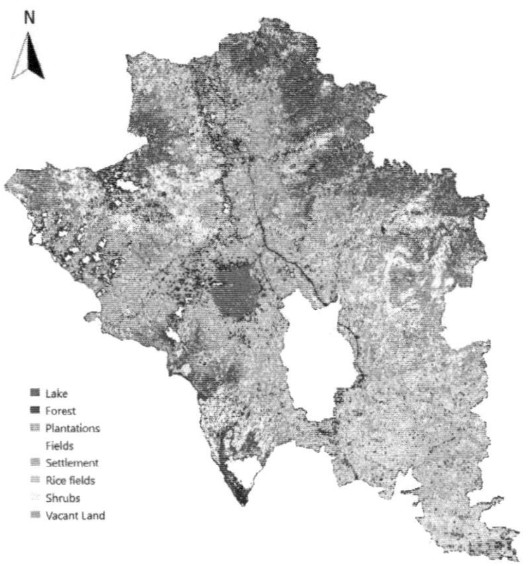


**Figure 7.** Land cover in study area by Landsat 8 image processing (supervised classification with maximum likelihood)

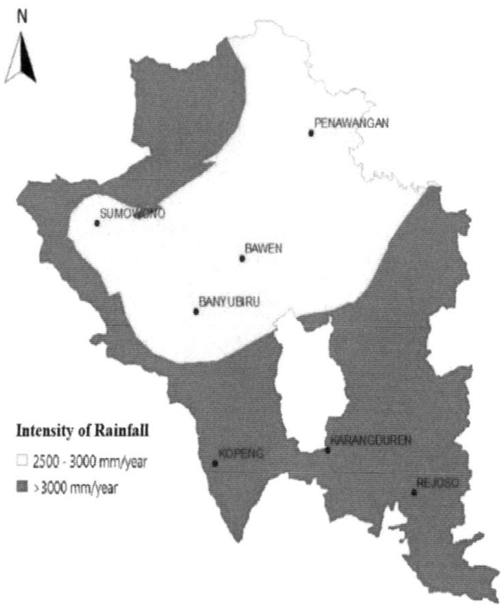


**Figure 8.** Intensity of rainfall in study area by IDW processing.

### *4.6. Analysis of Environmental Vurnerability in Potential Landslide Areas*

Potential of Landslide disaster in study area shows that Semarang Regency has 3 categories of the landslide are less prone, prone and extremely prone. The classification based on the regulations from Directorate of Volcanology and Disaster Mitigation, 2004. The calculation that uses to obtain the values of landslide prone areas using scoring and weighting by each parameters, which has been described earlier. Rainfall parameter is given the greatest weight (30 %), while for geological factors and soil type given weight (20 %), and for land cover and slope parameters given weight (15%). The value of landslide prone level in this study is acquired by summing the multiplication of weights with the scores in each parameters. Semarang Regency have the minimum value of landslide is 1.6 and the maximum value is 4.3. In this study, that cumulative score is classified in 3 categories are less prone ($\leq 2.5$); prone ($\geq 2.6$ and $\leq 3.6$); and extremely prone ($\geq 3.7$). Distribution of landslide prone areas in Semarang Regencycan be known by that classified. Figure 9 shows that landslide prone areas in Semarang Regency, which is dominated by prone areas about 791.27 km$^2$. Extremely prone areas spread in several locations, such as Suruh District, Pabelan District, Bancak District, Ungaran Timur District, and Bergas District. Some extremely prone areas are located at production forest land cover. The highest cumulative score is located in Bancak District. The result has been validated with data from Regional Disaster Management Agency (BPBD) in 2016, that data shows of the most landslide recorded in some location with prone category areas in Banyubiru District and Jambu District. While, period January until Mei in 2017, the amount of landslide incident has increased (154 incidents of 208 total accumulation disasters), if compared with landslide incident in 2016. Most of the incident is located in prone and extremely prone areas, that spread in several districts are Ungaran Timur District, Ungaran Barat District, Sumowono District, Banyubiru District, Suruh District, Getasan District and Bringin District.

Value of Environmental Vulnerability Index in Landslide areas is calculated by equation 4 in the methods section. Previously, its necessary to classified areas that belong to environmental vulnerability areas. That area consists of conservation forests (1472.29 Ha); conservation areas (1404.82 Ha), swamps (1351.87 Ha) and shrubs (12382.71 Ha), which can be seen in Figure 10. Delineation of that area based on the land cover by Landsat 8 image processing and refers to region map from Forest Ministry. Accumulation score of environmental vulnerability index in study areas is moderate value, that means environment condition must be considered, such as vegetation as ground cover and many others aspects. Some conservation forest and conservation areas are located in prone landslide areas, such as in Getasan District, Ungaran Barat District, and Sumowono District. Based on environment factors, that districts must be considered because of recorded data landslide incident in period January until Mei (2017) shown 16 landslide incidents located in Getasan Districs, 18 landslide incident in Ungaran Barat District, and 18 landslide incident in Somowono District. The most amount landslide incident in 2017 is located in Ungaran Timur District, but the environmental vulnerability score is smaller because some areas consist of production forest and shrubs. Vegetation density by NDVI algorithm processing in conservation areas in Getasan Districts about (0.030 - 0.844), while in conservation forest about (0.045 - 0.849). The range of NDVI value shows that density level, in conservation areas and conservation forest have rarely until high density level. Based on The Ministry of Forestry Republic of Indonesia, (2003), divided by density level are rarely (-1.0 – 0.32); moderate (0.32 – 0.42) and high (0.42 – 1) [6]. The rare vegetation density areas both in conservation areas and conservation forest should be done preventive efforts to minimize or reduce the risk of landslide disaster.

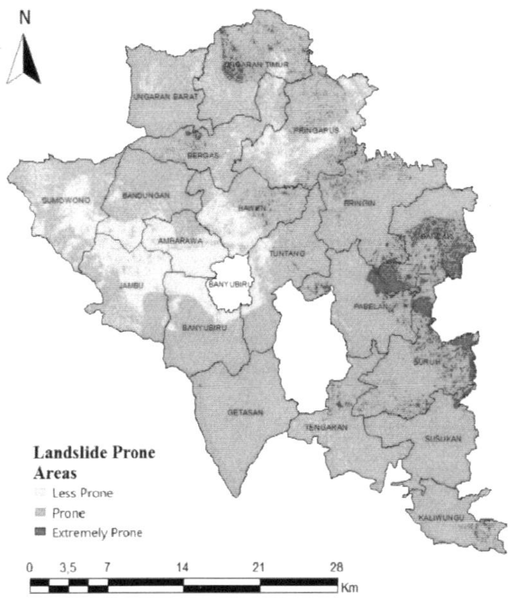


**Figure 9.** Landslide Prone Areas in Semarang Regency based on the regulations from Directorate of Volcanology and Disaster Mitigation.

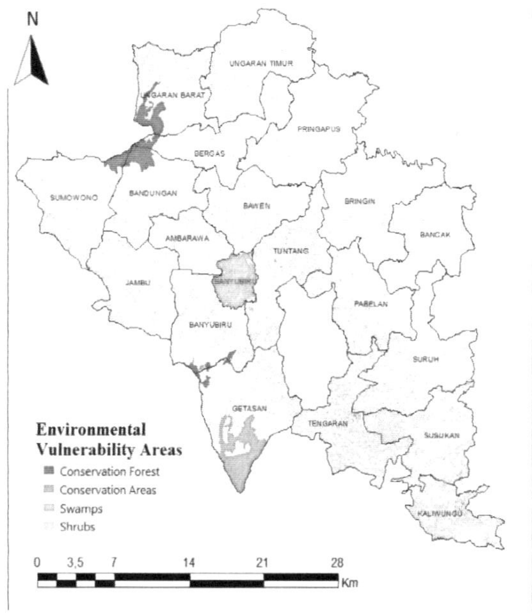


**Figure 10.** Environmental vulnerability areas in Semarang Regency

## 5. Conclusions

Mapping of the potential landslide areas is needed to reduce the risk of disaster. Remote sensing and geographic information system to be used in this study to mapping the landslide areas and analysis of environmental vulnerability. The parameters to mapping landslide areas based on land cover, rainfall, slope, geological factors and soil types. The high human activity caused some changes in land cover, so to obtain actual land cover in study areas using remote sensing is based on Landsat 8 imagery processing. Scoring and weighting method is a spatial analysis which utilizes geographic information system. Values of scoring and weighting are based on the relevant literature review and regulations such as the regulations from Directorate of Volcanology and Disaster Mitigation to mapping accurate landslide areas in Semarang Regency. The result of landslide prone areas that be analyzed to know about the environmental susceptibility in landslide areas. Based on Perka BNPB No. 2/2012, indicators of environmental vulnerability consist of forests, shrubs and swamps parameters in the study area. Semarang Regency have the minimum value of landslide is 1.6 and the maximum value is 4.3, which is dominated by landslide prone areas about 791.27 $km^2$.

Accumulation score of environmental vulnerability index in study areas is moderate value, that means environment condition must be considered, such as vegetation as ground cover and many others aspects. Some conservation forest and conservation areas are located in prone landslide areas, such as in Getasan District, Ungaran Barat District, and Sumowono District. Vegetation density by NDVI algorithm processing in conservation areas in Getasan Districts about (0.030 - 0.844), while in conservation forest about (0.045 - 0.849). The range of NDVI value shows that density level, in conservation areas and conservation forest have rarely until high density level. The rare vegetation density areas both in conservation areas and conservation forest should be done preventive efforts to minimize or reduce the risk of landslide disaster.

## References

[1] Department of the Interior U.S. Geological Survey. 2016. *Landsat 8 (L8) Data Users Handbook. Version 2*. Sioux Falls, South Dakota.
[2] Lillesand, T.M., Kiefer, R.W., and Chipman J.W.2004. *Remote Sensing and Image Interpretation*. Fifth Edition. New york : John Wiley & Sons.
*[3]* National Disaster Management Agency (BNPB) Decree. No 2/2012 about *General Guidelines for Disaster Risk Assessment.*
[4] Pramono, G. H. 2008. Akurasi Metode IDW dan Kriging Untuk Interpolasi Sebaran Sedimen Tersuspensi. *Forum Geografi*, Vol. **22**, No. 1, Juli 2008: 97 – 11.
[5] Taufik, M., Utama, W., and Firdaus, H.S. 2013. Environmental Baseline Imaging For Geothermal. Bandung : Proceedings, 2nd ITB Geothermal Workshop.
[6] The Ministry of Forestry Republic of Indonesia. 2003
[7] Urban Land Use Plan Semarang Regency 2011 – 2031

# Social Toponymy of Urban Ecotourism to Inventory Potential Tourism in Dolly Based on WebGIS

**B M Sukojo*, A K Asri and C A Swastyastu**

[1]Departement of Geomatics Engineering, Faculty of Civil Engineering and Planning, Sepuluh Nopember Institute of Technology (ITS) Surabaya, Indonesia

bangunms@gmail.com

**Abstract**. Before closed, Dolly is the largest localization area in Southeast Asia, located in Surabaya, East Java. After closing in 2014, to change Dolly into a region that has a positive image, the government of Surabaya City plans to develop Dolly becomes an Urban Ecotourism area. To support the program, various training activities have been conducted. In addition to, Dolly development is also encouraging the development of supporting facilities such as an information system that can describe the new Dolly. Knowing this issue, it is necessary to conduct a social toponymy survey that can support the availability of geospatial information. The method used in this research is field surveying and interview based on toponymy. The concept of standardization is verified the elements that have been standardized, propose that not have been standardized, and identify the potential of Dolly. The results of toponymy survey were analyzed using four geographic approaches (onomastics, spatial, ecological, and geographic) to determine the description of the area, naming criteria, and existing tourism potential. In addition to these four approaches, to describe how potential development of Urban Ecotourism in Dolly, also used qualitative-descriptive analysis techniques. The results of this research, in addition to information about history and gazetteer, also presented information about Urban Ecotourism in Dolly. From the identification, it is known that there are some villages in Dolly that have been developed into tourist areas with various characteristics, such as: "Kampung Orumi" which is famous with seaweed drink, "Kampung Samijali" with Samiler crackers, "Kampung Batik" with the design of Jarak and Semanggi Leaves Typical of Surabaya, and "Kampung Telur Asin" with various food made from salted eggs. And in the end, to present information about Dolly, in this research will also be built WebGIS of Dolly Ecotourism.

**Keywords:** Dolly, Geographic approaches, Social Toponymy, Urban Ecotourism, WebGIS

## 1. Introduction

Dolly is the largest localization area in Southeast Asia, larger than Phatpong localization in Bangkok and Geylang localization in Singapore. There is even a controversy where Gang Dolly was incorporated into one of Surabaya's tourist destinations for foreign tourists [1]. Based on statistical data after the declaration of dissolution (per July 2014) from 15 RW located in Putat Jaya Urban Village, not all of its territory become the location of prostitution. Only a few RTs, including 5 RWs, belong to the prostitution center, usually called Dolly and Jarak. In addition, in 1 RW only a few RTs are included in the scope of localization and the rest are only ordinary households. It is known that the number of prostitutes in the area Dolly and Jarak reached 1021 people with a total pimps reached 230 people.

After the dissolution launched by the Surabaya City Government, various trainings were conducted to provide skills to the residents so that they no longer rely on the existence of localization in the area [2]. In other words, the provision of skills training is aimed at starting a new business and meeting the daily needs with decent work.

To transform Dolly's image into a positive value area, it has planned Dolly's ex-localization area as an urban tourism area. There are several types of collaborative activities with Non-Governmental Organizations to develop Dolly region resources such as: i) training of samiler raw materials, ii) development of slamer dryer technology without solar energy, iii) seaweed processing training, iv) procurement of production expired test, V) business development Chatering Dolly (Sukojo, 2017). Development of Dolly region through various activities, encouraging the development of supporting facilities one of which is the existence of an information system. Where the development of information systems presented in the form of websites can be a media publication of the results of community production in the field of culinary, handicrafts or other micro businesses. And given the location of the target that Dolly as an ex-lokalisasi area has a special philosophy and history, then the development and packing of information areas to be widely known as a positive area should be done optimally.

In this research-based devotion, the method used is the Toponimi Survey. The concept that is done is the verification of the elements that have been standardized and give the proposed standardization of elements that have not been standardized and make inventory of existing potential, especially tourism potential. Basically, toponimi surveys can support the availability of basic geospatial information such as national gasetirs and also the provision of thematic geospatial information such as the inventory of tourism support elements in the Dolly region. In addition, toponimi is also required in development and activities related to the aspect of a country's sovereign defense [3].

The idea used in this devotion is the development of a Web-based Geographic Information System or so-called WebGIS. Where this system can store, repair, update, manage, manipulate, integrate and display existing tourism potential data in a website-based geographic information. In addition, there are several methods of approach in the analysis phase of the approach onomastics, spatial, ecological, economic, social and cultural to examine the tourism potential in the region Dolly.

## 2. Materials and Methods

### *2.1. Research Site*

**Figure 1.** Dolly Ex-Localization Map

The research area is located in the area that is the location of this research-based service is the Dolly area, Putat Jaya Village, Sawahan District, Surabaya, East Java.

### *2.2. Tools and Data Acquisition*

In this research used spatial and non-spatial data. For spatial data include RBI Map 1608-414 (Wonokromo) scale 1:25.000, SPOT 6 corrected image, administrative vector data of Dolly, and landmark coordinates that recorded by GPS Handheld. While for non - spatial data include:

IOP Conf. Series: Earth and Environmental Science **98** (2017) 012014 doi:10.1088/1755-1315/98/1/012014

a. Primary data, i.e.: topographic elements of the landmark and initial toponymy data
b. Secondary data, i.e.: physical data (topography, hydrography, geology, climatology) and non-physical data (demographic, social, economic and cultural)
c. Documentation of topographic elements.

In order to analyze the image for making village map, this research utilizes ArcGIS 10.3.

*2.3. Methods*

*2.3.1. Toponymy Survey.* The concepts of standardization in toponymy survey are verify the elements that have been standardized, propose the elements that not have been standardized, identify the potentials of village, and visual observation. To get detail information, also doing interview to the village heads, youth leaders, theologian, village elders and local people in Dolly. It aims to know the general conditions, history, social conditions, economic conditions, cultural, and local wisdom.

*2.3.2. Inventory of Topographic Elements.* In toponymy rules, to inventory topographic elements are used some survey forms based on standardization of topographic names [2]. The forms are Form A, B, C, D, and toponymy data collection form. The results of the inventory recorded on the D form, are then used as the basis for identification naming criteria of geographical elements

*2.3.3. Identification of Historical and Urban Ecotourism Potentials.* History and potentials of research area can be discovered by doing visual observation and conducting interviews to the surrounding villagers. The research has also identified about a landmark in Dolly that shows the Ecotourism area with various SMEs (Small and Medium Enterprises).

## 3. Results and discussion

*3.1. Naming Characteristics of Topographic Elements*

Based on field survey results, 49 landmarks or topographic elements have been identified, including: 19 portals, 13 trade location, 6 government facilities, 8 public facilities, 2 mosques and 1 old tomb. For the topographic elements nomenclature was based on several criterias. Characteristics of topographical elements nomenclature can't be separated from the cultural and linguistic influences. Those characteristics are based on name, script and pronunciation using Javanese. Topographic elements principle uses local names in order to preserve cultural wisdom, respect the tribes, religions, races and classes. The explanation of the nomenclature in Dolly:

*a. Naming of Villages*

- Based on Main Product, i.e.: Kampung Samijali, Kampung Orumi, Kampung Batik, Kampung Telur Asin, and Kampung Tempe.
- Based on the names of plant, i.e.: Jarak Village, and Putat Jaya Village
- Based on Names of Figure i.e.: Gang Dolly

*b. Naming of SMEs (Small and Medium Enterprises)*

- Based on address of production i.e.: Alpujabar (Alamat Putat Jaya Barat), Pujaa (Putat Jaya Gang 2A)
- Based on design of product i.e.: Jarak Arum, Canting Surya
- Based on name of creator: Tempe Bang Jarwo
- Based on community: KSM Kawan Kami
- Based on raw materials of product, i.e.: Samijali (Samiler Jarak-Dolly), and Orumi (Olahan Rumput Alami)

*3.2. The History of Dolly*

Dolly, a localization area that has existed since the 19th century was once proclaimed as one of the tourist destinations of Surabaya. Judging by its history, Dolly, who was once the largest prostitution area in Southeast Asia was founded with the aim of providing beautiful women to serve the Dutch soldiers.

There are various stories about who the main founder of this lokalisasi, ranging from the emergence of the name of Dolly Van de Mart who is a woman of Dutch descent to the name of Dolly Khavit who is a Philippines-Javanese woman. At first localization Dolly itself is divided into two areas, namely Putat Jaya and Jarak - Dolly. Once the Jarak is a tomb of China.

However, knowing that the lease contract of Chinese burial ground is exhausted, then the area is used as a dispute so that the tomb that has been completed the contract can be evicted. After carried out eviction, putat jaya area, Jarak and Dolly own erected cottage houses where doing prostitution activities. It can be said that the origin of the dolly and putut jaya spacecraft basically from the beginning has been built with the objective of localization area transfer from Jagir commercial sex worker.

*3.3. SMEs in Urban Ecotourism Area of Dolly*

After the closure of Dolly localization in 2014, Surabaya City Government began to conduct various activities of community empowerment, especially the community around the ex-localization. This empowerment is a government effort to provide employment so that people who used to rely heavily on localization as a source of income, can now conduct various business activities independently. There are a wide variety of businesses in every village that began to be occupied by the community and society Putat Jaya Dolly, among others. There are the dolly tourist villages that call “Kampung”.

a. Kampung Samijali

**Figure 2.** Kampung Samijali

Samiler Jarak-Dolly or Samijali is usually known as a snack specialties Dolly exclusive localization. SMEs Samijali which was established since April 2015 is one of the efforts of Non-Governmental Organizations namely the Painting of Hope Movement. Chaired by Mrs. Dwi and assisted by 10 employees, this SME has been producing processed samiler crackers with 4 flavor variants ie Samijali original flavor, baberque, balado, and cheese. The shape of small square-cut crackers is one of the innovations to make these snacks are not easy to crumble. In addition, the model of packing crackers are quite modern, making this product more selling than the samiler crackers that are usually sold in the market. Currently, for the samijali marketing area itself has spread in Surabaya area such as in Siola, Town Hall, Kertajaya, Kutai, Rumah Kukus Surabaya, and Dolly creative house. As for the production house itself is located in Kampung Samijali, precisely in Putat Jaya Gang 4A RT 13 / RW 03.

**Figure 3.** Samijali Product

b. Kampung Orumi

**Figure 4.** Kampung Orumi

In addition to snack crackers, one of the micro businesses that have been running under the guidance of the NGO Movement of Painting Hope is Orumi's SME. Orumi or what can be called Natural Seaweed Processing is one of the jelly products typical of Dolly ex-lokalisasi made from Seaweed. There are 4 variants of flavors produced are lemon oro flavor, lychees, strawberries, and green tea with the price of each package of 10 thousand. Early innovation of SMEs established since 2016 is to compensate for the existence of a typical food product Dolly Samijali which has existed since 2015. SMEs incorporated in the business group INAKOM is located in Kampung Orumi, precisely in Putat Jaya Barat Gang III A RT 03 / RW 03. Chaired by Mrs. Hanani and assisted by 4 employees, this Orumi product has been marketed around Surabaya, either through exhibitions or independent promotion.

**Figure 5.** Orumi Product

IOP Conf. Series: Earth and Environmental Science **98** (2017) 012014 doi:10.1088/1755-1315/98/1/012014

c. Kampung Gotong Royong dan Kreatif

**Figure 6.** Kampung Kreatif dan Gotong Royong

In this village there are some production that is managed by society, among others:

- SMEs Pujaa

  One of the existing SMEs in Kampung Gotong Royong and Kreatif is Pujaa SME. Named Pujaa because this SME is located in Putat Jaya Gang 2A. SMEs that began inaugurated in January 2017 is under the auspices of the Office of Food Security and chaired by Mr. Nirwono. Together with 11 employees, SMEs produce a variety of processed duck eggs such as salted eggs and also botok salted eggs. In addition, these SMEs also produce pure cow's milk and soy milk. To improve the quality of production, employees are also active in food processing training as well as entrepreneurship training conducted by the Food Security Agency, BPS, and Surabaya Community Training Board. And for now, the marketing area of production is around Surabaya such as in the Department of Population, Surabaya Zoo, and also producer for Soto Wawan.

**Figure 7.** SMEs Pujaa Product – Salted Egg

- KSM Kawan Kami

  In addition to SMEs Pujaa, the business group producing duck eggs in Kampung Gotong Royong and Kreatif is a Self-Help Group (KSM) of Our Friends. Since its founding in 2014, this business group headed by Mrs. Salamah has produced a variety of duck egg preparations, such as Salted Eggs, Salted Eggs Botok, and Pepes. In addition, to increase the diversity of production, KSM Kawan Kami also produces Wet Cakes, Pastries and Milk Sticks. Armed with the training of panga and entrepreneurship training organized by various institutions such as Community Training Board, Agriculture Agency, and Tristar, a business group consisting of 4 employees has marketed its products in various exhibition activities and marketed independently

**Figure 8.** KSM Kawan Kami Product – Various Preparation of Salted Egg

- Sablon Indana Collection

  In Kampung Gotong Royong and Kreatif there is also one micro business that serves penyablonan service, namely SME Sablon Indana Collection. The business started since 2014 is an independent business headed by Mas Arif and assisted by 6 employees. Although still standing for about 3 years, the marketing area of screen printing services is quite wide, in addition to local Surabaya, there are also customers who come from Nusa Tenggara, Makassar, and Sorong, Papua. In addition to serving screen printing services, employees of Indana Collection is also active role to train entrepreneurial candidates who want to open a business in the field penyablonan KASA Surabaya.

d. Kampung Batik

**Figure 9.** Kampung Batik

In this village there are some production of batik, that call "Batik Dolly", among others:

- Canting Surya Batik

  Canting Surya is one of the SMEs who are members of Dolly Batik business group and located in Putat Jaya Barat Gang VI B No. 6 RT 02 / RW 11. As the name implies, batik is produced by SMEs has a character that is typical in terms of art "Cantinganya". Still maintaining detailed and fine dash styles, supported by the use of natural colors are the hallmarks of Canting Surya batik. Established since 2014 under the guidance of the Department of Industry and Trade Surabaya, currently SME Canting Surya chaired by Mrs. Ike Setiowati and has 4 employees. Armed with the skills of the training results of the Community Training Board and the Department of Trade and Industry, SMEs is able to showcase and market its products through various activities such as the Colombo Plan held on 19-24 Sept 2016. In addition, improving the quality and quantity, making the sale value of batik produced has a fairly high price ranging from 250 thousand to 10 million. This depends on the complexity of batik design demanded by the buyer.

**Figure 10.** Canting Surya Batik

- Jarak Arum Batik

  In addition to Canting Surya, there are other SMEs who are members of the business group Batik Dolly, the Jarak Arum. As the name implies, the batik produced by Jarak Arum SMEs leaves no Jarak Leaf images with additional Clover Leaves in each design. Not only that, Jarak Leaf also characterizes that the batik is produced from the surrounding Jarak area precisely in Kelurahan Putat Jaya Timur Gang 2A RT 05 / RW 03. Currently, SMEs under the guidance of the Department of Industry and Commerce Surabaya is chaired by Mrs. Umi Ningsih and has 8 employees. To train batik skills, employees are also active in training activities conducted by the Department of Industry and Trade of Surabaya City. And until now, SMEs established since 2014 is able to produce various types of Dolly-Surabaya typical batik with selling from 200 thousand to 5 million.

**Figure 11.** Jarak Arum Batik

- Alpujabar Batik

  In contrast to the naming of other dolly batik SMEs that exhibit the character of batik being produced, the name "Alpujabar" on this SME is based on the address of the batik produced, "Alpujabar = Address Putat Barat". This is quite interesting, because behind the business name, indirectly shows the identity where the batik is coming from. Similar to other Dolly batik SMEs, the Alpujabar Batik, headed by Mas Tris also has its own characteristics. One characteristic in addition to the Jarak leaf and clover shapes typical of Surabaya, in the design of batiknya always displays icons that describe the city of Surabaya as the image of Suramadu Bridge, the Sun at Kenjeran, Tugu Pahlawan, Kapal Selam Monument and much more. One design that can also be said "Dolly Banget" is a batik design themed "Butterfly Night". From the design can be described briefly Dolly picture of the past. Since its establishment in 2015, Alpujabar SME under the guidance of the Department of Industry and Trade is still managed independently with its production house in Putat Jaya Barat Gang IX B No. 31. In addition, to market its products, Alpujabar SME has been active in sharing exhibition activities held by Surabaya City Government. The price of batik produced ranged between 200 thousand to 2.5 million.

**Figure 12.** Alpujabar Batik

e. Kampung Tempe Bang Jarwo

**Figure 13.** Tempe Bang Jarwo

The disbanding of Dolly 2014, in addition to closing community employment which is highly dependent on localization has also opened up community opportunities to create various innovations, and one of them is tempe production business. Tempe Bang Jarwo is one of the typical Dolly tempe produced by a former thug with his wife who used to work as a commercial sex worker. This is quite interesting considering the business innovation initiated along with the NGO of the Painting of Hope Movement, it has been able to avoid economic imbalances after the disbanding of Dolly. Currently there are several different productions of tempe bang jarwo like raw tempe with various shape innovation, dry tempe sambel, and also tempe nugget. In addition, for marketing tempe bang jarwo produced in the area of Kupang Gunung Gang Tembusan is done by going around using a bicycle. In addition, to expand its marketing, Bang Jarwo and his wife (Mbak Rina) also actively participate in business exhibition activities held by Surabaya City Government.

## 4. Conclusions

The conclusions of this research has been identified that there are some villages in Dolly that have been developed into tourist areas with various characteristics, such as: "Kampung Orumi" which is famous with seaweed drink, "Kampung Samijali" with Samiler crackers, "Kampung Batik" with the design of Jarak and Semanggi Leaves Typical of Surabaya, and "Kampung Telur Asin" with various food made from salted eggs.

## References

[1] Purnomo T and A Siregar 1982 *Dolly: Membedah Dunia Pelacuran Surabaya*. Jakarta: Grafiti pers, 1982) pp 33-34.
[2] Sari D F and M F Ma'ruf 2016 *Proses Pemberdayaan Bagi Warga Terdampak Penutupan Lokalisasi Dolly di Kelurahan Putat Jaya Kecamatan Sawahan Surabaya*. (Surabaya: Prodi S1 Ilmu Administrasi Negara, UNESA).
[3] Sukojo B M 2009 *Buku Ajar Toponimi* (Surabaya: Jurusan Teknik Geomatika, FTSP, Institut Teknologi Sepuluh Nopember) pp 23 – 29.

[4] Bakosurtanal 2009 *Survei dan Pemetaan Nusantara* (Jakarta: Bakosurtanal).
[5] Dahlia N R 2015 *Evaluasi Usaha Kecil Dan Menengah Menggunakan Sistem Informasi Geografis Di Kota Surabaya* (Surabaya: Institut Teknologi Sepuluh Nopember).
[6] Fathoni A 2014 *Pendekatan geografi (Keruangan, Ekologi dan Komplek Wilayah)* (www.zonasiswa.com/2014/06/pendekatan-geografi-keruangan-ekologi) Dikunjungi pada tanggal 24 Oktober 2016
[7] Herliningsih I 2013 *Gasetir sebagai Bagian Kekayaan Budaya Bangsa* (Bogor: Kepala Bidang Toponim, Pusat Pemetaan Rupabumi dan Toponimi – BIG)
[8] Kertasasmita G 1997 *Pembangunan untuk Rakyat, Memadukan Pertumbuhan dan Pemerataan* (Jakarta: Cides).
[9] Masyarakat. (Bandung: Rafika Aditama)
[10] Sukojo B M 2013 *Draft: Pedoman Teknis Inventarisasi dan Verifikasi Nama Rupabumi*(Jakarta: Dirjen PU, Direktorat Wilayah Administrasi dan Perbatasan).
[11] Yulius and H W L Salim 2013 *Inventarisasi Selat di Kabupaten Karimun Provinsi Kepulauan Riau Berdasarkan Kaidah Toponimi*. Jakarta: Pusat Litbang Sumberdaya Laut dan Pesisir, Badan Litbang KP-KKP..php>. Diakses tanggal 1 November 2016, jam 10.22.

IOP Conf. Series: Earth and Environmental Science **98** (2017) 012015 doi:10.1088/1755-1315/98/1/012015

# Multiscale Spatial Assessment of Determinant Factors of Land Use Change: Study at Urban Area of Yogyakarta

**Bowo Susilo**

Department of Geographic Information Science, Faculty of Geography, Universitas Gadjah Mada

bowosusilo@ugm.ac.id

**Abstract**. Studies of land use change have been undertaken by different researchers using various methods. Among those methods, modelling is widely utilized. Modelling land use change required several components remarked as model variables. Those represent any conditions or factors which considered relevant or have some degree of correlation to the changes of land use. Variables which have significant correlation to land use change are referred as determinant factors or driving forces. Those factors as well as changes of land use are distributed across space and therefore referred as spatial determinant factors. The main objective of the research was to examine land use change and its determinant factors. Area and location of land use change were analysed based on three different years of land use maps, which are 1993, 2000 and 2007. Spatial and temporal analysis were performed which emphasize to the influence of scale to both of analysis's. Urban area of Yogyakarta was selected as study area. Study area covered three different districts (kabupaten), involving 20 sub districts and totally consists of 74 villages. Result of this study shows that during 14 years periods (1993 to 2007), there were about 1,460 hectares of land use change had been taken place. Dominant type of land use change is agricultural to residential. The uses of different spatial and temporal scale in analysis were able to reveal different factors related to land use change. In general, factors influencing the quantities of land use change in the study area were population growth and the availability of land. The use of data with different spatial resolution can reveal the presence of various factors associated with the location of the change. Locations of land use change were influenced or determined by accessibility factors.

**Keywords**: multiscale spatial assessment, land use, Yogyakarta

## 1. Introduction

Studies on land use change have been conducted by the researcher along with the growing awareness to the importance of the phenomena. At the beginning of 1980s, land use and land cover change became a concern because of it has relation with global climate change by the carbon cycle [1, 5, 6, 7]. Changes in land use are also seen as a fundamental factor for the operation of the environmental and socio-economic systems, from local to global scale scale [2].

Land use change can be assessed using different methods. Among those methods, spatial modeling is widely utilized. Model is simplification of real-world phenomena [8, 11, 12]. Through a model or modeling, a complex phenomenon is represented using simplest manner i.e. a number of components referred to as modeling variables.

Variables used in the land use change model are representation of factors related to the phenomenon. These factors often referred to as determinant factors of land use change. Land use change is considered as a complex spatial phenomenon as a result of the interaction of various determinant factors. The physical condition of land, socio-economic, demographic and policy are factors that are generally supposed to be related to changes in land use [6]. These factors interact in the manner and intensity of which varies according to time and space. Study of the determinant factors therefore is necessary as a basis of modeling the spatial dynamics of land use.

Land use change is a spatial phenomenon as well as its determinant factors. Changes in land use occupy a certain location and occur in a certain period as well. Study of spatial phenomena requires appropriate methods that are known as spatial analysis. Spatial analysis, on the other hand, has a critical issue related to the use of different scale in analysis. The same phenomenon which is analyzed at different scales, very likely leads to a different conclusion. The problem is known as MAUP, which is an abbreviation of the modifiable area unit problem [4].

The objective of this study is to examine changes in land use and the factors associated with it. The most influential factor is referred to as determinant of change. The study was conducted through a spatial analysis by taking into account the effect of scale in the process of analysis. Urban area of Yogyakarta was selected as study area. Land use change in the study area were considered intensive, particularly from agriculture to other type of land use [9, 10].

## 2. Study Area

The study was conducted at the Urban area of Yogyakarta (Figure 1). This area covered whole area of Yogyakarta City and some parts of Sleman as well as Bantul Districts (kabupaten). Yogyakarta City consist of 16 sub districts (kecamatan). Parts of Sleman Districts used in the study, involved 3 sub districts namely Gamping, Mlati and Depok. Depok is quite well known for rapid development as well as for being the location of Universitas Gadjah Mada, one of the famous universities in Indonesia. Bantul Districts also involved 3 sub districts namely Kasihan, Sewon and Banguntapan. The entire study area, therefore involving 20 sub districts and totally consists of 74 villages (desa).

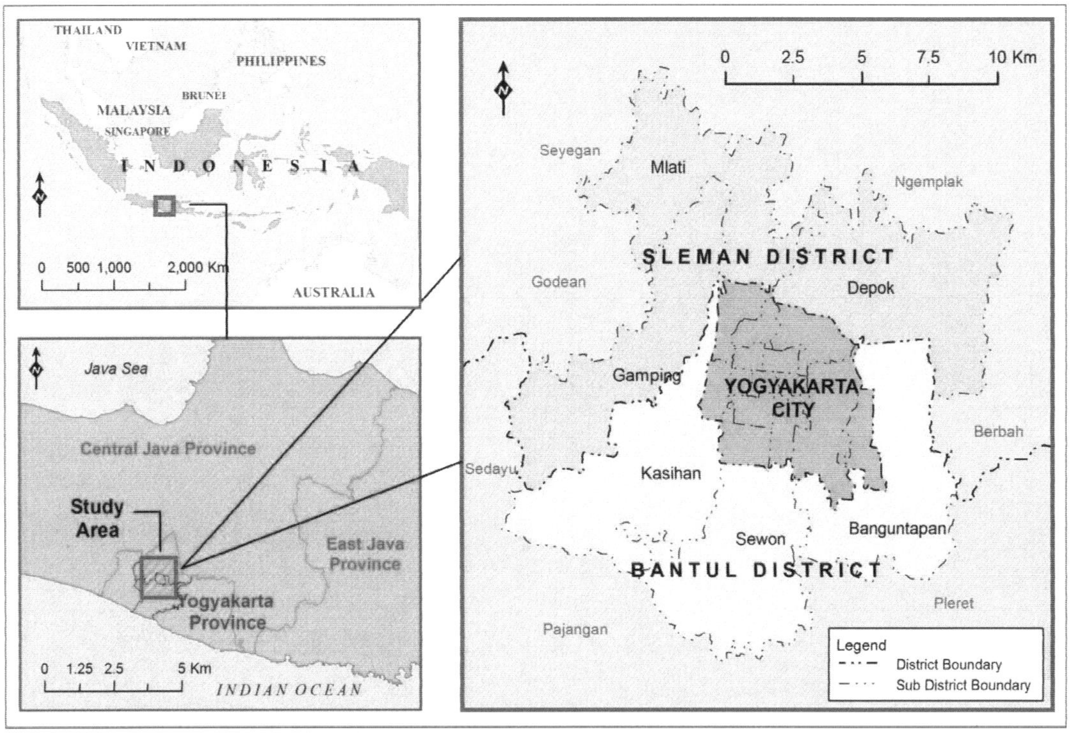


**Figure 1**. Study area

## 3. Methods

Determinant factors of land use changes in the study were assessed by means of mapping and spatial analysis. Mapping was conducted to identify spatial distributions of land use changes as well as presumably factors related to the land use changes. Spatial and statistical analyses were used to assess significant factors related to area and location of land use change.

### *3.1. Data*

Data used in this study include 1: 25.000 multi temporal land use maps, 1: 25.000 Indonesian topographic maps (RBI), demographic and socio-economic data. Multi temporal land use maps consisted of 1993, 2000 and 2007 maps respectively. Those maps originally derived from 1: 25.000 Indonesian topographic maps and updated periodically based on interpretation of remote sensing imagery.

Time series of demographic and socio-economic data, from year 1993 up to 2007, were obtained from secondary sources. Books namely Sleman District in Figures, Bantul District in Figures and Yogyakarta City in Figures were used as main source. Those books were published by Central Bureau of Statistics (Badan Pusat Statistik).

### *3.2. Mapping of Land Use Change as Dependent Variable*

Land use change maps were obtained by overlaying land use maps of 1993, 2000 and 2007. Those maps show the spatial distribution of land use change over the period of 1993 – 2000 and 2000-2007 (Figure 2). Land use changes were simplified in to two categories namely change and persistence. The location where the change occurs is referred to as change, while the location where the change does not occur is referred to as persistence.

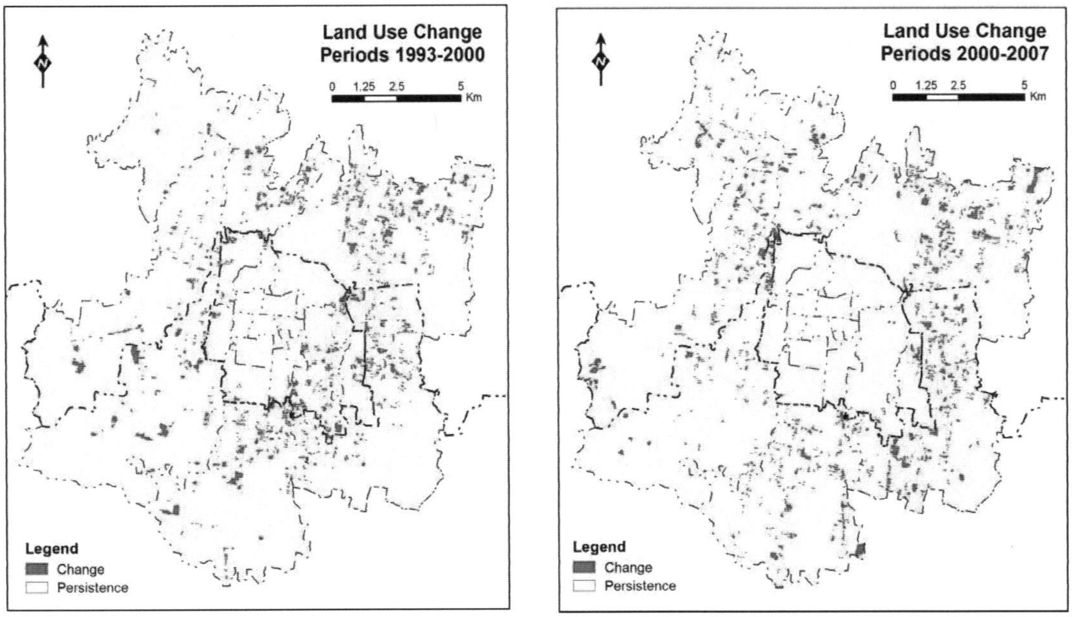


**Figure 2**. Spatial distribution of land use change 1993-2000 and 2000-2007

Quantities as well as spatial distribution of land use changes were vary across time periods. In the 1993-2000 periods, total area of land use change was 704.9 ha. Those changes were taken place in the city as well as in urban fringe area. In the urban fringe area, land use change mostly distributed in the northern and eastern part of study area. In the 2000-2007 periods, total area of land use change was 769.4 ha and mostly taken place in urban fringe area. Detail information about quantities and spatial distribution of land use change in the study area are shown in Table 1.

**Table 1**. Area of land use change by administrative division

| District | Sub district | Area of Land Use Change (Ha) | |
|---|---|---|---|
| | | 1993-2000 | 2000-2007 |
| Bantul | Banguntapan | 99,67 | 177,21 |
| | Kasihan | 83,39 | 60,31 |
| | Sewon | 108,08 | 109,88 |
| Sleman | Depok | 168,27 | 195,17 |
| | Gamping | 51,52 | 117,16 |
| | Mlati | 66,78 | 88,21 |
| Yogyakarta | Gondokusuman | 3,93 | |
| | Kotagede | 10,28 | 9,75 |
| | Mantrijeron | 3,43 | 0,81 |
| | Mergangsan | 8,05 | |
| | Tegalrejo | 14,56 | 2,58 |
| | Umbulharjo | 79,57 | 8,3 |
| | Wirobrajan | 7,33 | |
| Total | | 704,86 | 769,38 |

### *3.3. Mapping of Predetermining Factors as Independent Variable*

Any conditions or factors presumably related to land use change are referred to as predetermining factors. Identification of those factors was conducted based on literature and local knowledge about study area.

*3.3.1. Presumably Variable Related to Area of Land Use Change.* Factors presumably related to the quantity or area of land use change consisted of 6 (six) variables. Each variable was given a name and a specific code for purpose of analysis. Table 1 shows the presumably variables related to area of land use change. Those variables were used in analysis to assess determinant factors of area or quantity of land use change.

**Table 2.** Presumably variables related to area of land use change

| No | Variable Name | Variable Code | Description |
|---|---|---|---|
| 1 | Tegal | Var_A1 | Area and proportion of dry field |
| 2 | Kebun | Var_A2 | Area and proportion of mix garden |
| 3 | Sawah | Var_A3 | Area and proportion of rice field |
| 4 | Pdd | Var_A4 | Number of population |
| 5 | Tum_Pdd | Var_A5 | Rate of population growth |
| 6 | Kpdt | Var_A6 | Population density |

The first three variables were used to represent availability of land, which is land available for conversion. Previous analysis indicates that dominant type of land use change in the study area was conversion from agricultural land to settlement. Therefore, availability of agriculture land supposedly determines the quantity of land use change. The last three variables were used to represent demand for land. Quantity of land use change, particularly conversion from agricultural land to settlement, supposedly determines by increasing demand for residential use.

*3.3.2. Presumably Variable Related to Location of Land Use Change.* Factors presumably related to the location of land use change consisted of 7 (seven) variables. Each variable was given a name and a specific code for analysis. The first six variables (Var_L1 to Var_L6) were representing the accessibility of land. Accessibility can be expressed in terms of money, time, preference, effort or any concept of cost (Braimoh and Onishi, 2007). In this study, accessibility was measured in terms of distance to targets (government office, university, existing settlement and roads). The seventh variable (slope) represents topographic condition. Table 2 shows the presumably variables related to location of land use change.

**Table 3.** Presumably variables related to location of land use change

| No | Variable Name | Variable Code | Description |
|---|---|---|---|
| 1 | Jr_Kntr | Var_L1 | Distance to government office |
| 2 | Jr_kmps | Var_L2 | Distance to university |
| 3 | Jr_Bang | Var_L3 | Distance to existing settlement |
| 4 | Jr_JalAll | Var_L4 | Distance to roads |
| 5 | Jr_JalLok | Var_L5 | Distance to access roads (local roads) |
| 6 | Jr_JalUt | Var_L6 | Distance to main roads |
| 7 | Slope | Var_L7 | Slope |

Accessibility variables were obtained trough spatial analysis called distance analysis. This analysis used pixel, with a certain size, as a unit of measurement. Topographic variable (slope), derived from

DEM (Digital Elevation Model). All maps of presumably factors related to location of land use change are shown in Figure 3.

*3.4. Statistical Analysis Techniques*

Assessments of determinant factor were conducted using two statistical techniques namely regression and logistic regression analysis. Regression analysis was used to assess the determinant of area or quantities, while logistic regression analysis was used to assess the determinant of location.

*3.4.1. Multiple Regressions.* Area or quantities of land use change serve as dependent variable or response variable. Presumably factors serve as independent variable or predictor variable. Multiple regressions linking the dependent variable with several independent variables through an equation as shown in equation [1].

$$\boldsymbol{Y} = \alpha + \beta_1 X_1 + \beta_2 X_2, \beta_k X_k \quad [1]$$

The above equation linking the value of Y that is area of land use change with variables $X_1$, $X_2$ up to $X_k$. The value of $\alpha$ is a constant regression while $\beta_1$, $\beta_2$ and $\beta_k$ are regression coefficient for the predictor variables.

*3.4.2. Logistic Regression.* Location of land use change serves as dependent variable or response variable. Presumably factors serve as independent variable or predictor variable. Logistic regression analysis is basically the same as ordinary regression analysis. The difference lies in the value of the response or dependent variable. Instead of using the real value, logistic regression using logit function as the response variable. Using binary notation, the response variable values can be written with the number 1 and number 0. Number 1 represents the location where changes occur while the number 0 represents the location where the change does not occur. Logistic regression can be written in the form equation as shown in equation [2]

$$\text{Logit}\,(p_i) = \alpha + \beta_1 X_1 + \beta_2 X_2, \beta_k X_k \quad [2]$$

where,

$p_i$ : the probability of occurrence (change or no change)

$\alpha$ : regression constants

$\beta_1$ : coefficients of predictor variables 1

$X_{..k}$ : predictor variable (1, 2, ... $k$)

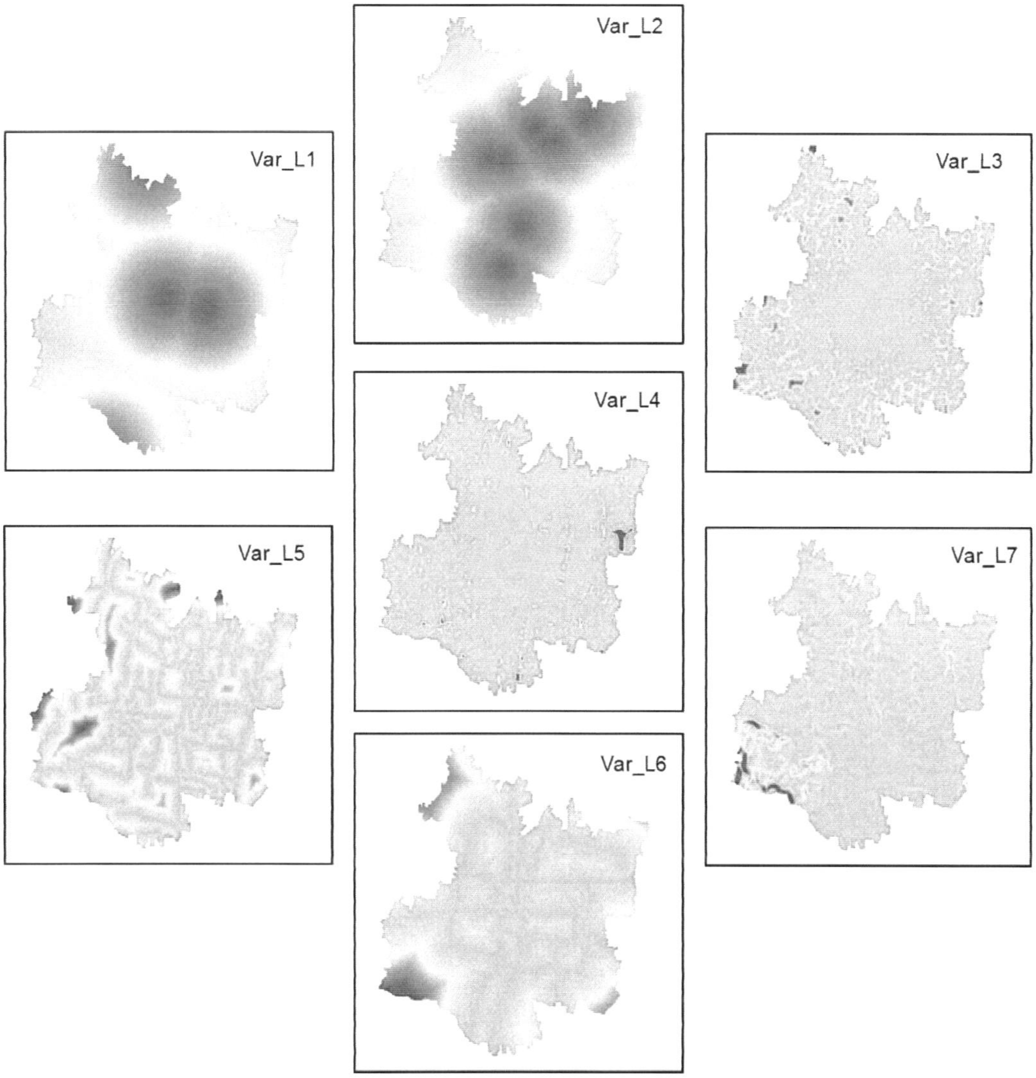


**Figure 3**. Maps of presumably variables related to location of land use change

*3.5. The Multi Scale Analysis*

Analysis of determinant factor using regression and logistic regression, as previously described, were performed on a different spatial and temporal scale. This was done to demonstrate what so called "multi scale effect" of performing spatial analysis. Multi scales in the context of spatial analysis were shown by using what is referred to as spatial extent and grain or resolution. Determinant factors of area or quantities of land use change were analysed using different spatial extent. Determinant factors of location of land use change were analysed using different grain or resolution. Figure 4 illustrates different spatial extent in the analysis of determinant factors related to area of land use change. Figure 5 illustrates different spatial resolution or spatial grain used in the analysis of determinant factors related to location of land use change.

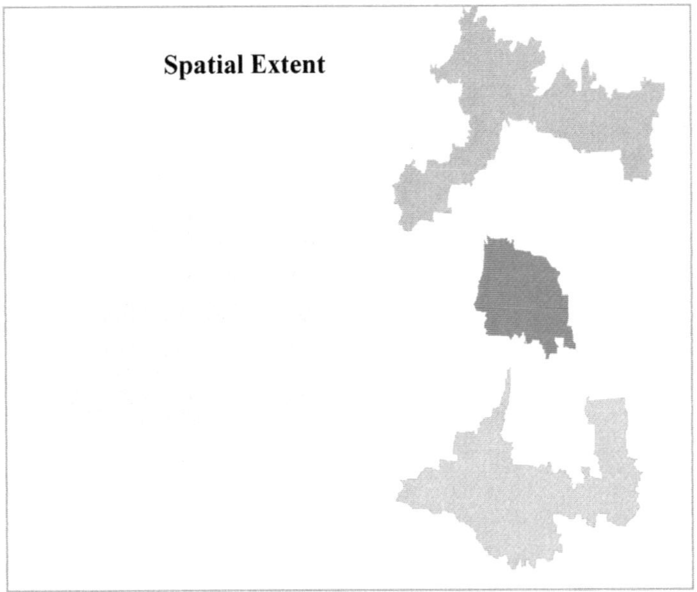


**Figure 4.** Three different spatial extent used in multi scale analysis

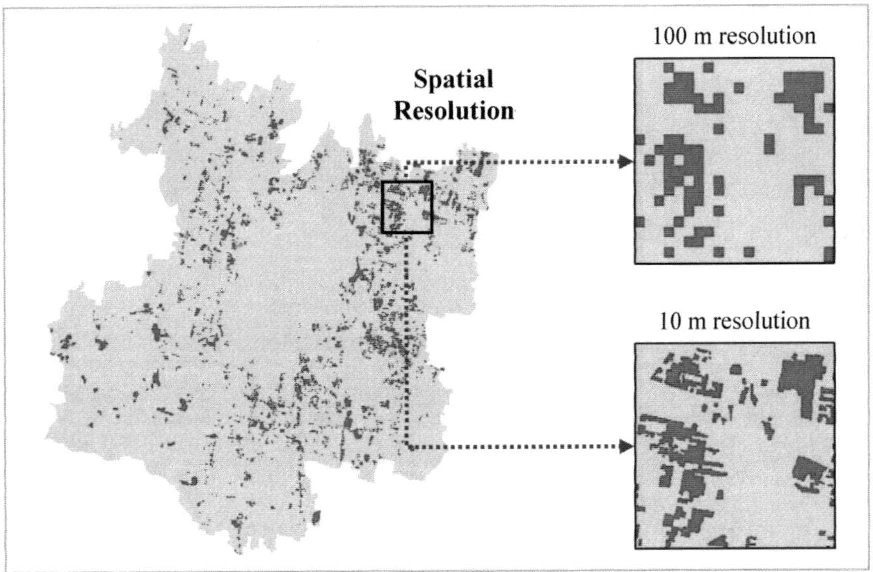


**Figure 5.** Two different spatial resolution used in multi scale analysis

## 4. Result and Discussion

### *4.1. Determinant Factors Related to Area of Land Use Change*

Statistical analyses show that area or quantity of land use change has significant correlation with several variables. As analyses performed using different spatial and temporal scales, different results were obtained. Table 4 show variables which have significant correlation to area of land use change. Those variables area referred as determinant factors of area of land use change. It should be noted that population growth variable (Var_A5), marked as significant variable in different scale of analyses.

**Table 4**. Significant variables related to area of land use change

| Analysis Extent (Spatial Scale) | Periods of Analysis (Temporal Scale) 1993 – 2000 | | 2000 – 2007 | | 1993 – 2007 | |
|---|---|---|---|---|---|---|
| | Variable | ß | Variable | ß | Variable | ß |
| Whole Area | Constant | -9.677 | Constant | -28.884 | Constant | -50.198 |
| | Var_A1 | .268 | Var_A3 | .100 | Var_A3 | .002 |
| | Var_A2 | -.098 | Var_A4 | .001 | Var_A4 | .129 |
| | Var_A3 | .031 | Var_A5 | 8.195 | Var_A5 | 18.806 |
| | Var_A4 | .001 | | | | |
| | Var_A5 | 3.595 | | | | |
| | R | .880 | R | .800 | R | .831 |
| | $R^2$ | .774 | $R^2$ | .640 | $R^2$ | .631 |
| Bantul  | Constant | -8.166 | Constant | -18.668 | Constant | -27.795 |
| | Var_A4 | .002 | Var_A3 | .088 | Var_A4 | .003 |
| | | | Var_A5 | 11.972 | Var_A5 | 15.310 |
| | R | .747 | R | .830 | R | .755 |
| | $R^2$ | .559 | $R^2$ | .689 | $R^2$ | .601 |
| Sleman 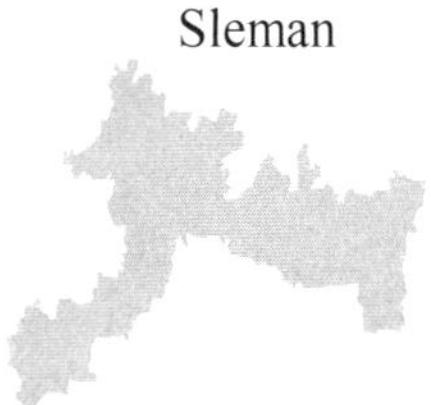 | Constant | 13.504 | Constant | -1.120 | Constant | 38.777 |
| | Var_A1 | .436 | Var_A1 | .489 | Var_A1 | 1.057 |
| | Var_A2 | -.197 | Var_A2 | -.302 | Var_A2 | -.495 |
| | | | Var_A3 | .106 | | |
| | R | .921 | R | .912 | R | .919 |
| | $R^2$ | .849 | $R^2$ | .832 | $R^2$ | .844 |
| Yogyakarta  | Constant | -.139 | Constant | .712 | Constant | -.627 |
| | Var_A1 | -.269 | Var_A1 | .470 | Var_A3 | .653 |
| | Var_A3 | .376 | | | | |
| | R | .925 | R | .837 | R | .902 |
| | $R^2$ | .855 | $R^2$ | .700 | $R^2$ | .813 |

### *4.2. Determinant Factors Related to Area of Land Use Change*

Statistical analyses show that location of land use change has significant correlation with several variables. Analyses were performed using two different spatial resolutions which are 100 meters and 10 meters. Spatial resolution reflects detail of data used in analysis. Some time, spatial data which has some level of detail is not available. For example, rather than using Ikonos or QuickBird imagery we use Landsat imagery as a source data for mapping and analysis the location of land use change. The use of less detail spatial data will affect the result of analysis. Table 4 show significant variables related to location of land use change.

**Table 5.** Significant variables related to location of land use change

| Resolution (Spatial Scale) | Periods of Analysis (Temporal Scale) 1993 – 2000 | | 2000 – 2007 | | 1993 – 2007 | |
|---|---|---|---|---|---|---|
| | Variable | Exp (B) | Variable | Exp (B) | Variable | Exp (B) |
| 10 m 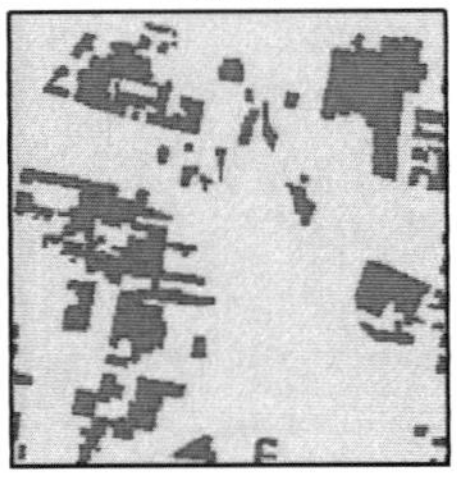 | Constant | 3.277 | Constant | .844 | Constant | 6.701 |
| | Var_L1 | 1.000 | Var_L2 | 1.000 | Var_L1 | 1.000 |
| | Var_L2 | 1.000 | Var_L3 | 1.000 | Var_L2 | 1.000 |
| | Var_L3 | 0.997 | Var_L4 | 0.989 | Var_L4 | 0.987 |
| | Var_L4 | 0.988 | Var_L5 | 0.999 | Var_L5 | 0.999 |
| | Var_L5 | 0.999 | Var_L6 | 0.999 | Var_L6 | 0.999 |
| | Var_L6 | 0.999 | Var_L7 | 0.917 | Var_L7 | 0.952 |
| | *Cox & Snell* $R^2$ | 0.112 | Cox & Snell $R^2$ | 0.077 | Cox & Snell $R^2$ | 0.200 |
| | Nagelkerke $R^2$ | 0.223 | Nagelkerke $R^2$ | 0.142 | Nagelkerke $R^2$ | 0.294 |
| 100 m 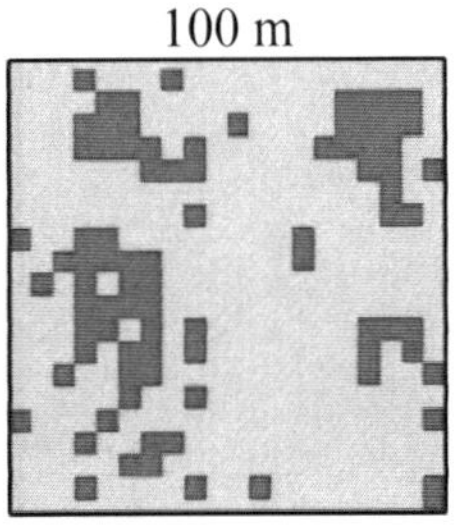 | Constant | 2.008 | Constant | 0.534 | Constant | 1.732 |
| | Var_L1 | 1.000 | Var_L1 | 1.000 | Var_L2 | 1.000 |
| | Var_L2 | 1.000 | Var_L2 | 1.000 | Var_L4 | 0.996 |
| | Var_L4 | 0.995 | Var_L4 | 0.997 | Var_L5 | 0.999 |
| | Var_L5 | 0.999 | Var_L5 | 0.999 | Var_L6 | 0.999 |
| | Var_L6 | 0.999 | Var_L6 | 0.999 | | |
| | Var_L7 | 1.139 | | | | |
| | Cox & Snell $R^2$ | 0.115 | Cox & Snell $R^2$ | 0.073 | Cox & Snell $R^2$ | 0.133 |
| | Nagelkerke $R^2$ | 0.226 | Nagelkerke $R^2$ | 0.136 | Nagelkerke $R^2$ | 0.205 |

## 5. Conclusions

Scale is a concept related to the use of specific dimensions to measure or assesses the phenomenon. Scale becomes an important aspect in conducting analysis particularly spatial analysis. This study clearly demonstrates the effect of spatial and temporal scale in the analysis of the determinant factors of land use change. The use of different analysis extents was able to reveal the different factors related to quantities of land use change. The use of data with different spatial resolution can reveal the presence of various factors associated with the location of land use change.

During the period of 14 years between 1993 and 2007 there were 1,460 ha of land use change occurred in the study area. Dominant type of land use change was agricultural land converted to residential use. Land use change in the study area has significant correlation with a number of factors. Number of factors as well as degree of correlation varies spatially and temporally. In general, factors influencing the quantities of land use change in the study area were population growth and the availability of land. Availability of land in this study refers to the presence of agriculture land. These two factors have significant influence in urban fringe areas, while in Yogyakarta City the only significant factor is availability of land. Locations of land use change were influenced or determined by accessibility factors. Accessibility of land in this study was indicate by it distance to roads and other facilities. The probability of land use change will take place in certain location, increase as distance to roads and other facilities decrease. Analysis of data with higher spatial resolution revealed that location of land use change has also significant correlation with topographic condition.

**Acknowledgement**

This research was carried out with funding support from The Faculty of Geography Universitas Gadjah Mada, through research grand for educational staff (2013). Process of writing this paper was supported by Bureau of Publisher and Publication (BPP) Universitas Gadjah Mada, trough incentive grant for writing scientific papers (2015). The authors wish gratitude to all parties for the assistance provided during the research.

**References**

[1] Anwar, Morshed, 2002, Land Use Change Dynamics: A Dynamic Spatial Simulation, *Thesis M.Sc.* AIT Bangkok Thailand.
[2] Aspinall,R.J and Hill, M.C. (Eds), 2008, *Landuse Change, Science, Policy and Management*, Taylor and Francis, New YorkMore references
[3] Braimoh, A.K., and Onishi ,T., 2007, Spatial determinants of urban land use change in Lagos, Nigeria, *Land Use Policy* Vol 24, 502-515.
[4] Heywood 1998 *Introduction to Geographical Information Systems*. New York: Addison Wesley Longman
[5] Lambin, E.F., Geist, H.J. and Lepers, E., 2003, Dynamics of Land-Use and Land-Cover Change in Tropical Region, *The Annual Review of Environment and Resources*, Vol. 28.
[6] Lambin, E.F., dan Geist, H.J. (lead authors), Ellis, E. (topic editor), 2007, *Causes of Land-Use and Land-Cover Change*, within Encyclopedia of Earth, internet access from http://www.eoearth.org
[7] Moran, E.F., Skole, D.L. and Turner, B.L., 2004, The Development of The International Land-Use And Land Cover Change (LUCC) Research Program and its Links to Nasa's Land-Cover And Land-Use Change (LUCC) Initiative, in *Land Change Science: Observing, Monitoring and Understanding Trajectories of Change on the Earth's Surface* (Eds: Gutman et al ), Kluwer Academic Publishers, Dordrecht, The Netherlands
[8] Sanders, Lena., 2007, *Model in Spatial Analysis*, ISTE Ltd London W1T 5DX UK
[9] Susilo, Bowo, 2013 Kajian Spasial Multi Skala Determinant Perubahan Penggunaan Lahan Di Sebagian Wilayah Daerah Istimewa Yogyakarta, *Research Report (unpublished)* Faculty of Geography Universitas Gadjah Mada
[10] Susilo, Bowo., 2016, Map Analysis and Spatial Statistic: Assessment of Spatial Variability of Agriculture Land Conversion at Urban Fringe Area of Yogyakarta, *IOP Conf. Series: Earth and Environmental Science* Vol 47, No 1
[11] Wainer,G.A., 2009, *Discrete-Event Modeling and Simulation : A Practitioner's Approach*, CRC Press New York
[12] Wainwright, J. and Mulligan, M. (Eds), 2004, *Environmental Modelling Finding Simplicity in Complexity*, John Wiley & Sons Ltd – England

# Estimation the Amount of Oil Palm Trees Production Using Remote Sensing Technique

**A C Fitrianto[1,2*], K Tokimatsu[1] and M Sufwandika[2]**

[1]Department of Transdisciplinary Science and Engineering, Tokyo Institute of Technology, 2-12-1 Oookayama - Meguro-ku -Tokyo
[2]Geospatial Information Agency (BIG), Jl. Raya Jakarta Bogor km. 46 Cibinong-Bogor – Indonesia

aangtimor@gmail.com

**Abstract**. Currently, fossil fuels were used as the main source of power supply to generate energy including electricity. Depletion in the amount of fossil fuels has been causing the increasing price of crude petroleum and the demand for alternative energy which is renewable and environment-friendly and it is defined from vegetable oils such palm oil, rapeseed and soybean. Indonesia known as the big palm oil producer which is the largest agricultural industry with total harvested oil palm area which is estimated grew until 8.9 million ha in 2015. On the other hand, lack of information about the age of oil palm trees and changes also their spatial distribution is mainly problem for energy planning. This research conducted to estimate fresh fruit bunch (FFB) of oil palm and their distribution using remote sensing technique. Cimulang oil palm plantation was choose as study area. First step, estimated the age of oil palm trees based on their canopy density as the result from Landsat 8 OLI analysis and classified into five class. From this result, we correlated oil palm age with their average FFB production per six months and classified into seed (0-3 years, 0kg), young (4-8 years, 68.77kg), teen (9-14 years, 109.08kg), and mature (14-25 years, 73.91kg). The result from satellite image analysis shows if Cimulang plantation area consist of teen old oil palm trees that it is covers around 81.5% of that area, followed by mature oil palm trees with 18.5% or corresponding to 100 hectares and have total production of FFB every six months around 7,974,787.24 kg.

**Keywords:** Remote sensing, GIS, Oil palm, FFB

## 1. Introduction

Currently, most countries are trying to find alternative energy resources which is renewable and environment-friendly to reducing their depending on fossil fuels such as petroleum, natural gas and coal as the main source to generate electricity. One of renewable energy that can be converted into electricity is biomass. Biomass energy or bio energy is the largest source of domestic renewable energy in Indonesia. Renewable energy such as biodiesel has the potential to replace petroleum-derived transportation fuel in the future. Biodiesel is defined as the mono-alkyl esters of long-chain fatty acids derived from vegetable oils such palm oil, rapeseed and soybean [1]. The energy output of oil palm is almost three times higher compared to soybean and rapeseed oil [2]. Moreover, another products are also produced from oil palm such as empty fruit bunch (EFB) fibers, shells, fronds, and trunk.

Palm oil is the largest agricultural industry in Indonesia. It means the potential sources for renewable energy especially from oil palm was abundant. Information about oil palm trees production and their spatial distribution very useful in organization or management of oil palm plantation and one of important variables affecting profit. On the other hand, lack of information about the amount of oil palm production and changes also their spatial distribution is mainly problem for energy planning. According to [3], collecting production data of oil palm trees especially on a large area or regional scale is time consuming and high costly. Remote sensing provides an effective and efficient approach for oil palm monitoring including the amount of oil palm production.

Our study, which is aimed at utilizing optical remote sensing data for oil palm plantation management including mapping, estimation of fruit bunch production and energy potential for electricity, has just been started and is in the preliminary study. In this paper, we focus on the amount of oil palm production based on the age of oil palm trees with Landsat 8 OLI. The study area is the oil palm plantations areas in Cimulang, Bogor Indonesia.

## 2. Methodology

This research was conducted in two step that are satellite image analysis using forest canopy density (FCD) formula to know relationships between canopy density with the age of oil palm trees, and the next step is estimate the amount of oil palm trees production. Generally, the methodology on this study can be seen on Figure 1.

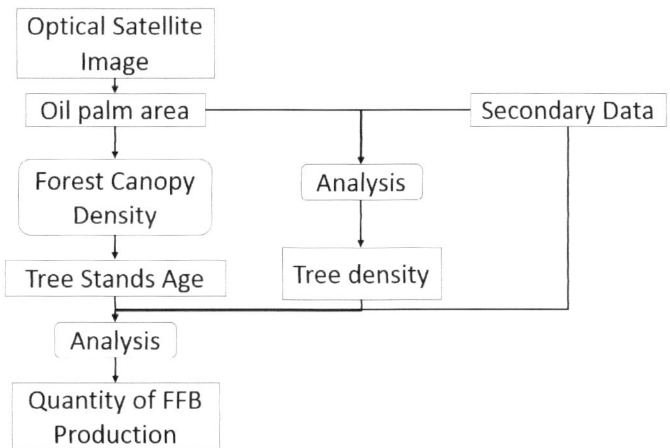


**Figure 1**. Methodology of Estimation the Amount of Oil Palm Trees Production

### *2.1. Forest Canopy Density (FCD)*

In this research, the age of oil palm trees is predicted using remote sensing technique based on canopy density of oil palm trees. Forest canopy density (FCD) utilized as essential method to estimate stand age. This model involves bio-spectral phenomenon modelling and analysis utilizing data derived from four indices that are Advanced Vegetation Index (AVI); Bare Soil Index (BSI); Shadow Index (SI); and Thermal Index (TI). The result of four indices integration in this modelling is percentage of canopy density in each pixel. Zero percent is no vegetation and 100 percent is very high density.

Advanced Vegetation Index (AVI) was used to measure green vegetation. In remote sensing field, healthy vegetation characterize by high absorption and low reflectance in visible region. AVI has been calculated using equation 1.

$$AVI = (B5 * (1 - B4) * (B5 - B4))^{1/3} \quad (1)$$

Similar with AVI, BI is normalized indices that used to separating vegetation and their background. This indices have opposite result with AVI because this is used to detected soil, it means near infrared wavelength will have low reflectance due to absorption by soil moisture. BSI has been calculated using equation 2.

$$BSI = \frac{(B6 + B4) - (B5 + B2)}{(B6 + B4) + (B5 + B2)} + 100\, x\, 100 \quad (2)$$

Canopy density also can detected by shadow characteristics that derived from spectral information and thermal information of the forest shadow. In remote sensing, shadow characteristics are defined by shadow index (SI) as spectral information because crown arrangement of oil palm trees will have shadow pattern that it affecting spectral responses. SI has been calculated using equation 3.

$$SI = ((1 - B2) * (1 - B3) * (1 - B4))^{1/3} \quad (3)$$

Thermal Index (TI) is used to check the black or shadow is real shadow not black soil. Land surface that was closed by shadow will have low temperature because leaf surface blocks and absorbs energy from the sun. TI has been calculated using equation 4 and 5.

$$L_I = M_I * Q_{cal} + A_I \quad (4)$$

$$T = K_2 / ln(\frac{K_1}{L_I} + 1) \quad (5)$$

FCD value was shown as percentage of canopy density. Integrated from four indices was used to calculated this model. AVI and BSI wan integrated and have result as vegetation density (VD). Processing method is using principal component analysis which is AVI and BSI have high negative correlation. Higher AVI value will have low value in BSI because canopy density or vegetation cover was high and no bare soil or open surface at there. After that, VD will set in the percentage scaling from zero percent to a hundred percent point.

Another calculation is scaled shadow index (SSI) that derived from integration of SI and TI. In areas where the SSI value is zero low, this corresponds with oil palm plantation that have the lowest shadow value or minimum canopy density, opposite with areas where the SSI value is 100 that have the high canopy density. VD and SSI will integrated to achieve FCD value with equation 6.

$$FCD = (VD * SSI + 1)^{1/2} - 1 \quad (6)$$

### *2.2. The Age of Oil Palm Tree*

Forest canopy density (FCD) was used to estimate oil palm trees age with assumption if older trees will have higher percentage of canopy density. The class number of stands age divided into 4 class, according to oil palm classification from Plantation Education Agency of Indonesia. Correlation between oil palm trees age and canopy density was shown in Table 1.

**Table 1.** Correlation of Stands Age and Canopy Density

| No | Class | Stands Age | % Canopy Density |
|---|---|---|---|
| 1 | Seed | 0 - 3 | < 10 |
| 2 | Young | 3 - 8 | 10 - 40 |
| 3 | Teen | 9-14 | 41 - 80 |
| 4 | Mature | 15 - 25 | > 80 |

Source: Plantation Education Agency, 2013 with modification

### *2.3. Oil Palm Trees Production*

Secondary data such as planting pattern was used to achieve trees density in every hectares. In this research we use assumption if planting pattern in research area using triangle pattern with 9 m distance between stands. According to Plantation Education Agency (2013), planting pattern with 9 m distance between stands and distance between rows is 7.8 m, in every hectares will have 143 trees.

Correlation of FFB production with stands age will show in Table 2. Average bunches or FFB in this table using 6 months because oil palm trees will harvest every 6 months. From this spatial data, overlay with oil palm trees stands age and FFB production for each trees will achieved the amount of FFB in every stands age in every hectares.

**Table 2.** Correlation of FFB Production and Stands Age

| No | Class | Stands Age | Average Bunch/trees (kg/6 months) |
|---|---|---|---|
| 1 | Seed | 0 - 3 | 0 |
| 2 | Young | 4 - 8 | 68.77 |
| 3 | Teen | 9-14 | 109.08 |
| 4 | Mature | 15 - 25 | 73.91 |

Source: Plantation Education Agency, 2013 with modification

## 3. Result and Discussion

### *3.1. Forest Canopy Density Model*

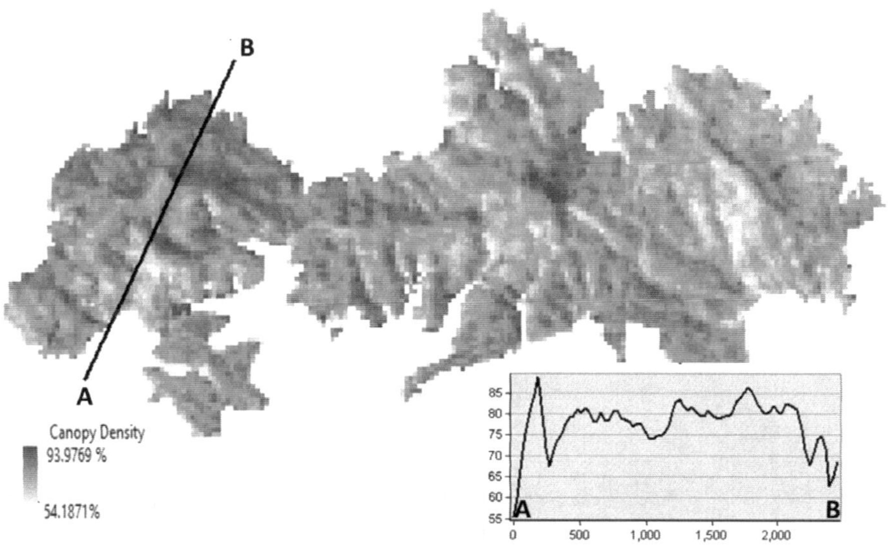


**Figure 2.** Canopy Density of Oil Palm Trees

Canopy density of oil palm trees in Cimulang plantation shows similar variation between 54%-93% with average density as 75%. Crops pattern and time of planting in this area that relatively same, indicated by similar variation of their canopy density. Furthermore, it is important to do other research in different area as a part to check the accuracy assessment for forest canopy density model to estimate the age of oil palm trees that have been used.

### *3.2. Oil Palm Trees Age*

Forest canopy density was used to estimate the age of oil palm trees with assumption older trees will have higher percentage of canopy density but using non linier assumption. Satellite image analysis shows the results for Cimulang area covers with FCD value ranges from 49.00 – 93.98 for each pixel.

These results was used to estimate oil palm trees age and from that value and based on Table 1, this area consist of two class of oil palm trees age that are teen with canopy density 41–80% and mature with canopy density more than 80%.

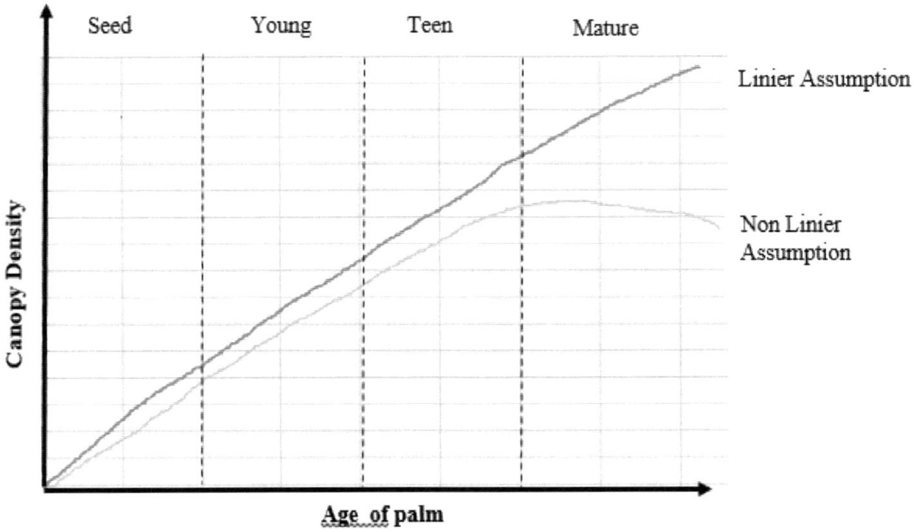


**Figure 3.** Correlation of Canopy Density with the Age of Oil Palm Trees

The result from satellite image analysis shows in Cimulang plantation teen old oil palm trees covers around 81.5% of that area, followed by mature oil palm trees with 18.5% or corresponding to 100 hectares. In this area we cannot found seed or young oil palm trees and it is means no new plantation in this area. Based on Table 2 about correlation of FFB production with the age of oil palm trees, we can calculate if Cimulang area have total production of FFB every six months around 7,974,787.24 kg.

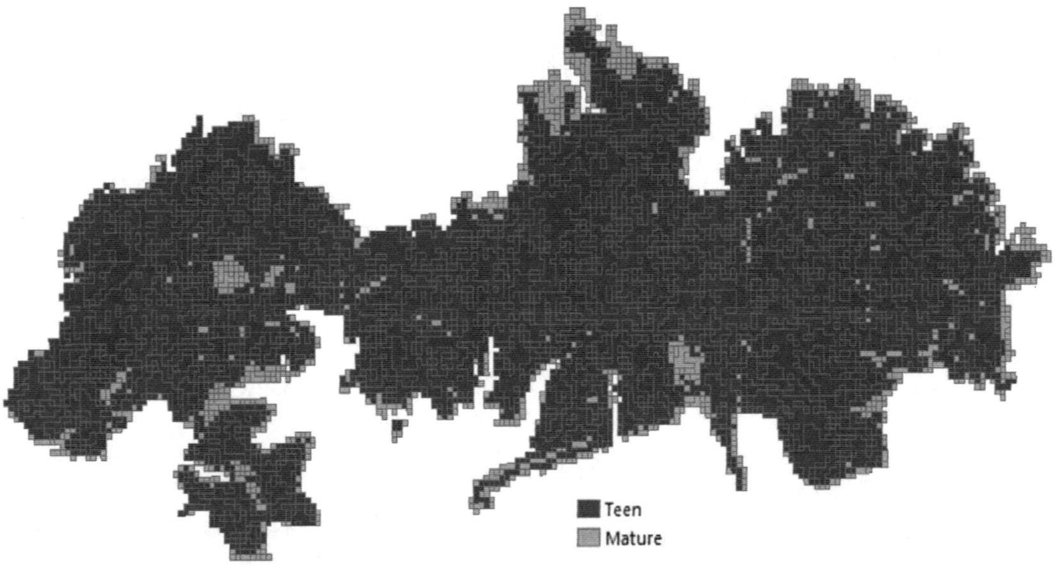


**Figure 4.** Class of oil palm trees age

In this research, we have assumption if some variable was equal for all area such as soil type, planting pattern, type of fertilizer, and another plantation management, because all of this variable have big correlation with canopy. Other plantation management have big correlation with canopy density is pruning tree branches. Many plantations have this policy to keep the amount of sunlight in that plantation and it will make the dense of oil palm trees canopy will different compare with natural oil palm trees.

The result in this paper is preliminary result, and in the future, field survey will be conducted as verification process for interpretation result. The age of oil palm trees is very important part for FFB production. The class number of stands age divided into 4 class and consist of seed, young, teen, and mature. From table 2, we can see if almost all of this class have small range of age difference such as seed class from 0 – 3 years old, young class from 4-8 years old, and teen from 9 – 14 years old. But this condition is different for mature class where this class have large range of age from 15 – 25 years old, this condition was happened because in this class was difficult or have problem to distinguish the mature (15-20 years) and old stands age (more than 20 years old). Characteristic from oil palm trees after 20 years old is different, the length of canopy or steam leaves increases is equal with growing phases, but the direction is different which steam leaves will grow with smaller angle and tend to lead downwards. Detailing the stands age class will make estimation of FFB production using remote sensing more accurate. Multi-temporal analysis can be used for detailing the stands age class and to know remote sensing capability to estimate the age of oil palm trees age and their production.

## 4. Conclusions

Forest canopy density can be used to estimate the age of oil palm trees with assumption older trees will have higher percentage of canopy density. Result from this research shows this method only can divide 4 class and consist of seed, young, teen, and mature caused by difficult to distinguish the mature (15-20 years) and old stands age (more than 20 years old). This problem was happened because characteristic from oil palm trees after 20 years old have steam leaves growing with smaller angle and tend to lead downwards.

Multi-temporal satellite image analysis of oil palm plantation area will be very useful for monitoring and detailing information of oil palm trees age particularly to distinguish the mature (15-20 years) and old stands age (more than 20 years old).

## References

[1] Tan KT, Lee KT, Mohamed AR and Bhatia S. Palm oil 2009 Addressing issues and towards sustainable development *Renew Sustain Energy Rev* **13** p 420–7.

[2] Sumathi S, Chai SP, Mohamed ARÃ et. A 2008 Utilization of oil palm as a source of renewable energy in Malaysia *Renew Sustain Energy Rev* **12** p 2404–21.

[3] Tan KP, Kanniah KD and Cracknell AP 2015 International Journal of Remote Use of UK-DMC 2 and ALOS PALSAR for studying the age of oil palm trees in southern peninsular Malaysia. p 37–41.

[4] Danoedoro P 2012 *Pengantar Penginderaan Jauh Digital* (Yogyakarta: ANDI).

[5] Banerjee, Krishnendu, S Panda, J Bandyopadhyay and M K Jain 2014 Forest Canopy Density Mapping Using Advance Geospatial Technique **1**(7) pp 358–63.

[6] S Darmawan, W Takeuchi, H Abidin, M N Rasid and M N Samat 2016 An Investigation of Age And Yield of Fresh Fruit Bunches of Oil Palms Based On Alos Palsar 2 *The 8th IGRSM International Conference and Exhibition on Geospatial & Remote Sensing* **5** pp 57–65.

[7] Teh, C B S 2016 Availability, Use, and Removal of Oil Palm Biomass in Indonesia *Report Prepared for the International Council on Clean Transportation* pp 1–39.

[8] Vadivelu, S, A Ahmad and Y H Choo 2014 Remote Sensing Techniques For Oil Palm Age Classification Using Landsat-5 TM Satellite **26**(4) p 1547–51.

[9] Z Azizi, et. a. 2008 Forest Canopy Density Estimating, Using Satellite Images *The International Archives of the Photogrammetry, Remote Sensing and Spatial Information Sciences* **XXXVII**

IOP Conf. Series: Earth and Environmental Science **98** (2017) 012016 doi:10.1088/1755-1315/98/1/012016

Part B8. Beijing p 1127–30.

# Unmanned Aerial Vehicle (UAV) Data Acquisition for Archaeological Site Identification and Mapping

**W Handayani[1*], E A Ayuningtyas[1], F S Candra R[2], B Arif S[3] and B Argadyanto[3]**

[1]Diploma Program of Remote Sensing and Geographic Information System, Vocational College, Universitas Gadjah Mada, Yogyakarta, Indonesia
[2]CV. Mitra Geotama, Yogyakarta, Indonesia
[3]CV. Systemaptic, Yogyakarta, Indonesia

warsini.handayani@ugm.ac.id

**Abstract**. Archaeological sites as part of human history and located around community are important to be preserved for connecting historical information from generation to generation. Mapping of archaeological sites can be done as one of preservation efforts. Yogyakarta has several archaeological sites such as Pleret Palace, the former royal palace of Mataram Islam in the 16th Century. Data limitations and the difficulty of reconstructing the site sketches into a map become obstacles in archaeological sites mapping. Unmanned Aerial Vehicle (UAV) can be an alternative of high-resolution spatial data acquisition for detail mapping, including archaeological sites mapping. This study aims to see how far the UAV acquisition results can be used for Archaeological Site mapping in Pleret Palace. Data acquisition using UAV generated to mosaic orthophoto, Digital Surface Model (DSM), and Digital Terrain Model (DTM). Archaeological sites identified using DTM and matched with site sketch made by Cultural Agency. From these data, it can be recognized some relics form, such as palace fortress, moats and canals, and also dikes of Segarayasa. This research is expected to be a reference in archaeological site mapping using detailed spatial data, especially UAV. Furthermore, it can be obtained archaeological site map close to real condition; as well as archaeological sites preservation in Indonesia.

**Keywords:** UAV, Identification and mapping, Archaeological Site, Pleret Palace

## 1. Introduction

### *1.1. Background and Research Problem*

Archaeological sites have an important role in part of human history and society. The previous culture of the community as the perpetrators of history as well as heritage buildings have an effect on the behaviour patterns of people living in the present. Therefore, archaeological sites are important to be preserved for connecting historical information from generation to generation [8]. Mapping of archaeological sites can be done as one of preservation efforts.

Yogyakarta has several archaeological sites such as Pleret Palace, the former royal palace of Mataram Islam in the 16th Century. This site is located in Pleret SubDistrict, Bantul District, special province of Yogyakarta. Based on existing historical records, this palace include area about 2,000 $m^2$ surrounded by

IOP Conf. Series: Earth and Environmental Science **98** (2017) 012017 doi:10.1088/1755-1315/98/1/012017

water buildings [4]. One of the privileges part of this palace is the existence of an artificial sea located behind the palace.

Archaeological site mapping as one of spatial data management effort in archaeological potential has been done and made the national archaeology maping system (SPAN) [12]. Each agency of heritage site management is expected to make detailed mapping for each research or field activity. However, archaeological sites mapping has not been optimal recently due to lack of data, resources, and existing site conditions which have changed from the original conditions. The condition of the Pleret Palace site is now quite worrying. Preservation is supposed to be contrary to current conditions. Land indicated as a relic of the palace has morphological changes by settlements development. This condition makes difficulty in identifying the palace sites by using remote sensing imagery, especially limitation in topographic information in order to archaeological sites identification. On the other hand, terrestrial mapping will be less effective due to a long time survey. Furthermore, spatial data preparation and site mapping become constrained.

Up to present, site reconstruction based on the Pleret Palace sketch (Figure 1) into the map (with more suitable geometry) was not yet possible. Special characteristic of Pleret Palace is the existence of some detail water building like dam and irrigation channel. This object identification requires more detailed and comprehensive data mainly related to existing topography of Pleret Palace site.

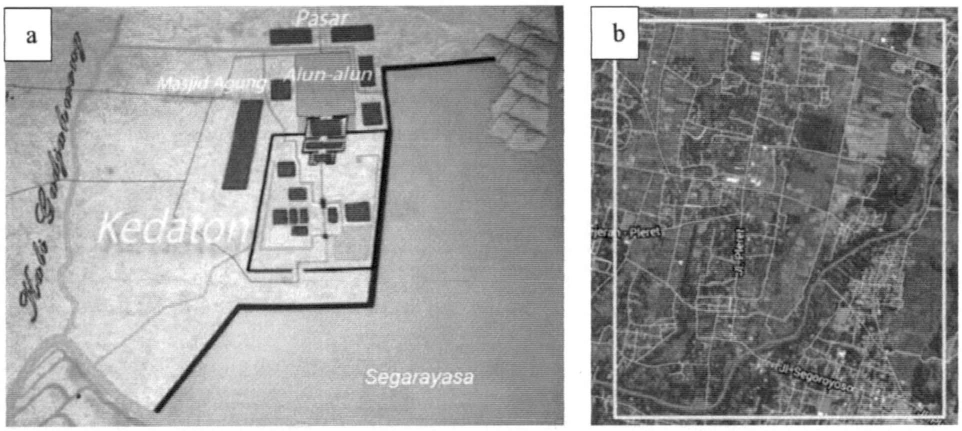


**Figure 1**. (a) Sketches of the ancient Pleret Palace [4], (b) environmental conditions at Kraton Pleret Site and surrounding which have been changed into settlements

Remote sensing technology is now widely developed by researchers for mapping and monitoring activities. One of technology has been developed is Unmanned Aerial Vehicle (UAV) as a platform for aerial photography which can produce orthophoto and digital elevation model (DEM) with high resolution [9][10][11]. It makes UAV become an effective, efficient, and economical method for detailed mapping and geographic phenomena monitoring [3][6]. By looking at these opportunities, the results of UAV can be an alternative of high-resolution spatial data acquisition for detail mapping, including archaeological sites mapping by recognizing and identifying objects in the image images and detailed topographic data as past relic based on the characteristics described in historical documents. Furthermore, archaeological site management can be helped with the archaeological site map which is closer to real condition, both in position and geometry aspect. Therefore, this research will help the archaeologist to do the archaeological sites mapping using UAV particularly Pleret Palace as the historical relic of Islamic Mataram Palace.

*1.2. Research Purposes*

2. This is a preliminary research in mapping of archaeological sites using UAV in Pleret Palace Site. The purposes of this research are: (1) perform data acquisition (aerial photography) using UAV to

generate detail spatial data at archaeological site of Pleret Palace, (2) identify and map Pleret Palace Site based on spatial data resulted from acquisition data using UAV.

## 2. Study Area

Study area of this research is site of Pleret Palace located in Pleret, Bantul. Yogyakarta. This area covers approximately 676.3 hectares (Figure 2). Pleret Archaeological site is one of the historical relics of the Islamic Mataram Palace that was built during the period of Susuhunan Amangkurat I (1646 - 1677). The palace is built with various elements, i.e. king's residence, fortress that surrounds the king's residence, great mosque (Masjid Agung), palace square, garden, water canal, artificial sea (Segarayasa), settlements, cemetery, hunting area (Krapyak), dikes and dam [2]. Based on historical records written by Van Goens [5] the form of the Pleret Palace was diamond with 2,256 $m^2$ area wide; different from others Javanese palace in general (square). This palace building was surrounded by water canal and artificial sea for security defense function.

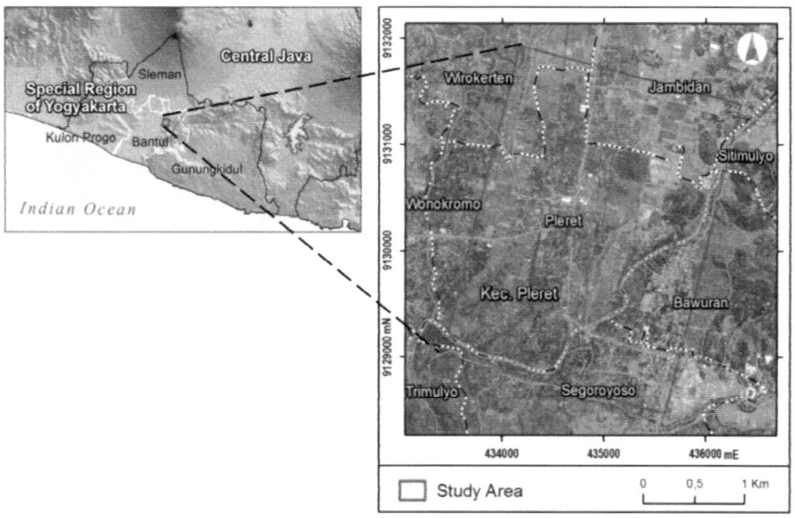


**Figure 2.** Map of Study Area.

## 3. Data, Instrument, and Method

### *3.1. Data and Instrument*

This reaseach collaborate spatial data and non-spatial data (ancient documents) for Pleret Site identification. Spatial data used in this research are base map (Rupabumi Indonesia Map) scale 1:25.000 including administrative boundary, roads, and hydrography layer as reference for relative orientation; and Pleret Palace sketch form Yogyakarta Tourism Agency. UAV fixed wing X8 is used aerial data acquisition platform, Sony A5000 camera as aerial photography sensor, and Geodetic GPS for control point measurement. Post processing aerial photos and control points is done using Agisoft PhotoScan software.

### *3.2. Method*

This research is conducted by following stages i.e. initial preparation and field observation, data acquisition using UAV, data processing, identification and interpretation of archaeological sites. Each stage describes as follow.

IOP Conf. Series: Earth and Environmental Science **98** (2017) 012017 doi:10.1088/1755-1315/98/1/012017

*3.2.1. Initial Preparation and Field Observation*

Initial stages of this research include the research method to the data processing technical, field observations, and aerial mission planning. In this stage also conducted data collection or Archaeological Site Pleret documents that have been inventoried by the Cultural Agency of Yogyakarta. This document is used as reference for site object recognition and identification in Pleret Archaeological Site at during field observation. Furthermore, flight plan and control points distribution data were design as data acquisition preparation, as well as equipment preparation for aerial photography mission and field survey.

*3.2.2. Data Acquisition using UAV*

Data acquisition using UAV is implemented using UAV fix wing and camera sensors. The aerial photography is carried out in accordance with flight plan. Simultaneously with aerial photography, premark of ground control point (GCP) and independent control point (ICP) was measured using Geodetic GPS. This control point which have x, y, and z value function as spatial reference. The results of this data acquisition are aerial photographs (singles) in the overall study area and control points measurement data. UAV flight at altitude ± 500 meters (agl) to produce a 10cm spatial photo resolution. Accuracy of aerial photographs depends on the number and distribution of GCP, so that the GCP evenly distributed throughout the area. ICP is used as a reference points for of aerial photography accuracy assessment. Control point is measured with Geodetic GPS with an accuracy limitation value at least 2 times smaller than aerial photography data resolution to be generated. Furthermore, these two data will be integrated to obtain spatial data that is aerial photography (mosaic orthophoto) and digital surface model (DSM).

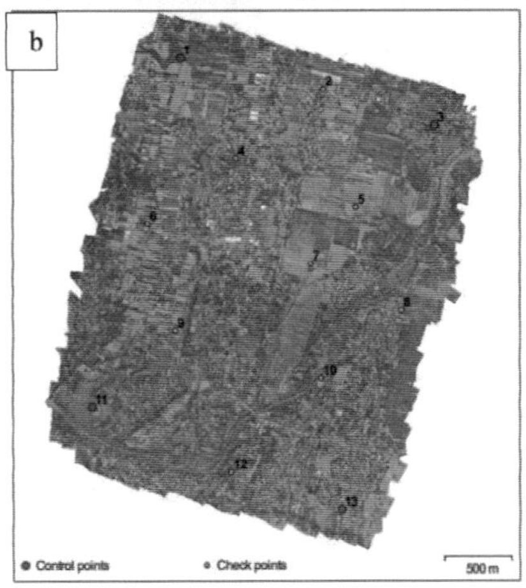

| NAME | X | Y | Z |
|---|---|---|---|
| 1 | 434.268,829 | 9.131.756,756 | 85,899 |
| 2 | 435.301,818 | 9.131.537,643 | 86,431 |
| 3 | 436.134,597 | 9.131.281,918 | 85,755 |
| 4 | 434.664,815 | 9.131.051,870 | 84,865 |
| 5 | 435.550,658 | 9.130.712,530 | 83,494 |
| 6 | 434.027,429 | 9.130.593,788 | 80,902 |
| 7 | 435.208,090 | 9.130.311,982 | 80,272 |
| 8 | 435.896,472 | 9.129.982,822 | 82,129 |
| 9 | 434.227,058 | 9.129.845,919 | 76,926 |
| 10 | 435.286,215 | 9.129.512,114 | 78,984 |
| 11 | 433.630,329 | 9.129.310,975 | 71,786 |
| 12 | 434.628,874 | 9.128.855,996 | 71,022 |
| 13 | 435.446,342 | 9.128.593,164 | 78,384 |

**Figure 3.** (a) Flight plan, preparation and aerial data acquisition, and UAV monitoring during flight mission, (b) GCP and ICP distribution and measurement

*3.2.3. Data Processing*

Data processing integrates aerial photographs of acquisition using UAV and control points measurement. Processing of UAV data into mosaic orthophoto and DSM is done by digital photogrammetric technique with block bundle adjustment principle. In this process all the measurement points and control point coordinates (GCP) are processed simultaneously with one adjustment process that ensures high accuracy on the results [1]. This process is using high difference between objects in aerial photography and generates tie points (Figure 4). The relatively high difference is combined with the elevation on each control point so that a height model is generated on the whole block of aerial photographs. In this process also determined the coordinate system of aerial photographs in accordance with the location of aerial photography and retrieval of control points.

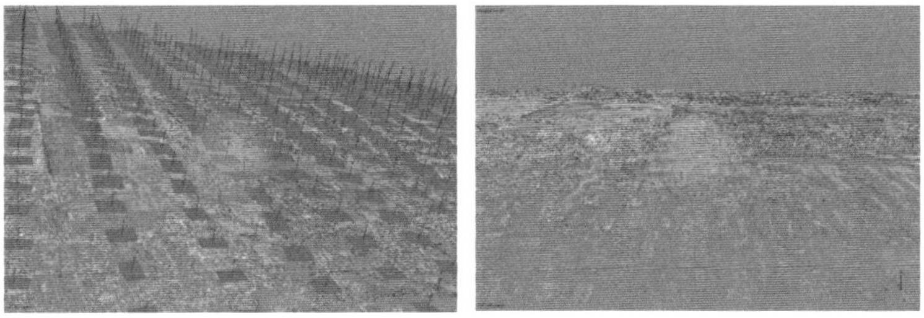

**Figure 4.** Determination of camera orientation and tie points on aerial photography

*3.2.4. Archaeological Site Identification and Mapping*

Identification is done by combining DSM, orthophoto, and sketches of the Pleret Palace including components, distribution, and dimensions of each site component. Visual interpretation of orthophoto and DSM is used in site component identification that was described in the historical document. Interpretation in this study is more focused on the relics that have micro morphology due to changes by human activities. These components include: palace wall walls, trenches and canals, embankment and dam of Segarayasa, as well as intact form of Segarayasa.

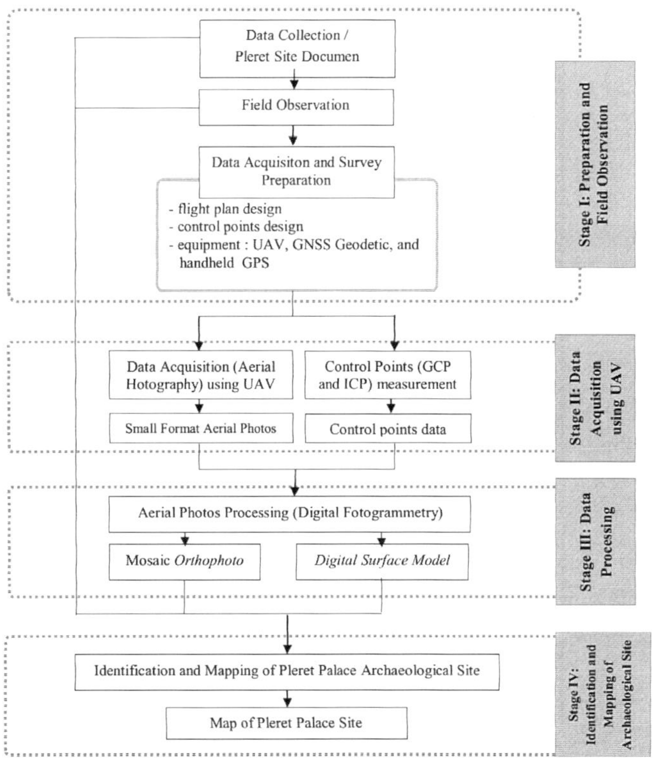


**Figure 5.** Flow Chart of Research Methods

## 4. Result and Discussion

### *4.1. Field Survey*

Field observation activities bring more detailed information about the Pleret Palace site components, i.e. great mosque of Pleret Palace, water canals, indication of palace square, indication of Segarayasa border, indication of Balekambang, indication of palace wall/fortress. The information is then used in Pleret Palace site component component identification from the aerial photography and DSM. Location observations and site descriptions is presented in Figure 6.

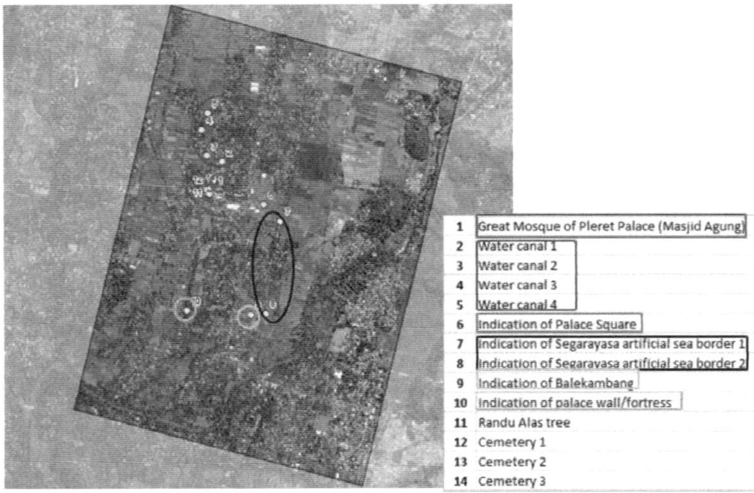


**Figure 6.** Field observation point and indication of site components

### *4.2. Aerial Photography*

The aerial photography resulted in 649 aerial photographs. Processing includes standardization of brightness, sharpness, and colour, and mosaic orthophoto processing. The orthophoto mosaic is generated from the input of a series of aerial photographs and 4 GCP. GCP serves as the coordinate point in the field on the georefencing process. Mosaic orthophoto processing is done with Agisoft PhotoScan software to obtain aerial photograph of the study area (Figure 7).

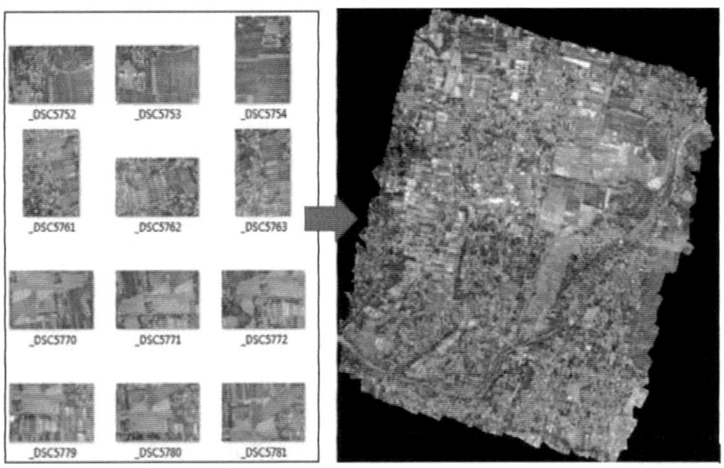


**Figure 7.** Aerial photographs and mosaic orthophoto resulted

Based on the accuracy assessment with 9 ICP, mosaic orthophoto result 0.23meter horizontal accuracy and 3.2 meters vertical accuracy. This accuracy conforms mapping specification at scale

1:2.500 for horizontal accuracy and scale 1:10.000 for vertical accuracy based on the regulation of Geospatial Information Agency's head (PERKA BIG) No. 15 year 2014 about Technical Guidance of Basic Map Accuracy. Mosaic orthophoto resulted have 7 cm spatial resolution; sufficient for detail mapping in Pleret Palace site. Detail of aerial photo is shown in Figure 8.

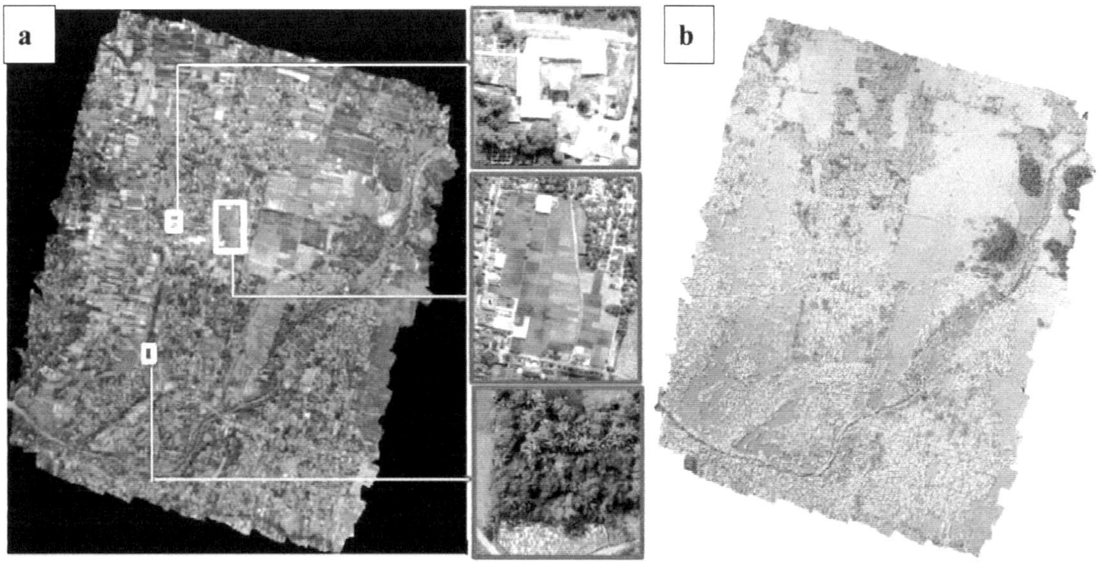


**Figure 8.** (a) Mosaic orthophoto resulted from aerial photography using UAV and detail object recorded, (b) digital surface model.

### *4.3. Archaeological Site Identification and Mapping*

The initial identification of Pleret Palace is done by referring to four main references: Pleret Palace map from de Graaf version, Pleret Palace map from Yogyakarta Cultural Agency, Historical Map of Trunojoyo Attack Scenario, and field observation results. In general, all maps show Pleret Palace location is on the northwest side of the Opak River, with geometry skewed to the northeast. There was road on the north side of the palace runs from north to south entering the palace. On the northwest side, there was a building of the Great Mosque and on the west there was a building of keputren (princess building complex). The maps also show the existence of water canals and dam that surround the palace building (Figure 9).

Based on the of the Pleret Palace components identification and existence interpretation, of similar phenomena on DSM, aerial photographs, field survey, and references map, there are several indications of the site of palace fortress, water canal, palace square, Balekambang, and embankment of Segarayasa. Interpretation of aerial photography to identify the Pleret Palace is done by using several interpretation keys such as shape, pattern, and association.

IOP Conf. Series: Earth and Environmental Science **98** (2017) 012017 doi:10.1088/1755-1315/98/1/012017

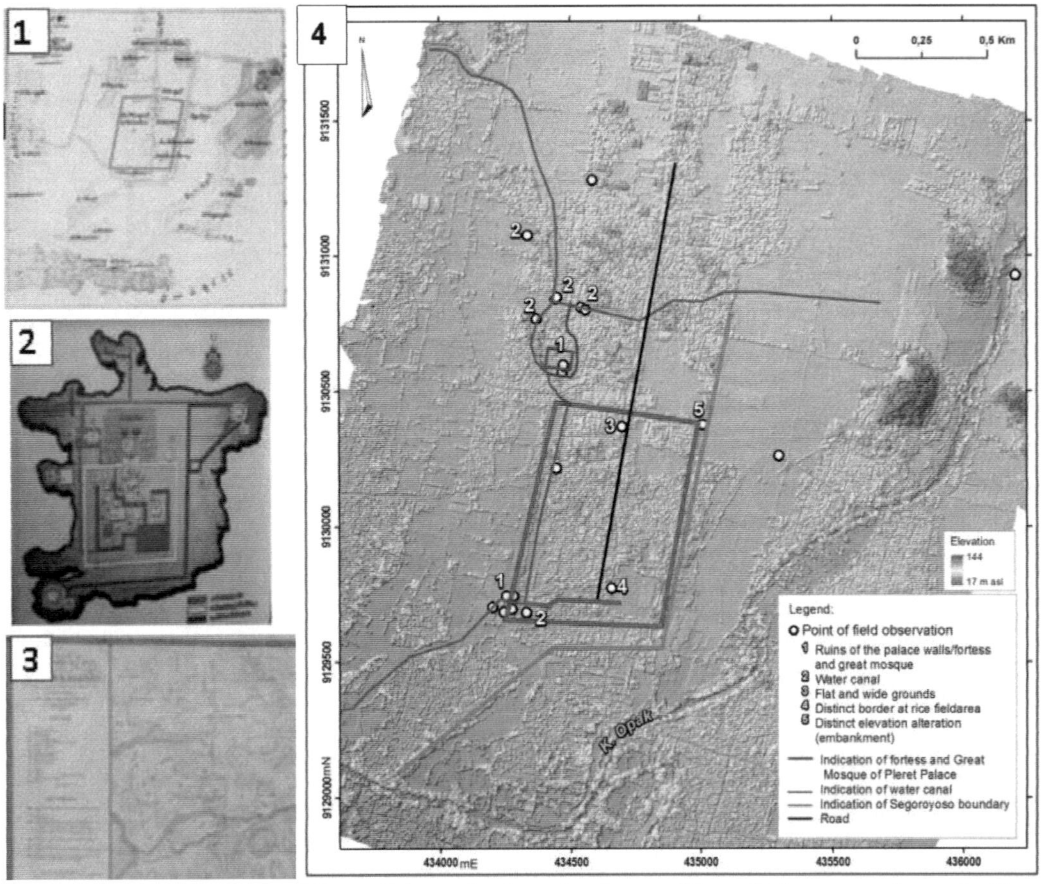


**Figure 9.** Identification of Pleret Palace site based refers to Pleret Palace map from de Graaf version, Pleret Palace map from Yogyakarta Cultural Agency, Historical Map of Trunojoyo Attack Scenario, and field observation results

On the west side there are indications of waterways that look concave along the field and there are bamboo and sugarcane plants that indicate the presence of natural water quantities. In the aerial image imagery, it can be recognize from the elongated area and lower altitude than the surrounding. This area has been utilized as a rice field whose pattern is different from the surrounding (Figure 10). The points that indicate the presence of drains in the field are combined with orthophoto and DSM object identification and connected to form a whole water canal.

**Figure 10.** Water canal and palace fortress identification and interpretation from DSM and orthophoto

The next identification is the situation of the Pleret palace consisting of plains and hills area boundary, Segarayasa embankment, fortress/wall, Great Mosque, palace square, and keputren. Objects depicted on the reference map can be identified using aerial photographs by looking at the patterns. For example, fortress identified by looking at the straightness pattern that has higher elevation than surrounding, grand mosque identified by delineating the excavation area of the grand mosque, keputren identified by delineating the area of 'settement (kampung) with toponimi keputren (from reference map), and Segarayasa embankment identified by straightness pattern on east and south side of fortress. It can be seen clearly pattern and distinct elevation from DSM that indicate fortress and embankment. Identification and nterpretation of Pleret Palace situation is shown in Figure 11.

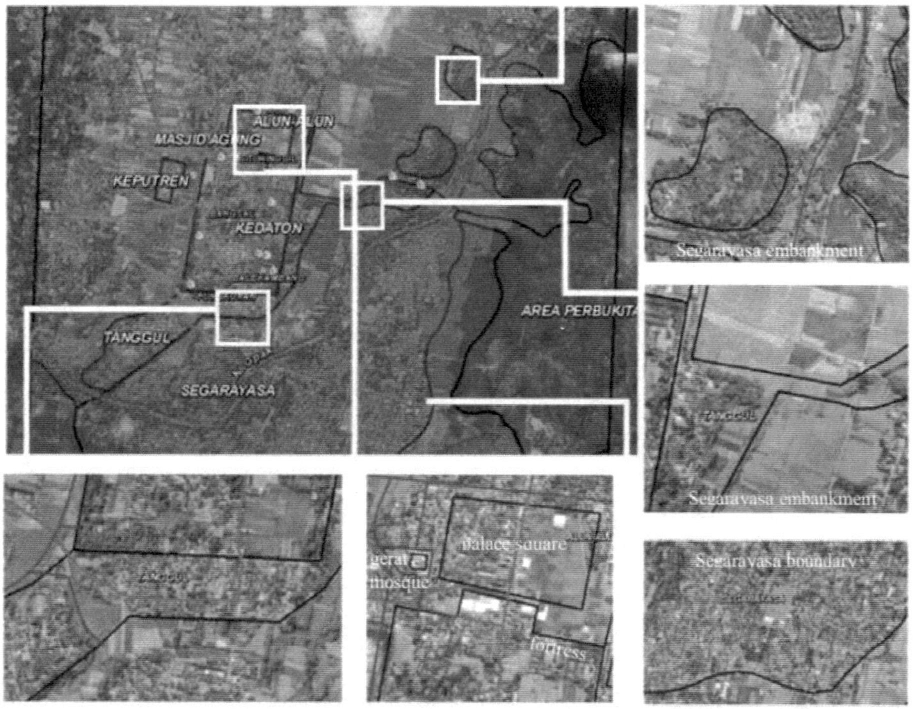


**Figure 11.** Identification and nterpretation of Pleret Palace situation including plains and hills area boundary, Segarayasa embankment, palace fortress/wall, Great Mosque, palace square, and keputren

The results of archaeological site identification in this research show that the existing phenomena from detail aerial photo and digital surface model can be the interpretation key of palace building/construction that exist in kingdom period. The existence of the palace, great mosque, Balekambang pond, Segarayasa embankment, Segarayasa artificial sea, road, and water canal built in the Pleret Palace can be identified and mapped by using orthophoto and DSM from data acquisition using UAV.

## 5. Conclusions

Data acquisition with UAV in Pleret Palace site can produce detail mosaic orthopoto and DSM that suitable in detail scale. These results can be used as input data for the identification of archaeological sites and recognize difference height as well as topographic patterns. Interpretation of aerial photography to identify the Pleret Palace is done by using several interpretation keys such as shape, pattern, and association. Based on the result, it can be identified some site element i.e. palace fortress, water canal, road, great mosque, palace square, Balekambang pond, and Segarayasa embankment, and Segarayasa artificial sea.

**Acknowledgment**
We would like to express my sincere gratitude to Research Directorate of Universitas Gadjah Mada who has provided research funding support through research scheme Capacity Improvement for Young Lecturer Researcher year 2017. Our sincere thanks also go to Lukman Ramadhani, S.Si. dan Edwin Seta Aji, S.Si as CV. Mitra Geotama team for technical data acquisition, Rhindang Bima Yudha, M.Ulil Azmi, Athif Fathul Hadi, Cahyo Anugra Ramdan, and Irfan Septian as research assisstants.

**References**

[1] Aber J S, Marzolff I, Ries J B 2010 *Small-Format Aerial Photography: Principles, Techniques and Geoscience Application* (The Netherlands: Elsevier).
[2] Adrisijanti I 2000 *Arkeologi Perkotaan Mataram Islam (Urban Archeology of Mataram Islam)* (Yogyakarta: Jendela).
[3] Comer D C and Harrowe M J 2013 *Mapping Archaeological Landscape from Space* (New York: Springer).
[4] Cultural Agency of Yogyakarta 2014 Menelusuri Jejak Kejayaan Kerajaan Mataram Islam di Pleret (Observation of the Glory Traces of the Islamic Mataram Kingdom at Pleret) (Video No. 6071/DVD/13/PA/12.2019/2014, launch on 5 Desember 2014).
[5] De Graaf H J 1987 *Disintregasi Mataram Dibawah Mangkurat I*: Terjemahan Bahasa Indonesia (Disintegration of Mataram under Mangkurat I: Indonesian Translation) (Jakarta: Pustaka Grafiti Press).
[6] Fallavollita P, Esposito S, Melis M G, Balsi M, Zappino L 2013 *UAS for Technology, New Perspective on Aerial Documentation* (Italy: University of Sassari).
[7] Geospatial Information Agency 2014 *The Regulation of Geospatial Information Agency's head (PERKA BIG) No. 15 year 2014 about Technical Guidance of Basic Map Accuracy* (Bogor: Badan Informasi Geospasial).
[8] Suprapta B 1996 *Pelestarian Benda Cagar Budaya* (Sejarah).
[9] Rokhmana C A 2006 Some Aspects on Developing a Truly Low-Cost Mapping System Using Handheld GPS and Camcorder *XXIII FIG Congress*, Munich.
[10] Rokhmana C A and Utomo S 2016 The Low-Cost UAV-Based Remote Sensing System Capabilities for Large Scale Cadaster Mapping. *IOP Conf. Series: Earth and Environmental Science* **47** (2016) 012005. doi:10.1088/1755-1315/47/1/012005
[11] Rosaji S C R, Handayani W, Nurteisa Y T, Suharyadi R, Widartono B S 2013. Aerial/Terrestrial Videography: Alternatif Teknologi Penginderaan Jauh untuk Survey dan Akuisisi data Spatial. *Simposium Nasional Sains Geoinformasi III – 2013*, ISBN 978-979-98521-4-4.
[12] Yuwono J S E 2010 *Sistem Pemetaan Arkeologi Nasional (SPAN): Strategi Integrasi dan Pengelolaan Data Spasial*

# Application of GIS for Assessment of Water Availability in the Cianten Watershed, West Java

**A A Mirrah and E Kusratmoko***

Department of Geography, Faculty of Mathematics and Natural Science, Universitas Indonesia, Depok 16424

eko.kusratmoko@sci.ui.ac.id

**Abstract.** Spatial information about the water availability in a region is very important in the management of sustainable water resources. Geographic information systems allows spatial information about water availability in watersheds to be assessed and monitored over time and within a predefined unit of space. This paper describes the application of GIS to assess the water availability and water demands spatially in the Cianten watershed, Bogor District, West Java. The availability of water for each sub-watershed was assessed using the water balance equation. Rainfall and temperature data for the 2007-2016 period, data of elevation, slope, soil type and land use for research area were used to calculate the water availabiliy, both annual and dry season. While the data on population size and land use were used to calculate water demands (domestic and non-domestic) for each sub-watershed based on the standards issued by the Ministry of Public Works and Agriculture. The analysis results using GIS platform shows a spatial variation of annual water availability and dry season. The water availability ranges from 9266 m3 / ha to 15,991 m3 / ha, while the dry season ranges from 2285 m3/ha to 4147 m3 / ha. Comparing water availability and water demands show that during the dry season most sub-watersheds in the study area experienced a high to low water deficit.

**Keywords:** SWAT Model, Water Availability, Water Balance, Water Demand

## 1. Introduction

Spatial information about the availability of water resources in a region is essential for sustainable management of water resources. Geographic information systems (GIS) facilitate the management of water resources more effectively and efficiently so as to enable spatial information about the availability of water resources in the watershed can be assessed and monitored over time and within a predefined unit space scale [1].

Definition of the water availability on earth space varies depending on the point of view of interest. Most studies and reviews on the subject of water availability focus on blue water or water available on the surface and subsurface of the earth [2]. [3] defined the availability of water as the amount of water expected to be continuously present at a given location (rivers, dams, lakes, reservoirs and other water structures) for a certain amount and within a certain period of time. In the hydrological cycle, the watershed is an open system in which the hydrological process takes place. The quantity of output (output) in the form of river flow discharge depends on the quantity of input (input), rainfall and watershed characteristics [4]. Thus, the water availability between watersheds will be different due to differences in climate factors, especially rainfall, and watershed characteristics, such as geological factors, topography, soil and land cover.

Spatial information on the water availability in a region, both regionally and locally, is essential to determine the potential of available water resources compared to the water demands of the area. The water availability index with an indicator of the comparison between human needs compared to water availability is one of the main parameters that is often used as a reference for addressing water and drinking water demands in an area [2]. In this study, we used GIS technology to assess water availability in the Cianten watershed, Bogor regency, West Java. The Cianten watershed is part of the Cisadane Basin located upstream. This watersheds was chosen because of the dynamic of land use changes, especially the increasingly widespread of built-up areas [5]. The problem of water shortage, especially in the dry season, is a classic problem that often occurs in most areas of the Cianten watershed. In this paper we describe how the spatial pattern of the water availability index in the Cianten watershed, especially during the dry season.

## 2. Methods

### *2.1 Study Area*

The Cianten watershed is a part of the Cisadane Basin located in upstream and geographically located at 6°28'45"- 6°43'0" South Latitude and 106°30'15"- 106°44'30" East longitude. The watershed area is 507.2 km$^2$ or 50,715.2 hectare and is administratively part of West Bogor Regency, West Java Province (see Figure 1).

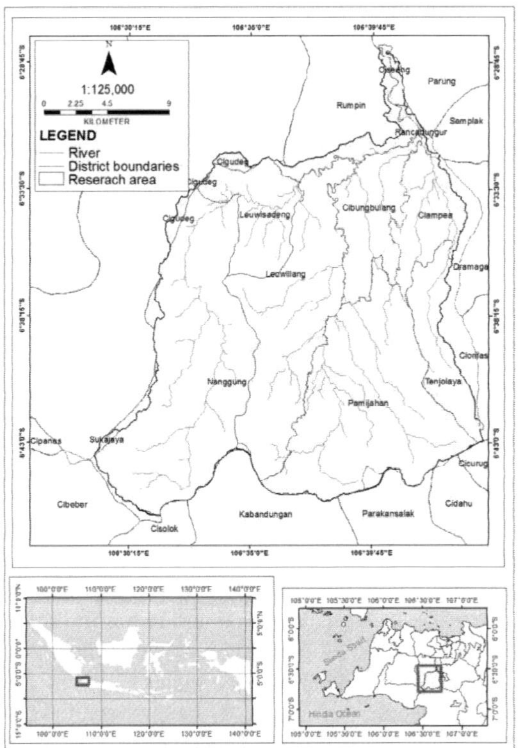


**Figure 1.** Map of the Cianten watershed, Bogor Regency, Province of West Java

### *2.2 Data and Materials*

The spatial and tabular data used and how to obtain it are described as in Table. 1. The GIS software used was ArcGis Ver. 10.1.

**Table 1**. Data collecting

| Data | Data acquisition |
|---|---|
| **DEM (*Digital Elevation Model*)** | DEM data derived by processing the SRTM data, which obtained from USGS Glovis. |
| **Land use** | Land use data derived from a digital Land use map in 2015 with a scale of 1: 100,000 issued by the Ministry of Land and Spatial Planning. |
| **Soil type** | The soil type data derived from a digital map of soil type with a scale of 1: 100.000 issued by Soil Research Institute, Ministry of Agriculture of the Republic of Indonesia. |
| **Climate (rainfall and temperature) Suhu)** | All data obtained from the Meteorology, Climatology and Geophysics Agency. The average monthly rainfall data for the period of 2007-2016 are available from 8 stations, namely Cianten Plantation, Leuwiliang, Kuripan, Dramaga, Rancabungur, Cigudeg, Kracak, and Cicurug. While the average data of monthly temperature for the same period obtained from Dramaga Climate Station. |
| **Drainage network** | Data were derived by the processing of the digital topography map with a scale of 1: 25.000 issued by Geospatial Information Agency (BIG). |
| **Total Population** | Secondary data obtained from the Central Bureau of Statistics, Bogor Regency, in 2016. |

*2.3. Data Processing and Analysis*

Water availability for each sub watershed were calculated by the equation [5]:

$$Q = (1 - C) \times (P - Ea) \times A \tag{1}$$

where Q is the water availability (mm / year), P is the average rainfall annual (mm / year), Ea is the actual evapotranspiration (mm / year), C is the runoff coefficient, and A is the watershed area (ha). The workflow to compute the water availability based on GIS technology is shown in Figure 2. The GIS database system for the Cianten watershed is built with layers of sub-watershed boundary, river network, slope, land use, rainfall and temperature.

Delineation process of the Cianten watershed to the sub-watershed unit is done using ArcSWAT software with ArcGIS 10.1 interface. Then, the hydrological unit for each sub-watershed were determined by the overlay processing of thematic layers of slope, land use and soil type. The classification for each thematic layer refers to the Cook method classification [7][8].

Spatial distribution of the annually and seasonally rainfall of the study area are calculated by Thiessen Polygon method based on the eight (8) available rainfall stations. Furthermore, the actual evapotranspiration is calculated using the Turk-Langbein equation [5]:

$$Ea = \frac{P}{(0.9 + \frac{P}{Eo})^2} \tag{2}$$

where Ea is the actual evapotranspiration, P is the average annual rainfall, and Eo is the potential evapotranspiration. The amount of Eo is calculated using the Langbein equation:

$$Eo = 300 + 25Ty + 0.05Ty^3 \tag{3}$$

where Ty is the annual temperature distribution. The water demands in a watershed are calculated based on the population's water demands and the water requirements for agriculture and crops. The population water demands in this study refer to the needs of the rural population of 60 liters/day per person [10]. While water requirements for agriculture and crops based on land use type [11] as follow, forest crops (1000 mm/ year per hectare), shrublands (1000), mixed garden (1000), non-irrigated field ("tegalan" in bahasa) (1350 mm), rice field with twice harvest (2400), rice field with one harvest (1200).

Population water demands (Wpo in $m^3$) in a watershed is calculated by the equation:

$$Wpo = n \times p \times \frac{q}{1000} \tag{4}$$

where n is the number of days, p is the number of population per village, and q equals 60 liter/day per person. Furthermore, the water requirement for each land use (Wlu in $m^3$) are calculated by the equation [1]:

$$Wlu = 10 \times n \times A \times r \tag{5}$$

where n is the number of days, A is the area of each type of land use and r is the water requirement for each type of land use. Therefore, the total water demands = Wpo + Wlu. We determined a water availability index (WInd) in a watershed with the basis of the comparison between total water demand to water availability,

$$WInd\ (\%) = \frac{WN}{W\ x\ 100\%} \tag{6}$$

Based on the value of WInd for each watershed, which can be positive (surplus) or negative (deficit), we classified the Wind, in five classes, is deficits or surplus: high, slightly high, medium, slightly low, and low.

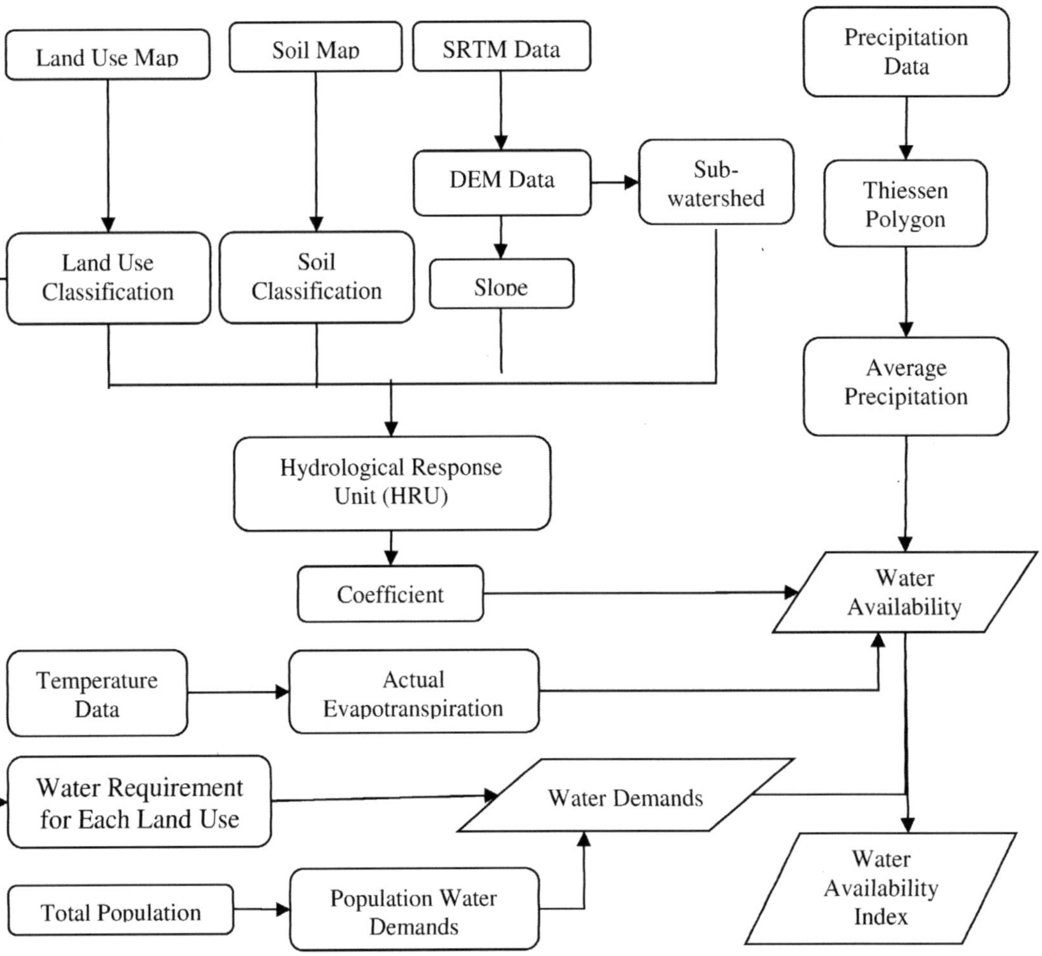


**Figure 2.** Workflow to calculate water availability based on GIS techology

## 3. Results and Discussion

### *3.1 Annual and Seasonal Water Availability*

The GIS modeling process produced two maps, ie a map of annual water availability and a map of water availability for the dry season in the Cianten watershed (see Figures 3a and 3b). The annual water availability varies with the range between 125,5 $10^6$ $m^3$ ( 3019 $m^3$/ha) to 6.1 $10^6$ $m^3$ (While in the dry season, its amount ranges from .67 $10^6$ $m^3$ to 30.1 $10^6$ $m^3$. The availability of water in the dry season is only a quarter of the annual water availability. The size of the watershed area affects directly the amount of the water availability, both annual and seasonal. . Statistically, the average annual water availability of Cianten watershed is 42.3 $10^6$ $m^3$ (10.1 $10^6$ $m^3$ in dry season) and with an inter-watershed variability of 84% and 83% respectively.

The amount of water availability in a watershed is divided by the watershed's area generated a water availability per unit area (hectare). The magnitude in the Cianten watershed is in the range of 9,266 to 15,991 $m^3$/ha for annual and 2285 to 3754 $m^3$/ha in the dry season. Compared with the total water availability in a watershed, the variability of water availability per unit area (hectare) is relatively smaller (15%), both annual and seasonal. This may provide a clue that the water availability in the Cianten watershed is spatially homogeneous. Spatially, regions (sub-watersheds) with the highest water availability tend to be found in regions with high rainfall (> 4000 mm) with land cover dominated by forest.

**Table 2**. Annual and seasonal water availability (WA) in the Cianten Watershed

| No | Sub-Watershed | Area (Ha) | WA in total area | | WA in ha area | |
|---|---|---|---|---|---|---|
| | | | Dry season (in $10^6$ $m^3$) | Annual (in $10^6$ $m^3$) | Dry season (m3/ha) | Annual (m3/ha) |
| **1** | Cisadane 1 | 740.5 | 1.69 | 6.7 | 2285 | 9266 |
| **2** | Cisadane 3 | 911.9 | 3.8 | 14 | 2608 | 9541 |
| **3** | Cisadane 2 | 642.8 | 1.67 | 6.1 | 3828 | 14201 |
| **4** | Cianten 2 | 1767.9 | 5.8 | 23.8 | 2997 | 12356 |
| **5** | Cikaniki 2 | 9955.0 | 30.1 | 125.5 | 3019 | 12611 |
| **6** | Ciaruteun | 2011.6 | 7.7 | 30.1 | 3835 | 14982 |
| **7** | Citeureup | 3019.2 | 7.4 | 30.5 | 2446 | 10110 |
| **8** | Cibungbulang | 2327.5 | 8.2 | 33.9 | 3260 | 13467 |
| **9** | Cinangneng | 2248.5 | 7.5 | 31.5 | 3355 | 14031 |
| **10** | Cigamea | 2237.1 | 6.9 | 29.4 | 3103 | 13153 |
| **11** | Cianten 1 | 7105.8 | 18.6 | 79.4 | 2623 | 11171 |
| **12** | Ciampea | 2820.4 | 10.1 | 42.1 | 3309 | 13791 |
| **13** | Cisarua | 2359.5 | 8.1 | 34.4 | 3421 | 14574 |
| **14** | Cipuraseda | 2455.9 | 7.9 | 33.6 | 3221 | 13670 |
| **15** | Cikaniki 1 | 7545.3 | 28.3 | 120.6 | 3754 | 15991 |
| **16** | Cikuluwung | 2566.1 | 8 | 35.6 | 3135 | 13856 |
| | Average | - | 10,11 | 42,33 | 3137 | 12923 |
| | Std Deviasi | - | 8,4 | 35,6 | 469,6 | 1966,1 |
| | Var (%) | - | 82,8 | 84,0 | 15,0 | 15,2 |
| | Total area | 50715.2 | 161,76 | 677.20 | 50198 | 206771 |

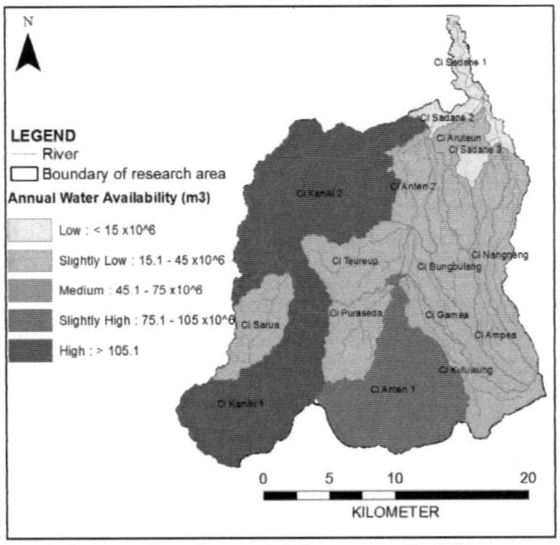


**Figure 3a.** Annual Water availability in Cianten watershed

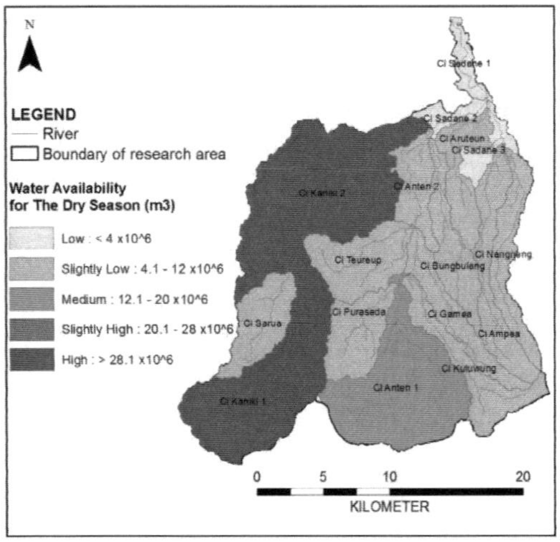


**Figure 3b.** Water availability for the dry season in Cianten watershed

*3.2. Water Availaibility Index*

We have calculated the total water demand for each sub-watershed in the Cianten watershed, both annual and dry season, and then we compared it with the water availability for each sub-catchment (see Figure 4 and Figure 5). Spatially, the distribution of WInd for each sub-watershed are shown in Figures 5a and 5b. It is clearly seen that based on the annual water availability index, most of the sub-watersheds (11 of 16 sub-watersheds) in the Cianten have surplus water resources. The sub-watersheds categorized deficit of water availability can be found only in the downstream area, ex. the Cisadane_2, Cinangneng. The average of water deficit is less than 15%.

The opposite happens during the dry season. Most sub-watersheds are characterized by the greater water demand more than its water availability. There are five sub-watersheds, ex. the Cisadane_1, Cisadane_2, Cianten_2, which the water deficit is classified by high (>30%). The sub-watersheds located in the downstream area. Only one watershed (the Cikaniki _1) is not water deficits.

In general, the results of this study indicate that the calculation of water availability based on annual rainfall data, especially in areas with high rainfall, does not guarantee a careful calculation of the index of water availability in a region. We need to consider the spatio-temporal factor of rainfall in the study area and the characteristics of the land use.

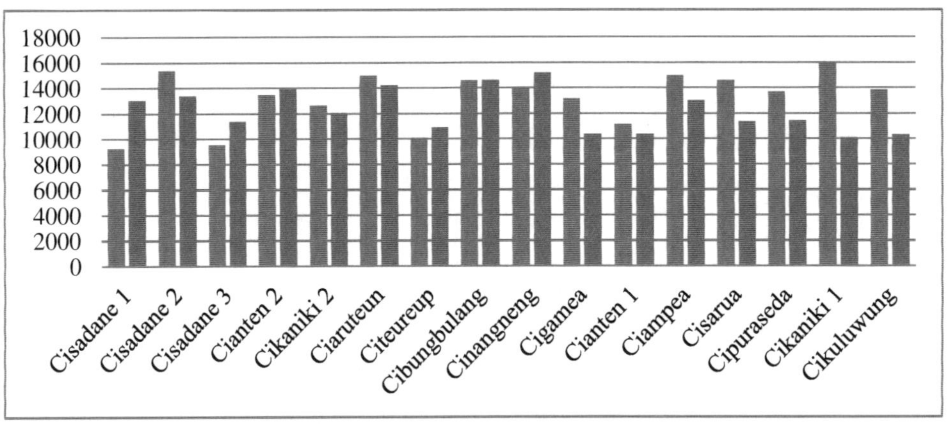


■ Water availability ($m^3$/ha) ■Water demand ($m^3$/ha)

**Figure 4.** Annual water availability and water demand in the Cianten watershed

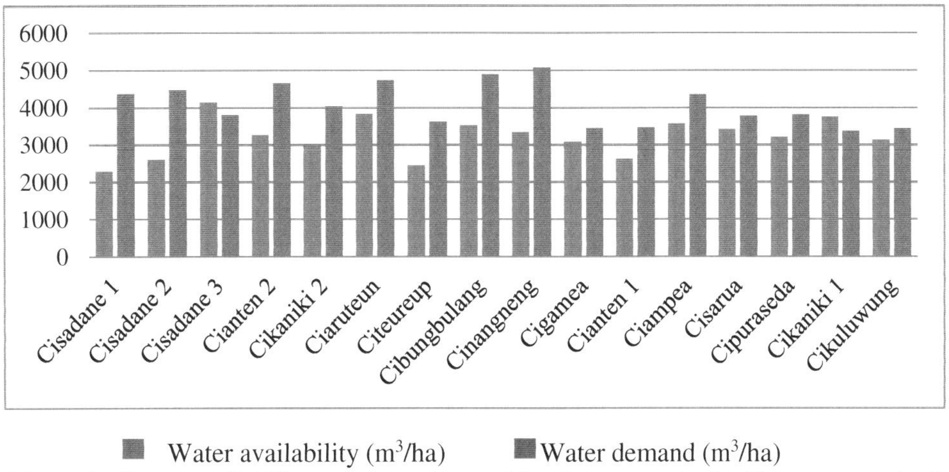


**Figure 5.** Water availability and water demand for dry season in the Cianten watershed

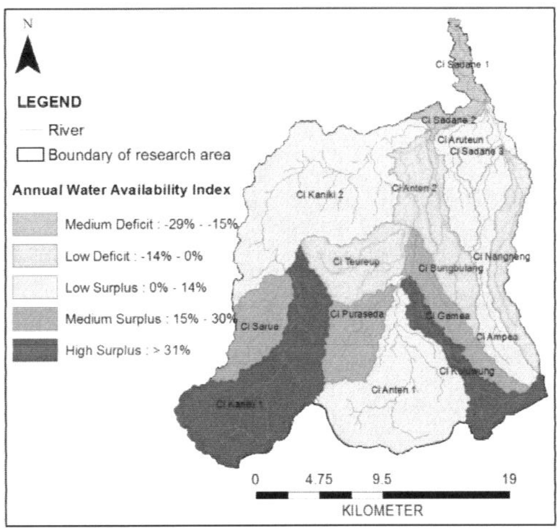


**Figure 6a**. Annual water availability index in the Cianten watershed

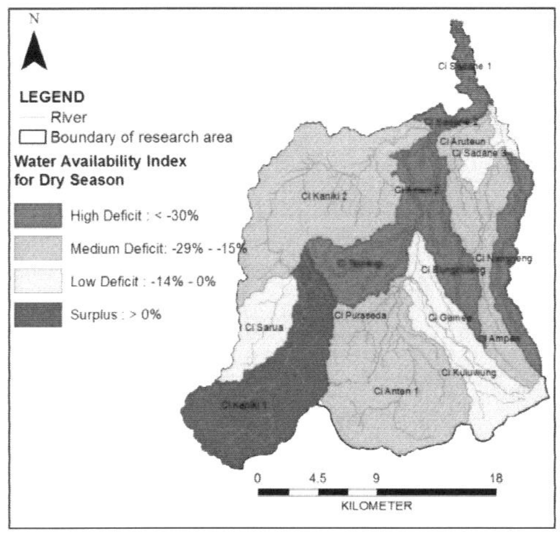


**Figure 6b**. Water availability index for dry season in the Cianten watershed

## 4. Conclusions

Based on the GIS modeling, this study has been able to calculate and assess the availability of water in the Cianten River Basin, Bogor, West Java, both spatially and temporally. The annual water availability index in the Cianten watershed shows that most of the sub-watersheds are classified as surplus. A factor of the high annual precipitation in the study area supports these results. However, an indication of water resource deficit in all sub-watersheds occurred during the dry season. The worst water deficit can be found especially in the sub-watersheds located downstream of the study area.

Further research of these findings can support sustainable management of water resources. For that, the verification process of the research results should be done through direct observation in the field. It is necessary to validate the results of this study.

## References

[1] Wilson John P, H Mitasova and Dawn J Wright 2000 Water Resource Applications of Geographic Information Systems *Journal of The Urban and Regional Information System Association.*

[2] Hui Xu and May Wu 2017 *Water Availability Indices – A Literature Review* Energy Systems Division, Argonne National Laboratory.

IOP Conf. Series: Earth and Environmental Science **98** (2017) 012018 doi:10.1088/1755-1315/98/1/012018

[3] Kodoatie Robert J dan Roestam Sjarief 2010 *Tata Ruang Air* (Yogyakarta: Andi Offset)
[4] Gregory K J and D E Walling 1973 *Drainage Basin form and process, a geomorphological approach* (London: Arnold).
[5] Trimarmanti T K E 2014 Evaluasi Perubahan Penggunaan Lahan Kecamatan Daerah Aliran Sungai Cisadane Kabupaten Bogor *Jurnal Wilayah dan Lingkungan* **2** pp 55-72.
[6] Maijerink A M J., et al. 1994 *Introduction of Geographical Information Systems for Practical Hydroogy* (Enschede: ITC).
[7] Kusratmoko E 2001 Curah Hujan dan Karakteristik Aliran Ci Liwung *Jurnal Geografi* **2** (2) pp 39-49.
[8] Ismail Arif, E Kusratmoko and Sobirin 2015 Pemodelan perubahan penggunaan lahan dan pengaruhnya terhadap koefisien aliran pada daerah tangkapan air waduk Darma, Kabupaten Kuningan, Jawa Barat *J. Ilmu Kehutanan Wanakarsa* **9** (2).
[9] Hudson N W 1993 Field measurement of soil erosion and runoff Food and Agriculture Organization of the United Nations Rome www.fao.org/docrep/T0848E/T0848E00.htm
[10] Badan Standarisasi Nasional 2002 *Penyusunan neraca Sumber Daya Bagian 1: Sumber Daya Air Spasial* SNI 19-6728.1-2002
[11] Dumairi 1992 *Ekonomika Sumber daya Air: Pengantar ke Hidronamika* (Yogyakarta: BPFE).

# Effect of Nock-Ten Tropical Cyclone on Atmospheric Condition and Distribution of Rainfall in Gorontalo, Ternate, and Sorong Regions

**A Lumbangaol[1*], Y R Serhalawan[2] and Endarwin[3]**

[1]Student, School of Meteorology Climatology and Geophysics
[2]Staff, Indonesian Agency for Meteorology Climatology and Geophysics
[3]Lecturer, School of Meteorology Climatology and Geophysics

adel.lumbangaol@gmail.com

**Abstract.** Nock-Ten Tropical Cyclone is an atmospheric phenomenon that has claimed many lives in the Philippines. This super-typhoon cyclone grows in the Western Pacific Ocean, North of Papua. With the area directly contiguous to the trajectory of Nock-Ten Tropical Cyclone growth, it is necessary to study about the growth activity of this tropical cyclones in Indonesia, especially in 3 different areas, namely Gorontalo, Ternate, and Sorong. This study was able to determine the impact of Nock-Ten Tropical Cyclone on atmospheric dynamics and rainfall growth distribution based on the stages of tropical cyclone development. The data used in this study include Himawari-8 IR channel satellite data to see the development stage and movement track of Tropical Cyclone Nock-Ten, rainfall data from TRMM 3B42RT satellite product to know the rain distribution in Gorontalo, Ternate, and Sorong, and reanalysis data from ECMWF such as wind direction and speed, vertical velocity, and relative vorticity to determine atmospheric conditions at the time of development of the Nock-Ten Tropical Cyclone. The results of data analysis processed using GrADS application showed the development stage of Nock-Ten Tropical Cyclone has effect of changes in atmospheric dynamics condition and wind direction pattern. In addition, tropical cyclones also contribute to very light to moderate scale intensity during the cycle period of tropical cyclone development in all three regions.

**Keywords:** Tropical Cyclone, Rainfall, TRMM, Satellite Imagery

## 1. Introduction

On 22 December 2016, tropical cyclone has been tracking in the Western Pacific Ocean region north of Papua. Nock-Ten tropical cyclone moves with maximum wind velocity reaches 93 km/h and triggers rain events in some areas. Generally, the formation of this cyclones occurs in areas with latitude positions above $10^{o}$ or ranges over a distance of 500 km from the equator. Indonesia territory is not a sufficiently effective area for the growth of tropical cyclones. However, due to Indonesia's geographical position adjacent to tropical cyclone trajectory areas, the potential for weather disturbance caused by tropical cyclones is quite influential in Indonesia. For that purpose, study conducted to discuss the impact of Nock-Ten tropical cyclone which occurred in the Gorontalo, Ternate and Sorong areas.

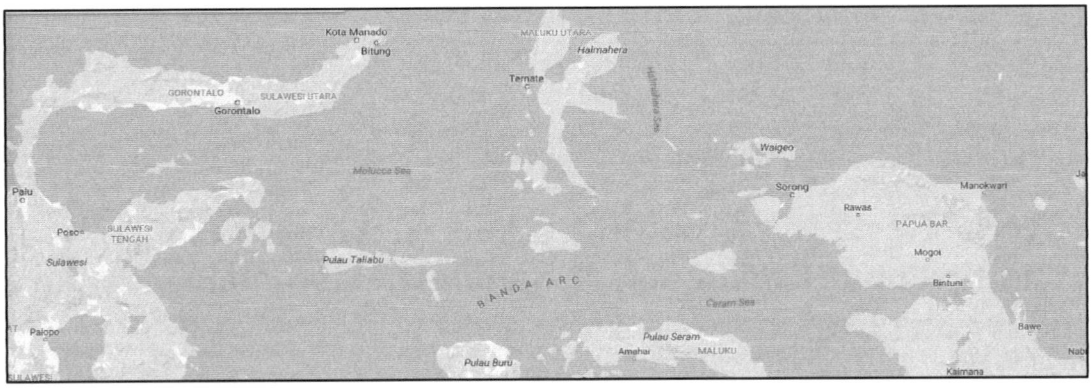


**Figure 1.** Location of Gorontalo, Ternate, and Sorong Areas located in the northern part of Indonesia *(Taken from Google Map)*

The purpose of this study is to determine the atmospheric condition that occur based on the growth stages and analyse the effect of the cyclones on the conditions of distribution of rainfall in the three areas. Tropical cyclone is a low-pressure system with mesoscale power plant in a synoptic-scale area which has a maximum wind velocity condition of 34 knots or more, which dominates in its central region with a life span of at least 6 hours and grows over the oceans [4]. The growth stage of tropical cyclone consists of four stages, namely the formation stage characterized by the presence of tropical disturbances and the convective cloud areas of Cumulonimbus clouds, immature stages with convective areas that form a circular belt with maximum wind speed increased to reach the gale force wind, ranging from ≥ 34 knots or 63 km/h, the mature stage is characterized by a stable tropical cyclone condition as well as the tropical depression stages show warm cyclone center begin to disappear, with maximum wind speed extending and widening away from the center of the cyclone. Nock-Ten tropical cyclone impact analysis needs to be done in order to know the atmospheric dynamics occurring in Gorontalo, Ternate, and Sorong regions through four growth stages cyclones, i.e. forming stage, immature stage, mature stage and tropical depression stage.

## 2. Data and Method

This research uses data of Himawari-8 satellite channel of IR channel with wavelength 6.7 µm -7.3 µm. The data taken are daily data based on classification of tropical cyclone stage development. The Himawari-8 satellite image has a 10-minute temporal resolution and spatial resolution in the range 0.5 - 2 km. This study also uses TRMM satellite data which is NASA product to determine the intensity of rainfall at the time of the incident on 21 December to 27 December 2016. Since December 1997, TRMM has observed rain with multiple sensors [7]. TRMM is the first meteorological satellite specially used to gauge tropical and subtropical precipitation quantificationally [2]. TRMM data accumulated daily rainfall used is TRMM 3B42RT with a resolution of space 0.25° and resolution time of 3 hours [5]. The next data used is ERA-Interim reanalysis data which is the ECMWF (The European Center for Medium-Range Weather Forecasts) output. ECMWF data are used in weather forecasts, weather parameter analysis, climate analysis, providing data for reforecast needs and multimodel data. This research data has a resolution 0.75° x 0.75° with weather parameters such as wind patterns and vertical velocity layer 925 mb, 850 mb, 700 mb, 500 mb, and 200 mb.

The method used in this research is the data processing of Himawari-8 satellite image using one of the SATAID (Satellite Animation and Interactive Diagnosis) variation program developed by JMA (Japan Meteorological Agency) in the form of GMSLPC. SATAID serves to display binary satellite data into images. SATAID can display satellite images and combine NWP (Numerical Weather Prediction) numerical weather data. NWP data is separate from satellite imagery data, also obtained from JMA in one package with satellite image data. In general, the processing steps in SATAID start with program, register image data in memory, display image data and advanced operation. This research also use ERA-

Interim reanalysis data processing and TRMM satellite data using GrADS (Grid Analysis and Display System) application. GrADS is a software of WMO recommendation (World Meteorological Organization) to display spatial weather parameters and visualize data into object of map and graph [3]. GrADS can manipulate and visualize geo-science data to be displayed in both graphics and flow lines [6].

## 3. Result and Analysis

### *3.1. Satellite Analysis Himawari-8 IR Channel*

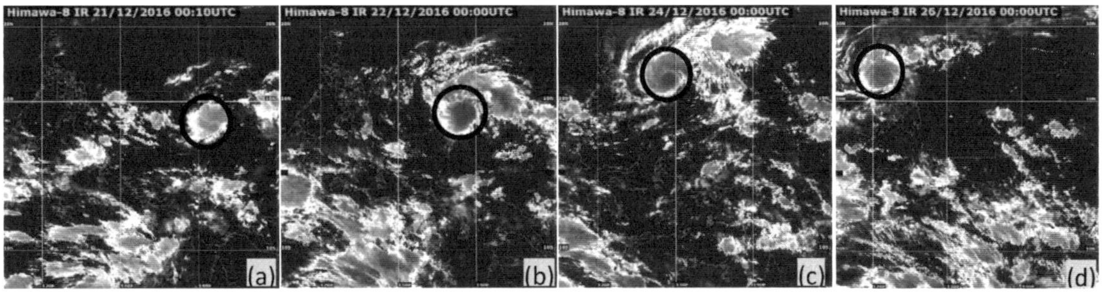


**Figure 2.** Satellite imageries shows the temporal change of Nock-Ten Tropical Cyclone movement track on **(a)** 21 December 2016, **(b)** 22 December 2016, **(c)** 24 December 2016, and **(d)** 26 December 2016 attached by black circle.

Himawari satellite IR channel imagery shows the temporal changes of Nock-Ten tropical cyclone movement track. Cloud conditions on 21 December 2016 show aggregate of convective clouds in the West Pacific Ocean region that marks the early stages of the cyclone. Then, the rotating cyclone moves and forms circulation also the presence of cloud circular belt indicates the start of cyclone eye formation on 22 December 2016 which shows the stage of cyclone intensification. On 24 December 2016, the cyclone reached a mature stage and moved towards Filipine. It is characterized by the thick convective wall of conventional around the center of the cyclone. On 26 December 2016, Nock-Ten tropical cyclone convective cloud area conditions gradually disappeared.

### *3.2. Satellite Analysis TRMM 3B42RT*

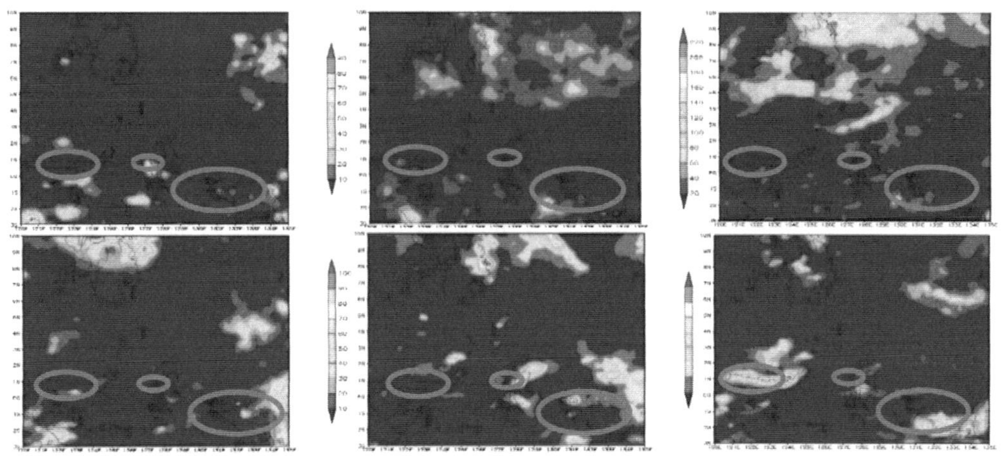

**Figure 3.** Accumulated rainfall conditions 1 Day through Satellite TRMM 3B42RT analysis ranging on 21 to 26 December 2016. Gorontalo, Ternate and Sorong areas marked by red circles.

From TRMM satellite image analysis results can be seen that on 21 December 2016, which is in the early stages of tropical cyclone formation, rainfall conditions in Gorontalo and Sorong areas generally only range ≤10 mm/day, except Ternate region has an intensity of 10-40 mm/day. On the next day, on 22 December 2016, tropical cyclones were immature and intensified moved towards the northeast of the Fhiliphina region. The rainfall potential of Gorontalo, Ternate and Sorong is generally only about ≤ 20 mm/day. The mature tropical cyclone stage begins on 24 December 2016. At this stage, the three regions generally have only a potential rainfall event of ≤10 mm/day. Furthermore, the depression occurred on 26 December 2016 with rainfall scale conditions in the three regions varying, ie 10-60 mm/day. From the TRMM analysis, it can be indicated at tropical depression, the three regions experienced an increased rainfall.

### *3.3. Surface Wind Analysis*

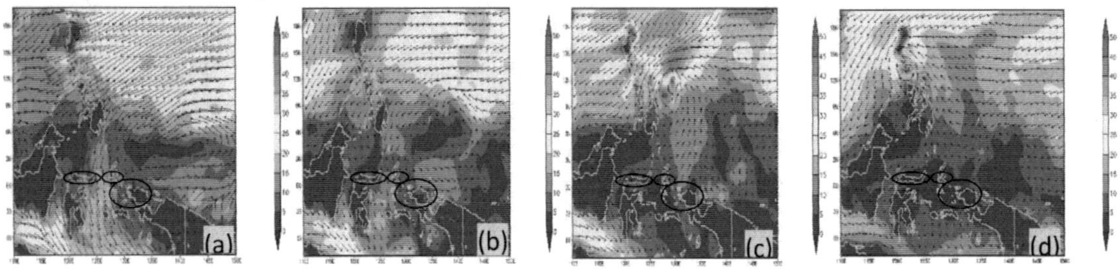


**Figure 4.** Surface Wind Conditions of GrADS analysis results on **(a)** 21 December 2016, **(b)** 22 December 2016, **(c)** 24 December 2016 and **(d)** 26 December 2016. Gorontalo, Ternate and Sorong regions are shown by black circles.

From the wind surface analysis, the cyclone initial stage of Nock-Ten formation, there is a cyclonic current turnover in the north of West Papua at West Pacific Ocean. This condition results in the formation of convective clouds. Wind condition conditions in parts of Gorontalo, Ternate and Sorong are generally influenced by cyclonic currents and have wind speeds ranging from 5-20 knots. In Ternate it was detected that shear line caused by cyclone generating conventive clouds. The shear line can be an important source of convection as is often associated with strong low-level moisture convergence [1]. The intensification stage shows the condition of the wind patterns in the Sorong area moving cyclonically toward the center of the cyclone with wind velocity between 5-15 knots. It also indicates shear line in all three areas. At a mature stage, air masses in all three regions are attracted toward the centre of the cyclone to the east of the Philippines. Conditions of wind patterns in the three regions generally move with of 5-15 knots velocity. These three areas generally have bright conditions. The depression stage shows the condition of wind patterns in Gorontalo and Ternate generally still influenced by cyclone with speed ≤5 knots. Sorong is generally no longer influenced by the cyclone wind conditions. Wind patterns move to south of Papua at speeds ranging from ≤5 knots.

### *3.4. Vertical Velocity Analysis*

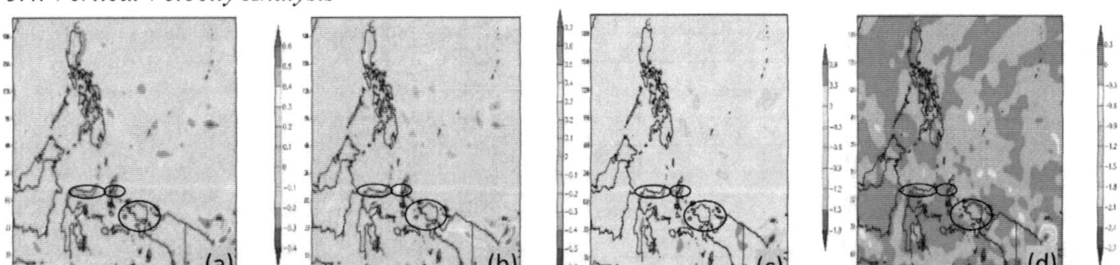


**Figure 5.** Vertical velocity GrADS analysis results on 21 December 2016 at layers **(a)** 925 mb, **(b)** 850 mb, **(c)** 700 mb and **(d)** 500 mb. The areas of Gorontalo, Ternate, and Sorong are shown in black circles.

IOP Conf. Series: Earth and Environmental Science **98** (2017) 012019 doi:10.1088/1755-1315/98/1/012019

At the first stage of Nock-ten Tropical cyclone shown by Figure 5, the GrADS analysis shows in Figure 4 the vertical velocity conditions in (a) 925 mb and (b) 850 mb layers ranging from (-0.2)-0.1 Pa $s^{-1}$ of air mass in the layer is quite low. The pattern indicated by layers (c) 700 mb and layer (d) 500 mb also shows the same value, ranging from (-0.3)-0.3 Pa $s^{-1}$. Generally, the areas of Gorontalo, Ternate and Sorong have the potential to increase the air mass is quite low and indicate the condition of atmospheric weather in the region is generally bright.

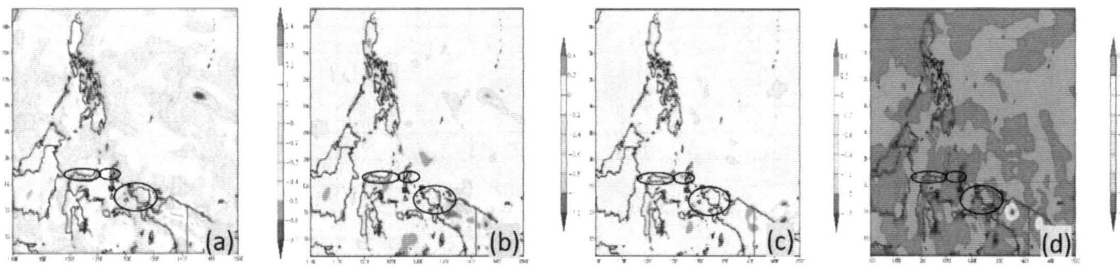


**Figure 6.** Vertical velocity conditions of GrADS analysis on 22 December 2016 at layers **(a)** 925 mb, **(b)** 850 mb, **(c)** 700 mb and **(d)** 500 mb. The areas of Gorontalo, Ternate, and Sorong are indicated by black circles.

The immature cyclone stage which can be seen from the Figure 6, shows the vertical velocity conditions in the three regions of the (a) 925 mb layer ranging from (-0.1)-0 $s^{-1}$ and layer (b) 850 mb to (-0.2)-0.2 Pa $s^{-1}$. The negative vertical velocity phase continues until the layer (c) is 700 mb and the layer (d) 500 mb, with each value indicating an average of (-0.3)-0.3 Pa $s^{-1}$ and the number (-0.5)-0 Pa $s^{-1}$. The pattern of vertical velocity change on a positive to negative scale indicates an indication of the increase in air masses in the region is high enough to indicate the potential for convective clouds in three areas.

The mature stage of the cyclones from Figure 7 shows the vertical velocity at (a) 925 mb layer ranging from 0.3-0 Pa $s^{-1}$. This indicates a positive vertical velocity condition and air mass decreased. It also occurs in layer (b) 850 mb which shows the value of 0.5-0 Pa $s^{-1}$. This is also supported by layer condition (c) 700 mb and (d) 500 mb, a decrease of air mass with vertical velocity $\geq 0$ Pa $s^{-1}$. This indicates the potential for the occurrence of convective clouds is quite low in Gorontalo, Ternate and Sorong.

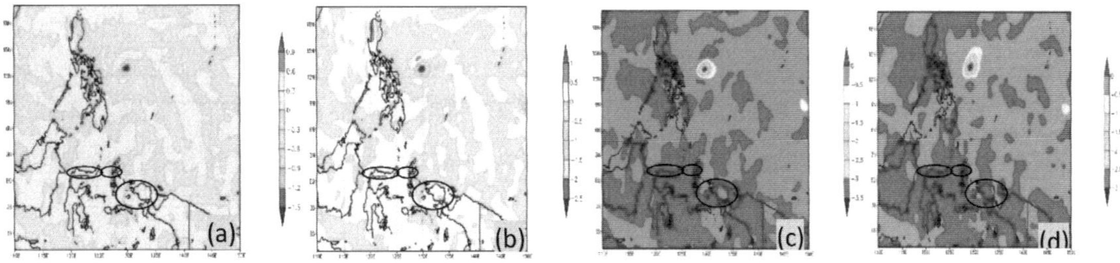


**Figure 7.** Vertical velocity conditions of GrADS analysis on 24 December 2016 at layers **(a)** 925 mb, **(b)** 850 mb, **(c)** 700 mb and **(d)** 500 mb. The areas of Gorontalo, Ternate, and Sorong are indicated by black circles

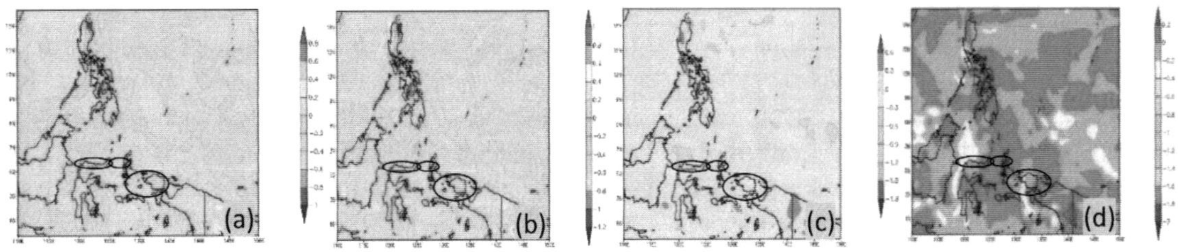


**Figure 8.** Vertical velocity conditions of GrADS analysis on 26th December 2016 at layers **(a)** 925 mb, **(b)** 850 mb, **(c)** 700 mb and **(d)** 500 mb. The areas of Gorontalo, Ternate, and Sorong are indicated by black circles.

At the tropical depression, the vertical wind velocity conditions of (a) 925 mb and (b) 850 mb range (-0.2)- 0.2 Pa $s^{-1}$. The potential for air mass increases is quite low in all three regions. The rising pattern continues to layer (c) 700 mb and layer (d) 500 mb with velocity ranging from (-0.3)-0.3 Pa $s^{-1}$ and (-0.4)-0 Pa $s^{-1}$. Parts of Sorong area are detected to have a vertical velocity of (-0.4) $s^{-1}$ indicating the potential for convective clouds to form. From the analysis result, it can be concluded that there is an increase of active air mass in Sorong city, while Gorontalo and Ternate areas generally have low rainfall potential.

### *3.5. Relative Vorticity Analysis*

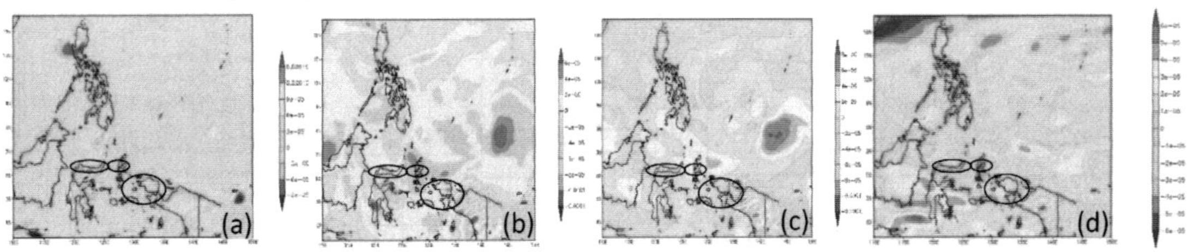


**Figure 9.** Relative Vorticity GrADS analysis results on 21 December 2016 at layers **(a)** 925 mb, **(b)** 850 mb, **(c)** 700 mb and **(d)** 200 mb. The areas of Gorontalo, Ternate, and Sorong are shown in black circles.

Based on the results of GrADS analysis in the early stages of Nock-Ten tropical cyclone formation, it can be seen that the average relative vorticity condition at 925 mb layer in Gorontalo, Ternate, and Sorong areas shows 3 x $10^{-5}$ $s^{-1}$ values. This indicates cyclonic wind conditions. Layers of 850 mb and 700 mb show the same relative vorticity values, ranging from 4 x $10^{-5}$ - (-2 x $10^{-5}$) $s^{-1}$. In the 200 mb layer, the relative vorticity values range from 2 x $10^{-5}$ - (-2 x $10^{-5}$) $s^{-1}$ and indicate that movement is generally anticyclonic. The analysis shows that the weather conditions in the three regions are generally clear.

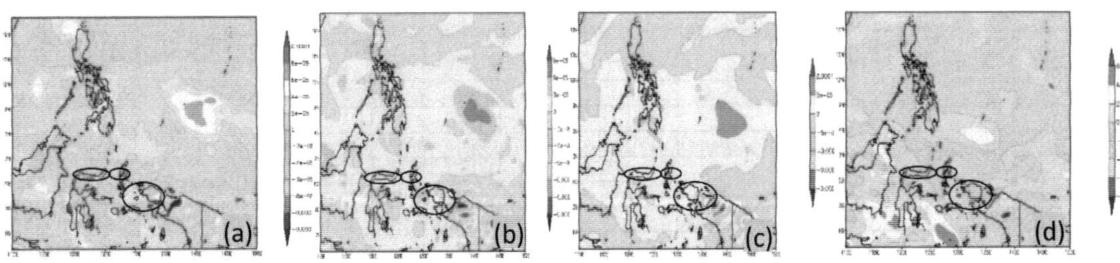


**Figure 10.** Relative Vorticity GrADS analysis results on 22 December 2016 at layers **(a)** 925 mb, **(b)** 850 mb, **(c)** 700 mb and **(d)** 200 mb. The areas of Gorontalo, Ternate, and Sorong are shown in black circles.

At the cyclone intensification stage, it can be seen in all three regions in 925 mb having an average relative vorticity value of 4 x $10^{-5}$-0 $s^{-1}$ except in the northern part of Sorong is -6 x $10^{-5}$ $s^{-1}$. This indicates a cyclonic condition. The cyclonic wind conditions are also supported on the 850 mb layer of 3 x $10^{-5}$-0 $s^{-1}$, except in northern Gorontalo and Sorong, and the 700 mb layer shows a value of 5 x $10^{-5}$ $s^{-1}$. The 200 mb layer shows a value of 2 x $10^{-5}$ - (-2 x $10^{-5}$) $s^{-1}$. This shows the condition of relative vortices at intensive stage is generally cyclonic.

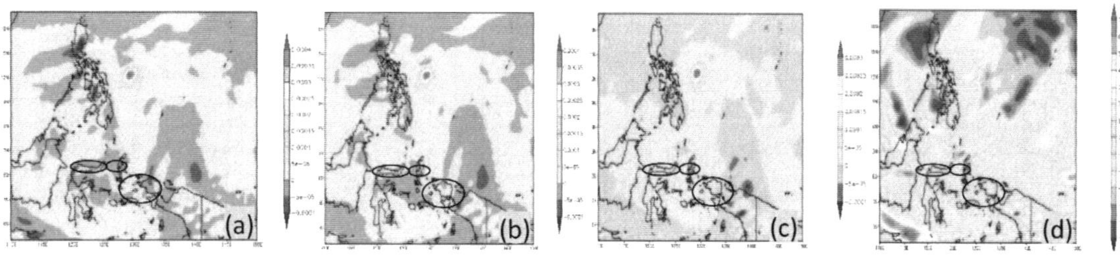


**Figure 11.** Relative Vorticity GrADS analysis results on 24 December 2016 at layers **(a)** 925 mb, **(b)** 850 mb, **(c)** 700 mb and **(d)** 200 mb. The areas of Gorontalo, Ternate, and Sorong are shown in black circles.

In the Nock-Ten tropical cyclone mature stage which shown in Figure 11, relative vorticity conditions at 925 mb and 850 mb layers show an average value of 5 x $10^{-5}$ - (-5 x $10^{-5}$) $s^{-1}$. At 700 mb layers, relative vorticity conditions in the three regions are positively 1 x $10^{-4}$ $s^{-1}$. This shows the existence of cyclonic movement in all three regions. Cyclones are also shown up to a layer of 200 mb indicating a range of reach (2 x $10^{-5}$) - 0 $s^{-1}$.

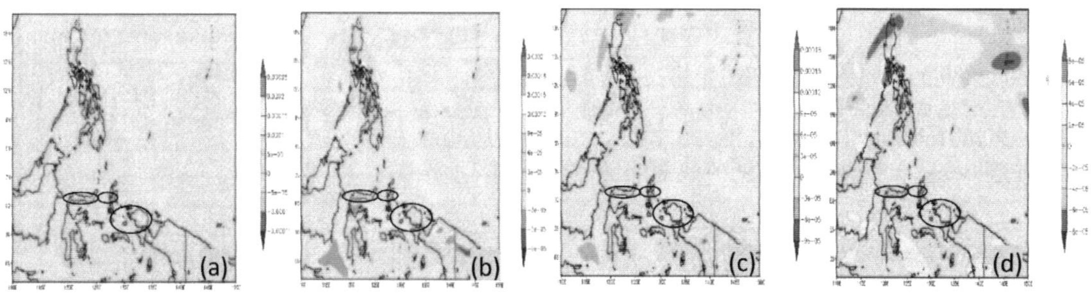


**Figure 12.** Relative Vorticity GrADS analysis results on 26 December 2016 at layers **(a)** 925 mb, **(b)** 850 mb, **(c)** 700 mb and **(d)** 200 mb. The areas of Gorontalo, Ternate, and Sorong are shown in black circles.

The tropical depression phase of Nock-Ten tropical cyclone is shown in Figure 12. At this stage, the relative vorticity condition indicates a positive value in the 925 mb layer, ie 5 x $10^{-5}$ $s^{-1}$, except for the Gorontalo region reaching 5 x $10^{-5}$ $s^{-1}$. This condition is also supported by layers of 850 mb and 700 mb, Sorong and Ternate regions show positive values about 6 x $10^{-5}$ -0 $s^{-1}$, except Gorontalo reach -3 x $10^{-5}$ $s^{-1}$. In the 200 mb layer, the relative vorticity conditions of the three regions show a value of 2 x $10^{-5}$ - (-2 x $10^{-5}$) $s^{-1}$. From the analysis results on 26 December 2016 it can be concluded that relative vorticity conditions in Sorong and Ternate move cyclonic, except Gorontalo region moves anticyclonic.

### *3.6. Rainfall Observation Analysis*

From the graph of rainfall accumulation, rainfall conditions detected in Gorontalo, Ternate and Sorong generally have the same pattern, except on 26 December having different intensity scale of rainfall. Maximum rainfall conditions generally occur on 26 December 2016, ie the tropical depression, except in the Gorontalo region experienced a decrease in intensity. The rainfall conditions during the early stage of cyclones in the three regions ranged from ≤ 10 mm/day. At the intensification stage, the three areas experienced an increase rainfall ranging from 15-35 mm/day, whereas mature stage generally low rainfall distribution ranges ≤ 10 mm/day.

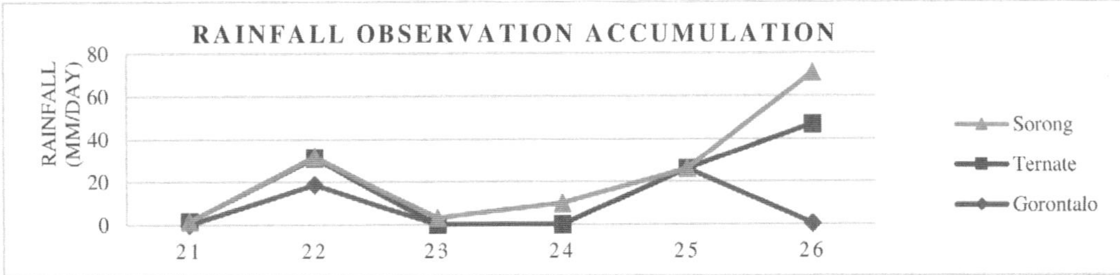


**Figure 13.** Average accumulated rainfall meteorological station of Gorontalo, Ternate, and Sorong on 21-26 December 2016.

## 4. Conclusions

Tropical Cyclone Nock-Ten events have an impact on atmospheric events in Gorontalo, Ternate and Sorong. Generally, at the early stage of cyclone growth, the three regions have fairly bright weather conditions with low rainfall intensity caused by the influence of warm air mass to the cyclone trajectory area and is supported by a low enough air mass potential increase and the anticyclonic movement. The intensification stage shows the three regions influenced by the cyclonic condition and the negative vertical velocity value and causing the shear line which contribute to slight-moderate scale rainfall. At the mature stage of tropical cyclones, the regions generally have bright conditions that are also supported by positive vertical velocity except in Sorong area which has slight intensity rain due to air masses in all three areas attracted towards the centre of the cyclone. In the tropical depression, the three regions have different scale of rainfall intensity. Seen from the wind pattern, the weather conditions of Gorontalo and Ternate areas are still affected by the cyclone, except for the Sorong region that is affected by the local scale factors indicated by negative value of vertical velocity. Generally, Gorontalo has clear weather condition which supported by anticyclonic condition in relative vorticity. Sorong and Ternate have moderate to heavy rain which supported by cyclonic movement.

## References

[1] Cabanerit J 2016 *Interaction between a shear line and the Inter Tropical Convergence Zone in the Northwestern Coast of Venezuela* (25 November 2010 Heavy Rainfall Event) (College Park, MD).
[2] Makmur E 2008 *Panduan Menggunakan GrADS untuk Pemula* (Jakarta: Pusat Klimatologi dan Kualitas Udara Badan Meteorologi dan Geofisika).
[3] Chen C, Yu Z, Li L and Yang C 2011 Adaptability Evaluation of TRMM Satellite Rainfall and its Application in the Dongjiang River Basin *Procedia Environmental Sciences* **10** pp 396-402.
[4] Ooyama K V 1982 Conceptual Evolution of the Theory and Modeling of the Tropical Cyclone *Journal of the Meteorological Society of Japan* **60** pp 369-380.
[5] Renggono F, Syaifullah M D 2011 Kajian Meteorologis Bencana Banjir Bandang Di Waisor Papua Barat *Jurnal Meteorologi dan Geofisika* **12** pp 33-41.
[6] Syifulllah M D and Nuryanto S 2016. Pemanfaatan Data Satelit GMS Multikanal untuk Informasi Perawanan dalam Rangka Mendukung Kegiatan Teknologi Modifikasi Cuaca *Jurnal Sains dan Teknologi Modifikasi Cuaca* **17** pp 49-57.
[7] Yokoyama C, Takayabu Y N 2008 A Statistical Study on Rain Characteristics of Tropical Cyclones Using TRMM Satellite Data *American Meteorological Society* **136** pp 3848-3862.

# Application of Landsat 8 OLI Image and Empirical Model for Water Trophic Status Identification of Riam Kanan Reservoir, Banjar, South Kalimantan

**A N Saputra[1*], P Danoedoro[2] and M Kamal[3]**

[1]Remote Sensing Masters Program, Faculty of Geography, Universitas Gadjah Mada, Yogyakarta, 55281
[2]Department of Geographic Information Science, Remote Sensing Program, Faculty of Geography, Universitas Gadjah Mada, Yogyakarta, 55281
[3]PUSPICS, Faculty of Geography, Universitas Gadjah Mada, Yogyakarta, 55281

aswin.geografi@gmail.com

**Abstract.** Remote sensing has a potential for observing, mapping and monitoring the quality of lake water. Riam Kanan is a reservoir which has a water resource from Riam Kanan River with the area width of its watershed about 1043 km$^2$. The accumulation of nutrient in this reservoir simultaneously deteriorates the condition of waters, which can cause an increasingly growth of harm *micro algae* or *Harmful Algal Blooms* (HABs). This research applied *Carlson's trophic status index* (CTSI) at Riam Kanan Reservoir using Landsat-8 OLI satellite image. The Landsat 8 OLI image was recorded on 14 August 2016 and was used in this research based on its surface reflectance values. The result of correlation test shows that band 3 of the image as coefficient of chlorophyll-a parameter, channel 2 as coefficient of phosphate, and band ratio of SDT as coefficient of SDT. Based on the result of modelling using CTSI, the majority scale of CTSI score at Riam Kanan Reservoir is between 60 to70 in medium eutrophic class. The class of medium eutrophic at Riam Kanan Reservoir potentially emerges the threat both of the improvement of water fertility and the reduction of water quality. Improvement of the fertility is apprehensive since it can trigger an explosion of micro algae which will endanger the ecological condition at the area of Riam Kanan Reservoir.

**Keywords:** Landsat-8 OLI, Eutrophic, Carlson's trophic status index, HABs

## 1. Introduction

A rising of eutrophication at Riam Kanan Reservoir waters will cause another effect that is the growth of the potential of HABs (Harmful Algal Blooms) [1]. It is because of the rate of eutrophication, which keeps on rising; make the growth of the micro algae type of *Microcystis sp.* also increases [2]. Eutrophication level that always keeps rising every year does not only increase the potential of HABs, but also causes the negative effect on ecological condition of reservoir, such as massive death of millions fish because of ammonia poisoning. That case is not added yet by potential of danger when HABs is happened. The danger which will be emerged is not only to the ecology of the reservoir but also to the humans around the reservoir who use it for their daily life. Thus, it is important to know the distribution level of eutrophication that potentially happens to HABs at Riam Kanan Reservoir waters.

An increasing of nutrient in the reservoir comes from some sources. Generally, the resource of the nutrient comes from inlet area of the reservoir, while the source of water of the reservoir comes from inlet area that consists of some stream branches. The potential source on increasing the nutrients and pollution in the reservoir are waste from the population, industry, agricultural, and livestock.

The use of remote sensing image to observe waters quality especially for eutrophication study and *Harmful Algal Blooms* (HABs) to map the distribution has been widely used [3][4][5][6]. The previous research concerning of HABs was done more on using image with higher spectral resolution on lower spatial resolution. Observation of waters land quality needs remote sensing image with bigger spatial resolution. The use of Landsat-8 OLI is tended to fulfil the need.

## 2. Methodology

This research was done at Riam Kanan Reservoir which has the wide 9.200 ha on flood water, and 1200 million $m^3$ volume on normal water. Riam Kanan Reservoir is a reservoir which has a water resource from Riam Kanan River with the wide of watershed is 1043 $km^2$. Riam Kanan Reservoir or Pangeran Mohammad Noor Reservoir is located in Aranio sub district, Banjar Regency, South Kalimantan. In this research, the data which was used was 1T level data that has been geometrically corrected automatically and the atmosphere disturbance has been also removed. Image conversion as the TOA reflectance value was done based on algorithm which was issued by United States Geological Survey (USGS), then was followed by relative atmospheric correction using the dark object subtraction (DOS) to generate image with surface reflectance value. The next process was image masking process which was limited to only cover land area and masking area based on the range of the pixel value on water and land as well by utilizing a range value of transformation of NDVI.

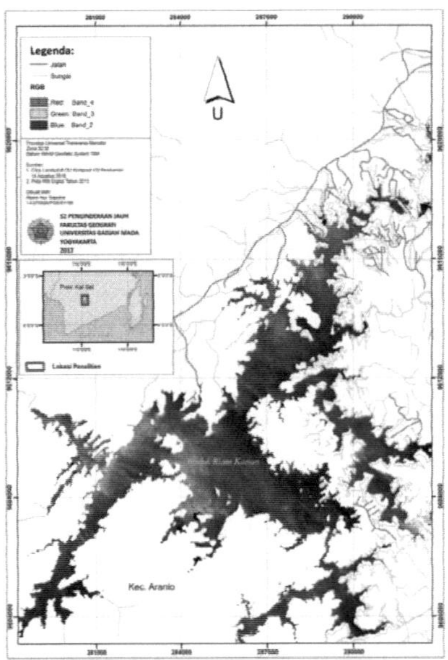


**Figure 1**. Research area

The result of image on masking then was being extracted by pre field primary data. Data extraction was done through the processing of digital image using spectral transformation with band ratio method. Band ratio method was used algorithm that has been generated by previous research on each parameter to be measured in the field which included transparency of Secchi Disk [7], that is:

$$\ln(SDT) = B1/B3(1.3083) + B1(-0.0631) + (-1.2937) \quad (1)$$

Clorophyll-a with algorithm [8] :

$$\ln(Chl) = 9.82(TM1-TM3)/TM2 \quad (2)$$

and algorithm for total of phosphor [9] :

$$\ln(TP) = -21.45(TM3/TM2)-14.42(TM1/TM3)+42.99(TM1)+27.1 \quad (3)$$

Where:
ln(SDT) = Natural logarithm from the depth of *Secchi Disk* transparency
ln(Chl) = Natural logarithm from chlorophyll-a
ln(TP) = Natural logarithm from total of phosphor
TM1 = Reflectance value band 1Landsat-5 TM
TM2 = Reflectance value band 2 Landsat-5 TM
TM3 = Reflectance value band 3 Landsat-5 TM

The result of transformation of band ratio would then be overlaid in which the result was used as a tentative map of the research and was used to determine the distribution of the sample. The sample was taken by stratified random sampling method based on classification of overlay result.

Furthermore, the score of content on each parameter in the field would be tested by accuracy test of the score from resulted image. Besides using the resulted image of band ratio between parameter of trophic status, image and real band were also included to be correlated to the score in the field. The highest input correlation then was tested by accuracy test of content score on each parameter in the field. Based on accuracy test above, then regression test was done between content score in the field and image score for spatial modelling to be done empirically. The image result of empirical modelling then was combined based on trophic status index by [10] so that it was obtained the trophic status at Riam Kanan Reservoir.

## 3. Results and Discussion

### *3.1. Statistical Analysis of Trophic Parameter*

Empirical modelling was done by using statistical analysis between the pixel value of image on a tentative map parameter and the result of field measurement. Statistical analysis was used linear regression analysis. The result of sampling in the field then tested in the laboratory for the parameters of total phosphor and chlorophyll-a. The result of parameter in the laboratory and in the field then was converted into natural logarithm. The distribution for each sample was tested using normality test of sampling then was correlated between sample score and pixel value by using SPSS software.

In the correlation which was done to chlorophyll-a, the most important predictor was on band 3 ($r = 0{,}46$) compared to alogarithm on initial estimation. Whereas for SDT, there were three predictors which have the same important correlation ($r = 0{,}28$); however, predictor alogarithm 2 was taken as an initial alogarithm in spectral transformation to estimate the initial distribution. On phosphate, the most influential predictor was on band 2 ($r = 0{,}48$) compared to alogarithm on initial estimation.

Accuracy test was also done between content score of chlorophyll-a (mg/L), phosphate (mg/L), and SDT (m) in the field and the predictor which has the highest correlation on each parameter. Based on the accuracy test, it is obtained that the score of chlorophyll-a RMSE is higher than the score of other RMSE.

**Table 1.** Table accuracy test of trophic parameter

| **Parameter** | **RMSE** |
|---|---|
| Chlorophyll-a | 0.011760836 |
| Phosphate | 0.529959 |
| Transparency of *Secchi Disk* | 1.13172707 |

Source: Analysis of Data (2016)

IOP Conf. Series: Earth and Environmental Science **98** (2017) 012020 doi:10.1088/1755-1315/98/1/012020

Then, on the result of linear regression statistical analysis of data showed that there was a low determinacy coefficient between the result of field data for chlorophyll-a and reflectance value on the green band of Landsat-8 OLI image. This was possible since water of reservoir has already mixed with the rainwater. In addition, it was also possible since the peak of productivity of micro algae in the water reservoir has already passed.

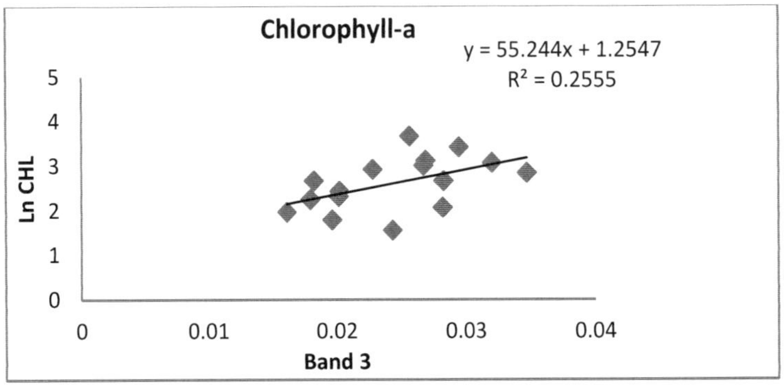


**Figure 2.** Graphic of linear regression of chlorophyll-a parameter
Source: Analysis of Data (2016)

Regression and correlation on phosphate on Figure 3 were compared to chlorophyll-a showed that there was an increasing of phosphorus content which was compared to a decreasing of chlorophyll-a. There was a possibility that the peak of micro algae productivity occurred between 14 August (image recorded) and 15 September (field measurement) then the number of micro algae declined. This possibility was strengthening by the result of a high regression (R) and determinacy ($R^2$).

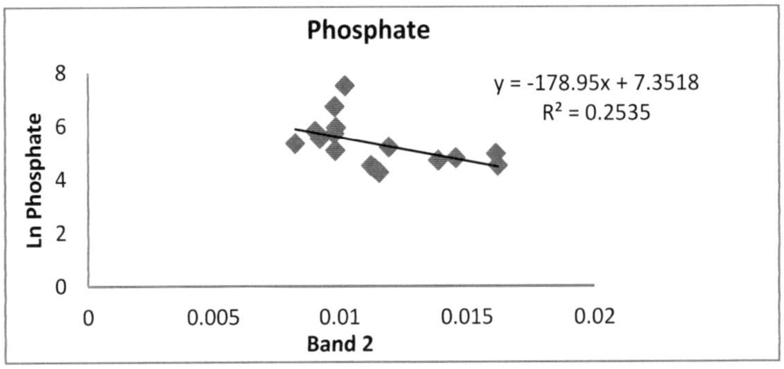


**Figure 3.** Graphic of linear regression of phosphate parameter
Source: Analysis of Data (2016)

The result of regression on SDT as shown in Figure 4 showed the level of sun translucency in water that relatively correlated with regression result and analysis of phosphate regression. Therefore, there was a possibility of low connection of reservoir permeability and the level of total phosphor in the water of reservoir.

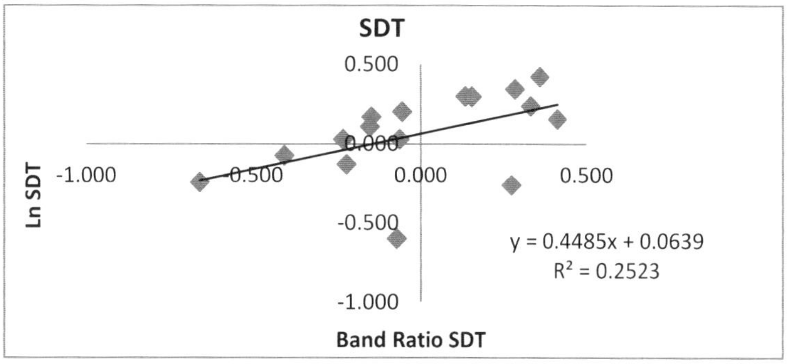


**Figure 4.** Graphic of linear regression of SDT parameter
Source: Analysis of Data (2016).

### *3.2. Analysis of Carlson Trophic Status*

The result of regression which was done on each parameter, then was being modelled for trophic status of reservoir water by using trophic status index by [10] that was developed based on 3 parameters, they are chlorophyll-a, total of phosphor, and transparency of *Secchi Disk* with the formulas as follows:

a. TSI for water transparency

TSI (SD) = 60-14,41 Ln *Secchi depth* (m) (4)

b. TSI for total of phosphor

TSI (TP) = 14,42 Ln *Total Phosporus* (µg/L) + 4,15 (5)

c. TSI for chlorophyll-a

TSI (CA) = 9,81 Ln *Chlorophyll a* (µg/L) + 30,6 (6)

*Carlson's Trophic State Index*

(CTSI) = [TSI (SD) + TSI(TP) + TSI(CA)]/3 (7)

The calculation of Carlson's trophic status index used the image of empirical modelling result to obtain the trophic status map of Riam Kanan Reservoir. The resulted map is the combination of 3 parameters, they are chlorophyll-a, phosphate, and SDT. The score which is obtained from the calculation index, results the majority of scale score between 60 and 70. The average scores which are obtained from the calculation are fit to trophic status class that are explained in Table 2.

**Table 2.** Trophic Classification Based on Carlson Trophic Index (1977)

| Score | Trophic Status | Explanation |
|---|---|---|
| <30 | Ultraoligotrophic | The fertility of waters is very low. The water is clear, has a high soluble oxygen concentration along year, and reaches hypo limnion zone. |
| 30 – 40 | Oligotrophic | The fertility of waters is low. The water is clear, and there might be anoxic limitation on hypo limnetic zone periodically (DO=0). |
| 40 – 50 | Mesotrophic | The fertility of waters is medium. The brightness of the water is in moderate condition, there is an increasing of anoxic changing in hypo limnetic zone, aesthetically it still supports for water sport activities. |

| | | |
|---|---|---|
| 50 – 60 | Light Eutrophic | The fertility of water is high. Decreasing in water brightness, hypo limnetic zone is anoxic, there are problems of water plant, only fish that can live in warm water, it supports water sport activities, but needs a treatment. |
| 60 – 70 | Medium Eutrophic | The fertility of water is high. Dominated by blue green algae, a clumping is occurred and water plant's problems are occured extensively. |
| 70 – 80 | Heavy Eutrophic | The fertility of water is high. Heavy *blooming algae* are occurred and water plant forms a layer, such as hyper eutrophic condition. |
| >80 | Hyper eutrophic | The fertility of water is high. Algae clumping are occurred, the death of fish is frequently occurred, water plant is slightly dominated by algae. |

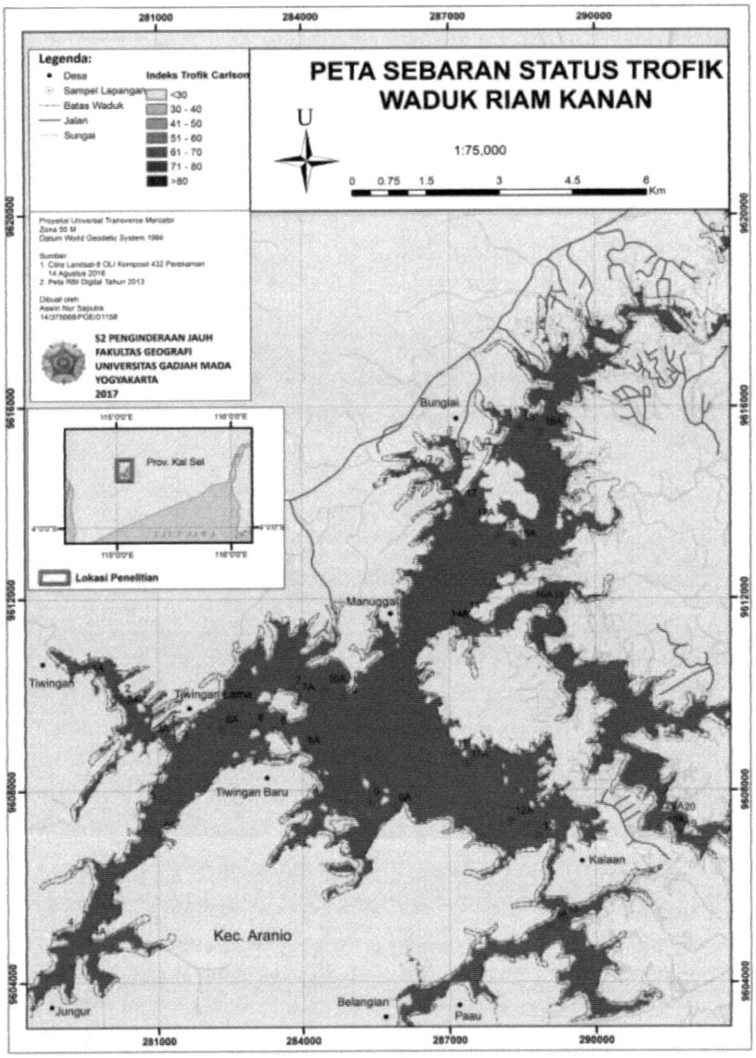


**Figure 5.** The Distribution of trophic status at Riam Kanan Reservoir

Based on the score that resulted by adjusting to the table of Carlson trophic classification, it is obtained that trophic status at Riam Kanan Reservoir is in medium trophic status. According to [10] area of water that has a medium trophic status is categorized as a high fertility area. This area of water has been dominated by blue green toxic micro algae. The dominated of micro algae has been very extensive and the clumping has been also occurred. The area of water in this condition has a high potential of the explosion of harmful micro algae population or *Harmful Algal Blooms* (HABs).

## 4. Conclusions

Based on the result of calculation by using Carlson's trophic status index, it is obtained that generally; the water condition at Riam Kanan Reservoir is in medium eutrophic status. Trophic status improvement which is occurred at Riam Kanan Reservoir is potentially cause a threat of water fertility improvement as well as water quality reduction. It is feared that the improvement of water fertility can trigger an explosion of micro algae population that will be very harmful to the ecological condition in the reservoir area.

Through the use of method and process which are used in the research, it is also known that the use of band ratio is not always has a high correlation on particular water area. The score of *at surface reflectance* on image has the similarity score to the score of trophic parameter content in the field. Therefore, the use of band ratio can be only used as an approach to the initial estimation of each parameter sampling in the field.

## References

[1] Effendi H 2003 *Telaah kualitas air: Bagi Pengelolaan Sumber Daya Dan Lingkungan Perairan* (Jakarta: Kanisius).

[2] Brahmana S S, Y Summarriani and F Ahmad 2010 Kualitas Air dan Eutrofikasi Waduk Riam Kanan di Kaliman Selatan *Prosiding Seminar Nasional Limnologi V* Puslitbang Sumber Daya Air, Bandung.

[3] Matthews M W, S Bernard and K Winter 2010 Remote sensing of cyanobacteria-dominant algal blooms and water quality parameters in Zeekoevlei, a small hypertrophic lake, using MERIS *Remote Sensing of Environment* **114**(9) pp 2070-2087.

[4] Stumpf R P and M C Tomlinson 2007 Remote sensing of harmful algal blooms In *Remote sensing of coastal aquatic environments* pp 277-296 Springer Netherlands.

[5] Kahru M, O P Savchuk and R Elmgren 2007 Satellite measurements of cyanobacterial bloom frequency in the Baltic Sea: interannual and spatial variability *Marine Ecology Progress Series* ***343*** pp 15-23.

[6] Barale V, J M Jaquet and M Ndiaye 2008 Algal blooming patterns and anomalies in the Mediterranean Sea as derived from the SeaWiFS data set (1998–2003) *Remote Sensing of Environment* **112** (8), pp 3300-3313.

[7] Kloiber S M, P L Brezonik, L G Olmanson and M E Bauer 2002 A procedure for regional lake water clarity assessment using Landsat multispectral data *Remote Sensing of Environment* **82**(1) pp 38-47.

[8] Brivio P A, C Giardino and E Zilioli 2001 Determination of Chlorophyll Concentration Changes in Lake Garda Using an Image-Based Radiative Transfer Code for Landsat TM Images *International Journal of Remote Sensing* **22**(2-3) pp 487-502.

[9] Wu C, J Wu, J Qi, L Zhang, H Huang, L Lou and Y Chen 2010 Empirical Estimation of Total Phosphorus Concentration in The Mainstream of The Qiantang River in China Using Landsat TM Data *International Journal of Remote Sensing* **31**(9) pp 2309-2324.

[10] Carlson R E 1977 *A Trophic State Index for Lakes* (Minneapolis: Limnology Research Center, University of Minnesota).

[11] Badan Pusat Statistik Kabupaten Banjar 2008 *Kabupaten Banjar Dalam Angka* Kabupaten Banjar.

[12] Department of Interior 2015 *Landsat 8 (L8) Data Users Handbook Version 1.0* (South Dakota: USGS).

[13] Dillon P J and Rigler F H 1974 The phosphorus-chlorophyll relationship in lakes *Limnol. Oceanogr* **19**(5) pp 767-773.
[14] Irianto W Eko and Anong Sudarna 1996 Karakteristik Beban Pencemaran Limbah Penduduk di Bandung dan Yogyakarta *Buletin PUSAIR, Media Kegiatan Penelitian Keairan (V)* No: 21 pp 15-35.
[15] OECD 1982 *Eutrophication of Waters: Monitoring, Assesment, and Control*. Organization for Economic and Co-operative Development, Paris, France.
[16] Prasad A G D and Siddaraju 2012 *Carlson's Trophic State Index for The Assessment of Trophic Status of Two Lakes in Mandya District. Pelagia Research Library,* Department of Environmental Sciences, University of Mysore, Karnataka, India, 3 pp 2992-2996.

# Flood Disaster Analysis Using Landsat-8 and SPOT-6 Imagery for Determination of Flooded Areas in Sampang, Madura

**B M Sukojo* and F Alfiansyah**

Departement of Geomatics Engineering, Faculty of Civil Engineering and Planning, Sepuluh Nopember Institute of Technology (ITS) Surabaya, Indonesia

bangunms@gmail.com

**Abstract**. Based on data of disaster which is defaced by Badan Penanggulangan Bencana Daerah (BPBD) of Sampang that in the period of 2015 - 2017 as many as 25 cases from 31 cases of disaster caused by flood disaster or 80.65% from total disaster. Therefore, the purpose of this research is to create a map of flood vulnerability in Sampang. From the vulnerability map, we can know the area with the impacted flood level in Sampang so that from the map of flood affected areas can be known the extent of the affected area in each class. In this study, two Landsat-8 and SPOT 6 data were used. For Landsat-8 imagery used for land cover on district level disaster level vulnerability maps, while high-resolution SPOT-6 images were used for land cover making maps of flood affected areas Sampang district. With the flood affected areas in this study, it is expected to be used as a determinant of flood affected areas in Sampang district. Based on data processing and analysis it is found that the highest impacted area is located in Sampang district with 12 cases of 17 cases of total flood disaster in Sampang district based on data from BPBD Kabupaten Sampang in 2016. There are 4 classes of flood affected areas in Sampang district i.e. not affected by 9039,540 ha, low impact 46262.881 ha, medium impact 43012.431 ha and high impact of 14009,760 ha.

**Keywords:** Flood, Landsat-8 Image, SPOT-6 Image, Remote Sensing, Geographic Information *System*

## 1. Introduction

Definition of Disaster according to Law Number 24 Year 2007 on Disaster Management mentions the definition of Disaster is an event or series of events that threaten and disrupt the life and livelihood of the community caused by both natural factors and / or non-alam factors or human factors resulting in the occurrence of human lives , Environmental damage, property loss, and psychological impact [1].

In Indonesia there are frequent disasters, one of which is the flood. Flood is an event or condition in which an area or land is inundated due to an increased water volume [2]. In general, floods occur in the rainy season. This is caused by the intensity of rain water is very high and lasted for a long time and the river cannot accommodate again so that the river has increased intensities water debit, consequently river water overflows and impacts on settlements or agricultural areas of citizens.

Sampang Regency is a district with high flood vulnerability. Every year floods occur not only during the rainy season but during high tide. According to the Regional Disaster Management Agency, one of the flood-prone areas in East Java. Floods that occur can cause harm to the community and local government.

The purpose of this research is to utilize Remote Sensing and Geographic Information System in determining flood vulnerability in Sampang Regency with the method used is scoring and weighting method. With the map of flood prone levels, it can be known areas with high levels of vulnerability. Referring to the disaster data of BPBD Kabupaten Sampang as a booster data, information obtained from the area affected by flood is mostly located in Sampang district. Furthermore, the analysis of flood-affected areas using high-resolution SPOT-6 images, rainfall data, altitude, slope type of soil and river buffer as a determinant variable of flood, expected to obtain the results of detail and accurate impact areas so that it can be used as the basis for determining the region Affected by the flood in Sampang district.

## 2. Materials and Methods

The location of this research is in Sampang Regency. Astronomically, Sampang District is located between 70 10 "- 70 20" South Latitude and 1130 13'- 1130 23 'East Longitude.

### *2.1. Data Used*

The data used in this research such are:

a. Landsat 8 OLI WRS imagery with *Surface Reflectance* correction date off acquisition 21th August 2016 path/raw 116/65. (Source: https://espa.cr.usgs.gov)
b. SPOT-6 ortho image with spatial resolution 1.5 m, district Sampang (Source: LAPAN)
c. Indonesian Hypsography maps scale 1:25.000 (Source: http://tanahair.indonesia.go.id)
d. Sampang Rainfall data on 2016 (Source: BMKG Karangploso Malang)
e. Soil Type maps Sampang scale 1:100.000 (Source: BPTP Karangploso Malang)
f. Administrative maps of Sampang (Source: Bappeda Kabupaten Sampang)
g. Data Flood Sampang (Source: BPBD Kabupaten Sampang)

1.*Equipment used*

Equipment used in this research such as:

1. Hardware
   a. *GPS Handheld*
   b. *UAV*
   c. Laptop
2. Software
   a. Satellite Imagery Processing Software
   b. Geospatial processing Software
   c. Word and number Processing software

*Data Collection Stages*: Data used in this research are Landsat 8 OLI WRS imagery with *Surface Reflectance* correction date off acquisition 21th August 2016 path/raw 116/65 from this data will be made a net design for measurement of GCP point in field and sampling location.

### *2.2. Stages of Data Processing*

*2.2.1 High Resolution Satellite Image Processing.* High resolution satellite imagery is done by making nets to find the SoF value, then test the accuracy with rectification process using image processing software to know the value of accuracy of image with RMSE requirement of ≤1,5 pixel.

*2.2.2 Land Use Classification.* Land use obtained from Landsat 8 and SPOT-6 image processing. The classification method used is the Supervised Maximum Likelihood Classification. To test the accuracy of the classification results used matrix confussion.

*2.2.3 Rainfall Data Processing Station.* The rainfall map is obtained from the interpolation of Rainfall using the Inverse Distance Weighted Method (IDW) then classified and scored the classification.

*2.2.4 Processing Altitude and Slope.* Tilt Height and slope maps are obtained from the interpolation of contour RBI data and then classified and scored

*2.2.5 Type of Soil.* From georeferenced soil map was to classify type of soil with digitation, after that score was given from classification data.

*2.2.6 River Buffer.* River Buffer map was obtained from buffer process, after that scoring is given.

*2.2.7 Scoring.* Scoring on each variable in this research, based on some references and previous research, but with many improvements like study area. There are table of scoring for each variable:

**Table 1.** Scoring Parameter Determination of Flooded Area

| **Parameter** | **Skor** | | | | | **Weight** |
|---|---|---|---|---|---|---|
| | 9 | 7 | 5 | 3 | 1 | |
| **Slope (%)** | 0-8 | 8-15 | 15-25 | 25-40 | >40 | 0,02 |
| **Height (m)** | 0 - 25 | 25 - 50 | 75 - 100 | 125-150 | >150 | 0,01 |
| **Rainfall(mm)** | 51-75, 76-100, 101-125, >125 | 26 –37, 38 - 50 | 13 – 25 | 6 - 12 | 0 - 5 | 0,02 |
| **Soil Texture** | Very Smooth | Smooth | Medium | Heavy | Very Heavy | 0,01 |
| **Rivers Buffer (m)** | | 0-50 | 50-100 | 100-150 | | 0,02 |
| **Land Use** | Water Area | Fields | Residence | Vegetation | Forrest | 0,02 |

Source: Reference and Previous Research

*2.2.8 Weighting and Overlay.* Determine flooded potention, is use that equation (Suhardiman 2010).

$$K = \Sigma_{i=1} W_i X_i \quad (1)$$

Explanation:
- K = potential
- Wi = weight for parameters i
- Xi = Score Class Parameter i

*2.2.9 Flood Vulnerability Mapping On Distric.* While creating map of flood vulnerability areas, must calculate interval for each class with calculation of total wight with the score of each flood causes variable. Referring to Kingma 1991 in Purnama, 2008, using with the formula

$$Ki = \frac{Xt - Xr}{k} \quad (2)$$

Explanation:
- Ki=Interval Class
- Xt= maximum value
- Xr=minimum value
- K=Number of classs

*2.2.10 Flood Vulnerability Mapping On SubDistric.* From the result of Flood Vulnerability Maps on distric Scale, it can be seen where the region with the highest vulnerability level. Using hight

IOP Conf. Series: Earth and Environmental Science **98** (2017) 012021 doi:10.1088/1755-1315/98/1/012021

resolution imagery SPOT-6, map can be made map of Flood-Affected Area in high risk areas with data from BPBD Sampang.

*2.2.11 Vegetation Indeks Map.* Vegetation Indeks Maps made from Normalized Difference Vegetation Index (NDVI) algoritm.

*2.2.12 Reflectant Value between Image Data Test.* Calculate reflectant value, with correlation test between image, Landsat-8 Imagery and SPOT-6 imagerywith ratio of NDVI value on each RGB value image.this is formmula for correlation according Sugiyono (2008) :

$$r = \frac{n\,(\sum xy) - (\sum x)(\sum y)}{\sqrt{[n(\sum x2) - (\sum x)^2\,][n(\sum y2) - (\sum x)^2\,]}} \quad (3)$$

Where :

- r = Correlation of variable
- x = independent variable
- y = dependent variable
- n = Number of observation

## 3. Results

### *3.1. RMS Error accuration*

To calculate RMSeror and SoF value, a net design and GCP distribution was made.

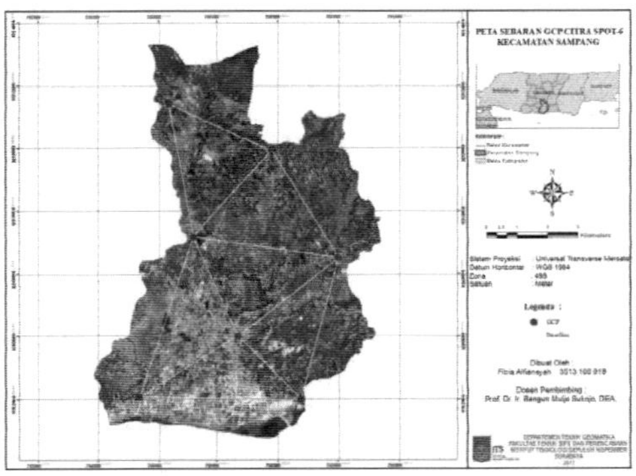


**Figure 1.** GCP distribution map SPOT-6 Imagery of Sub-distric Sampang

| | | |
|---|---|---|
| Sof Calculation: | | |
| Number of *baseline* | : 18 | |
| Number of Point | : 7 | |
| N_size | : Number of *baseline* x 3 | = 54 |
| N_parameter | : Number of point x 3 | = 21 |
| u | : N_size – N_parameter | = 33 |

$$SOF = \frac{Trace\ ([A^T].[A])^{-1}}{u} = 0{,}2242 \quad (4)$$

From the calculation of SoF value is 0.2242. Where the smaller number of network strength factors mentioned above, the better the network configuration [3].

GCP coordinate from measurement converted to point SRGI by using calculation of Least Square Adjustment then that coordinate used for rectification process using image processing software until got required RMSE <1,5 pixel, that is equal to 1,395 pixels. So, from these results, the image of SPOT-6 is acceptable.

### *3.2. Classification Accuration Test*

From result of classification with Supervised classification, obtained accuracy test with confusion matrix at Landsat 8 equal to 74.31%, while in SPOT-6 equal to 87.54%. The accuracy test in medium resolution image is ≥70% while in high resolution image is ≥85%. Thus, from the results of accuracy tests on Landsat 8 and SPOT-6 imagery was acceptable.

### *3.3. Flood Vulnerability Mapping On Distric Sampang*

The map of the flood vulnerability area in this study refers to research conducted by Primayuda, 2006 obtained from scoring, overlay and weighting of flood variables i.e. slope, altitude, rainfall, soil texture, river buffer and land cover. For the class used there are 4 classes. By using Landsat 8 image as land cover variable, we get the map of flood vulnerability as shown below:

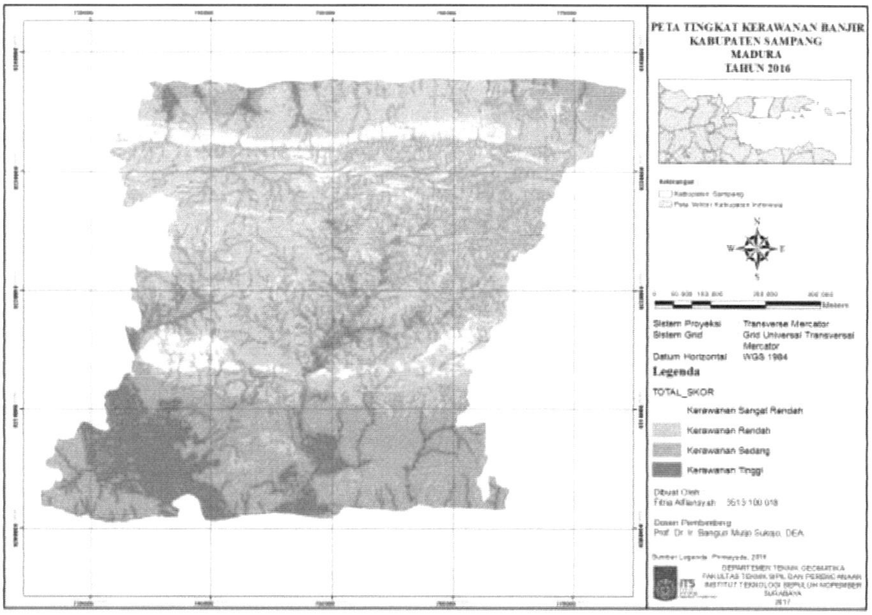


**Figure 2.** Flood Vulnerability Maps District Sampang

**Table 2.** Wide Flood Insectity of Sampang Regency

| No. | Class | Interval Class | Area (ha) | Area (%) |
|---|---|---|---|---|
| 1 | Very Low Vulnerability | 2,6 – 4,2 | 9039.540 | 7,39 |
| 2 | Low Vulnerability | 4,3 – 5,8 | 46262.881 | 37,82 |
| 3 | Medium Vulnerability | 5,8 – 7,4 | 53012.431 | 43,34 |
| 4 | High Vulnerability | 7,5 - 9 | 14009.760 | 11,45 |

According to disaster data of BPBD, the highest disaster-affected area is located in Sampang District with disaster data of 12 cases from 17 cases of flood disaster in 2016. In this study, the data is used as a reference to analyse the determination of flood affected areas.

## 4. Result of Flood Variable on Sampang District

In analysing the flood affected area in Sampang Sub district, 6 parameters of flooding area are determined as below.

### *4.1. Rainfall Station*

Rainfall data used the rainfall station data located in Sampang district in September 2016. From the interpolation results it can be seen that Sampang District has a very high rainfall. Classification and scoring refers to BMKG, 2013 with the results as in the following table:

**Table 3.** Precipitation

| No | Class (mm) | Information | Area (ha) | Area (%) |
|---|---|---|---|---|
| 1 | >125 | Heavy Rain | 7345.177 | 100 |

### *4.2. Hight and Slope on Sampang District*

The height affects the process of flooding. The lower the height of the area the greater the potential for flooding. Classification and scoring on the class height refers to previous research by [7] with the results as in the table below:

**Table 4.** Wide of Hight On Sampang Sub-District

| No | Class (m) | Area (ha) | Area (%) |
|---|---|---|---|
| 1 | 0 – 25 | 5766.369 | 78.44 |
| 2 | 26 – 50 | 1134.214 | 15.43 |
| 3 | 51 - 75 | 375.701 | 5.11 |
| 4 | 75 - 100 | 69.882 | 0.95 |
| 5 | >100 | 5.577 | 0.08 |

As for the slope class, Sampang district is also dominated by broad flat class of 83.3%. The results of the classification results are shown in the following table:

**Table 5.** Slope Area On Sampang

| No | Class (%) | Information | Area (ha) | Area (%) |
|---|---|---|---|---|
| 1 | 0 – 8 | Flat | 6160.357 | 83.80 |
| 2 | 8 – 15 | Choppy | 1062.089 | 14.45 |
| 3 | 15 - 25 | Wavy | 113.278 | 1.54 |
| 4 | 25 - 45 | Hilly | 14.895 | 0.20 |
| 5 | >45 | Steep | 1.018 | 0.01 |

### *4.3. Type of Soil*

With the result of classification of soil type map which refers to Primayuda, 2006 hence obtained of soil texture class shown in following table:

**Table 6.** Wide of Soiy Type on Sampang

| No | Class | Area (ha) | Area (%) |
|---|---|---|---|
| 1 | VerySmooth | 3393.06 | 46.21 |
| 2 | Smooth | 1033.32 | 14.07 |
| 3 | Medium | 382.64 | 5.21 |
| 4 | Heavy | 1033.32 | 14.07 |
| 5 | Very Heavy | 802.43 | 10.93 |

IOP Conf. Series: Earth and Environmental Science **98** (2017) 012021 doi:10.1088/1755-1315/98/1/012021

### *4.4. River Buffer*
River buffer criteria refer to Government Regulation of the Republic of Indonesia no. 38 Year 2011 About the River where, for the river line flood exposure at least 50 meters from the edge of the river water will be classification refers to Purnama (2006) with the results shown in the following table:

**Table 7.** Wide Of River Buffer on Sampang

| No | Class (m) | Area (ha) | Area (%) |
|---|---|---|---|
| **1** | 0-50 | 559.6570087 | 51.80 |
| **2** | 50-100 | 520.8443293 | 48.20 |

### *4.5. Land Use*
Land Use was derived from the classification result using the Supervised Maximum Likelihood classification method on SPOT-6 high-resolution satellite images. The classification refers to Primayuda, 2006 as the reference shown in the table below:

**Table 8.** Wide of Land Use on Sampang

| No | Class | Area (ha) | Area (%) |
|---|---|---|---|
| **1** | Water | 74.392 | 1.01 |
| **2** | Ponds | 550.948 | 7.49 |
| **3** | Rice Field | 2802.132 | 38.11 |
| **4** | Field | 1083.555 | 14.74 |
| **5** | Mangrove | 165.825 | 2.26 |
| **7** | Recidence | 1543.325 | 20.99 |
| **8** | Meadow | 38.823 | 0.53 |
| **9** | Garden | 1029.851 | 14.01 |
| **10** | Prod Forest | 63.175 | 0.86 |

### *4.6. Imapacted Flood Area Maps of Sampang*
Sampang's Flood Affected Area Maps was Map Sampang flood affected area is obtained from the overlay of variable rainfall, altitude, slope, rainfall, soil texture of river buffer and land cover. Referring to Primayuda, 2006 obtained classification of flood affected areas into 4 classes shown with the image below:

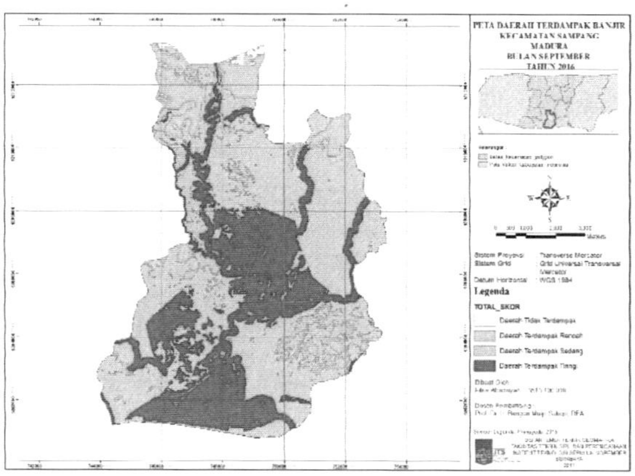


**Figure 3.** Sampang Flood Affected map on September 2016

From the results of the classification that has been done got the area of each affected area as the following table:

**Table 9.** Area Classification of Impacted Area of Sampang District Flood

| No. | Class | Interval Class | Area (ha) | Area (%) |
|---|---|---|---|---|
| 1 | Not Affected | 2,6 – 4,2 | 9039.540 | 7,39 |
| 2 | Low Affected | 4,3 – 5,8 | 46262.881 | 37,82 |
| 3 | Medium Affected | 5,8 – 7,4 | 53012.431 | 43,34 |
| 4 | High Risk | 7,5 - 9 | 14009.760 | 11,45 |

### *4.7. Vegetation Index Maps*

Vegetation Density Map is generated from NDVI algorithm which aims to determine the level of vegetation density. Referring to Regulation of the Minister of Forestry of the Republic of Indonesia No.P.12 / Menhut-II / 2012 resulted classification of NDVI into 5 classes, as shown below:

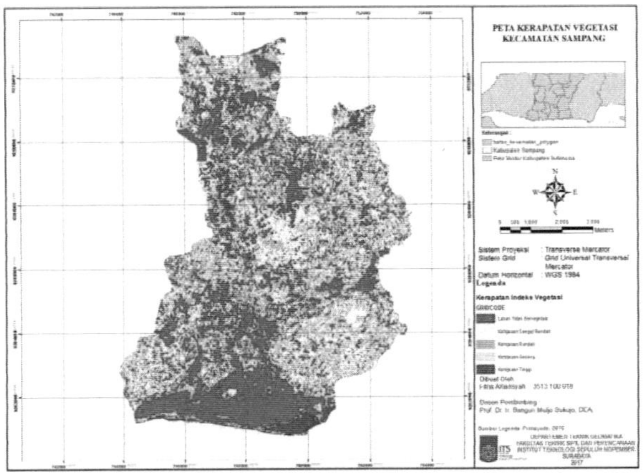


**Figure 4.** Vegetation Index maps

From the map, Sampang sub-district is dominated by non-vegetated land of 2297.55 ha. This can strengthen the causes of flooding in District Sampang due to lack of vegetated so that the lack of water absorption.

### *4.8. Reflectant Variable Data Result*

In this research, the test of reflectance value with correlation test on image, SPOT-6 image and Landsat -8 image by using ratio of NDVI value on each RGB value of image. The RGB value of the image is defined as the nearest wavelength in the two correlated images. It aims to find out how big the relation of reflectance value in the image data shown with the value of R that produced.

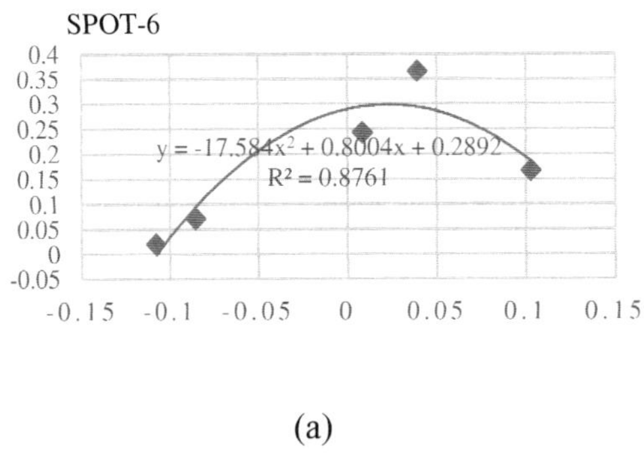


(a)

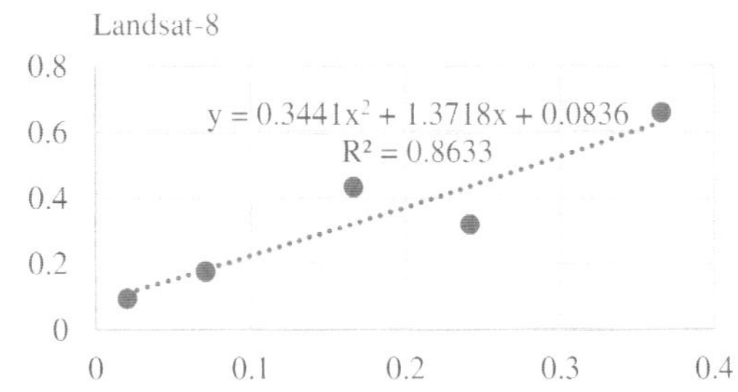


(b)

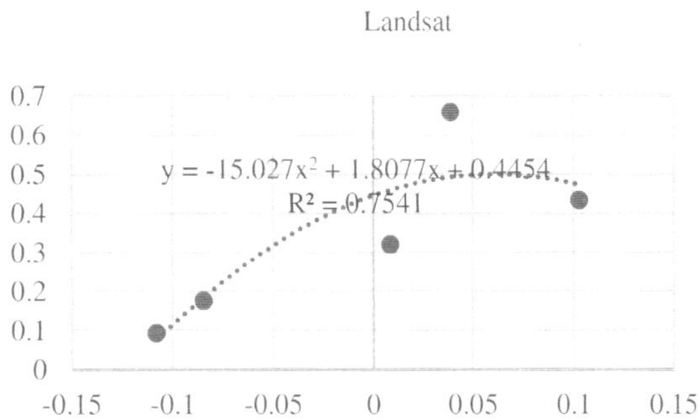


(c)

**Figure 5.** NDVI correlation value, (a) SPOT-6 imagery with Photo, (b) Landsat-8 with SPOT-6, (c) Landsat-8 with photo

From Figure 3.4 above shows that the result of correlation of NDVI ratio value in figure (a) gives R2 value equal to 0.8761 with R equal to 0.6764, figure (b) yields R2 value equal to 0.8633 with R equal to 0.9289, while in figure (c) With R of 0.7984.

Referring to Sugiyono, 2007 for interpretation of Correlation Coefficient (R) in the interval range 0.8 - 1 has a very strong relationship level. So, the strongest correlation result is shown in figure (b) that is correlation of NDVI value between Landsat-8 with SPOT-6 equal to 0.8633. This is because the wavelength of the RGB between SPOT-6 and Landsat-8 has an approximate wavelength.

## 5. Conclusions

Based on the processing and data analysis that has been done, here are the conclusions that the authors get in the author of this thesis.

1. From the classification result, got 4 class area flood vulnerability level in Sampang Regency that is very low, low, medium and very high.
2. Sampang District is a sub-district that has high flood vulnerability and one of the flood affected areas in accordance with disaster data BPBD Kabupaten Sampang. From the classification of the affected areas, the total area was not affected by 9039,540 ha, the low impact was 46262.881 ha, medium impact 53012.431 ha, and the high impact was 14009.760 ha.
3. From the results of analysis of each parameter in the flood affected areas in Sampang district, it can be seen that the most dominant parameters are rainfall, altitude and slope.

## 6. Recomendation

The advice given from the results of this research are:

1. Preferably in rectifying high-resolution images using GCP coordinate data obtained from GPS Geodetic, this is done so that the RMSE results are smaller.
2. Need for further research in the form of flood risk modeling in Sampang Regency that can be used to anticipate in flood prevention.

## References

[1] Undang-undang Nomor 24 Tahun 2007 Tentang Penanggulangan Bencana. Indonesia.
[2] Cahyono, A T 2011 Perencanaan Pengendalian Banjir Kali Kemuning, Sampang. Surabaya: Teknik Sipil FTSP-ITS.Osazuwa, I. B. 1988.
[3] Abidin, H Z, Jones A and Kahar J 2002 *Survei dengan GPS* (Jakarta: Pradnya Paramita).
[4] Badan Informasi Geospasial 2016 *Modul Validasi Peta Rencana Tata Ruang* Jakarta.
[5] Badan Meteorologi Klimatologi dan Geofisika (BKMG) 2013 *Analisa Hujan Mei 2013 Prakiraan Hujan Juli,Agustus, dan September 2013* Karangploso Malang.
[6] Menteri Kehutanan. 2012 Peraturan Menteri Kehutanan Tentang Perubahan Kedua Atas Peraturan Menteri Kehutanan Nomor P.32/Menhut-Ii/2009 Tentang Tata Cara Penyusunan Rencana Teknik Rehabilitasi Hutan dan Lahan Daerah Aliran Sungai (Rtk RHL-DAS) Jakarta.
[7] Purnama A 2008 Pemetaan Kawasan Rawan Banjir di Daerah Aliran Sungai Cisadane Menggunakan Sistem Informasi Geografis *skripsi* Bogor: Fakultas Kehutanan, Institut Pertanian.
[8] Primayuda A 2006 Pemetaan Daerah Bahaya dan Resiko Banjir Menggunakan Sistem Informasi Geografis: studi kasus Kabupaten Trenggalek, Jawa Timur *Tugas Akhir* Bogor: Fakultas Pertanian, Institut Pertanian Bogor
[9] Peraturan Pemerintahan Republik Indonesia No. 38 Tahun 2011 Tentang Sungai. Indonesia

IOP Conf. Series: Earth and Environmental Science **98** (2017) 012022 doi:10.1088/1755-1315/98/1/012022

# Remote Sensing and GIS Application for Sedimentation Modeling in Porong River Estuary as an Impact of Lapindo Mudflow, Sidoarjo

**A P Firdausy*** **and Sudaryatno**

Cartography and Remote Sensing, Faculty of Geography, Universitas Gadjah Mada Sekip Utara, Bulaksumur, Yogyakarta 55281

andikaputrif@gmail.com

**Abstract**. Lapindo mudflow that has occurred since May 2006 in Porong, Sidoarjo have raised thorny issue. One of the consequences that turned up physically was the alteration that has occurred in Porong River estuary. Porong River estuary has undergone drastic morphological change since 2006 to 2016 that became interesting phenomenon studied using remote sensing imagery. Landsat 7 ETM+ and Landsat 8 OLI/TIRS images were used to observe the changes that occurred in Porong River estuary. Porong River estuary had additional land area of 149.4 Ha in 2000-2006, 93.79 Ha in 2006-2010 and 100.24 Ha in 2010-2016. This addition was largely due to the change of land use in the sub-watershed, input from Lapindo mudflow, as well as human activities in the development of a new delta called Sarinah Island. There were several algorithms that were used in the calculation of the sediment concentration value (SSC), such as blue, green, near infrared band, also Doxaran, NDSSI, and the red/near infrared band ratio. However, the SSC of pixel value did not have strong correlation with the SSC field value. The dynamic of object and accuracy of the algorithm became the most influential factor that must be considered in calculating the value of SSC.

**Keywords**: Sediment, Estuary, Porong River, Landsat, Lapindo

## 1. Introduction

Sedimentation is a process that definitely occurs in every watershed. Sedimentation occurs because of the precipitate of erosion materials and material transport from previous processes from upstream to downstream of a watershed. The disaster was caused by PT Minarak Lapindo Jaya drilling resulted in a burst of hot mud that spurted up to now. This disaster has occurred for more than ten years since 26 May 2006 around Porong Sub- district, Sidoarjo Regency, East Java. As a result of this disaster, the surrounding area was drowned by continuous mud out of the earth.

The modeling of Lapindo mudflow has been done just after the first overflowed [1]. However, the dynamics of bursts that were constantly changing over time required continuously updated modeling as the mudflows addition, also the morphological and ecological changes of the estuary. The development of modeling was done to observe the sediment growing pattern as input in the downstream planning of the watershed as the reference for the future cases.

Monitoring the type, quantity, and spatial distribution of sedimentary minerals is crucial. Sediments might affect water quality and suitability for drinking water, tourism, and industrial purposes. Suspended sediment might block the transmission of solar radiation and reduce the photosynthesis of underwater vegetation and underlying phytoplankton. In fact, underwater vegetation and phytoplankton play a huge role in the food chain of aquatic ecosystems.

Research on sedimentation using Landsat 8 OLI/TIRS imagery combined with GIS-based modeling has not been done in Indonesia, especially which located in Porong River estuary. In addition, the influence of sediment input from outside the watershed will be interesting object to be studied, in this case a flow of overflow of Lapindo mud sedimentation.

The aim of this research is to find out the distribution and sediment concentration in Porong River estuary before and after Lapindo mudflow with Landsat 5 TM, Landsat 7 ETM+, and Landsat 8 OLI/TIRS. This study also aims to model changes in the pattern of sediment distribution in the estuary of Porong River in the period of 2000-2016 by using geographic information system.

## 2. Research Method

Landsat images were processed by performing geometric, radiometric, and atmospheric correction. Image to image geometric correction has been done image between one image and another image. Dark Object Substract method or adjusting the histogram has been done for atmospheric correction.
The corrected image has been chosen as the sampling basis for the fieldwork. Samples of estuary water which contain sediment were taken by transect method. Thus, it has been laboratory tested to determine the value of sediment concentration (SSC) in mg/l.

The laboratory result of SSC value were correlation and regression tested with image pixel value, resulted regression coefficient. These values were chosen as the benchmark for obtaining SSC values in other images in various years. Meanwhile, images from year 2000, 2006, and 2010 had been inter-band calibrated first before applying to calculate value of SSC. The development of Porong River estuary has been spatially mapped by overlaying the images based on the years required, such as images 2000 and 2006, 2006-2010, and 2010-2016.

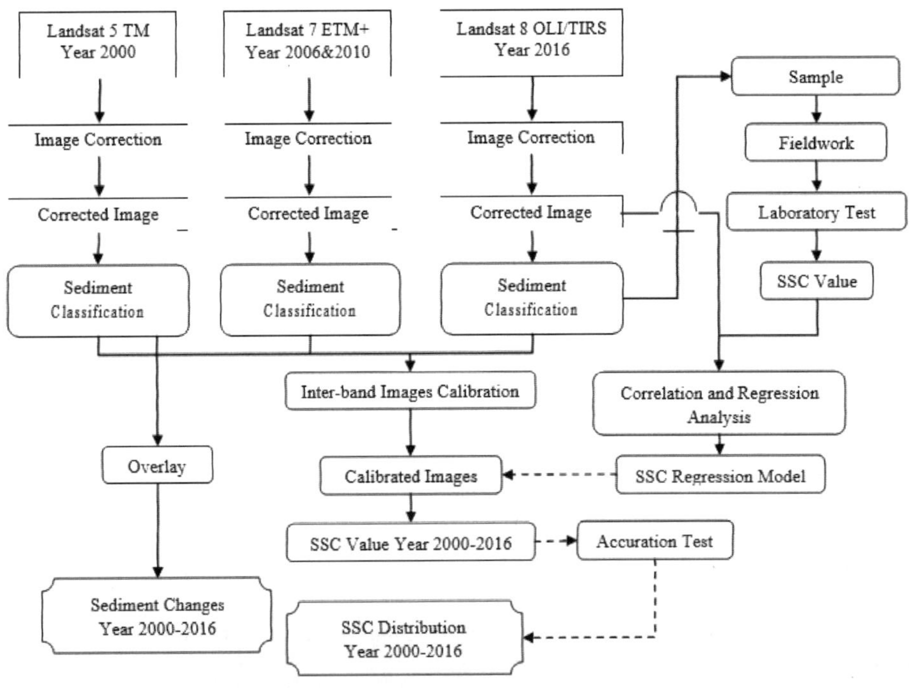


**Figure 1.** Flowchart of Method

## 3. Results and Discussion

Based on the results of field data retrieval, SSC values have been correlation and regression tested with the band values of the image. Correlation test has been used to find out the relationship between image pixel in each band and field SSC values. Several bands which have been used were blue, green, red, near infrared, and middle infrared. Meanwhile, band 1 in Landsat 8 or band blue stretching cannot be used because it would not be applied to Landsat images 5 and 7.

Based on the correlation test, it is known that each band has a relatively low correlation with the SSC value. The correlation reference is adjusted to the value of R (product moment) according to the number of samples. Thirty eight samples have been used in 5% significance resulted 0.320 for R value. The correlation degree has not been complied, so the band which has highest correlation degree was selected as the reference? Thus, it caused by the dynamic characteristic of sediment which has higher probability in both spectral and spatial biases. In addition, image spatial resolution can also affect the accuracy of research results.

Multitemporal images within spatial researches require inter-band calibration to align the spectral standard of image pixels. Thus, at different times, the same spectral response might have different recording objects, due to the different quality of satellites and recording conditions. Images calibration has been applied to tackle that problem.

Reference in calibration between image bands has been selected from one image that is considered to have appropriate resolution and is able to represent another image. Landsat 8 image has been used as a reference to calibrate the other images. This image has been chosen not because of the latest recording time, but it has been considered to choose the best spectral value among other images. In contrast, Landsat 7 image has a gap at its spectral value due to the offline of scan line corrector so as to interrupt the calibration process. Likewise with Landsat 5 image which has an inadequate spectral display and tends to blur.

The result of that calibration is applied to algorithms that have been built to obtain SSC values. The SSC value would represent the SSC value in the same location as the sample point at different times and images. The results of the best SSC values will be the basis for mapping the SSC values in the general assessment location during the appropriate recording year.

Based on the SSC values of the six algorithms, a correlation test has been applied between SSC pixel and SSC field to determine the strongest correlation. The strongest correlation algorithm is considered to represent the SSC value for mapping the SSC value classification. The correlation coefficient value of band 1 (blue) is 0.301895. The correlation coefficient value of band 2 (green) is 0.330227. The correlation coefficient value of band 4 (near infrared) is 0.04621. The correlation coefficient value of Doxaran band ratio is 0.093752. The correlation coefficient value of NDSSI band ratio is 0.060975. While the correlation coefficient of RIR band ratio is 0.086868. The coefficient value on green band is highest and more than R (n = 38) is 0.320. So by 2016, the algorithm of the green band is considered to be the most representative of SSC values. As for band ratio, Doxaran ratio has the highest coefficient value among others.

The results of SSC value algorithm implementation can be observed by several patterns. In contrast to the SSC value of 2016, the SSC value of 2000 on band 1 (blue) and band 2 (green) have the same characteristics, which are negative. This suggests that the algorithm did not succeed in building SSC values in 2000 because SSC is unlikely to be negative. In the other hand, SSC values based on band 4, Doxaran, NDSSI, and RIR ratios show a positive value. Although it is not definitely show the right value, but these four values are still appropriate if the value of SSC.

In general, the SSC value of field results has more varied distribution rather than the result of SSC logarithm. Thus, it could be happen because the SSC value of the algorithm result is mathematical so that the resulting value is not far from the normal regression line that had been formed. In contrast, the SSC value of field results is highly dynamic. Both have different angles of view so that high correlation and determination coefficients cannot always be found easily. Differences in the range of SSC values resulted in the analysis of the increase or decrease of SSC in each period cannot take place optimally.

The dynamics or changes in sedimentary patterns on the coast are a natural phenomenon occurring in an estuary or delta, including those occurring in the Porong River estuary. This change can occur due to several factors, such as the presence of sedimentary input from the river, the fluctuation of tidal and wave energy, and the presence of human influences. In certain circumstances, human activity may be the largest and fastest factor affecting the dynamics of the estuary or delta.

The dynamics of the estuary of Porong River are observed in three periods: 2000-2006, 2006-2010, and 2010-2016. In general, the changes that occur are the growing of coastal lines. Although there are some locations that suffered coastline retreat but the extent is not prominent. This change is natural and can happen anywhere. However, the presence of large additional sediment inputs from the Lapindo bursts since 2006 may have influenced the changes.

The change of estuary area of 2000-2006 was dominated by the change of territorial waters to the mainland in 2006, which was 1,494,364,624 m2 or equal to 149,4 Ha, as represented in Figure 2. This change is a considerable amount in coastal dynamics. This may be due to several factors such as significant land-use change and may also be derived from the influence of sediment input from the Lapindo mud since May 2006.

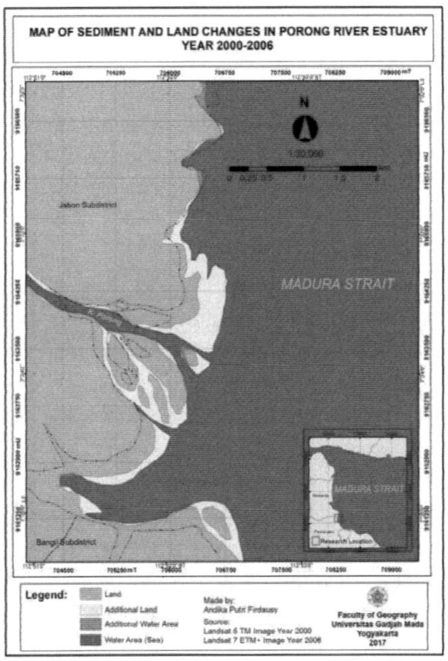


**Figure 2.** Map of Sediment and Land Changes Year 2000-2006

Changes in the area of estuary in 2006-2010 are more common in the newly formed delta or commonly called "Pulau Sarinah", as represented in Figure 3. The new area formed in the delta reaches tens of hectares. In addition to the influence of Lapindo mud input that flowed towards Kali Porong, the delta has undergone many changes due to efforts by the government to develop Sarinah Island. The delta area due to sediment deposition is increasing with the influence of human activity in it.

IOP Conf. Series: Earth and Environmental Science **98** (2017) 012022 doi:10.1088/1755-1315/98/1/012022

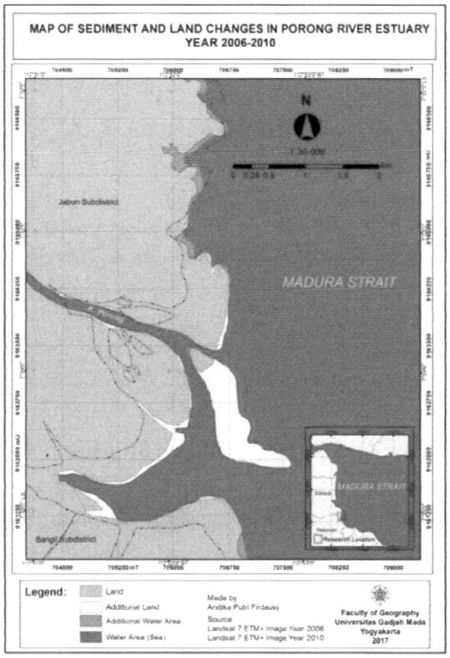


**Figure 3.** Map of Sediment and Land Changes Year 2006-2010

Changes in the Porong River estuary in 2010-2016 have increased again after in 2006-2010 is relatively lower than the year 2000-2006, which reached 100.24 Ha as represented in Figure 4. In these years the government has become more massive in handling the estuary of Porong River. The island of Sarinah that is formed today becomes a new icon that invites tourists. The island is used *wanamina* or silvo-fishery where it functions as mangrove forest on one side and pond on the other side.

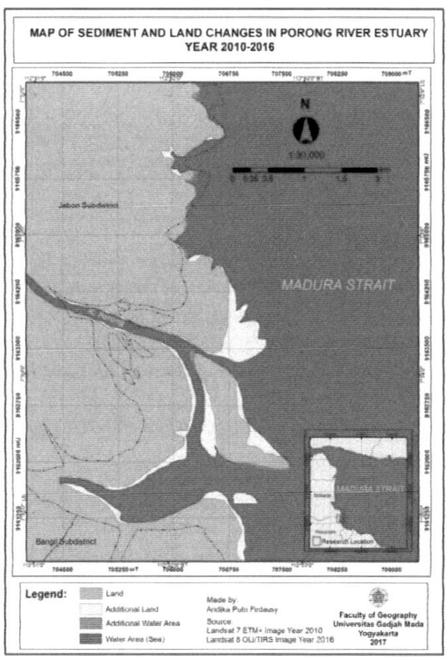


**Figure 4.** Map of Sediment and Land Changes Year 2010-2016

The addition of land area in the estuary of Porong River between 2000-2006, 2006-2010, and 2010-2016 has different patterns. The biggest change occurred between 2000 and 2006 with a total change of 149.4 Ha, then it changed in 2010-2016 amounted to 100.24 Ha, and the change in the year 2006-2010 amounted to 93.79 Ha. The change in 2000-2006 had the greatest change since land use change in early 2000 was considerable. This allows for more sediment input if the use of agricultural land is reduced.

## 4. Conclusion and Recommendation

The spatial distribution of sediment at Porong River estuary has undergone quite a drastic change since before the existence of Lapindo until afterwards. Some algorithms are used to calculate SSC values in 2000, 2006, and 2010 from the SSC value regression in 2016. However, the resulting SSC values did not have very strong correlation. Sediment concentration at Porong River estuary did not have a significant influence with Lapindo mud. In addition to the river flow is too far, also because the sediment concentration in the estuary is influenced by various factors that are dynamic in a short time, especially marine activities.

The development of sediment distribution and land area at Porong River estuary in 2000- 2006 was very large, as big as 149.4 Ha, followed by 100.24 Ha in 2010-2016, and 93,79 Ha in 2006-2010. Thus changes are affected by several factors such as land use change in the Porong River sub-watershed, input of Lapindo mud sediment, also marine activities. In addition, the government's efforts in developing the delta that has been formed also increase the area of the delta.

The large amount of sediment input from Lapindo mud which flow into the Porong River estuary resulted in the formation of a new delta, now called Sarinah Island. This new delta is one of the government programs to introduce mangrove and wanamina studies. Thus, it becomes a more separate value for the sustainability of ecosystems in the estuary of Porong River besides maintaining the balance of river flow and estuary, and so both government and the people should maintain it wisely.

## References

[1] Batubara B and P U Waluyo 2012 *Kronik Lumpur Lapindo Skandal Bencana Industri Pengeboran Migas di Sidoarjo* (Yogyakarta: INSIST).
[2] Arsyad S 2012 *Konservasi Tanah dan Air* (Bogor: IPB Press).
[3] Asdak C 1995 Hidrologi dan Pengelolaan Dearah Aliran Sungai (Yogyakarta: Gadjah Mada University Press).
[4] Chander et al 2009 Summary of Current Radiometric Calibration Coefficients for Landsat MSS, TM, ETM+, and EO-1 ALI sensors. *Remote Sensing of Environment* **113** pp 893-903.
[5] Danoedoro P 2012 *Pengantar Penginderaan Jauh Digital* (Yogyakarta: Andi).
[6] D Doxaran, J-M Froidefond and P Castaing 2002 A reflectance band ratio used to estimate suspended matter concentrations in sediment-dominated coastal waters *International Journal of Remote Sensing* **23**(23) pp 5079-5085.
[7] Jensen J R 2014 *Remote Sensing of The Environment: An Earth Resource Perspective* 2nd Edition (USA: Pearson).
[8] Kodoatie R J and S Roestam 2010 *Tata Ruang Air* (Yogyakarta: Penerbit Andi).
[9] Lillesand and Kiefer 1999 *Penginderaan Jauh dan Interpretasi Citra* Trans. Dulbahri. (Yogyakara: Gadjah Mada University Press).
[10] Ongkosongo Otto 2010 *Kuala, Muara Sungai, dan Delta* (Jakarta: Lembaga Ilmu Pengetahuan Indonesia).
[11] Sugiyono 2012 *Metode Penelitian Kuantitatif Kualitatif dan R&D* (Bandung: Alfabeta).

# Potential Land Mapping for Agricultural Extentification in Mengwi Sub-district to Support Food Balance in Badung Regency, Indonesia

**Ni Made Trigunasih[2*], Indayati Lanya[1], I.G.P Ratna Adi[1], JeremiaHutauruk[2], Feronika[2]**

[1]Study Program Agroecotechnology, Faculty of Agicultural, Udayana University, Indonesia
[2]Center for Spatial Data Infrastructure Development (PPIDS) UniversitasUdayana, Bali, Indonesia

tri5963@yahoo.com

**Abstract.** The availability of agricultural land for food crops, especially in Bali, is rapidly declining every year. The availability of rice fields in Badung regency, especially in Mengwi Sub-district until 2040 is no longer exist, this means that Mengwi Sub-district has lost the rice fields. The existence of land conversion will affect food availability for the country, so there will be food deficit. The food balance in Badung Regency in 2015 with Cultivation Index (IP) and initial productivity in each Sub-district showed a food deficit of 32,843.44 tons, then after increasing IP of 2,5 the productivity in Kecamatan Petang and Kuta at 7 tons / ha, and Abiansemal, Mengwi and North Kuta Sub-districts with 8 tons / ha which indicate a surplus in 2020 and 2030 respectively of 25,155.19 tons, and 3,401.79 tons. But in 2040 and 2050 there was a food deficiency of 18,434.78 tons and 11,824.82 tons respectively. Considering that productivity improvement efforts cannot rely solely on intensification approaches, but also need to be done with extensification or expansion of agricultural areas to support food production. This research was conducted in Mengwi Sub-district, Badung Regency. Mengwi Sub-district consists of 20 villages. The objectives of this research are: (1) to map potential land that can be converted to agricultural land of food crops, and (2) to know the amount of food demand to supply food balance in Badung Regency in 2040. Research methodology includes (1) preliminary study, (2) interpretation of satellite images, (3) mapping and measurement of land area, and (4) calculation of additional food availability. The results indicate that the potential land that can be converted to agricultural land for food crops is 132 ha, consists of 128.51 ha of mixed plantation and 3.49 ha of bare land/ bush. The result of additional land produced 1601.73 tons of rice that increased the food availability in Mengwi Sub-district to 45425.7 tons. The addition of surplus in 2040 in Mengwi sub-district amounted to 21669.23 tons but did not cover the food deficit in Badung regency in 2040.

**Keywords:** Extensification, Food Crop Agriculture, Food Balance

## 1. Introduction

The availability of agricultural land for food crops, especially in Bali, is rapidly declining every year. The availability of rice field in Badung regency, especially in Kuta Sub-district until 2040 is no longer exist, this means that Kuta has lost its traditional Subak rice field [1]. Subak is an irrigation organization in Bali that is very famous in the eyes of the world. The declining of the availability of land is not only caused by the increase in population, but also because of the conversion of cropland in relation to the widespread of urbanization [2]. The existence of land conversion will affect food availability for the country, so there will be food deficit. Food balance in Badung Regency in 2015 with Cultivation Index (IP) and initial productivity in each sub-district showed a food deficit of 32,843.44 tons, then after increased IP to 2,5 the productivity in Petang Sub-district and Kuta Sub-district for 7 tons / ha, and Abiansemal, Mengwi and Kuta Utara sub-districts of 8 tons / ha which indicate a surplus in 2020 and 2030 respectively of 25,155.19 tons, and 3,401.79 tons. But in 2040 and 2050 there was a food deficit of 18,434.78 tons and 11,824.82 tons respectively [1].

Avoiding the occurrence of food deficits, regional production capacity in the long run needs to be increased through intensification. Considering that productivity improvement efforts cannot rely solely on intensification approaches, but also need to be done with extensification or expansion of agricultural areas to support food production. However, it is not easy to establish areas that can be reserved for agricultural expansion due to the complexity of the problems encountered in land governance. The pressure of development and population also demands easier access to land, through the expansion of other food areas.

One of the efforts to increase land production in Badung Regency is by extensification of agriculture as it has done in Petang Sub-district of 100 ha. In this case extensification can be done through monitoring and mapping of potential land that can be converted to agricultural land for food crops such as: bare land, bush, swamp, moor, and mixed plantation. This mapping is based on remote sensing and Geographic Information System (GIS).

The aims of this research are (1) to map potential land that can be converted to agricultural land of food crops, (2) to know the amount of food demand to supply food balance in Badung regency in 2040.

## 2. Literature Review

### *2.1 Food Balance in Badung Regency*

Food balance in Badung Regency is calculated based on projection of population growth and projection of land function in every sub-district. Modeling by increasing IP or crop index of 2.5 and productivity up to 7-8 tons hence resulted food balance data of 2020, 2030, 2040 and 2050 shown in Table 1.[1]

**Table 1.** Food Balance Data in Badung Regency Year 2015, 2020, 2030, 2040 and 2050

| Years | Petang | Abianse-mal | Mengwi | Kuta Utara | Kuta | Kuta Selatan | Total | Information |
|---|---|---|---|---|---|---|---|---|
| 2015 | 196 | -371,13 | 778,06 | -8501,85 | -11022,8 | -13921,73 | -32843,44 | Defisit |
| 2020 | 6096,77 | 19533,37 | 30807,11 | -151,62 | -12286 | -18844,49 | 25155,19 | Surplus |
| 2030 | 5908,62 | 17076,07 | 25437,36 | -4803,49 | -15296,9 | -24919,87 | 3401,79 | Surplus |
| 2040 | 5679,73 | 14618,77 | 20067,5 | -9454,57 | -18161,6 | -31184,64 | -18434,78 | Defisit |
| 2050 | 5532,43 | 12161,47 | 14697,76 | -14105,77 | -21012,1 | -37310,18 | -11824,82 | Defisit |

### *2.2 Agricultural Extensification*

Potential land resources for agricultural expansion are possible. Nevertheless, taking into consideration: (a) tenure status, (b) administrative area (location), (c) availability of manpower, and (d) availability of infrastructure. For the procurement of inputs and distribution of agricultural output, and (e) the opportunity to be converted to agricultural land for food crops in relation to spatial planning (land use for settlement, urban development, forest conservation, etc.); then in the short - medium term that can be utilized for the expansion of agricultural areas is estimated at about 20-25 percent of the figure. But this number is still more than enough to meet the target of 2 million hectares. Especially for rice fields, it is estimated more than 650 thousand hectares (greater than the target of 250 thousand hectares of rice field expansion) [4].

*2.3 Geography Information System (GIS)*

Geography Information System (GIS) is one of many information technologies that have transformed the ways geographers conduct research and contribute to society. In the past two decades, these information technologies have had tremendous effects on research techniques specific to geography, as well as on the general ways in which scientists and scholars communicate and collaborate [3].

A geographic information system in the wider sense consists of software, data, people, and an organization in which it functions. In the narrow sense, we consider a GIS as a software system for which we discuss its architecture and functional components.According to the definition, a GIS always consists of modules for data capture (input) and preparation, data storage, data analysis, and presentation (output) of spatial data. For a particular GIS, each of these modules may provide many or only a few functions. However, if one of these functionswould becompletelymissing, thesystem should not be called a geographicinformation system. The samefunction may be offered by different modules: forinstance, data capture and data storage may have functions in common, and thesame holds for data preparation and data analysis.An explanation of the various functions of the four modules provide a functional description of a GIS. Here, we only briefly describe them in the followingsections. A more detailed treatment can be found in follow-up chapters.Beside data capture, storage and maintenance, analysis and output, geo-information processes also involve exchange of information, as well as issues ofembedding the spatial data handling process efficiently in the organizationalcontext. These issues define the context and rules according to which geo-information is acquired and processed [4].

*2.4 Extensification of Agriculture*

Decreasing of natural resources and the environmental impacts the food production per hectare. This objective translates to reducing external inputs, such as livestock densities while minimizing food loss [5]. Extensification programs should consider land suitability for agriculture. otherwise, it will face obstacles in its development [6].

In the agricultural development of Indonesia, agricultural extensification is the extension of agricultural areas to areas previously untapped by humans. The goal is to forest land, steppe meadow, peatland, or other forms of marginal land (marginalized). This term in Indonesian has no direct relationship to extensive agriculture; and in the international terminology the program is better known as agricultural (land) expansion ("expansion of agricultural land") [4].

Expansion of agricultural areas is necessary when available agricultural land is deemed incapable of supporting the expected supply of production (e.g. to provide food for the population of a region / country). The risks to be taken are the disruption of native natural ecosystems [7] and the potential for indigenous cultural destruction due to competition with outsiders [8].

## 3. Methodology

*3.1. Area of Study*

This research was conducted in Mengwi Subdistrict, Badung Regency. Mengwi sub-district consists of 20 villages, and the total area of rice field is covering 4366.03 ha [10].

*3.2 Tools and Material*

While the tool used are:

- *Computer*
- *Software (Qgis 2.10.1),*

The material used in this research are spatial data as follows:

- Quickbird satellite Image of Badung Regency, 2013
- Indonesia Topographic Map 2002

*3.3 Research Procedures.*

Phase of the research include:

*3.3.1. Preliminary studies*

Preliminary study is done through stages such as collecting and studying literature review from various sources. Collecting data used in research, identification of research area.

*3.3.2 Interpretation of satellite images*

Interpretation of orthoretification satellite images, compilation of reference coordinate system using Transverse Mercator (UTM) projection system, development of raster and vector data.

*3.3.3 Mapping and Measurement of Land Area*

Mapping is done with image interpretation and digitization on screen. Objects observed and mapped are bare land, shrubs, swamps, moorlands and mixed plantations. After mapping the area is calculated using GIS applications.

3.3.4 *Calculation of Increasing Food Availability*

Calculate the addition of food availability starting from the calculation of the addition of land area in 2040. The area of land projection in 2040 added with the area of potential land mapping. From the new land area obtained the availability of food is by multiplying the new land area with a maximum IP of 2.5 and maximum productivity of 8 tons / ha. Data of dry grain harvest is converted to dry grind mill multiplied by 0,86 and for Rice multiplied 0,63 of Daily Power Off (THL) - Agricultural Extension Aid (TBPP) of agricultural depart.

From the results of additional food availability reduced by the needs of residents in Mengwi District Year 2040, it will be seen the value of suplus or deficit. Then supply supply of food to scale badung district in 2040 to know the value of surplus or deficit.

## 4. Result and Discussion

The results of satellite image interpretation in Mengwi Sub-district obtained potential areas that can be converted to agricultural land for food crops, shown in Figure 1.

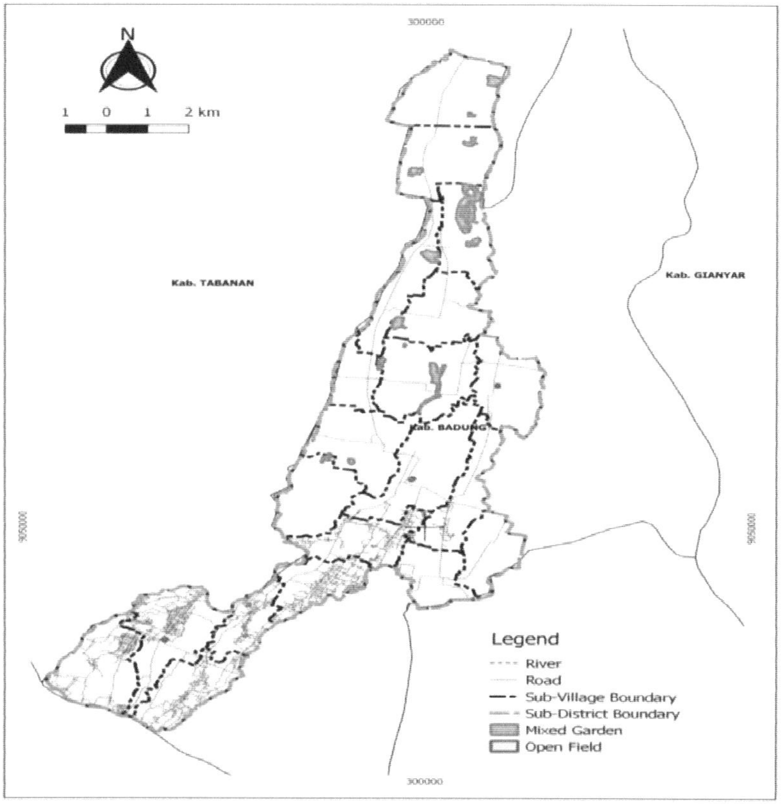


**Figure 1.** Potential Land Map

The above mapping results indicate the potential land of mixed plantation area of 128.51 ha and bare land / shrubs of 3.49 ha. Potential land in the form of dominant mixed plantation was found in Sobangan Village (50.07 ha), Gulingan Village (22.5 ha), and Sembung Village (11.97 ha) while potential land in the form of bareland/ bush was found in the Village Munggu (1.27 ha), Kapal Village (1.3 ha) and Desa Baha (0.92 ha). The result of bare land and mixed plantation are shown in figure 2 and figure 3.

**Figure 2.** Bareland

**Figure 3.** Mixed Plantation

Projection of Food Balance in Kecamatan Mengwi is calculated using projection of population growth and projection of land conversion to non agriculture. data on population projection in Badung Regency is shown in Table 2.

**Table 2.** Projection Data of Population of Sub-District in Badung Regency in 2020, 2040 and 2050

| No. | Districts | An Area ($km^2$) | Population *) Growth Rate | Population Densitiy (people/ $km^2$) | Total Population | | | | |
|---|---|---|---|---|---|---|---|---|---|
| | | | | | 2015 | 2020 | 2030 | 2040 | 2050 |
| 1 | Petang | 115,00 | 0,6 | 226,97 | 26.100 | 29.953 | 31.611 | 33.628 | 35.590 |
| 2 | Abiansemal | 69,01 | 2,2 | 412,85 | 90.600 | 98.809 | 116.627 | 134.445 | 152.265 |
| 3 | Mengwi | 82,00 | 3,14 | 1562,20 | 128.100 | 142.094 | 175.719 | 209.345 | 242.970 |
| 4 | Kuta Utara | 33,86 | 3,3 | 3.358,10 | 119.800 | 136.916 | 172.662 | 208.407 | 244.153 |
| 5 | Kuta | 17,52 | 2,8 | 5.576,48 | 97.700 | 109.805 | 134.923 | 160.042 | 185.161 |
| 6 | Kuta Selatan | 101,13 | 4,42 | 11.142,09 | 122.680 | 166060 | 219597 | 274803 | 328782 |

Description: *) = Central Bureau Statistics of Badung Regency [10]

**Table 3.** Projection Data of Rice Field Change in Badung Regency

| No. | Districts | An Area (ha) | Rate of Land Fuction (ha/y) | Paddy Field Area (ha) [1] | | | | |
|---|---|---|---|---|---|---|---|---|
| | | | | 2015* | 2020** | 2030** | 2040** | 2050** |
| 1 | Petang | 11.500 | 0 | 1.173,00 | 1173,00 | 1173,00 | 1173,00 | 117300 |
| 2 | Abiansemal | 6.901 | 4,033 | 2.868,67 | 2848,51 | 2808,18 | 2767,85 | 2727,52 |
| 3 | Mengwi | 8.200 | 14,40 | 4.420,03 | 4348,05 | 4204,08 | 4060,11 | 3916,14 |
| 4 | Kuta Utara | 3.386 | 5,51 | 1.452,90 | 1425,35 | 1370.25 | 1315,15 | 1260,05 |
| 5 | Kuta | 1.752 | 1,7 | 27,00 | 18,50 | 1,50 | 0 | 0 |
| 6 | Kuta Selatan | 10.113 | 0 | 0 | 0 | 0 | 0 | 0 |

Description: *) = Central Bureau Statistics of Badung Regency [10]

The relation of population, food requirement, land area and availability of food in Mengwi Sub-district with IP 2.5 and productivity of 8 ton / ha are presented in Figure 4.

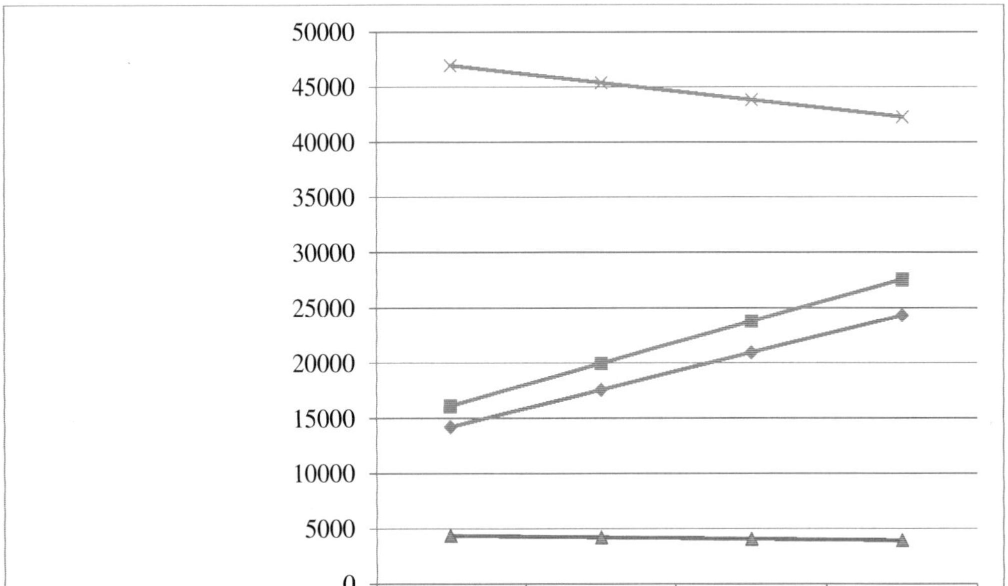


| | 2020 | 2030 | 2040 | 2050 |
|---|---|---|---|---|
| Jumlah Penduduk x 10 | 14209.4 | 17571.9 | 20934.5 | 24297 |
| Kebutuhan Pangan (ton) | 16124.83 | 19940.59 | 23756.47 | 27572.24 |
| Luas Lahan (ha) | 4348.05 | 4204.08 | 4060.11 | 3916.14 |
| Ketersediaan Pangan (ton) | 46931.94 | 45377.95 | 43823.97 | 42269.99 |

**Figure 4.**Graph of Food Balance Projection in Mengwi Sub-district

To increase availability in 2040 food in Kecamatan Mengwi from the addition of potential land area that can be converted to agricultural land of food crops, the calculation is as follows:

New Land Area = Land Area Year 2040 + Potential Land Area
= 4060,11 ha + 132 ha
= 4192.11 ha

Food Availability (GKP) = (New Land Area x 2.5 (IP)) x 8 (ton / ha)
= 4192.11 ha x 2.5, 8
= 83842.2

Food Availability (GKG) = 83842.2 x 0.86
= 72104.29

Availability of Food (Rice) = 72104.29 x 0.63
= 45425.7

From the result of the addition of potential land that can be converted as wetland (132 ha) then calculated the addition of food availability that is:

Availability of new food in 2040 - availability of old food year 2040 =45425.7 tons - 43823.97 tons = 1601.73 tons.

Projection of food balance in 2040 with productivity and maximal IP in Badung regency has a food deficit of 18434,78 tons [1], while from the addition of food availability in Mengwi District only 1601,73 tons. Therefore, this condition can not cover the food deficit condition of Badung Regency in 2040 but for Mengwi sub-district has a food surplus of 21669.23 tons, obtained from the reduction of food availability in Mengwi Subdistrict (45425,7 tons) reduced by food requirement in Mengwi Subdistrict 23756.47 tons.

With the extensification of agriculture in District Mengwi can increase the availability of food for 1601.73 so Mengwi Sub-district can supply food availability for other districts in Badung regency of the year. But for the entire Badung regency in 2040 still need food supplies from other regions to meet the needs of food.

## 5. Conclusions and Recommendations

### *5.1 Conclusions*

1. Mapping of potential land that can be converted to agricultural land of food crops obtained by 132 ha consists of 128.51 ha of mixed plantation/ garden and 3.49 ha of open land / shrubs.
2. From the results of additional land produced 1601.73 tons of rice that supplies food availability in Mengwi Sub-district to 45425.7 tons. With the addition of surplus in 2040 in Mengwi sub-district amounted to 21669.23 tons but did not cover the food deficit in Badung Regency in 2040.

### *5.2 Recomendations*

1. Agricultural extensification potential can be done in other sub-districts in Badung Regency such as Abiansemal and North Kuta Sub-districts.
2. Potential mapped potentials may be further adjusted to the Spatial Plans of Badung Regency Year 2013-2033, in particular additional lands in accordance with the zoning / designation of the agricultural crops area.

**Acknowledgments:**
Writers would like to thank the Rector of Universitas Udayana, Dean of Agriculture Faculty of Universitas Udayana, Head of Center for Spatial Data Infrastructure Development (PPIDS)

Universitas Udayana, fellow staf, researchers and professors at Universitas Udayana for all the support, comments and suggestions that makes this paper finished.

## References

[1] Trigunasih, N.M. 2017. Spatial Numerical Classification Model of Sustainable Agriculture Farm Area in Badung Regency and Denpasar City. *Dissertation.* Udayana University
[2] Pasandaran, E., N.Sutrisno, dan H.Purba. 2015.Land Expansion of Food Agriculture Building a Framework of Integration. Jakarta
[3] Fazal, S.2008.*Gis Basics.* New Delhi. New Age International (P) L.td
[4] de By, R. A. (*ed*), R. A. Knippers, M. J. C.Weir, Y.Georgiadou, M-J.Kraak, C. J. van Westen, Y. Sun. 2001. *Principles of Geographic Information Systems, R. A. de By (ed.) (ITC Educational Textbook Series; 1) Third edition*. The International Institute for Geo-Information Science and Earth Observation(ITC), Enschede, The Netherlands.
[5] Hans JM van Grinsven, Jan Willem Erisman, Wim de Vries, Henk Westhoek (2015). “Potensial of extensification of European agriculture for a more sustainable food system, focusing on nitrogen”. *Environ. Res. Lett. 10 (2015) 025002. Doi:10.1088/1748-9326/10/2/025002*
[6] Simanjuntak, Bistok Hasiolan. 2003. Land Sustability Evaluation for Agriculture. AGRIC Vol. 16 No.1 Juli 2003: 1-12
[7] Anomious. 80% of agricultural expansion since 1980 came at expense of forests. mogabay.com. Ed. 15-02-2009.
[8] Rivet J. Indonesia Aims to be World's Breadbasket. the Jakarta Globe. Edisi 22-02-2010.
[9] Feronika, I. Lanya, W. Nuarsa. 2017. Mapping of Subak Land Based on Remote Sensing and GIS in Mengwi, Badung Regency. *E-JurnalAgroekoteknologi Tropika Vol. 7 No.1.*
[10] Central Bureau of Statistics of Badung Regency. 2015. *Badung in Numbers*. Mangupura: Central Bureau of Statistics of Badung Regency

IOP Conf. Series: Earth and Environmental Science **98** (2017) 012024 doi:10.1088/1755-1315/98/1/012024

# Analysis of Subak Landuse Change Due to Tourism Accomodation Development in North Kuta Sub-district, Badung Regency, Indonesia.

**I. Lanya[1], I N Dibia[2], I W Diara[2], D G Suarjaya[2*]**

[1] Faculty of Agriculture, UniversitasUdayana, Bali, Indonesia
[2] Center for Spatial Data Infrastructure Development (PPIDS) Universitas Udayana, Bali, Indonesia

suarjayadewa27@gmail.com

**Abstract.** Subak is a traditional irrigation management system in Bali which governs the distribution of irrigation water into rice fields. The problems that occur in Subak in North Kuta Subdistrict is the high landuse change of rice field due to the rapid development of tourism and in the Region Spatial Plan is mostly designated as land for non agriculture. The aim of the research is to find out the Subak landuse change into tourism accommodation during the last five years, as well as knowing the area of Subak which located in the area of agricultural lands food crops related to landuse planning regulation of Badung Regency.The research methods included analysis of satellite imagery, literature studies, field survey, analysis of Subak landuse change, analysis of the subak in the area of the agricultural land food crop.
Subak landuse change into non rice fields in the North of Kuta as uch as 538.88 ha during the last five years. The change of rice fields into tourist accommodation as much as 92 accommodation. Subak which is within the area of farmland of food crops 548.35 ha and outside the area of farmland 459.04ha.

**Keywords:** Subak, landuse change, tourism accommodation, landfarm area.

## 1. Introduction

Subak is an organization farmers to management irrigation water that has a particular land area or have a source of water, have autonomous and Subak Temple. According to Regulation of Bali Province No. 2 the year 1972, irrigation system of Subak, defined as customary law society which has the characteristics of an agricultural-socio-religious, which is a gathering of farmers who manage their irrigation water in rice fields. The Council of the UNESCO in Russia at the end of June 2012 decided that organizations subak was accepted as World Cultural Heritage. Subak system different from other World Cultural Heritage because it is based on a concept or philosophy of Tri Hita Karana (THK) who has a meaning that is three-way to the happiness of life, consists of *parhyangan* (the harmonious relationship between man and God), *palemahan* (the harmonious relationship between man and natural environment) and *pawongan* (the harmonious relationship between human to human) [11].

Therefore the existence of Subak should be maintained and conserved. The existence of subak in North Kuta Sub-district are currently declining due to development of tourism.

North Kuta has total of 19 subak scattered in six villages namely Kerobokan as much one Subak, Kerobokan Kaja as much as two Subak, Kerobokan Kelod three Subak, Canggu four Subak Tibubeneng six subak, and three Subaks in Dalung. Subak landuse change is due to the development of tourism which provide a great influence towards landuse inNorth Kuta. Land for construction, tourism, infrastructure and other economic areas are generally derived from rice field conversion. In 2014, landuse change of Subak in North Kuta reached 123 ha [3]. The impact from Subak landuse change causing declining in rice productions and urban green space. Conversion of the landuse also caused by several factors including the regulation of the Regional Spatial Plan which sets the paddy fields is not entirely located on landfarm area [5].

This research aims to: (1) find out the landuse change subak in North Kuta and number of tourist accommodations using the ricefields over the last five years, (2) knowing Subak area which located inside and outside of landfarm area based on Regional Spatial Plan of Badung Regency.

## 2. Literature Review

Subak is the Organization of farmers who carry out irrigation areas, which each of its members has a certain area of rice fields, water springs, have autonomous and Subak Temple. Members of the subak irrigation water management is always based on the philosophical concept of Tri Hita Karana (THK). THK is one of life's philosophy that stems from the Hindu religion. Literally, the philosophy of Tri means three, Hita means happiness and Karana means cause. THK means three paths to happiness of life. THK philosophy consists of *Parhyangan* (the harmonious relationship between man and God), *Palemahan* (the harmonious relationship between man and the surrounding environment) and *Pawongan* (the harmonious relationship between man and other human beings). The implementation Subak in the form of the concept of *Parhyangan* can be seen by the existence of Pura Subak which is used by members of the Subak to perform worship to establish harmonious relationships with God Almighty. The concept of *Palemahan* can be seen by the existence of irrigation water management in the area of the rice field and the existence of elements of the rice fields. The concept of *Pawongan* can be seen from the organization of the autonomous nature of the Subak and the presence of Krama Subak (Subak member) [9].

Subak is one of the world's cultural heritage has been approved by UNESCO in June 2012. Subak has its own uniqueness that contrasts with other world cultural heritage. The uniqueness is meant that Subak as an organization of irrigation system management is a socio-cultural community, which in the Subak members achieved its goal has always been based on the concept of harmony and togetherness in accordance with the philosophy of THK as well as maintaining the balance of the environment [11]. Subak organisations based on socio-cultural community capable of providing the power of sustainability, because it was able to absorb the technologies that are being developed, able to adapt to the development of culture, simplicity the structure of the Organization and work system kooperative.
Subak organisations also have the weakness that is not able to withstand the intervention of outside parties like the foreign investors which led to occur over the function of the land from the agricultural land into non-farming. The ability of human resources limited is one of the weaknesses in this water organizations. This condition is caused due to the agricultural sector is one of the jobs that are less sought after by the public especially the young generation today. In General farmers still till it was over 50 years old. The low interest in the community in the field of work to cultivate the farms because it has a high risk of loss. Another disadvantage on water level limited land ownership.

Each year the land area of the water always decline. This is because landuse change is high especially in areas close to urban and tourist areas. According to data of the Central Bureau of Statistics, the Subak in district of North Kuta has changed into non farming as much as 123 ha in the year 2014. the conflict between agriculture and tourism continues, resulting in wetlands being constantly pressured by the interests of non-agricultural development, both for housing and for the

IOP Conf. Series: Earth and Environmental Science **98** (2017) 012024 doi:10.1088/1755-1315/98/1/012024

infrastructure of objects and tourist destinations. Conversion of agricultural land has occurred due to the supply of land in Bali [6].

Some of the factors that cause a decrease in Water sustainability, such as the lack of interest the younger generation to work in the agricultural sector, a decline in the quantity and quality of irrigation water, wetland constriction due to land conversion to other uses outside of agriculture. The competition is the utilization of agricultural land or water resources between agriculture and tourism. In the competition always surrendered agricultural sector under the pretext of defeating the more promising generate foreign exchange to finance the construction of Bali and Indonesia in General. Proven reduction of farmland and converted to non-agricultural very significant [12].

As for the effort that can be done to conserve water is to maintain the sustainability of the paddy fields in Bali, maintaining sustainability of the resource water for irrigation, maintaining the boundaries between the clear water, maintain subak organization system that is flexible, strengthen institutional subak as well as maintaining the concept of harmony in the togetherness and the mindset of farmers in managing irrigation systems in accordance with the philosophical concept of THK [11]. In addition, there needs to be support from the government in protecting the priority of food land to meet food security and food self-sufficiency [8].

## 3. Methodology

### *3.1. Area of Study*

Area of study is located in North Kuta Sub-district, Badung Regency, Bali Province, Indonesia. Geographically, North Kuta Sub-district located in 08°38'44,2"-08°43'32,6" South Latitude and 115°10'39,2"-115°38'42,3" East Longitude. North Kuta Sub-district bordering with Mengwi Sub-district in the north, Denpasar City in the east, Kuta Sub-District in the south and The Indonesian Ocean in the west.

### *3.2. Tools and Material*

The material used in this research are spatial data as follows:

- Quickbird Satellite Image of Badung Regency in 2012 [5]
- Google Satelite Image year 2017 from obtained from Openlayers Plugin in QGIS
- Map of Regional Spatial Plan of Badung Regency in 2013-2033 with scale1 : 50.000 [2]

While the tool used are:

- Computer to process the spatial data
- *QGIS* Software to perform geoprocessing analysis
- GPS (Global Positioning System) to determine the coordinate point of observation.
- Camera
- Stationary

### *3.3. Data collection*

Data collection method used in this research is primary data and secondary data method.

- Secondary data consists of Quickbird Satellite Imagery data of 2012 and Badung Regency Regional Spatial Plan Area Map Year 2013-2033.
- Primary data (direct data source in the field) obtained from observing and recording the point of tourism accommodation in North Kuta Subdistrict using GPS. Coordinate point data is then incorporated into QGIS Software. The process resulted in a tourism accommodation map in North Kuta Sub-district.

### *3.4. Data Analysis*

Data analysis in this research consist of two phase, which are

- Tourism accommodation analysis which come from Subak area during last five years and

- analysis of Subak land inside and outside of landfarm area based on Badung Regional Spatial Plan Map Year 2013-2033.

*A. Analysis of Tourism Accommodation Derived from Subak Land in the Past Five Years*
This analysis is augmented by inputting the Quickbird Satellite imagerry of 2012 into QGIS software. The next step is to delineate the rice field located in North Kuta Subdistrict. The process resulted in subak land in North Kuta Subdistrict 2012. The 2012 Subak field is then intersected with Subak data of 2017 [10]. The result of overlapping resulted in changes of subak use of rice fields into non-rice field l for the last five years. The land use change data is then integrated with tourism accommodation data in the North Kuta Subdistrict that has been obtained during the field survey. The process aims to obtain a percentage of tourism accommodation data derived from paddy fields from 2012 to 2017.

*B.Subak land analysis located inside and outside the Crop Farming Area*
Based on the Regional Spatial Plan Area of Badung Regency Year 2013-2033 in North Kuta Subdistrict is not found Green Open Space there is only landfarm area. The first step is to input the map of landfarm area of Badung Regency on the Regional Spatial Plan Area of Badung Regency in 2013-2033. Registration map of landfarm area to equalize with Subak Map of North Kuta Subdistrict 2017 [10]. Subak map is then arranged with map of landfarm area, analysis of spatial data matching result of digitized subak with digitized landfarm area result, so it can be determined the width of subak paddy field which is likely to be converted in North Kuta Subdistrict based on subak rice field area outside landfarm area and subak which is located inside the landfarm area.

## 4. Result and Discussion

### *4.1. Analysis of Tourism Accommodation Derived from Subak Land in the Past Five Years*

The intersect result result of Subak in 2012 with subak in 2017 resulting the Subak land use change in North Kuta Subdistrict. For the last five years Subak in North Kuta Sub-district experienced land conversion of 538.88 ha (35%). Subak land experienced the largest land conversion in each respective villages of Tibubeneng, Canggu, Kerobokan Kelod, Dalung, Kerobokan Kaja and Kerobokan. The use of Subak in North Kuta Subdistrict is transformed into gardens, vacant land, buildings and tourism accommodation. The increasing number of tourists North Kuta Subdistrict caused the rapid development of the villages so the agricultural land is decreasing. The rapid development of tourism poses a serious threat to Subak sustainability in Bali, especially in Subak at the tourism area with heterogeneous populations and many social economic facilities [1]. The tourism accomodation in North Kuta Subdistrict is presented in Figure 1.

Tourism Accommodation in North Kuta Subdistrict consists of three types i.e lodging, restaurant, and supporting facilities. Based on Figure 1, it can be seen that the pattern of wet land conversion to tourism accommodation in North Kuta Subdistrict is increasingly decreasing to the north. Kerobokan Kelod Village has the most tourism accommodation among other villages. The development of tourism accommodation in Kerobokan Kelod Village was earlier compared to other villages in Kuta Utara Sub-district. This is because Kerobokan Kelod Village adjacent to Legian and Seminyak whose tourism are well known.

Based on the results of the analysis of changes in Subak land use over the past five years, some of the development of tourism accommodation in North Kuta take advantage of Subak land. The number of tourism accommodation using Subak rice field is presented in Table 1.

Tourism accommodation using the most subak rice fieldis in Kerobokan Kelod Village. The number of tourism accommodation in Kerobokan Kelod using rice fields for the last five years is 33 inns and 19 restaurants. Figure 2 shows one of the tourism accommodation in Kerobokan Kelod Village which takes up rice fields. The development of tourism accommodation encourages non-agricultural development, either in the form of new tourism accommodation or other supporting facilities such as settlement. The number of supporting facilities in North Kuta Subdistrict that utilize Subak land as much as 15%. While the number of inns and food stalls or restaurants that use Subak rice fields as much as 32%. Development that uses Subak land causes irrigation channels to be cut off. Irrigation systems in the disrupted Subak cause the cultivation process is also hampered [12]. Conversion of land area of Subak becomes bigger due to disrupted of farming cultivation. Land conversion creates some impacts, not only the direct impact on agricultural production and other related variables but also on environmental aspect [4].

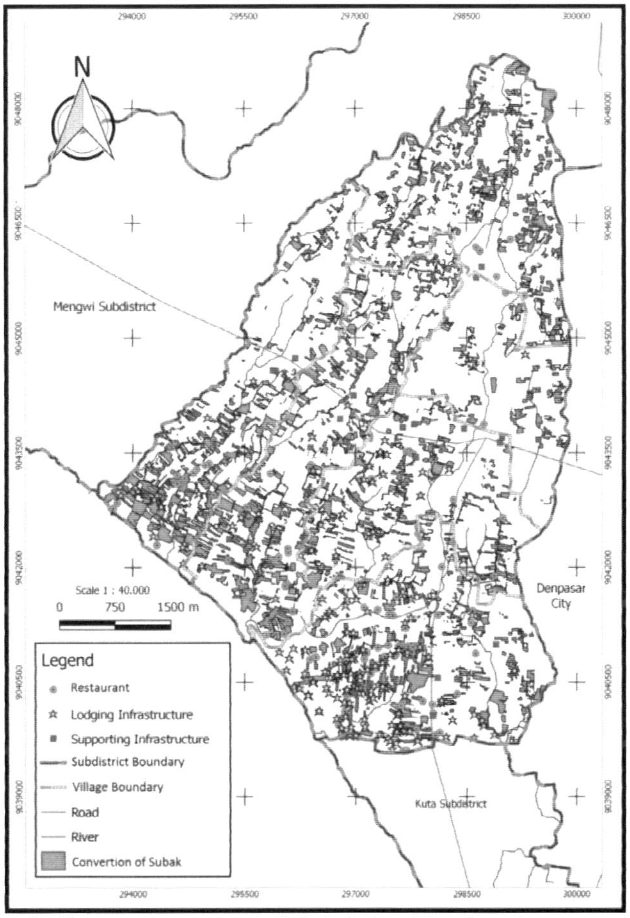


**Figure 1.** Changes in the Use of Subak into Tourism Accommodation

**Table1**. Tourism Accommodation in North Kuta Subdistrict

| No | Name of Village | Supporting Infrastructure | | Lodging Infrastrukture | | Restaurant | |
|---|---|---|---|---|---|---|---|
| | | Total | Using Paddy Fields | Total | Using Paddy Fields | Total | Using Paddy Fields |
| 1 | Dalung | 5 | 1 | 2 | 1 | 6 | 0 |
| 2 | Canggu | 3 | 0 | 16 | 11 | 13 | 4 |
| 3 | Tibubeneng | 4 | 1 | 10 | 2 | 17 | 9 |
| 4 | Kerobokan Kaja | 3 | 1 | 1 | 0 | 3 | 1 |
| 5 | Kerobokan | 6 | 1 | 22 | 6 | 14 | 2 |
| 6 | Kerobokan Kelod | 6 | 0 | 113 | 33 | 58 | 19 |

**Figure 2**. Tourism Accommodation Using Subak in Kerobokan Kelod

*4.2 Analysis of Subak Located Inside and Outside the Landfarm Area*

Subak spatial data in North Kuta Subdistrict intersected with landfarm area plan in Regional Spatial Plan Badung Regency Years 2013-2033 produces a stretch of Subak area inside and outside of landfarm area plan of Regional Spatial Plan of Badung Regency. The result is the area of Subak rice field located inside the landfarm area plan and outside the landfarm area plan. Table 2 shows the Subak area inside and outside of landfarm area in the Regional Spatial Plan Badung Regency. Subak which located in the landfarm area plan is a Subak rice field that can not be converted and planned as sustainable rice fields. Subak land area that is not in the landfarm area is Subak which most likely to be converted to be non rice field, so the possibility of landuse change is very big. Subak maps located inside and outside the landfarm area are presented in Figure 3.

**Table 2**. Subak Area Within and Outside The Lanfarm Area

| No | Name of Subak | Total Subak Area (ha) | Area of Subak (ha) | |
|---|---|---|---|---|
| | | | Inside of Landfarm Area | Outside of Landfarm Area |
| 1 | Subak Sebuah | 98,31 | 40,72 | 57,59 |
| 2 | Subak Tegal | 66,57 | 57,97 | 8,6 |
| 3 | Subak Muding | 38,52 | 8,14 | 30,38 |
| 4 | Subak Petitenget | 38,71 | 2 | 36,71 |
| 5 | Subak Kedampang | 76,47 | 52,02 | 24,45 |
| 6 | Subak Basangkasa | 72,63 | 42,76 | 29,87 |
| 7 | Subak Banjarsari | 31,42 | 2,45 | 28,97 |
| 8 | Subak Perancak | 35,38 | 4,88 | 30,5 |
| 9 | Subak Semat | 43,8 | 19,97 | 23,83 |
| 10 | Subak Bantan | 47,61 | 40,44 | 7,17 |
| 11 | Subak Daksina | 45,85 | 34,28 | 11,57 |
| 12 | Subak Dawas | 22,13 | 19,56 | 2,57 |

| | | | | |
|---|---|---|---|---|
| 13 | Subak Liplip | 25,57 | 19,62 | 5,95 |
| 14 | Subak Canggu | 74,64 | 34,54 | 40,1 |
| 15 | Subak Umaalas | 13,34 | 10,51 | 2,83 |
| 16 | Subak Umadesa | 77,49 | 62,22 | 15,27 |
| 17 | Subak Bernasi | 71,73 | 59,27 | 12,46 |
| 18 | Subak Saih | 82,96 | 19,49 | 63,47 |
| 19 | Subak Gaji | 28,27 | 17,5 | 10,77 |
| | **Total** | **1007,39** | **548,35** | **459,04** |

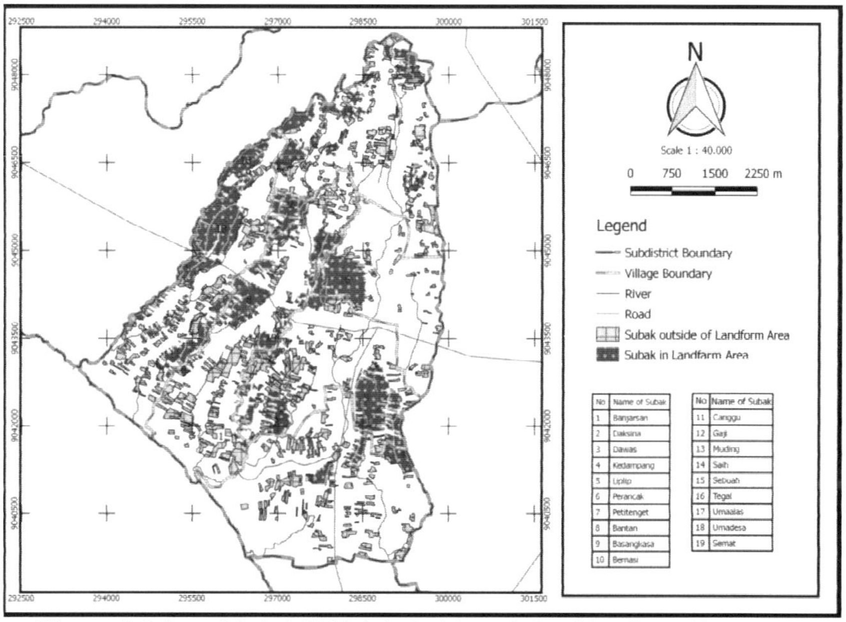


**Figure 3**. Map of Subak Which is Inside and Outside of Landfarm Area

Based on Figure 3 it can be seen that almost 50% of Subak area in Kecamatan Kuta Utara is located outside of landfarm area plan of Regional Spatial Plan Badung Regency. The subak area in North Kuta Subdistrict within landfarm area plan is 548.35 ha and the area of subak rice field outside ofland farm area plan is 459,04 ha. Highest Subak area which located inside the landfarm area plan is Subak Umadesa (62.22 ha) located in Canggu Village, North Kuta Subdistrict mean while highest Subak area located outside land farm area plan is Subak Saih (63.47 ha) located in Dalung Village, North Kuta District.

Subak area of more than 50% in the landfarm area plan are Subak Dawas (88%), Subak Tegal (87%), Subak Bantan (85%), Subak Bernasi (83%), Subak Umadesa (80%), Subak Umaalas (75%), Subak Kedampang (68%), Subak Gaji (62%) and Subak Basangkasa (59%). While subak more than 50% of land outside of the landfarm area plan are Subak Petitenget (95%), Subak Banjarsari (92%), Subak Perancak (86%), Subak Muding (79%), Subak Saih (77%), Subak Sebuah (59%), Subak Canggu (54%), Subak Semat (54%). Subak which is more than 50% of its land outside of the landfarm area is rice field located in the area close to Batu Belig Beach to Batu Mejan Beach (Echo Beach) and Dalung Permai Housing.

Land outside the landfarm area legally are possible to be converted to non-agricultural (settlements, roads, trade, services and tourism)according to its designation on he Regional Spatial Plan of Badung Regency. In contrast, the ricefields inside the landfarm area are not used for conversion. The area of non-rice field in landfarm area plan in North Kuta Subdistrict is 273,18 ha (33%) in total. Its means that maximum of 33% of existing rice fields are able to be converted into non-rice fields.

## 5. Conclusions

North Kuta Subistrict is an area of development of tourism in Badung regency. North Kuta Subdistrict experienced landuse change of 538.88 hectares over a period of 5 years from 2012 until 2017. Subak who experience land conversion to the use of tourism in general is on the tourist route Kuta - Tanah Lot. Tourism accomodation in North Kuta Sub-district that convert the Subak land for the last five years are as much as 92 accommodation (30.5%). Number of tourism accommodation in North Kuta Subdistrict which takes the most Subak land is in Kerobokan Kelod Village. The closer to the Kuta tourism center, the more rice fields that experience land conversion.

Subak rice field located inside landfarm area in North Kuta Subdistrict is 548.35 ha (54%) and outside of the landfarm area of 459.04 ha (46%). Land outside the landfarm area legally are able to converted into non-agricultural (settlements, roads, trade, services and tourism). In contrast, the paddy fields within landfarm area are sustainable rice fields that cannot be converted.

## Acknowledgments

Writers would like to thank the Rector of Universitas Udayana, Head of Center for Spatial Data Infrastructure Development (PPIDS) Universitas Udayana, Dean of Agriculture Faculty of Universitas Udayana, fellow staf, researchers and professors at Universitas Udayana for all the support, comments and suggestions that makes this paper finished.

## References

[1] Alit Artha W., I Wayan. Rachel P. Lorenzen, and Stephan Lorenzen .2006.The Past, Present And Future-Perspectives of Balinese Rice Farming. In Sumarno, Suparyono, Achmad M. Pagi, Made Oka Adnyana (Eds), *Rice Industry, Culture, and Environment, Book 1, Proceedings of the International Rice Conference*, September 12-14, 2005, Tabanan Bali. Indonesian Agency for Agricultural Research and Development in Cooperation with International Rice Research Institute.

[2] Regional Development Planning of Badung Regency. 2013. Regional Regulation of Badung Regency no. 26 Year 2013 About Spatial Planning of Badung Regency Year 2013-2033. Badung: Badung Regency Government.

[3] Central Bureau of Statistics of Badung Regency. 2016. *Badung in Figures Year 2015*. Badung: Badung Regency Government.

[4] Irene, Pinta Lizti. 2015. Farmland Conversion in Karawang, Indonesia: Discourse Analysis. *Dissertation Faculty of Bioscience Engineering, Universitiet Gent.*

[5] Lanya Indayati, N. Netera Subadiyasa, K. Sardiana, IGP RatnaAdi. 2015. Land Use Control Strategies to Reduce the Negative Impact of Tourism in Bali.*The final report MP3EI Research* (Udayana University, Bali Jimbaran Hill).

[6] Lanya Indayati and N. Netera Subadiyasa.2016.Role of Remote Sensing and Geographic Information System Mapping for Protected Areas Land Rice Field Subak, Buffer Zones, and Area Conversion (Case Studies In Gianyar Regency, Bali Province).Denpasar: Udayana University

[7] Nasoetion, L.I .2003. Conversion of Agricultural Land: Aspects of Law and its Implementation. Proceedings of The National Seminar on Multifunctional Agriculture and Land Conversion. Agency for Agricultural Research and Development. Jakarta: Ministry of Agriculture.

[8] Purbiyanti, Erni. 2013. Wetland Conversion in Indonesia : Determinant Factors, Impact on the National Food Availability, and Its Solutions. *Proceedings of 2013 Jakarta International Conference of Muslim Intelectuals*. Jakarta.

[9] Sriartha, I P., Sri Rum Giyarsih .2017. Subak Endurance in Facing External Development in South Bali, Indonesia. International Research Journal of Management, Iternational Research

Journal of Management, IT & Social Science Vol. 4 pages: 20-30.

[10] Suarjaya, D.G., I. Lanya, I G P Ratna Adi .2017. Aplication of Remote Sensing and GIS for Mapping and Information Resources in Wetland of Subak in Kuta and North Kuta, Badung Regency. *E-Jurnal Agroekoteknologi Tropika Vol. 6 No. 3.*

[11] Windia, W. and Wayan Alit Artha Wiguna .2013.*Subak as World Cultural Heritage*.Denpasar: Udayana University Press.

[12] Yuliana, E.D .2017. SUBAK: Traditional Irrigation Organization A Cultural Representation of Agriculture in Bali. *International Journal of Contemporary Research and Review Vol. 8 Issue 4 Page no. AG20179-20185.*

IOP Conf. Series: Earth and Environmental Science **98** (2017) 012025 doi:10.1088/1755-1315/98/1/012025

# Geothermal Potential Analysis Using Landsat 8 and Sentinel 2 (Case Study: Mount Ijen)

**B M Sukojo[1*] and R Mardiana [2]**

[12]Departement of Geomatics Engineering, Faculty of Civil Engineering and Planning, Sepuluh Nopember Institute of Technology (ITS) Surabaya, Indonesia

bangunms@gmail.com

**Abstract**. Geothermal energy is also a heat energy contained in the earth's internal. Indonesia has a total geothermal potential of around 27 GWe. The government is eager for the development of geothermal in Indonesia can run well so that geothermal can act as one of the pillars of national energy. However, the geothermal potential has not been fully utilized. One of the geothermal potention is Mount Ijen. Mount Ijen is a strato volcano that has a crater lake with a depth of about 190 m and has a very high degree of acidity and the volume of lake water is very large. With the abundance of potential geothermal potential in Indonesia, it is necessary to have an activity in the form of integrated geoscience studies to be able to maximize the potential content that exists in a geothermal area. One of the studies conducted is to do potential mapping. This research performs image data processing of Landsat 8, Sentinel 2, RBI Map, and preliminary survey data. This research carried out the Vegetation Index, surface temperature and altitude. The equipment used in this research includes image processing software, number processing software, GPS Handheld and Laptop. Surface Temperatures in the Mount Ijen have anomalies with large temperatures ranging between 18° C to 38° C. The best correlation value of altitude and ground surface temperature is -0.89 ie the correlation of January surface temperature. While the correlation value of Landsat 8 and Sentinel 2 vegetation index was 0.81. The land cover confidence matrix scored 80%. Land cover in the research area is dominated by forests by 35% of the research area. There is a potential area of geothermal potential is very high on Mount Ijen with an area of 39.43 hectares located in Wongsorejo District and adjacent to District Sempol.

**Keywords:** Geothermal, Landsat 8, Sentinel 2, Surface temperature, Vegetation index

## 1. Introduction

A total of 252 geothermal locations in Indonesia are scattered following the path of volcanic formation that stretches from Sumatra, Java, Nusa Tenggara, Sulawesi to Maluku. With a total potential of around 27 GWe [12]. Judging from the appearance of geothermal on the surface per unit area, Indonesia ranks fourth in the world, even in terms of high temperatures, is the second largest.

The government is eager for the development of geothermal in Indonesia can run well so that geothermal can act as one of the pillars of national energy. It is seen through the stipulation of Presidential Regulation no. 5 of 2006 on National Energy Policy (KEN). In the Presidential Regulation, the Government targets geothermal energy contribution in 2025 of 9500 MW [4]. However, geothermal potential in Indonesia is only 4% of potential that has been utilized [5].

One source of geothermal energy stored in Indonesia is located on Mount Ijen. Mount Ijen is located in District Licin and District Sempol, Banyuwangi and Bondowoso, East Java. The height of the crater lake of Mount Ijen reaches 2145 m and the edge of the crater reaches 2386 from sea level. Currently Ijen crater measuring about 1160 m x 1160 m, with a crater lake Ijen measuring about 910 m x 600 m and has a depth of up to 200 m [1].

Geothermal energy is one of the renewable energy that can be used as an alternative because the energy made from fossil raw materials increasingly rare. The energy is clean and is considered safer because it does not require large space for exploration and exploitation purposes. In addition to geothermal energy produces steam that can be mixed also into the air [1]. In addition, geothermal is a renewable, potentially enormous, state-controlled natural resource and plays an important role as a source of energy of choice in national energy diversity to support sustainable national development for the realization of people's welfare [8].

Utilization of geothermal energy is one of the government's efforts in the diversification of energy sources. Planned in 2010 has produced 3600 MW of electrical energy from geothermal heat. This policy attracts many foreign investors, even lately, the hotbed areas with a production potential of around 20 MW also attract investors [11].

With the abundance of potential geothermal potential in Indonesia, it is necessary to have an activity in the form of integrated geoscience studies to be able to maximize the potential content that exists in a geothermal area. One of the studies conducted is to do potential mapping. Where the usefulness of the potential map is to describe geospatial information from a geothermal region to determine the potential area of the geothermal content. To pursue these targets required mastery of technology and science management of geothermal energy in a relatively very short time

Current technological advances in the field of remote sensing should be put to good use. Utilizing remote sensing technology that has high observation accuracy and relatively low cost for broad unity, thus providing the possibility to integrate the level of accuracy and efficiency in the supply of geothermal data and information. Using Landsat 8 satellite images we can obtain the values of vegetation density and ground surface temperature of the area while using DEM from the RBI map is used to obtain the height of the land formed from the digital elevation model of an area and for the land cover map using the analysis of the Landsat image 8 and Sentinel 2.

## 2. Materials and Methods

### *2.1. Research Site*

The location of this study took the location of study Ijen volcano is a strato crater volcano located in the District Tetelan and District Sempol, Banyuwangi and Bondowoso, East Java. This mountain is about 33 km from Banyuwangi with geography position about 8 ° 3.5 'South and 114 ° 14.5' East:

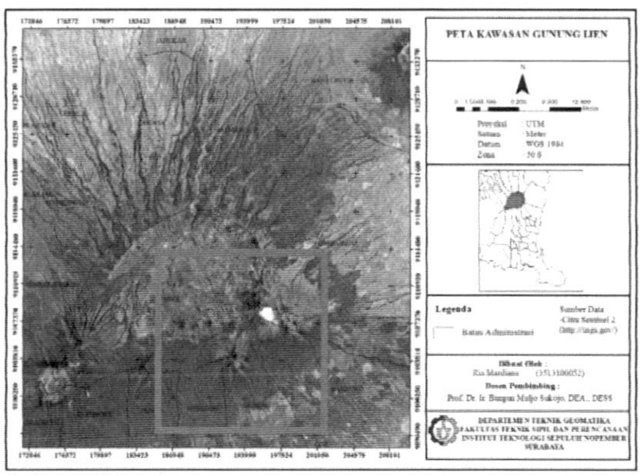


**Figure 1.** Research Site

### *2.2. Data and Equipment*

*2.2.1. Data.* The data used in this research are:

1. Landsat Image 8 January, May and August 2016 (http://usgs.gov/).
2. Geothermal preliminary survey data on Mount Ijen are geological data, geochemical data, and geophysical data (PT Medco Power Indonesia)
3. Sentinel Image 2nd of October 2015 (http://usgs.gov/).
4. Map RBI scale 1: 25.000 Sub Sempol and Tetelan (Department of Geomatics Engineering).

*2.2.2.Equipment.*Equipment used in this research are:

1. Hardware
   a. Laptop
   b. GPS Handheld
2. Software
   a. Image processing software
   b. Software calculatating data

### *2.3. Step to Processing Data*

The process of data processing performed can be seen will be explained as follows:

1. Cropping the image of Landsat 8 and Sentinel 2 in the desired area to facilitate data processing as it focuses more on the research area.
2. Geometric correction to coordinate the image in accordance with the coordinates of field conditions. This process is done by placing the Ground Control Point (GCP) point in the uncorrected image based on the image that has been corrected by RBI Map 1: 25000. The image can be said to be corrected if it meets the tolerance limit based on the value of RMS Error $\leq 1$ pixel and SOF $\leq 1$.
3. Make a radiometric correction to change the Digital Number (DN) value to Spectral Radiance. Radiometric correction aims to improve distortion due to atmospheric disturbances during image recording process. In addition, geometric correction is done so that the coordinates on the map are the same as the coordinates in the field.
4. To obtain land cover map on Sentinel 2 image, the method used is supervised classification with maximum likelihood method. Performed by determining the training area where a sample pixel area is defined as a certain type of land cover. The sampling area for this type of land cover is based on the interpretation of the image and the RBI map. After obtaining several samples representing the type of land cover in the study area, then classified by using the maximum likelihood algorithm.
5. Landsat 8 image processing to obtain vegetation density map (NDVI) and surface temperature map. The bands used to obtain surface temperature values are bands 10 and 11 while to get the value of vegetation density used band 4 and 5. Do also sentinel processing 2 to get the value of vegetation density using bands 8 and 4.
6. Processing DEM data from digitization the next RBI map of DEM data is converted to TIN surface format showing the surface model of the earth in the research area. So get the height in the area under study.
7. Digitizing geological maps to obtain geological data if a site is found fumarole and hot springs, it is undoubtedly underneath there is a geothermal source that makes the ground water temperature increase and of course only be in certain locations with typical geological conditions. Easy observation is to look for the existence of manifestation make it out surface soil as hot water so it can be used as an analysis when overlay.
8. Overlay all data then analyse geothermal potential.

IOP Conf. Series: Earth and Environmental Science **98** (2017) 012025 doi:10.1088/1755-1315/98/1/012025

## 3. Result and analysis

### *3.1. Geometric Corrections and Calculations of Strength of Figure*

The design of the control point nets used in the geometric correction of images as follows:

**Figure 2.** Design of Point Control

The calculation of SOF (Strength of Figure) of the net is:

Number of Baseline : 16
Point Count : 4
N Size : The number of Baseline x 3: 16 x 3 = 48
N Parameters : Number of Points x 3: 4 x 3 = 12
U : N Size - N Parameters: 48 - 12 = 36

$$\text{SOF: Trace } \frac{Trace\,[A^T x\, A]^{-1}}{u} : 0{,}088 \qquad (1)$$

### *3.2. GCP calculation and RMS Error Value*

The process of giving GCP to the image is done by using image data processing software, that is by using Image to map menu using RBI map. The geometric correction of Landsat 8 and Sentinel 2 satellite images is done using 5 selected GCP dots are the same object in the image and on the reference where the possibility of change is relatively slow (fixed) such as the peak of the mountain and the edge of the cliff. The RMS error value in Landsat 8 image is 0.0003.

| | Map X | Map Y | Image X | Image Y | Predict X | Predict Y | Error X | Error Y | RMS |
|---|---|---|---|---|---|---|---|---|---|
| #1+ | 195485.00 | 9108268.00 | 3417.59 | 1575.82 | 3417.6119 | 1575.8232 | 0.0219 | 0.0032 | 0.0222 |
| #2+ | 193457.00 | 9110910.00 | 3350.05 | 1487.70 | 3350.0201 | 1487.6956 | -0.0299 | -0.0044 | 0.0302 |
| #3+ | 192943.00 | 9103666.00 | 3332.95 | 1729.27 | 3332.9263 | 1729.2665 | -0.0237 | -0.0035 | 0.0239 |
| #4+ | 191647.00 | 9107351.00 | 3289.69 | 1606.36 | 3289.7220 | 1606.3647 | 0.0320 | 0.0047 | 0.0323 |
| #5+ | 189602.00 | 9110740.00 | 3221.55 | 1493.38 | 3221.5497 | 1493.3800 | -0.0003 | -0.0000 | 0.0003 |

**Figure 3.** RMS Error Value of Landsat Image 8.

The RMS error value on Sentinel 2 image is 0.0012

| | Map X | Map Y | Image X | Image Y | Predict X | Predict Y | Error X | Error Y | RMS |
|---|---|---|---|---|---|---|---|---|---|
| #1+ | 195485.00 | 9108268.00 | 3989.51 | 3072.57 | 3989.5925 | 3072.5964 | 0.0825 | 0.0264 | 0.0867 |
| #2+ | 193457.00 | 9110910.00 | 3790.67 | 2805.08 | 3790.5575 | 2805.0440 | -0.1125 | -0.0360 | 0.1181 |
| #3+ | 192943.00 | 9103666.00 | 3728.69 | 3529.40 | 3728.6008 | 3529.3714 | -0.0892 | -0.0286 | 0.0937 |
| #4+ | 191648.00 | 9107350.00 | 3604.29 | 3158.60 | 3604.4104 | 3158.6385 | 0.1204 | 0.0385 | 0.1264 |
| #5+ | 189602.00 | 9110741.00 | 3404.75 | 2816.48 | 3404.7488 | 2816.4796 | -0.0012 | -0.0004 | 0.0012 |

**Figure 4.** RMS Error Value Sentinel Image 2

### *3.3. Image Processing for Surface Temperatures*

The following is the result of surface temperature processing on the thermal band 10 on the three Landsat 8 imageries of 19 January 2016, 26 May 2016 and 30 August 2016.

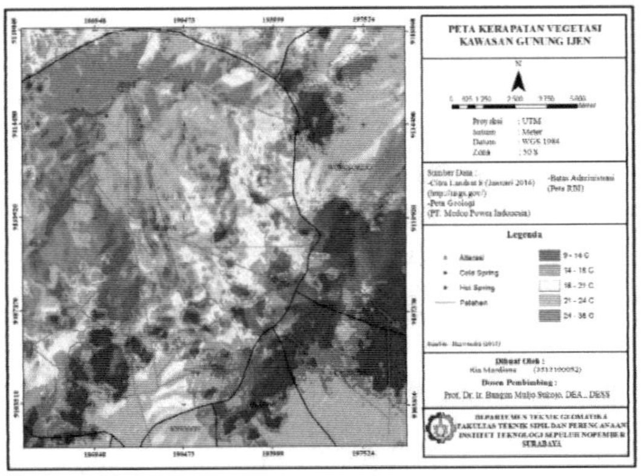


**Figure 5.** Image Processing Results of Land Surface Temperature January

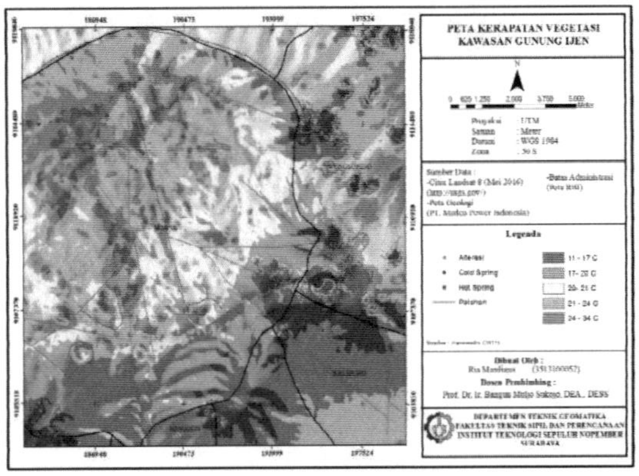


**Figure 6.** Image Processing Results of Land Surface Temperature May

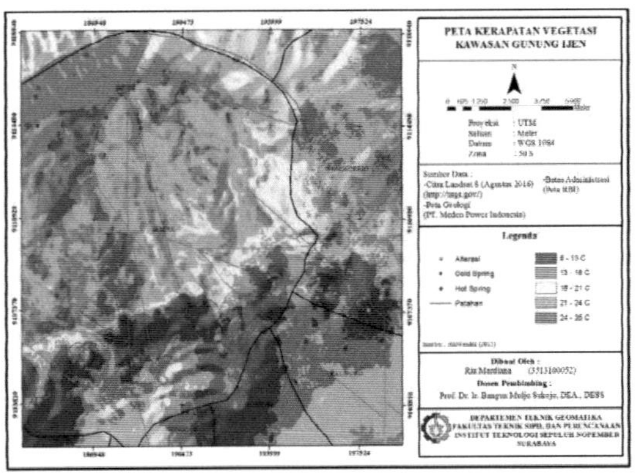


**Figure 7.** Image Processing Results of Land Surface Temperatures August

The Ijen mountain area has a surface temperature of 21 ° C - 34 ° C [3]. For the result of surface temperature of soil at data processing yield the lowest temperature that is 8 °C and highest temperature 38°C. The higher the ground surface temperature of an area, the higher the potential for geothermal in the area [14]. Here is a table of minimum and maximum temperature on surface temperature data processing using Landsat image 8.

**Table 1.** Surface Temperature

| Month | Low Temperature | High Temperature |
|---|---|---|
| January | 9 ° C | 38 ° C |
| May | 11° C | 34 ° C |
| August | 8° C | 35 ° C |

The results of all three thermal bands have differences. In January is a wet month so it can detect higher temperatures higher than in May and August, this is because in wet months the water content in the soil surface more so as to absorb more sunlight [6].

### *3.4. Land Elevation*

The results of processing for altitude of land based on the class height of land as follows:

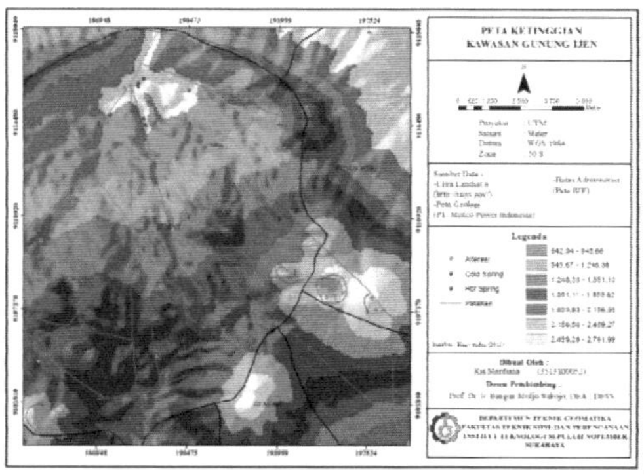


**Figure 8.** Map of Land Elevation

Then tested the correlation between elevation and ground surface temperature. Correlation analysis aims to determine the relationship between two or more variables. In this research, correlation analysis is used to find out the correlation of surface temperature with altitude. The following is the result of correlation of surface temperature with the height that will be displayed in the table

**Table 2.** Surface and Land elevation Correlation Results

| Month | Residu |
|---|---|
| January | -0,89 |
| May | -0,86 |
| August | -0,82 |

The result of land elevation correlation and surface temperature can be seen that the correlation between height and surface temperature has a very strong relationship with the best correlation that is the correlation of height and surface temperature of January is -0.89. The relationship between altitude and surface temperature is negative due to their unrelated relationship. The higher the land, the lower surface temperature will be lower [9]

### *3.5. Land Cover*

In this study the classes specified are shrubs, vacant land / grass, plantations, forests, settlements and craters. For accuracy of land cover processing is calculated using confusion matrix method, where the tolerance limit given is ≥ 80% [7].

IOP Conf. Series: Earth and Environmental Science **98** (2017) 012025 doi:10.1088/1755-1315/98/1/012025

Here is a land cover map of data processing:

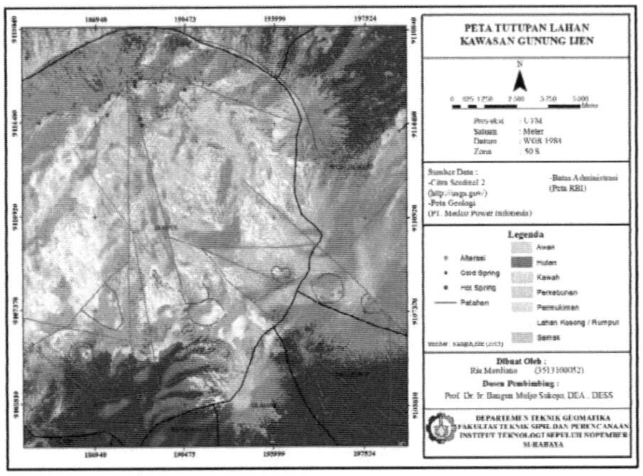


**Figure 9.** Map of land cover

To perform the accuracy test, it takes ground truth data as sample data taken from the location of sample points by 30 sample points. The result of the confusion matrix calculation is 80%. If the land cover is dominated by trees, the smaller the geothermal potential in the area [10]. The following is the area of each land cover.

**Table 3.** Results of Land Cover

| No | Land Cover | Luas (ha) | % |
|---|---|---|---|
| 1 | Farm | 6621,4923 | 21,72 |
| 2 | Forest | 10711,301 | 35,13 |
| 3 | Copse | 6896,3687 | 22,62 |
| 4 | Grass | 5744,9547 | 18,84 |
| 5 | Crater | 42,962237 | 0,14 |
| 6 | Settlement | 464,89779 | 1,52 |

Land cover in the research area is dominated by forests by 35% of the research area.

### *3.6. Vegetation Index*

The value of vegetation index is one indicator to determine the level of drought or greenness in an area. The band channel in this study can be from band 4 and band 5 from Landsat 8 image whereas for image Sentinel 2 use band 4 and band 8 by using NDVI algorithm (Normalized Difference Vegetation Index). The following is the result of the processing of vegetation density on Landsat 8:

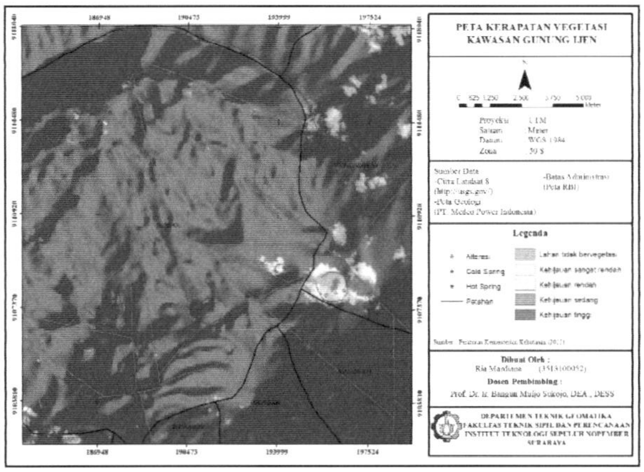


**Figure 10**. Maps of Vegetation Spread Using Landsat Image 8

The following is the result of the processing of vegetation density on Sentinel 2 image:

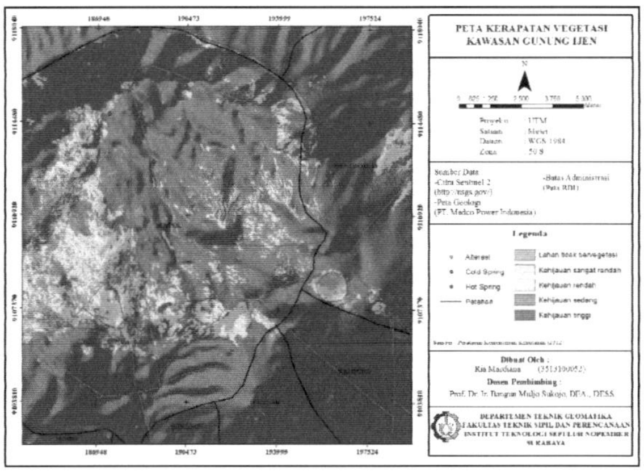


**Figure 11.** Maps of Vegetation Spread Using Sentinel Image 2

From the results of processing vegetation index in the area of Mount Ijen there are 5 classification of index level in the area of Mount Ijen. In Landsat 8 imagery, most of the research area is dominated by high greenish-ness and clouds are also defined with very low greenish density.

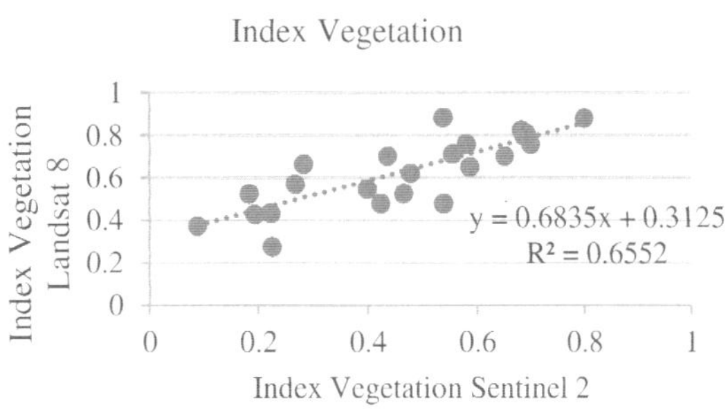


**Figure 12.** Correlation of vegetation index

From result of correlation of Landsat 8 and Sentinel 2 image vegetation density obtained residue equal to 0,80. This shows the correlation of the two relationships is very strong [10]. The higher the value of vegetation density, the lower the geothermal potential of the area [9].

IOP Conf. Series: Earth and Environmental Science **98** (2017) 012025 doi:10.1088/1755-1315/98/1/012025

### *3.7. Analysis of Geothermal Potential*

After all parameters in the determination of the geothermal potential area is processed then the entire parameter area is overlaid so that it produces a new area that is the geothermal potential area in the area of Mount Ijen as follows:

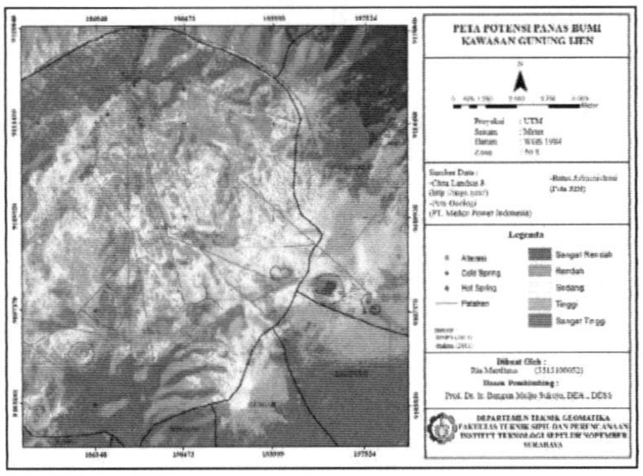


**Figure 13.** Geothermal Potential Map

Based on the results of classification on data processing in the area of Mount Ijen, then obtained the area of each class. The following is a table of geothermal potential in the area of Mount Ijen.

**Table 4.** Area of Geothermal Potential

| | Area (Ha) | % |
|---|---|---|
| Very low | 2017,31 | 6,23 |
| Low | 15724,19 | 48,62 |
| Medium | 10918,9 | 33,76 |
| High | 3638,02 | 11,25 |
| Very High | 39,43 | 0,12 |

The identified areas have very high geothermal potentials located on Mount Ijen located in Wongsorejo Subdistrict. The area has a high surface temperature on a high surface. Besides it has low vegetation index value and is in land cover in the form of grass, crater, settlement and copse.

## 4. Conclusions and suggestions

### *4.1. Conclusion*

Based on the results of this final project, it can be concluded as follows:

1. Based on satellite image processing can be concluded:
   - Surface temperatures in the area of Mount Ijen have anomalies with large temperatures ranging between 18° C to 38° C. With the correlation between surface temperature and altitude for each thermal band on Landsat 8 image produces a very close relationship with the highest value of -0.89.
   - The correlation of Landsat 8 and Sentinel 2 image vegetation densities is very strong with a residual value of 0.81.
   - The land cover confidence matrix receives a value of 80%. Land cover in the research area is dominated by forests by 35% of the research area
2. There is a potential area of geothermal potential is very high on Mount Ijen with an area of 39.43 hectares located in Wongsorejo sub-district and adjacent to District Sempol.

### *4.2. Suggestion*

The advice given from the results of this study are as follows:

1. We recommend using Landsat imagery that is a bit cloud because the sensor is not able to penetrate the cloud and can affect the results on the classification of vegetation density and temperature processing.
2. To generate the geothermal potential, further geophysical, geochemical and geological studies on geothermal potential points are required.

**References**

[1] Abidin H 2007 Karakteristik Deformasi Gunungapi Ijen dalam Periode 2002-2005 Hasil Estimasi Metode Survei GPS *PROC. ITB Sains & Tek* **39 A**(1&2) 2007 pp 1-22

[2] Cahyono A T 2011 *Perencanaan Pengendalian Banjir Kali Kemuning, Sampang* (Surabaya: Teknik Sipil FTSP-ITS).

[3] Azhari A, S Maryanto and A Rachmansyah 2016 *Identifkasi Struktur Geologi dan Pengaruhnya Terhadap Suhu Permukiman Tanah Berdasarkan Data Landsat 8 di Lapangan Panas Bumi Blawan. Jurnal Penginderaan Jauh* **13** pp 1-12.

[4] Ibrahim Herman Darnel 2009 *Mempercepat dan Menadikan Indonesia Negara yang Leading dalam Panasbumi di Dunia pada Periode Kepemimpinan Presiden SBY* Jakarta.

[5] Irfandy Ariono 2012 *Analisa Skema Bisnis Pengembangan dan Penentuan Harga Listrik Panas Bumi di Indonesia* (Jakarta : UI).

[6] Klimatologi 2009 *Klimatologi Suatu Pengantar* (Makassar :Universitas Hasanuddin).

[7] Prakosa D and A Wuryanta 2004 *Kajian Perubahan Penutupan Lahan dengan Menggunakan Landsat 7 ETM+ di Sub DAS Batanghari Bulu Tengah, Jambi.*

[8] Presiden 2003 *Undang-Undang Republik Indonesia Nomor 27 Tahun 2003 tentang Panas Bumi.*

[9] Saragih B, Y Prasetyo and B Sasmito 2015 *Identifikasi Manifestasi Panas Bumi dengan Memanfaatkan Kanal Thermal pada Citra Landsat (Studi Kasus: Kawasan Dieng).*

[10] Sugiyono 2007 *Metode Penelitian Administrasi* (Bandung: Alfabeta).

[11] Sutrisno 1995 Penguasaan Teknologi Energi Panasbumi Indonesia. Yogyakarta: *Seminar Nasional Teknologi Energi.*

[12] Wahyuningsih R 2005 Potensi Dan Wilayah Kerja Pertambangan Panas Bumi di Indonesia *Kolokium Hasil Lapangan.*

[13] Zaenuddin A et al Prakiraan bahaya letusan Gunung Api Ijen Jawa Timur *Jurnal Lingkungan dan Bencana Geologi* **3**(2) pp 109-132.

[14] Zhang N, Q Qin, L He and H Jiang 2012 Remote Sensing and GIS Based Geothermal Exploration in Southwest Tengchong, China *Geoscience and Remote Sensing Symposium (IGARSS) IEEE International* pp 5364-5367.

# Evaluation of Total Suspended Sediment (TSS) Distribution Using ASTER, ALOS, SPOT-4 Satellite Imagery in 2005-2012

**T Hariyanto[1*], T C Krisna[1], C B Pribadi[1], A Kurniawan[1], B M Sukojo[1] and M Taufik[1]**

Geomatics Engineering, Institut Teknologi Sepuluh Nopember

tgh_hary@yahoo.com

**Abstract**. Lapindo mud thrown to Porong River from September 27, 2006 brought an enormous impact to the environment and surrounding communities. This will exacerbate the damage Porong ecosystems, and pollute the Madura Strait and surrounding areas (Wibisono, 2006). Disposal of sludge in large quantities and continuously to Porong also indicated sedimentation resulted in Porong River, Porong River estuary and along coastal of Surabaya-Pasuruan. This is because the material sediment transport along water flow, and the influence of geographical conditions, and the waves of the sea water. Satellite image data used in this study is the ASTER in 2005-2008, ALOS/AVNIR-2 in 2010, and SPOT-4 years 2009.2011 and 2012. In the satellite image processing, for obtain the value of is used TSS algorithm of Jing Li (2008) for ASTER satellite imagery, algorithms of Hendrawan and Asai (2008) for the ALOS satellite imagery, and algorithm of Budiman (2004) for the SPOT-4 satellite imagery. TSS value of the image processing results then performed validation / test precision using reference data TSS In-Situ to obtain linear correlation (R2). R2 value was obtained is 0.854 in 2009, 0.761 in 2011, and 0712 in 2013. That indicates that the value of TSS in the field is proportional with the TSS value in image and has a very good correlation. The results show the value of TSS in the study area ranged from 25 until more than 150 mg/L and according to the results of the analysis showed an upward trend of TSS values over time. There are several locations that indicated experiencing severe sedimentation impacts such as in Porong River, Porong River Estuary, Alo River Estuary, and the surrounding area of the estuary. According to Government Regulation Number 82 in 2001, the maximum value of TSS in the river or water is must less than 50 mg/L and so the value of TSS in the study area is very improper that if allowed to continue may damage the ecosystem in the area. Results from this study is expected to be referenced in the rehabilitation, development, and utilization of the area in the research area and coastal zone management in the future.

**Keyword:** TSS, Sedimentation, Satellite imagery, Linear Correlation

## 1. Background

Lapindo mudflow was phenomenom of hot mud spurt at drilling site of Lapindo Brantas since May 27th 2006 in Porong District, Sidoarjo Regency, East Java Province. According to [9], the volume of mudflow reaches 50,000 - 126,000 m3 per day. On September 27th, 2006, the Government approved that mudflow was dumped into Porong River which will be flown into the sea to provide additional time in effort to prepare for others alternative.

The impact of the mud disposal into Porong River against the environment and society was enormous. It will be worsening damage of ecosystem in Porong River, contaminate of Madura Strait and its surrounding [9]. Sedimentation in Porong River, Porong estuary, and along the coast of Surabaya-Pasuruan caused by mud disposal in wavelengths that can be used to measure TSS values, have good spatial resolution and multitemporal.

Parameters which can be used to measure sedimentation were TSS (Total Suspended Sediment). Remote sensing methods using satellite imagery can be solution to do research about issue of sedimentation, because this method more efficient and more effective in large-scale research. One of many kind of satellite imagery which can be used to examine TSS such as ASTER, ALOS / AVNIR-2, and SPOT-4 because it has wavelengths which can be used to measure TSS values, have good spatial resolution and multitemporal.

Based on these issue, so it necessary to do sedimentation research using TSS parameters in estuary of Porong River using ASTER, ALOS / AVNIR-2, and SPOT-4 satellite imagery to provide information about value, distribution, sedimentation potential, and recommendations for coastal area management in these areas. The results of this study were expected to be used to optimize the results of sedimentation research using satellite imagery technology and can be useful for coastal area management in the future.

## 2. Methodology

### *2.1. Study Area*

Study area for this study is located in Porong River, Sidoarjo Regency, East Java, Indonesia with geographical location being on 7°10'20" south latitude - 7°36'00" south latitude and 112°34'52" east longitude - 112°54'36" east longitude.

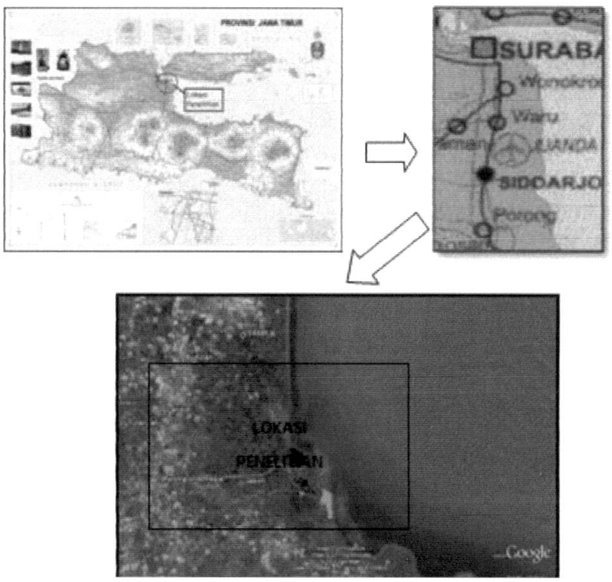


**Figure 1.** Study Area

### *2.2. Data Acquisition*

ASTER Satellite Imagery Level 1B data acquired since 2005 until 2008 which have 15 meters resolution, ALOS / AVNIR-2 level 1B data acquired at 2009 which has 10 meters resolution, and SPOT-4 level 2A data acquired at 2011 which has 20 meters resolution. Geometric correction process used Landsat-7 ETM Ortho data acquired at 2000 which has 14,5 meters resolution. The use of

IOP Conf. Series: Earth and Environmental Science **98** (2017) 012026 doi:10.1088/1755-1315/98/1/012026

satellite imagery due to ASTER, ALOS / AVNIR-2, and SPOT-4 satellite imagery suitable for the study of sedimentation and it has small aquatic areas (Kishino, 2005).

*2.3. Data Processing*

*2.3.1. Cropping Image.* Coverage area used administrative boundary of the village in some districts of Sidoarjo and Pasuruan, for coverage area to the sea used area of study as desired. In order to crop the coverage area, base data used RBI Map scale of 1: 25.000.

*2.3.2. Geometric correction.* The geometric platform of ASTER sensor has 10 control points which was provided with geographic coordinates reference of DEM data were obtained from GTOPO30 1997 data, so ASTER product has been corrected 3D (x, y, z) from the sensor platform and ASTER satellite imagery has a high geometric accuracy. However, there is one important step for geometric correction such as an angular rotation parameter of the sensor. Geometric accuracy of ASTER imagery reaches RMSerror <0.2 for scale of 1: 50,000. It's served as the basis that geometric correction of ASTER imagery up to the rotation only for the accuracy of mapping in this study. Whereas, geometric correction of ALOS / AVNIR-2 level IB and SPOT-4 level 2A using Landsat-7 ETM Ortho image as a reference. Geometric correction aims to reduce geometric errors by taking into account the RMS error must be less than or equal to one. Geometric correction was done by specifying GCP (Ground Control Point). GCP was performed using Landsat-7 ETM Ortho path / row: 118/065 acquisition of August 17, 2000 with a spatial resolution of 14.25 meters. In order to process of geometric correction, used polynomial method with linear interpolation and nearest neighbour resampling method. After determining GCP point then the next step is to calculate the SoF (Strength of Figure) of the GCP point distribution net.

*2.3.3. Masking Image (Separation of Land and Sea).* Masking process was done by looking for the range of maximum and minimum brightness values on the object. Decreasing the brightness value occurs at near infrared wavelengths about 1.4 - 2.7 μm, it is not the case in soil objects (land). In these channels the electromagnetic wave radiation does not penetrate the water body (Lillesand / Kiefer, 1997). Bands with these wavelengths produce very dark brightness values in the image display, so that can be seen the boundary between the waters and the mainland. For ASTER image, masking process using band 3. It was calculated using the following algorithm:

$$\text{if i1<30 then i2 else null} \quad (1)$$

*2.3.4. Conversion of DN to reflectance for ASTER Imagery*

The radian data ($L_{rad}$) were calculated using the following equation:

$$L_{rad} = (DN-1) \text{ x Unit Conversion Coefficient (UCC)} \quad (2)$$

(i.e. Abrams and Hook 1998)

The reflectance data (RTOA) were calculated using the following equation:

$$R = (pi \times L_{rad} \times d^2)/(ESUN_i \times \cos(z)) \quad (3)$$

(i.e. Abrams and Hook 1998)

Where,

Z = $90^0$ - solar elevation angle (5)

ESUN = Solar Atmospheric Irradiance

The count of DN multiplier variables were calculated using the following equation:

$$R= ((DN-1) \times UCC \times pi \times d^2) / (ESUN_i \times \cos(z) \quad (4)$$
$$a = (UCC \times pi \times d^2) / (ESUN_i \times \cos(z)(7)$$

where, a is the multiplier variable (DN-1).

*2.3.5. DN to Reflectance Conversion for SPOT-4 Imagery*

The equations used to change the value of DN into the reflectance value are:

- Digital Number (DN) to Spectral Radiance

$$L\lambda = DN / (G\lambda * A\lambda) + B\lambda \quad (5)$$

- Spectral Radiance to Reflectance

$$\rho_p = \frac{\pi * L_\lambda * d^2}{ESUN_\lambda * \cos \theta_s} \quad (6)$$

Where,
L = spectral radiance at aperture sensor (mWcm-2. $ster^{-1}$. $\mu m^{-1}$).
DN = Digital Number
Gλ = Gain
Aλ = Absolute Correction
Bλ = Bias
P = unitless planetary reflectance at the satellite (on a scale of 0-1).
Π = 3.141593

*2.3.6. TSS algorithm*

- Jing Li (2008) algorithm for ASTER imagery
  Formula Jing Li was a formula which uses Multi Band in its calculations, in this formula uses reflectance band 1 and band 2 for ASTER image. The Value of Suspended Sediment Concentration (SSC) / TSS (Li et al., 2008) using the following model:

$$X=[R_w(\lambda_{550})+R_w(\lambda_{670})]*[R_w(\lambda_{670})/R_w(\lambda_{550})] \quad (7)$$

(Han et al., 2006)

where,
Rw (λi) is the reflectance value of the precipitate with units of wavelength nm.
Rw (λ550) is suitable for detecting high reflections against chlorophyll.
Rw (λ670) is used to detect high absorption of chlorophyll.
[Rw (λ670) / Rw (λ550)] can reduce chlorophyll disorder.

Jing Li proposes the following equation:

$$\text{Log}_{10}\ S\ (mg/l) = 0.892+6.2244*X \quad (8)$$

(Li et al., 2008)

Where S is the value of sediment

$$X = [Rw\ (tm1) + Rw\ (tm2)] * [Rw\ (tm2) / Rw\ (tm1)] \quad (9)$$

Then the formula was converted to the following equation:

$$S = 10^{(0,892 + 6.2244*X)} \quad (10)$$

- Hendrawan and Asai Algorithm (2008) for ALOS / AVNIR-2 imagery
  This equation was applied for ALOS / AVNIRR-2 satellite imagery, using a combination of digital number bands 1, 2, and 3. The algorithm equations used to obtain TSS (mg/L) values are as follows:

$$TSS\ (mg/l) = -1.315*b1 + 2.371*b2 + 0.791*b3 + 9.649 \quad (11)$$

  Where,
  b1 = Digital Number Band 1
  b2 = Digital Number Band 2
  b3 = Digital Number Band 3

- Budhiman Algorithm (2004) for SPOT-4 image
  This equation was based on regression result and refers to [2] algorithm. This equation was applied for SPOT-4 satellite images, using the Reflectance band 2. The following equation:

$$TSS\ (mg/l) = 7{,}9038*\exp(23{,}942*\text{red band}) \quad (12)$$

  Where,
  Red band = reflectance band 2

*2.3.7. Linear Regression and Correlation.* Correlation aims to looking for relationship value between suspended sediment with reflectance band, suspended sediment of each algorithm at each image. From the three values, the regression correlation test was calculated using value of ground truth, and the algorithm with the corresponding value (R2)> 70% was obtained. Furthermore, from the strongly correlated canals a model was prepared through regression analysis, which is an analysis which allows to predict the values of non-free variable of one or more non-free variables. In this study, the sediment value was suspended as non-free variable (Y), based on the reflectance value as known free variable (X).

*2.3.8. Ground truth.* It's aims to determine the conditions of sedimentation in the field and the condition of waters in general. This field data was used as a basis in interpretation of satellite imagery which represents the area.

*2.3.9. Accuracy Test.* The accuracy test in the research was conducting the field check and the measurement of some point (sample area) chosen from each class classification. Purwadhi (2001) in Pahlevi explained that the accuracy of the results was greater than 70% so the classification results are said to have better accuracy.

*2.3.10. Classification.* Classification is the process of grouping pixels into a particular class. In this research, the selected classification method is to classify spectral values using the tool band threshold to ROI on the ROI tools menu. The classification results are:

1. Class 1: classification of sediment values in the range of ≤20 mg / l.
2. Class 2: classification of sediment values in the range of 25-≤50 mg / l.
3. Class 3: classification of sediment values in the range of 50-≤75 mg / l.
4. Class 4: classification of sediment values in the range of 75-≤100 mg / l.
5. Class 5: classification of sediment values in the range of 100> 125 mg / l.
6. Class 6: classification of sediment values in the range of 125> 150 mg / l.
7. Class 7: classification of sediment values in the range> 150 mg / l.

## 3. Result and Discussion

### *3.1. $RMS_{Error}$ of Geometric Correction*

$RMS_{Error}$ of ASTER amount <0.3 pixel, SPOT amount 0.4 pixel, and ALOS amount 0.2 pixel. It can be concluded that the result has fulfilled the RMS Error value for geometric correction of image that must be <1 pixel.

### *3.2. Accuracy Test*

The accuracy test was calculated using In-Situ data from sampling of water in field (2009) and using Water Checker Troll (2011). The linear correlation between TSS field data and TSS satellite imagery on June 11, 2009 has a value of linear correlation ($R^2$) amount 0.84. Whereas, the value of linear correlation ($R^2$) on July 23, 2011 obtained is 0.98. The closer the value of one, can be interpreted that the linear correlation is better and can be interpreted that the value of TSS in the image can describe real TSS value in field.

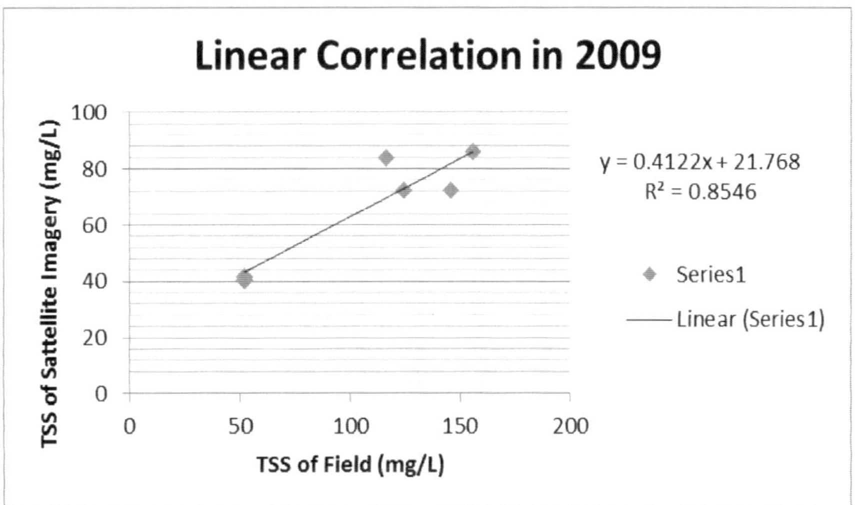


**Figure 2.** Linear Correlation in 2009

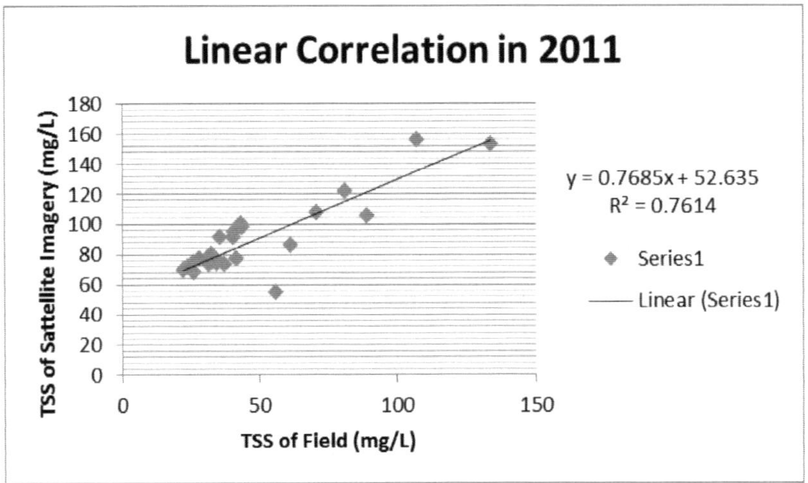


**Figure 3.** Linear Correlation in 2011

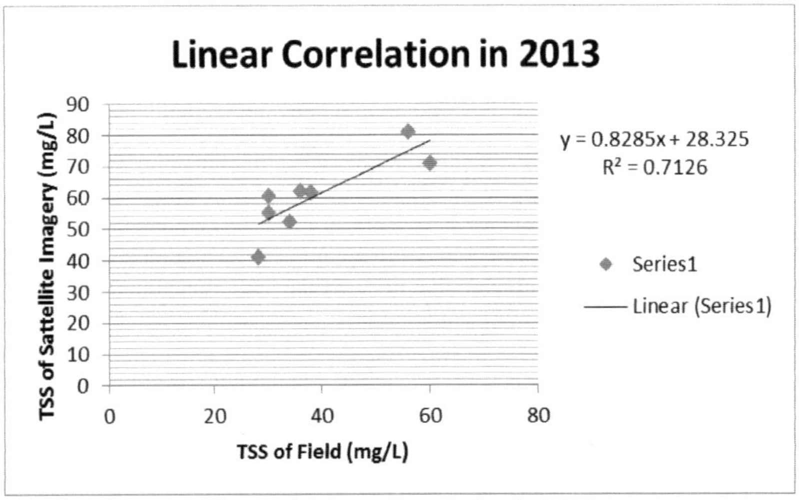


**Figure 4.** Linear Correlation in 2013

## 4. TSS Distribution Area

**Table 1.** TSS Distribution Area (km$^2$)

| CLASS (mg/L) | ASTER 2005 | ASTER 2006 | ASTER 2007 | ASTER 2008 | SPOT 2009 | ALOS 2010 | SPOT 2011 | SPOT 2012 |
|---|---|---|---|---|---|---|---|---|
| **0-25** | 0.000675 | 0 | 0.031725 | 160.107 | 0 | 15.374 | 0.0342 | 0.0477 |
| **25-50** | 108.149 | 181.703 | 143.868 | 54.073 | 106.73 | 33.445 | 104.222 | 127.093 |
| **50-75** | 67.043 | 38.936 | 27.302 | 3.472 | 60.538 | 47.923 | 31.459 | 22.254 |
| **75-100** | 22.532 | 0.3438 | 10.271 | 1.335 | 31.856 | 75.007 | 53.032 | 28.471 |
| **100-125** | 21.059 | 0.001575 | 8.418 | 0.6759 | 14.177 | 34.745 | 20.113 | 22.045 |
| **125-150** | 2.203 | 0.00045 | 3.808 | 0.245475 | 7.997 | 9.256 | 5.863 | 11.014 |
| **>150** | 0.0297 | 0.000225 | 10.417 | 0.37395 | 1.73 | 0.652292 | 5.314 | 8.496 |

Table 1 show us that the area distribution of TSS in each year fluctuated up and down from year to year. So, it cannot be generalized that the broad value of TSS distribution always increase or always decreases from year to year.

**Table 2.** TSS Distribution Area in 2005 and 2006 ($km^2$)

| CLASS (mg/L) | ASTER 2005 | ASTER 2006 |
|---|---|---|
| **0-25** | 0.000675 | 0 |
| **25-50** | 108.149 | 181.703 |
| **50-75** | 67.043 | 38.936 |
| **75-100** | 22.532 | 0.3438 |
| **100-125** | 21.059 | 0.001575 |
| **125-150** | 2.203 | 0.00045 |
| **>150** | 0.0297 | 0.000225 |

Table 2 show us that TSS area occurred before mudflow to Porong River. The Lapindo mud disaster occurred since early 2006 and the mud began to be flowed to Porong River since 26 September 2006. From the data can be described that the area of TSS value generally decrease.

**Table 3**. TSS Distribution Area in 2006 and 2007 ($km^2$)

| CLASS (mg/L) | ASTER 2006 | ASTER 2007 |
|---|---|---|
| **0-25** | 0 | 0.031725 |
| **25-50** | 181.703 | 143.868 |
| **50-75** | 38.936 | 27.302 |
| **75-100** | 0.3438 | 10.271 |
| **100-125** | 0.001575 | 8.418 |
| **125-150** | 0.00045 | 3.808 |
| **>150** | 0.000225 | 10.417 |

Table 3 show us that increase of TSS value before and after the disposal of Lapindo mud into Porong River. From the data can be described that the area increased TSS value.

There was a significant increase in TSS values > 75 mg / L. It indicates that the disposal of mud to Porong River influences the value of TSS in Porong River and its surroundings.

**Table 4.** TSS Distribution Area in 2007 and 2008 ($km^2$)

| CLASS (mg/L) | ASTER 2007 | ASTER 2008 |
|---|---|---|
| **0-25** | 0.031725 | 160.107 |
| **25-50** | 143.868 | 54.073 |
| **50-75** | 27.302 | 3.472 |
| **75-100** | 10.271 | 1.335 |
| **100-125** | 8.418 | 0.6759 |
| **125-150** | 3.808 | 0.245475 |
| **>150** | 10.417 | 0.37395 |

Table 4 show us about comparison area TSS of each class in 2007 and 2008. From the data can be described that the area to not so much changed.

IOP Conf. Series: Earth and Environmental Science **98** (2017) 012026 doi:10.1088/1755-1315/98/1/012026

**Table 5.** TSS Distribution Area in 2008 and 2009 ($km^2$)

| CLASS (mg/L) | ASTER 2008 | SPOT 2009 |
|---|---|---|
| **0-25** | 160.107 | 0 |
| **25-50** | 54.073 | 106.73 |
| **50-75** | 3.472 | 60.538 |
| **75-100** | 1.335 | 31.856 |
| **100-125** | 0.6759 | 14.177 |
| **125-150** | 0.245475 | 7.997 |
| **>150** | 0.37395 | 1.73 |

Table 5 show us about comparison area TSS of each class in 2008 and 2009. From the data can be described that the all area TSS value generaly increased.

**Table 6.** TSS Distribution Area in 2009 and 2010 ($km^2$)

| CLASS (mg/L) | SPOT 2009 | ALOS 2010 |
|---|---|---|
| **0-25** | 0 | 15.374 |
| **25-50** | 106.73 | 33.445 |
| **50-75** | 60.538 | 47.923 |
| **75-100** | 31.856 | 75.007 |
| **100-125** | 14.177 | 34.745 |
| **125-150** | 7.997 | 9.256 |
| **>150** | 1.73 | 0.652292 |

**Table 7.** TSS Distribution Area in 2010 and 2011 ($km^2$)

| CLASS (mg/L) | ALOS 2010 | SPOT 2011 |
|---|---|---|
| **0-25** | 15.374 | 0.0342 |
| **25-50** | 33.445 | 104.222 |
| **50-75** | 47.923 | 31.459 |
| **75-100** | 75.007 | 53.032 |
| **100-125** | 34.745 | 20.113 |
| **125-150** | 9.256 | 5.863 |
| **>150** | 0.652292 | 5.314 |

**Table 8.** TSS Distribution Area in 2011 and 2012 ($km^2$)

| CLASS (mg/L) | SPOT 2011 | SPOT 2012 |
|---|---|---|
| **0-25** | 0.0342 | 0.0477 |
| **25-50** | 104.222 | 127.093 |
| **50-75** | 31.459 | 22.254 |
| **75-100** | 53.032 | 28.471 |
| **100-125** | 20.113 | 22.045 |
| **125-150** | 5.863 | 11.014 |
| **>150** | 5.314 | 8.496 |

Meanwhile, the comparison between the value and distribution of TSS in 2005 representing the value before the occurrence of Lapindo mudflow and TSS in 2012 representing the current condition can be seen in the following table:

**Table 9.** TSS Distribution Area in 2005 and 2012 ($km^2$)

| CLASS (mg/L) | ASTER 2005 | SPOT 2012 |
|---|---|---|
| **0-25** | 0.000675 | 0.0477 |
| **25-50** | 108.149 | 127.093 |
| **50-75** | 67.043 | 22.254 |
| **75-100** | 22.532 | 28.471 |
| **100-125** | 21.059 | 22.045 |
| **125-150** | 2.203 | 11.014 |
| **>150** | 0.0297 | 8.496 |

Table 9 show us about comparison area TSS of each class in 2005 and 2012. From the data can be described that all areas TSS value increased. The increase in terms of value and TSS distribution area indicated that the water quality is getting worse with the higher sedimentation potential. It also illustrates that Lapindo Mud Disposal event has a negative impact on the increase of TSS value in the area of research location.

## 5. Conclusions

The result for this research indicate that TSS value obtained from image processing has good correlation in describing field TSS value, there is a high increase of TSS value in 2005 and 2012 which indicates the influence of the Lapindo mudflow to Porong River, the value of TSS in the majority of research areas has exceeded TSS value threshold set by the Ministry of Environment <70 mg / L while in the study area it can reach > 150 mg / L which can be fatal to ecosystems in research areas such as death of aquatic biota, and severe sedimentation, there is an area indicated sedimentation will occur continuously indicated by the high value of TSS.

## References

[1] Bellote R T 2007 *Sediment Transport on A river-Dominated Shallow Water Shelf : Atchafalaya Bay Region* (Louisisana: Louisana State University).
[2] Budhiman S 2004 Mapping TSM Concentrations From Multisensor Satelite Images in Turbid Tropical Coastal Waters of Mahakam Delta Indonesia *Msc. Thesis* Endesche : ITC.
[3] Fang L G 2009 *Use of Reflectance Ratios as a Proxy for Coastal Water Constituent Monitoring in the Pearl River Estuary* ISSN 1424-8220.
[4] Guzman V R and F G Santaella 2009 Using MODIS 250 m Imagery to Estimate Total Suspended Sediment in a Tropical Open Bay. *International Journal of System Applications, Engineering & Development* **3**(1).
[5] Hariyanto T 2008 Pemetaan Semburan Lumpur Panas Lapindo. Surabaya: *Simposium Nasional Mitigasi Bencana Karena Cuaca Himpunan Mahasiswa Geomatika*.
[6] Lillesand T M and Kiefer R W 1990 *Remote Sensing and Image Interpretation* (New York: John Willey&Son.Inc).
[7] Lillesand T M, Kiefer R W and Chipman J W 2004 *Remote Sensing And Image Interpretation* Fifth Edition (New York : John Wiley & Sons).
[8] Sukresno B *Investigation of Total Suspended Matter in Porong Region Using Aqua-MODIS Satellite Data and Numerical Model.*
[9] Wibisono 2006 *Tragedi Lumpur Lapindo* <URL: http://agorsiloku.wordpress.com/2006/10/11/tragedi-lumpur-lapindo/>. Dikunjungi pada

IOP Conf. Series: Earth and Environmental Science **98** (2017) 012026 doi:10.1088/1755-1315/98/1/012026

tanggal 25 Oktober, jam 18.43.

## Appendix

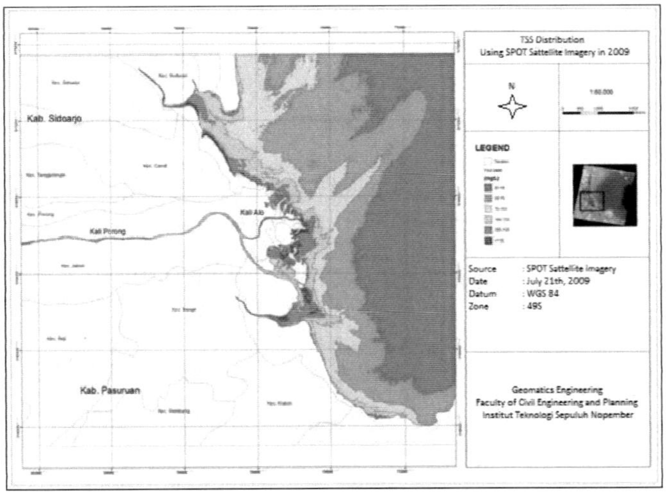


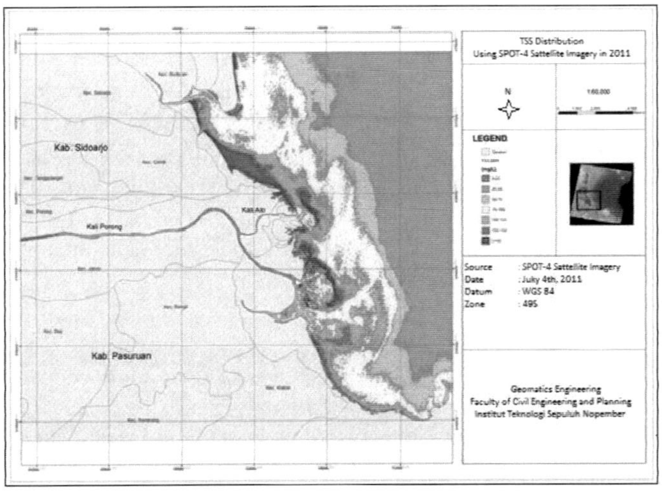

IOP Conf. Series: Earth and Environmental Science **98** (2017) 012026 doi:10.1088/1755-1315/98/1/012026

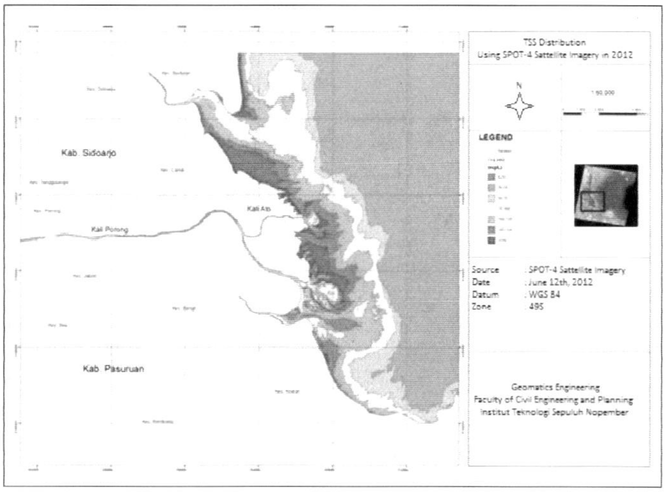

# Analysis of Artificial Neural Network in Erosion Modeling: A Case Study of Serang Watershed

**N Arif [1,2*], P Danoedoro [3], Hartono [3]**
[1]Doctoral Program in Faculty of Geography, Universitas Gadjah Mada, Sekip Utara Bulaksumur, Yogyakarta, 55281, Indonesia
[2]Faculty of Science and Technology, Universitas Muhammadiyah Gorontalo, Pentadio Timur, Gorontalo, 96181, Indonesia
[3]Faculty of Geography, Universitas Gadjah Mada, Sekip Utara Bulaksumur, Yogyakarta, 55281, Indonesia

nursida.arif@um-gorontalo.ac.id

**Abstract**. Erosion modeling is an important measuring tool for both land users and decision makers to evaluate land cultivation and thus it is necessary to have a model to represent the actual reality. Erosion models are a complex model because of uncertainty data with different sources and processing procedures. Artificial neural networks can be relied on for complex and non-linear data processing such as erosion data. The main difficulty in artificial neural network training is the determination of the value of each network input parameters, i.e. hidden layer, momentum, learning rate, momentum, and RMS. This study tested the capability of artificial neural network application in the prediction of erosion risk with some input parameters through multiple simulations to get good classification results. The model was implemented in Serang Watershed, Kulonprogo, Yogyakarta which is one of the critical potential watersheds in Indonesia. The simulation results showed the number of iterations that gave a significant effect on the accuracy compared to other parameters. A small number of iterations can produce good accuracy if the combination of other parameters was right. In this case, one hidden layer was sufficient to produce good accuracy. The highest training accuracy achieved in this study was 99.32%, occurred in ANN 14 simulation with combination of network input parameters of 1 HL; LR 0.01; M 0.5; RMS 0.0001, and the number of iterations of 15000. The ANN training accuracy was not influenced by the number of channels, namely input dataset (erosion factors) as well as data dimensions, rather it was determined by changes in network parameters.

**Keywords**: neural network, erosion, watershed

## 1. Introduction

Soil erosion is a process of soil displacement from one place to another, which becomes an indicator of declining quality of land, thus affecting the sustainability of agricultural production globally [1-4]. Prediction of erosion-prone area is important for evaluation of land use and to support decisions in making regulations on land management. The problem is there is high uncertainty of spatial data,

IOP Conf. Series: Earth and Environmental Science **98** (2017) 012027 doi:10.1088/1755-1315/98/1/012027

greatly affecting the accuracy of prediction result, including erosion prediction. Spatial data is very complex and contains a lot of uncertainty, as evident in the nature, collection, and analysis technique of the data.

Problems in the classification of non-linear spatial data with dimension of big data can be solved by artificial neural network approach [5]. The artificial neural network is one machine learning that is widely used in solving various problems of classification and prediction. An artificial neural network is a process of training input data to identify an output pattern, i.e. the target that has been determined by its ability to learn. Study of erosion of artificial neural network has been performed by several researchers by erosion-prone area prediction modeling with good resulting accuracy [6-7]. [8] Perform similar research, which is mapping landslide-prone area using artificial neural network with accuracy of prediction up to 82,92 %. The weakness of artificial neural network is incomprehensible supporting theoretical evidence in selecting a combination of parameter value of ANN, including ANN structure selected to produce effective results, making ANN difficult to apply [8-9]. The current research tested several parameter combinations to produce good accuracy. The algorithm used for execution was backpropagation. There are many artificial neural network algorithms that can be used in the training process. Backpropagation algorithm in multilayer perceptron (MLP) network is one that is widely utilized for complex problems. The previous researcher [10] compare the guided algorithms: Backpropagation, Self-Organizing Mapping and Simulated Annealing using Landsat TM imagery in the classification of land cover conclude that the backpropagation method obtains the most accurate precision compared to other algorithms. Some other researchers also prove that backpropagation algorithm is reliable in the classification of spatial data with good accuracy [8,11-13].

Backpropagation algorithm uses error output to change the loading values backward. To fix this error, forward propagation phase must be performed in advance [14]. In addition to write algorithm selection, network operation is likewise influenced by the pattern of data input and the value of the network parameters, hence that the network can perform the function of generalization and memorization well. In this paper, the focus of study was comparing the combination of various input parameters of artificial neural network to get good classification results.

## 2. Method

The entire experiment was performed on Windows 8 (64-bit) operating system with i3 processor and 2 GB RAM. Artificial neural network training was performed on IDRISI using MLP (multilayer perceptron) tool. The training data consisted of 9 layers which are the factors of erosion control, i.e. erosivity (R), erodibility (K), length and slope (LS), land cover management (C), factors supporting land conservation practice (P) and four additional layers of remote sensing data, i.e. SPOT 5 multispectral imagery with four bands/channels that provided different spectral reflectance values for each land cover/land use. The neural network architecture in this modeling is as shown in Figure 1. Total data collected was 53 samples of which 30 samples were used as training data and 23 samples as test data for verification. The experiment conducted with 23 simulations by changing the values of network parameters, i.e. learning rate (LR), momentum (M), number of hidden layers (HL), number of iterations (I) and RMS Error (Table 1)

The training results were evaluated using an overall accuracy calculation. One that is widely used in accuracy calculation of the classification is by preparing classification error matrix. The correlation between known reference data and classification results can be compared with error matrix. Overall accuracy was computed by dividing the number of right pixels (the number of main diagonal) with the total number of pixels in the error matrix [15].

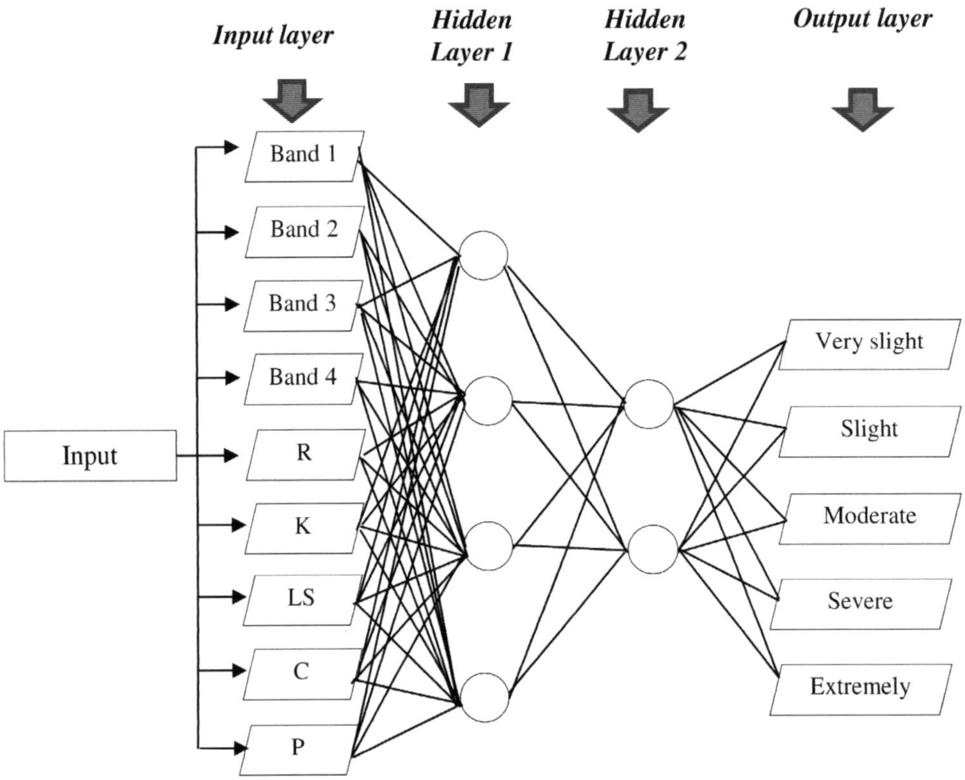


**Figure 1**. Architecture of artificial neural network for erosion (Band 1, band 2, band 3, band 4, band 5 = SPOT 5 satallite image), R = erosivity data image, K = erodibility data image, LS = length and slope steepness data image; c = vegetation cover management data image; P = supporting land conservation practice data image

**Table 1**. Parameters of network training structure

| Name | HL | LR | M | RMS | I | Name | HL | LR | M | RMS | I |
|---|---|---|---|---|---|---|---|---|---|---|---|
| ANN 1 | 1 | 0.001 | 0.5 | 0.0001 | 5000 | ANN 13 | 1 | 0.001 | 0.5 | 0.0001 | 25000 |
| ANN 2 | 1 | 0.001 | 0.5 | 0.0001 | 10000 | ANN 14 | 1 | 0.01 | 0.5 | 0.0001 | 15000 |
| ANN 3 | 1 | 0.001 | 0.5 | 0.0001 | 15000 | ANN 15 | 2 | 0.01 | 0.9 | 0.001 | 10000 |
| ANN 4 | 2 | 0.001 | 0.5 | 0.0001 | 5000 | ANN 16 | 2 | 0.001 | 0.9 | 0.0001 | 15000 |
| ANN 5 | 2 | 0.001 | 0.5 | 0.0001 | 15000 | ANN 17 | 1 | 0.001 | 0.9 | 0.0001 | 5000 |
| ANN 6 | 1 | 0.001 | 0.9 | 0.0001 | 15000 | ANN 18 | 1 | 0.001 | 0.5 | 0.0001 | 20000 |
| ANN 7 | 2 | 0.4 | 0.5 | 0.0001 | 5000 | ANN 19 | 2 | 0.001 | 0.5 | 0.0001 | 20000 |
| ANN 8 | 1 | 0.01 | 0.5 | 0.001 | 5000 | ANN 20 | 1 | 0.001 | 0.9 | 0.0001 | 20000 |
| ANN 9 | 1 | 0.01 | 0.5 | 0.001 | 10000 | ANN 21 | 2 | 0.001 | 0.9 | 0.0001 | 20000 |
| ANN 10 | 2 | 0.01 | 0.5 | 0.001 | 5000 | ANN 22 | 2 | 0.001 | 0.9 | 0.0001 | 25000 |
| ANN 11 | 2 | 0.01 | 0.5 | 0.001 | 10000 | ANN 23 | 1 | 0.001 | 0.9 | 0.0001 | 30000 |
| ANN 12 | 2 | 0.01 | 0.4 | 0.001 | 10000 | | | | | | |

## 3. Experimental Results and Discussion

Table 2 shows the highest training accuracy achieved in this study was 99.32%, occurred in ANN 14 simulation with combination of network input parameters of 1 HL; LR 0.01; M 0.5; RMS 0.0001, and the number of iterations of 15000. The previous researcher [16] have proved the accuracy of training reaches 91.9% when the iterations at 15000. The experimental results showed there was a simulation with low accuracy and some simulations that cannot be executed by artificial neural network because the combination of network input parameters was not considered ideal in the simulations (Table 2)

**Table 2**. Results of simulations with error/low accuracy

| Name | HL | LR | M | RMS | I | Overall accuracy (%) |
|---|---|---|---|---|---|---|
| ANN 4 | 2 | 0.001 | 0.5 | 0.0001 | 5000 | *error* |
| ANN 5 | 2 | 0.001 | 0.5 | 0.0001 | 15000 | 48.69 |
| ANN 7 | 2 | 0.4 | 0.5 | 0.0001 | 5000 | *error* |
| ANN 15 | 2 | 0.01 | 0.9 | 0.001 | 10000 | *error* |

Table 2 shows ANN 4 resulted in an error or "out of memory" whose results were not as expected. This condition occurred in the simulation with 2 HL, but cannot be said that only the hidden layer that affected on accuracy. The following are the effects of each network parameter against the level of accuracy. The first parameter, i.e. **iteration**, was the parameter that had great effect on accuracy. It can be seen from multiple simulations that had the same network parameter values, only differed in the value of iterations (Table 3)

**Table 3**. The comparison of accuracy with different iteration

| Name | HL | LR | M | RMS | Iteration | Overall accuracy (%) |
|---|---|---|---|---|---|---|
| ANN 4 | 2 | 0.001 | 0.5 | 0.0001 | 5000 | *error* |
| ANN 5 | 2 | 0.001 | 0.5 | 0.0001 | 15000 | 48,69 |
| ANN 8 | 1 | 0.01 | 0.5 | 0.001 | 5000 | 98,58 |
| ANN 9 | 1 | 0.01 | 0.5 | 0.001 | 10000 | 95,8 |
| ANN 10 | 2 | 0.01 | 0.5 | 0.001 | 5000 | 98,37 |
| ANN 11 | 2 | 0.01 | 0.5 | 0.001 | 10000 | 99,05 |
| ANN 21 | 2 | 0.001 | 0.9 | 0.0001 | 20000 | 98,78 |
| ANN 22 | 2 | 0.001 | 0.9 | 0.0001 | 25000 | 98,06 |

Table 3 shows overtraining occurs in ANN 8 and ANN 9 simulations. ANN 8 with 5000 iterations had reached 98.58% accuracy, but when added by iterations in ANN 9 to 15000, the accuracy value was decreased. Basically, there was no provision or previous study mentioned the best number of iterations. The initial determination of the number of iterations will be a reference for the next simulation only when the first simulation is carried out. Thus, the thing to be considered when determining the number of iterations is the network is not stuck in overtraining, rather than what are the numbers of iteration that must be carried out in every training. Because the data pattern in each case was different, the ideal treatment or artificial neural network architecture will also be different. Overall, from 23 simulations that were carried out, the best iteration was 15000 since it occurred in some simulations. The optimal number of iterations was also strongly affected by the combination with other parameters, such as momentum and RMS [13]. The simulations had been carried out with two different momentum values tested in different iterations (Table 4).

**Table 4**. The comparison of accuracy in different iteration and momentum

| Name | HL | LR | M | RMS | I | Overall accuracy (%) |
|---|---|---|---|---|---|---|
| ANN 1 | 1 | 0.001 | 0.5 | 0.0001 | 5000 | 81,81 |
| ANN 2 | 1 | 0.001 | 0.5 | 0.0001 | 10000 | 89,8 |
| ANN 3 | 1 | 0.001 | 0.5 | 0.0001 | 15000 | 94,58 |
| ANN 13 | 1 | 0.001 | 0.5 | 0.0001 | 25000 | 95,48 |
| ANN 18 | 1 | 0.001 | 0.5 | 0.0001 | 20000 | 96,16 |
| ANN 17 | 1 | 0.001 | 0.9 | 0.0001 | 5000 | 96,34 |
| ANN 6 | 1 | 0.001 | 0.9 | 0.0001 | 15000 | 98,51 |
| ANN 20 | 1 | 0.001 | 0.9 | 0.0001 | 20000 | 98,6 |
| ANN 23 | 1 | 0.001 | 0.9 | 0.0001 | 30000 | 99.23 |

The changes of iteration carried out in the values of momentum in 0.5 and 0.9 showed M value = 0.5 higher accuracy by the addition of the number of iterations. The researcher performed the addition because of seeing the pattern of increased accuracy when the number of iterations added by the assumption of low accuracy because of the number of training was less (under training). An overview of accuracy changes at each iteration with the values of M = 0.5 and M = 0.9 is presented in Figure 2.

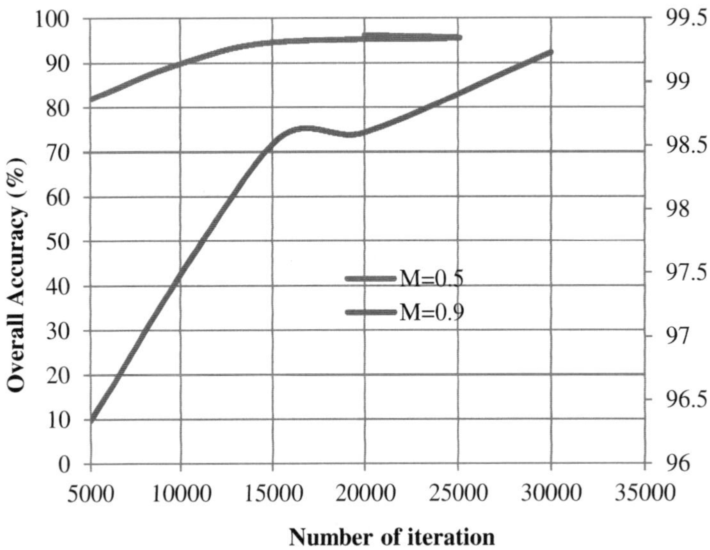


**Figure 2.**The correlation between iteration, momentum, and accuracy

The second parameter, i.e. **learning rate**, affected on the intensity of the training process. Learning rate becomes a problem when the number of training is too much, so that the system is not able to generalize well. Learning rate that is too high requires a longer time for continuous training, so that optimal convergence is not achieved [17]. Lower learning rate guarantees a decrease in gradient, but can increase the number of iterations. In the simulations carried out in the study, the most optimal learning rates producing the best accuracy was 0.001 and 0.01, as on previous research conducted by [13]. The correlation with accuracy and time was inversely related. The lower the value of learning rate, the higher the accuracy, and the longer the computation time. Otherwise, the higher the value of learning rate, the lower the accuracy, and the faster the computation time. The value of learning rate that was higher had been tested in the simulation conducted at ANN 7, i.e. 0.4 (Table 2), but the results of training cannot be executed. In contrast to those produced by [16], learning rate of 0.4 produces an accuracy of 75%. Generally, the value of LR been ranging from 0.001 to 1 [18]. The study had been attempted with LR 0001; 0.01 was quite optimal to use. Both values have also been proven by [13]. His study can yield the highest accuracy, but also can yield low accuracy if other input parameters are not correct.

The third parameter, i.e. **hidden layer**, functions as a place where algorithm improves the load of input before comes to the output layer. ANN has knowledge rules that are difficult to formulate as seeing how much each of the parameters affect the accuracy, including hidden layer. Some simulations with 1 HL had very good accuracy, but contrarily, some simulations with 2 HL had better accuracy. Some of the following simulation shows the difference in accuracy between 1 HL and 2 HL (Table 5)

**Table 5**. The comparison of accuracy in different hidden layer

| Name | HL | LR | M | RMS | I | Overall accuracy (%) |
|---|---|---|---|---|---|---|
| ANN 1 | 1 | 0.001 | 0.5 | 0.0001 | 5000 | 81,81 |
| ANN 4 | 2 | 0.001 | 0.5 | 0.0001 | 5000 | *error* |
| ANN 3 | 1 | 0.001 | 0.5 | 0.0001 | 15000 | 94,58 |
| ANN 5 | 2 | 0.001 | 0.5 | 0.0001 | 15000 | 48,69 |
| ANN 8 | 1 | 0.01 | 0.5 | 0.001 | 5000 | 98,58 |
| ANN 10 | 2 | 0.01 | 0.5 | 0.001 | 5000 | 98,37 |
| ANN 9 | 1 | 0.01 | 0.5 | 0.001 | 10000 | 95,8 |
| ANN 11 | 2 | 0.01 | 0.5 | 0.001 | 10000 | 99,05 |
| ANN 18 | 1 | 0.001 | 0.5 | 0.0001 | 20000 | 96,16 |
| ANN 19 | 2 | 0.001 | 0.5 | 0.0001 | 20000 | 74,72 |
| ANN 20 | 1 | 0.001 | 0.9 | 0.0001 | 20000 | 98,6 |
| ANN 21 | 2 | 0.001 | 0.9 | 0.0001 | 20000 | 98,78 |

Based on Table 5 above, the combination of parameters of LR 0.001; M 0.5; RMS 0.0001 would be optimal with 1 HL although changes were made to the number of iterations as in simulations ANN 1 and ANN 4, ANN 3 and ANN 5, ANN 18 and ANN 19. When the momentum parameter and the value of learning rate were changed, the resulting accuracies on some combinations with 2 HL were optimal, and in other combinations with 1 HL were sufficient. These conditions occurred in the simulations ANN 9 and ANN 11, ANN 20, and ANN 21. Overall, the best accuracy occurred with 1 HL. Some simulations with 2 HL achieved the highest accuracy as in ANN 11 and ANN 21, but accuracy with 1 HL on the same simulations and parameters had reached an accuracy of > 90%. So, it can be said that one hidden layer was sufficient to solve the problems with the characteristics of the data in this study.

[14] supports the above facts that 1 HL has been already quite optimized to use, because the more the number of HL, the lower the computing process; even in some cases, the computer cannot process them as in ANN 4 and ANN 5.

The fourth parameter, i.e. **momentum**, functions to accelerate the convergence of the algorithm by forcing the load change process to continue to move, so that not being trapped in a local minimum, resulting in the training process to be faster. However, the training time in this study was not recorded since not been the focus of study. The difference in accuracy with different momentum values is presented in Table 6.

**Table 6**. The Comparison between Accuracy and Different Momentum Values

| Name | HL | LR | M | RMS | I | Overall accuracy (%) |
|---|---|---|---|---|---|---|
| ANN 1 | 1 | 0.001 | 0.5 | 0.0001 | 5000 | 81,81 |
| ANN 17 | 1 | 0.001 | 0.9 | 0.0001 | 5000 | **96,34** |
| ANN 3 | 1 | 0.001 | 0.5 | 0.0001 | 15000 | 94,58 |
| ANN 6 | 1 | 0.001 | 0.9 | 0.0001 | 15000 | **98,51** |
| ANN 11 | 2 | 0.01 | 0.5 | 0.001 | 10000 | 99,05 |
| ANN 12 | 2 | 0.01 | 0.4 | 0.001 | 10000 | **99,23** |
| ANN 18 | 1 | 0.001 | 0.5 | 0.0001 | 20000 | 96,16 |
| ANN 20 | 1 | 0.001 | 0.9 | 0.0001 | 20000 | **98,6** |
| ANN 19 | 2 | 0.001 | 0.5 | 0.0001 | 20000 | 74,72 |
| ANN 21 | 2 | 0.001 | 0.9 | 0.0001 | 20000 | **98,78** |

Based on the observation on the results of simulations, the high value of learning rate required lower momentum, otherwise low value of learning rate required higher value of momentum to achieve optimal effectiveness and convergence during the training process. It can be seen from the simulation results of ANN 6, ANN 20, and ANN 21, in which the momentum, value of 0.9 in learning rate value of 0.001 was more optimal than the simulation using a momentum value of 0.5. While [11] give a rather good accuracy on the momentum value of 0.5.

The fifth parameter that affected the network was **root mean square (RMS) error**. RMS is the root mean square that occurs between the network output and target output. This study used two RMS values in simulations, i.e. 0.001 and 0.0001. From the test results, the highest accuracy on each combination occurred in RMS value of 0.0001. The analysis results of the entire simulation performed showed both the RMS values (0.001 and 0.0001) likely to yield better accuracy if the values of other network parameters (learning rate, hidden layers, iteration) were right through trial-and-error. The comparison of the accuracy, resulting in both RMS values is presented in the following Table 7

**Table 7**. The Comparison of Accuracy in Different RMS

| Name | HL | LR | M | RMS | I | Overall accuracy (%) |
|---|---|---|---|---|---|---|
| ANN 9 | 1 | 0.01 | 0.5 | 0.001 | 10000 | 95,80 |
| ANN 14 | 1 | 0.01 | 0.5 | 0.0001 | 15000 | 99,32 |

The number of iterations has a correlation with the value of the RMS. The small RMS error requires many iterations. Otherwise, the great RMS error requires fewer iterations, as proven by [13]. The same conclusion obtained in this study as described in the graphic of the results of simulations ANN 1 and ANN 2 (Figure 3).

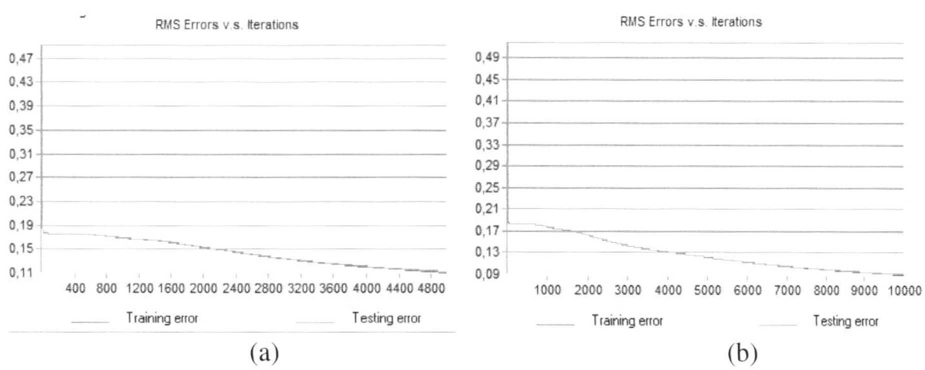


**Figure 3**. The Correlation between RMS Error and Iteration (a)ANN 1; (b) ANN 2

Test results on 23 simulations showed the highest accuracy ANN 14 simulation with overall accuracy reaching 99,32 %. This resulting accuracy is higher than that produced by [19] in the land use classification using artificial neural networks which resulted in the training accuracy by 95%. An image of classificationof ANN 14 is shown in Figure 4 where distribution of erosion class in the research site was dominated by the following erosion classes, namely very severe erosion (34.22%) spreading across most of the area of Kokap Sub-district, Girimulyo Sub-district, and some of the area of Pengasih Sub-district; severe erosion (20.75%) spreading all over Panjatan Sub-district, Pengasih Sub-district, and Nanggulan Sub-district; moderate erosion (29.78%) spreading across Pengasih Sub-district, some of the area in Wates Sub-district, and Panjatan Sub-district; as well as slight erosion (10.81 %) and very slight erosion (4.46%) spreading all over Temon Sub-district and Wates Sub-district. The percentages of distribution for each erosion class are shown in the histogram presented in Figure 5

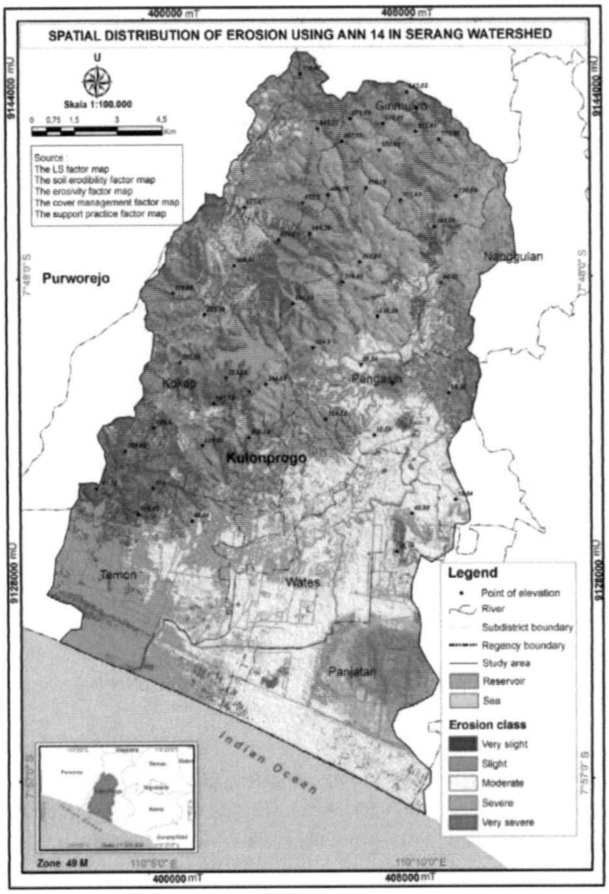


**Figure 4**. Map of spatial distribution of erosion (Analysis, 2016)

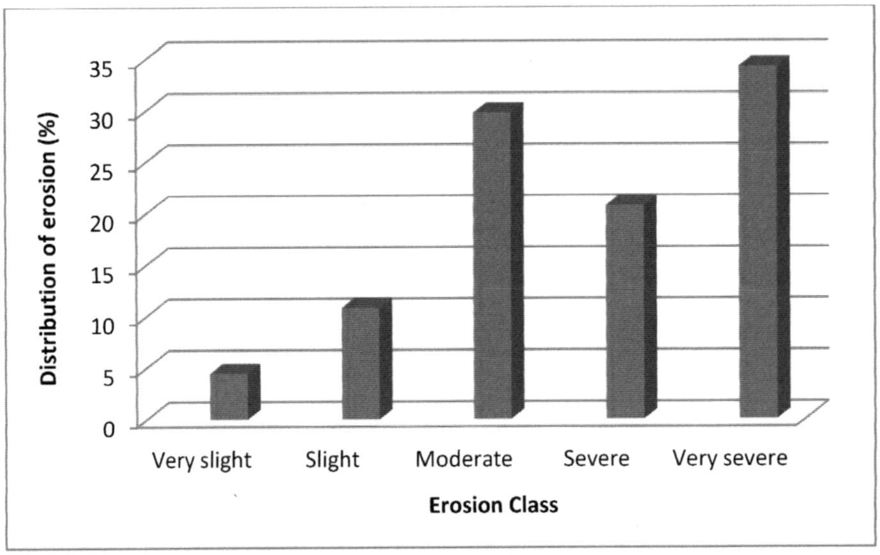


**Figure 5**. The Histogram for the Distribution of Erosion Classes According to the Results of the ANN 14 (Analysis, 2016)

## 4. Conclusion

The artificial neural network is effectively used for the classification of spatial data, although trained by the small number of samples. The complexity of data input can be understood well by the network while training with the accuracy of 99.32%. The most influential parameter in this study was the number of iterations although generally, the right combination among network parameters was also an important concern. The parameter of hidden layer did not have a significant effect on the level of accuracy.

## Acknowledgement

The authors would like to thank National Institute of Aeronautics and Space, Indonesia, Meteorological Climate and Geophysics Agency for providing the data. Financial support from the Ministry of Research, Technology and Higher Education of the Republic of Indonesia, which has provided postgraduate scholarship in Universitas Gadjah Mada, Yogyakarta and funding this Doctoral Dissertation Skim.

## References

[1]. Arsyad S 1989 *Konservasi Tanah dan Air* Penerbit IPB Press Bogor
[2]. Bouaziz M, Leidig M, Gloaguen R 2011 Optimal Parameter Selection for Qualitative Regional Erosion Risk Monitoring: A Remote Sensing Study of SE Ethiopia. *Geoscience Frontiers* **2** (2) pp 237-245
[3]. Morgan R P C 1995 Soil Erosion and Conservation, 2nd Edition. Longman Group, Ltd., London, 198 p.
[4]. Parveen R, Kumar U 2012 Integrated Approach of Universal Soil Loss Equation (USLE) and Geographical Information System (GIS) for Soil Loss Risk Assessment in Upper South Koel Basin, Kharkhand. *J.Geographic Information System* **4** pp 588-596
[5]. Melchiorre C, Matteucci A, Azzoni A, Zanchi 2008 Artificial Neural Networks and Cluster Analysis in Landslide Susceptibility Zonation *Geomorphology* **94** pp 379-400
[6]. De la Rosa D, de la Mayol, F Lozano S 1999 An Expert System/Neural Network Model (impelERO) for Evaluating Agricultural Soil Erosion in Andalucia Region, Southern Spain. Agriculture, *Ecosystem and Environment* **73** pp 211-226
[7]. Kim M, Gilley J E 2008 Artificial Neural Network Estimation of Soil Erosion and Nutrient Concentrations in Runoff From Land Application Areas *Comput Electron Agric.* doi://10.1016/j.compag.2008.05.021
[8]. Pradhan B, Lee S, Buchroitner M F 2010 A GIS-Based Backpropagation Neural Network Model And Its Cross-Application and Validation for Landslide for Susceptibility Analyses. Computers *Environment and Urban Systems* **34** pp 216-235
[9]. Liu Q, Wu G, Chen J, Zhou G 2012 Interpretation Artificial Neural Network in Remote Sensing Image Classification. *IEEE*
[10]. Yuan H, C F Van Der Wiele, S Khoram 2009 An Automatad Artificial Neural Network System for Land Use/Land Cover Classification from Landsat TM Imagery *Remote Sens.* **1** (3) pp 243-265
[11]. Song H, Xu, Ma Y, Li G 2013 Classification of ETM+ Remote Sensing Image Based on Hybrid Algorithm of Genetic Algorithm and Backpropagation Neural Network *Hindawi Publishing Corporation, Mathematical Problems Engineering*
[12]. Conforti M, Pascale S, Robustelli G, Sdao F 2014 Evaluation of Prediction Capability of The Artificial Neural Networks for Mapping Landslide Susceptibility in The Turbolo River Catchment (northern Calabria, Italy) *Catena* **113** pp 236-250

[13]. Arif N, Danoedoro P 2013 An Analyze of Backpropagation Neural Network in The Identification of Critical Land Based on Alos Imagery *Proc. Int. Conf. on 34$^{th}$ Asian Conference on Remote Sensing*, Volume **1** pp 589-593
[14]. Fausett L 1994 Fundamentals Of Neural Networks: Architectures, Algorithms, and Applications, Prentice-Hall, New Jersey
[15]. Congalton R G, Green K 2009 Assessing the Accuracy of Remotely Sensed Data: Principles and Practices CRC Press
[16]. Chen Y R, Chen J W, Hsieh S C, Ni P N 2009 The Application of Remote Sensing Technology to The Interpretation of LU for Rainfall – Induced Landslide Based on Genetic Algorithm and Artificial Neural Network *IEEE Journal of Selected Topics in Applied Earth Observations and remote Sensing*, Vol **2** No 2 pp 87-95
[17]. Wilson W R, Martinez T R 2001 The Need for Small Learning Rate on Large Problems. *Proceedings of The International Journal Conference on Netherlands* pp 115 – 119
[18]. Purnomo M H, Kurniawan A 2006 Supervised Neural Networks, Graha Ilmu, Yogyakarta, 2006 (Text in Indonesia)
[19]. Chen Y R, Chen J W, Hsieh S C, Ni P N 2013 The Application of Remote Sensing Technology to The Interpretation of LU for Rainfall – Induced Landslide Based on Genetic Algorithm and Artificial Neural Network *IEEE Earth Observations and remote Sensing*, Vol **2** (2) pp 87-95

# Determination of the Priority Aspect in Process of Determining Urban Residential Area

**Cherie Bhekti Pribadi[1*], Teguh Hariyanto[1]**

[1]Geomatics Engineering, Institut Teknologi Sepuluh Nopember, Surabaya, Indonesia

cherie_b@geodesy.its.ac.id

**Abstract**. Recently, the population growth is increasingly crowded in urban area and as a source of land supply for housing development into the needs of each individual becomes increasingly limited. the home is one of the basic rights of the people and therefore every citizen has the right to reside and got a good environment and healthy. Besides the house is also a basic human need to enhance the dignity, the dignity, quality of life and livelihood, as well as personal self-reflection in order to improve the standard of living, as well as the formation of character, character and personality of the nation. The construction of such housing should always take into consideration many aspects in order to create a balanced environment not only based on the suitability of land but also based on the aspects that exist in surrounding. Therefore, this study will be to analyze the priority aspect in process of determining urban residential area using Analytical Hierarchy Process Method. This method aims to determine priority aspect were used based on primary data collection form of questionnaires and interviews to respondents who expert in the field.

**Keywords :** Aspect, AHP, Urban Area, Residental Area

## 1. Background

Population growth is increasingly crowded in urban area every time especially in major city in Indonesia such as Surabaya. It plays important rule as a source of land supply for residential area development into the needs of each individual becomes increasingly limited. Residential area is the one of the basic rights of the people, therefore every citizen has the right to reside and got a good environment and healthy. Besides the house is also a basic human need to enhance the dignity, the dignity, quality of life and livelihood, as well as personal self-reflection in order to improve the standard of living, as well as the formation of character, character and personality of the nation. The construction of such housing should always take into consideration many aspects in order to create a balanced environment not only based on the suitability of land but also based on the aspects that exist in surrounding. The transformation of land use in urban areas is often carried out in order to meet the housing needs for people are increasing.

Residential area development should meet the basic requirements for urban development such as accessibility, compatibility, flexibility, and ecology. The designated residential area is an area intended for residential or residential environments and places of activity which can support the life and livelihoods [1]. The construction of such housing should always take into consideration environmental aspects and development sites in accordance with the spatial plan to create a balanced environment and the provision of facilities and infrastructure necessary to develop an efficient and integrated way, directed, planned and ongoing / continuous.

Therefore, this study will be to analyze the priority aspect in process of determining urban residential area using Analytical Hierarchy Process Method. This method aims to determine priority aspect were used based on primary data collection form of questionnaires and interviews to respondents who expert in the field.

## 2. Methodology

### 2.1. *Study Area*

Study area for this study is located in West Surabaya, Surabaya city, East Java, Indonesia with geographical location being on 7°13'22",238 south latitude - 7°14'44",476 south latitude and 112°35'29",605 east longitude - 112°43'3",973 east longitude.

### 2.2. *Data Processing*

In making policy on the determination of the location of residential area development, we used three criteria in this study such as criteria of social, economy, and accessibility. The factors which influence the determination of the location of residential area development used as sub-criteria, where the similar subcriteria were grouped in the same criteria. The Analytical Hierarchy Process (AHP) method aims to determine the priority level of the alternative location of residential area development by determining the weighting of each criterion and subcriteria[2]. The value of the priority scale used in this study was derived from primary data collection in the form of questionnaires and interviews to some respondents who are experts in social, economics, accessibility, population, education facilities, employment, land / building price, population income, transportation, electricity network, and water distribution at Lakarsantri, Sukomanunggal, and Sambikerep.

There were some steps in processing data. First, to obtain the priority scale value which will be used for pairwise comparison matrix, then for the criteria and subcriteria priority scale were calculated geometric averages to level out the pairwise results because the assessment involves more than one person. Key informants are members who reflect the interests of people who know a lot about the culture, are able and willing to share their knowledge [6]. Geometric averages are calculated using the following equation:

$$G= \sqrt{(n\&X1.X2.X3 \ldots\ldots\ldots\ldots\ldots\ldots Xn)} \quad (1)$$

Where,

G = geometric average
X1, X2, …., Xn = assessment to-1,2,…………,n
n = number of assessment

Second, these priority scale results as a pairwise comparison matrix of criteria and sub criteria, then normalized to standardize every element of eigenvector. So, we can get the priority weight of each criterion and sub criteria. Before normalizing in pairwise comparison matrix, the columns of each criterion/sub criteria are summed to use as a calculation in the next step by using the following equation:

$$Sj = \sum_{i=1}^{n} a_{ij} \quad (2)$$

Where,

Sj = Total Value of Each Column
$aij$ = Matrix i to j
n = Number of Criterion [3]

Normalizing in pairwise comparison matrix were calculated by using the following equation:

$$V_{ij} = \frac{a_{ij}}{s_j} \quad (3)$$

Where,
Vij = Normalized Value of Matric I to j
Sj = Total Value of Each Column
*aij* = Matrix i to j [3]

The priority weight of each criterion was obtained from summation of each line of the normalization of pairwise comparison matrix, then it was divided by the number of criterion as in the following equation:

$$w_i = \frac{\sum_{j=1}^{n} v_{ij}}{n} \quad (4)$$

Where,
Wij = Priority Weight
Vij = Normalized Value of Matric i to j
n = Number of Criterion [4]

The maximum eigen value (λmax) was calculated to measure the consistency of a matrix by using the following equation:

$$\lambda_{max} = \frac{1}{n} \sum_{i=1}^{n} \frac{[Aw]i}{w_i} \quad (5)$$

Where,
Λmax = Maximum Eigen Value
A = Eigenvalue
*wi* = Priority Weight
n = Number of Criterion [4]

The consistency index (CI) aims to provide information about logical consistency of pairwise comparison. When CI = 0.0, there were a logical inconsistency between pairwise comparison, or the assessment was considered 100% consistent.

After maximum eigen value (λmax) was obtained, then consistency index value calculated by using the following equation:

$$CI = \frac{\lambda_{max} - n}{n-1} \quad (6)$$

Where,
CI = Consistency Index
λmax = Maximum Eigen Value
n = Number of Criterion [5]

The Consistency Ratio (CR) aims to measure how a consistent assessment relative to a large sample of random purely assessments. After the consistency index value (CI) was obtained, then the value of consistency ratio calculated by using the following equation:

$$CR = \frac{CI}{RI} \quad (7)$$

Where,
CI = Consistency Index
CR = Consistency Ratio
RI = Random Index

## 3. Results and Discussion

The weighting factor calculation is performed on each criterion and sub criteria as follows:

### *3.1. Criteria*

Primary data / quantitative data obtained from eight respondents to the assessment among criteria first calculated average by using geometric mean that aims to flatten the results of pairwise comparisons. The results of the priority scale of the questionnaire of respondents as well as the results of the calculation of geometric mean is contained in the following table.

**Table 1.** Average geometric criteria

| District | Social-Economic | Social-Accessibility | Economic-Accessibility |
|---|---|---|---|
| Lakarsantri District | 1/6 | 5 | 1 |
| Lidah wetan Subdistrict | 7 | 5 | 1/5 |
| Sambikerep District | 1/5 | 1/9 | 1/7 |
| Lontar Subdistrict | 4 | 1 | 3 |
| Sambikerep Subdistrict | 1/3 | 1/5 | 1/3 |
| Sukomanunggal District | 4 | 1/5 | 5 |
| Sonokwijenan Subdistrict | 1 | 5 | 5 |
| Sukomanunggal Subdistrict | 1/7 | 1/9 | 1/9 |
| **Average Geometric** | 0.80581 | 0.70601 | 0.72854 |

Based on table 1, the average geometric value between social to economic criteria amount 0.80581, among social to accessibility criteria amount 0.70601, and between economic criteria against accessibility criteria of 0.72854. The geometric value of the value aij = 1 for i = j, pairwise comparison matrix can be seen in the following table.

**Table 2.** Pairwise comparison matrix of criteria

| Criteria | Social | Economic | Accessibility |
|---|---|---|---|
| Social | 1 | 0.80581 | 0.70601 |
| Economic | 1.24098 | 1 | 0.72854 |
| Accessibility | 1.41641 | 1.37261 | 1 |
| Total | 3.65739 | 3.17842 | 2.43455 |

Based on table 2, total value of each criteria shows the total value on the social criteria amount 3.65739, economic criteria amount 3.17842, and accessibility criteria amount 2.43455. Then the normalization of pairwise comparison matrix which aims to determine the relative importance of the various criteria computed was calculated using the formula in equation 3. Normalization of pairwise comparison matrix can be seen in the following table.

**Table 3.** Normalization of pairwise comparison matrices criteria

| Criteria | Social | Economic | Accessibility |
|---|---|---|---|
| Social | 0.27342 | 0.25353 | 0.29000 |
| Economic | 0.33931 | 0.31462 | 0.29925 |
| Accessibility | 0.38727 | 0.43185 | 0.41075 |
| Total | 1 | 1 | 1 |

To obtain the priority weight of each criteria, the result of normalization was calculated using the formula in equation 4. So, the priority weight of each criterion can be seen in the following table.

**Table 4.** Priority weight criteria

| Criteria | Total | Weight | Priority |
|---|---|---|---|
| Social | 0.81694 | 0.27231 | 3 |
| Economic | 0.95318 | 0.31773 | 2 |
| Accessibility | 1.22988 | 0.40996 | 1 |
| Total | 3 | 1 | |

Based on table 4, the priority of each criteria used in the process of determining the location of residential area development in West Surabaya were dominated by the highest accessibility criteria amount 0.40996, The weight of the obtained must be determined the ratio of consistency of the opinions of various respondents in this study to measure the quality of priority vector, if the value of consistency ratio less than 0.1, the opinion of respondents which were considered inconsistent were acceptable. Whereas, if the value of consistency ratio is more than 0.1, the opinion of the respondent which were considered inconsistent were not acceptable and must be done by taking re-questionnaire to determine the weight so that the value of consistency ratio is in accordance with the specified. Eigenvalue maximal ($\lambda max$) in these criteria amount 3.00378. Consistent index ($CI$) in these criteria amount 0.00189. Ratio of consistency ($CR$) in these criteria amount 0.00326.

### *3.2. Social Criteria*

The results of the priority scale of the results of questionnaires of respondents as well as the results of the calculation of average geometric mean were contained in the following table.

**Table 5.** Average geometric social sub criteria

| District | Population-Educational | Population – Health Facility | Educational-Health Facility |
|---|---|---|---|
| Lakarsantri District | 1 | 1 | 1 |
| Lidah wetan Subdistrict | 5 | 7 | 3 |
| Sambikerep District | 1 | 3 | 3 |
| Lontar Subdistrict | 7 | 4 | 1/9 |
| Sambikerep Subdistrict | 3 | 1/5 | 1/5 |
| Sukomanunggal District | 3 | 3 | 1 |
| Sonokwijenan Subdistrict | 1 | 1/5 | 1 |
| Sukomanunggal Subdistrict | 1/81 | 1/6 | 9 |
| **Average Geometric** | 1.58271 | 1.06700 | 1.07624 |

Based on Table 5, the average geometric value between sub criteria of population condition to educational amount 1.58271, Pairwise comparison matrix can be seen in the following table.

**Table 6**. Pairwise comparison matrix of social sub criteria

| Subcriteria | Population | Educational | Health Facility |
|---|---|---|---|
| Population | 1 | 1.58271 | 1.06700 |
| Educational | 0.63183 | 1 | 1.07624 |
| Health Facility | 0.93721 | 0.92916 | 1 |
| Total | 2.56903 | 3.51188 | 3.14324 |

Based on Table 6, total value of each sub criteria shows that total value in the sub criteria of population amount 2.56903, sub criteria of educational facility amount 3,51188, and sub criteria of educational facility amount 3.14324. Normalization of pairwise comparison matrix can be seen in the following table.

**Table 7.** Normalization of pairwise comparison matrix of social sub criteria

| Subcriteria | Population | Educational | Health Facility |
|---|---|---|---|
| Population | 0.38925 | 0.45068 | 0.33946 |
| Educational | 0.24594 | 0.28475 | 0.34240 |
| Health Facility | 0.36481 | 0.26458 | 0.31814 |
| Total | 1 | 1 | |

**Tabel 8.** Weight of subcriteria social priority

| Sub criteria | Total | Weight | Priority |
|---|---|---|---|
| Population | 1.17938 | 0.39313 | 1 |
| Educational | 0.87309 | 0.29103 | 3 |
| Health Facility | 0.94753 | 0.31584 | 2 |
| Total | 3 | 1 | |

Based on Table 8, the priority of each sub criteria in the social criteria used in the process of determining the location of residential area in West Surabaya were dominated by the population sub criteria amount 0.3931. Eigenvalue maximal ($\lambda max$) in this sub criteria amount 3.02479. Consistent index ($CI$) in these criteria amount 0.01239. Ratio of consistency ($CR$) in these criteria amount 0.02137.

*3.3. Economic Sub criteria*

The results of the priority scale of the results of questionnaires of respondents as well as the results of the calculation of average geometric mean were contained in the following table.

**Table 9.** Average geometric economic sub criteria

| District | Labor-Land Value | Labor-Income Level | Land Value – Income Level |
|---|---|---|---|
| Lakarsantri District | 2 | 9 | 2 |
| Lidah wetan Subdistrict | 1/9 | 7 | 9 |
| Sambikerep District | 0.2 | 1/3 | 3 |
| Lontar Subdistrict | 1/3 | 5 | 1 |
| Sambikerep Subdistrict | 7 | 1/7 | 1/7 |
| Sukomanunggal District | 7 | 1 | 3 |
| Sonokwijenan Subdistrict | 9 | 1 | 0.2 |
| Sukomanunggal Subdistrict | 1/9 | 1/9 | 9 |
| **Average Geometric** | 0.96075 | 1.06594 | 1.59390 |

Based on Table 9, the average geometric value between sub criteria of labor to land value amount 0.96075, between sub criteria of labor to income level amount 1.06594, and sub criteria of land value to income level amount 1.59390. Pairwise comparison matrix can be seen in the following table.

**Table 10.** Pairwise comparison matrix of economic sub criteria

| sub criteria | Labor | Land Value | Income Level |
|---|---|---|---|
| Labor | 1 | 0.96075 | 1.06594 |
| Land Value | 1.04085 | 1 | 1.59390 |
| Income Level | 0.93814 | 0.62739 | 1 |
| Total | 2.97899 | 2.58814 | 3.65984 |

Based on Table 10, total value of each sub criteria shows that total value of the sub criteria of labor amount 2.97899, the sub criteria of land value amount 2.58814, and sub criteria of income level amount 3.65984. Normalization of pairwise comparison matrix can be seen in the following table.

**Table 11.** Normalization of pairwise comparison matrix of economic sub criteria

| sub criteria | Labor | Land Value | Income Level |
|---|---|---|---|
| Labor | 0.33568 | 0.37121 | 0.29125 |
| Land Value | 0.34940 | 0.38638 | 0.43551 |
| Income Level | 0.31492 | 0.24241 | 0.27324 |
| Total | 1 | 1 | 1 |

**Table 12.** Weight of economic sub criteria priority

| sub criteria | Total | Weight | Priority |
|---|---|---|---|
| Labor | 0.99815 | 0.33272 | 2 |
| Land Value | 1.17129 | 0.39043 | 1 |
| Income Level | 0.83057 | 0.27686 | 3 |
| Total | 3 | 1 | |

Based on Table 12, the major priority of each sub criteria in the economic criteria used in the process of determining the location of residential area development in West Surabaya highest dominated by sub criteria of land value amount 0.39043. Eigenvalue maximal ($\lambda max$) in this sub criteria amount 3.01489. Consistent index ($CI$) in these criteria amount 0.00744. Ratio of consistency ($CR$) in these criteria amount 0.01284.

### *3.4. Accessibility sub criteria*

The results of the priority scale of the results of questionnaires of respondents as well as the results of the calculation of average geometric were contained in the following table.

**Table 13.** Average geometric accessibility sub criteria

| District | Public Transportation -Electric Network | Public Transportation –Water Distribution | Electric Network - Water Distribution |
|---|---|---|---|
| Lakarsantri District | 1 | 1 | 1 |
| Lidah wetan Subdistrict | 7 | 1 | 7 |
| Sambikerep District | 1/3 | 3 | 5 |
| Lontar Subdistrict | 1 | 1 | 1 |
| Sambikerep Subdistrict | 7 | 7 | 1/7 |
| Sukomanunggal District | 7 | 1 | 1/6 |
| Sonokwijenan Subdistrict | 1 | 1 | 3 |
| Sukomanunggal Subdistrict | 1/9 | 9 | 9 |
| **Average Geometric** | 1.37401 | 1.92556 | 1.47578 |

Based on Table 13, the average geometric value between sub criteria of public transportation to electric network amount 1.37401, between sub criteria of public transportation to water distribution amount 1.92556, and between sub criteria of electric network to water distribution amount 1.47578. Pairwise comparison matrix can be seen in the following table.

**Table 14.** Pairwise comparison matrix of accessibility sub criteria

| **sub criteria** | **Public Transportation** | **Electric Network** | **Water Distribution** |
|---|---|---|---|
| Public Transportation | 1 | 1.37401 | 1.92556 |
| Electric Network | 0.72779 | 1 | 1.47578 |
| Water Distribution | 0.51933 | 0.67761 | 1 |
| Total | 2.24712 | 3.05162 | 4.40135 |

Based on Table 14, the total value of each sub criteria shows that total value of the sub criteria of public transportation amount 2.24712, the sub criteria of electric network amount 3.05162, and the sub criteria of water distribution amount 4.40135. Normalization of pairwise comparison matrix can be seen in the following table.

**Table 15.** Normalization of pairwise comparison matrix of accessibility sub criteria

| **sub criteria** | **Public Transportation** | **Electric Network** | **Water Distribution** |
|---|---|---|---|
| Public Transportation | 0.44501 | 0.45026 | 0.43749 |
| Electric Network | 0.32388 | 0.32769 | 0.33530 |
| Water Distribution | 0.23111 | 0.22205 | 0.22720 |
| Total | 1 | 1 | 1 |

**Table 16.** Weight of accessibility sub criteria priority

| sub criteria | Total | Weight | Priority |
|---|---|---|---|
| Public Transportation | 1.33276 | 0.44425 | 1 |
| Electric Network | 0.98688 | 0.32896 | 2 |
| Water Distribution | 0.68036 | 0.22679 | 3 |
| Total | 3 | 1 | |

Based on Table 16, the priority of each sub criteria in accessibility criteria used in the process of determining the location of residential area development in West Surabaya highest dominated by public transportation sub criteria amount 0.44425. Eigenvalue maximal ($\lambda max$) in this sub criteria amount 3.00378. Consistent index ($CI$) in these criteria amount 0.00189. Ratio of consistency ($CR$) in this criteria amount 0.00326.

## 4. Conclusion

The result for this research indicate that the priority aspect of each criteria used in the process of determining the location of residential area development in West Surabaya were dominated by the highest accessibility criteria amount 0.40996. The other hand, we used economic criteria as the second priority aspect with the weight amount 0.31773, and the third priority was the social criterion with the weight amount 0.27231.

## References

[1] Republic of Indonesia 2007 *Regulation of Minister of Public Work No. 41/PRT/M/2007 about Guidelines for Cultivation Area Technical Criteria* Jakarta
[2] Saaty T L 1980 *The Analytic Hierarchy Process Planning, Piority Setting, Resource Allocation* (New York: McGraw-Hill Companies)
[3] Riyanto A, I M A Anthara 2008 Determination of Priority for Selection of Gravel Pump Component Using Analytic Hierarchy Process *National Conference of Information Technology 2008* (SNATI 2008) pp A1-A6.
[4] Cabala P 2010 *Using the Analytiv Hierarchy Process in Evaluating Decision Alternatives* Department of Management Process Cracow University of Econimics, pp 5-23
[5] Alonso J A 2006 Consistency in The Analytic Hierarchy Process: A New Approach *International Journal of Uncertainty* **14**(4) pp 445−459
[6] Dolores Ma, C Tongco 2007 Purposive Sampling as a Tool for Informant Selection *a Journal of Plants, People, and Applied Research* pp 147-158

# Physical Land Suitability for Civet Arabica Coffee: Case Study of Bandung and West Bandung Regencies, Indonesia

**E Chairani[1*], J Supriatna[2], R Koestoer[3], and M Moeliono[4]**

[1]School of Environmental Science, University of Indonesia, Jakarta, Indonesia
[2]Faculty of Mathematic and Natural Sciences, University of Indonesia, Depok, Indonesia
[3]School of Environmental Science, University of Indonesia, Jakarta, Indonesia
[4]Consultative Group of International Agricultural Research, Bogor, Indonesia

ellyna.chairani@ui.ac.id

**Abstract**. Indonesia has been widely known as the best Arabica coffee (*Coffea arabica*) producer, in terms of both aspects, quality and number of product. Currently, its production, however, declines to the 3rd rank internationally. Issues emerged in the coffee cultivation are: land unsuitability, low quality of seeds, and poor management. Among Arabica coffee types, wild civet coffee is the most expensive one and harvested from the coffee beans which have been digested naturally. The study aims to determine the physical suitability of land as well as the constraints related to land for civet Arabica coffee in selected study cases, e.g., Bandung and Bandung Barat. The research methods employ multi-criteria analysis, and combined with weighted overlaying techniques for mapping. The criteria include temperature, rainfall, humidity, duration of dry season, slope, altitude, type of soil, soil texture, and erosion potential. Parameters of civet (*Paradoxurus hermaphroditus*) are land use, altitude, and temperature. Local policy strongly supports the extensive management for land and the increase of coffee export. Processing data involved matching the comparison between guideline requirements for the land suitability classes, characteristics of Arabica coffee and civet habitat. The results covered the profile suitable land of the civet *Arabica coffee* in the study areas.

**Keywords**: land suitability, civet arabica coffee, Bandung

## 1. Introduction

Coffee cultivation was introduced to Java in 1696. It spread to other Indonesian regions due to greater demand and geographically suitable for coffee [1, 2, 3]. The total area of coffee plantation is 1,230,495 ha. However, the Country dominance as the major coffee exporter continues to decline [2, 3 ,4]. National productivity was only 761 kg/ha in 2015 compared to the 2nd biggest Vietnam that reached 2000 kg/ha. The main challenges of national coffee are low productivity and poor quality. Classic obstacles include: largely (96.07%) cultivated by smallholder farmers; and nearly 25% of the existing area are not productive nor classified as suitable [5], [6]. Global Arabica consumption amounts to 70%. Conversely, of the national coffee export, Arabica simply shares 15% [4,5].

Coffee plantation occupies 12,000 ha in Bandung district [7]; and 1,406 ha in West Bandung [8]. The District governments implement policy to increase the export and expanding coffee farming. However, research on the suitable area for sustainably coffee farming is inadequate [9]. Belongs to

*Rubiaceae* family, *Coffea arabica* is considerably to have higher quality and therefore used for most premium coffee [10]. Key factors to have optimum coffee yield includes physical environment, practices of agricultural management, and genetic resources [11]. In Indonesia, optimum growing conditions are characterized by an altitude of 700-2.000 m a.s.l., annual rainfall 1.500-2.500 mm, temperature of 15-25$^{o}$ C, 3 months of dry season per year, and fine-textured soils as well [13].

Luxury (wild) civet Arabica coffee (kopi luwak) involves Asian palm civet (luwak pandan or luwak) digesting Arabica coffee cherries and mixing with enzymes passing through the digestive tract before its feces are collected. Digestive systems of civet give kopi luwak a uniquely rich aroma and smoothly rounded flavour. The quality of the coffee is better as the civet naturally selects the ripest cherries [14, 15, 16]. Luwak are native to tropical jungles and rainforests, mostly in southern China, northern Himalayas, southern India, south-east Asia [17]. Taxonomically, civet is in the Family *Viverridae* [18, 19]. Civets are highly active nocturnal animals, arboreal, mostly solitary and territorial by nature. Living in a variety of habitats as long as the food and areas they can rest in are available. A single civet needs an average of 17 km$^2$ territory and elevation range extend up to about 2.234 m a.s.l [17,20].

There are various approaches to study land suitability. This research employs GIS approach on the basis of multi criteria rankings and weights assigned to variable that affect coffee growing and production [21, 22, 23, 24, 25]; and similar approach used to analyze civet habitat suitability. The multi-criteria analysis framework integrates nine physical land constraints for Arabica coffee including temperature, rainfall, humidity, dry season, slope, altitude, type of soil, soil texture, and erosion. As for luwak, the habitat characteristics consist of land use, altitude, and temperature. The study goals are: 1) to analyze land suitability of Arabica coffee; (2) to analyze habitat suitability of luwak; (3) to analyze suitability of civet Arabica coffee; (4) to provide maps of area expansion of civet Arabica coffee.

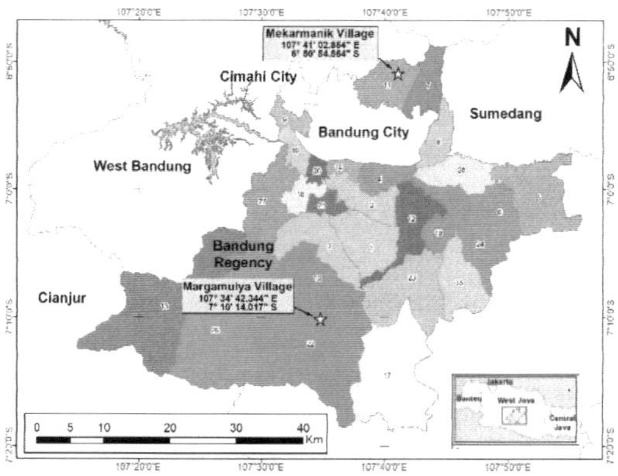


**Figure 1.** Administrative Region of Bandung Regency
Source: Thematic Map of Spatial Plan 2007-2027

| | | | | | |
|---|---|---|---|---|---|
| 1 | Arjasari. | 10 | Cimaung. | 19 | Margaasih. |
| 2 | Baleendah. | 11 | Cimenyan. | 20 | Margahayu |
| 3 | Banjaran | 12 | Ciparay. | 21 | Pameungpeuk |
| 4 | Bojongsoang | 13 | Ciwidey. | 22 | Pengalengan |
| 5 | Cicalengka | 14 | Dayeuhkolot. | 23 | Pacet |
| 6 | Cikangkung | 15 | Ibun. | 24 | Paseh |
| 7 | Cilengkrang | 16 | Katapang. | 25 | Pasirjambu |
| 8 | Cileunyi. | 17 | Kertasari. | 26 | Rancaekek |
| 9 | Cimahi Selatan | 18 | Majalaya | 27 | Soreang |

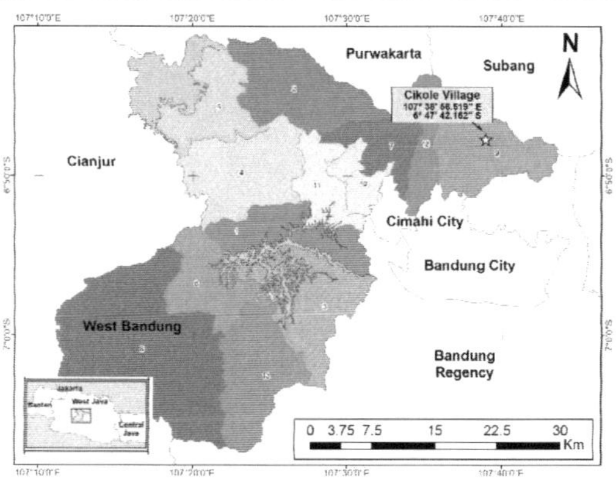


**Figure 2.** Administrative Region of West Bandung Regency
Source: Thematic Map of Spatial Plan 2009-2029

| | | | |
|---|---|---|---|
| 1 | Batujajar | 7 | Cisarua |
| 2 | Cikalong Wetan | 8 | Gununghalu |
| 3 | Cililin | 9 | Lembang |
| 4 | Cipatat | 10 | Ngamprah |
| 5 | Cipendeuy | 11 | Padalarang |
| 6 | Cipongkor | 12 | Parompong |
| | | 13 | Sindangkerta |

IOP Conf. Series: Earth and Environmental Science **98** (2017) 012029 doi:10.1088/1755-1315/98/1/012029

## 2. Study Area

Study area of the two districts can be seen in Figures 1-2. Figure 1 shows that Bandung Regency is located between latitudes of 6°41` and 7°19 South and longitudes of 107°22` and 108°5` East. Figure 2 shows Regency of West Bandung is located by latitude 6°53' South and longitude 107°25' East. The total area of districts Bandung and West Bandung covers 1.762,40 km$^2$ and 1.305,77 km$^2$. It is inhibited by 3.534.111 (2015); and 1.614.495 (2013). The main source of income in the study areas is industry and agriculture; and agriculture respectively [8, 9].

## 3. Data and Methodology

### *3.1. Data*

The study uses datasets as summarized in the Table 1. Digitized and tabulated data were obtained from regional (district) spatial plan, including thematic maps of slope, elevation, soil type, temperature, rainfall, humidity, duration of dry month, slope, altitude, type of soil, soil texture, erosion, protected area and land use (scale 1:250.000). Data on micro climate were assembled from Meteorological, Climatological and Geophysical Agency (central and local stations). Data on existing coffee farms was acquired from survey and primary data/interview during field works. The expert opinion ratings were used for determining variable weighting for multi criteria modelling.

**Table 1.** Data sets used in the study

| Data set | Year | Source |
|---|---|---|
| Climate: rainfall, temperature, relative humidity, duration of dry months | 2014-2015 (in excel) | Meteorological, Climatological and Geophysical Agency (Central and local stations) |
| GIS thematic maps at a 1:250.000 scale of elevation, slope, soil type and texture. | 2007-2027 (Bandung District)<br>2009-2029<br>(W. Bandung District) | 1. Bandung Regency Spatial Plan [25]<br>2. West Bandung Regency Spatial Plan [26] |
| Land use/cover maps, protected area | 2007 | Indonesian base map |
| Village maps | 2016 | Potensi Desa, BPS [27] |
| Erosion potential | 2015 | Calculated using the method of USLE [28] |
| Existing Arabica coffee cultivation | Survey, interview during field works 2016 | Survey, interview during field work |

### *3.2. Methodology*

The methodology used for this study integrates steps as follow:

- **Collecting thematic maps** of elevation, slope, soil type, soil texture that were obtained from geo-data of regional spatial plan. Reclassification of classes with respect to Arabica coffee requirement using the spatial analyst tool in ArcGIS 10.3 (2013); **Generating maps of isohyet (climate parameters) using method of interpolating in the** ArcGIS 10.3 (2013) software include rainfall, temperature, period of dry month, relative humidity [29]; **Generating map of erosion potential** by applying USLE methods [35] and raster calculation of weighted overlay; **Generating map of land use cover and protected area** was interpolated from Indonesian base map 2007.
- **Multi Criteria Analysis:** weighting is aimed to express the importance of each criteria relative to other criteria that influences the growth rate and coffee yield [29, 30] and civet habitat. The essential criteria were selected and weighted on the basis of expert opinions, scientific references, and

IOP Conf. Series: Earth and Environmental Science **98** (2017) 012029 doi:10.1088/1755-1315/98/1/012029

farmer's experience. The classes of suitability for each criterion was defined and these classes were used as basis for constructing the individual criteria map. Referring to FAO 1976 Framework for land suitability classification, the classes consist of highly suitable (S1), moderate suitable (S2), marginally suitable (S3), not suitable (N) [13, 21, 22]. Table 2 shows criteria and weight for Arabica suitability and Table 3 presents criteria and weight for Asian palm civet suitability.

**Table 2.** Criteria and weight for Arabica coffee suitability

| Parameter | Weight (%) | Suitability Class | | | |
|---|---|---|---|---|---|
| | | Highly suitable/S1 | Moderately suitable/S2 | Marginally suitable/S3 | Not suitable/ N |
| | | 4,00-3,25 | 3,24-2,50 | 2,51-1,75 | 1,74-0 |
| Temperature (°C) | 0,114 | 16 – 22 | 15-16; 22-24 | 14-15; 24-26 | < 14; > 26 |
| Rainfall (mm/year) | 0,114 | 1.200 - 1.800 | 1.000 - 1.200 1.800 - 2.000 | 2.000-3.000 800-1.000 | > 3.000; < 800 |
| Dry season month) | 0,114 | 1 – 4 | < 1; 4 - 5 | 5-6 | > 6 |
| Relative humidity % | 0,114 | 40 – 70 | 30-40; 70-80 | 20-30;80-90 | < 20; > 90 |
| Elevation (m.a.s.l.) | 0,136 | 1.100 - 1.750 | 1.750 - 2.000 | 700 – 1.100 | >2.000; <100 |
| Slope (%) | 0,114 | < 8 | 8-16 | 16-30;16-50 | > 30 ;> 50 |
| Erosion potential | 0,091 | Very low | Low-medium | Heavy | Very heavy |
| Soil type | 0,091 | Andosol, Latosol | Organosol, top soil | Regosol | Organosol,podzol Podsolik, Aluvial peaty, Laterit Litosol,Grumosol Mediteran,Renzi-na, Hidromorf |
| Soil texture | 0,114 | fine | Moderately fine-medium | Medium-Moderately coarse | Coarse |

Source: Analysis, June 2017 and [13]

**Table 3.** Criteria and weight of civet habitat suitability

| Parameter | Weight | Highly suitable/ S1 | Moderately suitable/ S2 | Marginally suitable/ S3 | Not suitable/ N |
|---|---|---|---|---|---|
| | | **4,00-3,25** | **3,24-2,50** | **2,51-1,75** | **1,74-0** |
| Land use cover | **0.4** | Primary & protected forests | Secondary forest, agricultural cultivation, settlement area | fields, moor, shrubs | Paddys field, rivers, lakes |
| Elevation (m.a.s.l) | **0.3** | 0-2500 | | | >2500 |
| Temperature (°C) | **0.3** | 18-20; 30-36 | | | <18; > 36 |

Source: Analysis, June 2017

- **Cross tabulation and overlay the maps** of Arabica coffee and civet habitat suitability to create the maps of civet Arabica coffee as can be seen in Table 4.

**Table 4.** Cross tabulation for civet Arabica coffee suitability

| | | Asian Palm Civet Habitat | | | |
|---|---|---|---|---|---|
| | | Not Suitable | Marginally Suitable | Moderately Suitable | Highly suitable |
| Arabica coffee | Not Suitable | Not Suitable | Not Suitable | Not Suitable | Not Suitable |
| | Marginally Suitable | Not Suitable | Marginally Suitable | Marginally Suitable | Marginally Suitable |
| | Moderately Suitable | Not Suitable | Marginally Suitable | Moderately Suitable | Moderately Suitable |
| | Highly suitable | Not Suitable | Marginally Suitable | Moderately Suitable | Highly suitable |

Source: Analysis, June 2017

- **Validation:** suitability maps for Arabica coffee is validated by overlaying them with the 3 (three) existing coffee farms in the field works in villages of Mekarmanik, Margamulya and Cikole areas.

## 4. Results and Discussion

The overall study results show that Arabica is more suited to be grown in Bandung as more than three fourth of the areas are in the class of highly and moderately suitable. In comparison to habitat suitability of civet, most of the areas (more than 90%) is classified as highly and moderately suitable; and none of the areas that are classified unsuitable.

### *4.1. Distribution of Arabica coffee land suitability*

Based on the criteria selected and weighted scheme, equation to find the suitable class of farming area as follows:

$$Arabica\ Coffee\ Land\ Suitability\ (SKA) = 0{,}114\,t + 0{,}114\,r + 0{,}114\,d + 0{,}114h + 0{,}136\,a + 0{,}114\,s + 0{,}091\,e + 0{,}091\,l + 0{,}114\,tx \quad (4.1)$$

| | | |
|---|---|---|
| t = temperature | r = Rainfall | d = Dry season |
| h = Relative humidity | a = Elevation | s = Slope |
| l = Soil type | tx= Soil texture | e = Erosion potential |

#### *4.1.1 Arabica coffee suitability in Bandung Regency*

The study site is located geographically in a mountainous region (500-1,800 m.a.s.l) part of the eastern and southern part of the rocky volcanic Bandung basin. Lying on the upper stream area of the Citarum watershed, the rivers of Citarik, Cikeruh, Cirasea, Cisangkuy, Ciwidey dan Cimahi bring a lot of benefits to local people as a source of clean water and agriculture. Annual rainfall in the area ranges from 1,500-4,000 mm/year and the relative humidity from 70 to 78%. During dry season (May-October) the rainfall average is under 2,000 mm. Land sloping towards south is steeper (more than 45%). This condition results in soil susceptible to erosion and avalanche. In addition, land use change and bad attitude of the society towards environment have placed the region as the fourth of disaster-prone in the Country [30].

The study results specify that the suitable zones (highly and moderately) for Arabica coffee growth and production in the region cover an area of 13,688.43 ha (7.73%) and 124,447.1 ha (70.25%) respectively. The remaining portion (38,998.35 ha or 22.01%; and 13.12 ha or 0.01%) is marginally suitable and not suitable respectively. Of the total sub districts (27), Pengalengan sub district provides the largest highly suitable area for the coffee farming (10.802,5 ha). On the contrary, very small area in sub districts of Cicalengka, Cileunyi and Rancaekek areas are classified unsuitable (3,37 ha, 1,75 ha, dan 8,01 ha, respectively). Map of land suitability for coffee growing in the area is shown in Figure 3.

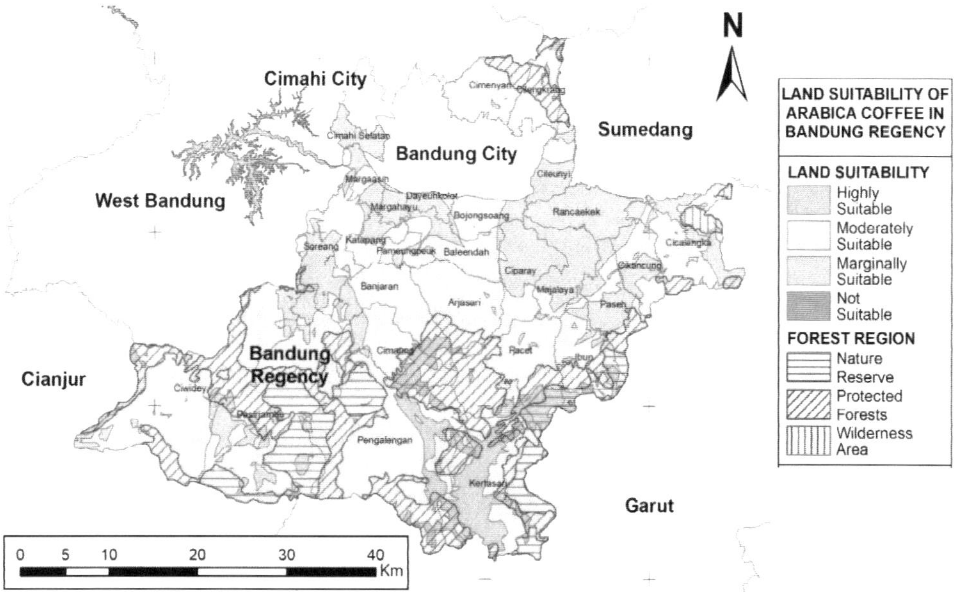


**Figure 3**. Land suitability map of Arabica coffee in Bandung Regency
Source: Analysis, June 2017

### *4.1.2 Arabica coffee suitability in West Bandung Regency*

The morphology of West Bandung area consists of plains, ramps, hills and mountains with altitude ranges of 500-2,000 m.a.s.l. Of the 1.305,77 ha area, 46,68% (59.614,15 ha) is located at an elevation of 500-1000 m.a.s.l; and the rest is at an elevation of 1500-2000 m.a.s.l. In terms of annual rainfall, the area is varying (1500-3.500 mm/year) with the highest amount is in Cikalong Wetan and Cipeundey. The majority of the area has slope more than 40%, and the steepest area can be found at Gununghalu sub district. Batujajar area slope, however, is only 0-8% (Figure 4).

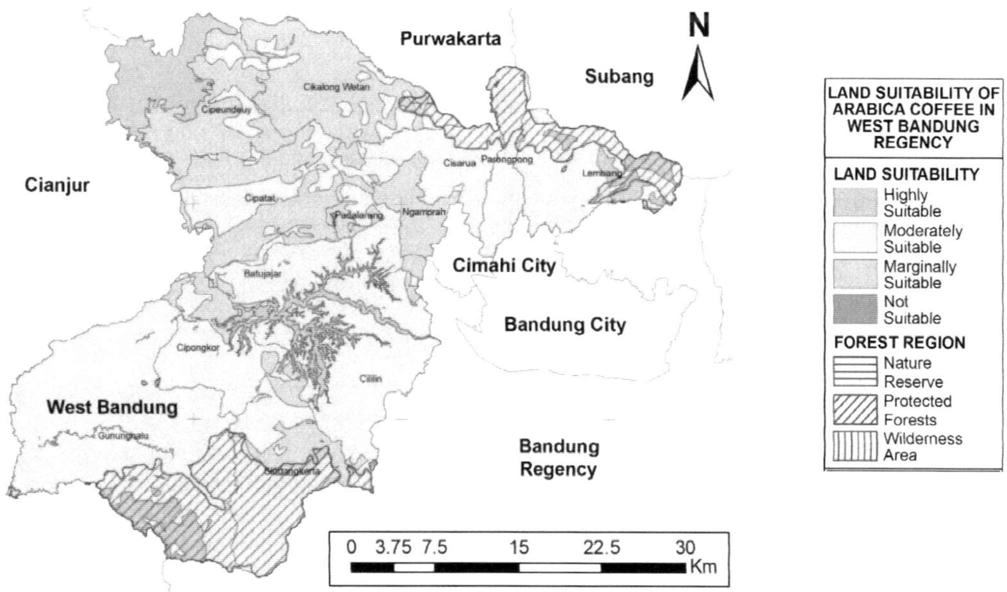


**Figure 4**. Land suitability map of Arabica coffee in West Bandung Regency
Source: Analysis, June 2017

IOP Conf. Series: Earth and Environmental Science **98** (2017) 012029 doi:10.1088/1755-1315/98/1/012029

The results revealed that 3.46% (4,296.83ha) of the region area was identified as the most suitable for Arabica coffee; 85,778.60 ha (55%) is moderately suitable area, the rest portion is 27.43% and 7.69 ha or 0.01% as marginally suitable and not suitable respectively. Four out of 13 sub districts in the area have few unsuitability lands for coffee, i.e., Cikalong Wetan (0,76 ha); Cipatat (0,02ha); Cipendeuy (20,39ha); and Gunung Halu (7,07ha). The total area should be deducted by nature reserve area (5,24 ha) and protected forest (21,152.9 ha) of which area prohibited for cultivating activities.

The overall result of land suitability analysis for Arabica coffee in the two regions is summarized and presented in Table 5.

**Table 5.** Land suitability for Arabica coffee in study areas

| Suitability Classification/ | Highly suitable/ S1 | Moderately Suitable/ S2 | Marginally Suitable/ S3 | Not Suitable/ N | Total Area (ha) |
|---|---|---|---|---|---|
| **Bandung** | | | | | |
| Arabica coffee | 13,688 | 124,447 | 38,998 | 13 | 177,146 |
| % | 7.73% | 70.25% | 22.01% | 0.01% | 100% |
| Nature reserve | 1,761 | 12,929 | 683 | 0 | 15,373 |
| Protected forest | 4,080 | 31,881 | 1,730 | 0 | 37,691 |
| Wilderness area | 0 | 974 | 42 | 0 | 1,016 |
| **West Bandung** | | | | | |
| Arabica coffee | 4,297 | 85,779 | 34,043 | 7 | 124,126 |
| % | 3.46% | 69.11% | 27.43% | 0.01% | 100.00% |
| Nature reserve | 0 | 5 | 0 | 0 | 5 |
| Protected forest | 3,529 | 16,730 | 894 | 0 | 21,153 |

Source: Analysis, June 2017

Physical characteristics frequently used for Arabica suitability analysis and evaluation are topography, climate and soil [31]. In this study, it was expanded to be slope, altitude, type and texture of soil, temperature, humidity, the length of dry season, and potential erosion. Altitude is limiting barrier factor for the coffee to grow. The plants are vulnerable from pest and parasites at the elevation of lower than 800 m.a.s.l, and limit by frost which generally occurs at elevations of higher than 1500 m.a.s.l. [31].

On the other hand, coffee productivity is significantly attributable to climate factor [31]. Arabica relishes some altitude-influenced temperatures of 18-24°C. The role of temperature and duration of sun light can be seen from the formation of the primary branch until the coffee tree flowering. The more sun exposure, the faster the formation of flowers and fruits. Temperature does not directly affect the development of plants but rather its resistance to disease attacks. In practices, coffee farming is commonly combined with shade trees. The benefit of shade trees along with elevation factors, soil types and climates is to control pests and tree diseases because shade trees affect the temperature and humidity and moisture of micro soils [32, 33, 34]. In addition, shade tree plays the role in coffee growth, productivity and flavor [35].

Coffee plantation also requires rainfall between 2000-3000 mm/year for the development of Arabica beans, sensitive to humidity and demand fertile organic sufficiently soil that has good drainage [36, 37]. Most of the soil type and structure in the study areas support the Arabica requirements to grow, particularly in Pengalengan area. As explained above, productivity is more influenced by environment and cultivation techniques, while for bean quality is more influenced by climate factor, that is rainfall. In the rainy season, the size of coffee bean formed is greater [37, 38].

IOP Conf. Series: Earth and Environmental Science **98** (2017) 012029 doi:10.1088/1755-1315/98/1/012029

### *4.2. Distribution of Asian palm civet habitat suitability*
Based on criteria selected and weighted scheme, the equation to find the suitable class of luwak habitat area as follows:

$$Asian\ palm\ civet\ habitat\ suitability\ (SHL) = 0{,}4\,pl + 0{,}3\,a + 0{,}3t \qquad (4.2)$$

pl = land use   t = temperature   a = elevation

#### *4.2.1 Asian palm civet habitat suitability in Bandung Regency*
The study result clearly demonstrates the habitat suitability distribution for luwak in Bandung. As can be seen in Figure 6, the majority of the region are highly suitable/S1 and moderately suitable/S2 as civet habitat, covering an area of 49,337 ha (27.85%) and 118,356 ha (66.81%) respectively. The remaining 9453 ha (5.34%) is classified as marginally suitable/S3; and no area that is unsuitable for luwak. In the highly suitable class, the area includes nature reserve (1,761 ha) and protected forest (4,080 ha). Similarly, the area of moderately suitable involves protected area (31,881 ha) and wilderness area (974 ha); and 42 ha of suaka margasatwa should be included as marginally suitable. The map of suitability distribution for luwak habitat is shown in Figure 5.

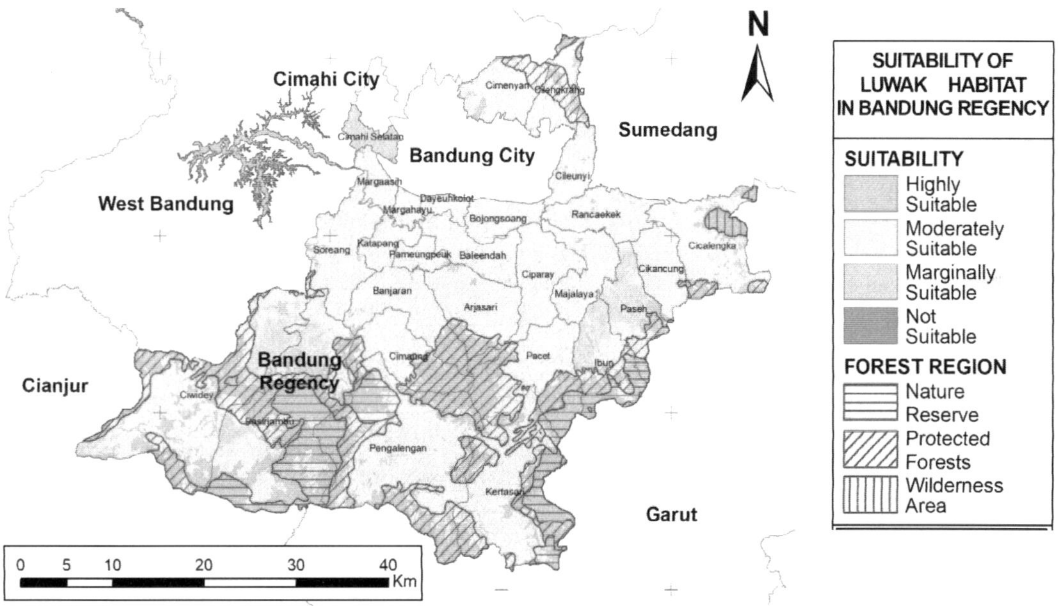


**Figure 5.** Habitat suitability map of Asian Palm Civet in Bandung Regency
Source: Analysis, June 2017

#### *4.2.2. Asian palm civet habitat suitability in West Bandung Regency*
In West Bandung, highly and moderately suitable cover the areas of 21.85 % (27,122 ha) and 96,945 ha (78.10%) as respectively. Of the total 124,126 ha, only 57 ha (0.05%) is categorized as marginally suitable. Contrasting to the other region, nature reserve provides only 5 ha that moderately suitable for luwak, while in protected forest has classes of highly suitable (3,529 ha), moderately suitable (16,730 ha) and marginally suitable (894 ha). Figure 6 illustrates the habitat suitability of civet in this area.

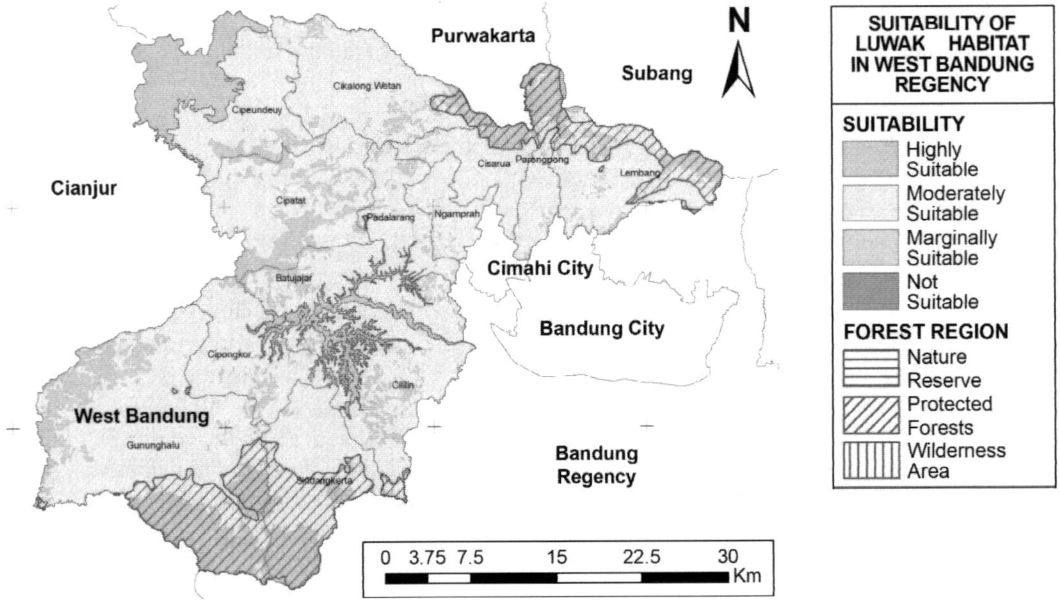


**Figure 6.** Habitat suitability map of Asian Palm Civet in West Bandung Regency
Source: Analysis, June 2017

Table 6 presents the summary of the result for analysis for civet habitat suitability in the two regencies.

**Table 6.** Habitat Suitability of Asian Palm Civet in Bandung and West Bandung Regencies

| Suitability Classification/ | Highly suitable/ S1 | Moderately Suitable/ S2 | Marginally Suitable/ S3 | Not Suitable/ N | Total Area (ha) |
|---|---|---|---|---|---|
| **Bandung** | | | | | |
| Civet habitat | 49,337 | 118,356 | 9,453 | 0 | 177,146 |
| % | 27.85% | 66.81% | 5.34% | 0.00% | 100% |
| Nature reserve | 1,761 | 12,929 | 683 | 0 | 15,373 |
| Protected forest | 4,080 | 31,881 | 1,730 | 0 | 37,691 |
| Wilderness area | 0 | 974 | 42 | 0 | 1,016 |
| **West Bandung** | | | | | |
| Civet habitat | 27,122 | 96,945 | 57 | 2 | 124,126 |
| % | 21.85% | 78.10% | 0.05% | 0.00% | 100.00% |
| Nature reserve | 0 | 5 | 0 | 0 | 5 |
| Protected forest | 3,529 | 16,730 | 894 | 0 | 21,153 |

Source: Analysis, June 2017

Luwak (pandan) belong to the family of *Viverridae* are known as nocturnal, arboreal and mostly solitary animal. Being native to temperate and tropical forest within and around Asia, however, civet can survive in diverse terrestrial habitat (0-2,234 m.a.s.l), e.g., primary and secondary forests, shrubs, parks, settlement area, or plantations. Their activity is determined by the obtainable food and presence of the areas such as tree hollow and dense foliage where they can rest. Males need a larger territory than females with the range size is 5-20 km$^2$. In addition, it is recorded that daily range of males reach 17 km$^2$ whereas females up to 1.6 km$^2$ for females. To mark the territory, they have perineal scent gland [14, 17, 20].

The research location is dominated by areas with altitudes less than 2000 m.a.s.l whose land-cover varies from agriculture, forests and settlements. The areas provide plenty of food to enable the surviving luwaks. The results of the habitat suitability analysis distinguished the study area in the class of S1, S2, and S). It should, however, take into consideration that civets are extremely adaptable and opportunistic (omnivore, eating whatever is available).

### *4.3. Distribution of kopi luwak Arabica land suitability*

The final cross-tabulating and overlaying of Arabica and civet suitability results in the map of civet Arabica coffee in the two areas.

#### *4.3.1 Distribution of kopi luwak Arabica land suitability in Bandung Regency*

The study area of Bandung is classified as being moderately suitable (71.62%, or 126,877 ha) for civet Arabica coffee. This finding is in agreement with the result of Arabica coffee and civet habitat suitability. The areas mapped as moderately suitable are having necessary requirements, e.g., elevation, temperature, rainfall, relative humidity, soil texture, and land cover with minor limitations. The map of the analysis result is presented in Figure 7.

**Figure 7.** Civet Arabica coffee suitability map of Bandung
Source: Analysis, June 2017

#### *4.3.2 Civet Arabica coffee land suitability in West Bandung District*

In West Bandung, result shows that an area of 4,296.83 ha (3.46%) is highly suitable for kopi luwak. The majority area (69.10% or 85,775.84 ha) is moderately suitable; and marginally suitable amounted to 34,043.46 ha (27.43%). The rest (9.77 ha or 0.01%) is classified as unsuitable due to of un-correctable criteria include slope and elevation (Figure 8).

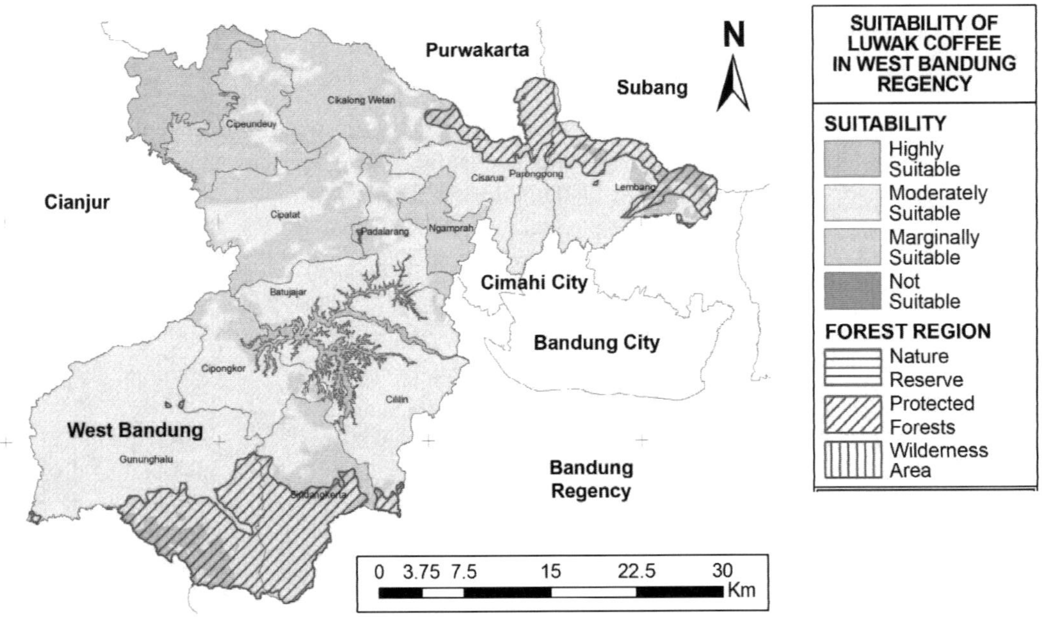


**Figure 8.** Civet Arabica coffee suitability map of West Bandung
Source: Analysis, June 2017

To sum up the result of study, Table 7 presents the distribution kopi luwak suitability in both areas.

**Table 7.** Civet Arabica Coffee Suitability in Bandung and West Bandung Regencies

| **Suitability classification** | **Highly suitable** | **Moderately suitable** | **Marginally suitable** | **Not suitable** | **Total area (ha)** |
|---|---|---|---|---|---|
| Bandung District | | | | | |
| **Civet Arabica coffee** | 6,408.10 | 126,877.30 | 43,848.45 | 13.12 | 177,147 |
| % | 3.62% | 71.62% | 24.75% | 0.01% | 100% |
| **Nature reserve** | 1,760.78 | 12,928.52 | 682.95 | 0.00 | 15,372.25 |
| **Protected forest** | 4,080.43 | 31,880.86 | 1,729.66 | 0.00 | 37,690.94 |
| **Wilderness area** | 0.00 | 974.28 | 42.02 | 0.00 | 1016.31 |
| West Bandung | | | | | |
| **Civet Arabica coffee** | 4,296.826 | 85,775.84 | 34,043.46 | 9.77 | 124,125.90 |
| % | 3.46% | 69.10% | 27.43% | 0.01% | 100.00% |
| **Nature reserve** | 0.00 | 5.24 | 0.00 | 0.00 | 5.24 |
| **Protected forest** | 3,529.15 | 16,729.82 | 893.93 | 0.00 | 2,1152.90 |

Source: Analysis, June 2017

### *4.4 Validation*

Validation was conducted by overlaying suitability maps of civet habitat, Arabica coffee and the existing coffee farms as surveyed in the field works. The suitable areas are identified as suitable coincide with most Arabica coffee farming at Mekarmanik and Margamulya villages of Bandung District; and Cikole village in West Bandung. The result of validation is presented in Table 8.

**Table 8.** Validation of civet Arabica coffee suitability

| Suitability class/ Village | Nature Preserve | Protected Forest | Non-Conservation | Total | % hectare |
|---|---|---|---|---|---|
| **Margamulya** | | | | | |
| Moderately Suitable/S2 | 111.10 | 252.76 | 415.06 | 778.93 | 68.04 |
| Highly Suitable/S1 | 0.00 | 329.45 | 36.37 | 365.82 | 31.96 |
| Total | 111.10 | 582.21 | 451.44 | 1144.75 | |
| % | 9.71 | 50.86 | 39.44 | | |
| **Mekarmanik** | | | | | |
| Moderately Suitable/S2 | | 507.13 | 793.53 | 1300.66 | 99.98 |
| Marginally suitable/S3 | | 0.00 | 0.32 | 0.32 | 0.02 |
| Total | | 507.13 | 793.85 | 1300.98 | |
| % | | 38.98 | 61.02 | | |
| **Cikole** | | | | | |
| Moderately Suitable | | | 177.6344 | 177.6344 | 60.30 |
| Highly Suitable | | | 116.9468 | 116.9468 | 39.70 |
| Total | | | 294.58 | 294.58 | |
| % | | | 100 | | |

Source: Analysis, June 2017

Figure 9 shows that 68.04% and 31.96% areas in Margamulya are categorized as moderately suitable and highly suitable respectively. However, only 39,44% of the Arabica suitable area is allowed to be cultivated (non-conservation). The rest is classified as nature preserve (9.71%) and protected forest (50,86%).

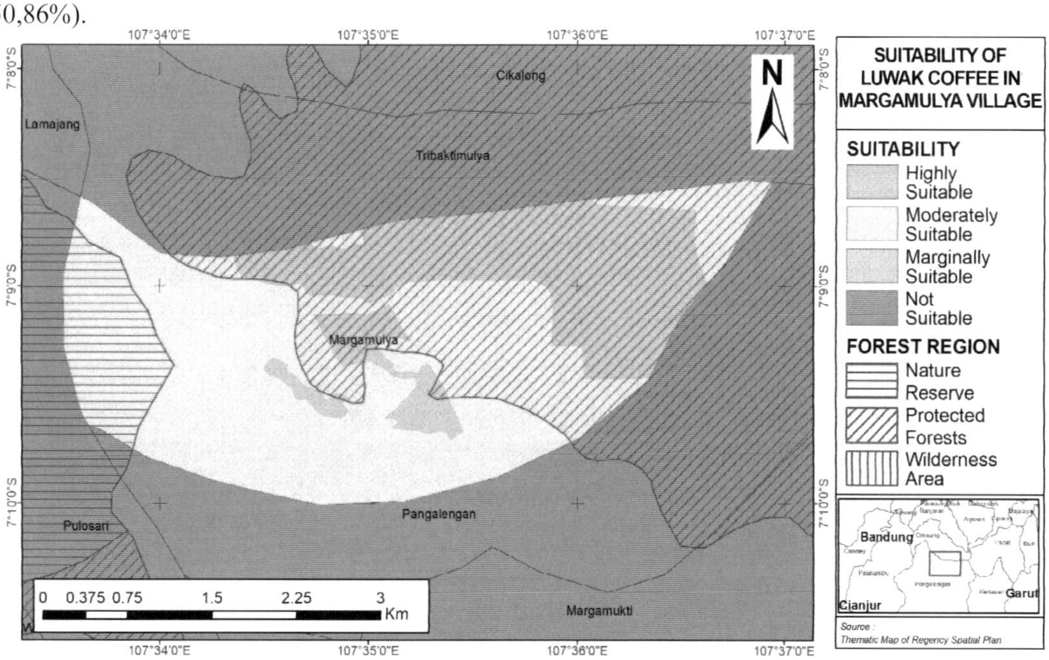


**Figure 9.** Validation of civet Arabica coffee suitability maps in Margamulya coffee farming area
Source: Analysis, June 2017

Similar to Mekarmanik area, the majority (99.98%) of the coffee farming is in moderately suitable class, and the rest (0,32 ha) is marginally suitable (Figure 10).

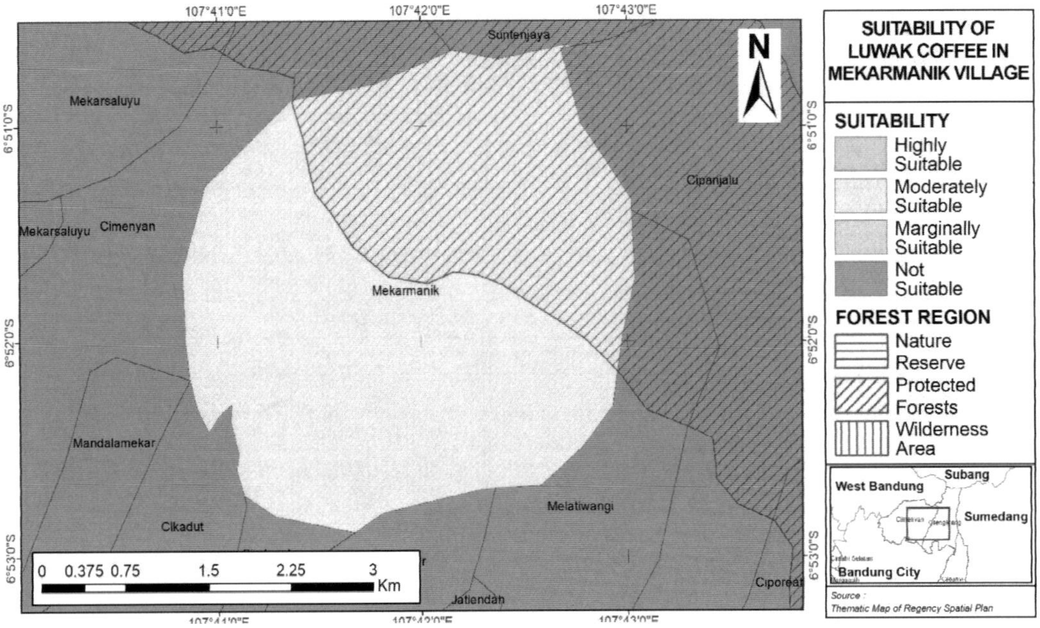


**Figure 10.** Validation of civet Arabica coffee suitability maps in Mekarmanik coffee farming area
Source: Analysis, June 2017

Whereas the whole area in Cikole village is suitable for Arabica farming, in which 60,3% is moderately suitable and 39,7% is highly suitable (Figure 11).

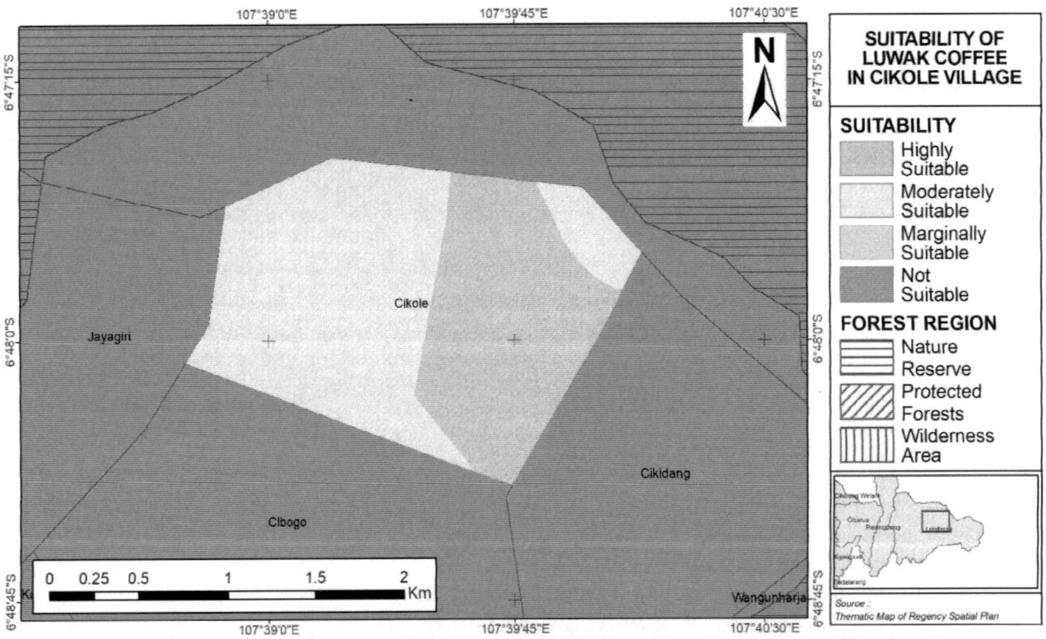


**Figure 11**. Validation of civet Arabica coffee suitability maps in Cikole coffee farming area
Source: Analysis, June 2017

## 5. Conclusions and Recommendation

This study was undertaken as a means of enhancing productivity of civet Arabica coffee in study area. Results show that the total suitable area for Civet Arabica coffee production West the Bandung is 69,808.46 ha, whereas in Bandung is amounted to 31,984.22 ha. Field works validated the Arabica suitability maps which indicate accuracy of the model developed. The maps provide spatial information to farmers, related government institutions and private sectors as their guidance in selecting the right areas of Arabica growing and kopi luwak yielding.

### References

[1] Wild A 2005 *Coffee, a Dark History* (New York: W.W. Norton & Company)
[2] Indarto P 2014 *The Road to Java Coffee* (Jakarta: Asosiasi Eksportir Kopi Indonesia)
[3] Hoffman J 2014 *The World Atlas of Coffee: From Beans to Brewing-Coffees Explored, Explained and Enjoyed* (Canada: Firefly Octopus Publishing Group Ltd. Ontario)
[4] International Coffee Organization (ICO) Coffee Statistic 2014
[5] Asosiasi Eksportir Kopi Indonesia (AEKI) 2016 Konsumsi Kopi di Pasar Domestik Naik Pesat. http://www.aeki-aice.org/news/1/laporan- bpp-aeki--2014/56.
[6] Direktorat Jenderal Perkebunan 2016 Produksi Kopi (Jakarta: DJK, Kementerian Pertanian)
[7] Kabupaten Bandung Dalam Angka 2014 http://bandungkab.bps.go.id/ index.php/publikasi
[8] Kabupaten Bangli Dalam Angka 2014 http://banglikab.bps.go.id/ v2/index/
[9] Jonasson, O 2017 *Agricultural Geography*. www.jstror.org/stable /10.2307.4996431.
[10] Integrated Taxonomy Information Report IT IS 2016 https://www.itis.gov/servlet/SingleRpt/SingleRpt?search_topic=TSN&search_calue=35190#
[11] Bosselman AS, Dons K, Oberthur T, Sminth OC, Herman,U 2009 The influence of shade trees on coffee quality in small coffee agroforestry system in Southern Columbia *Agriculture, Ecosystems & Environment 129*(1-3) pp 253-260 Harvard.
[12] Ridley F V 2010 Department of Biological and Biomedical Sciences University of Durham. The past and future climatic suitability of arabica coffee (Coffea arabica L.) in East Africa *Agriculture, Ecosystems & Environment 80*(1–2).
[13] Djaenudin D, Marwan H, Subagjo H, Hidayat A 2011 *Petunjuk Teknis Evaluasi Lahan Untuk Komoditas Pertanian* Balai Besar Penelitian dan Pengembangan Sumberdaya Lahan Pertanian Badan Penelitian dan Pengembangan Pertanian Bogor
[14] Jumhawan U, Putri S P, Yusianto, Marwani E, Bamba T, Fukusaki E 2013 Selection of discriminant markers for authentication of Asian palm civet coffee (Kopi Luwak): a metabolomics approach *J Agric Food Chem 2013*(33) pp 7994-8001.
[15] Marcone, F M 2004a Composition and properties of Indonesian palm civet coffee (Kopi Luwak) and Ethiopian civet coffee *Food Research International* 37 pp 901-912.
[16] Marcone, FM 2004b The science behind luwak coffee: an Analysis of the World's Rarest and Most Expensive Coffee *Annals of Improbable Research* University of Guelph Ontario Canada 1-2.
[17] Duckworth, J 1997 Ranging Behaviour, Actity, Habitat Use and Morphology of the Malay Civet (*Viverra tangalunga*) on Peninsular Malaysia and Comparison with Studies on Borneo and Sulawesi *Mamalian Biology* pp 437-447.
[18] Nowak R M, Paradiso J L 1983 *Walker's Mammals of the World $4^{th}$ Edition* The John Hopkins University Press. Baltimore and London
[19] Grzimek B, Schlager D, Olendorf 2004 *Grzimek's Animal Life Encyclopedia* Detroit: Gale.
[20] Duckworth J, Widmann P, Custodio J, Gonzalez A, Jennings G, Veron 2011 Paradoxurus hermaphroditus. The UN Red list of Threatened species http://www.uicnredlist.org/apps/redlist/details/416930.
[21] FAO 1976 *Land evaluation towards a revised framework*, Food and Agriculture Organization of the United Nations, Rome, Italy.
[22] FAO 2007 *Land Evaluation Towards a Revised Framework* (Rome, Italy: Food and Agriculture

Organization of the United Nation)
[23] Malczewski J 1999 *GIS and Multi Criteria decision Analysis USA and Canada* (USA: John Wiley & Sons)
[24] Karim Z 2012 Land Suitability Analysis for Coffee Cultivation Using Multi Criteria EvaluatiApproach and GIS. *European Journal of Experimental Biologi 4*(**3**) pp 639-648.
[25] Rencana Tata Ruang Kabupaten Bandung Tahun 2007-2027
[26] Rencana Tata Ruang Kabupaten Bandung Barat Tahun 2009-2029
[27] Potensi Desa Tahun 2016
[28] Wischmeier W H, Smith D D 1978 *Predicting Rainfall Erosion Losses—A Guide to Conservation Planning. U.S. Department of Agriculture, Agriculture Handbook* No. 537
[29] ArcGIS spatial analyst 2013 www.esri.com/software/arcgis/extensions/spatialanalyst/
[30] Badan Penanggulanan Bencana Daerah Kabupaten Bandung, 2016
[31] Pohlan H A J, Janssens M J 2010 Growth and production of coffee. Encyclopaedia of life support system (AOLSS) Vol III Oxford UK
[32] Castro L M 2010 *Shade Coffee Plantations A Sustainable Way to Protect Wildlife* US. Fish and Wildlife Service.
[33] Bote A D and Struik P C 2011 Effects of Shade on Growth, Production and Quality of Coffe (*Coffea arabica*) in Ethiopia. *Journal of Horticulture and Forestry 3*(**11**) pp 336-341.
[34] Schroth G, Tscharntke T, Clough Y, Bhagwat S, Buchori D, Faust H 2000 Multifunctional shade-tree management in tropical agroforestry landscape. *Journal of Applied Ecology.*
[35] Bufo R, Diwew CC 2004 Flavor: an overview. *Journal Flavor and Fragnance* 1999-104. Manitoba.
[36] Asosiasi Eksportir Kopi Indonesia AEKI 2012 *Produsen Kopi Kesulitan Memenuhi Permintaan Kopi yang Meningka*t http://www.aeki-aice.org/link_aeki.html
[37] Erwiyono R, Yacob RY, Usmad 2012 Pengaruh pola curah hujan terhadap produksi kopi (Banyuwangi) *Pusat Penelitian Kopi dan Kakao Indonesia.*
[38] Pujiyanto 2007 Heading Toward Sustainable Coffee Production 23(1) *Warta Pusat Penelitian Kopi dan Kakao Indonesia* (23)1—10.

# A Spatial Approach to Identify Slum Areas in East Wara Sub-Districts, South Sulawesi

**W Anurogo[1*], M Z Lubis[1], D S Pamungkas[2], Hartono[3], F M Ibrahim[3]**

[1] Geomatics Engineering, Politeknik Negeri Batam, Batam 29461, Indonesia
[2] Electrical Engineering, Politeknik Negeri Batam, Batam 29461, Indonesia
[3]Faculty of Geography, Universitas Gadjah Mada, Sekip Utara, 55281 Yogyakarta, Indonesia

wenang@polibatam.ac.id

**Abstract.** Spatial approach is one of the main approaches of geography, its analysis emphasizes the existence of space that serves to accommodate human activities. The dynamic development of the city area brings many impacts to the urban community's own life patterns. The development of the city center which is the center of economic activity becomes the attraction for the community that can bring influence to the high flow of labor both from within the city itself and from outside the city area, thus causing the high flow of urbanization. Urbanization has caused an explosion in urban population and one implication is the occurrence of labor-clumping in major cities in Indonesia. Another impact of the high urbanization flow of cities is the problem of urban settlements. The more populations that come in the city, the worse the quality of the existing settlements in the city if not managed properly. This study aims to determine the location of slum areas in East Wara Sub-Districts using remote sensing technology tools and Geographic Information System (GIS). Parameters used to identify slum areas partially extracted using remote sensing data and for parameters that cannot be extracted using remote sensing data, information obtained from field surveys with information retrieval based on reference data. Analysis results for slum settlements taken from the parameters indicate that the East Wara Sub-District has the largest slum areas located in Pontap village. The village of Pontap has two classes of slums that are very shabby and slums. Slum classes are also in Surutangga Village. The result of the analysis shows that the slum settlement area has 46,324 Ha, which is only located in Pontap Village, whereas for the slum class are found in some villages of Pontap and Surutangga Urban Village, there are 37.797 Ha area. The class of slum settlement areas has the largest proportion of the area among other classes in East Wara Subdistrict. The class of slum settlement areas has an area of 74,481 Ha. This class is located in Kelurahan Salekoe and Kelurahan Benteng. The less grungy and quite shabby class is the rest which each has an area of 29,144 Ha and 18,228 Ha. There is quite a slum class in Kelurahan Ponjale and Less Slum Class Available in Malatunrung Urban Village.

**Keywords:** Spatial Approach, Remote Sensing, GIS, Slum Area

## 1. Introduction

Spatial approach is one of the main approaches of geography, its analysis emphasizes the existence of space that serves to accommodate human activities [1]. The dynamic development of the city area brings many impacts to the urban community's own life patterns [2] The development of the city center which is the center of economic activity becomes the attraction for the community that can bring influence to the high flow of labour [3] both from within the city itself and from outside the city area, thus causing the high flow of urbanization. This is what causes the regional gap in Indonesia [4]. Urbanization has caused an explosion in urban population and one implication is the occurrence of labor-clumping in major cities in Indonesia. Another impact of the high urbanization flow of cities is the problem of urban settlements. The more populations that come in the city, the worse the quality of the existing settlements in the city if not managed properly [5]. Concentrated urbanization as described above, in addition to harming also have advantages. Some of the advantages are infrastructure equipment for the modernization of the cost becomes cheaper and faster economic development [6].

The high number of residents in the center of the city requires the fulfilment of the need for settlements that are habitable, especially to accommodate the urbanists whose work is concentrated in the trade and services sector in the commercial area in the city center. The availability of complete facilities and infrastructure in the center of the city has attracted people to settle in the area [7]. They need more shelter around the commercial area of the city, this is also possible because they approach the trading center to open a business by utilizing the crowd and crowded visitors who come to the shopping centers in the city. In addition, other reasons for the community are interested in residing in the vicinity of the downtown area as it facilitates workplace reach for those working in the city center, as well as meeting the needs of the residential communities that work in Central Business District (CBD) area of the city [8]. The availability of complete facilities and infrastructure in the city center is also the main attraction of the people to live in the area. A large number of immigrants in the area requires land to be settled [9]. The availability of available land does not match the number of people who come to the cause of the number of slums.

The urban deficiency with proper urban planning and management system, in anticipation of population growth with various motives and diversity, seems to be the main cause that triggers the emergence of settlement problems [10]. Fulfilment of the needs of infrastructure and settlement facilities both in terms of housing and neighbour-hoods affordable and decent settlement has not been fully provided by the community itself and the government. Thus, the carrying capacity of the existing settlement infrastructure and facilities begins to decline and will ultimately contribute to slum settlement.

Slum areas are defined as a residential area or non-residential area which is used as a residence whose buildings are substandard or inadequate inhabited by densely populated poor. The real area is not designated as a residential area in many large cities, by poor, low-income and non-permanent residents being resettled for residents, such as riverbanks, by railroads, the vacant land around factories or downtown, and under the bridge. Slums are inadequate settlements because they do not meet the requirements for both technical and non-technical occupancy [11]. A slum can be regarded as the embodiment of poverty, because in general in slums the poor life and many of us meet in urban areas. Poverty is one of the causes of the emergence of slums in urban areas. Basically, poverty can be overcome with high economic growth and equity, increased employment and the income of the poor and the improvement of basic services for the poor and the development of poverty reduction institutions. Improvement of basic services can be realized by increasing water supply, sanitation, provision and improvement of housing and residential environment in general [12].

Slums can be found in many major cities in the world. Slums are generally associated with high levels of poverty and unemployment. Slums can also be a source of social problems such as crime, drugs, and alcohol. In many poor countries, slums are also the center of health problems because of unhygienic conditions [13]. In slum areas, especially in poor countries, people live in very close proximity, making it very difficult to pass vehicles like ambulances and firefighters. The lack of garbage disposal services also resulted in the piles of garbage [14].

## 2. Methods

Research method is a step that must be done to complete the research objectives to be achieved. This research conducted to identify slums in East Wara Sub-District Palopo City. The research area in East Wara Sub-district Palopo City is geographically located between 2°53'15 "- 3°04'08" South Latitude and 120°03'10 "- 120°14'34" East Longitude. Palopo City as an autonomous region resulting from the expansion of Luwu regency, with boundaries:

a. Regency of Luwu is adjacent to Regency of Luwu in the north

b. East with the Bay of Bone
c. Southern borders with Bua District of Luwu Regency
d. West side bordering Tondon Nanggala Sub-district of Tana Toraja Regency.

The total area of Palopo City administration is about 247.52 square kilometers or equal to 0.39% of the total area of South Sulawesi Province. Administratively Palopo City is divided into 9 districts and 48 urban villages. The Most of the East Wara Sub-District Palopo City which is the study area of research is a lowland area, in accordance with its existence as an area located on the coast. The location of the research is shown in figure 1. The method used in this study descriptive qualitative through spatial level is supported by questionnaires (learning structured and closed) to analyze the spread of slums in East Wara Sub-District Palopo City. The spatial approach will later be combined with data from relevant agencies relating to this research. In addition, it will be taken data by questionnaire by doing random sampling. Random Sampling is a sample taken from a population and each member of the population has equal opportunity to be selected as a sample [15]. This research uses remote sensing data tool to get various parameters which influence to slum settlement. Remote sensing data used in this study is the QuickBird high-resolution data. These high-resolution data are used to get parameters that affect slum settlements. The parameters affecting the slums are shown in table 1.

**Table 1**. Parameters affecting slums area

| No | Variables | Information | Indicator |
|---|---|---|---|
| 1 | Basic Services | Sanitation | |
| | | Source of Clean Water | |
| | | Waste Management System | |
| | | Fire Safety Condition | |
| | | Electric network | |
| 2 | Physical Components | Home Physical Condition | Roof |
| | | | floor |
| | | | Wall |
| | | | Size of the house |
| | | Population Conditions | Building Density per Block |
| | | | Number of Houses per Hectare |
| | | | Tree Protector |
| | | | Entrance area |
| | | | Road surface conditions |
| 3 | Rural services | Government Facilities | |
| | | Education Facilities | |
| | | Health Facilities | |

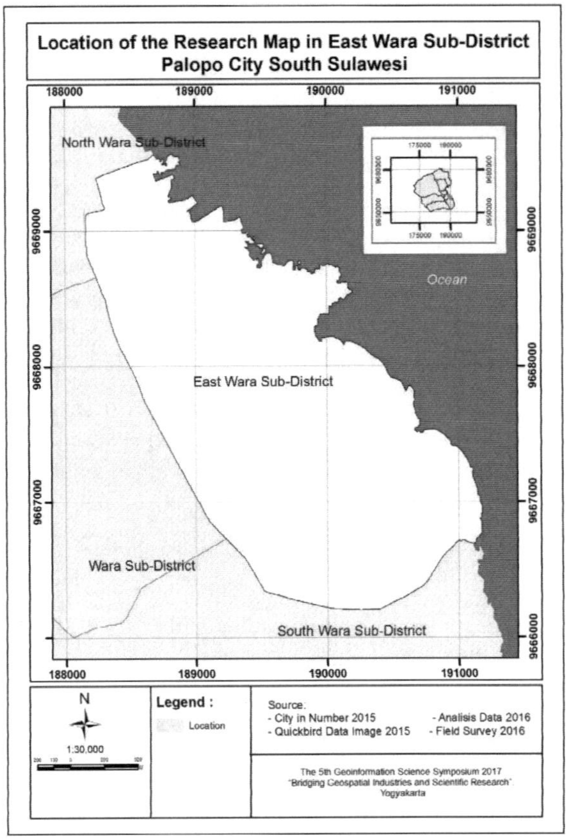


**Figure 1**. Location of the research

The stages undertaken to carry out this research consists of 3 stages of a pre-field stage, field stage, and post-field stage. In the pre-field activity consists of 3 parts, among others, data collection, retrieval of land cover information and parameters that can be retrieved information from visual interpretation and the determination of samples used to retrieve information directly in the field of information that cannot be extracted from remote sensing data, simultaneously validating the extracted data from the Quickbird image. the research stages are shown in figure 2.

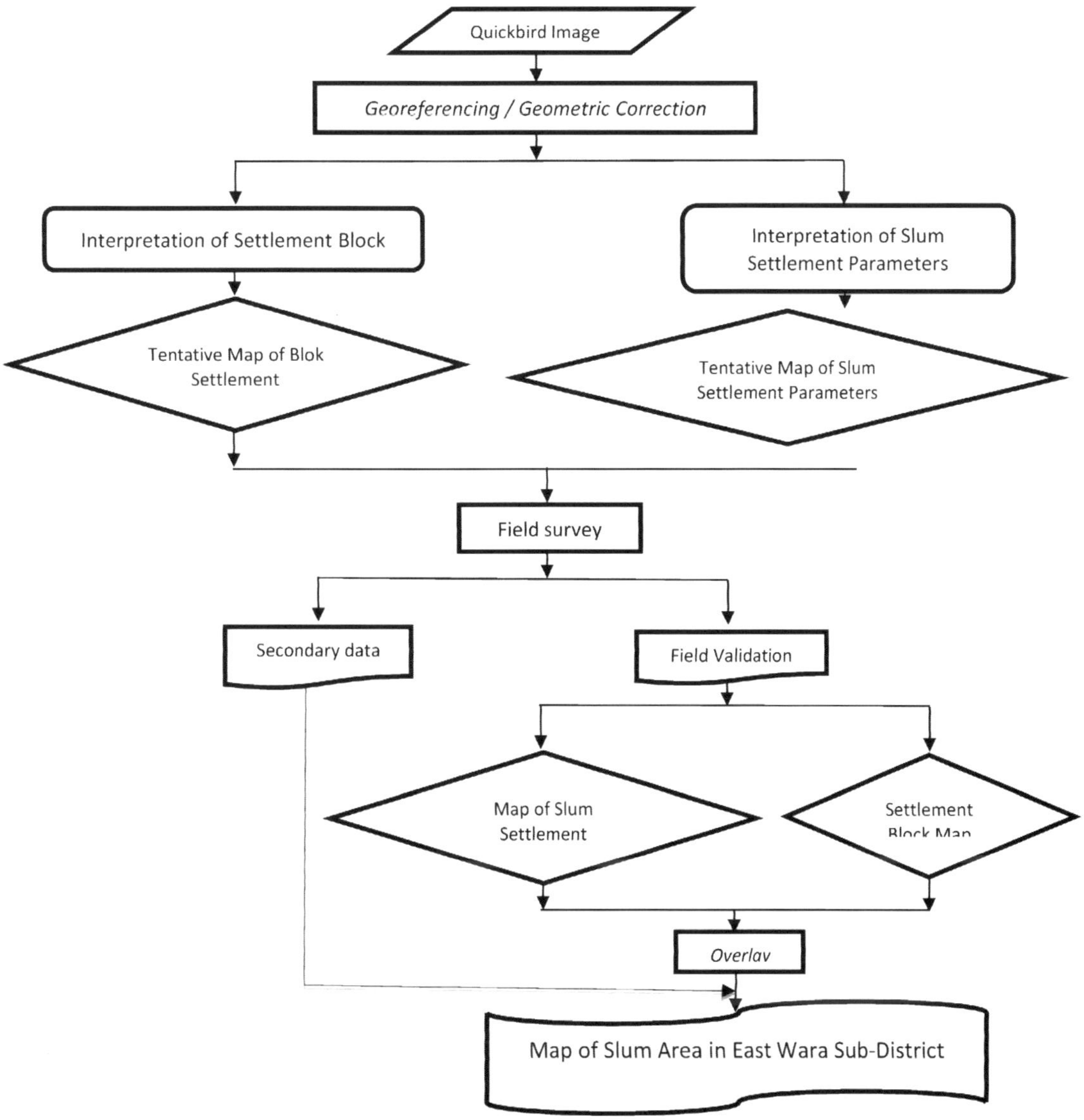


**Figure 2.** The research stages

In the pre-field activity consists of 3 parts, among others, data collection, reduction of land cover information and parameters that can be retrieved from the visual interpretation of visual images and the determination of samples used to retrieve information directly in the field of information that cannot be extracted from remote sensing data as well as data validation of extraction results from Quickbird images. Data collection phase consists of primary and secondary data collection process such as acquisition of google earth image, data collection of population density, education level data collection, health data collection and population economic data collection.

The phase of reduction of land cover information from google earth is done by identifying the land cover visually which will be classified into 2 classes namely settlement class and non-settlement class. The last stage is the determination of field samples. In the determination of this field sample using a random sampling method that provides equal opportunities or opportunities to all members of the population to be sampled. It is expected that the samples obtained are representative samples.

Samples taken in this study represent the conditions of the slum settlement in the study area and the community. The research took a sample of 50 people with random sampling method.

Fieldwork is aimed at obtaining information on slum settlement parameters in the study area. Data measured during field activities provide physical conditions of the cultivation, social and economic conditions of the population. The way of data collection in the field is done by interview and questionnaire. Interviews were conducted to obtain information such as air circulation, solar lighting, health, garbage, flood frequency and drainage. While the questionnaire to obtain information in the form of education level, monthly income, work. The resulting analysis was conducted descriptively qualitatively based on the parameters of slum areas and public perception of slums. This parameter analysis is performed to identify the slum classes that exist in the study area based on predetermined parameters. The analysis of people's perceptions of slums is based on the results of the questionnaire and the interviews by comparing them with the slum-dwelling classes that had been previously produced.

A field survey was undertaken to obtain additional information to supplement the parameter data as well as to retrieve slum area information based on community perceptions. The field survey was conducted by visiting relevant agencies to obtain additional data, cross-check interpretation, and interviews or interviews with respondents.

Data collection methods in this study using a qualitative approach, by providing questionnaires to 60 respondents who are residents living in the settlement District East Wara Palopo City. Determination of the number of respondents was taken based on the interpretation of settlement blocks created as well as checks the interpretation of the settlement area. The result of the interpretation is that the residential area of Wara Timur sub-district is divided into 60 blocks of settlements with the assumption that each of the settlement blocks represents one class of area.

Additional information obtained in this field survey for additional information on physical parameters as well as for community perceptions. Additional physical information sought is information that cannot be derived from visual interpretations such as Sanitation, Clean Water Source, Waste Management System, Electricity and Fire Conditioning Conditions. Additional information will be used as additional input to determine the slum areas in the study area.

Poor environmental sanitation affects the sustainability of the existing environment. The lack of knowledge and understanding of the importance of healthy sanitation in healthy settlements is evident from the community's less friendly behaviour. This is marked by the existence of some people who do unhealthy lifestyles such as using the river as a means of toilets and clean water for the necessities of life, and the habit of disposing of household waste directly into the river that has potential as the cause of the spread of disease outbreaks.

Source of clean water is indirectly related to environmental sanitation. Poor sanitation causes the source of clean water to be poor or polluted causing the environment to become visibly shabby. People in the study area use a lot of water obtained from middlemen around "*the tengkulak*" or from PDAM because sanitation in their environment is bad so pollute the clean water source that is there. The emergence of slums in urban areas is a problem that is often faced by a number of major cities in Indonesia. The lack of provision of facilities and infrastructure in slums is generally dilator by the problems of the legality of settlements, resulting in the decreasing quality of the settlement environment. For example, with the unavailability of garbage facilities, the community will tend to contaminate the settlements with garbage so that waste generation will be piled up in every corner of the settlement.

Waste that is under the house, not continuously cleaned by residents or residents. Factors of altitude under the house, the physical properties of muddy soil and the back of the soil under a short time by the garbage that drifts the tide, which is the factor that inhibits the motivation of citizens to clean it up. The process of cleaning garbage under the house will generally be done if the individual or residents have sufficient time and energy, hot weather that supports and river water in certain receding conditions.

The fire safety conditions in the study area do not meet the standards. Fire safety contained in the research area is only water from the water source that is around so it is still said to be inadequate.

## 3. Results and Discussion

The data used to determine the distribution and extent of the settlement blocks that serve as the basis or mapping unit to find slum areas found in East Wara Sub-District Palopo City is Quickbird image data. The data is then done equation coordinates with coordinates that exist in the field with georeferencing process or better known as a geometric correction. Other data used in this research is public perception data about slum areas. The data is obtained from field surveys and/or data from relevant agencies in accordance with the theme of the study conducted. The geometric correction performed on this data is non-systematic geometric correction performed using a ground control point (GCP). Non-systematic geometric correction uses first-order polynomial algorithms because the physiographic conditions of the region are mostly terrain. Non-systematic geometric correction is performed by an image to image registration with reference to georeferencing image map using base map of RBI scale of 1: 25.000 which has been corrected, with Universal Transverse Mercator (UTM) projection system, WGS datum 1984. The study area included in zone 51 M. Zone M is south of the equator.

The data tapping activity is done by digitizing on the screen to restrict the existing settlement blocks in the research area. In the interpretation of the settlement block is done based on the physical appearance of the environment where the object identifier is the parameter of the quality of the settlement environment, such as the density of the building, the layout of the building, the width of the entrance, the condition of the road surface of the settlement, the road protector tree. Interpretation of each parameter is done by referring to the element or key interpretation. The location map of the settlement is obtained from the land use interpretation. The location of residential area is shown in figure 3.

Interpretation of the quality parameters of the settlement of the image is done by first doing some things as follows:

a. Determining the boundaries of settlements and non-settlement and distinguish between settlements as a residence with buildings as other functions such as offices and education.
b. Mapping the boundaries of residential neighbour-hood units based on roadblocks as well as differences in settlement characteristics that distinguish the research area.

The parameters used to assess excessive slum areas that can be extracted using high resolution remote sensing data are:

a. Density of Settlements
   Settlement density data can easily be known through high-resolution imagery. The density of home settlements assessed is the relative density, ie based on the density of buildings in a settlement block. In determining the units of settlement units (settlement blocks) are measured qualitatively based on the level of uniformity. Areas that have a relatively homogeneous density will be included in the same unit of settlement. the density classification of the settlement is then given the weight of the scale and classified into 3 classes of good, medium, and bad. Classification of weighted density of settlements is shown in table 2

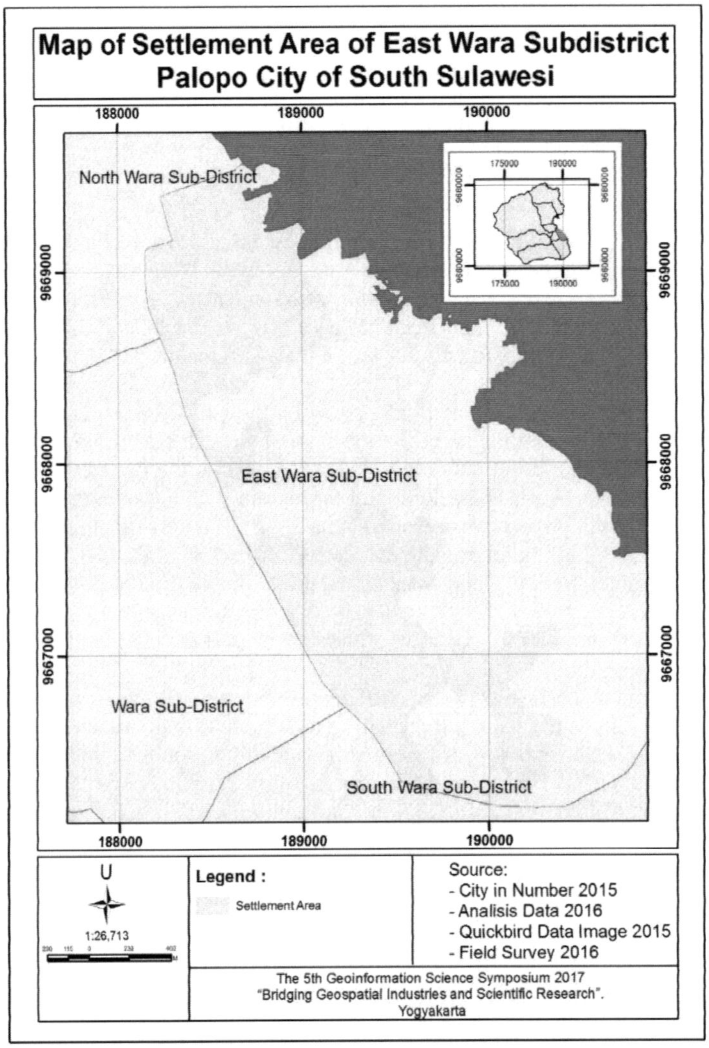


**Figure 3**. Settlement Area of East Wara Sub-district

**Table 2**. Classification of Weighted Density of Settlements

| Criteria | Classification | Score |
|---|---|---|
| Rarely (density less than 40%) | Good | 3 |
| Medium (density between 40% -60%) | Medium | 2 |
| High (over 60% density) | Bad | 1 |

The results of the analysis indicate that the density of settlements with bad grades is located in Pontap Village while for the density of settlements with good value is found in the village of Salekoe and Benteng. The result of analysis shows that residential area in study area based on visual interpretation has an area of 197.4507 ha. The interpretation results are then made per block of settlements used to be the smallest mapping unit as the basis for sampling the field.

b. Settlement Layout Patterns

Assessment of settlement layout patterns related to slums can be seen from the regularity of the location, and the size of the building. Building settlements that have relatively the same size and its location follow a certain pattern, then the building will be grouped in units of the same settlement units. The layout of the settlement is calculated by comparing the number of structured buildings with the number of buildings in the settlement blocks. Weight classification level of settlement locations is shown in table 3.

**Table 3.** Weight Classification Level of Settlement Locations

| Criteria | Classification | Score |
|---|---|---|
| More than 50% buildings in a regularly organized settlement unit | Good | 3 |
| 25-50% buildings in a regularly arranged settlement unit | Medium | 2 |
| less than 25% buildings in a regularly organized settlement unit | Bad | 1 |

c. Street Shade Trees

This road shade trees referred to as shading driveways and being on either side of the driveway on a settlement block. It also can work to reduce pollution caused by motor vehicle fumes.

**Table 4.** Street Shade Trees Classification

| Criteria | Classification | Score |
|---|---|---|
| More than 50% of existing access roads in residential units on both sides have a protective tree | Good | 3 |
| 25-50% of existing access roads in residential units on both sides have a protective tree | Medium | 2 |
| less than 25% of existing access roads in residential units on both sides have a protective tree | Bad | 1 |

d. Width of Entrance Road

The width of the entrance of the settlement is defined as a road that connects the neighbourhood road with its main road. Assessment of these parameters is intended to determine the ease of transportation to and from the settlement blocks concerned. With the spatial resolution possessed by the remote sensing data used, the different paths between the segments with each other can easily be distinguished. Weighting classification street width is shown in table 5.

**Table 5.** Street Shade Trees Classification

| Criteria | Classification | Score |
|---|---|---|
| average entrance width of more than 6 m | Good | 3 |
| average entrance width is between 3m-4m | Medium | 2 |
| average entrance width of less than 4 m | Bad | 1 |

e. Road Surface Condition
What is meant by the entrance is the road connecting the neighbour-hood road with the main road. The condition of the inlet surface is the hardening of the road surface distinguished by the hardening material based on a percentage of the asphalt roadway condition or cemented to the entire road. How to interpret it by noticing the hue on the object being observed. Classification of surface road conditions is shown in table 6.

**Table 6.** Classification of Surface Road Conditions

| Criteria | Classification | Score |
|---|---|---|
| more than 50% of the length of existing driveways in hardened residential units | Good | 3 |
| Between 25-50% of the length of existing driveways in hardened residential units | Medium | 2 |
| less than 25% of the length of existing driveways in hardened residential units | Bad | 1 |

Assessment of the quality class of settlements is done after the completion of all parameters of the quality of the settlement environment is completed in the input in the attribute table. The environmental quality class determinants are based on the total score. The total score is obtained from the sum and multiplication of each determinant parameter with the weighting factor. The parameters are then given the weighting of each value, then overlay or overlapping analysis to determine the slum areas. The classification of slum settlements is divided into five classes so that from the result of the overlay result, it can be seen that the settlement area that has the value of all the most influential parameters (the smallest value) is included in the bad slum class, and vice versa. According to the survey results of the questionnaire distribution and after analysed using the analysis method, obtained the result that the number of occupants has a low influence on the formation of slum areas in this study area. In reality in the study area most of each house with an area of not more than 36 m2 inhabited by 5 people. From the minimum standard of Dirjen Cipta Karya this is considered not ideal because it can be assumed that 1 bedroom is utilized by 2-3 people, also of course with the limited area the needs of the spaces are not fulfilled and cannot be organized properly. In this case, the number of occupants in a house can take part in shaping the slums of a region. if the house's number of inhabitants increases often the homeowner adds or expands their building regardless of the applicable rules like ROI Line (boundary of land) or KDB (building coefficient), allowable outbreaks. If such a situation occurs in 35% of the area of course will give the impression of slum because of the developing area without good planning.

However, according to the public perception, the house is considered to have qualified provided that they can take shelter and can rest is considered sufficient. Especially for those who open the home industry crackers, the more residents either brothers or others who live in a house will benefit them because of their energy needed to help their business.

From the results of questionnaires conducted on the characteristics of incomes, the results obtained that this factor has a strong influence on the formation of slum conditions in East Wara Sub-District. Economically, with relatively low economic conditions, it is possible the ability of residents to realize the improvement of their residential environment is not possible. The need for sustainability, such as clothing and food, is the top priority of residents in allocating expenditures from the funding they earn.

From the results of the questionnaire analysis conducted on the status of residential ownership obtained results that this factor has a strong enough influence on the formation of slum areas. According to the community if a house with rent status most residents do not care about the conditions or conditions of occupancy that they rent. So, if there is damage to the dwelling, the residents do not care about it because it felt it was not his responsibility. The longer the occupancy is getting worse because it is not maintained. The results of field observations show that the number of immigrants in this area caused the number of residential homes that are rented, this would further increase the level of the slum area.

The long-lived factor is a factor that strongly affects the occurrence of slum area, this is the result of the analysis conducted on the perception of the population about the long-lived factors for residents. This is closely related to the previous factor of occupancy status factors, where many of the people living with rent status are factors that have a strong influence on slums. If those who rent a house for a long time or more than 5 years of age will not neglect their shelter maintenance, but many also rent a room for only a short time, usually their rental rate is per day. So, of course, that occupy the room or rental house is changing or not fixed. This is what causes the area to be a slum. The results of the analysis for slum settlements taken from the parameters show that the district of East Wara has the largest slum area located in Pontap village. The village of Pontap has two classes of slums: very slums and slums. Slum classes are also located in Surutangga Village. The result of the analysis shows that the slum settlement area has 46,324 Ha, which is only located in Pontap Village, while for the slum class is found in some villages of Pontap and Surutangga Subdistrict, there is 37.797 Ha area.

The class of slum settlement areas has the largest proportion of the area among other classes in East Wara Subdistrict. The class of slum settlement areas has an area of 74,481 Ha. This class is located in Kelurahan Salekoe and Kelurahan Benteng. The less grungy and quite shabby class is the remainder which each has an area of 29,144 Ha and 18,228 Ha. Quite slum class is found in Kelurahan Ponjale and Less Slum Class Available in Malatunrung Urban Village. Map of Slum Residential Area, East Wara Subdistrict, Palopo Municipality, South Sulawesi is shown in figure 4.

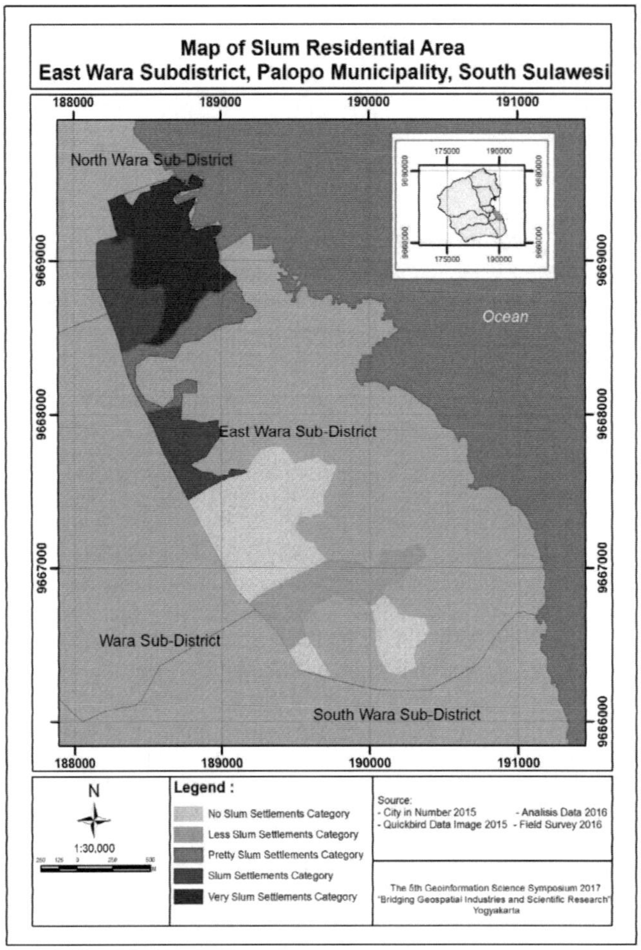


**Figure 4**. Map of Slum Residential Area

## 4. Conclusions

Analysis results for slum settlements taken from the parameters show that the district of East Wara has the largest slum area located in Pontap village. The village of Pontap has two classes of slums: very slums and slums. Slum classes are also located in Surutangga Village. The class of slum settlement areas has the largest proportion of the area among other classes in East Wara Subdistrict. The class of slum settlement areas has an area of 74,481 Ha. This class is located in Kelurahan Salekoe and Kelurahan Benteng. The less grungy and quite shabby class is the remainder which each has an area of 29,144 Ha and 18,228 Ha. Quite slum class is found in Kelurahan Ponjale and Less Slum Class Available in Malatunrung Urban Village.

## References

[1] Wesnawa, I. (2010). Dinamika Pemanfaatan Ruang Berbasis Kearifan Lokal di Kabupaten Buleleng Provinsi Bali.

[2] Muta'ali, L. (2002). Pola Perkembangan Karakteristik Kekotaan Pada Desa-Desa Di Propinsi Daerah Istimewa Yogyakarta. Majalah Geografi Indonesia, 16(2002).

[3] Rachmawati, R., Rijanta, R., & Subanu, L. P. (2004). Peranan Kampus Sebagai Pemicu Urbanisasi Spasial di Pinggiran Kota Yogyakarta. Majalah Geografi Indonesia, 18(2004).

[4] Wilonoyudho, S. (2009). Kesenjangan dalam Pembangunan Kewilayahan.

[5] Farizki, M., & Anurogo, W. (2017). Pemetaan kualitas permukiman dengan menggunakan penginderaan jauh dan SIG di kecamatan Batam kota, Batam. Majalah Geografi Indonesia, 31(1), 39-45.

[6] Atmajaya, J., & Jamli, A. (2014). Kontribusi Infrastruktur terhadap Pertumbuhan Ekonomi Sumatera Selatan (Doctoral dissertation, Universitas Gadjah Mada).

[7] Pigawati, B., & Rudiarto, I. (2011, December). Penggunaan Citra Satelit untuk Kajian Perkembangan Kawasan Permukiman Di Kota Semarang. In Forum Geografi (Vol. 25, No. 2, pp. 140-151).

[8] Anurogo, W., Lubis, M. Z., Hartono, H., Pamungkas, D. S., & Dilaga, A. P. Kajian Ketahanan Kawasan Wisata Berbasis Masyarakat Dalam Penguatan Ekonomi Lokal Serta Pelestarian Sumberdaya Kebudayaan Kawasan Kotagede Yogyakarta. Jurnal Ketahanan Nasional, 23(2), 114-136.

[9] Manning, C., & Effendi, T. N. (1985). Urbanisasi, pengangguran dan sektor informal di kota. Diterbitkan untuk Yayasan Obor Indonesia dan Pusat Penelitian dan Studi Kependudukan UGM [oleh] Gramedia.

[10] Campbell, S., & Fainstein, S. S. (2003). Readings in Planning Theory (Studies in Urban & Social Change).

[11] Berger, S. G., de Pee, S., Bloem, M. W., Halati, S., & Semba, R. D. (2008). Malnutrition and morbidity among children not reached by the national vitamin A capsule programme in urban slum areas of Indonesia. Public Health, 122(4), 371-378.

[12] Abate, G., Kogi-Makau, W., & Muroki, N. M. (2017). Child-feeding practices as predictors of nutritional status of children in a slum area in Addis Ababa, Ethiopia. The Ethiopian Journal of Health Development (EJHD), 13(3).

[13] Kanbur, R., & Sumner, A. (2012). Poor countries or poor people? Development assistance and the new geography of global poverty. Journal of International Development, 24(6), 686-695.

[14] Aziz, T. A., & Shawket, I. M. (2011). New strategy of upgrading slum areas in developing countries using vernacular trends to achieve a sustainable housing development. Energy Procedia, 6, 228-235.

[15] Stehman, S. V. (1992). Comparison of systematic and random sampling for estimating the accuracy of maps generated from remotely sensed data. Photogrammetric Engineering and Remote Sensing, 58(9), 1343-1350.

# Contextualizing Cave Maps as Geospatial Information: Case Study of Indonesia

**H Reinhart**

Master of Environmental Management, Graduate School of Universitas Gadjah Mada, Indonesia

hilary.reinhart@mail.ugm.ac.id

**Abstract.** Caves are the result of solution processes. Because they are happened from geochemical and tectonic activity, they can be considered as geosphere phenomena. As one of the geosphere phenomena, especially at karst landform, caves have spatial dimensions and aspects. Cave's utilizations and developments are increasing in many sectors such as hydrology, earth science, and tourism industry. However, spatial aspects of caves are poorly concerned dues to the lack of recognition toward cave maps. Many stakeholders have not known significances and importance of cave maps in determining development of a cave. Less information can be considered as the cause. Therefore, it is strongly necessary to put cave maps into the right context in order to make stakeholders realize the significance of it. Also, cave maps will be officially regarded as tools related to policy, development, and conservation act of caves hence they will have regulation in the usages and applications. This paper aims to make the contextualization of cave maps toward legal act. The act which is used is Act Number 4 Year 2011 About Geospatial Information. The contextualization is done by scrutinizing every articles and clauses related to cave maps and seek the contextual elements from both of them. The results are that cave maps can be regarded as geospatial information and classified as thematic geospatial information. The usages of them can be regulated through the Act Number 4 Year 2011. The regulations comprised by data acquisition, database, authorities, surveyor, and the obligation of providing cave maps in planning cave's development and the environment surrounding.

**Keywords:** cave, cave maps, geospatial information, regulation

## 1. Introduction

Caves are one of the typical features of karst region [1]. The definition of caves for the non-scientific purpose is a cavity or void space with enough size which human can enter [2]. That definition is non-scientific because it doesn't consider the genetic aspect, loose, and contains ambiguity. Scientific definition of caves was defined by White (1988) as "*Natural cavity in a rock which acts as a conduit for water flow between input points, such as streamsinks, and output points, such as springs or seeps*". Caves were formed under control of geosphere dynamics and processes such as tectonic activities and solution process involving reactions of water, air, and rocks [3]. Because it is the result of the process and dynamics caves are one of the significance features of earth and geosphere. The significance of caves relies on their values comprised by scientific, economical, to socio-cultural values. Scientific values of caves are related with the new discoveries and their relation with other sciences [4][5].

Utilization of caves to make profit and income is the economic value of caves. Socio-cultural values regard the relation of human and caves.

With its fragile environment and very high values, caves must be protected and managed satisfactorily [6]. However, science solely will not be adequate in management and conservation act. A more formidable and stronger instrument must be put into account. Regulation, institution, and governing aspects as that stronger instruments have to be applied to ensure the preservation of caves.

The management of caves also must be put above the pillars of spatiality as one of the foundations of environmental management. As physical entities on geosphere, caves have all of the location, size, dimension, and boundaries. Those aspects embed caves into spatial dimension. Caves are spatial therefore they contain geographical dimension and thus, can be taken into GIS science [7]. GIS science and the geographical consideration of caves are a fruitful tools and paradigms in establishing a far better understanding of cave itself. The better understanding will yield a more exact and precise management strategies.

The manifestation of amalgamation between geographic-spatial and the managerial aspects of caves values and resources is the cave maps and Cave GIS. Cave maps visualize the cave with geographic-spatial discipline by its exact measurements. Utilization of cave maps into management of caves can be based on ecological and ecosystem [8][9][10], tourism [11], and archaeology [12][13]. Cave GIS is very helpful in assessing and determine the characteristics, fragility, and management strategy in the surface area of karst with its ability to visualize the spatial distribution of cave entrance as the one of the surface features [14][15][16]. Cave maps and Cave GIS both are the speleo-spatial aspect of speleology.

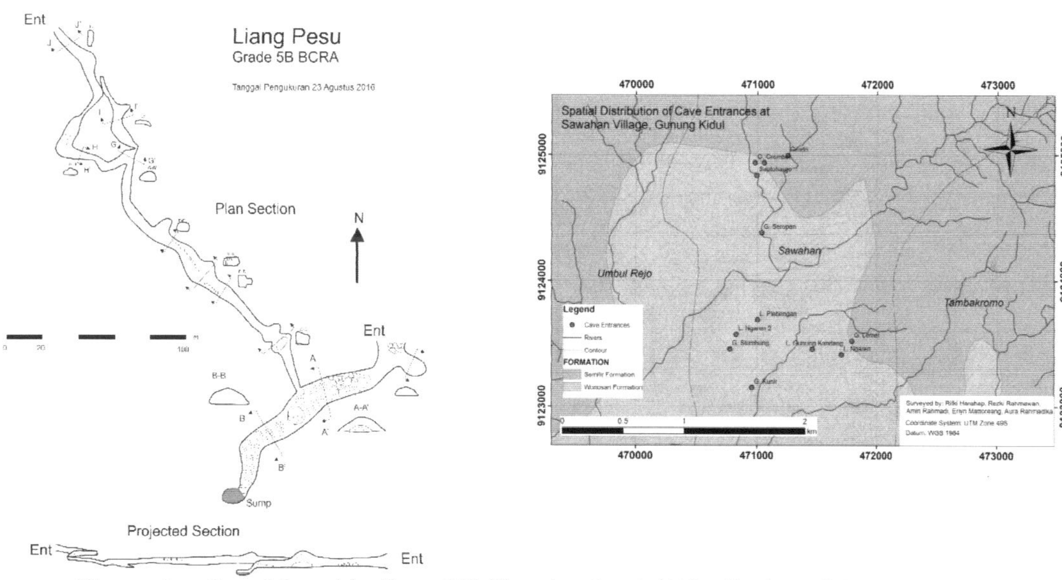


**Figure 1**.a. Cave Maps 1.b. Cave GIS Showing Spatial Distribution of Cave Entrances

Scientific application must be entailed by regulation. Practices of karst management in Indonesia frequently do not involve the speleo-spatial elements, whereas the speleo-spatial element is crucial. The lack of knowledge and information of cave cartography and GIS can be considered as the cause. Another reason is because of the karst management has not included geospatial aspects into its elements. To formulate an exact management system, speleo-spatial elements must be contextualized into geospatial aspect. To contextualize the speleo-spatial aspect means to put its scientific result in the regulation and juridical context. It is essential to examine regulations of karst and management and the geospatial one then merge them into more contextual. A comparison also needed to be made between what is written on the regulation and what science has provided. The comparison will enrich both side of management.

IOP Conf. Series: Earth and Environmental Science **98** (2017) 012031 doi:10.1088/1755-1315/98/1/012031

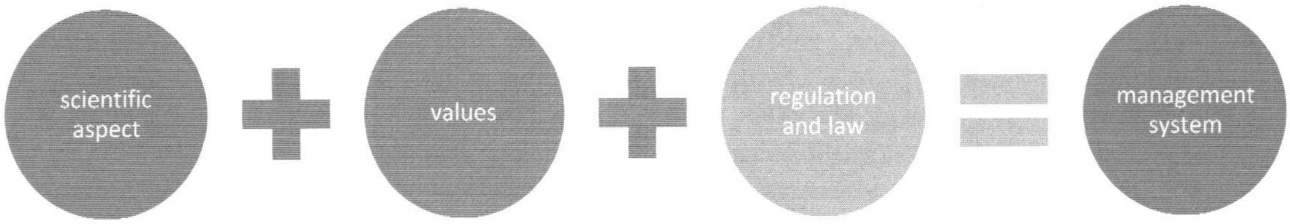


**Figure 2**. Basic Management System

2. **Geospatial Information**

In Indonesia, geospatial information can be divided into two: Base Geospatial Information (Informasi Geospasial Dasar (IGD)) and Thematic Geospatial Information (Informasi Geospasial Tematik(IGT)). The primary regulation on these geospatial information is the Act Number 4/2011 (Undang-undang Nomor 4/2011) Concerning Geospatial Information. This juridical paper regulates and becomes the main reference on the following and derivative regulation. Geospatial information is defined at Article 1. Geospatial itself is defined as "spatial aspects points to locations, lies, and positions an object or phenomena below, on, or above earth surface and represented in particular coordinate system." Geospatial aspect is expressed on geospatial data.

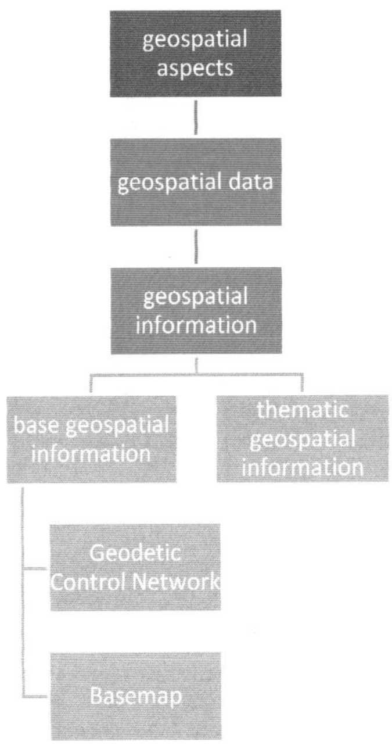


**Figure 3**. Schema of Geospatial Information

Geospatial information is derived from geospatial data. Article 1 defines geospatial information as "geospatial data which have been processed thus can be used as tools in formulating policies, taking decisions, and/or executing actions which are related with geospatial". Geospatial information is divided into Base Geospatial Information which is a Geospatial Information containing object which is directly visible or measured by physical appearance on earth surface and steady in a relatively long-time. Another Geospatial Information is the Thematic Geospatial Information which is "Geospatial Information which represents one or more certain theme which is created refers to Base Geospatial Information". A more complete elaboration of Base Geospatial Information can be found on the Part 2 of Act Article 5-18. Article 5 states that Base Geospatial Information consists of Geodetic Control Network and Base map which is comprised by three maps of Indonesia's Topographic Maps (Peta Rupabumi Indonesia), Indonesia Coastal Environment Maps (Peta Lingkungan Laut Indonesia), and National Marine Environment Maps (Peta Lingkungan Laut Nasional).

It can be extracted from the definition above that geospatial aspect has these characteristics: (i) location and position (ii) positioned below, on, or above earth surface, and (iii) has particular coordinate system. These three elements are the path to understand the context of caves as geospatial data and information.

## 3. Karst Management

The national regulation about the management of karst is the Minister of Energy and Mineral Resources of the Republic of Indonesia Decree Number 17/2012 (Keputusan Menteri Energi dan Sumber Daya Mineral Republik Indonesia Nomor 17 Tahun 2012) about Establishment of Karst Landscape Area. This decree regulates the definition of karst and its status of conservation or preservation. Karst, according to this decree must have permanent spring, karst hill, doline, uvala, polje and/or lake as exokarst features (Article 4 (5)) and Underground River and speleothem as endokarst features (Article 4(6)). Caves do not have their own regulation. They are still clung to the Decree Number 17/2012 and considered as endokarst features. Cave maps are not even included in the decree.

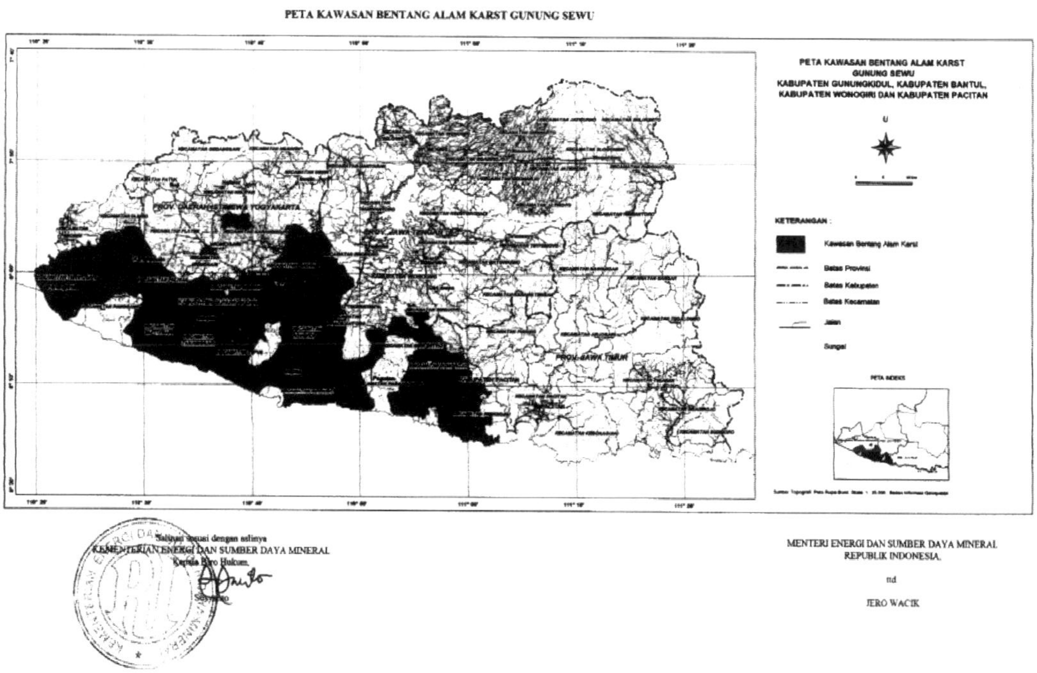


**Figure 4.** Maps of Established Karst Landscape Area's Status. Source: Decree of Energy and Mineral Resources Number 3045 Year 2014.

To establish a status of karst landscape area, all of those features must be inventoried and reported completed with the explanation and elaboration (Article 9(2)). This report is obliged to have maps on it (Article 9(1)). The authority to establish the status is on the minister of energy and mineral resources with motion and consideration of the status are given by the regency government and the Head of National Board of Geology (Article 11). The status of karst area is very crucial because area of karst is the geological reserve as written on Government Decree Number 26/2008 About National Land Use Plan (Article 60(2)). As geological reserve, it has a strict zoning and very limited activities. Activities which are allowed limited only due to the research and scientific purpose and tourism development (Article 104).

## 4. Cave Maps and Geospatial Information

Cave maps had gone through a very long period of time and history since 14$^{th}$ century where the first cave map was created [17][18]. Cave map contains cartographic elements on its layout and drawn using terrestrial survey on cave. The survey is done to measure the dimension and size of the cave thus it can be projected on a media [19].

As it contains cartographic elements, cave maps are presenting the situation inside the cave by graphic. Caves, as a geosphere feature, have precise location and position which makes them geospatially be. Without the element of location and position, caves will be difficultly understood. To contextualize this in juridical context, cave maps must be analysed and compared by the geospatial aspects at Act 4/2011.

Geospatial, according to the act must fulfil these criteria: (i) position and location, (ii) below, on, and above, earth surface and (iii) has particular coordinate system. Caves, therefore, has geospatial aspects because they have certain position and location. They are also on the earth surface and below the earth surface. Caves, in the cave maps or Cave GIS, when they are overlaid with some base maps in the drawing, will be set on certain coordinate system. If the caves are inventoried and recorded, they are transformed from phenomena into raw data. It brings caves as geospatial data because every data which satisfies the geospatial criteria will be considered as geospatial data.

Caves as geospatial data are a series of record and inventories. This data are spatial data because they have both location and position and the attribute. If the spatial data were projected on the media with cartographic elements and disciplines, then a map will be created. Position of certain points or data must be terrestrially measured by their length and direction. Inside the cave, passages are measured by this method because passages are invisible from earth surface. In cave cartography, those points on the passaged are called as stations. Every station is related and locked to another station and eventually attached with the ground reference point (GCP). The GCP is usually the cave entrances because the cave entrances are the only connection between the surface and the sub-surface features.

Next process is to advance the data into information. Caves have a very significance role on providing ecosystem services [20]. They keep the water as one of the the karst aquifers [21] and are the habitat of bats which act as pollinator or pest control. From the ecosystem services they provide, it is urged and very relevant to put them in the consideration on karst management. Thus, data of the caves can be used to help formulating relevance regulation and directing the management of karst.

In case of Indonesia, presence of cave entrances obviously and boldly signifies area of karst. Not only has the entrance, the passage also indicated the presence of karst. It is stated at decree 17/2012 where the Underground River can be categorized as the passages for the passages are formed due to the activity of Underground River. As the data, the entrances and the passages must be included into the report in establishing karst's status. Activity report on karst inventory must attach maps with coordinate of the exokarst and endokarst elements (Article 9(2)). To achieve that, spatial data must be considered. By using the data as the consideration of karst status, it means the data has been used in implementing regulation and making the decision. Caves maps and Cave GIS are now geospatial information.

Advancing again, cave maps as the geospatial information must be derived and categorized whether it is base geospatial information or thematic geospatial information. From the Part 2 Act 4/2011, base geospatial information is limited only to the Geodetic Control Network and the Base Maps with their derivations. It is clear then, that cave maps cannot be classified as the base geospatial information. Thus, the only possibility is to classify the cave maps as the thematic geospatial information. Thematic geospatial information must have these two conditions which are: (i) has certain theme and (ii) refers to base geospatial information. Certain theme is very loose concept. The theme of cave maps is relative due to its usage whether it is tourism, scientific, or ecological. To satisfy the second condition, a cave maps must be referred or overlaid with the base maps. At this case, to make it is easier, is referred to topographic maps rather than Geodetic Control Network. Nevertheless, joining the cave survey to Geodetic Control Network is highly recommended.

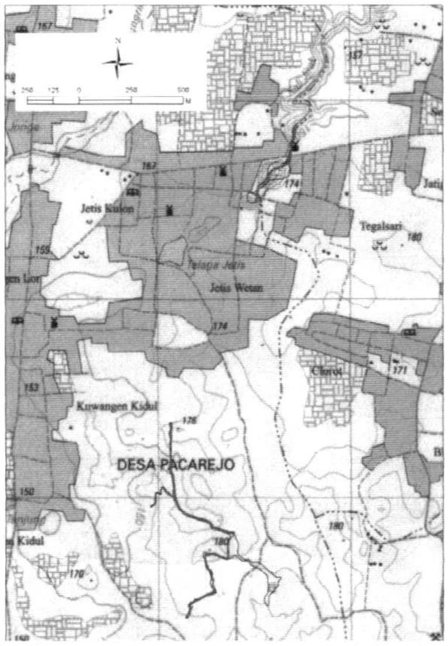


**Figure 5.** Overlaid Topographic Maps as Base Geospatial Information with Cave Passages

## 5. Management of Data and Challenges

After categorizing cave maps and Cave GIS as the thematic geospatial information, it is important to review how the management of thematic geospatial information is. Thematic geospatial information can be run and organized by Government Institution, local Government, and everybody (Article 23 (1)). Regardless the government, everybody may organize the thematic geospatial information for private purpose (Article 23(4)). It means that institution and body such as speleological club, *mapala*, or speleological association are legal in organizing and creating cave maps or Cave GIS as the part of thematic geospatial information.

At the Part 5 which regulates the organization of Geospatial Information, it is written at Article 25 that organization of Geospatial Information can be done through: (i) gathering geospatial data, (ii) processing geospatial data and geospatial information, (iii) archiving and securing the geospatial data and geospatial information, (iv) dissemination of geospatial data and geospatial information, (v) usage of geospatial data and geospatial information. This statement must be noticed by every speleologist and caver who runs for geospatial data.

The challenges on organizing and running the geospatial data and information is that cave surveyor has not been considered profession. It will be a violation regarding Article 55 Act 4/2011 which states that geospatial information must be organized by a competent and certified institution. Meanwhile, caver and speleologist in Indonesia are hardly to be certified. Lack of knowledge will lead to the lack of motivation. This can be considered as another challenge where the dissemination of the cave cartography is relatively low. Standardization is also poorly developed since the application and the production of cave maps is small number.

## 6. Conclusion

It can be concluded that both cave maps and Cave GIS are Geospatial Information regarding the Act Number 4 Year 2011 About Geospatial Information. The organization of them must obey the legal regulation.

**Acknowledgements**

This paper is one of the efforts to establish and encourage cave cartography become science and decision support. Author would like to give full appreciation to Stasiun Nol Festival as the only event which support that support and all person involved within.

## References

[1] White B W 1988 *Geomorphology and Hydrology of Karst Terrains* (New York: Oxford University Press)
[2] Klimchouk A Caves in Gunn, J (eds) 2004 Encyclopedia of Karst and Caves (Taylor & Francis) p 417-421.
[3] Ford D & Williams P 2007 *Karst Hydrogeology and Geomorphology* (England: British library)
[4] Palmer A N 2007 Cave geology and speleogenesis over the past 65 years: role of the National Speleological Society in advancing the science *Journal of Cave and Karst Studies* **69** p 3–12.
[5] White B W 2007 A brief history of karst hydrogeology: contributions of the NSS *Journal of Cave and Karst Studies* **69** p 13–26.
[6] Gillieson D S Management of Caves in van Beynen, P. (ed.) 2011 *Karst Management* (Springer)
[7] Szukalski B 2002 Introduction to cave and karst GIS *Journal of Cave and Karst Studies* **64** p 3.
[8] Wynne J J and Pleytez W 2005 Sensitive ecological areas and species inventory of Actun Chapat Cave, Vaca Plateau, Belize *Journal of Cave and Karst Studies* **67** p 148–157.
[9] Reinhart H, Fathur F, Erviana A, Arifta Y, Trias Y, Harahap R, Tamimi S, Mattoreang E, Rahmawan R, Hakiki N, Mashlikhah Y E, Kuncoro W, Fauzan R, The Application of a Cave Ecosystem Profile to Anjani Cave in Karst Jonggrangan, Central Java, Indonesia in Moore K., White S. (Eds), 2017. *Proceedings of the 17th International Congress of Speleology* **1** July 22–28, Sydney, NSW Australia pp 179-183
[10] Lavoie K H, Huang A, Huang C. Microhabitat Distribution Of Invertebrates In White Cave, Mammoth Cave National Park, Usa in Moore K, White S (Eds), 2017 *Proceedings of the 17th International Congress of Speleology* **1** July 22–28, Sydney, NSW Australia pp 88-89
[11] Lobo H A S, Trajano E, Marinho M A, Bichuette M E, Scaleante J A B, Scaleante O A F, Rocha B N, Laterza, F V 2013 Projection of tourist scenarios onto fragility maps: Framework for determination of provisional tourist carrying capacity in a Brazilian show cave. *Tourism Management* **35** pp 234-243
[12] Wibisino M W A Speleo-Archaeology Study Of Kali Banjar Underground River: Approach On The Research Of Site Formation Procces in Moore K., White S. (Eds), 2017. *Proceedings of the 17th International Congress of Speleology* **1**, July 22–28, Sydney, NSW Australia pp 30-32
[13] O'Connor S, Barham A, Aplin, L, Maloney T 2017 Cave stratigraphies and cave breccias: Implications for sediment accumulation and removal models and interpreting the record of human occupation *Journal of Archaeological Science* **77** pp 143-159
[14] Florea L and Paylor R L, Simpson, L, & Gulley J 2002 Karst GIS advances in Kentucky *Journal of Cave and Karst Studies* **64** pp 58-62.
[15] Komac M. and Urbanc J. 2012 Assessment of Spatial Properties of Karst Areas on a Regional Scale using GIS and Statistics – The Case of Slovenia *Journal of Cave and Karst Studies* **74** pp 251-262.
[16] Haryono E, Reinhart H, Effendi S, Irianto P A, Fathoni W, Ardi, T E, Putera A, Setiawan, T, Dayatullah M, Rinding Y, Sulistiono A. Cave Exploration in Sangkulirang-Mangkalihat Peninsula, East Kalimantan-Indonesia in Moore K., White S. (Eds), 2017. *Proceedings of the 17th International Congress of Speleology* **1**, July 22–28, Sydney, NSW Australia pp 344-350

[17] Mattes J 2015 Underground fieldwork – A cultural and social history of cave cartography and surveiing instruments in the 19th and at the beginning of the 20th century. *International Journal of Speleology* **44** 3 pp 251-266. Tampa, FL (USA) ISSN 0392-6672 http://dx.doi.org/10.5038/1827-806X.44.3.4

[18] Chirol B 2017 The oldest cave maps in the world. in Moore K., White S. (Eds), 2017. *Proceedings of the 17th International Congress of Speleology,* July 22–28, Sydney, NSW Australia pp 417

[19] Laksamana E.E 2016 *Stasiun Nol: Teknik-Teknik Pemetaan Dan Survey Hidrologi Gua vol. 2* (Yogyakarta: Megalith Books dan Acintyacunyata Speleological Club)

[20] Medellin R A, Wiederholt R, Lopez-Hoffman L 2017 Conservation relevance of bat caves for biodiversity and ecosystem services *Biological Conservation* **211** pp 45–50

[21] Worthington S Management of Carbonate Aquifers in van Beynen, P. (ed.) 2011 *Karst Management* (Springer)

IOP Conf. Series: Earth and Environmental Science **98** (2017) 012032 doi:10.1088/1755-1315/98/1/012032

# Spatial Point Data Analysis of Geolocated Tweets in the First Day of Eid Al-Fitr 2017 in Java Island

**T W Wibowo**

Faculty of Geography, Universitas Gadjah Mada

totok.wahyu@ugm.ac.id

**Abstract**. Eid Al-Fitr is a worldwide Muslim feast day, which in Indonesia generally accompanied by tradition of going home (mudik). The demographic patterns at the time of the holiday are generally shifted, in which some urban residents will travel to their hometowns. The impact of this shifting is that there is a quite massive mobility of the population, which is generally accompanied by traffic congestion. The presence of location sensors on smartphone devices, open the opportunity to map the movement of the population in realtime or near-realtime. Especially now that social media applications have been integrated with the capability to include location information. One of the popular social media applications in Indonesia is Twitter, which provides microblogging facilities to its users. This study aims to analyze the pattern of Geolocated Tweets data uploaded by Twitter users on the first day of Eid Al-Fitr (1 Syawal 1438H). Geolocated Tweets data mining is done by using Streaming API (Application Programming Interface) and Python programming language. There are 13,224 Geolocated Tweets points obtained at the location of the study. Various point data analysis techniques applied to the data have been collected, such as density analysis, pattern analysis, and proximity analysis. In general, active Twitter users are dominated by residents in major cities, such as Jakarta, Bandung, Surabaya, Yogyakarta, Surakarta and Semarang. The results of the analysis can be used to determine whether the Geolocated Tweets data mined by the Streaming API method can be used to represent the movement of the population when mudik.

**Keywords:** Text data mining, Twitter, Geolocated Tweet, Spatial point data analysis

## 1. Introduction

Population mapping is one challenging subject, since the object is moving continuously and dynamically. Traditional mapping technique such as choropleth map, is now still popularly used since it's conception in 1826 by Baron Pierre Charles Dupin [1]. The usage of administrative boundary in the choropleth map sometimes can be misleading, therefore a new technique was initiated by George Julius Poulett Scrope and Henry Drury Harness [2]. The new technique was then popularized by using a Russian term "dasymetric" which means density measurement [2][3]. Generally, the dasymetric mapping is used to estimate the population by using one or several ancillary data, such as street, address, parcel distribution, and limiting weight [4]. Modifications were made by several researchers to improve the original binary idea of settlement and non-settlement. Instead of using binary logic in the original idea of dasymetric map, several researchers have tried to modify the mapping unit, such as dividing the low and high population density of the settlement [5][6] and population density fraction [7]. Nevertheless, the mapping techniques discussed above still relying on population data availability, which is also relatively immobile.

The world we know today is a very different place than before with the emergence of internet. Data is now rich, hence there is a new concept called big data. It is called big data because of the high velocity, volume and variety of the data [8]. Since big data is having its unique characteristics, the means of processing will also be special. Among other factors that influencing rich data environment, social media is gaining the most popularity in society.

By definition social media is online platforms that allow users to create and exchange content [9]. The data and/or information sharing in social media is made easier by the increasing number of smartphone user. Many social media applications have the option to share user's location (geolocation), and smartphone has enable user to activate the feature conveniently. Geolocation data, which is designed primarily for mobile devices, is mainly used for geo-positioning, geocoding, and geotagging [10]. This geolocated data will enriched the geospatial data sources, which can be used further in spatial analysis especially in spatial distribution of population [11].

Twitter can provide its user to activated geolocated features to generate Geolocated Tweets. The existence of social media such as Twitter can provide opportunity to explore public opinion, social interaction, identity and cultural values [12]. Even though there are several assumptions that considered Twitter Data as "data-light", many researchers have proved its potentially rich data [11][12]. As of 2016, Indonesia was ranked 3rd in active Twitter user, just behind United States and India [13]. As Twitter is a microblogging site, there is limitation in the number of characters that a user can post in a single tweet. This fact actually can inhibit the attribute data analysis, particularly in thematic related study [11]. For the last reason, this research only aims to map the distribution of population in Java Island in the first day of Eid Al-Fitr.

The majority of population in Java Island is moslem, and each year there is mudik tradition. Mudik can be interpreted as someone that is going home for a relatively short time after they settled in big city, mostly because they are working in big city. However, people that are conducting mudik usually will not stay in their hometown for a long time. This tradition can be seen as social, cultural, and/or psychological phenomena. This paper will examine the usage of Geolocated Tweets data, to provide the population movement during the first day of Eid Al-Fitr, in which many people committing mudik tradition. Generally, mudik is characterized by rapid and massive population movement, in a relatively short time. This will be an ideal time to test Geolocated Tweets data to get the knowledge of population movement.

## 2. Research Methods

### *2.1. Research Area*

Java is the most populated island in the world. Based on 2010 census data, there are 106.45 million inhabitants in an island that has an area of 128,297 km$^2$. Generally, the population movement in Mudik tradition, is always focusing on the migrants in the capital/main cities back to their hometown. The most highlighted movement when mudik is migration from Jakarta to other cities in Java. Indonesian Government highly anticipated mudik activity in Java, since the risk of traffic accident usually rising. There are two main road that can be used by people, the north path (Jalur Pantai Utara) and the south path (Jalur Lintas Selatan). Mostly the north path is more congested than the south path (Figure 1). Not only by bus or car, mudik are also done by other transportation mode as well such as airplane, train, and even motorcycle. However, the people are usually not stay for long, as they have to go to work in their current residence.

### *2.2. Data*

Twitter allows user to access their data through Application Programming Interface (API), one of which can be done by using streaming method. Typically there are three kinds of Streaming API, namely the

Public Stream, the User Stream, and Site Stream. Public Stream will allow the user's application to get public data on Twitter, while User Stream enable user to stream Twitter data in real-time. Site Stream allow the user to monitor real-time Twitter feeds for a large number of users. The data used in this study were collected using Streaming API method, which employs Python Script to run the data mining [14]. This method dictate the user's web server to always stay connected to Twitter in order to get real-time Tweets [15]. Once the connection is interupted, then the requested data will be forfeited. However, there is limitation on how much the user can stream the data, which is only 1% of the data pool each time request is made.

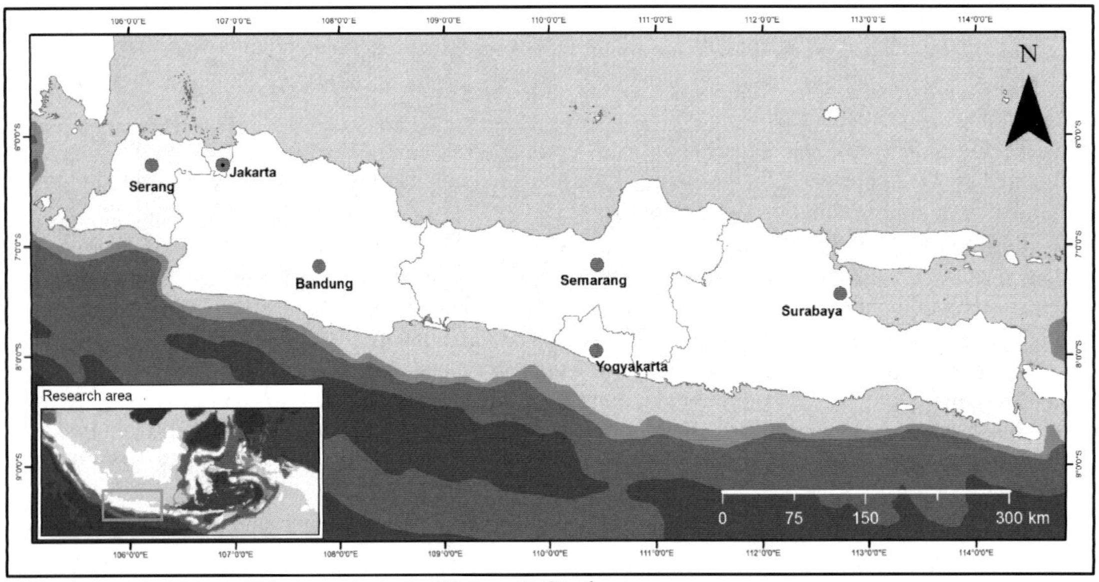


**Figure 1**. Study area

### 2.3. *Data Analysis*

As mentioned before that big data needs to be processed with special technique. Though the data used in this study is relatively small to be called big data, it still needs a way to harness insight throuh spatial analysis. In this study there are four analysis conducted to the data, which are distance analysis, density analysis, pattern analysis, and proximity analysis. All point data analysis used in this research aim to generalize the abundant points and get a new perception about the data.

*2.3.1. Global Density.* Global density was the first measure used in this study. It is simple yet can be used to get the first impression of the data. The density can be calculated by dividing the number of Tweets with it's respective administrative area. In this research, provincial area based on Minister of Internal Affairs Regulation No. 66/2011 was used to calculate the global density of Tweets.

*2.3.2. Quadrat Density and Point Density.* Regular tessellation (square and hexagonal) was used as a based for quadrat count method, in which the number of point data is summed up according to it's respective area [16]. Both tessellation polygons were created with an area of 1 km2 in each polygon. Points that are located within a polygon were then calculated, and presented in choropleth map. However, density calculation has difficulty in considering the definition of the study area [16]. The area for density calculation can be changed easily, but determining which is more representative area can be a challenging task. Therefore, point density analysis were done to provides an alternative of density calculation, as it offers different area determination of the previous two methods. Point density would produce a raster based density surface, in which the density calculation based on the points located within each cell and it's neighbour cell.

*2.3.3. Variance-Mean Ratio (VMR).* The quantitative pattern measurement of point data could give the illustration of how the data is distributed, evenly spaced, random, or clustered. VMR calculates the ratio between variance and mean of quadrat count density, in a Poisson distributed data. VMR greater than 1.0 indicates a tendency toward clustering in the pattern, and a VMR less than 1.0 is indicates an evenly spaced arrangement. However, this approach is considered as an unreliable unless the volume of the data is very large and high intensity of events per quadrat [16].

*2.3.4. Average Nearest Neighbour (ANN).* Empirical point patterns, though studied for many reasons, is mainly expected to reveal the process that is responsible for creating the pattern [17]. Moreover, study of point pattern will inform us about fundamental relationships in space or about possible causes for the observed patterns [18]. The second order effects was detected using pattern analysis through Average Nearest Neighbour (ANN) analysis. It calculates a nearest neighbor index based on the average distance from each points to its nearest neighboring points. The index is the result of ratio between observed distance and the hypothetical data [19]. The following formula were used to calculate ANN:

$$ANN = \frac{\overline{D}_O}{\overline{D}_E} \quad (1)$$

where $\overline{D}_O$ is the observed mean distance between each points and it's nearest neighbour:

$$\overline{D}_O = \frac{\sum_{i=1}^{n} d_i}{n} \quad (2)$$

and $\overline{D}_E$ is the expected observed mean distance between each point and its nearest neighbour:

$$\overline{D}_E = \frac{0.5}{\sqrt{n/A}} \quad (3)$$

*2.3.5. Tweets Proximity Index (TPI).* To determine the influence of a city on the data, an index that combines the normalization of data number with the average distance was built. Point distance analysis is used as basic analysis, in which the result can be used to calculate the number of points and also providing the average distance from source point. Among all cities in Java, 35 were selected as qualitatively assessed using the raw Tweets data. The city center were used in the calculation of Tweets Proximity Index (TPI), acted as the representation of the center of activity in the city. Search radius in point distance analysis can be set subjectively in regards characterisics found in study area. TPI can be calculated using the following formula:

$$TPI = N + D \quad (4)$$

where normalization of data number (N) can be calculated based on data in each city (Xc) as follows:

$$N = \frac{X_c - X_{min}}{X_{max} - X_{min}} \quad (5)$$

and the average distance index (D) is the inversed value of average distance divided ($\bar{x}$) by search radius (r):

$$D = 1 - \left(\frac{\bar{x}}{r}\right) \quad (6)$$

The summary of methods used in this study is illustrated in Figure 2.

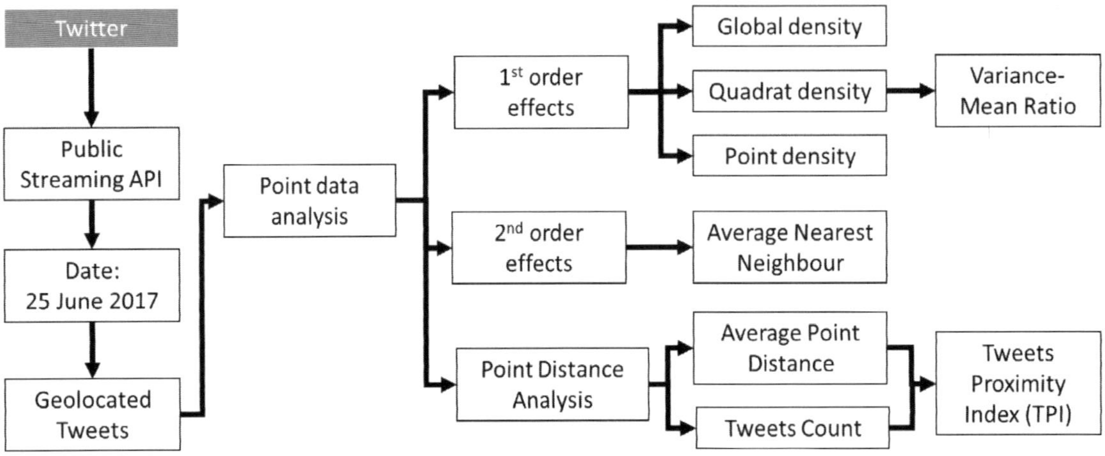


**Figure 2**. Research methods

## 3. Results and Discussion

### *3.1. Spatial Point Analysis*

By using Public Stream API, 13,224 Tweets were collected during the first day of Eid Al-Fitr in the research area. The data were then imported in GIS Software as point data, showing exactly where the Tweets has been made (Figure 3). General distribution of the points, showed that major cities residents still dominating in the study area.

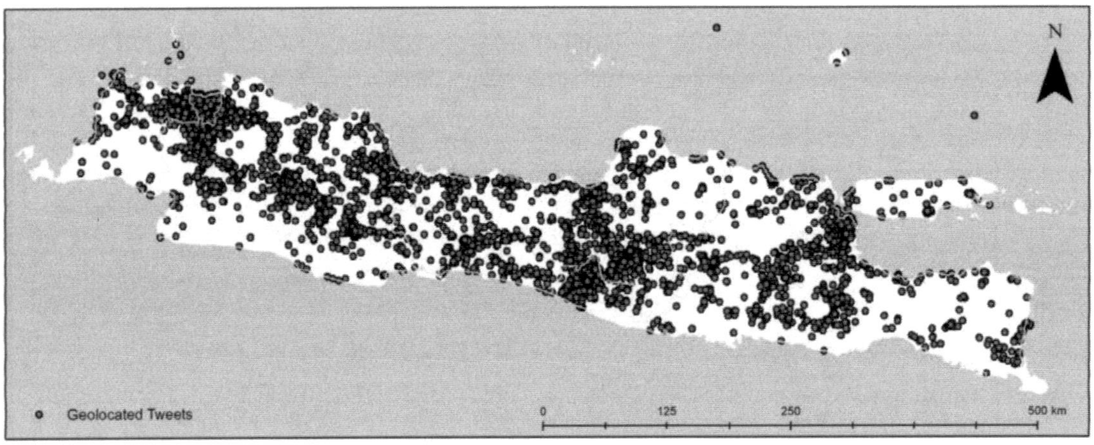


**Figure 3**. Geolocated Tweets acquired in 25th June 2017

Global density of the points showed that DKI Jakarta has the highest point density which reached 4.33 Tweets/km2, meanwhile East Java has the lowest (0.04 Tweets/km$^2$). However, the global density might have a bias, since the area of each provinces is not the same. Among all provinces in study area, DKI Jakarta has the smallest area, whilst West Java is the largest. Table 1 provide the calculation of global density analysis.

IOP Conf. Series: Earth and Environmental Science **98** (2017) 012032 doi:10.1088/1755-1315/98/1/012032

**Table 1**. Global density

| No | Province | Tweets Count | Area ($km^2$) | Global Density (Tweets/ $km^2$) |
|---|---|---|---|---|
| 1. | Banten | 1,173 | 9662.92 | 0.12 |
| 2. | DI Yogyakarta | 765 | 3133.15 | 0.24 |
| 3. | East Java | 1,875 | 47799.75 | 0.04 |
| 4. | Central Java | 1,871 | 32800.69 | 0.06 |
| 5. | West Java | 4,667 | 35377.76 | 0.13 |
| 6. | DKI Jakarta | 2,873 | 664.01 | 4.33 |

Quadrat density provides a more balanced density calculation by using tessellation area, since the density is calculated using the same area. The result of square and hexagon tessellation quadrat density shows similar pattern, with the latter having higher maximum density (Figure 4a & Figure 4b). The shape of hexagon made it possible to have a higher point density, since it is the most effective tessellation shape. However, the pattern shown by both tessellations is quite the same.

The result from point density analysis shows a wider area of density compared to quadrat density. Addition of search area parameter in point density analysis makes the probability of a point entering a calculation area becomes higher. Jakarta and it's vicinity still filled with high density Tweets, and so does other provincial capital and main cities (Figure 4c). From three calculated density analysis, it can be concluded that the Tweets still dominated by Twitter users in provincial capital and several main cities.

Further analysis using average point distance method was calculated to determine which provincial capital holds the most Tweets. Two search radius were used to determine the average distance of points from city center. The lower the average distance, it means that there are a lot of Tweets that posted arround the city center. In the first search radius (25 km), Yogyakarta has the lowest average distance points (Figure 5). But the second search radius (50 km), shows that Tweets in Bandung has the lowest distance. Although has a high density, evidently the Tweets that were made in Jakarta has the lowest average distance.

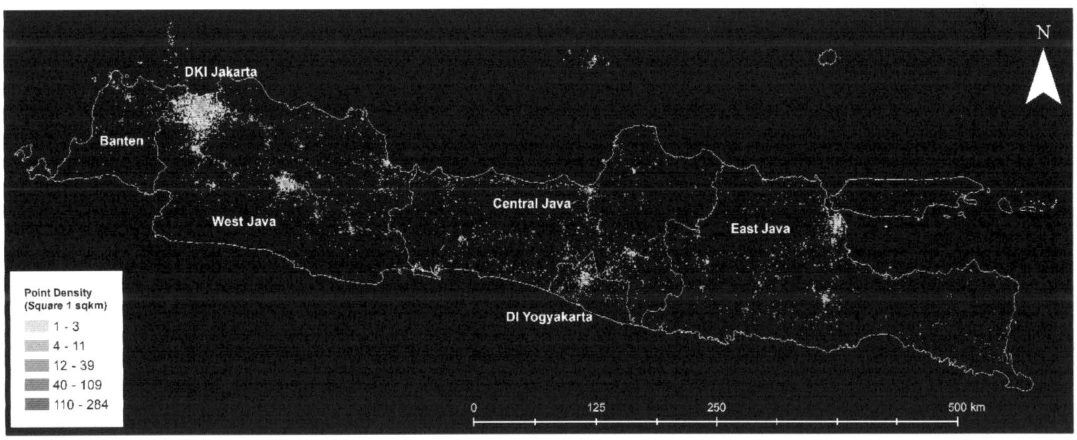

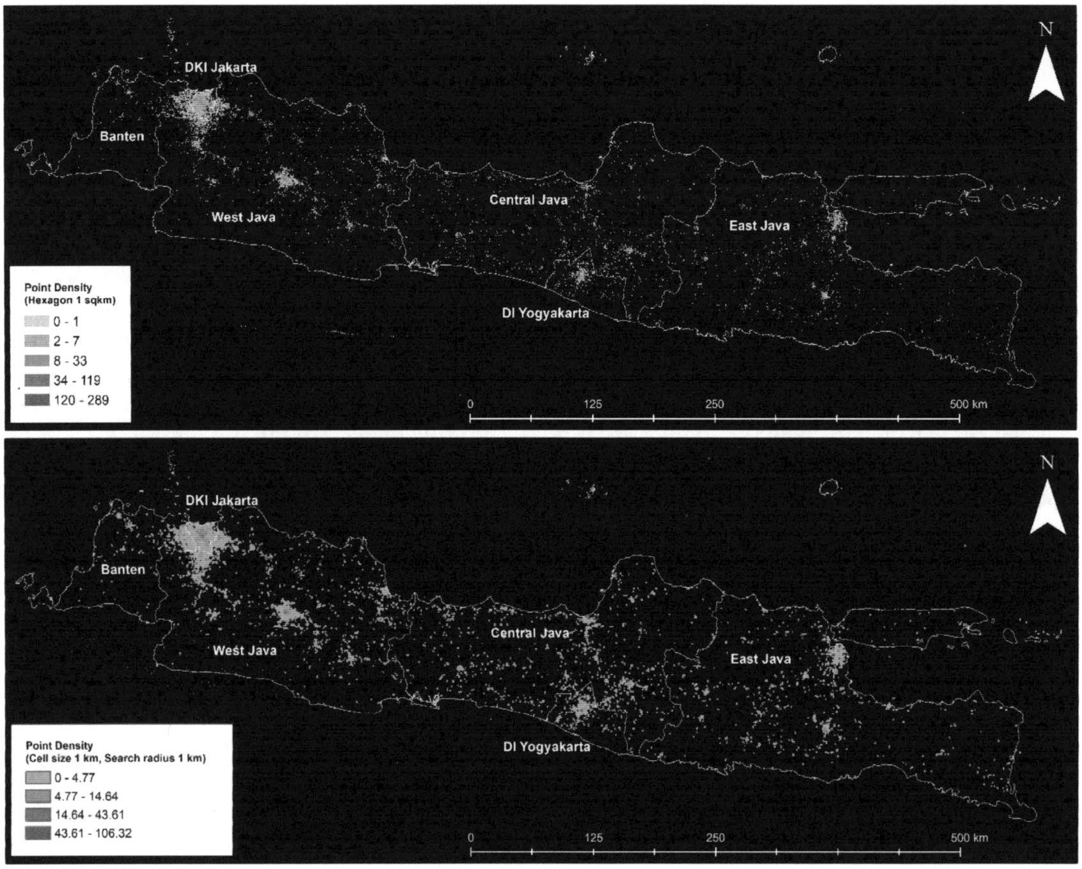


**Figure 4**. a) Quadrat density – square; b) Quadrat density – hexagon; and c) Point density

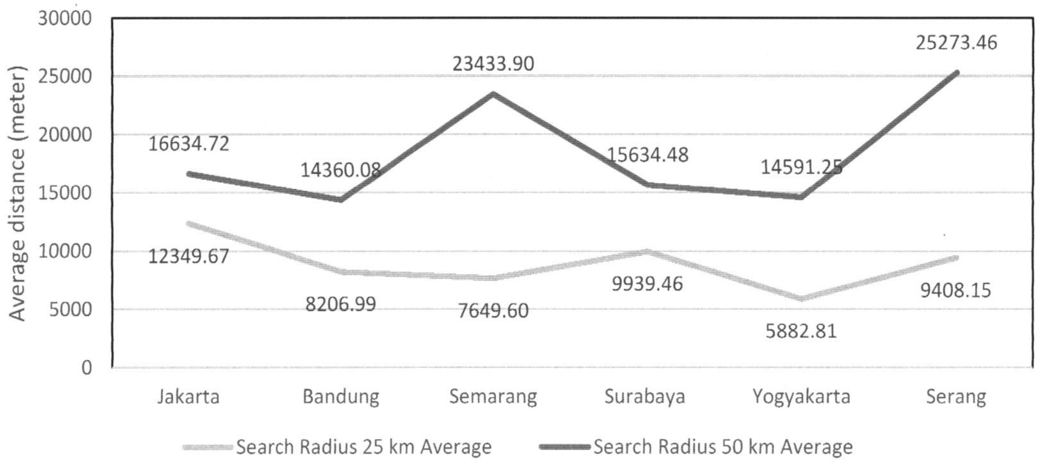


**Figure 5**. Average point distance to provincial capital

In order to obtain the quantitative measurement of pattern, Variance to Mean Ratio and Average Nearest Neighbour were performed. VMR was performed by using the same tessellations structure as quadrat density calculations. Both VMR analysis were yielded a relatively high above the threshold of an evenly distributed event, in which the threshold is 1 [16]. The findings of VMR calculations was confirmed by Average Nearest Neighbour analysis. Given the z-score of -169.05, there is a less than 1% likelihood that this clustered pattern could be the result of random chance. Table 2 summarize the VMR analysis on both tessellations, meanwhile Figure 6 shows the result of ANN analysis.

**Table 2**. Summary of VMR analysis

| No | VMR Analysis | Tessellations | |
|---|---|---|---|
| | | Square (1 $km^2$) | Hexagon (1 $km^2$) |
| 1 | Number of Tweets | 13,224 | 13,224 |
| 2 | Number of tessellations | 326,508 | 330,948 |
| 3 | Mean quadrat count | 0.04 | 0.04 |
| 4 | Observed variance | 0.86 | 0.88 |
| **VMR** | | **21.62** | **22.08** |

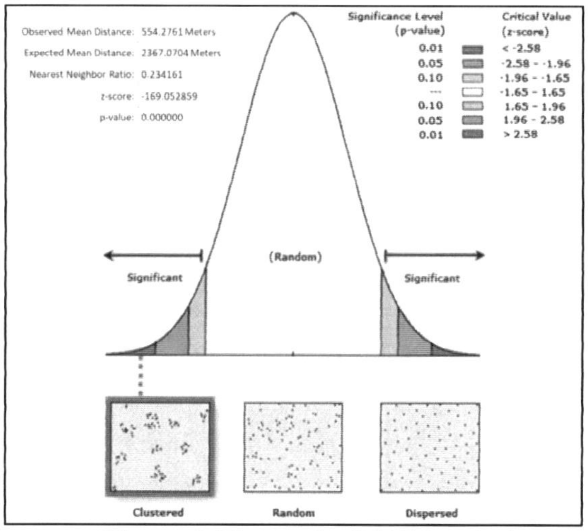


**Figure 6**. The result of ANN analysis

### *3.2. First and Second Order Impact*

From various analyzes that have been done, it is known that the characteristics of tweet data collected is clustered and located arround several main cities. The average distance from provincial capital is also relatively close. Based on the evidence, it can be drawn that the main city people still dominate active users Twitter who is also happened to be active (Tweeted) at the time of Eid-Al Fitr. Presumption elaborated previously in the background, in this case is not proven, since evidently there are much geolocated Tweets in main cities (especially Jakarta) instead of destination city of mudik. This leads to another assumption that the first order impact is stronger than the second order, which means that infrastructure and supportive environment to access the internet still plays a large role in social media activity particularly Twitter. With far better internet facilities and infrastructure in main cities, the citizen can easily become active in social media. Furthermore, Twitter users who usually make Geolocated Tweets, do not migrate even though the mudik tradition is going on.

The findings also indicated that there are differences between citizen and netizen. The purpose of this study aims to connect the movement of citizen and the Geolocated Tweets. However, apparently not all citizen are also active netizen, particularly who use Twitter. As a result, the data is showing a clustered pattern in which many netizen actively using Twitter during the first day of Eid Al-Fitr. A more elaborated study is needed to explain this behaviour.

During the first day of Eid-Al Fitr in Indonesia, Muslims have a tradition to apologize to each other. Not only verbally, apologies are also often delivered through social media. The majority of Tweets mentioned Eid Al-Fitr or something that are still associated. It means that Geolocated Tweets data

IOP Conf. Series: Earth and Environmental Science **98** (2017) 012032 doi:10.1088/1755-1315/98/1/012032

acquired from Public Streaming API provides the content of user's Tweets, in which can provide many applications regarding thematic mapping.

### *3.3. Tweet Proximity Index*

The Tweet Proximity Index (TPI) was developed and used to determine which city has the highest influence in Geolocated Tweets. This index was based upon two approaches, which are average distance and the counts of Tweets. The smaller average distance indicates the stronger impact of a city, on the other hand the less Tweet counts indicates the less impact of a city. For the purpose of TPI calculation, 35 cities were selected by considering the point density.

Table 3 shows TPI calculation using search radius of 25 km. Megapolitan cities, that consist of Jakarta, Depok, South Tangerang, Bekasi, and Tangerang, were still dominating the top list of TPI, only Bogor that placed outside the top 10. Despite having a slightly significant number of Tweets, the average distance of Tweets in Bogor is relatively distant. The usage of circular shape search radius might not be an advantage for an elongated shape city like Bogor, so that the TPI was low. In this case an ellipsoid search radius is more suitable, but there must be a consideration of the angle of the axes.

Jember unexpectedly, despite having the least number of Tweets (within a radius of 25 km), was able to occupy the eighth rank because it has the closest average tweet distances. This shows that Jember residents mostly make Tweets around the city center, which can be attributed to the completeness of facilities and infrastructure that centered arround city center as has been alluded to earlier. Cilacap, Pekalongan, and Madiun have the similar characteristic to Jember, in which they are only having small number of Tweets (below 100) but the average distance is relatively close.

**Table 3**. TPI calculation using search radius of 25 km

| No | City | Average Distance | Tweet Counts | TPI | No | City | Average Distance | Tweet Counts | TPI |
|---|---|---|---|---|---|---|---|---|---|
| 1 | Jakarta | 12.35 | 4358 | 1.51 | 19 | Banyuwangi | 8.73 | 41 | 0.65 |
| 2 | South Tangerang | 16.80 | 3954 | 1.23 | 20 | Serang | 9.41 | 146 | 0.65 |
| 3 | Bekasi | 15.84 | 3373 | 1.14 | 21 | Tulungagung | 9.76 | 73 | 0.62 |
| 4 | Depok | 17.08 | 3491 | 1.12 | 22 | Tegal | 10.11 | 85 | 0.61 |
| 5 | Bandung | 8.21 | 1557 | 1.02 | 23 | Purwokerto | 10.37 | 124 | 0.61 |
| 6 | Tangerang | 16.89 | 2797 | 0.96 | 24 | Kediri | 10.55 | 103 | 0.59 |
| 7 | Yogyakarta | 5.88 | 753 | 0.93 | 25 | Garut | 11.60 | 183 | 0.57 |
| 8 | Jember | 4.88 | 35 | 0.80 | 26 | Kebumen | 11.50 | 76 | 0.55 |
| 9 | Malang | 7.35 | 293 | 0.77 | 27 | Sukabumi | 12.41 | 184 | 0.54 |
| 10 | Semarang | 7.65 | 259 | 0.75 | 28 | Karawang | 12.78 | 236 | 0.54 |
| 11 | Surabaya | 9.94 | 596 | 0.73 | 29 | Sidoarjo | 15.03 | 592 | 0.53 |
| 12 | Cilacap | 7.72 | 45 | 0.69 | 30 | Kuningan | 12.97 | 142 | 0.51 |
| 13 | Surakarta | 9.57 | 326 | 0.68 | 31 | Ungaran | 15.26 | 338 | 0.46 |
| 14 | Pekalongan | 8.11 | 73 | 0.68 | 32 | Salatiga | 14.43 | 160 | 0.45 |
| 15 | Tasikmalaya | 8.98 | 197 | 0.68 | 33 | Mojokerto | 14.28 | 91 | 0.44 |
| 16 | Madiun | 8.46 | 97 | 0.68 | 34 | Klaten | 17.13 | 351 | 0.39 |
| 17 | Bogor | 12.57 | 785 | 0.67 | 35 | Magelang | 16.40 | 185 | 0.38 |
| 18 | Cirebon | 9.27 | 213 | 0.67 | | | | | |

## 4. Conclusions

At least for the time being, the abundance of Twitter Data couldn't represent the movement of people during mudik. Apparently there is still a big differences between citizen and netizen, in this case netizen who posted Geolocated Tweets. Twitter user, especially those who Tweet with Geolocated features, still dominated by Metropolitan/Urban/Downtown resident. However, further analysis is needed to elaborate this fact. Not only assess the data spatially, but also collecting Twitter's user opinion regarding their behaviour of using Twitter. On the other side, there is a huge possibility that Twitter Data can be used in attribute data analysis through it's user Tweet, although a robust data cleaning will be needed.

**References**

[1] Friendly M and Denis D J 2001 *Milestones in the history of thematic cartography, statistical graphics, and data visualization* [Online]. Available: http://www.datavis.ca/milestones/index.php?page=home.
[2] Mennis J 2009 Dasymetric Mapping for Estimating Population in Small Areas *Geography Compass* **3**(2) pp 727-745.
[3] Wright J. K 1936 A method of mapping densities of population with Cape Cod as an example *Geographical Review* **26** pp 103-110.
[4] Tapp A F 2010 Areal Interpolation and Dasymetric Mapping Methods Using Local Ancillary Data Sources *Cartography and Geographic Information Science* **37**(3) pp 215-228.
[5] Mennis J 2003 Generating surface models of population using dasymetric mapping *The Professional Geographer* **55**(1) pp 31-42.
[6] Weischelbaum J, Petrini-Monteferri F, Papathoma M, Wagner M and Hackner N 2005 Sharpening census information in GIS to meet real-world conditions: the case for Earth Observation *WIT Transactions on Ecology and The Environment* **84** pp 143-154.
[7] Steiner P and Paulus G 2007 *Dasymetric mapping for public health planning* Aalborg.
[8] Kaisler S, Armour F, Espinosa J A and Money W 2013 *Big data: Issues and challenges moving forward* Hawaii.
[9] Gandomi A and Haider M 2015 Beyond the hype: Big data concepts, methods, and analytics *International Journal of Information Management* **35** pp 137-144.
[10] Ranitovic J, Nikolic V and Velmirovic L 2015 *Geolocation-based social networks and mapping in emergencies.*
[11] Mislove A, Lehmann S, Ahn Y, Onnela J and Rosenquist N 2012 Understanding the Demographics of Twitter Users in *Fifth International AAAI Conference on Weblogs and Social Media.*
[12] Sloan L, Morgan J, Burnap P and Williams M 2015 Who Tweets? Deriving the Demographic Characteristics of Age, Occupation and Social Class from Twitter User Meta-Data *PLoSONE* **10**(3) pp 1-20.
[13] Statista 2016 *Statista* [Online]. Available: https://www.statista.com/statistics/242606/number-of-active-twitter-users-in-selected-countries/. [Accessed 12 08 2017].
[14] Klinger R 2015 *The Geography of Tweets: Reading Tweets with QGIS* [Online]. Available: http://www.digital-geography.com/the-geography-of-tweets-reading-tweets-with-qgis/#.WM75ODt97IU. [Accessed 13 08 2017].
[15] Reifman J 2014 *Building With the Twitter API: Using Real-Time Streams* [Online]. Available: https://code.tutsplus.com/tutorials/building-with-the-twitter-api-using-real-time-streams--cms-22194. [Accessed 12 08 2017].
[16] O'Sullivan D and Unwin D J 2010 *Geographic Information Analysis, 2nd ed* (New Jersey: John Wiley & Sons, Inc).
[17] Okabe A, Boots B, Sugihara K and Chiu S N 2000 *Spatial Tessellations: Concepts and Applications of Voronoi Diagrams, 2nd ed* (Chichester: John Wiley & Sons Ltd).
[18] Rogerson P 2006 *Statistical Methods for Geography, 2nd ed* (London: Sage).
[19] Mitchell A 2005 *The ESRI Guide to GIS Analysis, Volume 2: Spatial Measurements and Statistics*, Esri Press.
[20] Langford M and Unwin D 1994 Generating and mapping population density surface within a geographical information system *The Cartographic Journal* **31** pp 21-26.

# Utilization of ALOS PALSAR-2 Data for Mangrove Detection Using OBIA Method Approach

**N Anggraini* and A Julzarika**

Remote Sensing Applications Center, Indonesian National Insitute of Aeronautics and Space (LAPAN), Jl. Kalisari No. 8, Pekayon, Pasar Rebo, Jakarta 17310, Indonesia.

nanin.anggraini@lapan.go.id

**Abstract**. Mangroves have an important role for climate change mitigation. This is because mangroves have high carbon stock potential. The ability of mangroves to absorb carbon is very high and it is estimated that the mangrove carbon stock reaches 1023 Mg C. The current problem is the area of mangrove forest is decreasing due to land conversion. One technology that can be used to detect changes in the area of mangrove forest is by utilizing ALOS PALSAR-2 satellite imagery. The purpose of this research is to detect mangrove forest area from ALOS PALSAR-2 data by using object-based image analysis (OBIA) method. The location of the study is Taman Nasional Sembilang in Banyuasin Regency of South Sumatra. The data used are ALOS PALSAR-2 dualpolarization (HH and HV), recording year 2015. The calculation of mangrove forest area in Sembilang National Park has ~ 82% accuracy. The results of this study can be used for various applications and mapping activities.

**Keyword**: mangrove, ALOS PALSAR-2, OBIA

## 1. Introduction

[1] defines mangroves as typical coastal vegetation along tropical and sub-tropical sheltered beaches, influenced by tidal sea water, and able to adapt in brackish waters. One of the important functions of mangrove forests is as carbon reserves to reduce greenhouse gas emissions. [2] stated that the potential of carbon stock from mangrove is 1023 Mg C per hectare. The current problem is the degradation of mangrove forests. [3] released information that in 2005, the area of mangrove forest in Indonesia reached 3,062,300 ha and continues to decrease to date. Mangrove forests are much reduced due to land conversion into settlements, ponds, agriculture, industry, and so forth. Therefore, it is necessary to identify mangrove forest changes. One of the technologies that can be used is by utilizing satellite images of Advanced Land Observing Satellite-Phased Array Type L-Band Synthetic Aperture Radar (ALOS PALSAR) 2.

PALSAR-2 is an instrument of ALOS 2 with a frequency of L-Band (1.2 GHz) which has the primary objective to monitor disasters at any time and weather conditions. The main missions of ALOS PALSAR-2 are for disaster monitoring, monitoring of global environmental issues (deforestation, polar ice caps reduction), and for economic and food issues. Compared to previous generations, the ability of ALOS PALSAR-2 is better than basic satellite capacity, resolution, channel, and temporal resolution. The resolution of ALOS PALSAR-2 is 1x3 m, whereas in PALSAR the previous generation is 10 m. Temporary resolution is also getting faster, ie 14 days. ALOS PALSAR-2 has three modes, Spotlight, Strip Map, ScanSAR. ALOS PALSAR-2 has a single polarization (HH, HV, VH, VV), dual (HH + HV, VH + VV), and full polarimetry (HH + HV + VH + VV) [4].

Research by using ALOS PALSAR-2 has been done, among others [5] using ALOS PALSAR-2 for damage assessment of buildings caused by earthquake, [6] identify the location and characteristics of mangrove species in Hai Phong Vietnam . [7] used ALOS PALSAR-2 to calculate the stem volume of a forest area, [8] using ALOS PALSAR-2 for rapid response to floods that occurred in the Kanto and Tohoku areas.

One technique that can be used for image identification is object-based image analysis (OBIA). OBIA is a classification technique that not only classifies objects based on pixel hue or texture, but based on unity of objects. Simply stated, OBIA is a classification that not only uses spectral values but also considers the spatial aspects of objects [9]. [10] states that object-based classification is able to define object classes based on spectral and spatial aspects at once. The OBIA technique has been widely used for the identification of objects from images, among which are tidal site location maps [11], mapping of geomorphological zone of coral reef ecosystem [12], detection of land use change [13,14], mapping of mangrove composition [15], and karst morphology [16]. The purpose of this research is to detect mangrove forest area from ALOS PALSAR-2 data by using object-based image analysis (OBIA) method. It is hoped that the results of this research can be used for application and mapping activities.

## 2. Material and Methods

### *2.1. Study Area*

The research location is Taman Nasional Sembilang (TNS) located in Banyuasin Regency, South Sumatera Province which geographically located at 1040 11'- 1040 94 'East Longitude and 10 53'-20 27' South Latitude (Figure 1). TNS region has estuarine habitat because many have estuaries of ± 70 rivers with the largest river is Sungai Sembilang. All the river flows to the South China Sea and Bangka Strait. The mangrove forest area in TNS is 78,597.55 ha with eight (8) mangrove species, including *Excoecaria agallocha, Rhizophora apiculata, Avicennia alba, Rhizophora mucronata, Avicennia officinalis, Brugueira gymnorrhiza, Xylocarpus granatum and Nypa fruticans* [17].

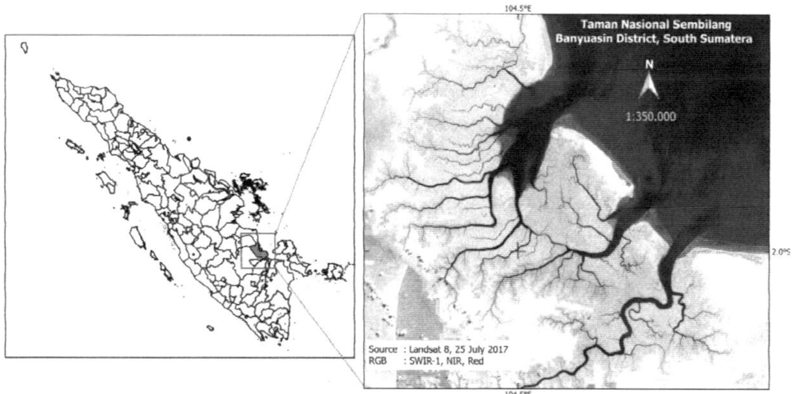


**Figure 1.** TNS from Landsat 8 image.

### *2.2. Materials*

The data used in this study is the image of mosaic ALOS PALSAR-2 in 2015, with dual SAR polarization mode HH and HV. Besides ALOS PALSAR-2 data, data from One Map Mangrove Nasional is used as mangrove location validation.

### 2.3. *Methods*

The initial process is geometric correction which refers to ALOS PALSAR-2 metadata. The geometric correction method used is a rigorous satellite sensor with 1.96 sigma accuracy. Digital Number (DN) is converted to sigma-zero normalization value using equation:

$$\sigma^{o}[db] = 10.\ \mathrm{Log}_{10}(DN)^2 + CF \quad (1)$$

where:

| | | |
|---|---|---|
| $\sigma^{o}$ | : | backscattering coefficient |
| DN | : | Digital number |
| CF | : | Calibration Factor (-83 db) |

The next step is a filtering that aims to reduce the speckle (noise). The filters used in this study were Goldstein and Low Pass filters. After getting the image free of noise, the next process is classification with OBIA segmentation approach. In the process of segmentation, the example object used for each class is 60. Segmentation using the multiresolution method with parameters used are scale, shape, compactness.

## 3. Results and Discussion

Filtering results by using Goldstein and Low Pass filter, able to remove speckle (noise) on ALOS PALSAR-2 image. To get the real view, a channel combination process (RGB) is performed. The combination used is Red: HH, Green: HV, Blue: (HH + HV)/3. Visually, mangrove looks darker than other vegetation objects, this is influenced by mangrove growing places in the form of muddy areas. Water in muddy areas causes a darker look. This is because water absorbs more electromagnetic waves than other objects. In addition to mangrove and water, other visible objects are non mangrove vegetation and open land. Non mangrove vegetation looks brighter and open field looks purplish. To clarify mangrove location in ALOS PALSAR-2 data, transect is done to see the DN profile of the object. Transects are performed at three locations that are considered to represent objects in the image. The RGB combination and object transect locations are shown in Figure 2.

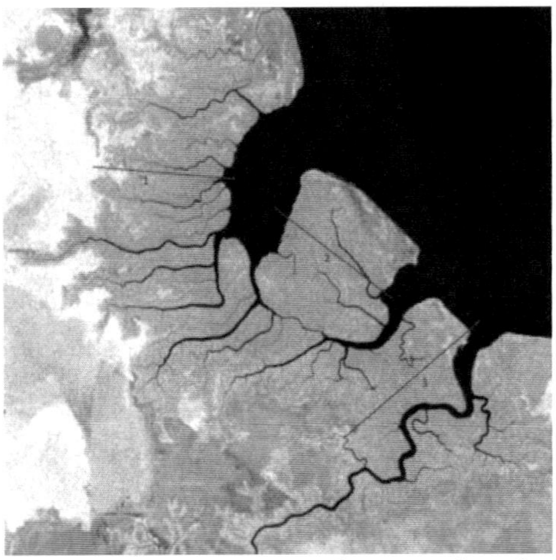

**Figure 2.** RGB combination and transect locations.

The transect results show different patterns for each object. Of the three transect sites, non mangrove vegetation has a higher DN profile than mangrove and open land, while water has a very low profile (Figure 3). This is due to the water having properties to absorb electromagnetic waves and slightly reflected compared to other objects.

Results transect helps to determine the location of objects that were sampled for segmentation. Segmentation is done to classify objects in the image into three classes, namely mangrove, non mangrove, and water. In the process of segmentation, each object is represented by 60 homogeneous samples. The segmentation used in the research is the multiresolution method, which minimizes the average heterogeneity of the object of an image at a certain resolution. This segmentation process uses the parameters of scale, shape, and compactness.

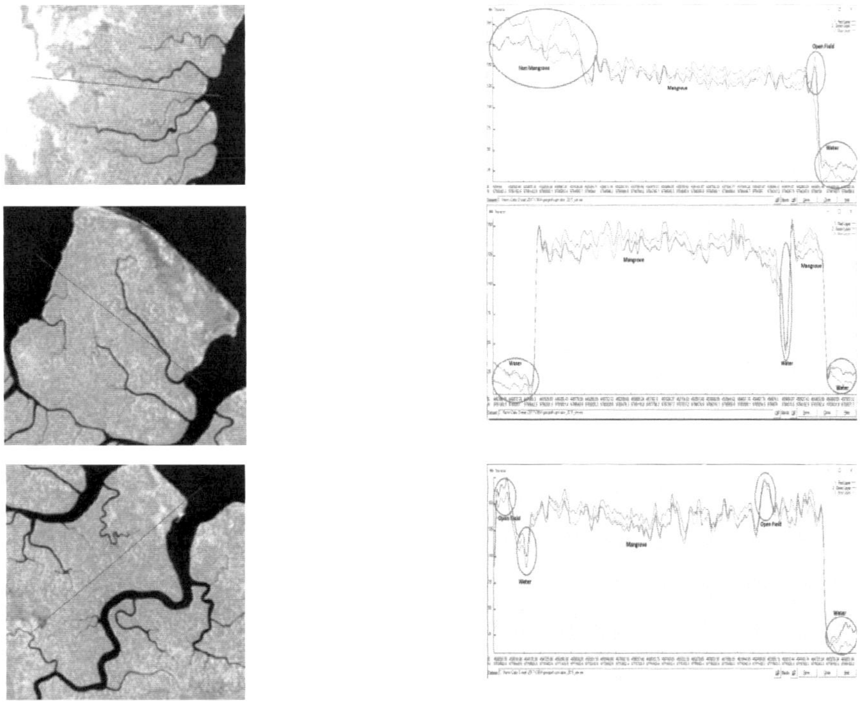

**Figure 3.** DN profile at transect locations

The value of each parameter will affect the segmentation result. The scale parameter is the term for the determination of the value of the maximum heterogeneity allowed on an object and its value is directly proportional to the size of the object. Needed modification on the scale parameters, the greater the value of the scale parameters then the homogeneous object will be less so that the results will be more coarse segmentation and the resulting object has a larger size.

The shape parameter influences the color parameter value so it must be adjusted to the study done. The larger the shape parameter values, then the segmentation process will be affected by the texture. The compactness parameters are used to separate compact objects from non-compact ones. The greater the value of the compactness parameters used then the resulting object form will be more compact [18]. In this study, the values used for each parameter are scale 30; shape 0,5; and compactness 0.5.

Prior to the homogeneity of classification according to the object, the boundary between the land with the water looks very clear, but the object is still difficult to distinguish. Classification is done to separate between mangrove, open land, non-mangrove vegetation, and water. Open land and non-mangrove vegetation are described as non mangrove so that the final result of segmentation is three classes ie mangrove, non mangrove, and water. To clarify the three objects of study, then the smoothing process has been done. This is shown in Figure 4.

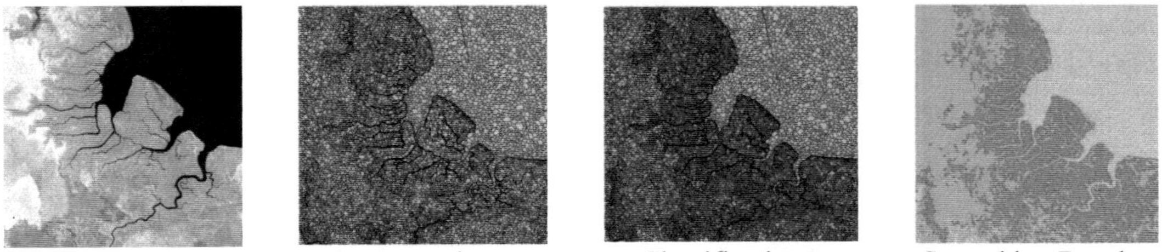

ALOS PALSAR-2 | Segmentation Result | Classification Result | Smoothing Result

**Figure 4.** Segmentation, Classification, and Smoothing Results

The smoothing results from the segmentation give a clearer picture of the object. Mangrove was seen around the coastal area and a small portion at a location some distance to the mainland. Vegetation on the land is partly described as a mangrove, this is possible because of the similarity to the homogeneity of the object image. In addition to mangroves, non mangrove classes are also seen in TNS areas. Non mangrove class besides non mangrove vegetation and open land, in this segmentation result, part of river flow is described as non mangrove (Figure 5).

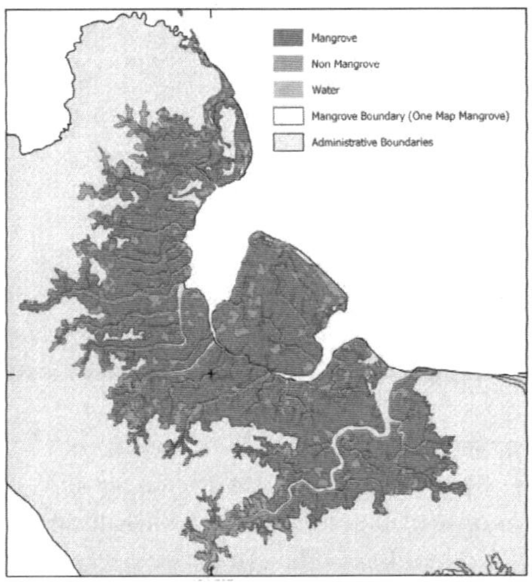


**Figure 5.** TNS Land cover

To validate the existence of mangrove, overlay done between mangrove from segmentation result with mangrove vector from One Map Mangrove National activity. The overlay result shows that mangrove from segmentation result is 82% according to One Map Mangrove vector. The area of mangrove forest of TNS segmentation is 73,854 ha while mangrove from One Map Mangrove activity has wide 90,085 ha (Figure 6).

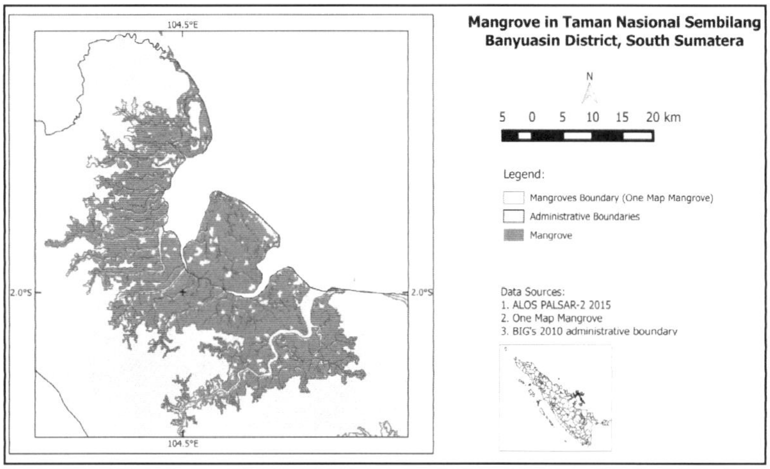


**Figure 6**. Mangrove in TNS from segmentation

## 4. Conclusions

ALOS PALSAR-2 ScanSAR mode can be used for object detection one of them is mangrove forest. By using OBIA segmentation method, mangrove forest in TNS can be detected clearly. Based on the result of overlay between mangrove result of segmentation with mangrove from One Map Mangrove National activity has 82% accuracy. Further research is needed to obtain a combination of scale, shape, and compactness parameters suitable for mangroves.

## Acknowledgments

Authors would like to thank Mukhoriyah, S.T., M.Si and Putri Widya Budiarti, S.T who have helped in the process of data processing on the writing of this paper.

## References

[1] SNI 7717: 2011. *Survei dan Pemetaan Mangrove*. Badan Standarisasi Nasional. 2011

[2] Daniel C Donato, J Boone Kauffman, Daniel Murdiyarso, Sofyan Kurnianto, Melanie Stidham, Markku Kanninen 2011 Mangroves among the most carbon-rich forests in the tropics *Nature Geoscience 2011*(**4**) pp 293–297. doi:10.1038/ngeo1123

[3] Food and Agriculture Organization of the United Nations (FAO) *The World's Mangroves 1980–2005*. FAO Forestry Paper 153. 2007: 24.

[4] PASCO. PALSAR-2: Phased Array type L-band Synthethic Apeture Radar [Internet]. 2014. [cited 2017 Agust 28]. Available from: http://en.alos-pasco.com/alos-2/palsar-2/.

[5] Bai, Yanbing. Bruno Adriano, Erick Mas, Hideomi Gokon, Shunichi Kashimura. Object Based Building Damage Assessment Methodology Using Only Post Event ALOS-2/PALSAR-2 Dual Polarimetric SAR Intensity Images *JDR 2017 12*(**2**) pp 259-271. doi: 10.20965/jdr.2017.p0259.

[6] Pham, T.D., Kunihiko Yashino. Characterization of mangrove species using ALOS-2 PALSAR in Hai Phong city, Vietnam. 8th IGRSM International Conference and Exhibition on Remote Sensing & GIS (IGRSM 2016). 2016. *IOP Publishing IOP Conf. Series: Earth and Environmental Science 37*(**2016**) 012036 doi:10.1088/1755-1315/37/1/012036

[7] Fransson JES, Maurizio Santoro, Jorgen Wallerman, Henrik J Persson, ...... & Lars M H Ulander 2016 Estimation of forest stem volume using ALOS-2 PALSAR-2 satellite images *In: 2016 IEEE International Geoscience & Remote Sensing Symposium: proceedings*. New Jersey: 2016 pp 5327-5330.

[8] Rimba A B, Fusanori Miura 2017 Evaluating the Extraction Approaches of Flood Extended

Area by Using ALOS-2/PALSAR-2 Images as a Rapid Response to Flood Disaster *Journal of Geoscience and Environment Protection 2017*(**5**) pp 40-61

[9] Bakar Abu. *OBIA (Object Base Image Analysis)*. 2014. [cited 2017 Sep 5]. Available from: http://www.citrasatelit.com/obia/.

[10] Danoedoro Projo 2012 *Pengantar Penginderaan Jauh Digital* (Yogyakarta: Penerbit ANDI)

[11] Hurd J D, D L Civco, M S Gilmore, S Prisloe, E H Wilson. Tidal Wetland Classification from Landsat Imagery Using an Integrated Pixel-Based and Object-Based Classification Approach. *In ASPRS 2006 Annual Conference, Reno, Nevada, USA*, May 1–5, 11 p. [Internet]. 2006. [cited 2017 Sep 5]. Available from http://www.ecognition.com/sites/default/files/171_asprs2006_0063.pdf.

[12] Anggoro Ari, Vicentius P Sireger, Syamsul B Agus 2015 Pemetaan Zona Geomorfologi Ekosistem Terumbu Karang Menggunakan Metode Obia, Studi Kasus di Pulau Pari. *Jurnal Penginderaan Jauh 12*(**1**) pp 1-12.

[13] Wibowo T S & R Suharyadi 2016 Aplikasi Object-Based Image Analysis (OBIA) untuk Deteksi Perubahan Penggunaan Lahan Menggunakan Citra ALOS AVNIR-2 *Jurnal Bumi Indonesia* 1(**3**) pp 130-138.

[14] Sari N M, Doni Kushardono 2014 Klasifikasi Penutup Lahan Berbasis Obyek Pada Data Foto Uav Untuk Mendukung Penyediaan Informasi Penginderaan Jauh Skala Rinci *Jurnal Penginderaan* Jauh 11(**2**) pp 114-127

[15] Kamal M, Stuart Phinn, Kasper Johansen 2015 Object-Based Approach for Multi-Scale Mangrove Composition Mapping Using Multi-Resolution Image Datasets *Remote Sens* **7** pp 4753-4783; doi:10.3390/rs70404753

[16] Zylshal, Projo Danoedoro, Eko Haryono 2013 An Object Based Image Analysis Approach To Semi-Automated Karst Morphology Extraction *Proceeding of Asian Conference on Remote Sensing* 2013 doi: 10.13140/RG.2.1.2243.1609.

[17] Theresia 2016 Pengelolaan Ekosistem Mangrove Di Taman Nasional Sembilang Kabupaten Banyuasin Provinsi Sumatera Selatan *Thesis* (Bogor: Sekolah Pascasarjana. Institut Pertanian Bogor)

[18] eCognition. *eCognition® Developer 8.7 Reference Book*. Trimble Germany GmbH. 2011.

# Urban forest topographical mapping using UAV LIDAR

**Iqbal Putut Ash Shidiq[1], Adi Wibowo[1*], Eko Kusratmoko[1], Satria Indratmoko[1], Ronni Ardhianto[2], Budi Prasetyo Nugroho[2]**

[1]Department of Geography, Faculty of Mathematics and Natural Sciences, Universitas Indonesia, Kampus Depok
[2]PT. Pangripta Geomatika Indonesia

adi.w@sci.ui.ac.id

**Abstract**. Topographical data is highly needed by many parties, such as government institution, mining companies and agricultural sectors. It is not just about the precision, the acquisition time and data processing are also carefully considered. In relation with forest management, a high accuracy topographic map is necessary for planning, close monitoring and evaluating forest changes. One of the solution to quickly and precisely mapped topography is using remote sensing system. In this study, we test high-resolution data using Light Detection and Ranging (LiDAR) collected from unmanned aerial vehicles (UAV) to map topography and differentiate vegetation classes based on height in urban forest area of University of Indonesia (UI). The semi-automatic and manual classifications were applied to divide point clouds into two main classes, namely ground and vegetation. There were 15,806,380 point clouds obtained during the post-process, in which 2.39% of it were detected as ground.

**Keywords:** topography, DEM, vegetation, drone, LIDAR

## 1. Introduction

Spatial data has an important role nowadays. Accurate and timely spatial databases are needed for a proper planning, evaluation and monitoring process in many applications [1]. An example of spatial database is topographical data, which is highly needed by many parties such as government, mining companies and agricultural sector.

One of the solution to quickly and precisely mapped topography is using remote sensing system, specifically LiDAR. LiDAR or Light Detection and Ranging is an active remote sensing system, capable of sensing objects by illuminating them with a pulsed laser light and measuring the reflection with a sensor. High accuracy data can be achieved because this system used light energy that moved in constant speed to measure distance between the LiDAR and target. Similar with other radar system technology, LiDAR also used several shorter wavelengths typically in the ultraviolet, visible or near infrared range [2].

In relation with forest management, a high accuracy topographic map is necessary for planning, close monitoring and evaluating forest changes. LiDAR system has been widely used for various application, such as vertical forest structure analysis [3], trees characterization [4], forest fire management [5], timber volume estimate [6] and many more. Recently, more studies are conducted in this area because of technological development. UAV system presents as an alternative to the more-conventional airborne or satellite remote sensing system [7] [8].

In this study, we test high-resolution data using Light Detection and Ranging (LiDAR) collected from unmanned aerial vehicles (UAV) to map topography and differentiate vegetation classes based on height in urban forest area of University of Indonesia (UI). With this study, we are hoping to know the capability of UAV-based LiDAR to create elevation data in such a heterogenic urban forest area.

## 2. Methodology

### *2.1. Study area*

This study was conducted in urban forest area of Universitas Indonesia campus in Depok, West Java. The area comprised 90 hectares, planted with various kinds of trees such as Teak, Meranti, Coffee, Rubber and Matoa (Figure 1). Besides its function as water infiltration area, UI urban forest also being utilized for student's research area.

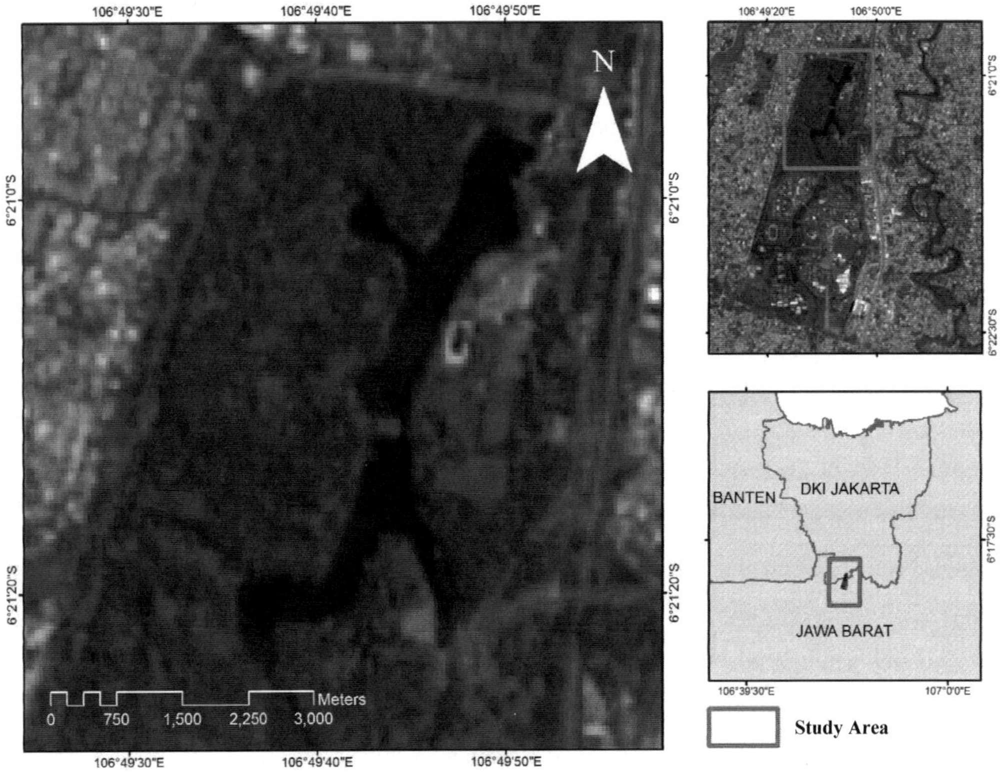


**Figure 1.** Location of the study area

### *2.2. Data acquisition*

*2.2.1. Initial planning.* This study used Yellowscan Mapper, a lightweight laser scanner (2.1 kg with battery included) with 10 cm on its precision and 15 cm on its XY accuracies. The scanner is suitable for various works, such as environmental research, engineering [9], archaeology, forestry [10], topography and geology (www.yellowscan.fr). The laser scanner was mounted on DJI Matrice 600, a six-rotor flying platform controlled by ground operator. A flight plan was designed based on elevation data and grid. It was necessary due to the complexity of the terrain and its cover. For the flight plan, elevation data was collected form Google Earth application, topographic map and SRTM (Shuttle Radar Topographic Mission). From there, a DEM data was generated and utilized for determining the flight plan. A grid was also necessarily created in order to determine the flying path. In this process, we also decided the flying altitude and proportion for grid sidelap. Different flying altitudes were applied by considering the density of vegetation cover (Table 1). Final step in this initial planning was focused on drawing the flight plan using Google Earth application. This process was finalized by inserting the previously created AOI (area of interest) and the grid. These information were later on saved in ".KML" format and ran using Autopilot application on iPad device (Figure 2).

**Table 1.** Different sidelap proportion for different flying altitude.

| Flying altitude (m) | Swath width (m) | Distance between Path 1 and Path 2 | |
|---|---|---|---|
| | | Sidelap 50% (m) | Sidelap 40% (m) |
| 100 | 130 | 65 | 78 |
| 90 | 117 | 58.5 | 70.2 |
| 80 | 104 | 52 | 62.4 |
| 70 | 91 | 45.5 | 54.6 |
| 60 | 78 | 39 | 46.8 |

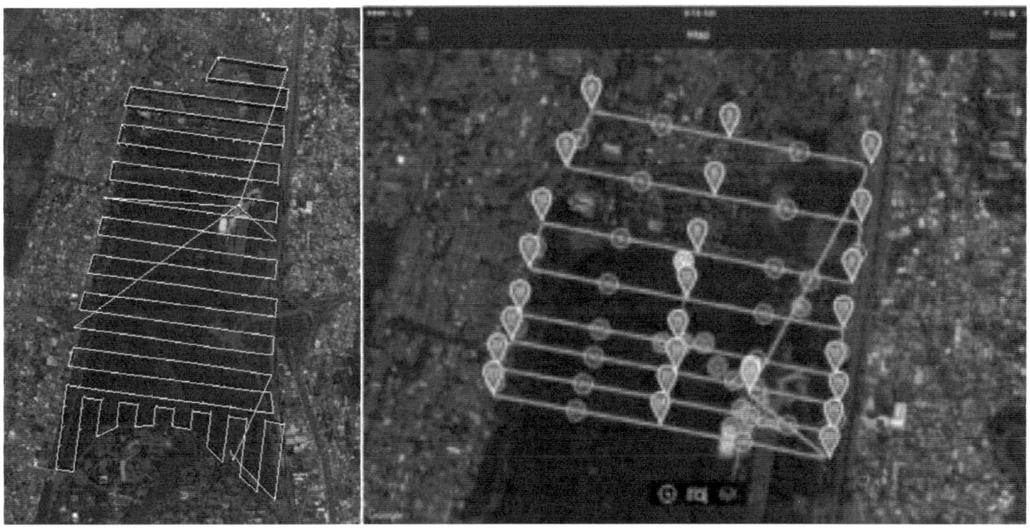

**Figure 2.** UAV flight plan

*2.2.2. Trajectory and GPS data processing.* A trajectory is flight plan data, consist of GPS and LiDAR data. The trajectories were processed using QGIS software equipped with Yellowscan plugin. This process includes determining the coordinate system, selecting trajectory line to be used and storing LiDAR data storage in ".LAS" format. A GPS post-processing were also done in this step, using Precise Point Positioning web application from Natural Resources Canada (https://webapp.geod.nrcan.gc.ca).

*2.2.3. Point cloud matching.* LiDAR data in ".LAS" format were further processed using Terrasolid application. Two main processes here, were flight line matching and point cloud classification. LiDAR data from previous process were not yet geometrically corrected. The differences in X, Y, Z position between each flight line were causing gaps and affecting the validity of the information. In this step, we did surface-to-surface matching by correcting each point cloud and flight line and aligning them in the same X, Y, and Z references (Figure 3).

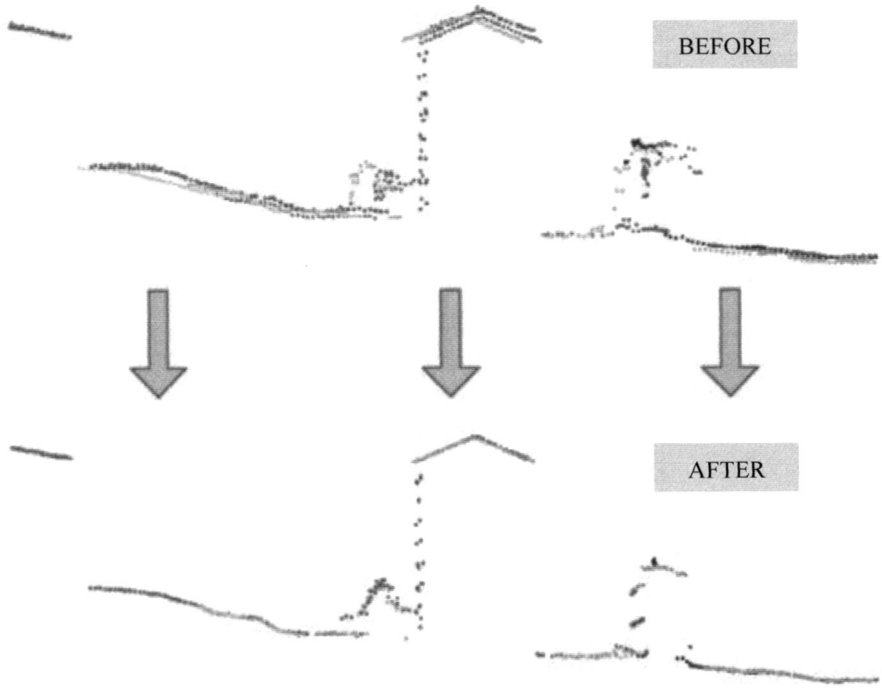


**Figure 3.** Illustration on flight line and point clouds matching

## 3. Result and Discussion

### *3.1. LiDAR data acquisition*

This study performed three times LiDAR data acquisitions. Each process was defined as different flight plan. The detail of data acquisition process was explained in Table 2 below.

**Table 2.** Flight plan characteristics in this study.

| Flight plan | Velocity | Altitude | Sidelap |
|---|---|---|---|
| F1 | 8 m/s | 90 m | 40-70% |
| F2 | 8 m/s | 70 m | 40-70% |
| F3 | 8 m/s | 90 m | 40-70% |

### *3.2. Point cloud classification*

LiDAR data in the form of point cloud were still unidentified when it was first time generated. The data will be well understood when the point cloud has been identified based on the relationship one to another. The filtering process was needed to segregate the point clouds for earth surface objects and point clouds for ground. Each point cloud will be classified into the same class with consideration on the similar nature, area, distribution and character of the object. In this study, the point clouds will mainly classify into vegetation and ground. There are also subclasses for vegetation such as high vegetation, medium vegetation and low vegetation. The classification process was applied using semi-automatic and manual methods. In semi-automatic, point clouds were classified into ground and vegetation based on controlled parameters, whilst the manual method was applied on the point clouds with the wrong class.

To classify ground, this study used ground filtering technique. The result of this process are two classes, namely ground and non-ground. The segregation was based on four physical parameters of ground points, such as lowest elevation, ground surface steepness, ground surface elevation difference and ground surface homogeneity [11]. On the other, vegetation was classified into three classes. It is based on the height of point clouds from the ground. Criteria for vegetation classification is shown in Table 3 below.

In total, there are 15,806,380 point clouds collected in this study (Figure 4). For ground class, there are 378,149 points or about 2.39% from the total point clouds. The rest were classified as non-ground or vegetation (Figure 5).

**Table 3.** Vegetation classification parameters [12].

| Vegetation class | Height from the ground (m) |
|---|---|
| Low | 0.10 – 0.50 |
| Medium | 0.51 – 1.00 |
| High | > 2.00 |

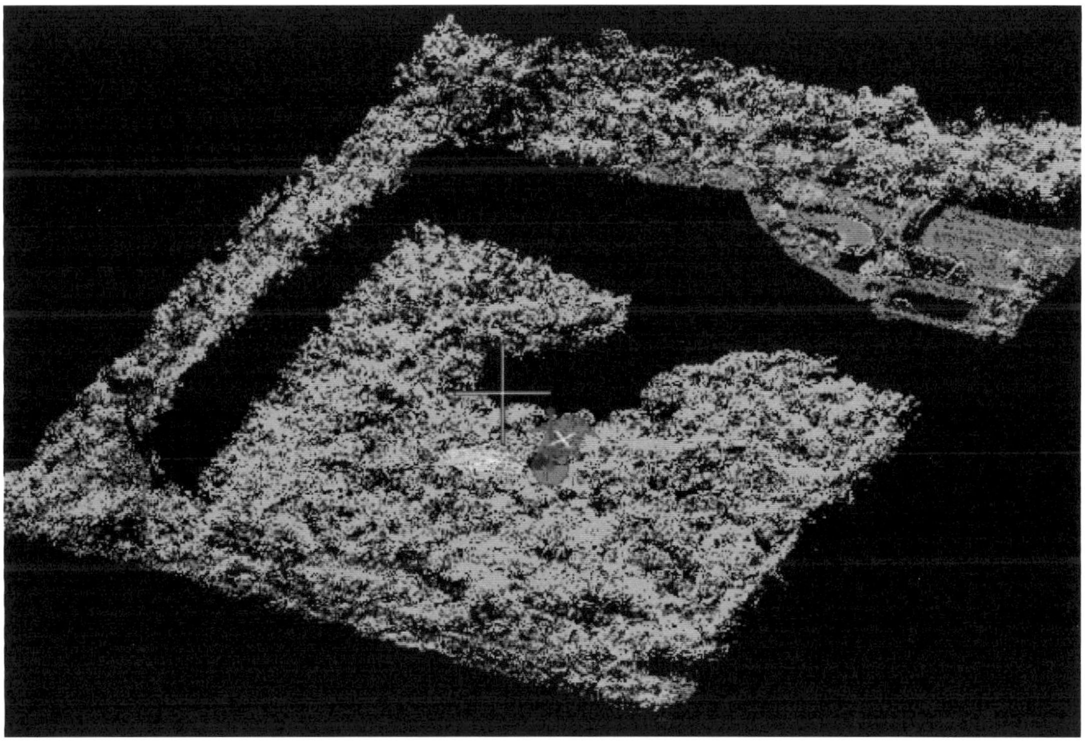

**Figure 4.** Collected point clouds

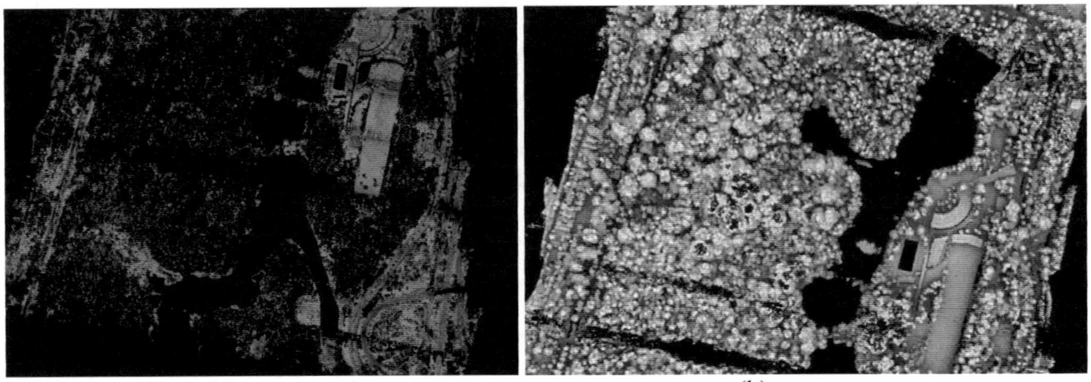

(a) (b)

**Figure 5.** The result from (a) ground classification and (b) non-ground classification

*3.3. Digital Terrain Model (DTM)*

After ground and non-ground object were successfully segregated, a topographical data in the form of DTM were able to be generated. DTM were created solely based on the ground class, whilst non-ground object can be used for creating Digital Surface Model (DSM) data. A brief looks on generated DTM are shown by Figure 5 and Figure 6 below. Based on DTM data, the elevation in our study area ranges from 46.8 to 68.5 meter above sea level.

**Figure 6.** The resulted Digital Terrain Model (3D view).

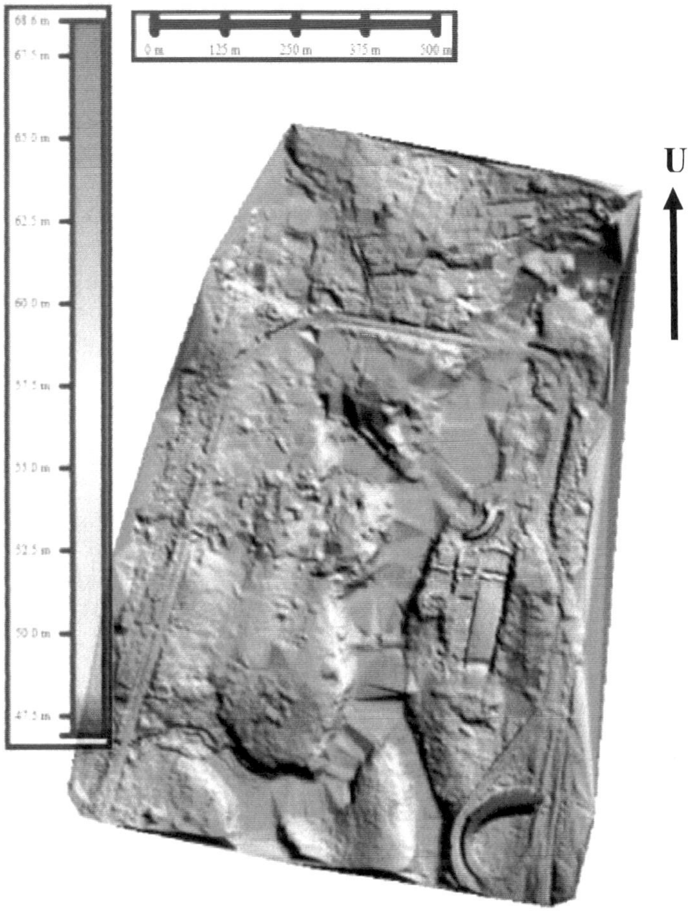


**Figure 7.** The resulted Digital Terrain Model (2D view)

## 4. Conclusions

This study shows the capability of UAV-based LiDAR to create topographical data and elevation map. The system is able to map the terrain of urban forest area in both densely vegetated area or less dense vegetated area. Due to the absence of elevation data from terrestrial survey, this study is not able to assess the accuracy of produced elevation map. It is the limitation to this study, however potentially addressed from the future studies.

## References

[1] Foody G, Fritz S, Fonte C C, Bastin L, Olteanu-Raimond A-M, Mooney P, See L, Antoniou V, Liu H-Y, Minghini M and Vatseva R 2017 *Mapping and the Citizen Sensor*, ed G Foody *et al* (London: Ubiquity Press) p 398

[2] Sumerling G 2011 *Lidar analysis in ArcGIS 10 for forestry applications* vol 1 (Adelaide: ESRI Inc) p 53

[3] Zimble D A, Evans D L, Carlson G C, Parker R C, Grado S C and Gerard P D 2003 *Remote Sens. Environ.* **87** 171

[4] Wulder M A, Bater C W, Coops N C, Hilker T and White J C 2008 *For. Chron.* **84** 807

[5] Koetz B, Morsdorf F, Van der Linden S, Curt T and Allgöwer B 2008 *Forest Ecol. Manag.* **256** 263

[6] Naesset E 1997 *Remote Sens. Environ.* **61** 246

IOP Conf. Series: Earth and Environmental Science **98** (2017) 012034 doi:10.1088/1755-1315/98/1/012034

[7] Fardusi M J, Chianucci F and Barbati A 2017 *Annals of Silvicultural Research* **41** 3
[8] Goodbody T R, Coops N C, Marshall P L, Tompalski P and Crawford P 2017 *For. Chron.* **93** 71
[9] Bakuła K, Ostrowski W, Szender M, Plutecki W, Salach A and Górski K 2016 *International Archives of the Photogrammetry, Remote Sensing & Spatial Information Sciences* **41**
[10] Sačkov I, Santopuoli G, Bucha T, Lasserre B and Marchetti M 2016 *Forests* **7** 307
[11] Meng X., Currit N and Zhao K 2010 *Remote Sens.* **2** 833
[12] Gallant A L 2015 *Remote Sens.***7** 10938

# Important Value of Economic Potency Mangrove Using NDVI Satellite High Resolution Image To Support Eco Tourism Of Pamurbaya Area (Case Study: East Cost of Surabaya)

**B M Sukojo* , H Hidayat and D Ratnasari**

Departement of Geomatics Engineering, Faculty of Civil Engineering and Planning, Sepuluh Nopember Institute of Technology (ITS) Surabaya, Indonesia

bangunms@gmail.com

**Abstract.** Indonesia is a vast maritime country; many mangrove conservations is found around coastal areas of Indonesia. Mangroves support the life of a large number of animal species by providing breeding, spawning and feeding. Mangrove forests as one of the unique ecosystems are potential natural resources, supporting the diversity of flora and fauna of terrestrial aquatic communities that directly or indirectly play an important role for human life in economic, social and environmental terms. East Coast Surabaya is an area with the most extensive and diverse mangrove ecosystems along the coast of Surabaya. Currently Pamurbaya used as a recreational object or nature tourism called eco tours. Utilization of mangrove ecosystem as a place of this eco tour bring positive impact on economic potency around pamurbaya area. So, to know the value of the economic potential of mangrove ecosystems for support of nature tourism Pamurbaya region needs to study mapping mangrove ecosystem conditions in the East Coast area of Surabaya. Mapping of mangrove conditions can use remote sensing technology by utilizing satellite image data with high resolution. Data used for mapping mangrove ecosystem conditions on the east coast of Surabaya are high resolution satellite image data of Pleiades 1A and field observation data such as Ground Control Point, soil spectral parameters and water quality. From satellite image data will be classification of mangrove vegetation canopy classification using NDVI vegetation index method using algorithm formula which then will be tested correlation with field observation data on reflectant value of field and water quality parameter. The purpose of this research is to know the condition of mangrove ecosystem to know the economic potential of mangrove ecosystem in supporting Pamurbaya nature tourism. The expected result of this research is the existence of basic geospatial information in the form of mangrove ecosystem condition map. So that can be used as decision makers to find out how big economic potential of mangrove ecosystem in supporting nature tourism Pamurbaya beside that can be used as reference research related mapping, conservation and development of mangrove ecosystem.

**Keywords**: Correlation, NDVI, Mangrove, Remote Sensing

## 1. Introduction

Indonesia became the country with the most extensive mangrove forest in the world. According to [13] in 2005 Indonesia has 3 million ha of mangrove forest. Mangrove is a complex ecosystem consisting of flora and fauna in coastal areas or sea tidal boundaries. Mangrove is an influential vegetation in coastal ecosystems. Because Indonesia is a vast maritime country, so much found mangrove conservation around the coastal areas of Indonesia.

Mangrove is a plant community or an individual plant species that forms a community in tidal areas. Mangrove forests are a type of forest that is naturally affected by tidal seawater, flooded during high tides and free from puddles during low tides. The mangrove ecosystem is a system consisting of biotic and abiotic environments interacting within a mangrove habitat. The mangrove forest ecosystem is complex and dynamic, but labile. Complex because of its ecosystem in addition there is mangrove vegetation, is also a habitat of various animals and aquatic biota. Dynamic because the mangrove forest can grow and develop and continue to succeed in accordance with changes in its natural place of growth [16].

Mangrove forest management as a form of conservation of natural resources and coastal waters resources is necessary considering the mangrove forest ecosystem functions physically to keep the coastline stable, protecting the beach from erosion (abrasion), storm and wave dampers, sediment catchers, Rahmawaty in Suci, (2011), while the biological mangrove functions according to [3] as a spawning or shrimp area for shrimp, crabs, shellfish and others, as the area for shelter, nesting and breeding.In addition to ecologically functioning mangroves also function socioeconomically, Rahmawaty in Suci (2011) mangroves as a source of livelihood, production of various forest products such as wood, charcoal, medicine, building materials and handicrafts, natural attractions.

Mangrove ecosystems are found throughout the coastal city of Surabaya, one of which is the East Coast of Surabaya (Pamurbaya) which is a protected area. Pamurbaya is known as the remaining green open space and a fortress to protect Surabaya from the threat of abrasion, sea water intrusion, and land subsidence. Besides, Pamurbaya is also used as a recreational object of nature tourism or so-called eco tour. Utilization of mangrove ecosystem as a place of eco this tour brings a positive impact on the economic potential around the pamurbaya region. To know the economic potential value of mangrove ecosystem in supporting eco-tourism Pamurbaya area can be done by mapping mangrove ecosystem condition in East Coast area of Surabaya by using remote sensing technology that is by utilizing satellite image data of high resolution.

Mangrove ecosystem is one of the objects that can be identified by using remote sensing technology. The geographical location of the mangrove ecosystem located in the land and sea transition areas provides a distinctive recording effect when compared to other land vegetation objects. The recording effect is closely related to the spectral characteristics of the mangrove ecosystem, to the extent that identification requires a separate transformation. In general, for vegetation detection transformation used vegetation index [8].

In this study will map the condition of mangrove ecosystem by studying the condition of mangrove ecosystem vegetation based on Normalized Difference Vegetation Index (NDVI) value using remote sensing technique using satellite imagery. Also in this study the NDVI image value will be validated with field data. From the results of this study is expected to provide a spatial information about the condition of mangrove ecosystems used to determine the economic potential value of mangrove ecosystems in supporting eco-tourism Pamurbaya region. In addition, this research is expected to be used as reference material for research related to the field of mapping, conservation and development of mangrove ecosystem.

## 2. Materials and Methods

The location of this study included in the land suitability area for mangrove conservation, geographically located at coordinates 7°15'19,60 "LS - 7°17'13,25" LS 112°48'35,69 "BT - 112°48 ' 40.72 "east of East Coast Surabaya.

### 2.1. Data Used

The data used in this research are Pleiades 1B satellite image 2015, Map of RBI 1: 25,000, GCP checking data field, Test data sample parameters.

IOP Conf. Series: Earth and Environmental Science **98** (2017) 012035 doi:10.1088/1755-1315/98/1/012035

*2.1.1 Data Collection Stages*

Data used in this research are Pleiades 1B image 2015 from this data will be made a net design for measurement of GCP point in field and sampling location.

*2.2. Stages of Data Processing*

*2.2.1 Geometric Correction*

Geometric correction aims to reduce geometric errors resulting in geometric corrected images. This technique is required to determine the position of the image in accordance with the coordinates of the map or the coordinates of the earth. In this research geometric correction on Aster and Landsat-8 image is done with vector map of RBI 1: 25.000 by taking GCP point coordinates in same area with image. While geometric correction on Pleiades 1B image is done by making the design of the net on the image then do the measurement of GCP point field. Geometric correction is correct if the RMS Error value is less or equal to one pixel [24], it means the image has been corrected geometrically.

*2.2.2 Radiometric Correction*

Radiometric corrections include radiometric calibration and atmospheric correction. The radiometric calibration converts the DN (Digital Number) to Reflectance i.e. the process of changing the Pixel value to an image originally worth DN in the change to Reflectance. Then the atmospheric correction by the method of Gain of Set.

*2.2.3 Cropping*

Cropping is done to facilitate the image processing, and only covers the research area. So, minimize the image area to be processed. Cutting by area or area coverage of coastal mangrove ecosystem of North Surabaya.

*2.2.4 Vegetation Index*

Vegetation index indices vegetation index is done by entering the vegetation index algorithm. In this study using the algorithm Normalized Difference Vegetation Index (NDVI). NDVI algorithm as follows:

$$\text{NDVI} = \frac{NIR\text{-}RED}{NIR+RED} \tag{1}$$

Information :
NIR = Near infrared spectral band value
RED = The value of the red spectral band
the process of incorporating this vegetation index algorithm will produce an image spectral value between -1 to 1.

*2.2.5 The Claasification of Terelia*

Furthermore, the classification of terelia by entering the value range of each type of vegetation health.

**Table 1.** The Classification of Terelia

| Class | NDVI Range | Satisfaction Level |
|---|---|---|
| 1 | 0 s.d 0,32 | Avicennia |
| 2 | 0,32 s.d 0,42 | Bruguera |
| 3 | >0,42 s.d 1 | Rhizopora |

Field observations included taking Parameter Samples (water and soil), Samples of parameters taken are soil and water physically and chemically. Chemically sample of water and soil parameters is done by laboratory test to know PH and salinity, then result of correlation test between index value of vegetation (reflectan) with value of parameter test result. From the tabulation data will be calculated correlation value.

*2.2.6 Correlation Test*
Correlation test is done on parameter sample and Spectral Ground. Correlation test is done by connecting the value of test result of parameter sample and Spectral Ground value with vegetation index value in the image so it can know the relation form. If the PH value is high, Salinity is low, with a high vegetation index value then it has a positive relationship and vice versa.

*2.2.7 Analysis*
At this stage we analysed the correlation value between vegetation index value with parameter sample test in the field. Furthermore, carried out the final stage of research, namely the preparation and writing of the report and its map information. Map obtained in this research is map of mangrove ecosystem scale 1: 5.000.

## 3. Results

*3.1 Image Improvement by Image Fusion Data Method (Pansharpening) on Pleiades image*
The incorporation of low resolution multispectral images and high resolution panchromatic images is of great importance for remote sensing and mapping applications. Usually the panchromatic image has a wide wavelength ranging from the visible spectrum and near infrared, whereas the multispectral image only includes a narrow spectrum. However, multispectral images have more than three spectral bands, while the panchromatic image has only one spectral band. So by combining the two images are obtained multispectral image that has a high spectral resolution and high spatial information as well. The image integration in this research is done by using Gram-Schmidt transformation method, where each image is multispectral and georefenced choreographic image, so firstly done the co-registration of the images. Figure 1. is a view of the merging of Multispectral and Panchromatic images of Pleiades 1B. Here is a view of the image of Satellite image of Pleiades 1B

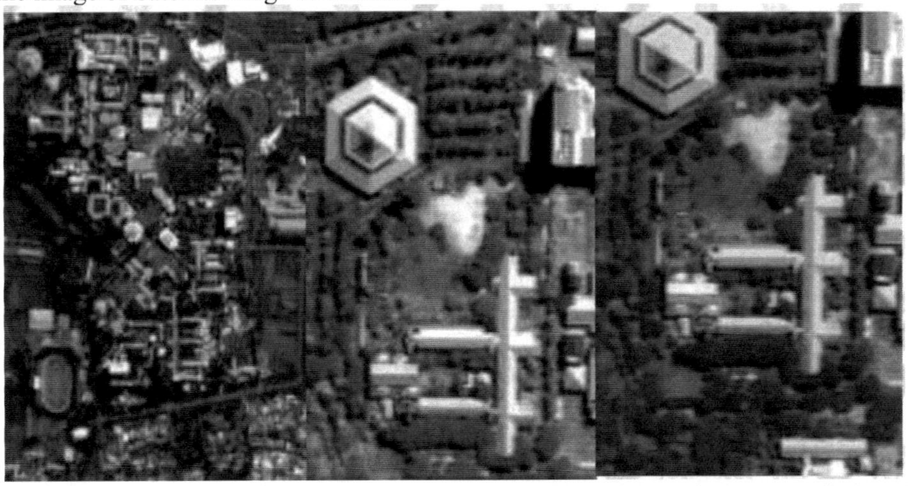

**Figure 1.** (a) Multispectral Images of Pleiades 1B, (b) Panchromatic Image Peliades 1B, (c) Pleiades 1B Image Fusion Result

*3.2 Determination of the Number and Distribution of Soil Control Points*
The Soil Control Point (GCP) should be planned on a specific object, clearly visible in the image and easily recognizable in the field or object on the surface of the earth which changes relatively slowly /

permanently, for example: roads, street corners, areas of striking colors, buildings / monuments that are easily identifiable or recognizable.

**Figure 2**. Location Selection of GCP Placement in Building Corner

**Figure 3.** Location Selection at Ujung Jembatan

In the process of image rectification process, it is needed the right number of ground control points, so that the implementation of measurement in the field is not too long (Yudha, 2015). The number of ground control points is cultivated according to need, with considerations such as area size, mapping area relief condition, cost, and tools to be used for the measurement to be carried out efficiently.

The determination of the GCP point is also attempted to spread to the outermost position of the image to be performed koreksigeometrik. In this study, the number of GCP used is 8 points. Here is the GCP point spread:

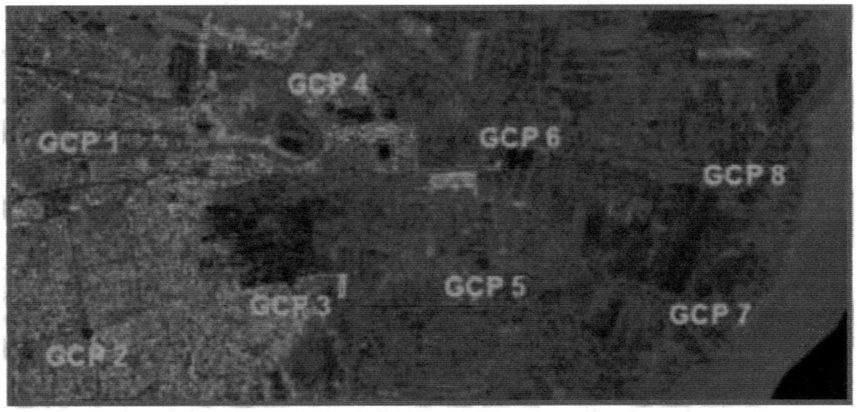


**Figure 4.** Spread of GCP Point on Pleiades 1B Image

Based on Figure 4. Control Point Land / GCP in this study amounted to 8 GCP, and there are 13 baselines.

**Table 2.** Length between GCP

| Baseline | Length (km) |
|---|---|
| GCP 1 - GCP 2 | 1,311 |
| GCP 1 - GCP 3 | 1,916 |
| GCP 1 - GCP 4 | 2,069 |
| GCP 2 - GCP 3 | 1,705 |
| GCP 4 - GCP 3 | 1,393 |
| GCP 4 - GCP 5 | 1,753 |
| GCP 4 - GCP 6 | 1,590 |
| GCP 3 - GCP 5 | 1,581 |
| GCP 6 – GCP 5 | 0,873 |
| GCP 6 - GCP 7 | 1,825 |
| GCP 6 - GCP 8 | 1,835 |
| GCP 5 - GCP 7 | 1,797 |
| GCP 8 - GCP 7 | 0,796 |

Based on Table 2. the distance between GCP is 0.873-2.069 km, can be seen if the points used are in accordance with the standard distance between points used by the BIG, where 3-4 km for a relatively flat area and 2-3 km for hilly areas, and study areas in the area of the village of Wonorejo, Surabaya, which includes a relatively flat area.

*3.3 Calculation of Strength of the Net (Strength of Figure / SoF)*

To determine the position of the ground control point (GCP) for the purpose of geometric correction required the planning and calculation of strength of the net (Strength of Figure / SoF). SoF (Strength of Figure) is a geometric power (form) of a triangular circuit that determines the spread of errors in network flattening. The geometric strength of a good triangular web is mirrored by the small price of the SoF and will ensure thoroughness of the whole jarring. The smaller the power factorof the net, the better the net configuration will be, and vice versa. Here is a net design and SoF calculation on Pleiades 1B image.

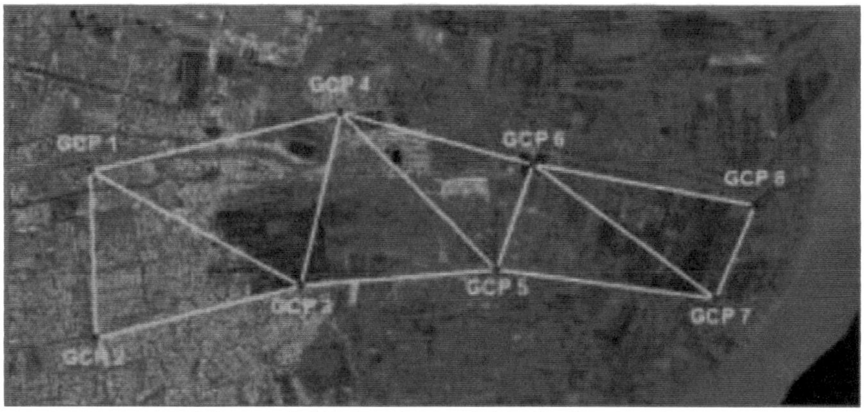


**Figure 5**. Net Design on Pleiades Image 1B

SoF calculation using parameter alignment method. Based on the design of nets in figure 5. and the value of SoF is 0.5938, it can be said if the calculation of the SoF meets the tolerance of ≤ 1. In this case, the strength factor of the net value of the smaller the better the network configuration and vice versa ( Anjasmara, 2005). The matrix of the SoF calculation can be seen in Appendix 1.

*3.4 Ground Control Point (GCP) Measurement Results*
GCP was measured in the field using the Topcon Hyper Pro Geodetic GPS tool with a static method with 40 minutes of observation. Here are the coordinate results obtained from field measurements.

**Table 3.** Ground Control Point Data (GCP) Field Measurement Results

| **Point** | ***GROUND CONTROL POINT*(GCP)** | |
|---|---|---|
| | **X (m)** | **Y (m)** |
| GCP 1 | 697694.529 | 9191696.257 |
| GCP 2 | 697740.773 | 9190389.81 |
| GCP 3 | 699407.414 | 9190800.332 |
| GCP 4 | 699721.094 | 9192151.699 |
| GCP 5 | 701088.897 | 9190613.219 |
| GCP 6 | 701281.548 | 9191748.465 |
| GCP 7 | 702795.57 | 9190703.69 |
| GCP 8 | 703107.998 | 9191437.327 |

The measurement data in Table 3. is the result of retrieving coordinate data which has been done post processing using Topcon Tools software v.7.5.1.

*3.5 Independent Check Point (ICP) Measurement Results*
The ICP point is used to test the accuracy of the geometric accuracy of the corrected image. The number of ICP points dimabil by 13 points, whose distribution is adjusted to the standard distribution of ICP point by BIG (Geospatial Information Agency) on Spatial Plan Validation module. Here is a picture of the sample point depth taken:

**Figure 6.** Distribution of Test Point Accuracy in Pleiades Image 1B Rectification Result

ICP was measured in field using Topcon Hiper Pro GPS tool with static rapid method with observation for 10-20 minutes and radial geometry. Here are the coordinate results obtained from field measurements:

**Table 4.** Independent Check Point (ICP) Results of Measurement Reports

| **Point** | **INDEPENDENT CHECK POINT (ICP)** | |
|---|---|---|
| | **X (m)** | **Y (m)** |
| ICP 1 | 697,911,493 | 9,190,670,503 |
| ICP 2 | 698,867,089 | 9,190,861,748 |
| ICP 3 | 699,812,476 | 9,190,842,260 |
| ICP 4 | 700,801,105 | 9,190,906,203 |
| ICP 5 | 701,461,131 | 9,190,963,780 |
| ICP 6 | 702,220,809 | 9,190,787,287 |
| ICP 7 | 702,887,550 | 9,191,368,168 |
| ICP 8 | 702,118,056 | 9,191,249,111 |
| ICP 9 | 701,042,913 | 9,191,675,572 |
| ICP 10 | 699,612,522 | 9,191,669,395 |
| ICP 11 | 699,021,604 | 9,191,585,024 |
| ICP 12 | 698,044,366 | 9,191,413,773 |
| ICP 13 | 697,546,610 | 9,190,950,773 |

*3.6 Coordinate Point Control on the Image*

The control point coordinates (x, y) in the satellite imagery are used for the transformation process during geometric correction. Here are the control point coordinates (x, y) on satellite imagery:

**Table 5**. Image Coordinate Data of Pleiades 1B

| Point | CONTROL POINT | |
|---|---|---|
| | X (Pixel) | Y (Pixel) |
| GCP 1 | 15133,5 | 32135,0 |
| GCP 2 | 15237,0 | 34758,5 |
| GCP 3 | 18554,5 | 33906,0 |
| GCP 4 | 19168,0 | 31186,0 |
| GCP 5 | 21906,5 | 34254,0 |
| GCP 6 | 22280,0 | 31970,0 |
| GCP 7 | 25306,0 | 34042,5 |
| GCP 8 | 25922,5 | 32563,5 |

*3.7 Results of Image Rectification*

To find out how big the value of GCP error, it must be done image rectification process. Things to consider when performing the image rectification process is at the time of determining the GCP point of the image, try the dots to be determined on the image in accordance with the points measured in the field so that will get a minimum error value. The rectification process uses order 1 and 2nd order polynomials. This is applied equally to both imagery, so that the comparison can be known. At least 1st order polynomial required 3 GCP and for polynomial 2nd order requires 6 GCP minimum, while for 3rd order polynomial required minimum 10 GCP. Therefore, because the number of GCP used in this study is 8 GCP, so the most appropriate is the order 1 and 2nd order polynomial methods. In addition to the number of control points available, there are also other factors that influence the selection of order, such as the topographic state of the region as well as errors or distortions on the image to be rectified (Yudha, 2015). To know the accuracy of the image rectification, it can be seen from RMSE (Root Mean Square Error) value per pixel unit in image [24]. RMSE shows the degree of error by comparing GCP coordinates of transformation with actual coordinates in this case ie the coordinates of GPS measurement results. The following is the result of the Pleiades 1B image rectification process using the 1st order polynomial method and the 2nd order:

**Table 6**. RMSE Value of Geometric Correction Result of Satellite Images of Pleiades 1B

| Point | RMSE (Pixel) | |
|---|---|---|
| | Polinomial | Polinomial |
| | Orde 1 | Orde 2 |
| GCP 1 | 0,328 | 0,231 |
| GCP 2 | 0,309 | 0,136 |
| GCP 3 | 0,293 | 0,173 |
| GCP 4 | 0,215 | 0,143 |
| GCP 5 | 0,524 | 0,218 |
| GCP 6 | 0,205 | 0,151 |
| GCP 7 | 0,344 | 0,215 |
| GCP 8 | 0,356 | 0,242 |
| **Total** | 2,575 | 1,508 |
| **Mean of RMSE (Pixel)** | 0,322 | 0,188 |

In Table 6. The average RMSE value for Pleiades 1B with 8 GCP images is 0.322 pixels for 1st order polynomial method and 0.188 pixel for 2nd order polynomial method. Thus, the Pleiades 1B image rectification process is acceptable because it meets the given tolerance of ≤ 1 as suggested.

*3.8 Geometric Accuracy Test*

The geometric accuracy test is obtained by finding the residual measurement data in the field which is the coordinate point of the Geodetic GPS measurement and the coordinate point of the rectified satellite image. The geometric accuracy test is performed on Pleiades 1B satellite images rectified in each method, ie 1st order and 1st order polynomial methods. Here is a list of ICP coordinates obtained from the image interpretation:

**Table 7.** Coordinate Data of ICP (Interpretation) on Satellite Images of Pleiades 1B

| **Point** | **ICP COORDINATIC POINT (INTERPRETATION)** | | | |
|---|---|---|---|---|
| | Image of Polynomial Recapitulation Result of Order 1 Image | | Image of Result of Polynomial Recondition of Order 2 | |
| | **X (m)** | **Y (m)** | **X (m)** | **Y (m)** |
| ICP 1 | 697911,8 | 9190669,9 | 697911,9 | 9190670,0 |
| ICP 2 | 698867,4 | 9190862,0 | 698867,4 | 9190862,0 |
| ICP 3 | 699812,7 | 9190842,4 | 699812,8 | 9190842,0 |
| ICP 4 | 700801,7 | 9190907,1 | 700801,4 | 9190905,4 |
| ICP 5 | 701460,5 | 9190964,2 | 701461,3 | 9190963,9 |
| ICP 6 | 702219,3 | 9190788,0 | 702220,2 | 9190787,0 |
| ICP 7 | 702886,5 | 9191368,5 | 702886,5 | 9191368,4 |
| ICP 8 | 702118,1 | 9191249,9 | 702118,2 | 9191249,5 |
| ICP 9 | 701042,8 | 9191675,5 | 701043,3 | 9191675,6 |
| ICP 10 | 699612,3 | 9191669,5 | 699612,8 | 9191669,5 |
| ICP 11 | 698043,9 | 9191413,9 | 698043,9 | 9191413,9 |
| ICP 12 | 697545,9 | 9190950,5 | 697546,0 | 9190949,9 |
| ICP 13 | 697911,8 | 9190669,9 | 697911,9 | 9190670,0 |

After obtained image image rectification image, then next is done RMSE calculation, to know how big shift of coordinate of point (residue) in that image. In RMSE calculations for geometric precision tests, 12 points are used because ICP 11 points can not be used because the results do not meet the tolerance. At the point of ICP 11 there is an error in the measurement, where there is no epoch with receiver 1, so the measurement of ICP 11 is absolute, which according to [1] absolute method is not in high accuracy (precision position up to several meters), while required for accuracy testing This geometric is less than 0.5 meters accurate, since the spatial resolution of the Pleiades 1B satellite image is 0.5 meters. The following is the result of the processing of the geometric accuracy test of Pleiades 1B satellite images.

**Table 8.** Geometric Accuracy Test Results Satellite Image Pleiades 1B Rectification Results

| POINT | GEOMETRIC RESEARCH TEST POINT | | | |
|---|---|---|---|---|
| | Image of Polynomial Recapitulation Result of Order 1 | | Image of Result of Polynomial Recondition of Order 2 | |
| | **Residue X (m)** | **Residue Y(m)** | **Residue X(m)** | **Residue Y(m)** |
| ICP 1 | 0,310 | 0,573 | 0,393 | 0,479 |
| ICP 2 | 0,285 | 0,243 | 0,263 | 0,236 |
| ICP 3 | 0,177 | 0,126 | 0,366 | 0,305 |
| ICP 4 | 0,562 | 0,874 | 0,265 | 0,815 |
| ICP 5 | 0,650 | 0,447 | 0,130 | 0,130 |
| ICP 6 | 1,501 | 0,727 | 0,623 | 0,286 |
| ICP 7 | 1,068 | 0,343 | 1,032 | 0,212 |
| ICP 8 | 0,002 | 0,752 | 0,134 | 0,436 |
| ICP 9 | 0,079 | 0,089 | 0,372 | 0,008 |
| ICP 10 | 0,187 | 0,064 | 0,303 | 0,057 |
| ICP 12 | 0,482 | 0,169 | 0,505 | 0,175 |
| ICP 13 | 0,681 | 0,301 | 0,656 | 0,885 |
| **RMSE (m)** | 0,806 | | 0,647 | |

Based on the calculation of RMSE geometric accuracy test of satellite image of Pleiades 1B result of rectification in Table 8. The value of RMSE 0,806 on the image of the 1st order polynomial rectification and on the image of the 2nd order polynomial rectification is 0,647 meter.

*3.9 Image Feasibility Analysis*

Based on the Technical Specification of 2014 Country Map Map by BIG, village maps can be made on a scale of 1: 2,500, 1: 5,000 and 1: 10,000. For the requirements of horizontal geometry accuracy that must be met in the process of making a village map that meets the standard accuracy of the map in accordance with Perka BIG No. 15 Year 2014.

**Table 9.** Horizontal Village Map
(Source: PerKa BIG No. 15 Year 2014)

| **No.** | **Scale** | **No. Scale (CE90 in m)** | | |
|---|---|---|---|---|
| | | **Class 1** | **Class 2** | **Class 3** |
| 1 | 1 : 10.000 | 2 | 3 | 4 |
| 2 | 1 : 5.000 | 1 | 1,5 | 2,5 |
| 3 | 1 : 2.500 | 0,5 | 0,75 | 1,25 |

Based on Table 9. The high resolution satellite imagery used for village map design is said to meet the standard accuracy of the village map if the horizontal accuracy is ≤ 5 meters (grade 3).

**Table 10.** Results Calculation of Horizontal Accuracy on Satellite Images of Pleiades 1B

| Results Calculation of HORIZONTAL ACCURACY (m) (CE90 = 1,5175 x RMSE) | | |
|---|---|---|
| | **Polynomial Order 1** | **Polynomial Order 2** |
| PLEIADES 1B | 1,222 | 0,982 |

Therefore, based on the multiplication of RMSE value with the coefficient of accuracy (1,5175), the horizontal accuracy of the 1B poliomial image is 1,222 meters. The 2nd horizontal polynomial method is 0.982 meters. Based on Table 10. the image feasibility as the basis for making the village map, Pleiades 1B satellite image meets the requirements of making a 1: 2.500, 1: 5.000 and 1: 10.000 village map.

*3.10 Correlation Test*

After processing NDVI and soil and water parameters test, the correlation test is done.

**Table 11.** Soil Analysis Data

| **Point** | **Coordinate** | | **NDVI** | **N** | **P** | **K** | **pH** | **Salinity (ppt)** |
|---|---|---|---|---|---|---|---|---|
| | **X** | **Y** | | **%N** | **%P** | **(ppm)** | | |
| **R1** | 683,759.63 | 9,204,052.30 | 0.7595 | 1.08 | 0.21 | 0.026 | 6.9 | 4.43 |
| **R2** | 683,772.81 | 9,204,058.97 | 0.7089 | 0.61 | 0.2 | 0.019 | 7.15 | 0.29 |
| **R3** | 683,787.79 | 9,204,064.44 | 0.7165 | 0.65 | 0.26 | 0.021 | 6.9 | 1.31 |

**Table 12.** Data Analysis Water Table

| **Point** | **Coordinate** | | **NDVI** | **pH** | **Salinity (ppt)** |
|---|---|---|---|---|---|
| | **X** | **Y** | | | |
| **R1** | 683,759.63 | 9,204,052.30 | 0.7595 | 7.2 | 0.21 |
| **R2** | 683,772.81 | 9,204,058.97 | 0.7089 | 7.45 | 0.81 |
| **R3** | 683,787.79 | 9,204,064.44 | 0.7165 | 7.35 | 1.46 |

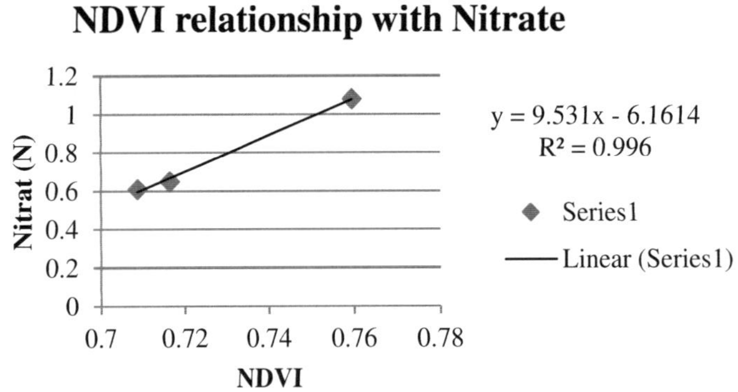


**Figure 7.** NDVI relationship with Nitrate

IOP Conf. Series: Earth and Environmental Science **98** (2017) 012035 doi:10.1088/1755-1315/98/1/012035

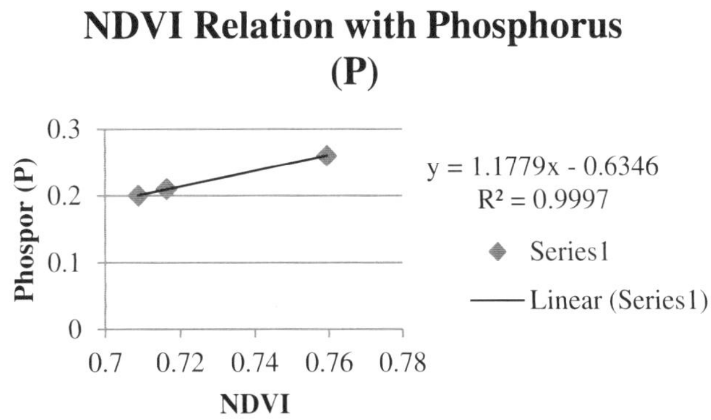


**Figure 8.** NDVI Relationship with Phosphor (P)

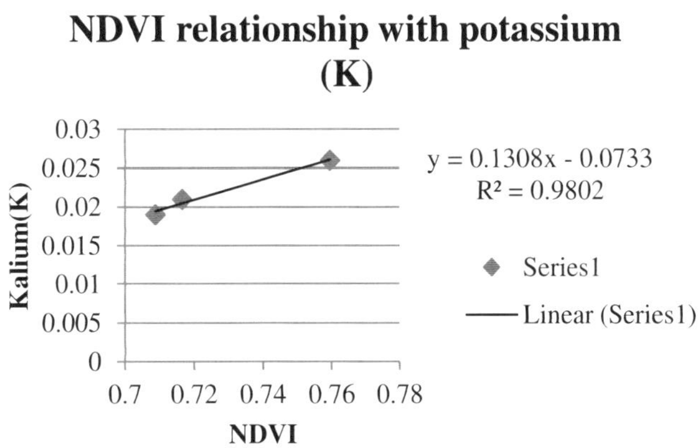


**Figure 9.** NDVI relationship with potassium

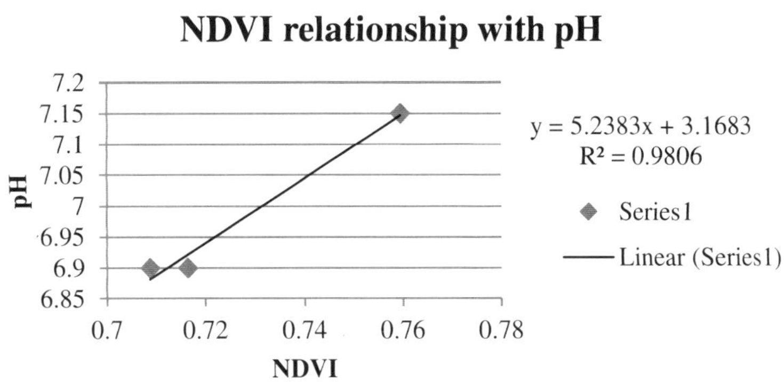


**Figure 10.** NDVI relationship with pH

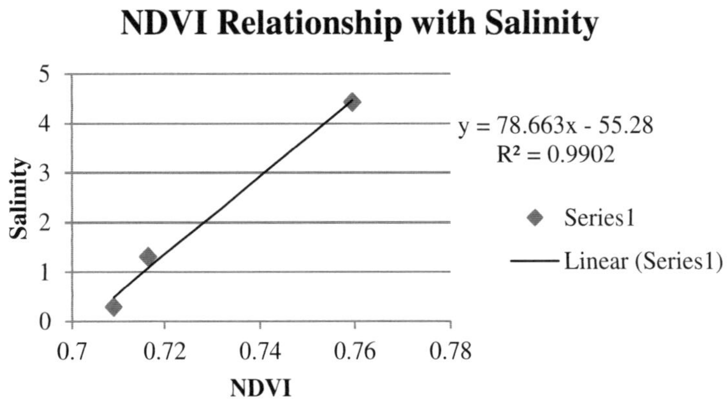


**Figure 11.** NDVI Relationship with Salinity

The results of the correlation test between NDVI with Nitrate, Phosphorus, Potassium, pH, High Salinity with value (R) of 0.99, 0.99, 0.99, 0.99.0.99 this shows that there is a strong relationship between NDVI value and NPK nutrient, pH , salinity. Correlation coefficient marked positive mean relation of NDVI value with one way nutrient parameter, if NDVI value is higher then NPK, pH, salinity is also higher because mangrove ecosystem requires NPK nutrient for photosynthesis process. Plants containing N, P, K, pH, enough salinity leaves will be greener. This greener leaf causes a higher reflectant value in the image.

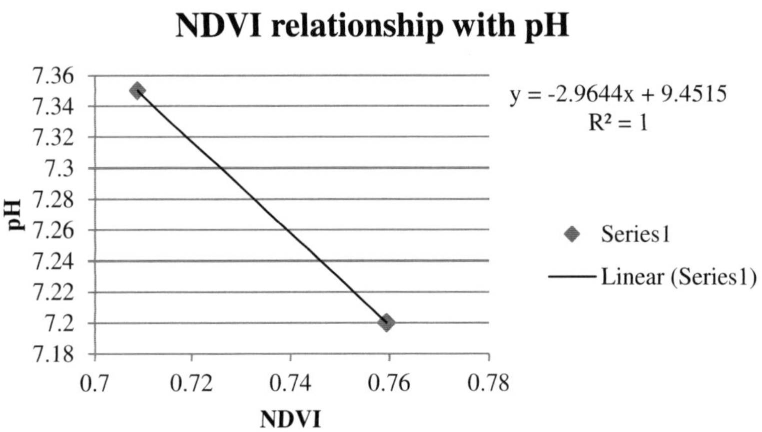


**Figure 12.** NDVI relationship with pH

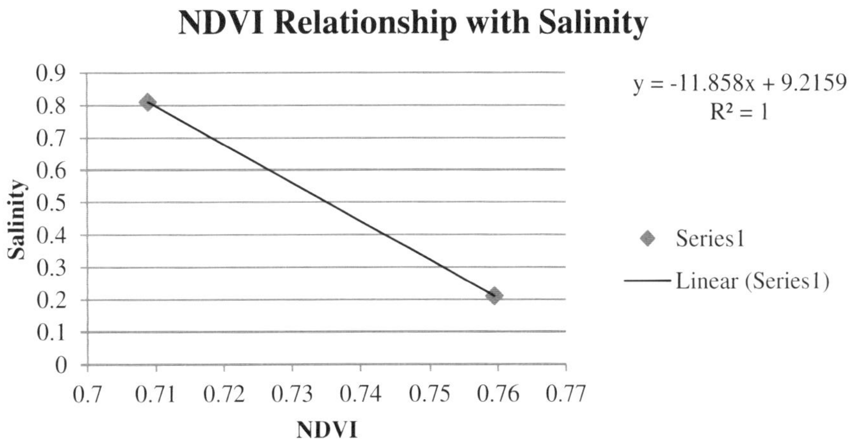


**Figure 13.** NDVI Relationship with Salinity

The NDVI correlations with water parameters include correlations with very strong relationships with a value (R) of 0.99, 0.99, respectively. The correlation coefficient of NDVI with pH parameter and positive salinity means the higher the NDVI value, the higher the pH value and the water salinity. So the parameters are closely related to the mangrove ecosystem is pH and Salinity.

## 4. Discussion

*4.1 Result of Rectification*

Based on result of rectification, in Pleiades 1B image with order 1 polynomial method got RMSE value 0,322 pixel and with method of polynomial of order 2 got value RMSE 0,188 pixel. whereas the RMSE value per largest point is 0.524 on Pleiades 1B. In the 2nd order polynomial method the RMSE value per the smallest point of 0.136 on the image leiades 1B. While the RMSE value per largest point of 0.242 on the image of Pleiades 1B. Based on the results of processing, the RMSE image of Pleiades 1B better methods used for geometric correction of images of Pleiades 1B are 2nd order polynomial methods.

*4.2 RMSE based on The Geometrical Accuracy*

Based on the geometric accuracy test using ICP points, the RMSE value in Pleiades 1B image is 0.806 meters on the rectification image of 1st order polynomial method. While the rectification image of the 2nd order polynomial method is obtained RMSE in Pleiades image 1B 0,647 meter.

*4.3 Result of Rectification*

For the feasibility of high-resolution imagery for making village maps, Pleiades 1B satellite images meet the requirements of making a village map with a scale of 1: 2.500, 1: 5.000 and 1: 10.000.

*4.4 The Correlation Test*

The results of the correlation test between NDVI with Nitrate, Phosphorus, Potassium, pH, High Salinity with value (R) of 0.99, 0.99, 0.99, 0.99.0.99 this indicates that there is a strong relationship between NDVI value and NPK , pH, salinity. Correlation coefficient marked positive mean relation of NDVI value with one way nutrient parameter, if NDVI value is higher than NPK, pH, salinity is also higher because mangrove ecosystem requires NPK nutrient for photosynthesis process. Plants containing N, P, K, pH, enough salinity leaves will be greener. This greener leaf causes a higher reflectance value in the image.

### *4.5 The NDVI Correlation with Parameters*

The NDVI correlation with water parameters includes correlation with very strong relation with value (R) of 0.99, 0.99, respectively. The correlation coefficient of NDVI with pH parameter and positive salinity means the higher the NDVI value, the higher the pH value and the water salinity. So the parameters are closely related to the mangrove ecosystem is pH and Salinity.

## 5. Conclusion

Based on geometric (area) and radiometric (classification) image processing for Pleiades 1B image with scale 1: 5000 obtained mangrove species dominated by *Rhizophora sp.* With an area of 966,45 Ha. The value of economic potency of *Rhizopora sp* Rp. 773.160.000/year with assuming a 25-year cutting cycle.

## References

[1] Abidin H Z 2002 *Survei dengan GPS* (Jakarta: Pradnya Paramita).
[2] Abrams Michael, S Hook and B Ramachandran 2002 *ASTER user Handbook Version 2* Jet Propulsion Laboratory 4800.
[3] Arifin 2003 *Hutan Mangrove Fungsi dan Manfaatnya* (Yogyakarta: Kanisius).
[4] Badan Informasi Geospasial 2014 *Peraturan Kepala Badan Informasi Geospasial Nomor 15 Tahun 2014 Tentang Pedoman Teknis Ketelitian Peta Dasar* Cibinong-Bogor.
[5] Badan Pusat Statistik 2015 *Kecamatan Bulak Dalam Angka 2014* (Surabaya: Badan Pusat Statistik).
[6] Badan Perencanaan dan Pembangunan Kota *Rencana Detail Tata Ruang (RDTR) Kota Surabaya 2010* Pemerintah Kota Surabaya Surabaya.
[7] Campbell J B and R H Wynne 2011 *Introduction to Remote Sensing (5th Ed.)* (New York The Guilford Press)
[8] Danoedoro Projo 1996 *Pengolahan Citra Digital Teori dan aplikasinya dalam bidang Penginderaan Jauh* (Yogyakarta: Fakultas Geografi UGM).
[9] Departemen Kehutanan 1999 *Undang – undang kehutanan Pasal 1 Ayat (1) Nomor 41.*
[10] Departemen Kehutanan 2003 *Buku Panduan Rehabilitasi Lahan dan Perhutanan Sosial* Penerbit Badan Penelitian dan Pengembangan Kehutanan (Jakarta: Departemen Kehutanan)
[11] Ecoton 1998 *Panduan Pengenalan Ekosistem Mangrove Pantai Timur Surabaya* (Surabaya: Ekoton).
[12] Fandeli C and Mukhlison 2000 *Pengusahaan Ekowisata* (Yogyakarta: UGM).
[13] FAO 2007 *The World's Mangroves 1980–2005* Forest Resources Assessment Working Paper No. 153 Food and Agriculture Organization of The United Nations, Rome.
[14] Hakim Luckhman 2004 *Dasar Dasar Ekowisata* (Malang: Bayumedia Publishing).
[15] Indica Mangifera 2010 *Perubahan Luasan Mangrove dengan Teknik Penginderaan Jauh di Taman Nasional Sembilang Kabupaten Banyuasin Propinsi Sumatera Selatan* (Palembang: Teknik Kelautan FMIPA Universitas Sriwijaya).
[16] Kusmana C 1995 Manajemen Hutan Mangrove di Indonesia *Prosiding Simposium Penerapan Ekolabel di Hutan Produksi Jakarta*, 10-12 Agustus 1995.
[17] Kovacs J M *et al* 2009 Evaluating The Condition of a Mangrove Forest of The Mexican Pacific Based on an Estimated Leaf Area Index mapping Approach *Environ Monit Assess* **157** pp 137-149.
[18] Kusumaningrum T E 2013 *Analisa Kesehatan Vegetasi Mangrove Berdasarkan Nilai Normalized Differencce Vegetation Index (NDVI) Menggunakan Citra Alos* (Surabaya: Teknik Geomatika FTSP-ITS).
[19] Landgrebe D 2003 *Signal Theory Methods In Multispectral Remote Sensing* (New Jersey: John Willey & Sons Inc).
[20] LAPAN 2015 *Spesifikasi Citra Satelit Pleiades* http://pusfatekgan.lapan.go.id/wp-content/uploads/2015/02/Informasi-Satelit-PL.pdf. (diakses November 09, 2016).
[21] Liang S 2004 *Quantitative Remote Sensing of LandSurface* (New Jersey: John Willey & Sons

IOP Conf. Series: Earth and Environmental Science **98** (2017) 012035 doi:10.1088/1755-1315/98/1/012035

Inc).
[22] Lillesand T M, Kiefer R W and Chipman J W 2004 *Remote Sensing and Image Interpretation* Fifth Edition (New York: John Wiley & Sons).
[23] Moko G I 2011 *Evaluasi Perubahan Tutupan Lahan Wilayah Perairan Pesisir Surabaya Timur Sidoarjo Dengan Menggunakan Citra Satelit Multitemporal* (Surabaya: Teknik Geomatika FTSP-ITS).
[24] Purwadhi S H 2001 *Interpretasi Citra Digital* (Jakarta: Grasindo).
[25] Sugiyono 2008 *Metode Penelitian Kunatitatif Kualitatif dan R&D (*Bandung: Alfabeta).
[26] Sumich J L 1992 *An Introduction to The Biology of Marine Life* Fifth Edition (New York: Wm. C. Brown Publishers).
[27] USGS 2015 <URL: http://Landsat.usgs.gov/band_designations_ Landsat_satellites.php>. Diakses tanggal 1 November 2016, jam 10.22.

IOP Conf. Series: Earth and Environmental Science **98** (2017) 012036 doi:10.1088/1755-1315/98/1/012036

# Analysis of Water Availability for Domestic Needs in Denpasar City

**Wiyanti*, T B Kusmiyarti, N M Trigunasih and N Juwita**

Center for Spatial Data Infrastructure Development (PPIDS) Universitas Udayana, Bali, Indonesia

wiyanti1259@yahoo.com

**Abstract.** This research aims to know the power carrying capacity of water for domestic needs in Denpasar City, Indonesia. Primary data collection techniques and field research using interviews to the people who directly need water for domestic purposes. Determination of the respondents was conducted by the method of purposive random sampling. The number of respondents that were used in this study was 100 respondents. The percentage of the number of communities that the fulfilment of the needs of the water comes from the PDAM as much as 38%, the well bore as much as 44% and 18% of the community's daily water needs fulfilment comes from the well dig. Most of the drinking water needs of the community in the Denpasar city, use bottled water. Water potential of Denpasar city run off coefficient calculation is 52.35 million $m^3$/year. Water needs for domestic purposes of Denpasar city of 65,469,173 m3/year. A comparison between the availability of water by domestic needs is. 0.9, which means the Denpasar city was included in the area that has water deficit.

**Keywords:** Water availability, Domestic needs, Denpasar

## 1. Introduction

The high rate of population growth, especially in urban areas will have an impact on the increase of water needs. The main human living needs come from the availability of adequate water. Fulfilling the need for water is one of them is taken from available water. Water conditions on the other hand are increasingly alarming, due to factors such as overcrowding, overfishing, damaged water catchment areas due to land conversion and environmental destructive activities. Water use utilization activities need to pay attention to the carrying capacity.

Water that is used continuously and excessively to meet the needs of the community will cause a decrease in quantity, so it is necessary to do water management steps to be used in a long time. Data / information on the need and potential of water resources is needed for the analysis of the fulfilment of domestic needs in Denpasar City. The purpose of this research is to analyses the potential of water availability for domestic water needs in Denpasar City.

Denpasar city, is the capital city and the centre of economic and service activities in Bali Province. It has a unique culture and its economic development relies on tourism and services. Since late 1970s, its urban population and economic development have grown fast resulting in fast growing settlements and rapid need for infrastructure and facilities [1]. This condition makes the population needs for water increased due to increasing of population and infrastructure and facilities.

IOP Conf. Series: Earth and Environmental Science **98** (2017) 012036 doi:10.1088/1755-1315/98/1/012036

## 2. Methodology

### *2.1. Area of Study*

Area of study is located in Denpasar City, Bali Province, Indonesia. Geographically the research area is located between 08°36'56" – 08°42'23" south latitude and 115°10'23"–115°16'27" east longitude. The research location showed in Figure 1.

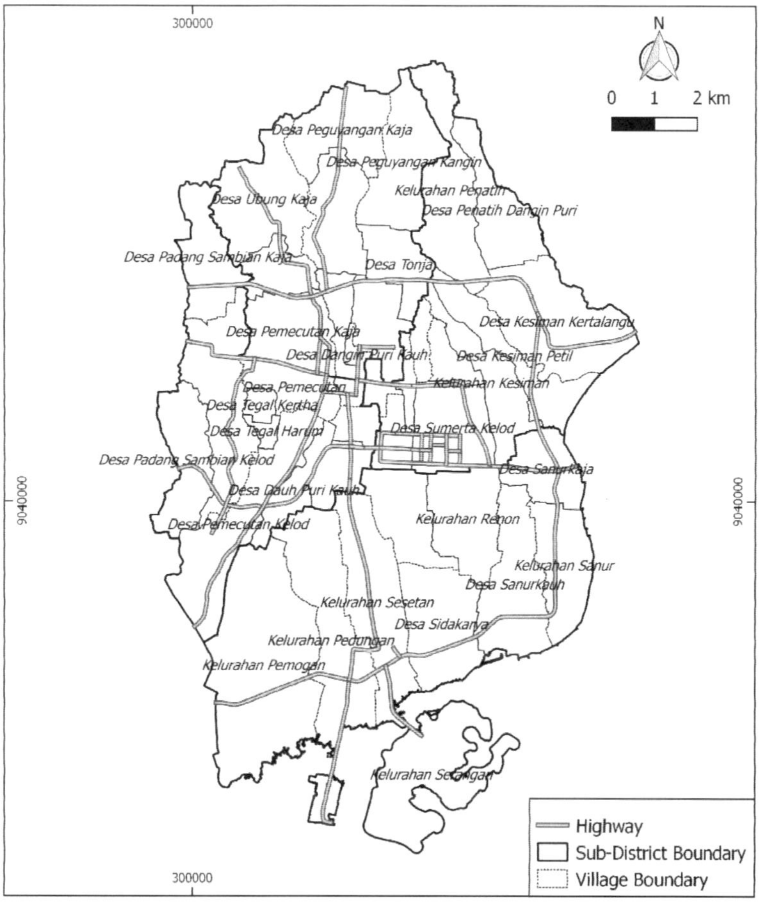


**Figure 1.** Research Location

### *2.2. Tools and Material*

Materials for this study includes

- Population data of Denpasar City of 2016 from Bali Provincial Statistics Bureau 2015,
- Water potential data from Public Works Agency (PU) ) and the Office of Energy and Mineral Resources (ESDM) and
- Data on the production and distribution of water from the regional drinking water company (PDAM) of Denpasar City.

Tools used in this research are:

- GPS (Global Positioning System) to determine coordinate point,
- Camera for field documentation, stationery to record the fieldata and
- A set of computer for the processing and presentation of data.

### *2.3. Data collection*

Data collection method used in this research is primary data and secondary data method.

- Secondary data is obtained through literature review, reports and previous research results. The secondary data used in this study were obtained from Department of Mineral Resources (ESDM) of Bali Province, and the data of water production and water distribution of the drinking water company (PDAM) of Denpasar City.
- Primary data collection was obtained through field observation and interview. Field observation was held to re-check the calculation of water availability and to recalculate and re-check the domestic water needs for population in Denpasar City. Determination of respondents to be interviewed by purposive sampling method. Sampling by purposive method by taking samples based on the will of researchers based on certain criteria (area, population density, informal settlements).

*2.4. Data Analysis*

Data analysis includes the estimation of water availability and the water needs for domestic use.

*2.4.1. Estimation of Water Availability.* Estimation of water availability were determined using the runoff coefficient method based on landuse information and annual rainfall data as noted in Equation 1.

$$SA = 10 \times C \times R \times A \quad (1)$$
$$C = \Sigma (Ci \times Ai) / \Sigma Ai$$
$$R = \Sigma (Ri \times Ai) / \Sigma Ai$$

Annotation:
C : weighted runoff coefficient
Ci : landuse i runoff coefficient
Ai : landuse i area (ha)
R : average annual rainfall (mm/yr)
SA : water availability ($m^3$/yr)
A : area (ha)
10 : conversion factor from ha to $m^3$.

*2.4.2. Weighted Runoff Coefficient.* The value of C is the average value of the runoff coefficient based on the type of landuse. The runoff coefficient relates to the type of landuse and the ability of landuses to infiltrate the surface water. The type of landuse and the runoff coefficient value are presented in Table 1 [2].

**Table 1.** C Value

| No | Landuse | Ci |
|---|---|---|
| 1 | Building | 0.9 |
| 2 | Industry | 0.,9 |
| 3 | Bareland | 0.4 |
| 4 | Housing | 0.7 |
| 5 | Port | 0.7 |
| 6 | Trade and services | 0.9 |
| 7 | Ricefields | 0.15 |
| 8 | Garden | 0.25 |
| 9 | Fishpond | 0,05 |

*2.4.3. Average Annual Rainfall.* The average annual rainfall is calculated using Thiessen's Polygon method. The Thiessen method is based on the assumption that measured amounts at any station can be applied halfway to the next station in any direction, which means that for any point rainfall is equal to the observed rainfall at the closest gauge. The weights of the rain gauges are computed by their relative areas, which are estimated with the Thiessen polygon network. The polygons are formed by the perpendicular bisectors of the lines joining nearby stations. The area of each polygon is used to weight the rainfall amount of the station in the center of the polygon. Figure 2 shows the Thiessen's Polygon [3].

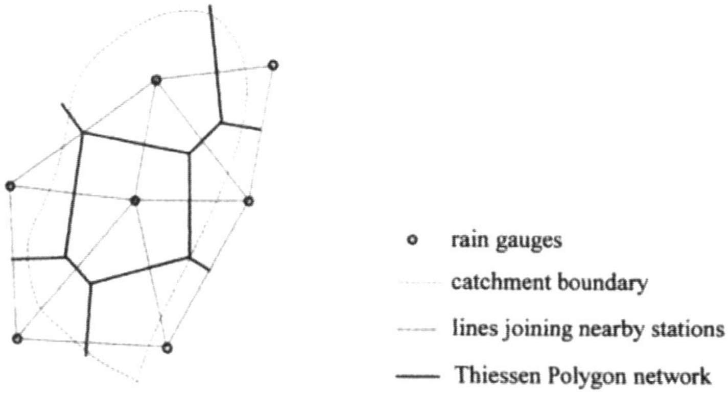


**Figure 2.** Thiessen Polygon Method.

*2.4.4. Domestic Water Needs.* The water needs are differing for each area. The higher the number of population, the higher the level of water demand. Water requirements standard based on number of population presented in Table 2 [4].

**Table 2.** Standard for Domestic Water Use

| Category | Number of Population | Water Use (L/people/day) |
|---|---|---|
| Metropolitan City | > 1.000.000 | 150 |
| Big City | 500.000 – 1.000.000 | 135 |
| Medium City | 100.000 – 500.000 | 120 |
| Small City | 20.000 – 100.000 | 105 |
| Village/ Rustic | < 20.000 | 82,5 |

## 3. Result and Discussion

### *3.1. Domestic Water Requirements*

The population of Denpasar City in 2016 is 896,838 people with population growth rate of 1.27 [4]. Based on the population of Denpasar City, then Denpasar City has not been included in the category of Metropolitan City, but included in the category of Large City, where the population is less than 1 million people. If the water requirement of each individual in Denpasar City is about 200 liters per person, then the total water requirement of Denpasar City is 179,367,600 $m^3$ / day, while for water requirement per year is 65,469,173,760 $m^3$.

### *3.2. Water Potential*

Water potential of Denpasar can be known from the calculation of water availability with the runoff coefficient method and water basin. The calculation of runoff coefficient method is based on runoff coefficient, annual rainfall, and the total landuse area. Total landuse area is 18,895.80 ha consist of building, industry, bare land, housing, port, trade and services, rice fields, garden and fishpond.

IOP Conf. Series: Earth and Environmental Science **98** (2017) 012036 doi:10.1088/1755-1315/98/1/012036

The rainfall intensity is calculated using Thiessen's polygon with five rainfall station i.e Sanglah, Peguyangan Kaja, Suwung Kangin, Pedungan and Sumerta. The areal rainfall in Denpasar City is 1989.51 mm/ yr. Total water potential from runoff coefficient method is 7,512,264,192.90 $m^3$ / year. Table 2 shows the areal rainfall using Thiessen's Polygon Method and Table 3 shows the calculation of runoff coefficient method.

**Table 2**. Areal Rainfall using Thiessen's Polygon Method

| **Stasiun** | **x** | **y** | **R** | **A (Ha)** | **A*R** |
|---|---|---|---|---|---|
| Sanglah | 115.21 | -8.68 | 1817.18 | 2910.76 | 5289374.86 |
| Peguyangan Kaja | 115.61 | -9.51 | 1952.32 | 1875.25 | 3661088.08 |
| Suwung Kangin | 115.37 | -9.64 | 1982.26 | 1894.43 | 3755252.81 |
| Pedungan | 116.08 | -9.05 | 1928.96 | 1745.87 | 3367713.40 |
| Sumerta | 116.08 | -9.55 | 1879.22 | 3939.10 | 7402435.50 |
| Total | | | | 12365.41 | 23475864.65 |
| **Areal Rainfal** | | | | | **1989.51 mm** |

**Table 3.** Runoff Coefficient Calculation

| **No** | **Landuse** | **Area (ha)** | **Ci** | **Area * Ci** |
|---|---|---|---|---|
| 1 | Building | 3,031.26 | 0.9 | 2728.134 |
| 2 | Industry | 6.44 | 0.,9 | 5.796 |
| 3 | Bareland | 83,237.57 | 0.4 | 33295.028 |
| 4 | Housing | 31,271.40 | 0.7 | 21889.98 |
| 5 | Port | 52.60 | 0.7 | 36.82 |
| 6 | Trade and services | 1,117.61 | 0.9 | 1005.849 |
| 7 | Ricefields | 40,120.04 | 0.15 | 6018.006 |
| 8 | Garden | 18.97 | 0.25 | 4.7425 |
| 9 | Fishpond | 39.91 | 0,05 | 1.9955 |
| | **Total** | **158,895.80** | | **64,986.351** |
| | **C** | **2.44** | | |
| | **R** | **1898.51 mm** | | |
| | **SA** | **7,512,264,192.90 m³/yr** | | |

*3.3. Water Availability for Domestic Requirements*

The carrying capacity of a region's water resources shows the comparison between the need and availability of water, or it can be defined as the maximum capacity of the region in the provision of water for a population of a certain number with its various activities. The water carrying capacity is obtained from the comparison between the availability of water (SA) and the water requirement (DA). If SA> DA then the carrying capacity is stated surplus, if SA < DA, then the water carrying capacity is expressed deficit (exceeded). Denpasar water potential based on calculation of runoff coefficient is respectively 7,512,264,192.90 $m^3$ / year.

The water requirement for domestic needs of Denpasar City is 65,469,173,760 $m^3$ / year. Comparison between the availability of water to the domestic needs and the people of Denpasar City based on the calculation of the water carrying capacity of 0.11, which means that the city of Denpasar is included in areas experiencing water deficit.

IOP Conf. Series: Earth and Environmental Science **98** (2017) 012036 doi:10.1088/1755-1315/98/1/012036

## 4. Conclussions

The water potential of Denpasar City is based on the calculation of runoff coefficient and water basin respectively for 13.75 million $m^3$ / year and 52.35 million $m^3$ / year. The number of domestic water needs of Denpasar City is 65.469.173 $m^3$ / year, while for non-domestic purposes of 1.482.410,87 $m^3$ / year. Water carrying capacity index is 0.11, which means Denpasar is included in areas with water deficit.

## Acknowledgments

Writers would like to thank the Rector of Universitas Udayana, Dean of Agriculture Faculty of Universitas Udayana, Head of Center for Spatial Data Infrastructure Development (PPIDS) Universitas Udayana, fellow staf, researchers and professors at Universitas Udayana for all the support, comments and suggestions that makes this paper finished.

## References

[1] Winarso H, I Dharmapatni and M Chen 2016 Community Empowerement for Succesfull Lad Consolidation, Case Study: Banda Aceh and Denpasar, Indonesia. Paper prepared for presentation at the 2016 World Bank Conference on Land and Poverty. The World Bank - Washington DC, March 14-18, 2016
[2] Asdak C 2010 *Hidrologi dan Pengelolaan Daerah Aliran Sungai* Cetakan Kelima (Yogyakarta: Gadjah Mada University Press).
[3] Schumann A H 1998 Thiessen Polygon. *Encyclopedia of Earth Science* pp 648-649.
[4] Triatmodjo B 2008 *Hidrologi Terapan* (Yogyakarta: Beta Offset).

# Benthic Habitat Mapping by Combining Lyzenga's Optical Model and Relative Water Depth Model in Lintea Island, Southeast Sulawesi

**M Hafizt[1*], M D M Manessa[2], N S Adi[3], B Prayudha[1]**

[1]Research Center for Oceanography LIPI, Indonesia
[2]Geodesy Department, Pakuan University, Indonesia
[3]Research Centre for Coastal and Maritime Resources, The Ministry of Maritime Affairs and Fisheries, Indonesia

muhammadhafizt@gmail.com

**Abstract**. Benthic habitat mapping using satellite data is one challenging task for practitioners and academician as benthic objects are covered by light-attenuating water column obscuring object discrimination. One common method to reduce this water-column effect is by using depth-invariant index (DII) image. However, the application of the correction in shallow coastal areas is challenging as a dark object such as seagrass could have a very low pixel value, preventing its reliable identification and classification. This limitation can be solved by specifically applying a classification process to areas with different water depth levels. The water depth level can be extracted from satellite imagery using Relative Water Depth Index (RWDI). This study proposed a new approach to improve the mapping accuracy, particularly for benthic dark objects by combining the DII of Lyzenga's water column correction method and the RWDI of Stumpt's method. This research was conducted in Lintea Island which has a high variation of benthic cover using Sentinel-2A imagery. To assess the effectiveness of the proposed new approach for benthic habitat mapping two different classification procedures are implemented. The first procedure is the commonly applied method in benthic habitat mapping where DII image is used as input data to all coastal area for image classification process regardless of depth variation. The second procedure is the proposed new approach where its initial step begins with the separation of the study area into shallow and deep waters using the RWDI image. Shallow area was then classified using the sunglint-corrected image as input data and the deep area was classified using DII image as input data. The final classification maps of those two areas were merged as a single benthic habitat map. A confusion matrix was then applied to evaluate the mapping accuracy of the final map. The result shows that the new proposed mapping approach can be used to map all benthic objects in all depth ranges and shows a better accuracy compared to that of classification map produced using only with DII.

**Keywords**: Lyzenga, benthic habitat, mapping, Sentinel-2A, water depth

## 1. Introduction

Lyzenga, through his published work in 1978, was one of the pioneers introducing benthic habitat mapping technique using multispectral remote sensing data [1]. Since the targeted object in benthic habitat mapping is located underwater Lyzenga [1] proposed a model to reduce the effect of light attenuation in the water column called as the depth-invariant index (henceforth named "DII") which can be used to improve the accuracy of benthic classification. Using the model deeper objects of benthic habitat becomes more discernible and the accuracy of benthic habitat map is improved [2].

In 1985, the Lyzenga's technique [1] was tested on a commercial multispectral image of Landsat MSS [3]. Since then a number of publications [2–18] have tested the technique on various types of shallow water environment using various imageries with overall satisfactory accuracies. However, several publications found that the Lyzenga's model still has a few weaknesses. Manessa et al. [9] reveal that the Lyzenga method is sensitive to a complex shallow water environment and is not capable to completely remove the light attenuation effect, especially in the bands with longer wavelengths. Also, the DII method becomes less effective with increasing depth, or in other word, the Lyzenga's model is not working in deep areas [10]. Moreover, Hafizt & Danoedoro [11] and Hafizt et al. [12] found that when the Lyzenga's model is applied to shallow or over exposed area, pixel values on the DII image sometimes show negative values due to poor sample selection which causes the pixels become unclassified. It would be a problem if the unclassified class is fully covered by seagrass which is exposed at low tide condition, meaning that the information of seagrass cover could be lost.

This study aims to optimize information in benthic habitat mapping using satellite imagery by proposing a new approach combining Lyzenga's water column correction and relative water depth index model. Based on the review of the previous studies, the approach presented here attempts to apply different techniques to different water type in term of its depth variation for classifying benthic habitat and then combine the final benthic habitat maps. The approach is to optimize the use of the DII technique for deeper water and the use of sun-glint corrected image for shallower water.

This study uses Sentinel 2A image which has blue band image that is an important requirement in benthic habitat mapping. Moreover, the Sentinel 2A image is classified as a high-resolution image that suitable to map benthic habitat cover in the study area which has a high variation of benthic habitat cover. This study was conducted in Lintea Island, part of Wakatobi Islands, Southeast Sulawesi. In addition to propose a new approach to produce a more accurate benthic habitat map, the result of this study is also expected to increase the insights of academics and practitioners about coral mapping using Sentinel image.

## 2. Method

### *2.1. Image data and Study Area*

Sentinel 2A level 1C satellite imagery of study area was acquired on August 17, 2016. Level 1C means the satellite imagery has been corrected at Top-of-Atmosphere reflectance in cartography geometry [4]. The study was utilized imagery of 10-meter spatial resolution that consist of three visible bands (B2: blue band, B3: green band, B4: red band) as well as one near infrared band (B8: NIR band) which are needed for image corrections. The study area, Lintea Island, is located in the north part of Wakatobi Islands, Southeast Sulawesi. Figure 1 shows the satellite imagery of study area in natural colour composite.

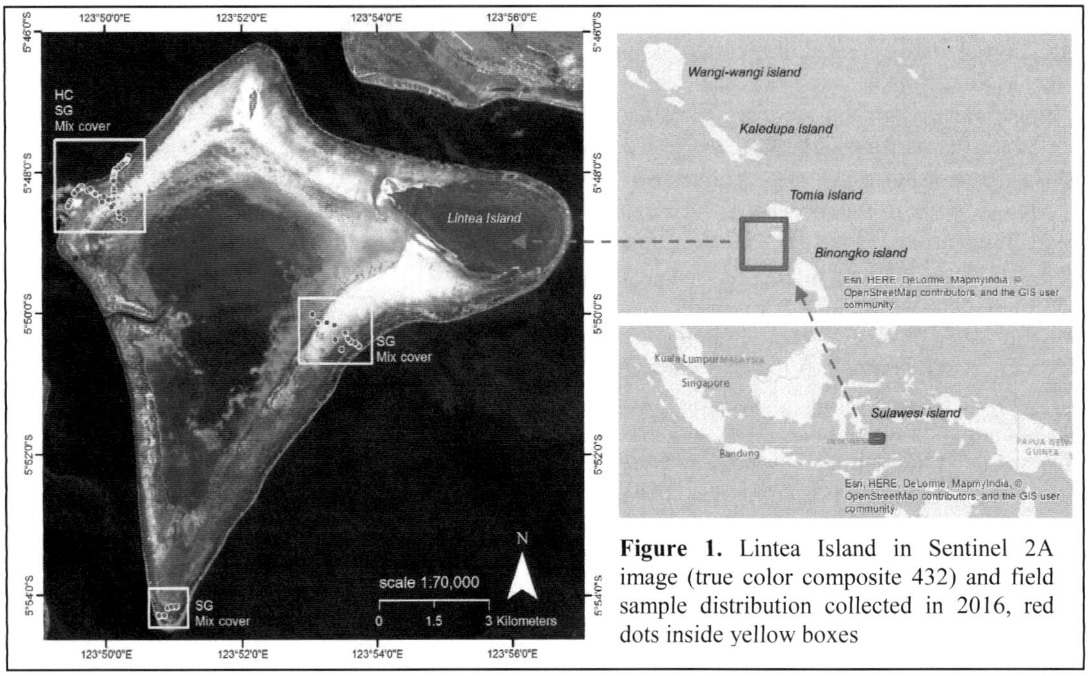


**Figure 1.** Lintea Island in Sentinel 2A image (true color composite 432) and field sample distribution collected in 2016, red dots inside yellow boxes

*2.2. Image processing*

Image processing consist of image correction, modeling and classification, the image correction itself consists of sunglint and water column corrections. Sunglint correction is useful for reducing glint effect on water surface due to sunlight reflection. We used an equation developed by Hedley at.al. [13,14] which uses deep water pixels as dark object to create offset value and use it on his equation for sunglint removal. Selecting dark object for creating offset values is similar concept to atmospheric correction using Dark Object Correction [15] method, therefore atmospheric correction is not applied to avoid double offset values creating negative values in shallower area. The sunglint equation applied to Sentinel 2A bands can be seen in Table 1.

**Table 1.** The equations for correcting sunglint effect, created from 131510 samples Region of Interest (ROI) in visible bands and Near-Infrared band with determination coefficient ($R^2$) relatively high.

| **Visible Bands** | **Determination coefficient ($R^2$)** | **Equation of sunglint correction** |
|---|---|---|
| Blue band (B2) | 0.8688 | B2-(1.0241*(B8-115)) |
| Green band (B3) | 0.8773 | B3-(1.0851*(B8-115)) |
| Red band (B4) | 0.8630 | B4-(0.9186*(B8-115)) |

The result of sunglint correction then becomes input data for water column correction step. The aim of this correction is to reduce water column attenuation effects causing misclassification in the image classification process and would decrease the accuracy level of benthic habitat map. The water column correction is performed using Depth Invariant Index (DII) model [16], modified by Lyzenga et.al. [17]. The method used pixel values of sand samples on different depth to calculate several parameters required to create DII equation, namely variance of each visible bands, covariance between visible bands, attenuation coefficient (a) and attenuation ratio (ki/kj). The parameters of DII can be seen in Table 2.

**Table 2.** Parameters of the DII model, created from 739 samples ROI where the variance values of blue, green and red band are 0.027999, 0.072238, and 0.081889 respectively.

| Band combination | covariance | attenuation coefficient (a) | attenuation ratio ($K_i/k_j$) | DII equations |
|---|---|---|---|---|
| Blue and green | 0.044200 | -0.500445 | 0.617788 | (alog(B2))-(0.6178*(alog(B3))) |
| Blue and red | 0.044109 | -0.610872 | 0.560949 | (alog(B2))-(0.5610*(alog(B4))) |
| Green and red | 0.067810 | -0.071161 | 0.931368 | (alog(B3))-(0.9314*(alog(B4))) |

Image modelling is performed to generate depth information from the image which is used to develop benthic habitat map. We use Relative Water Depth Index (RWDI) model developed by Stumpf et.al [18]. The model is available in image processing software and applied to Sentinel 2A image. The model output is an image showing depth information in indexed values for each pixel. The depth image is then classified as shallow and deep areas based on index value and is used as input data for classification process.

The corrected images (sunglint and DII images) were used as input data in the classification process using ISO-DATA method. The method is categorized as unsupervised classification approach meaning that spectral classes are generated automatically by computer. Our role is only determining minimum and maximum classes and number of iteration. We applied two models for the classification step for the first model (Model 1) we classified all pixels of the entire reef area using DII image as input data. For the second model (Model 2) we first classified the reef areas into shallow and deep reef areas. The shallow area is then classified using sunglint image and deep area is classified using the DII image. The accuracy level of our proposed model is assessed using confusion matrix table for its overall, producer and user accuracies [19]. The steps of image processing method which are used on each model in this research can be seen in Table 3.

**Table 3.** The steps of image processing method which are used on each model in this research

| Treatment descriptions | Model 1 | Model 2 (shallow area) | Model 2 (deep area) | Relative Water Depth Index Model (RWDI) |
|---|---|---|---|---|
| Using Sentinel 2A, level 1C | * | * | * | * |
| Atmospheric correction (software automatically) | | | | * |
| Sunglint correction | * | * | * | |
| Water column correction | * | * | * | |
| Density slice image | | | | * |
| ISO-DATA classification | * | * | * | |
| Accuracy assessment | * | * | * | |

### *2.3. Sampling method*

Sea truth for validating the resulting benthic habitat map is conducted using photo transect method developed by Roelfsema et.al. [20]. The samples are photos of bottom cover and its georeferenced information on each photo. All objects on the photos were then classified into benthic habitat classes according to previous research [21–24]. We used the field sample data to calculate the accuracy level of benthic habitat map and to validate the image classification result. Information on the accuracy level, calculated using confusion matrix table [22,25,26], is needed to measure the improvement of our proposed model presented here (Table 4).

**Table 4.** Confusion matrix table for calculating overall accuracy of benthic habitat map, user and producer accuracy

| | Cells | A | B | C | D | E | F | G |
|---|---|---|---|---|---|---|---|---|
| | | | | Ground truth sample | | | | |
| | 1 | Object classes | Class 1 | Class 2 | Class 3 | total | *user accuracy (%)* | *error commission* |
| Classification result | 2 | Class 1 | | | | =B2+C2+D2 | =(B2/E2)*100 | =100-F2 |
| | 3 | Class 2 | | | | =B3+C3+D3 | =(C3/E3)*100 | =100-F3 |
| | 4 | Class 3 | | | | =B4+C4+D4 | =(D4/E4)*100 | =100-F4 |
| | 5 | Total | =B2+B3+B4 | =C2+C3+C4 | =D2+D3+D4 | =E2+E3+E4 | | |
| | 6 | *producer accuracy (%)* | =(B2/B5)×100 | =(C3/C5)*100 | =(D4/D5)*100 | *Overall accuracy (%)* | = ((B2+C3+D4)/E5)*100 | |
| | 7 | *error omission* | =100-B6 | =100-C6 | =100-D6 | | | |

## 3. Result and Discussion

### *3.1. Samples for sea truth*

The samples used in this research are 56 samples in total, consisting of field samples collected in 2016 and additional samples collected through the image by interpretation. We added more samples because the reef area is too shallow and it is difficult to reach the inner area. We added samples which are easier to be recognized from image data i.e. sand (SD) and seagrass (SG Hi and SG Sp). In general, all samples can be classified into four classes in the field, abbreviated as HC, SG, SD, and Mix Cover. Using georeferenced photos obtained from the photo transect all classed can be detailed into sub-classes (Table 5). The dominant class is hard coral (15 samples), whereas the other classes are almost equal in number, about 7 samples in average. The samples collected are sufficient to calculate the accuracy level of benthic habitat map.

**Table 5.** Detailed classes of field samples (sea truth classes)

| General classes | Detail Classes | Information | Number of classes |
|---|---|---|---|
| HC | HC | Hard coral/healthy coral | 15 |
| SG | SG HI | seagrass high density | 8 |
| | SG SP | Seagrass sparse density, only mix with carbonate sand | 7 |
| SD | SD | Sand (carbonate sand) | 6 |
| MIX Cover | MIX SG BS | Mix cover seagrass and bare substratum (rubble and rock) | 7 |
| | MIX HC AL | Mix cover hard coral and algae (green algae or brown algae) | 6 |
| | MIX HC BS | Mix cover hard coral and bare substratum (sand, rubble, rock) | 7 |
| **Total** | | | **56** |

### *3.2. Image Processing*

The Image processing step is begun with image correction consisting of sunglint and water column correction (DII). The results of both corrections are shown in Figure 2, where the applied corrections transform the visualization of the image from its original image to sunglint image and finally to DII image. The sunglint correction reduces the glint effects on surface water, sharpens the contrast features on reef area, and eliminates land area. Moreover, after water column correction is applied deep area is clearly exposed but it fails in the shallow area. The unexpected result is found in shallow water area as where negative pixels values are observed after the water column correction is applied. This is due to higher pixel values for longer wavelength bands causing negative values in the DII image (Table 2). Since, Lyzenga [1] assumed that reflected radiance/reflectance in water area is exponentially linear with depth and the attenuation coefficient, then in shallow depth the attenuation effect is relatively small compared with deep area and also the attenuation coefficient is larger in longer wavelength [9].

IOP Conf. Series: Earth and Environmental Science **98** (2017) 012037 doi:10.1088/1755-1315/98/1/012037

This fact became the reason why in shallow area the longer wavelength bands (i.e. green and red band) will have higher value compared with the deep area. To solve this issue, shallow and deep area are separated using the RWDI model which generates depth information in index values and can be used to classify benthic object according to its depth zones.

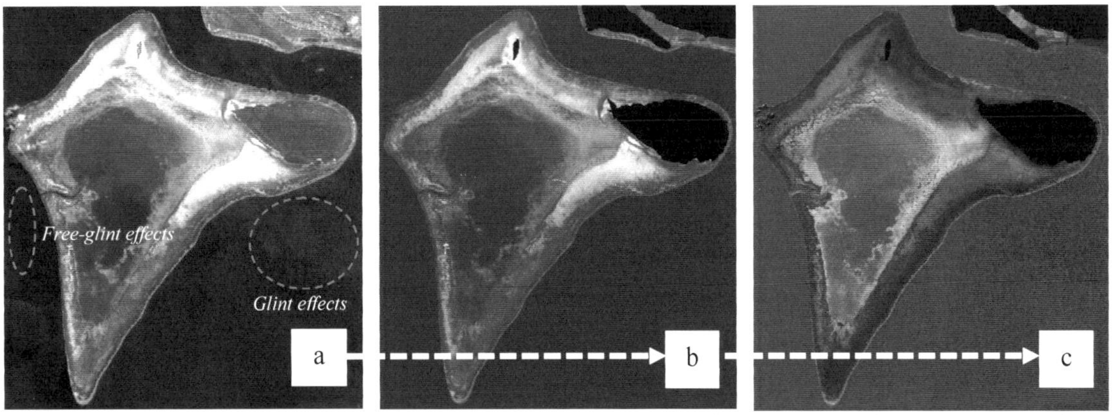


**Figure 2**. Transformation of image visualization, begin with original image (a) then become sunglint corrected image (b) and continue become DII image (c) as a final result in correction process

The RWDI image result can be seen in Figure 3a where the output image is a single band displayed in grayscale representing the depth of waters in index values. Slicing index values of RWDI image is implemented to obtain shallow and deep area (Figure 3b) where small and high values of the index show shallow and deep areas, respectively. Shallow area is in the range of 0.0725 to 0.15 (cyan color) and deep area is in the range of 0.01 to 0.0725 (blue color). Moreover, seagrass is classified as a personal class because the area can be delineated using index value 0 to 0.01 (green color). The classified areas convert into masker imageries which used in classification process afterwards.

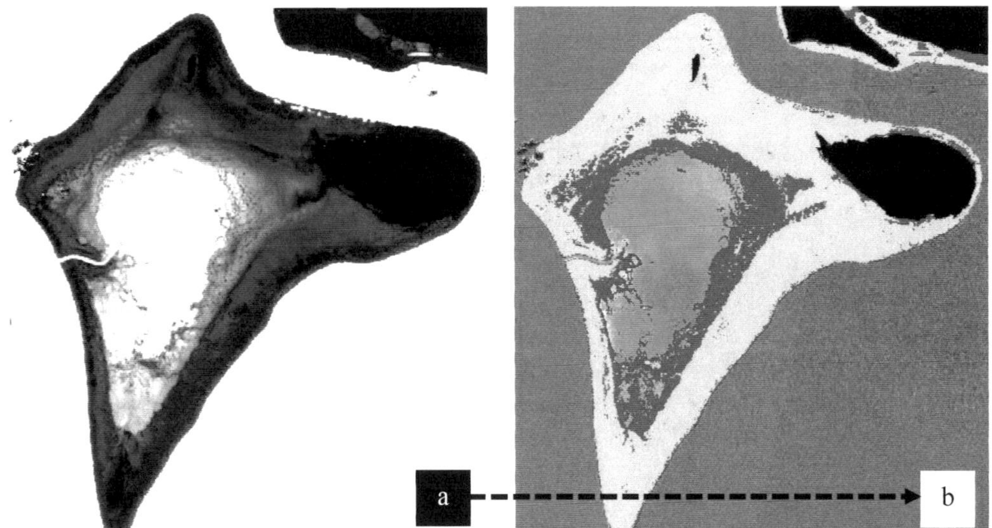


**Figure 3**. The greyscale band which is result of RWDI model (a) than convert into depth zones (b) using density slice classification, shallow area: cyan, deep area: blue, and seagrass area: green.

### *3.3. Image classification*

The important part of the proposed new mapping approach is in the classification process, where the two classifications models are used, as described in section 2.2. Figure 4 shows the classification result of model 1 (without depth data) and compared with the model 2 (the proposed new approach) whereas the number class of the classification results are 7 and 17 classes respectively. More detailed classes in Model 2 is caused we used two images (sunglint and DII) in two different area (Shallow and Deep). The shallow area using sunglint image produced 10 classes and deep area which was used DII image produced 7 classes then combined the all class into total 17 classes of benthic habitat map.

The classes in Model 1 and Model 2 are spectral classes (Figure 4) which were validated according to benthic habitat classes which has created from sea truth samples (Table 5). The seven classes in Model 1 almost named according to the classes in Table 5, however not detailed to Mix cover classes (only Mix SG BS). It is obviously different to Model 2 which very detailed to carbonate sand cover (SD), hard coral cover (HC) and Mix Cover which can represented Table 4 classes. The classes in Model 2 then generalized into seven classes of sea truth (Table 5) in order to be assessed the accuracy level.

Although the benthic habitat classes created from sea truth samples was only seven classes, the variation of benthic habitat cover on the field was more than that (17 classes). The un-matching class between the sea truth classes and model 2 classes happened because the limitation in identifying benthic cover from photo samples or thoroughly on the field. So that, we must take more sample on the field to validate the undefined classes of Model 2 map to make more detailed benthic habitat map.

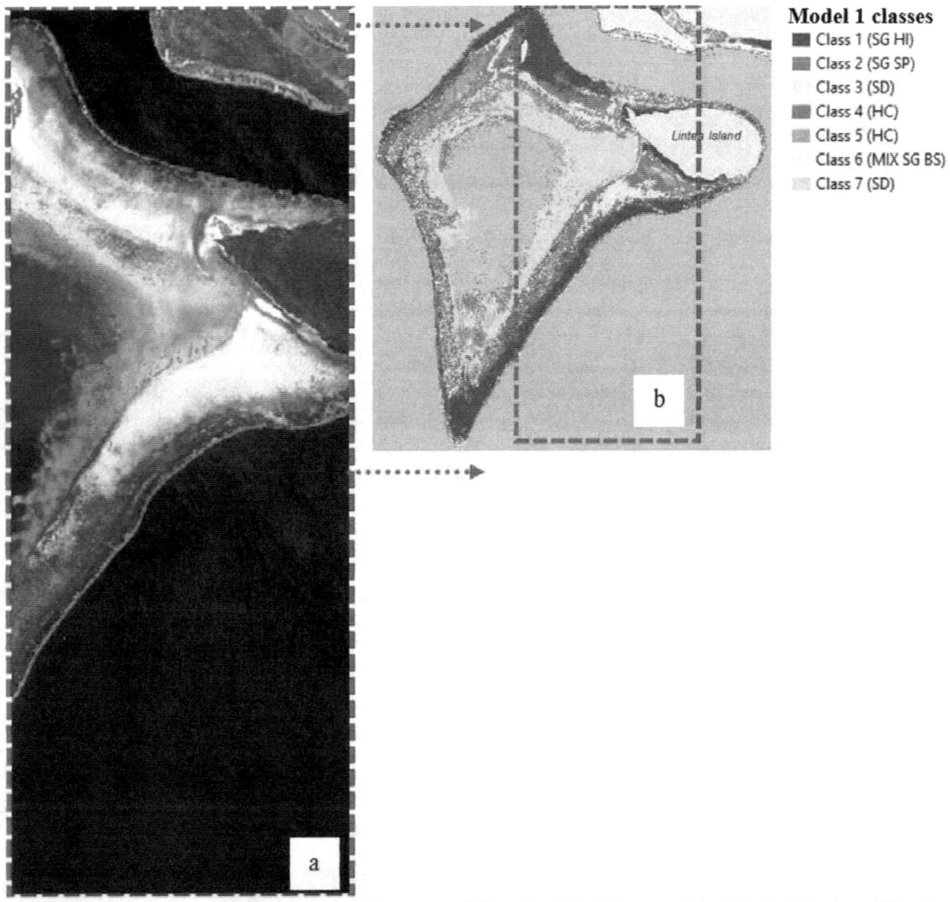


**Figure 4**. Classification results, original image of Sentinel 2A image (a), Model 1 classification result (b), Model 2 classification result (c) where S and D initial means Shallow and Deep Zone.

### *3.4. Result analysis*

Based on benthic habitat sample distribution of both models, model 2 shows a better accuracy, because the shallow water area shows dominance with sand and mixed sand with other objects (mixed cover), this fact fit the image visual appearance. Moreover, coral cover class in model 2 is concentrated in slope area, a transition zone from shallow to deep area which is naturally covered by coral reef. On the other hand, model 1 shows that most of shallow area is covered by wide area of high dense seagrass and coral reef. Interestingly, based on image visualization the sand shows as the dominant object, while the image classification result shows the opposite fact.

**Figure 5.** Overall accuracy three different input image on classification process

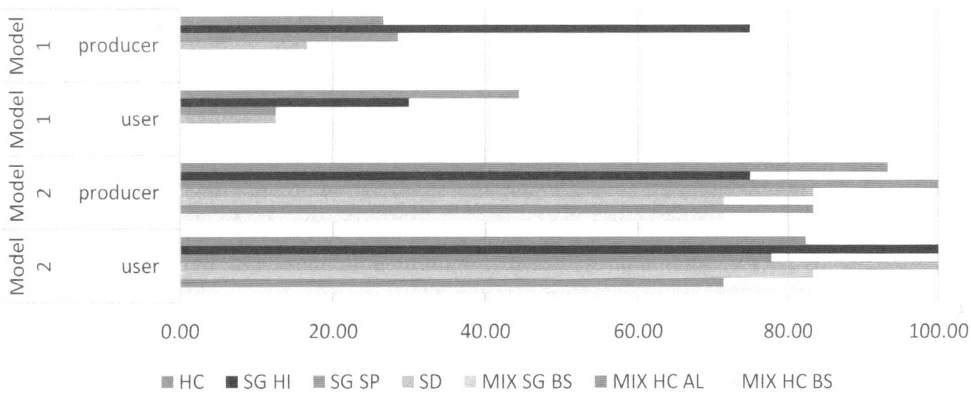


**Figure 6.** Accuracy assessments for two evaluated models. model 1: DII image as input, and model 2: Combination of DII and RDWII image as input

The confusion matrix shows the accuracy comparison of both methods in more detail and is summarized as Figure 5 and 6. The accuracies of the two models are compared and evaluated. The result shows that the proposed new approach (model 2) shows superiority (83.93 % overall accuracy) over model 1 (23.21 % overall accuracy). The cause of low accuracy in model 2 is due to misclassification of classified image, mix classes samples on shallow area classified as seagrass cover, so that the mix classes (Mix SG BS, Mix HC AL, and Mix HC BS) is zero producer and user accuracy. Meanwhile, the high accuracy in model 2 is due to all classes obtained from ISODATA classification is able to accommodate the classification scheme created from sea truth sample (Table 5). Figure 6 shows that all benthic classes of Model 2 have high producer and user accuracy (more than 60%), that qualifies the limit of benthic habitat mapping standard [27].

## 4. Conclusions

This paper introduces a new approach of benthic habitat mapping by combining the result of water column correction (DII image) and water relative index extraction (RDWII image) using Sentinel 2 image of Lintea Island. Since shallow water area is under exposure after the water column correction, this study proposed (model 2) to separately classify the target mapping area in each depth zone as shallow and deep water. By adding the depth information (model 2), the accuracy of the final benthic habitat map was found to improve, with overall accuracy of 83% or almost three time higher compared with model 1 model (common model), respectively. The new approach also has another improvement: benthic objects can be classified into more detailed classes. The improvement is due to smaller area coverage for classification process enabling the identification and clustering of high pixel variation in smaller area (shallow and deep water), as shown in this study. The proposed new method is applicable in practice, as the depth information is extracted from the same image, meaning that no extra data such as in situ depth data are needed.

**Acknowledgment**
Field data in this study provided from Reef Health Monitoring activity in the year of 2015 - 2017. The activity conducted by Research Center for Oceanography, Indonesian Institute of Sciences as a part of COREMAP-CTI projects funded by World Bank. The authors also would like to thank Ir. Suyarso for contributing the data.

**References**
[1] Lyzenga D R 1978 Passive remote sensing techniques for mapping water depth and bottom features. *Appl. Opt.* **17** 379–83
[2] Wicaksono P 2010 *Integrated Model of Water Column Correction Technique for Improving Satellite-Based Benthic Habitat Mapping. A Case Study on Part of Karimunjawa Islands, Indonesia* (Gadjah Mada Univresity)
[3] Mumby P J, Green E P, Edwards A J and Clark C D Coral reef habitat mapping: how much detail can remote sensing provide?
[4] John R, Vinluan N, Don J and De Alban T Evaluation of LANDSAT 7 ETM+ Data for Coastal Habitat Assessment in Support of Fisheries Management
[5] Andréfouët S, Kramer P, Torres-Pulliza D, Joyce K E, Hochberg E J, Garza-Pérez R, Mumby P J, Riegl B, Yamano H, White W H, Zubia M, Brock J C, Phinn S R, Naseer A, Hatcher B G and Muller-Karger F E 2003 Multi-Site Evaluation of IKONOS Data for Classification of Tropical Coral Reef Environments *Remote Sens. Environ.* **88** 128–43
[6] Philipson P and Lindell T 2003 Can Coral Reefs be Monitored from Space? *AMBIO A J. Hum. Environ.* **32** 586–93
[7] Vanderstraete T, Goossens R and Ghabour T K 2004 Coral Reef Habitat Mapping in The Red Sea (Hurghada, Egypt) Based on Remote Sensing *EARSeL eProceedings 3*
[8] Marlina N 2004 *The Application of QuickBird and Multi-temporal Landsat TM Data for Coral Reef Habitat Mapping. Case Study: Derawan Island, East Kalimantan, Indonesia* (The Netherlands: ITC)
[9] Manessa M D M, Haidar M, Budhiman S, Winarso G, Kanno A, Sagawa T and Sekine M 2016 Evaluating The Performance of Lyzenga's Water Column Correction in Case-1 Coral Reef Water Using a Simulated Wolrdview-2 *IOP Conf. Ser. Earth Environ. Sci.* **47**
[10] Jensen J 2007 *Remote Sensing of The Environment: an Earth Resource Perspective* (United States of America: Pearson Prentice Hall)
[11] Hafizt M, Danoedoro P and Prayudha B 2016 Study of Shallow Water Ecosystem Based on Spatial Using Worldview-2 Imagery (Case Study: Kemujan Island, Karimunjawa Islands) *Seminar Nasional Penginderaan Jauh* (Depok, Indonesia)
[12] Hafizt M, Iswari M Y and Prayudha B 2017 Assessment of Landsat-8 Classification Method for Benthic Habitat Mapping in Padaido Islands, Papua *Oseanologi dan Limnol. di Indones.* **2** 1–13
[13] Kay S, Hedley J D and Lavender S 2009 Sun Glint Correction of High and Low Spatial Resolution Images of Aquatic Scenes: a Review of Methods for Visible and Near-Infrared Wavelengths *Remote Sens.* **1** 697–730
[14] Anggoro A, Siregar V P and Agus S B 2016 The Effect of Sunglint on Benthic Habitats Mapping in Pari Island Using Worldview-2 Imagery *Procedia Environmental Sciences* vol

33pp 487–95
[15] Wicaksono P and Hafizt M 2013 The Effectiveness of Various Dark Targets for Dark Subtract Atmospheric Correction Method: Case Study on Vegetation Index for Mangrove Carbon Stock Mapping *Asian Conf. Remote Sens.*
[16] Green E, Mumby P, Edwards A and Clark C 2000 *Remote Sensing Handbook for Tropical Coastal Management* ed A J Edwards (Paris: The United Nations Educational, Scientific and Cultural Organization)
[17] Lyzenga D R 1981 Remote Sensing of Bottom Reflectance And Water Attenuation Parameters in Shallow Water Using Aircraft and Landsat Data
[18] Stumpf R P, Holderied K, Spring S and Sinclair M 2003 Determination of Water Depth with High-Resolution Satellite Imagery Over Variable Bottom Types *Limnol. Oceanogr.* **48** 547–56
[19] Danoedoro P 2012 *Pengantar Penginderaan Jauh Digital* (Yogyakarta: Andi Offset)
[20] Roelfsema C and Phinn S 2009 *A Manual for Conducting Georeferenced Photo Transects Surveys to Assess the Benthos of Coral Reef and Seagrass Habitats* (Brisbane: Centre for Remote Sensing & Spatial Information Science School of Geography, Planning & Environmental Management University of Queensland)
[21] Goodman J, Purkis S and Phinn S 2013 *Coral Reef Remote Sensing: A Guide for Mapping, Monitoring and Management* (London: Springer)
[22] Phinn S, Roelfsema C and Mumby P 2012 Multi-Scale, Object-Based Image Analysis for Mapping Geomorphic and Ecological Zones on Coral Reefs *Int. J. Remote Sens.* **33** 3768–97
[23] Leon J, Phinn S, Woodroffe C, Hamylton S, Roelfsema C and Saunders M 2012 Data Fusion for Mapping Coral Reef Geomorphic Zones: Possibilities and Limitations *GEOBIA* (Rio de Janeiro) p 261
[24] Hedley J, Roelfsema C, Koetz B and Phinn S 2012 Capability of The Sentinel 2 Mission for Tropical Coral Reef Mapping and Coral Bleaching Detection *Remote Sens. Environ.* **120** 145–55
[25] USGS 2015 *LANDSAT 8 (L8) DATA USERS HANDBOOK* ed J 2015 (Department of the Interior U.S. Geological Survey)
[26] Jensen J R 2005 *Introductory Digital Image Processing: A Remote Sensing Perspective* (Prentice Hall)
[27] Peraturan Kepala Badan Informasi Geospasial Nomor 8 Tahun 2014 and BIG 2014 *Pedoman Teknis Pengumpulan dan Pengolahan Data Geospasial Habitat Dasar Perairan Laut Dangkal* ed B I Geospasial (Jakarta)

# Sea Surface Temperature Dynamics in Indonesia

**D W Kusuma[1*], A Murdimanto[2], L Y Aden[2], B Sukresno[1], D Jatisworo[1] and R Hanintyo[1]**

[1]Researcher, Institute for Marine Research and Observation, Jembrana, Bali, Indonesia
[2]Map Analyst, Institute for Marine Research and Observation, Jembrana, Bali, Indonesia

dennywijaya_k@yahoo.com

**Abstract.** Indonesia as a country that has large marine area reserves the large potential of marine resources. One of parameters to identify marine resources is sea surface temperature (SST). Dynamics of SST in Indonesia is influenced by global phenomena, such as ENSO and IOD. This paper aims to understand the spatio-temporal distribution and statistical attribute of SST in Indonesia between 2007 – 2016, respond of SST in Indonesia against ENSO and IOD, and comparison between two infrared sensors for SST measurement. Level 3 Aqua/Terra MODIS monthly dataset are used to obtain spatio-temporal distribution and statistical attribute of SST. Dipole Mode Index and TAO buoy are used to assess the respond of SST in Indonesia against ENSO and IOD. AHI Himawari-8/9 monthly-composite dataset is used to compare SST measured from Aqua/Terra MODIS. From 2007 to 2016, in Indonesia SST is ranging between 27.91 to 30.46 °C. The coldest SST occurred in August (2007, 2012, 2015) while the warmest SST occurred in April (2010 and 2016). Strong correlation has been identified between the SST and IOD (especially in western part of Indonesia) and between the SST and ENSO (especially in WPPNRI 573 and 718). There is no significant difference of SST measurement result between Aqua/Terra MODIS and Himawari-8/9.

**Keywords:** Sea Surface Temperature (SST), SST dynamic, Aqua/Terra MODIS, Himawari 8/9

## 1. Introduction

The oceans have large heat-storage play an important role in climate system [1]. Composed of more than 3,000 islands, lying between two ocean basins (Indian Ocean and Pacific Ocean) and two continents (Asia and Australia), also with a cumulative costal perimeter that is more than twice Earth's Circumference, Indonesia plays the important role in Earth climate, it also acts as an oceanic connection between warm pools of Pacific and Indian Oceans. The SST variations within the Indonesian region are a consequence of complex physiography and connectivity between the two major oceans [2]. The main factors influence the SST variability in Indonesia are ENSO (El Niño Southern Oscillation), Monsoon, IOD (Indian Ocean Dipole) [3][4][5]. One of leading mode in tropical climate variability is ENSO, characterized by warm SST for positive El Niño in eastern tropical Pacific Ocean and cold SST for negative La Niña [6]. Another mode that influences Indonesia is IOD, which is coupled ocean-atmosphere mode with positive phase characterized by warm SST over Eastern Indian Ocean, and negative phase by cold SST over Eastern Indian Ocean [4].

The Sea Surface Temperature (SST) is a result when ocean and the atmosphere interact via turbulent and radiative energy exchange in the sea surface, then energy fluxes in turn depend on a single oceanic quantity [1]. SST is one of important variables in climate observation database [7][8]. In fisheries, SST also plays an important factor to fish for feeding, spawning, and migrating. For example, spawning according [9] the common to all Tunas species is the relationship between spawning activity and SST more than about 24 °C, because of their mode of reproduction, repetitive broadcast spawners, and tunas must have very high lifetime fecundities to be successful. For small pelagic fish, SST also plays important role during spawning, and at the development and survival of the eggs and larvae, as well as influencing distribution, aggregation, migration and schooling behaviour of juveniles and adults [10][11][12]. SST also has been used to investigate productive frontal zones related to feeding ground of fishes [13]. In Sri Lanka high season or high frequencies of Yellowfin Tuna catch were obtained in the areas where SST varied primarily in 28 – 30 °C [14].

SST measurement from satellite is performed by utilizing the infrared and microwave channel. Infrared measurement can present highly accurate SST measurement (~0.5 K) when the atmosphere is clear while microwave measurement is less affected by clouds [15]. Several sensor instruments have been launched and operated to measure SST. Some examples for infrared sensor are **AVHRR** (Advanced Very High-Resolution Radiometer) sensor carried by NOAA satellite uses its infrared channel (3.7, 11, and 12 μm) [16]; **MODIS** (Moderate Resolution Imaging Spectroradiometer) sensor on Aqua/Terra satellite uses its mid (3-4 μm) and far (11-12 μm) infrared channel [17]; and **AHI** (Advanced Himawari Imager) carried by Himawari-8/9 satellite uses its infrared channel on 3.9, 8.6, 10.4, 11.2, and 12.4 μm [18]. Several examples of passive microwave channel used to derive SST are **SMMR** (Scanning Multi-Channel Microwave Radiometer) carried on Nimbus-7 and Seasat satellite uses its microwave channel ranging from 6.6-37 GHz [19] and **AMSR** (Advanced Microwave Scanning Radiometer) with its 6.9-89 GHz frequency of microwave channel carried on ADEOS-II satellite [20].

Indonesian Fisheries Management Area (Wilayah Pengelolaan Perikanan Negara Republik Indonesia, abbreviated as WPPNRI) is fisheries management area for fish catchment, fish cultivation, conservation, research, and fisheries development that cover internal sea, island sea, marine territorial, additional zone, and exclusive economy zone of Indonesia [21]. It consists of 11 areas, shown as yellow area in Figure 1 and its coverage area is shown in Table 1.

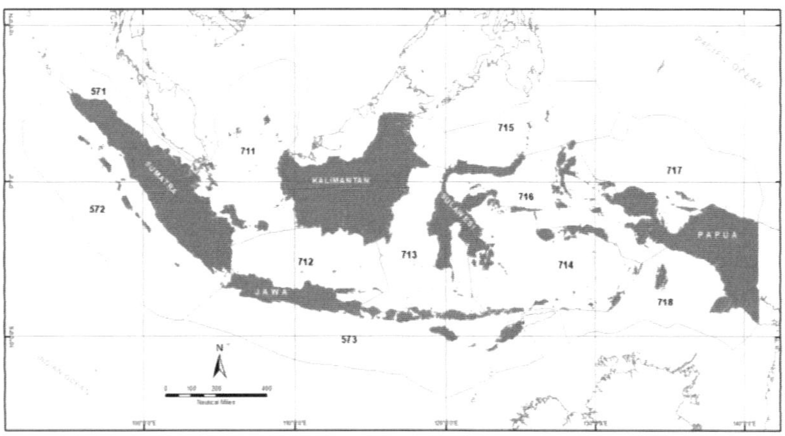

**Figure 1.** WPPNRI boundaries.

**Table 1.** WPPNRI division.

| WPPNRI | Coverage Area |
|---|---|
| WPPNRI 571 | Malaka Strait and Andaman Sea |
| WPPNRI 572 | Indian Ocean of western of Sumatra and Sunda Strait |
| WPPNRI 573 | Indian Ocean of southern of Java to Nusa Tenggara, Savu Sea, and western of Timor Sea |
| WPPNRI 711 | Karimata Strait, Natuna Sea, and South China Sea |
| WPPNRI 712 | Java Sea |
| WPPNRI 713 | Makassar Strait, Gulf of Bone, Flores Sea, and Bali Sea |
| WPPNRI 714 | Gulf of Tolo and Banda Sea |
| WPPNRI 715 | Gulf of Tomini, Maluku Sea, Halmahera Sea, Seram Sea, and Gulf of Berau |
| WPPNRI 716 | Sulawesi Sea and northern of Halmahera Island |
| WPPNRI 717 | Gulf of Cendrawasih and Pacific Ocean |
| WPPNRI 718 | Aru Sea, Arafuru Sea, and eastern of Timor Sea |

Although there is statement that SST in Indonesia generally range at 28 – 31 °C [22], this range may be different as an impact of climate change. This paper aims to 1) assess the spatio-temporal distribution and statistical attribute of SST in WPP-NRI; 2) assess respond of SST in WPPNRI to ENSO and IOD; and 3) compare performance of Aqua/Terra MODIS and Himawari-8/9 in SST detection.

## 2. Data and Methods

MODIS is a sensor attached in Aqua/Terra satellite. Both satellite have sun-synchronous orbit that view the entire Earth's surface every 1-2 days, acquiring data in 36 spectral bands, where Aqua passes from south to north across the equator in the afternoon (1.30 pm) and Terra passes from north to south across the equator in the morning (10.30 am) [23]. To determine the SST, MODIS uses certain bands (shown in Table 2) that are chosen based on particular aspect of the atmospheric total column transmissivity in each part of the mid- and far- infrared spectrum [17].

**Table 2.** Bands for MODIS Infrared SST Determination.

| Band Number | Band Center (μm) | Bandwidth (μm) |
|---|---|---|
| 20 | 3.750 | 0.1800 |
| 22 | 3.959 | 0.0594 |
| 23 | 4.050 | 0.0608 |
| 31 | 11.030 | 0.5000 |
| 32 | 12.020 | 0.5000 |

Based on mid- and far- infrared spectrum, two ocean skin SST products are produced. The first product (SST) is produced from band 31 and 32 while the second product (SST4) is produced from band 22 and 23. Algorithms to obtain those products are respectively shown in Equation (1) and (2). By using these algorithms, the standard deviation of MODIS Aqua/Terra is ranged between 0.42 – 0.51 K by comparing the satellite SST dataset to sub-surface buoy [24].

$$SST = c_1 + c_2 * T_{31} + c_3 * (T_{31} - T_{32}) * T_{sfc} + c_4 * (\sec(\theta) - 1) * (T_{31} - T_{32}) \quad (1)$$

$$SST4 = c_1 + c_2 * T_{22} + c_3 * (T_{22} - T_{23}) + c_4 * (\sec(\theta) - 1) \quad (2)$$

where

| | | |
|---|---|---|
| *SST* | : | retrieved surface temperature |
| $T_{22}$, $T_{23}$, $T_{31}$, and $T_{32}$ | : | brightness temperature from band 22, 23, 31, and 32 respectively |
| $T_{sfc}$ | : | estimate of the surface temperature, based on NOAA/NCEP weekly or daily blended Reynolds 0.25° spatial resolution product |

| | | |
|---|---|---|
| $\theta$ | : | satellite zenith angle |
| $c_1$, $c_2$, $c_3$, and $c_4$ | : | algorithm coefficients for band 31 and 32, derived using collocated in situ buoy SST and satellite observations |

Himawari-8/9 satellite is located at 140.7° E and observes from 80° E to 160° W between 60° N and 60° S. Advance Himawari Imager (AHI) sensor that has 16 spectral bands from visible to infrared wavelengths is attached to the satellite. The infrared bands located at 3.9, 8.8, 10.4, 11.2, and 12.4 μm are used for SST retrieval. Japan Aerospace Exploration Agency (JAXA) produces Himawari-8/9 skin SST from 2 spectral band combinations. The first spectral band combination is 10.4, 11.2, and 8.6 μm and the second spectral band combination is 10.4, 11.2, and 3.9 μm. The result of SST Himawari-8/9 data is obtained from 2 steps. Step one is calculating algorithm SST by using Equation (3) below:

$$T^{S}_{\lambda_{i,x}} = B^{-1}_{\lambda_i}(\bar{I}^{S}_{\lambda_i} + \Delta I^{S}_{\lambda_i}) \quad (3)$$

where:

| | | |
|---|---|---|
| $T^{S}_{\lambda i,.x}$ | : | SST for the $\lambda_i$ band |
| $x$ | : | Vector of orthogonal components |
| $B$ | : | inverse of the Planck function |

Step two is cloud screening by using threshold-based cloud screening with Boolean values and Bayesian method with probabilities. Himawari-8/9 SST has been validated by comparing its data with the nearest drifting and moored buoy data point that was located and observed within 3 km and 3 hours respectively. SST data with more than 0.3 cloud probabilities are rejected, as well as buoy data when the difference exceeded 4 K from Himawari-8/9 SST. The comparison showed that Himawari-8/9 SST data has root mean square difference ~0.59 K against the buoy data [18].

To obtain the environmental data for oceanographic prediction and coastal dynamics monitoring, Institute for Marine Research and Observations have deployed buoys in several marine areas in Indonesia. It also can be used to validate data measured from satellite such as sea surface temperature [25]. SST data in December 2016 from buoys located in Gulf of Tomini and Savu Sea have been extracted as a comparison against SST measured from Aqua/Terra MODIS.

Dipole Mode Index (DMI) is anomalous SST gradient between the western equatorial Indian Ocean and the south eastern equatorial Indian Ocean to represent the intensity of IOD. Positive DMI (positive IOD) results lower temperature in eastern Indian Ocean while negative DMI (negative IOD) results the opposite condition [4]. Comparison between DMI and SST in Indonesia within period 2007 – 2016 is performed to obtain the respond of SST against the IOD phenomenon.

The TAO/TRITON array is developed and supported by National Oceanic and Atmospheric Administration, USA and Japan Agency for Marine-Earth Science and Technology. It was designed as instrument to have better understanding on climate variations prediction related to ENSO. Located in 140° - 180° E and 95° - 180° W, the instrument can provide various information, i.e. wind and SST. Information from TAO buoy is observed and compared to SST in Indonesia to obtain the respond of SST against the ENSO phenomenon.

Aqua/Terra MODIS dataset from 2007 to 2016 are analysed to obtain spatio-temporal distribution and statistical attribute of SST in WWPNRI. Level 3 monthly dataset in 4 km spatial resolution are used to minimize the influence of cloud-cover that often occurred over the Indonesian region. Hourly dataset of Himawari-8/9 close to acquisition hour of Aqua MODIS (1.30 pm) and Terra MODIS (10.30 am) are selected to be composited and compared to monthly dataset of SST Aqua/Terra MODIS to obtain statistical comparison that consists of mean difference test and standard deviation difference test.

## 3. Result and Discussions

### *3.1. Spatio-temporal distribution and statistical attributes of SST*

The analysis for study area consists of SST monthly average for each WPPNRI and SST monthly average for whole WPPNRI during 2007 - 2016. SST monthly average for each WPPNRI are then classified into three classes of SST: cold (26 – 27 °C), moderate (28 – 29 °C), and warm (30 – 31 °C).

Monthly average SST during 2007 – 2016 in WPPNRI ranging from 26.04 °C (occurs in August in WPPNRI 718) to 30.66 °C (occurs in May in WPPNRI 571 and 711). Figure 2 shows that in January – February all regions in Indonesia are in moderate SST except in Natuna Sea (WPPNRI 711) that has cold SST because of the impact of ocean phenomenon named “cold tongue”. This phenomenon exists because of the influence of winter monsoon that occurred along the Vietnam coast transports through South China Sea [26]. It occurs from November to March and disappears in April [27]. However, previous study showed that the coldest SST of cold tongue phenomenon is around 298 K or 24 °C [28].

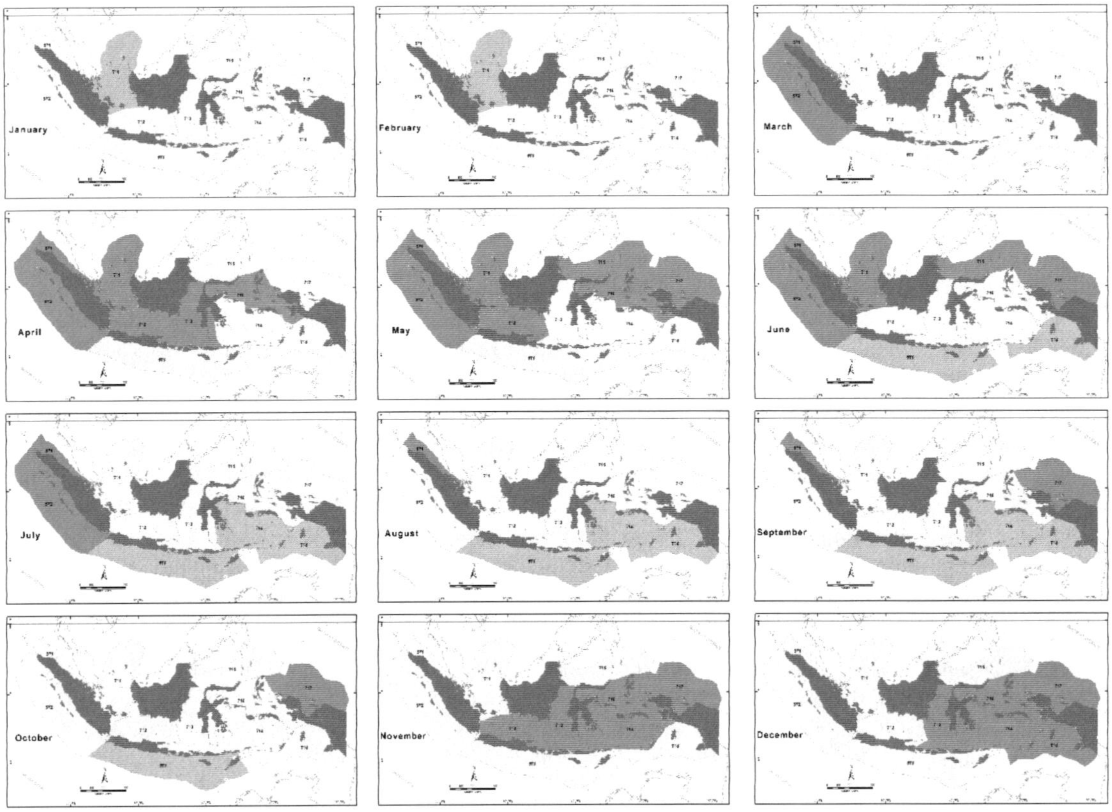


**Figure 2.** Monthly average of SST data (2007 – 2016). Green for cold SST, yellow for moderate SST, and red for warm SST.

In March the warm SST occurs in WPPNRI 571 and 572 from Indian Ocean, then continues in April where the warm SST expands to WPPNRI 711, 712, 713, and 716 while the moderate SST appears in WPNRI 573, 714, 715, 717, and 718. This situation of warm and moderate SST continues in May, but some area become warmer (WPPNRI 715 and 717) and cooler (WPPNRI 713).

The upwelling, indicated by cold SST, occurs in WPPNRI 573 and WPNNRI 718 in June, expands to WPPNRI 714 in July – September, then shrinks to only in WPPNRI 573 in October, then disappears in November. This phenomenon happened during the southeast (SE) monsoon, when southeast wind flows from Australia to generate upwelling along the Java – Sumatera coasts [29][30]. The warm and moderate SST occur again in all WPPNRI in November to December with most WPPNRI are in warm SST which indicates rainy season in Indonesia region.

For whole WPPNRI, monthly average of SST ranging from 27.91 °C to 30.46 °C. This range is similar with previous study stated that SST in Indonesia range from 28 – 31 °C [22]. During 2007 – 2016, the coldest SST often occurred in August (2007, 2012, 2015) at around 27.9 °C while the warmest SST occurred in April (2010 and 2016) at 30.46 °C. Figure 3 shows graph of SST monthly average for whole WPPNRI from 2007 to 2016 with blue line symbolized the SST and orange dot symbolized the trendline of SST during 2007 – 2016. In general, there is an average 0.8 °C increase from January 2007 to December 2016.

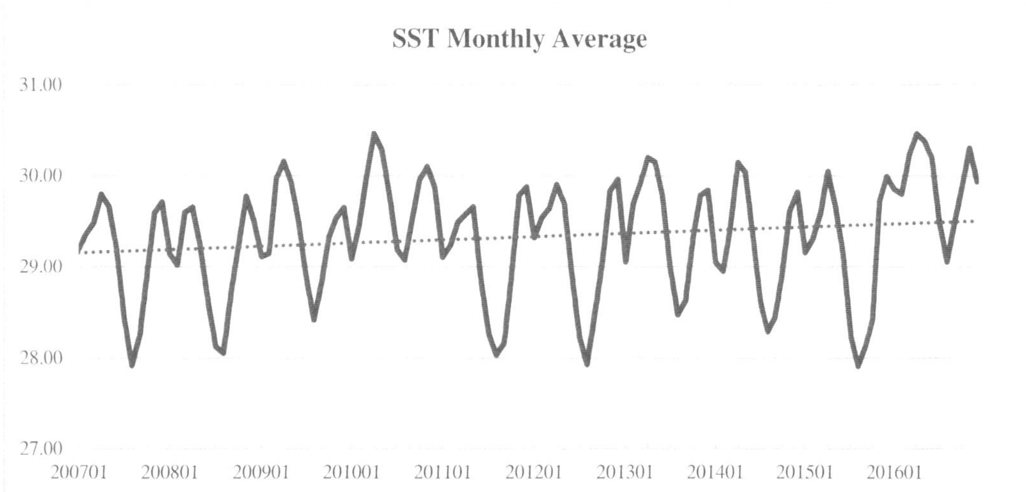


**Figure 3.** Graph of SST monthly average from 2007 to 2016.

Four buoys deployed by Institute for Marine Research and Observation located in Savu Sea and Gulf of Tomini that obtain various information such as SST [25] show similar result compared with SST measured from MODIS. Difference between both SST are -0.28 – 0.63 °C with negative sign shows MODIS SST is lower than buoy SST. Figure 4 shows location of buoys and Table 3 shows comparison between SST measured from buoys and MODIS.

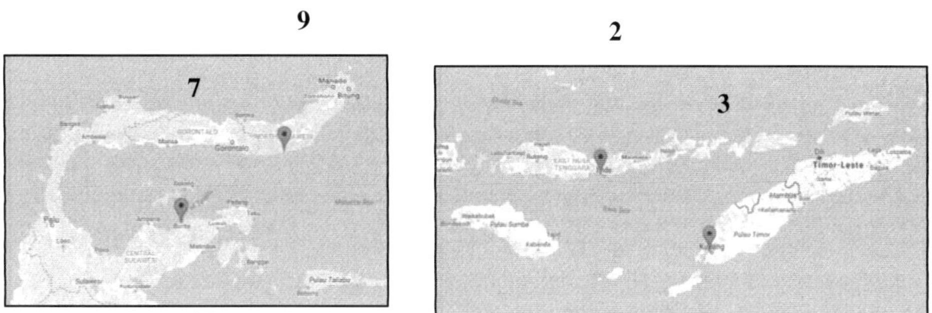


(a) (b)

**Figure 4.** Location of buoys in (a) Gulf of Tomini and (b) Savu Sea.

IOP Conf. Series: Earth and Environmental Science **98** (2017) 012038 doi:10.1088/1755-1315/98/1/012038

**Table 3.** Comparison of SST measured from buoy and MODIS in December 2016.

| Location | Unit No. | SST (in °C) | | |
|---|---|---|---|---|
| | | Buoy | MODIS | Difference |
| Savu Sea | 2 | 29.61 | 30.24 | 0.63 |
| (WPPNRI 573) | 3 | 30.52 | | -0.28 |
| Gulf of Tomini | 7 | 30.24 | 30.17 | -0.07 |
| (WPPNRI 716) | 9 | 29.89 | | 0.28 |

*3.2. SST respond to ENSO and IOD*

The geographic location of Indonesian Sea is influenced by two oceans system, IOD for Indian Ocean and ENSO for Pacific Ocean, both of system occur in Indonesia region. IOD impacts western part of Indonesia and ENSO impacts eastern part of Indonesia. The IOD influence in Indonesia sea is clearly shown during upwelling period in WPPNRI 573, when IOD+ occur (June – October) the SST in western part of Sumatera and southern part of Java become colder. The correlation between IOD+ and SST in Indonesia is shown in figure 5.

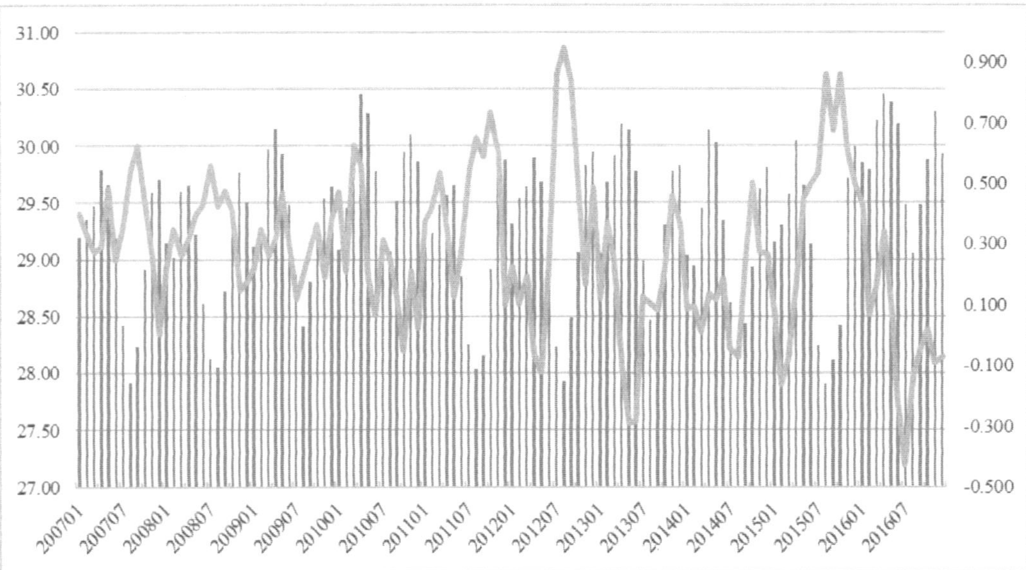


**Figure 5.** Correlation between Dipole Mode Index (orange line) and SST (blue bar).

In August 2012, low SST occurred as an impact of IOD+ during that period. This event can be identified by negative (cold) SST anomaly in the tropical eastern Indian Ocean and positive (warm) SST anomaly in the tropical Indian Ocean [4], as shown in figure 6. IOD+ occurrences (Table 4) also impacted SST in September 2007 (SST 28.23 °C) and July 2008 (28.12 °C). The IOD is measured by an index that is the difference between SST anomalies in western part of Indian Ocean and the eastern part called Dipole Mode Index (DMI), the index has two signs of negative and positive [4].

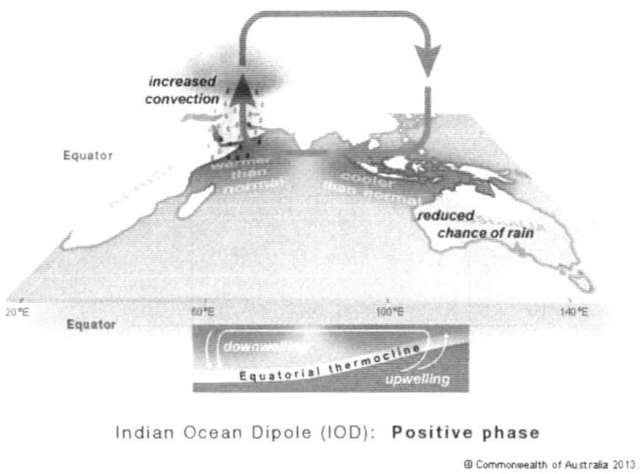


**Figure 6.** Illustration of IOD+ [31]

**Table 4.** The peak amplitude of IOD event in 1982 – 2012 [32]

| | Year | IOD Index | Peak Month |
|---|---|---|---|
| 1 | 1997 | 4.71447 | November |
| 2 | 1994 | 4.45823 | October |
| 3 | 2006 | 2.95482 | October |
| 4 | 2012 | 2.23291 | August |
| 5 | 2008 | 1.78743 | July |
| 6 | 1991 | 1.58422 | May |
| 7 | 1982 | 1.57511 | June |
| 8 | 2007 | 1.55126 | September |
| 9 | 2003 | 1.52663 | June |

The ENSO correlation with Indonesia Sea is shown by Tropical Atmosphere Ocean (TAO) buoy SST image data compare with Himawari-8/9 satellite data. Figure 7 shows that during La Niña in Pacific Ocean, the warm SST from Pacific direction is westward and enter the Indonesia water through WPPNRI 717 with moderate SST, then become warmer in April – December. The warm SST distribution also flows through the southern part of Indonesia via Makassar Strait and Lifamatola, this distribution known as Indonesia Through Flow (ITF).

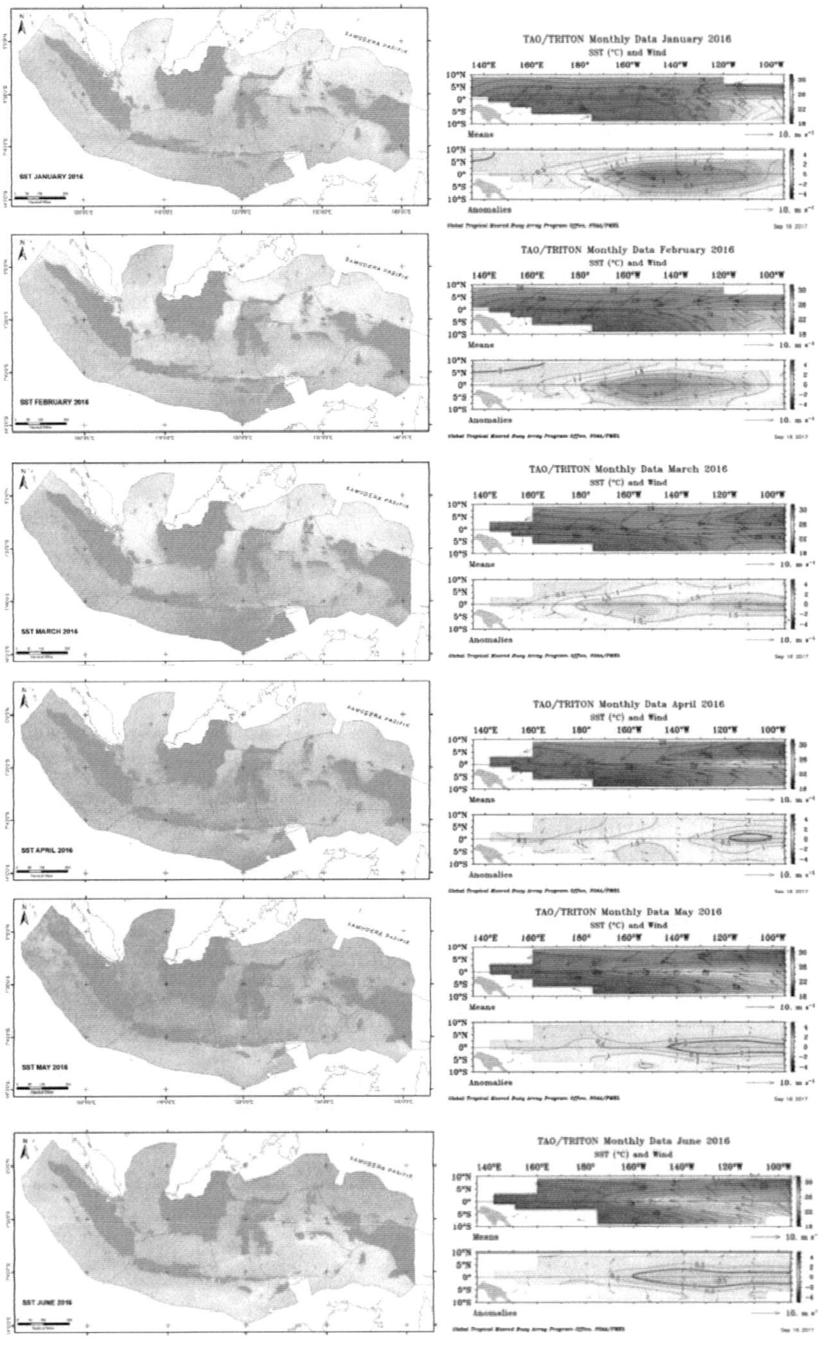
SAMUDERA PASIFIK
SST JANUARY 2016
TAO/TRITON Monthly Data January 2016
SST (°C) and Wind
Means
Anomalies
SST FEBRUARY 2016
TAO/TRITON Monthly Data February 2016
SST (°C) and Wind
Means
Anomalies
SST MARCH 2016
TAO/TRITON Monthly Data March 2016
SST (°C) and Wind
Means
Anomalies
SST APRIL 2016
TAO/TRITON Monthly Data April 2016
SST (°C) and Wind
Means
Anomalies
SST MAY 2016
TAO/TRITON Monthly Data May 2016
SST (°C) and Wind
Means
Anomalies
SST JUNE 2016
TAO/TRITON Monthly Data June 2016
SST (°C) and Wind
Means
Anomalies

**(a)**

IOP Conf. Series: Earth and Environmental Science **98** (2017) 012038 doi:10.1088/1755-1315/98/1/012038

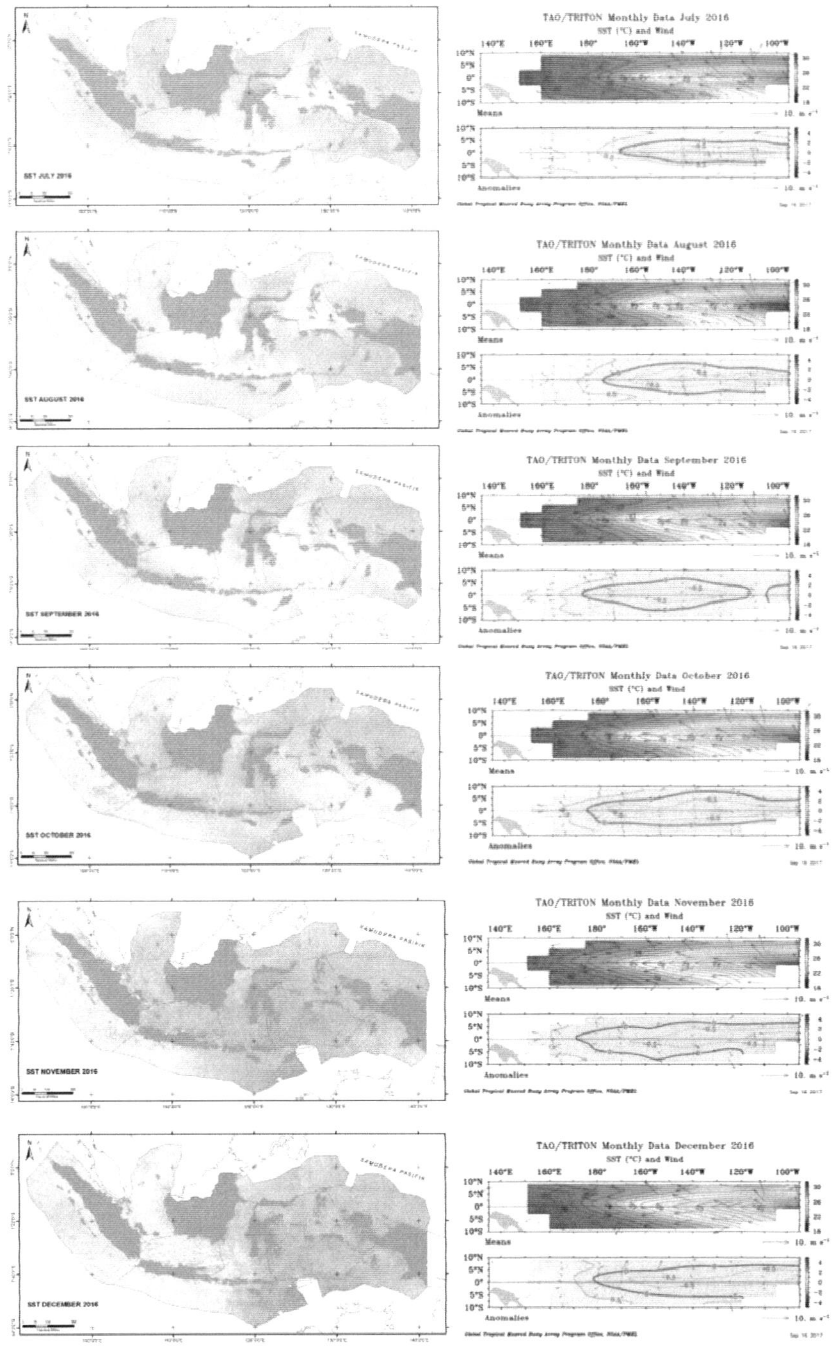


**(b)**

**Figure 7.** The correlation between SST from Himawari-8/9 satellite and TAO buoy measurement (a) January – June 2016; and (b) July – December 2016.

IOP Conf. Series: Earth and Environmental Science **98** (2017) 012038 doi:10.1088/1755-1315/98/1/012038

*3.3. Performance comparison between Aqua/Terra MODIS and Himawari-8/9*

Statistical comparisons of monthly dataset, mean difference test and standard deviation test, are conducted to compare monthly SST measurement results in 2016 from Aqua/Terra MODIS and Himawari-8/9. Both comparisons show that there is no significant difference of SST measurement results from Aqua/Terra MODIS and Himawari-8/9. Mean difference between both satellites is ranging from -0.19 to 0.19 °C, while standard deviation difference is ranging from 0.01 to 0.27 (see Table 5). Minus sign indicates that Aqua/Terra MODIS has lower SST value than Himawari-8/9. This could be possible because of monthly composite of Himawari-8/9 data is performed to selected data that close to Aqua/Terra MODIS acquisition. However, some conditions have not considered for this comparison. Inclusion of differences in spatial (Aqua/Terra MODIS 4 km – Himawari-8/9 2 km) and temporal (Aqua/Terra MODIS monthly – Himawari-8/9 monthly composite from hourly data) resolution between those data in future research could present different results.

**Table 5.** Comparison of SST measurement results (unit in °C).

| Month | Mean Difference Test | | | Standard Deviation Test | | |
|---|---|---|---|---|---|---|
| | **MODIS** | **Himawari** | **Difference** | **MODIS** | **Himawari** | **Difference** |
| January | 29.85 | 29.89 | -0.04 | 0.80 | 0.67 | 0.13 |
| February | 29.80 | 29.72 | 0.07 | 1.09 | 0.88 | 0.21 |
| March | 30.23 | 30.15 | 0.08 | 1.01 | 0.90 | 0.12 |
| April | 30.46 | 30.39 | 0.06 | 0.61 | 0.46 | 0.15 |
| May | 30.38 | 30.57 | -0.19 | 0.64 | 0.42 | 0.23 |
| June | 30.20 | 30.02 | 0.18 | 0.78 | 0.55 | 0.23 |
| July | 29.49 | 29.30 | 0.19 | 1.18 | 0.94 | 0.25 |
| August | 29.05 | 28.89 | 0.16 | 1.20 | 0.95 | 0.25 |
| September | 29.49 | 29.34 | 0.14 | 0.99 | 0.72 | 0.27 |
| October | 29.87 | 29.75 | 0.12 | 0.82 | 0.67 | 0.15 |
| November | 30.30 | 30.19 | 0.11 | 0.82 | 0.66 | 0.15 |
| December | 29.93 | 29.82 | 0.11 | 0.91 | 0.89 | 0.01 |
| | | ***Mean*** | 0.08 | | ***Mean*** | 0.18 |

## 4. Conclusions and Discussions

Cold tongue phenomenon occurred in WPPNRI 711 during January and February when the influence of winter monsoon that occurred along the Vietnam coast flows through South China Sea to this area. Upwelling phenomenon, indicated by cold SST, occurred in WPPNRI 573 and WPNNRI 718 in June, expands to WPPNRI 714 in July – September, then shrinks to only in WPPNRI 573 in October, then disappears in November.

During 2007 – 2016, SST within WPPNRI ranging from 27.91 to 30.46 °C. The coldest SST occurred in August (2007, 2012, 2015) while the warmest SST occurred in April (2010 and 2016). Generally, it is found that SST within WPPNRI has been increased 0.8 °C during the last ten years (2007 – 2016).

There is strong influence from IOD to SST in Indonesia, where cold SST will be found on the same period as IOD+. This influence can be found in August 2012 and 2015 where respectively DMI shows 0.948 and 0.863 and SST shows 27.93 and 27.91 °C.

ENSO has influenced SST dynamics on eastern of Indonesia, especially in WPPNRI 717. During La Niña in Pacific Ocean, the warm SST from Pacific flows westward and enter the Indonesia water through WPPNRI 717 to WPPNRI 573 with moderate SST, then become warmer in April – December.

In the future the SST become the important information for indicate the environment change in Indonesia sea, therefore the better spatial resolution and temporal resolution is needed. AHI Himawari-

8/9 that has hourly temporal resolution and 2 km spatial resolution offers similar data as Aqua/Terra MODIS. The mean differences and standard deviation test between Aqua/Terra MODIS and Himawari-8/9 data show no significant difference. SST measurement result from both satellites have mean difference of 0.08 °C and standard deviation difference of 0.18 °C.

**Acknowledgments**
This research is a part of bigger research under and supported by Institute for Marine Research and Observation, Ministry of Marine Affairs and Fisheries of Indonesia.

**References**

[1] Deser Clara, Michael A Alexander, Shang Ping Xie and Adam S Philips 2010 *Sea Surface Temperature Variability: Patterns and Mechanisms*. 10.1146/annurev-marine-129408-151453.
[2] Qu Tangdong, Yan Du, Jane Strachan, Gary Meyers, and Julia Slingo 2005 Sea Surface Temperature and its Variability in the Indonesian Region *Journal of Oceanography Society* **18** (4).
[3] Wyrtki 1962 The upwelling in the region between Java and Australia during south-east monsoon *Marine and Freshwater Research* **13**(3) pp 217-225.
[4] Saji N H, B N Goswami, P N Vinayachandran and T Yamagata 1999 A dipole mode in the tropical Indian Ocean *Nature* **401**(6751) pp 360-363.
[5] Susanto R D, A L Gordon and Q Zheng 2001 Upwelling along the coasts of java and Sumatera and its relation to ENSO *Geophysical Research Letters* **28**(5) pp 1599-1602.
[6] Chiang J C H and A H Sobel 2002 Tropical tropospheric temperature variations caused by ENSO and their influence on remote tropical *J. Climate* **15** pp 2616 – 2631.
[7] Kaplan Alexey, Mark A Cane, Yochanan Kushnir, Amy C Clement, M Benno Blumenthal, and Balaji Rajagopalan 1998 Analyses of global sea surface temperature 1856 – 1991 *Journal of Geophysical Research* **103** **(**C9) pp 18567 – 18589.
[8] Banzon Viva, Thomas M. Smith, Toshio Mike Chin, Chunying Liu, and William Hankins 2016 A long-term record of blended satellite and in situ sea-surface temperature for climate monitoring, modelling, and environmental studies *Earth Syst. Sci. Data* **8** pp 165 – 176.
[9] Schaefer K M 2001 *Reproductive biology of yellow fin tunas. In: Block BA, Steven ED (eds) Tuna: Physiology, ecology, and evolution* (California: Academic Press) pp 225 – 270.
[10] Laevastu, T and Hayes M L 1981 *Fisheries Oceanography and Ecology*. 199p., Fishing News (books), Oxford, U.K. ISBN: 9780852381175.
[11] Sund, P N, Balcburn, M and Wiliams, F 1981 Tuna and Their Environment in the Pacific Ocean: A Review. *Oceanography*, "Oceanography and Marine Biology: An Annual Review **19** pp 443-512.
[12] Gordoa A, M Maso and Voges L 2000 Satellites and Fisheries: The Namibian hake, A case study in D. Halpern. Satellite *Oceanography and Society*, pp. 193 – 205, Elsevier, Amsterdam, The Netherlands. DOI: 10.1016/So422-9894(00)8011-2.
[13] Lu H-J, K-T Lee, H-L Lin and C H Liao 2001 Saptio temporal distribution of yellow fin tuna Thunnus albacares and big eye tuna Thunnus obesus in Tropical Pacific Ocean in relation to large scale temperature fluctuation during ENSO episodes *Fish. SCi* **67** pp 1046 – 1052.
[14] Rajapaksha J K, Nishida T, and Samarakoon L 2010 *Environmental preferences of yellowfin tuna (Thunnus albacores) in the northeast Indian Ocean: an application of remote sensing data to longline catches*. IOTC-2010-WPTT-43.
[15] Njoku E G, T P Barnett, R M Laurs and A C Vastano 1985 Advances in Satellite Sea Surface Temperature Measurement and Oceanographic Applications *Journal of Geophysical Research* **90** (C6) pp 11573-11586.
[16] Robel J and A Graumann 2014 *NOAA KLM User's Guide: with NOAA-N, N Prime, and MetOp Supplements* National Oceanic and Atmospheric Administration.
[17] Brown O B and Minnet P J 1999 *MODIS Infrared Sea Surface Temperature Algorithm – Algorithm*

*Theoretical Basis Document Version 2.0* (Miami: University of Miami).

[18] Kurihara Y, H Murakami and M Kachi 2016 Sea surface temperature from the new Japanese geostationary meteorological Himawari-8 satellite *Geophys. Res. Lett* **43** pp 1234–1240, doi:10.1002/2015GL067159.

[19] Gloersen P and F T Barath 1977 A Scanning Multichannel Microwave Radiometer for Nimbus-G and SeaSat-A *IEEE Journal of Oceanic Engineering* **2** pp 172-178.

[20] Imaoka K, T Sezai, T Takeshima, T Kawanishi, A Shibata 2002 Instrument characteristics and calibration of AMSR and AMSR-E *Proceedings of IGARSS 2002*. Toronto. Canada.

[21] Ministry for Marine Affair and Fisheries (MMAF) 2014 *Peraturan Menteri Kelautan dan Perikanan No. 18/PERMEN-KP/2014*.

[22] Nontji A 1993 *Laut Nusantara* (Jakarta: Penerbit Djambatan).

[23] MODIS Specifications. Accessed on September 14, 2017 from https://modis.gsfc.nasa.gov/about/specifications.php

[24] Kilpatrick K A, Podesta G, Walsh S, Williams E, Halliwell V, Szczodrak M, Brown O B, Minnet P J and Evans R 2015 A decade of sea surface temperature from MODIS *Remote Sensing of Environment* **165** pp 27-41. http://dx.doi.org/10.1016/j.rse.2015.04.023.

[25] Institute for Marine Research and Observation 2016 *Final Report of Ocean Observation System Development Project* Ministry of Marine Affairs and Fisheries, Indonesia.

[26] Stommel H 1948 The westward intensification of wind-driven ocean currents *Trans. Am. Geophys. Union* **99** pp 202 – 206.

[27] Chen J M, Chang C-P and Li T 2003 Annual cycle of the South China sea surface temperature using NCEP/NCAR reanalysis *Journal of Meteorological Society of Japan* **81** pp 879 – 884.

[28] Koseki S, Tieh-Yong Koh, Chee-Kiat Teo 2013 Effects of the cold tongue in the South China Sea on the monsoon, diurnal cycle and rainfall in Maritime Continent *Q.J.R. Meteorol.Soc* **139** pp 1566 – 1582.

[29] Clarke A J and X Liu 1993 Observations and dynamics of semiannual and annual sea levels near the eastern equatorial Indian Ocean boundary *Journal of Oceanography* **23** pp 386 – 399.

[30] Sprintall, J, A L Gordon, R Murtugudde and R D Susanto 2000 A semi-annual Indian Ocean forced Kelvin wave observed in the Indonesian Seas *J. Goephy. Res* **105** pp 17217 – 17230.

[31] Bureau of Meteorology Australian Government. Indian Ocean influences on Australian climate. Accessed on September 14, 2017 from http://www.bom.gov.au/climate/iod/#tabs=Indian-Ocean-climate-drivers

[32] Mo Lan 2013 *Analysis of 2012 Indian Ocean Dipole Behavior* Accessed on September 14, 2017 from https://www.s.u-tokyo.ac.jp/en/utrip/archive/2013/pdf/05MoLan.pdf

# Multispectral Resampling of Seagrass Species Spectra: WorldView-2, Quickbird, Sentinel-2A, ASTER VNIR, and Landsat 8 OLI

**Pramaditya Wicaksono[1*], Ignatius Salivian Wisnu Kumara[2], Muhammad Kamal[3], Muhammad Afif Fauzan[2], Zhafirah Zhafarina[2], Dwi Agus Nurswantoro[4], Rifka Noviaris Yogyantoro[2]**

[1] Remote Sensing Laboratory, Department of Geographic Information Science, Faculty of Geography, Universitas Gadjah Mada, Yogyakarta, Indonesia 55281
[2] Cartography and Remote Sensing, Department of Geographic Information Science, Faculty of Geography, Universitas Gadjah Mada, Yogyakarta, Indonesia 55281
[3] PUSPICS, Faculty of Geography, Universitas Gadjah Mada, Yogyakarta 55281
[4] Diving Unit, Universitas Gadjah Mada, Yogyakarta, Indonesia 55281

prama.wicaksono@geo.ugm.ac.id

**Abstract**. Although spectrally different, seagrass species may not be able to be mapped from multispectral remote sensing images due to the limitation of their spectral resolution. Therefore, it is important to quantitatively assess the possibility of mapping seagrass species using multispectral images by resampling seagrass species spectra to multispectral bands. Seagrass species spectra were measured on harvested seagrass leaves. Spectral resolution of multispectral images used in this research was adopted from WorldView-2, Quickbird, Sentinel-2A, ASTER VNIR, and Landsat 8 OLI. These images are widely available and can be a good representative and baseline for previous or future remote sensing images. Seagrass species considered in this research are *Enhalus acoroides* (Ea), *Thalassodendron ciliatum* (Tc), *Thalassia hemprichii* (Th), *Cymodocea rotundata* (Cr), *Cymodocea serrulata* (Cs), *Halodule uninervis* (Hu), *Halodule pinifolia* (Hp), *Syringodum isoetifolium* (Si), *Halophila ovalis* (Ho), and *Halophila minor* (Hm). Multispectral resampling analysis indicate that the resampled spectra exhibit similar shape and pattern with the original spectra but less precise, and they lose the unique absorption feature of seagrass species. Relying on spectral bands alone, multispectral image is not effective in mapping these seagrass species individually, which is shown by the poor and inconsistent result of Spectral Angle Mapper (SAM) classification technique in classifying seagrass species using seagrass species spectra as pure endmember. Only Sentinel-2A produced acceptable classification result using SAM.

**Keywords**: seagrass, species, resampling, spectral resolution, multispectral

## 1. Introduction

Presently, multispectral image is still the primary source of remote sensing data for seagrass mapping [1]. However, the limited number of spectral band of multispectral image limit the possibility of accurate seagrass mapping, especially when mapping their species composition. Mapping seagrass species is a challenging task, either using multispectral or hyperspectral image [2]. Sub-pixel species composition mixing adds the complexity of benthic habitat mapping using remote sensing image

[3][4], and this includes seagrass species mapping. Several seagrass species may present together and create mix-species seagrass beds, for instance, smaller species such as *Cymodocea rotundata* (Cr) frequently present under bigger species such as *Enhalus acoroides* (Ea) [5].

Seagrass species is spectrally-different [6][7][8]. However, the difference is primarily not in the spectra shape but in the absolute reflectance value [8]. This becoming an issue for spectral-based algorithm such as Spectral Angle Mapper (SAM), which utilized spectra shape instead of the variation in the absolute reflectance [9]. Furthermore, the limited band of multispectral image simplifies the spectra shape of seagrass species and generalizes the unique absorption feature of different seagrass species. Therefore, it is important to quantitatively assess the possibility of mapping seagrass species using multispectral images by resampling seagrass species spectra to multispectral bands. In addition, it is also important to assess to which extent seagrass species spectra can be used as the input endmember for SAM algorithm for seagrass species classification using multispectral image.

## 2. Study Area and Image Data

This study was conducted in Karimunjawa Islands National Park, Jepara Regency, Central Java. Astronomically, the National Park is located between 110°23'32" – 110°31'29" E and 05°45'06" – 05°54'59" S. Karimunjawa main islands consist of land and open water areas about 7.645 ha, especially 3.097 ha for shallow sea water area. Figure 1 shows the location study area. Shoreline typology of Karimunjawa islands consist of sandy shore and mangrove habitat. Shallow water area up to the depth of 20 m consist of seagrass, coral reef, and bare substratum. Seagrass can be found mainly in optically-shallow water near the shoreline, with *Enhalus acoroides* (Ea), *Cymodocea rotundata* (Cr), and *Thalassia hemprichii* (Th) as the most dominant species. Based on Karimunjawa Islands National Park report, there are nine seagrass species found in Karimunjawa Islands [10].

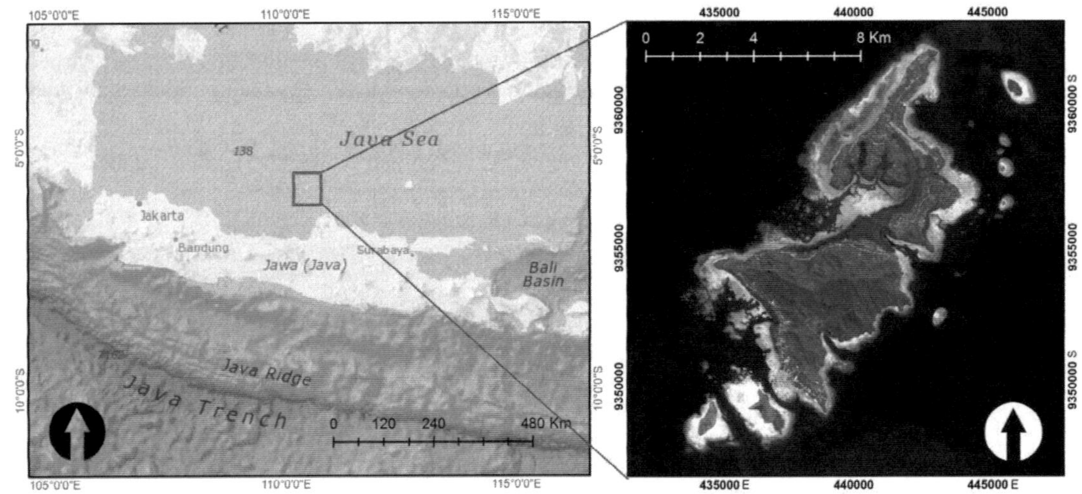


**Figure 1.** Sentinel-2A image of Karimunjawa and Kemujan Island in study area. Small islands surrounding these two main Islands are also visible.

This study used five multispectral images that have been frequently used for remote sensing-based seagrass mapping [1][2][5][11][12]. These images are WorldView-2, Quickbird, Sentinel-2A, ASTER VNIR, and Landsat 8 OLI. These images representing remote sensing data with different spatial resolution that still able to detect seagrass patch. Spectral bands specification of these images is provided in Table 1. Atmospheric correction was performed using Dark-Object Subtraction (DOS) method as described in [13] for WorldView-2, ASTER VNIR and Landsat 8 OLI, while Fast Line-of-Sight Atmospheric Analysis of Hypercubes (FLAASH) method was applied to Quickbird image.

Sentinel-2A was atmospherically-corrected using Sen2Cor module. Afterward, sunglint correction was performed on all images using method developed by [14].

**Table 1.** Spectral band specification of multispectral images used in this research.

| Image | Acquisition Date (dd/mm/yyyy) | Bands (Wavelength in nm) | Central Wavelength (nm) | Pixel Size (at nadir) | Radiometric resolution |
|---|---|---|---|---|---|
| WorldView-2 | 24/05/2012 | Coastal (400 - 452)<br>Blue (448 – 510)<br>Green (518 – 586)<br>Yellow (590 – 630)<br>Red (632 – 692)<br>RedEdge (706 – 746)<br>NIR1 (772 – 890)<br>NIR2 (866 – 954) | 427.0<br>278.0<br>546.0<br>608.0<br>659.0<br>724.0<br>831.0<br>915.0 | 1.84 m | 11 bits |
| Quickbird | 23/08/2004 | Blue (430 – 545)<br>Green (466 – 620)<br>Red (590 – 710)<br>NIR (715 – 918) | 477.2<br>555.7<br>658.1<br>803.1 | 2.44 m | 11 bits |
| Sentinel-2A MSI | 19/05/2017 | Blue (458 – 523)<br>Green (543 – 578)<br>Red (650 – 680)<br>NIR (785 – 900) | 490.0<br>560.0<br>665.0<br>842.0 | 10 m | 12 bits |
| ASTER VNIR | 26/08/2004 | 1 (520 – 600)<br>2 (630 – 690)<br>3N (780 – 860) | 556.0<br>661.0<br>807.0 | 15 m | 8 bits |
| Landsat 8 OLI | 18/05/2017 | Coastal (435 – 451)<br>Blue (441 – 515)<br>Green (525 – 600)<br>Red (630 – 680)<br>NIR (845 – 885) | 442.9<br>482.1<br>561.4<br>654.6<br>864.6 | 30 m | 16 bits |

## 3. Method

### *3.1. Sample Collection and List of Species*

Seagrass species sample collection was done by using harvest method in seagrass bed of optically-shallow water area. Seagrass leaves of different species were harvested in a healthy and fresh conditions, and were put into non-translucent bottle and were placed inside cool box until spectra measurement was performed. This was done to prevent structural and leaf pigment damages due to the delay between harvesting time and spectra measurement (6 hours differences in average for all species). List of seagrass species collected from the field is listed on Table 2.

**Table 2.** List of seagrass species collected from Karimunjawa Islands

| No | Species | Abbreviation |
|---|---|---|
| 1 | *Enhalus acoroides* | *Ea* |
| 2 | *Thalassia hemprichii* | *Th* |
| 3 | *Cymodocea rotundata* | *Cr* |
| 4 | *Cymodocea serrulata* | *Cs* |
| 5 | *Halophila ovalis* | *Ho* |
| 6 | *Halophila minor* | *Hm* |
| 7 | *Halodule uninervis* | *Hu* |
| 8 | *Halodule pinifolia* | *Hp* |
| 9 | *Thalassodendron ciliatum*[a] | *Tc* |
| 10 | *Syringodum isoetifolium* | *Si* |

[a]*Tc* was collected from Nusa Lembongan Island as accuracy control

### *3.2. Spectra Measurement*

Spectra measurement was done above water surface on harvested seagrass leaves. Each species leaves stored in bottle were grouped until covering an area with minimum size of 25 $mm^2$. The spectra measurement instrument is the field spectrometer Jaz EL-350 VIS/NIR with 350 – 1100 nm wavelength range (effective at 400 – 900 nm). The white and dark reference was calculated by measuring the reflectance of white reference (Spectralon® WS-1-SL) and dark reference (performed by covering the collimating lens with its cap). Spectrometer default field-of-view is 14.25°, hence, to prevent mixing with background spectral responses, the spectra measurement was conducted very close to the seagrass leaves. The setting for spectra measurement, including Boxcar width, Scans-to-average, and Electric dark correction were similar with the setting published in [8] and [15]. We also added the spectra measurement of *Tc* as a control in the spectral-based classification. *Tc* spectra were measured from samples collected in Nusa Lembongan Island. Each species was measured several times and the average spectra were calculated. The output of spectra was recorded as .jaz format data and was processed in spreadsheet after being converted to ASCII format.

### *3.3. Spectra Resampling*

The seagrass species spectra were resampled based on the spectral resolution of multispectral images. Resample procedure can be performed in several ways, such as using image containing central wavelength of each band, ASCII format file that contains central wavelength, or pre-defined filter function provided by image processing software. In this research, we used pre-defined filter function for each multispectral image.

### *3.4. Spectra Analysis (Tukey-Test)*

Post-Hoc Analysis Tukey-test commonly used to assess the statistical difference between data groups. Tukey-test represents the differences between groups statistics more explicitly. Therefore, this method can be applied on spectra data to identify if seagrass species can be spectrally distinguished. Tukey-test analysis procedure was divided into three steps, i.e., one-way ANOVA calculation, yardstick of honestly significant difference (HSD) calculation, and pairwise comparisons (Tukey matrix calculation). Each seagrass species spectra data between 400 – 900 nm was treated as individual group. Therefore, there are 10 groups of seagrass species as listed in Table 2.

One-way ANOVA test was performed to gain average value of entire reflectance's value in each group ($\bar{X}$), number of treatments or groups ($K$), and number of spectral data or wavelength within a group ($N$). $SS_{wi}$ and $df_{wi}$ value were required to calculate the mean square error within groups ($MS_{wi}$) using equation (1).

$$MS_{wi} = \frac{SS_{wi}}{df_{wi}} \tag{1}$$

$SS_{wi}$ is sum of square within group and $df_{wi}$ is degree of freedom within groups from compared spectral groups. These two values were generated by one-way ANOVA calculation in spreadsheet.

Furthermore, $MS_{wi}$ value was used to calculate the yardstick HSD value for determining the significantly difference margin. Yardstick calculation require $MS_{wi}$, $q$-value, and $N$. The $q$-value was obtained using table of critical values for studentized range with $\alpha = 0.05$ (required $df_{wi}$ and $K$ values). In addition, the Tukey-test can be continued if the $P$-value is less than 0.05. The yardstick HSD value was calculated using equation (2).

$$HSD = q\left(\frac{MS_{wi}}{N}\right)^{1/2} \tag{2}$$

Pairwise comparisons principle was used to test the mean of each species pair. Average value of each seagrass species obtained from one-way ANOVA was calculated for each absolute difference (e.g., $|Ea - Th|$; $|Ea - Cr|$; and $|Th - Cr|$). A pairwise comparison matrix was created for easier and structured difference calculation. Difference value of each seagrass species pair group was compared with the yardstick HSD value. If the difference value of pair group greater than the yardstick HSD value, then the two seagrass species spectra is honestly significantly different.

### *3.5. Spectral Angle Mapper (SAM)*

SAM classification technique was applied on multispectral images to map seagrass species distribution. This technique is a physical-based spectral classification. SAM algorithm requires resampled spectra as pure endmember to determine spectra similarity between reference spectra and target spectra (pixel) [16]. The difference in the angle of radian between those spectra was used as threshold to classify image pixels to particular seagrass species class. SAM is insensitive to illumination and albedo effect, therefore noise such as atmospheric path radiance does not impact on classification process [9].

The radian angle threshold assigned for each image in SAM classification is 0.1; 0.4; 0.7; and 1.0. The threshold was set from the narrowest to the widest possible angle to identify the trend of seagrass species spatial distribution classified by SAM classification technique. Wider angle of radian threshold gives more opportunities for pixels to be classified, and vice versa.

## 4. Result and Discussion

### *4.1. Spectra of Seagrass Species*

The spectra of seagrass species are shown in Figure 2. The shape of all seagrass species is relatively similar. There are two minor peaks identified around 600 nm and 635 nm for most seagrass species. *Ea* and *Ho* have higher reflectance up to 30% and 25% across visible wavelengths, meanwhile the rest are below 18%. *Ea* reflectance across the visible wavelengths are almost twice than other species, but *Ho* reflectance only higher than another species (except *Ea*) in green to red region of wavelengths. *Cs* and *Si* have the lowest reflectance in almost all visible wavelengths. However, *Hu* has the lowest reflectance of all in the red band region.

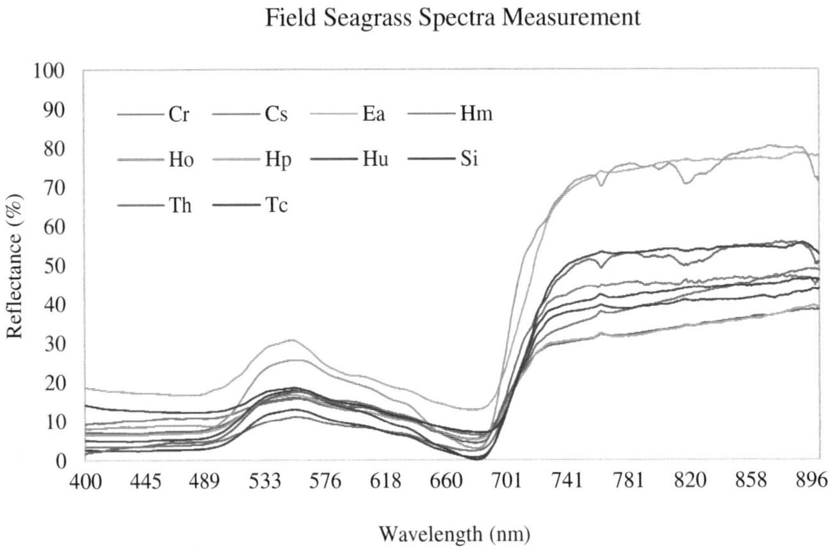


**Figure 2.** Spectra of ten seagrass species measured by field spectrometer

All of these seagrass species spectra are similar to the spectra of healthy vegetation leaf. The major peak at ±550 nm indicates the domination of chlorophyll pigments. Since all species exhibit similar absorption band locations, but different in the strength of absorption, the pigment composition may be similar. However, the concentration of pigments between species is definitely unique.

### *4.2. Resampling Result of Species Spectra*

Resampling result according to multispectral images is shown on Figure 3. The species spectra pattern of each image corresponds to the number and center wavelength of each band. Generally, compared to the field seagrass spectra (Figure 2), all resampled spectra are less precise and more generalized but still retain the same pattern of the original spectra i.e. *Ea* and *Ho* have higher reflectance than other species. It is more difficult to differentiate between species based on the resampled shape, which is caused by the limited band number and wider bandwidth. The peak reflectance of resampled spectra is still at 540 – 560 nm (green band) and the lowest reflectance is located between 650 – 670 nm (red band). The absence of blue band in ASTER VNIR image made the species spectra even harder to be differentiated. These resampled spectra were used as pure endmember for SAM classification.

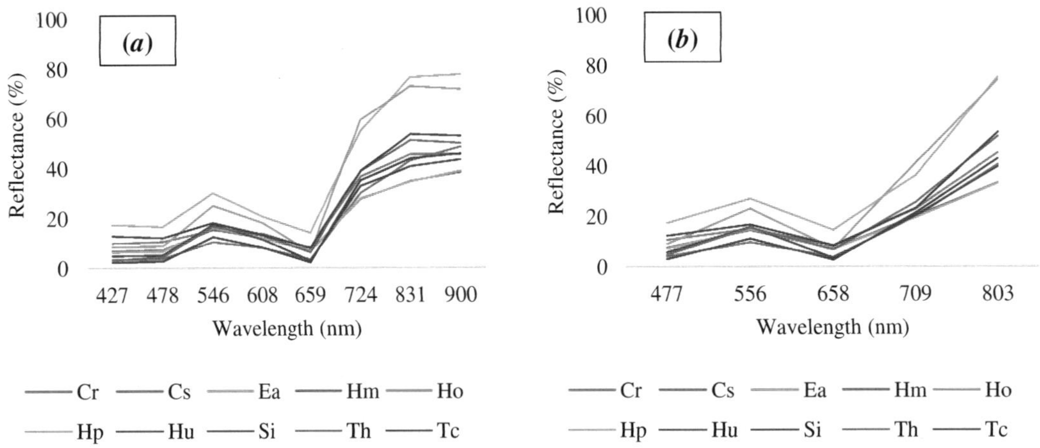

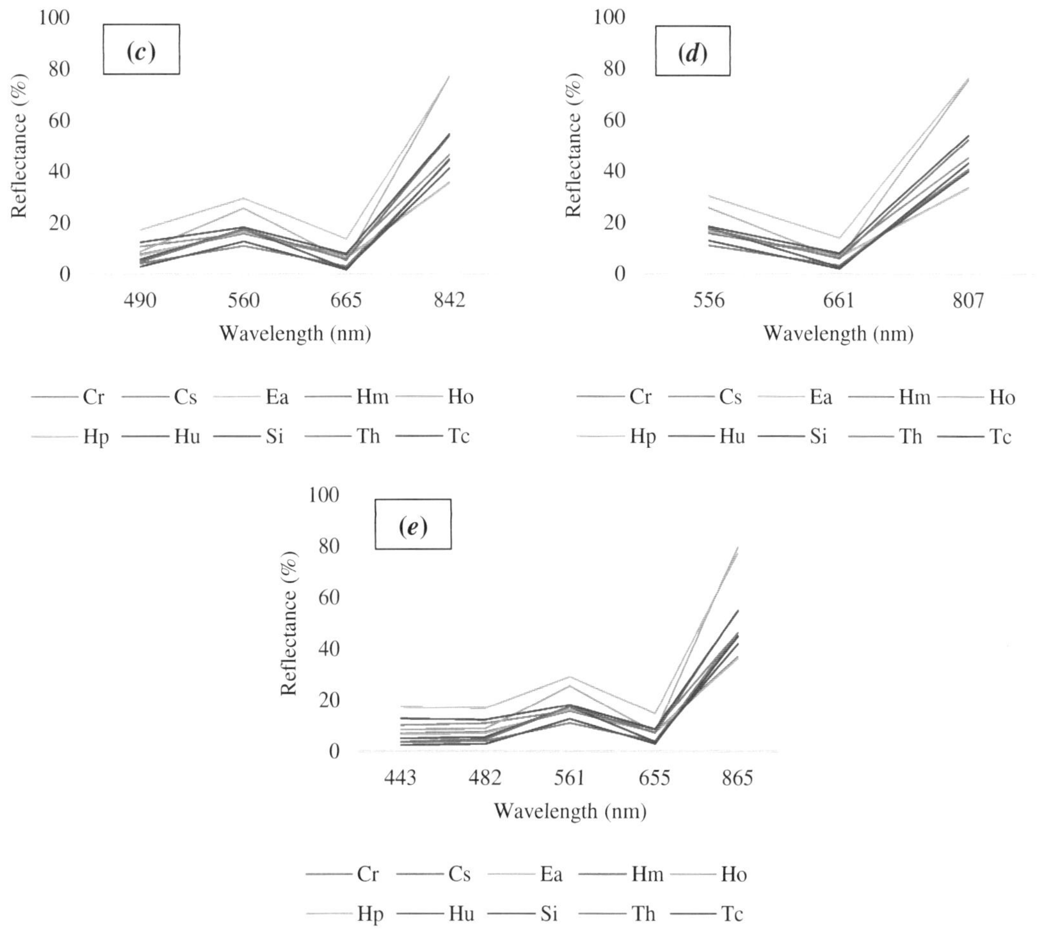


**Figure 3.** Seagrass species spectra resampled to multispectral images (*a*) WorldView-2, (*b*) Quickbird, (*c*) Sentinel-2A, (*d*) ASTER VNIR, and (*e*) Landsat 8 OLI

### 4.3. *Tukey-test Result*

Tukey-test analysis showed that most of seagrass species were spectrally-different. From 45 pairs of seagrass species, 34 pairs clearly different and the rest 11 pairs are similar. These results refer to the calculation of species pair difference and yardstick HSD value. Through one-way ANOVA calculation, $MS_{wi}$ value of 384.305 was obtained. Then, based on the original field spectra, the $N$ value between 400 – 900 nm is 1,457 for each group. Table of critical values gave $q$-value at 4.47 ($\alpha = 0.05$). Since the $P$-value is less than 0.05, yardstick HSD can be calculated and the value of 2.29 was obtained. This yardstick value was used as margin in pairwise comparison of Tukey-test calculation (Table 3).

Table 3 is a 2-dimensional Tukey-test matrix of seagrass species pairwise comparisons. Spectral average value was used to assess the absolute difference. Green boxes on Tukey-test matrix show the honestly significant different of seagrass species pair. For example, *Ea – Th* pair difference is 15.25, which is clearly higher than the yardstick. Oppositely, *Hu – Cs* pair difference is 0.94, far lower the yardstick value. *Ea*, *Ho*, and *Tc* are completely different from other species spectrally. Thereafter, followed by *Th* and *Cr* that which is similar to each other. *Cs, Hu, Hm, Hp*, and *Si* could not be differentiated due to the very low value of pair difference, which is between 0.5 – 1.0. Hence, these species are not spectrally-distinguishable.

**Table 3.** Pairwise comparison of seagrass species spectra in Tukey-test Matrix

| Species | Spectral average ($\bar{X}$) | Difference | | | | | | | | |
|---|---|---|---|---|---|---|---|---|---|---|
| | | *Ea* | *Th* | *Cr* | *Cs* | *Ho* | *Hu* | *Hm* | *Hp* | *Tc* |
| *Ea* | 41.27 | | | | | | | | | |
| *Th* | 26.01 | 15.25 | | | | | | | | |
| *Cr* | 24.85 | 16.41 | 1.16 | | | | | | | |
| *Cs* | 20.12 | 21.14 | 5.89 | 4.72 | | | | | | |
| *Ho* | 38.12 | 3.14 | 12.10 | 13.27 | 17.99 | | | | | |
| *Hu* | 21.06 | 20.20 | 4.94 | 3.78 | 0.94 | 17.05 | | | | |
| *Hm* | 19.58 | 21.68 | 6.43 | 5.27 | 0.54 | 18.54 | 1.48 | | | |
| *Hp* | 19.46 | 21.80 | 6.55 | 5.38 | 0.66 | 18.66 | 1.60 | 0.11 | | |
| *Tc* | 28.48 | 12.78 | 2.47 | 3.63 | 8.36 | 9.63 | 7.42 | 8.90 | 9.02 | |
| *Si* | 20.69 | 20.57 | 5.32 | 4.15 | 0.57 | 17.42 | 0.37 | 1.11 | 1.23 | 7.79 |

*Green box indicates the difference between two species greater than the yardstick HSD value = 2.29

### *4.4. SAM Result*

The application of SAM classification on each image produced different result. Generally, almost all pixels were misclassified (Figure 4). In the classified image, several species such as *Ea*, *Cr*, *Hp*, and even the unlikely *Tc* dominate the seagrass area. All images, especially WorldView-2 and ASTER VNIR images failed to classify any seagrass species using smaller angle thresholds (<0.7). SAM threshold of 0.1 produced high number of unclassified pixels due to the wide angle between reference spectra and target spectra. Pixels started to be classified at 0.4 and 0.7 angle threshold, except for WorldView-2 image. Landsat 8 OLI pixels near the shoreline stayed unclassified despite the threshold, which is suspected due to the mixing between seagrass and both emerged and submerged substrates.

WorldView-2 image only classified *Hp* and *Hm* species, but only on some pixels near the shoreline. WorldView-2 result is surprisingly poor as *Hp* and *Hm* in fact is the most difficult species to be found in the study area. Quickbird image only classified *Cr* species, yet other species were unclassified. Sentinel-2A image produced better seagrass species distribution and the only one to map *Ea* correctly along water near the shoreline. Other classified species are *Cr* and small area of *Tc*. *Cr* distribution may be overestimated but still acceptable as *Th* and *Cr* are mixing strongly in the study area and their spectra based on Tukey-test is similar. Similar to WorldView-2 classified image, ASTER VNIR image was unable to classify seagrass species at 0.1, 0.4, and 0.7 thresholds, and at 1.0 threshold most pixels were classified as *Hp*, which is highly unlikely. Other species were also classified, this includes *Cr*, *Cs*, *Hm*, and *Ea*. Landsat 8 OLI image produced consistent result at the 0.1, 0.4, 0.7, or 1.0 angle thresholds and produced similar results to Sentinel-2A but failed to classify *Ea* near the shoreline and produced more misclassification of *Tc* in the seaward margin of seagrass habitat. Quickbird and Landsat 8 OLI results are almost similar, where *Cr* and *Tc* dominates the classified image. In the actual condition, *Cr* may dominate some places in the eastern coast, but *Tc* was not found across the study area.

Analyzing all SAM results, there are much of inconsistency in the seagrass species spatial distribution between images. Different images classified different species with different spatial distribution. This indicates that multispectral images with limited number of bands do not provide sufficient spectral information for the spectral-based classification of seagrass species. Interestingly, image with four bands had less confusion in classifying seagrass species based on their spectra. Although may not be fully accurate, the spatial distribution of seagrass species is consistent between image and is more acceptable than from WorldView-2 that has more bands, for instance, in the distribution of *Ea* and *Cr*. ASTER VNIR produced the least reliable result, which is understandable as it has the smallest number of bands. Nevertheless, this needs to be confirmed in other areas whether the pattern of result is consistent. In addition, this highlight the importance of tuning the threshold for SAM classification algorithm to produce more accurate classification result, by not only using the default value (0.1) or similar value for all class (set the threshold value similarly for all class), but using unique value for different seagrass species based on the abundance and probability of each species encounter in the study area.

Seagrass species spectra (Figure 2 and Figure 3) and Tukey-test result (Table 3) have opposite characteristics. The Tukey-test shows that seagrass species are spectrally-different. However, the classification of field spectra and the resampled spectra using SAM indicated otherwise. Future research should investigate more appropriate classification technique needed for better classification result, considering the SAM classification technique could not differentiate spectra with similar shape. Furthermore, hyperspectral image should also be tested, with the expectation that its narrow wavelength and more number of water penetration bands may improve the classification result.

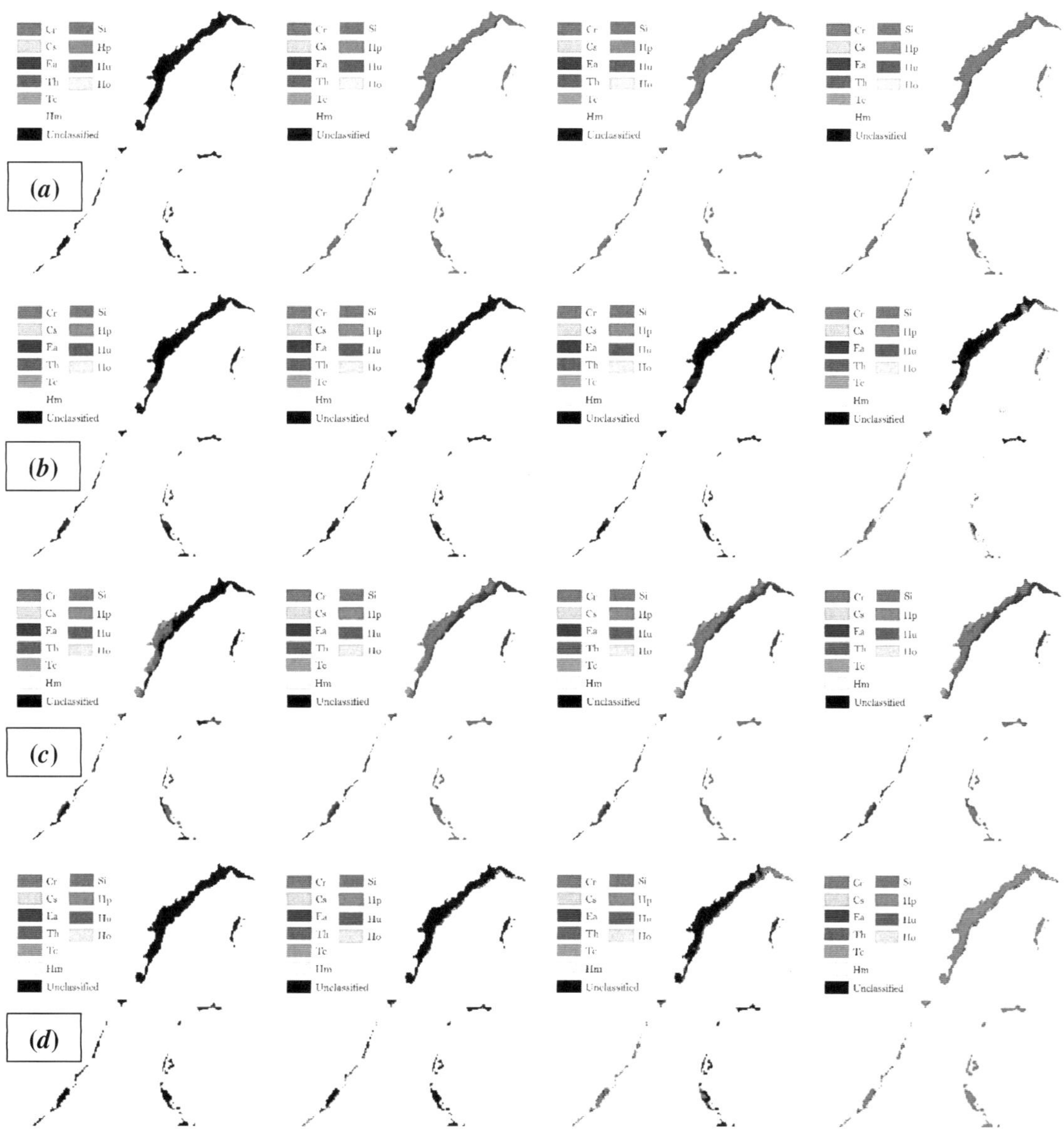

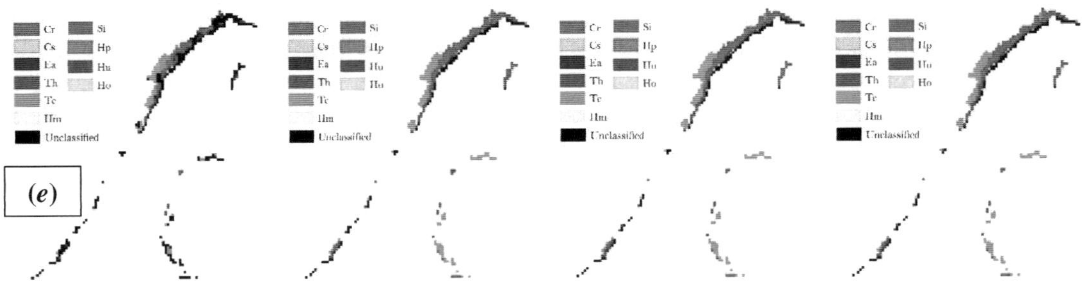


**Figure 4.** SAM classification of (*a*) WorldView-2 with angle 0.1; 0.4; 0.7; and 1.0, (*b*) Quickbird with angle 0.1; 0.4; 0.7; and 1.0, (*c*) Sentinel-2A with angle 0.1; 0.4; 0.7; and 1.0, (*d*) ASTER VNIR with angle 0.1; 0.4; 0.7; and 1.0, and (*e*) Landsat 8 OLI with angle 0.1; 0.4; 0.7; and 1.0

## 5. Conclusions

Multispectral resampling of seagrass species spectra shows that the resampled spectra exhibit similar shape and pattern with the original spectra, but less precise and lose the unique absorption feature of seagrass species. Although spectrally-different based on spectra measurement, the difference in seagrass species spectra is not in the shape but in the absolute reflectance value. As a consequence, using seagrass species spectra as pure endmember for SAM classification is unrealistic, as the algorithm normalizes the variation in absolute value and only sensitive to the shape of spectra. The classification results between images are inconsistent, and as a consequence, no agreement between results, which indicates the ineffective use of SAM algorithm.

**Acknowledgements**

This research was funded by "Direktorat Riset dan Pengabdian Masyarakat – Direktorat Jenderal Penguatan Riset dan Pengembangan – Kementerian Riset, Teknologi, dan Pendidikan Tinggi Republik Indonesia" through Penelitian Fundamental Research Scheme, Grant Number 2232UN1-P.III/DIT-LIT/LT/2017.

## References

[1] Hossain M S, Bujang J S, Zakaria M H, and Hashim M 2015 The application of remote sensing to seagrass ecosystems: an overview and future research prospects *Int. J. of Remote Sensing 36*(**1**) pp 61-113 doi: http://dx.doi.org/10.1080/01431161.2014.9

[2] Phinn S R, Roelfsema C M, Brando V, and Anstee J 2008 Mapping seagrass species, cover and biomass in shallow waters: An assessment of satellite multi-spectral and airborne hyper-spectral imaging systems in Moreton Bay (Australia) *Remote Sensing of Environment 112* pp 3413–3425 doi: http://dx.doi.org/10.1016/j.rse.2007.09.017

[3] Hedley J D, Roelfsema C M, Phinn S R, and Mumby P J 2012 Environmental and sensor limitations in optical remote sensing of coral reefs: implications for monitoring and sensor design *Remote Sensing 4* pp 271-302 doi: http://dx.doi.org/10.3390/rs4010271

[4] Joyce K E, Phinn S R, and Roelfsema C M 2013 Live coral cover index testing and application with hyperspectral airborne image data *Remote Sensing 5*(**11**) pp 6116-6137 doi: http://dx.doi.org/10.3390/rs5116116

[5] Wicaksono P and Hafizt M 2013 Mapping seagrass from space: Addressing the complexity of seagrass LAI mapping *European Journal of Remote Sensing 46* pp 18-39 doi: http://dx.doi.org/10.5721/EuJRS20134602

[6] Fyfe S K 2003 Spatial and temporal variation in spectral reflectance: Are seagrass species spectrally distinct? *Limnology and Oceanography 48*(1, part 2) pp 464–479 doi: http://dx.doi.org/10.4319/lo.2003.48.1_part_2.0464

[7] Thorhaug A, Richardson A D, and Berlyn G P 2007 Spectral reflectance of the seagrasses: Thalassia testudinum, Halodule wrightii, Syringodium filiforme and five marine algae. *Int. J. of Remote Sensing 28*(**7**) pp 1487 – 1501 doi: http://dx.doi.org/10.1080/01431160600954662

[8] Wicaksono P and Kamal M 2017 Spectral response of healthy and damaged leaves of tropical seagrass *Enhalus acoroides, Thalassia hemprichii*, and *Cymodocea rotundata* In *Proceeding of SPIE Remote Sensing 2017* (Warsaw: SPIE)

[9] Kutser T, Vahtmae E, and Mersamaa L 2006 Spectral library of macroalgae and benthic substrates in Estonian coastal waters *Proc. Estonian Acad. Sci. Biol. Ecol. 55*(**4**) pp 329-340

[10] Nababan M G, Munasik, Yulianto I, Kartawijaya T, Prasetia R, Ardiwijaya R L, . . . Syaifudin Y 2010 *Status Ekosistem di Taman Nasional Karimunjawa: 2010* (Bogor: Wildlife Conservation Society-Indonesia Program)

[11] Topouzelis K, Spondylidis S C, Papakonstantinou A, and Soulakellis N 2016 The use of Sentinel-2 imagery for seagrass mapping: Kalloni Gulf (Lesvos Island, Greece) case study *Proc. SPIE 9688, Fourth International Conference on Remote Sensing and Geoinformation of the Environment (RSCy2016), 96881F* SPIE. doi: http://dx.doi.org/10.1117/12.2242887

[12] Fauzan M A, Kumara I S W, Yogyantoro R N, Suwardhana S W, Fadhilah N, Nurmalasari I, . . and Wicaksono P (In Press) Assessing the capability of Sentinel-2 data for seagrass cover mapping in Jerowaru, East Lombok *Indonesian Journal of Geography*

[13] Green E P, Mumby P J, Edwards A J, and Clark C D 2000 *Remote Sensing Handbook for Tropical Coastal Management - Coastal Management Sourcebooks 3* (A. J. Edwards, Ed.) (Paris: UNESCO)

[14] Hedley J D, Harborne A R, and Mumby P J 2005 Simple and robust removal of sunglint for mapping shallow-water benthos *International Journal of Remote Sensing 26*(**10**) pp 2107-2112 doi: http://dx.doi.org/10.1080/01431160500034086

[15] Wicaksono P 2017 Karakterisasi respon spektral spesies lamun *Enhalus acoroides* dan *Cymodocea rotundata* di Pulau Karimunjawa *Globe 19*(**1**) pp 1-10 doi: http://dx.doi.org/10.24895/MIG.2017.19-1.%25x

[16] Kruse F, Lefkoff A B, Heidebrecht K B, Shapiro A T, Barloon P J, and Goetz A F 1993 The spectral image processing system (SIPS) - Interactive visualization and analysis of imaging spectrometer data *Remote Sensing of the Environment 44* pp 145-163 doi: http://doi.org/10.1016/0034-4257(93)90013-N

# Identifying Distribution of Chlorophyll-a Concentration Using Landsat 8 OLI on Marine Waters Area of Cirebon

**Gilang Buditama[1*], Astrid Damayanti[2], Tjiong Giok Pin[2]**

[1]Bachelor Program of Geography, Universitas Indonesia, Depok – Indonesia

[2]Department of Geography, Faculty of Mathematic and Natural Science, Universitas Indonesia, Depok – Indonesia

buditama.gilang@gmail.com

**Abstract.** Phytoplankton is a microscopic plant that has a function to produces oxygen and organic substances. It also plays a role as a main producer in the food chain of a marine ecosystem. Chlorophyll-a is a colour pigment most common in phytoplankton species, so that the concentration level of chlorophyll-a can be used as an indicator of the abundance of phytoplankton cells, and as a reference for predicting organic potency in the aquatic area. This research discusses about the spatial and temporal distribution of chlorophyll-a and its correlation with salinity and total suspended solid (TSS), in the seawaters of Cirebon, West Java. The goal of this research is to be a source of information for fishermen, and other stakeholders whose related subjects in the field of marine and fisheries to predict fertile water regions and can also be used as an indicator in discovering potential areas to catch pelagic fish in Cirebon seawaters. Chlorophyll-a concentration, salinity, and TSS are identified using remote sensing data such as Landsat-8 OLI multi temporal images according to dry and wet month parameters in the 2014-2015. The results of the processed image are then validated between in-situ measurements in the field and remote sensing imagery at the same time. This research utilizes descriptive analysis, and statistics with spatial approach. The results of the research show that temporally, chlorophyll-a levels have a tendency to be higher in wet months compared to dry months, while chlorophyll-a is higher in areas near the coastline compared to open sea areas. The distribution of chlorophyll-a concentration is affected by salinity and TSS distribution, where salinity has the negative relationship and TSS has the positive relationship with chlorophyll-a distribution.

**Keywords:** Landsat 8, chlorophyll-a, Cirebon

## 1. Introduction

Phytoplankton is a microscopic plant that lives floating in the water, be it in estuary waters, coast, or sea. Phytoplankton is also an autotrophic organism that produces food through photosynthesis process. Therefore, phytoplankton has an important role as a producer in marine ecosystems cycle [1, 2]. Chlorophyll-a is a colour pigment most common in phytoplankton species, so that the existence of phytoplankton can be identified by looking at the concentration of chlorophyll-a in water bodies. Therefore, concentration levels of chlorophyll-a are able to indicate the abundance of phytoplankton and primary productivity in a waters area [3].

IOP Conf. Series: Earth and Environmental Science **98** (2017) 012040 doi:10.1088/1755-1315/98/1/012040

The abundance of phytoplankton and chlorophyll-a can experience change according to seasons. This is caused by the variability of nutrients between wet season and dry season. In the wet season of Cirebon, tends to have high rainfall which causes an abundance of river nutrients compared to dry season which has low rainfall. The abundance of nutrient supply will cause the tendency of high concentration of chlorophyll-a in water bodies to rely on oceanographic conditions in the region [4].

Oceanographic conditions can affect the distribution of phytoplankton and chlorophyll-a spatially, like salinity and Total Suspended Solid (TSS). Salinity as a physical parameter in seawater that states the level of dissolved salt in water can cause osmotic pressure between the body of the organism and seawater, so that the higher the level of salt will lessen the existence of phytoplankton [5]. TSS as a representation of the level of turbidity can affect the productivity of seawaters as an effect of the disrupted light penetration inside seawater, where the light is needed by phytoplankton to photosynthesize [6]. Chlorophyll-a, salinity, and TSS can be detected using remote sensing technology.

Remote sensing is a technology which can monitor the sea appearance through satellites, where from the satellite creates images consist of several spectral band sensors. One of the images which has said properties are Landsat 8 OLI Imagery [7]. Ocean colour sensors on Landsat-8 OLI can be processed using certain algorithms and software therefore this method can identify spatial information related to the distribution of chlorophyll-a concentrations [8]. This statement is also supported by Snyder et al. [9] in his research to map potential of oyster cultivation based on chlorophyll-a concentration, sea surface temperature, and sea surface turbidity using Landsat-8 OLI Imagery.

This research explains the spatial and temporal distribution of Chlorophyll-a. The results of this research are expected to be used by the stakeholders whose activities concern the field of marine and fisheries as a source of information in predicting fertile seawaters. This research can also be used as an indicator to map pelagic fish catching potential areas in Cirebon seawater according to concentration of chlorophyll-a distribution conditions in seawater.

## 2. Methodology

The study area of this research, is done in parts of seawater that exist in the coast of a part of a district that is in the city and regency administrative border of Cirebon next to Java Sea. The districts mentioned in this research, are Mundu, Astanajapura, and Pangenan district in Cirebon regency along with Kejaksan district, and Lemahwungkuk in Cirebon city. The research area is in the coordinate location of 108°34’00”-108°39’30” E and 6°41’30”-6°47’00” S.

The variables used in this study are the distribution of chlorophyll-a concentration, salinity, and TSS. These three variables are obtained from several data which are rainfall data and reflectance data on Landsat 8 OLI images. Materials needed to gather data are Landsat 8 OLI images and in-situ samples from the three variables. The rainfall data used to determine wet months and dry months, are rainfall >200 mm/month are categorised as wet month and rainfall <100 mm/month are categorised as dry month [10]. Landsat 8 OLI images are used as a representation of the distribution of chlorophyll-a variables, salinity, and TSS. In-situ data samples are used to validate images that have been processed.

The processing of Landsat 8 OLI imagery (Figure 1) are done using ENVI 5.1 and ArcGIS 10.1 software. ENVI 5.1 is used to process radiometric correction and calibration, cropping, and algorithm input. Radiometric calibration is used to convert image pixels into reflectance value. The radiometric corrections that are done are sun angle corrections to fix error reflectance values caused by sun positioning [11]. Cropping is done to minimize the image coverage according to the study area. Algorithm inputs are done to convert reflectance values of images into chlorophyll-a, salinity, and TSS values using a certain equation. The data processing uses ENVI 5.1 software to produce three raster data, which are raster with the three variable values. After the three data of raster are produced, ArcGIS 10.1 is used to process said data for image validation and classification. A detailed explanation regarding validation is provided in the next paragraph. Classification is done to divide three variable values into four or five classes to simplify the spatial analysis of the three variables. Data processing using ArcGIS 10.1 software will produce the chlorophyll-a distribution map, salinity map, and TSS map.

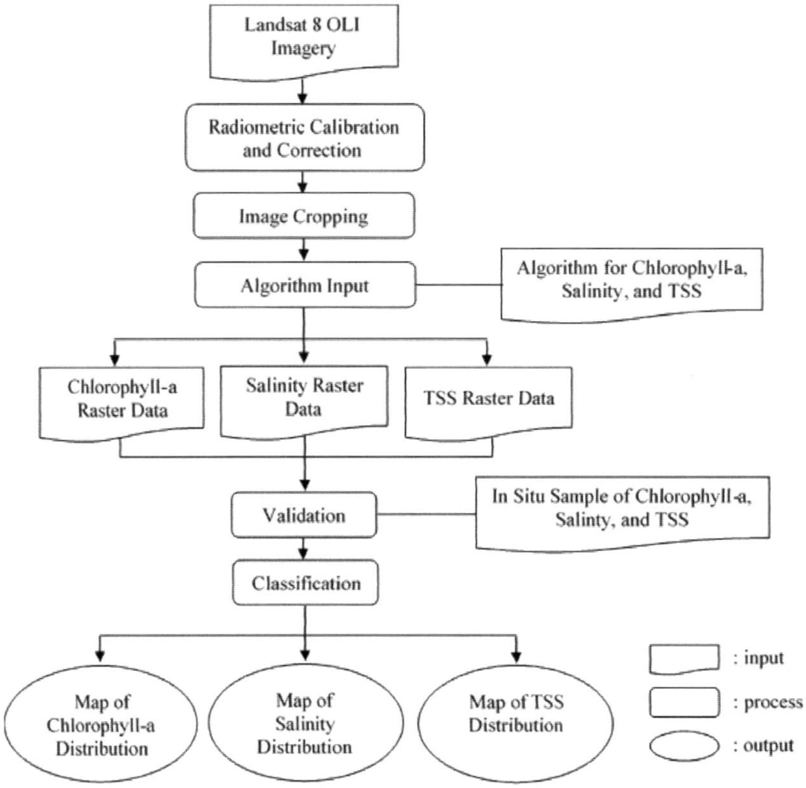


**Figure 1.** Landsat 8 OLI Data Processing Workflow

Landsat 8 OLI imagery data that have been corrected and calibrated are then converted into chlorophyll-a concentration values, salinity, and TSS. There are three algorithms which are used to identify the concentration of chlorophyll-a, salinity, and TSS. Chlorophyll-a concentration using algorithm of Wibowo et al. [12], salinity using the algorithm of Supriatna et al. [13], and TSS using the algorithm of Budhiman [14], where these three researches have done the mapping of variables related to Indonesian seawaters. These three algorithms can be seen in equations 3.1, 3.2, and 3.3.

The imagery that has been processed into the three variables, can then be validated by counting the Root Mean Square Error (RMSE) between specified sample data from image processing with in-situ measurements which obtained by field survey. Acquiring sample points are based on distance from coastline each 600 meters until it reaches the distance approximated about 6 km as many as 20 points (Figure 2). RMSE uses the equation 3.4 below.

$$Log\ Chl = (2,41*B4/B3)+0,187 \quad (3.1)$$

$$Sln = 29,983+165,047(B2)-260,227(B3)+2,609(B4) \quad (3.2)$$

$$TSS = 7,9038*exp(23,942*B4) \quad (3.3)$$

$$RMSE = \sqrt{\frac{\sum(z_i-z_j)^2}{n}} \quad (3.4)$$

| Notes: | | | | |
|---|---|---|---|---|
| Chl | = Chlorophyll-a Concentration ($mg/m^3$) | RMSE | = Root Mean Square Error |
| Sln | = Salinity (ppt) | Zi | = imagery data |
| TSS | = Total Suspended Solid (mg/L) | Zj | = in-situ data |
| B(n) | = Landsat 8 OLI canal reflectance value on n order band | n | = samples count |

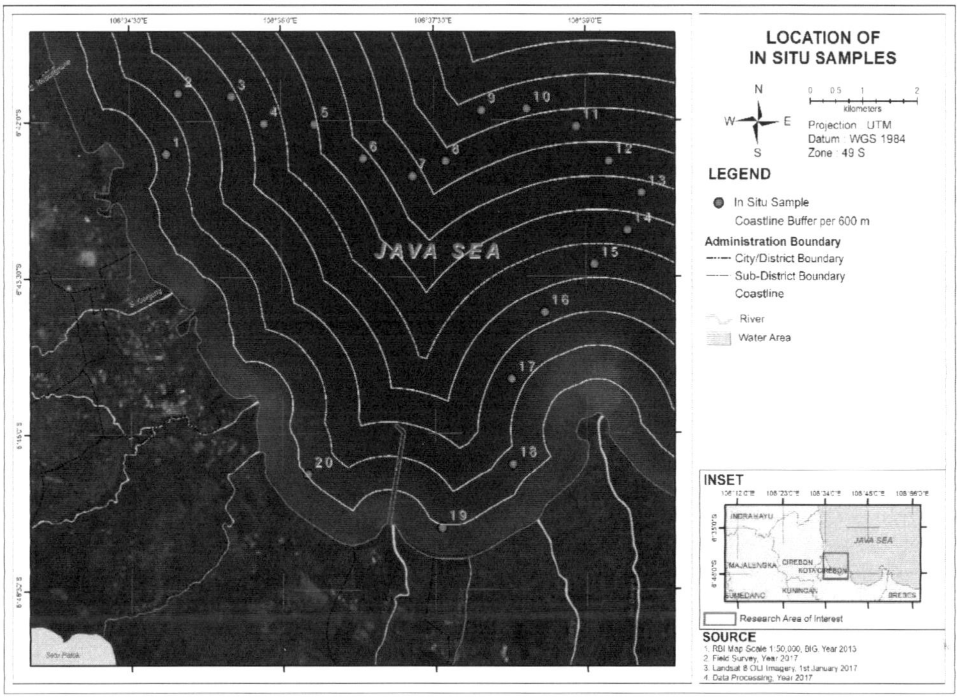


**Figure 2.** In-situ Samples Location

Data analysis used in this research are spatial analysis and descriptive statistics. Spatial analysis is done by overlaying the three map variables. Statistical analysis is done with testing the relationship between the sample values of the three variables during wet season and dry season with regression method. Temporally, chlorophyll-a concentrations are discussed according to wet month and dry month on the year 2014-2015. In regard to spatial distribution, the concentration of chlorophyll-a is discussed with salinity variables and TSS to discover its relationship.

## 3. Result and Discussion

### *3.1. Rainfall Condition of Cirebon*

Cirebon regency and city which is in the coastal area including inside the tropical climate and has a monsoon rainfall pattern caused by wind movement. Rainfall season usually occurs on the months October till May, while dry season falls on the months of June-September. Badan Meterorologi Klimatologi dan Geofisika (BMKG) data (Table 1) shows that the count of monthly rainfall in the year 2014-2015 has a range of 0-710,4 mm/month.

**Table 1.** Rainfall Data in Cirebon Region

| Month | Rainfall in 2014 (mm/month) | Rainfall in 2015 (mm/month) |
|---|---|---|
| January | 710.4 | 276 |
| February | 319.7 | 383.1 |
| March | 243.4 | 107.9 |
| April | 216 | 48 |
| May | 85 | 139.6 |
| June | 169.5 | 10 |
| July | 62 | 0 |
| August | 0 | 0 |
| September | 0 | 0 |
| October | 37 | 0 |
| November | 66.7 | 30.6 |
| December | 459.4 | 119.5 |

**Notes:**
Wet Month
Dry Month

According to Oldeman classification [10], there are 5 wet months and 6 dry months in the year 2014, while in the year 2015 there are 2 wet months and 7 dry months. This research uses four Landsat 8 OLI imagery data, among others acquisition of date 9 January 2014, 6 September 2014, 28 January 2015, and 9 September 2015. The selection of this data is based on the conditions of the image that is avoided from the amount of cloud cover and as a representation of the wet months and dry months on each year.

### *3.2. Data Validation*

The data which is measured to validate Landsat 8 OLI processed data are the in-situ value of chlorophyll-a concentrations, salinity, and TSS which are obtained from the field survey on the date 7 April 2017 and pixel values from Landsat 8 OLI imagery data processing of Landsat 8 OLI acquisition of date 1 January 2017.

**Table 2.** Comparison of In-situ and Imagery Processed Data

| Sample | Coordinates | | Chlorophyll-a (mg/m3) | | Salinity (ppt) | | TSS (mg/L) | |
|---|---|---|---|---|---|---|---|---|
| | Lat | Long | in situ | image | in situ | image | in situ | image |
| 1 | -6.7052 | 108.5811 | 0.77 | 0.59 | 26 | 24 | 40.50 | 39.39 |
| 2 | -6.6956 | 108.5831 | 0.77 | 0.57 | 29 | 27 | 28.50 | 32.19 |
| 3 | -6.6961 | 108.5921 | 0.49 | 0.55 | 27 | 27 | 26.25 | 30.43 |
| 4 | -6.7004 | 108.5974 | 0.66 | 0.55 | 28 | 28 | 22.25 | 26.53 |
| 5 | -6.7006 | 108.6056 | 0.70 | 0.56 | 26 | 29 | 31.75 | 26.31 |
| 6 | -6.7060 | 108.6135 | 0.41 | 0.54 | 28 | 29 | 26.25 | 24.03 |
| 7 | -6.7089 | 108.6215 | 0.35 | 0.53 | 29 | 29 | 25.75 | 23.83 |
| 8 | -6.7065 | 108.6270 | 0.46 | 0.53 | 28 | 29 | 31.50 | 24.07 |
| 9 | -6.6984 | 108.6329 | 0.55 | 0.53 | 28 | 29 | 27.75 | 24.38 |
| 10 | -6.6982 | 108.6404 | 0.33 | 0.53 | 25 | 28 | 23.75 | 25.94 |
| 11 | -6.7010 | 108.6486 | 0.37 | 0.56 | 26 | 28 | 27.25 | 28.94 |
| 12 | -6.7065 | 108.6540 | 0.36 | 0.50 | 25 | 28 | 31.50 | 24.92 |
| 13 | -6.7116 | 108.6594 | 0.50 | 0.50 | 25 | 27 | 32.50 | 27.31 |
| 14 | -6.7176 | 108.6571 | 0.64 | 0.54 | 24 | 27 | 33.00 | 29.83 |
| 15 | -6.7231 | 108.6515 | 0.59 | 0.51 | 23 | 26 | 28.50 | 31.03 |
| 16 | -6.7307 | 108.6433 | 0.69 | 0.55 | 23 | 25 | 41.75 | 37.59 |
| 17 | -6.7414 | 108.6377 | 0.69 | 0.47 | 26 | 27 | 25.50 | 25.87 |
| 18 | -6.7550 | 108.6379 | 0.43 | 0.48 | 20 | 22 | 37.50 | 38.29 |
| 19 | -6.7651 | 108.6262 | 0.95 | 0.61 | 21 | 19 | 68.75 | 61.20 |
| 20 | -6.7565 | 108.6042 | 0.93 | 0.60 | 19 | 20 | 65.50 | 58.21 |
| RMSE | | | 0.168 | | 2 | | 4.35 | |

The results of validation calculation by comparing in-situ values with image processing results show that RMSE values (Table 2) as large as 0,168 for chlorophyll-a, two for salinity, and 4,35 for TSS. According to the RMSE values on each variable this shows that there exists a positive correlation between in-situ data with remote sensing image, so that the algorithms which are used on the image can be used to represent the actual condition of chlorophyll-a concentration, salinity, and TSS distribution in the study area.

### *3.3. Chlorophyll-a Distribution*

Values of chlorophyll-a concentration in the study area, are according to Landsat-8 OLI imagery processing, that has a range value of 0.35-1.1 mg/m$^3$. Averages of chlorophyll-a concentration on 9 January 2014 (Wet 1) are 0.60 mg/m$^3$, on 6 September 2014 (Dry 1) with the value of 0.54 mg/m$^3$, on 28 January 2015 (Wet 2) with the value of 0.61 mg/m$^3$, and on 9 September 2015 (Dry 2) with the value of 0.55 mg/m$^3$. This shows that the average values of chlorophyll-a concentrations tend to grow higher in wet month compared to dry month.

Distribution of chlorophyll-a in the study area varies temporally as well as spatially. Temporally, the area of chlorophyll-a classification in wet months has a higher score compared to dry months (Figure 3). This is shown by the wet months which are dominated by the chlorophyll-a values of 0.5-0.65 mg/m$^3$ with the area of 60.20% on the date 9 January 2014 and the area 81.52% on the date 28 January 2015, while the dry month is dominated by chlorophyll-a 0.35-0.5 mg/m$^3$ with an area of 61.76% on the date 6 September 2014 and the area of 59.94% on 9 September 2015. Chlorophyll-a with the highest classification of 0.85-1.1 mg/m$^3$ is only available in wet months such as 9 January 2014 with the area of 6.22% and 28 January 2015 with an area of 0.82%, while on dry months there are almost none. Temporal variations are caused by rainfall which affects the runoff from river flow as a main nutrient supplier for the phytoplankton. Wet months which have higher rainfall compared to dry months cause the high amount of freshwater poured out into the sea which have salt water, so that the concentration of chlorophyll-a becomes higher. According to the results of this temporal analysis, it can be shown that phytoplankton in this study area has a higher abundance on wet months than dry months.

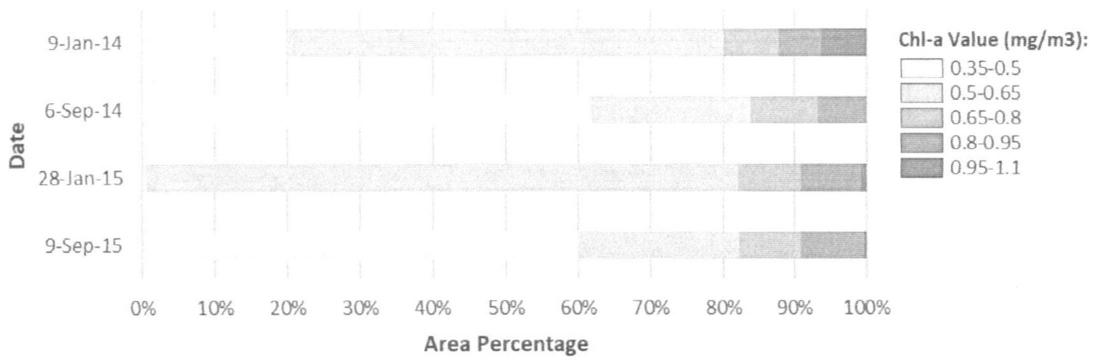


**Figure 3.** Graph for Area Percentage of Chlorophyll-a Classification

The distribution of chlorophyll-a concentration on study area is spatially tend to have significant differences based on distance from coastline. Chlorophyll-a has higher value on the area near of coastline and estuary zone, while it has getting lower towards the open sea area. This statement is based on sample data of number 1, 2, 18, 19, and 20 which are the nearest samples with the coastline and estuary zone. These five samples have highest average chlorophyll-a concentration level with pixel value of each 0.589, 0.592, 0.599, 0.752, and 0.829 mg/m$^3$. These values are higher than other samples pixel value which are located on further marine water area. This dominancy of chlorophyll-a concentration spatial pattern caused by mainland influence which has more nutrient supply brought by runoff of river flow.

The nutrient mentioned can be formed by many causes, be it caused by agricultural activities, domestic waste, industry waste, and many other activities. This chlorophyll-a concentration pattern means that phytoplankton abundance has higher amount if it near of coastline of estuary zone, be compared to far on the marine waters area.

On the map of chlorophyll-a concentration distribution (Figure 4) it can also be seen that the distribution has similar patterns between wet months and dry months on the year 2014 as well as 2015, however there is a significant difference on the level of concentration. Chlorophyll-a levels on the dry months have a tendency to be lower compared to the wet months. This difference in chlorophyll-a levels can be seen on the outer regions of the coastline which are dominated by the concentration 0.35-0.5 mg/m3, however on the wet months this region tends to change into higher levels of concentration 0.5-0.65 mg/$m^3$.

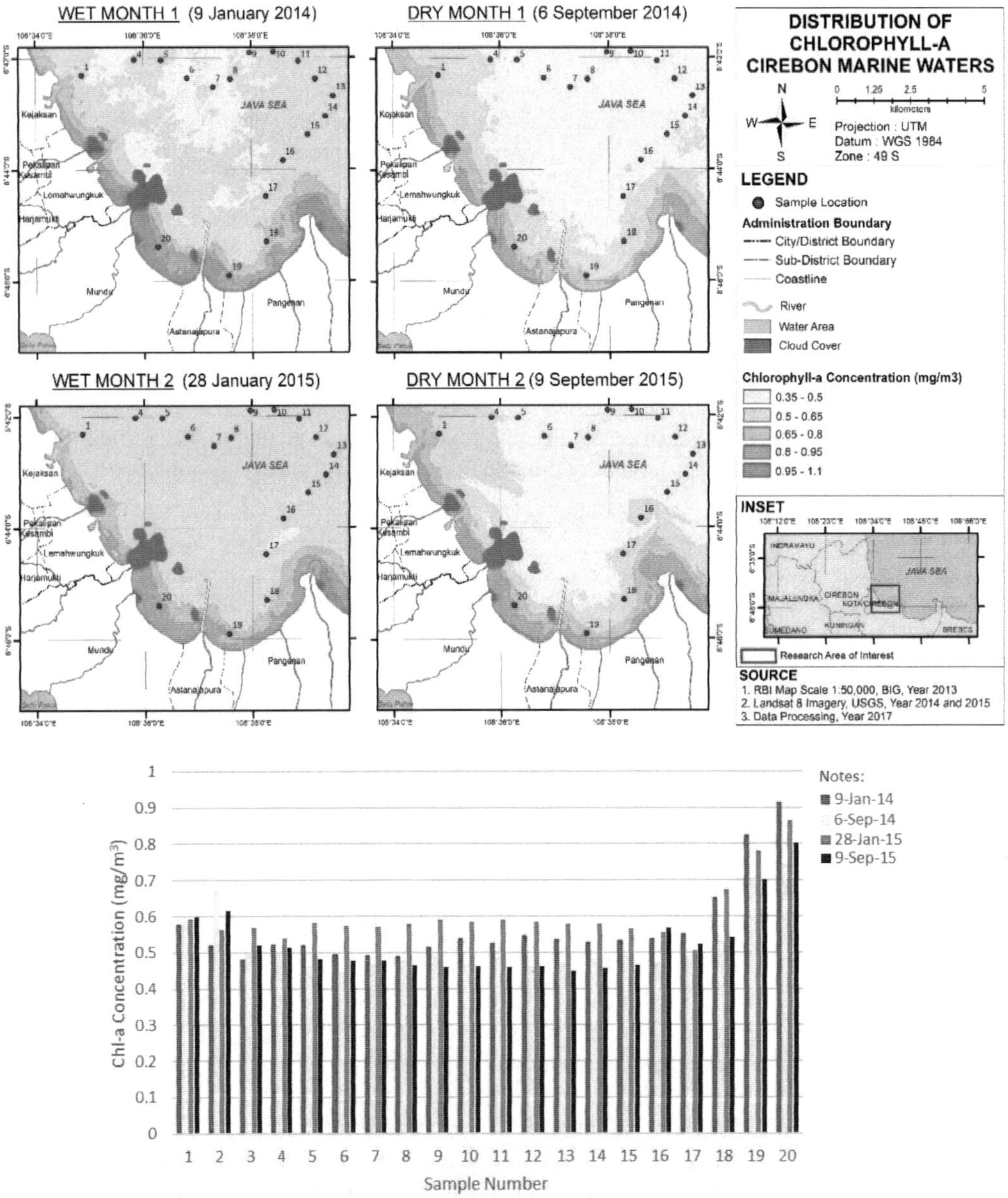


**Figure 4.** Map of Chlorophyll-a Distribution and Its Value on each Samples

### *3.4. Relationship of Chlorophyll-a with Salinity and TSS*

The result of linear regression test between chlorophyll-a concentration and salinity (Figure 5a and 5b) shows that salinity variable affects the distribution of chlorophyll-a concentration whether on wet or dry month in study area. The regression result on both wet and dry month shows significance value of 0.000, which is less than significance level of 5% or 0.05. It is means that $H_0$ rejected so that salinity significantly affects chlorophyll-a on wet and dry month. Salinity on wet month has strong relationship as indicated by R value of 0.852 and $R^2$ value of 72.6%. These values are higher when it compared with dry month condition which has R value of 0.798 and $R^2$ value of 63.8%. The regression model on wet month shows the equation of Y=1,148-0.022X, which means that in each addition on salinity with value of 1 ppt can reduce the level of chlorophyll-a concentration in amount of 0.022 mg/m$^3$. This equation model has a little bit different when it compared on dry month condition which has equation of Y=1,236-0,027X, which means that in each 1 ppt value of salinity will reduces 0.027 mg/m$^3$ of chlorophyll-a concentration. Both on these wet and dry month equation shows that salinity has a negative or reversed linear relationship with chlorophyll-a on both season.

The other linear regression is also test the relationship between chlorophyll-a concentration and TSS (Figure 5c and 5d) which shows that TSS affects the distribution of chlorophyll-a concentration on both wet and dry month in study area. It has same value of significance value from salinity test, which shows value of 0.000. This means there is significant influence of TSS variable into chlorophyll-a distribution. This result supported by R value of 0.902 on wet month and 0.948 on dry month, and also $R^2$ value of 81.3% on wet month and 90% on dry month. These values show that TSS has higher influence on dry month than on wet month. The regression model shows the equation of Y=0,341+0,005X on wet month and Y=0,293+0,004X on dry month. It means that in each addition of 1 mg/L of TSS can increase the amount of chlorophyll-a concentration with value of 0.005 on wet month and 0.004 on dry month. These equations also show that TSS has positive or direct linear relationship with chlorophyll-a on both season.

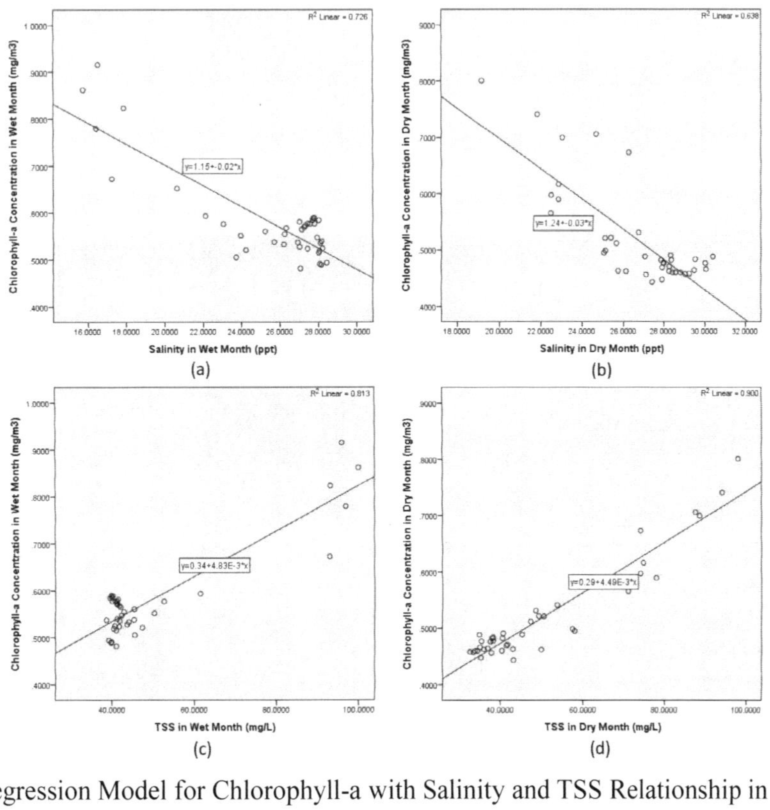


**Figure 5.** Regression Model for Chlorophyll-a with Salinity and TSS Relationship in Each Season

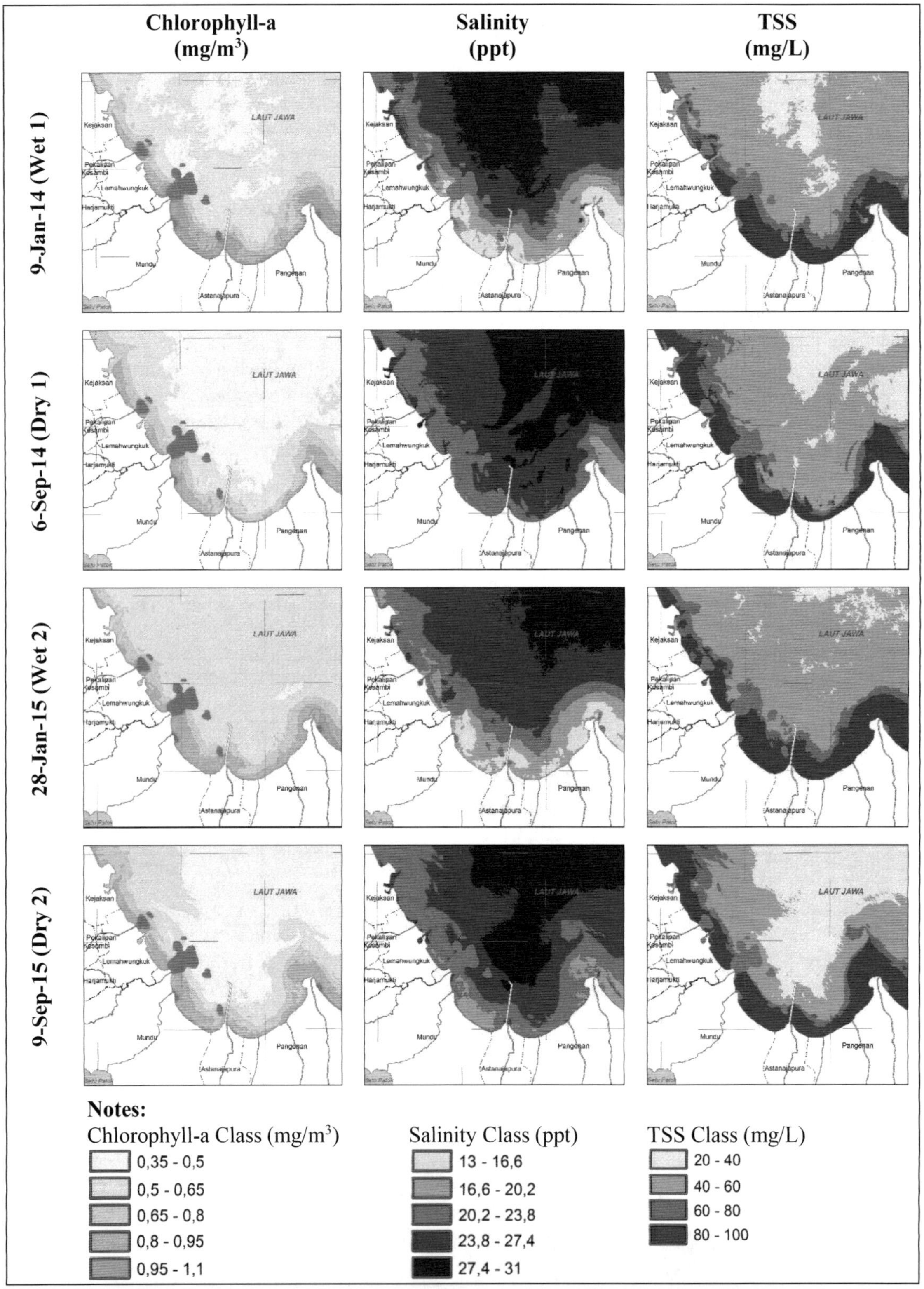


**Figure 6.** Spatial Comparison of Chlorophyll-a, Salinity, and TSS Distribution

Spatial appearance of chlorophyll-a concentration and salinity on each month (Figure 6) which obtained by Landsat 8 OLI processed data has a similar statement with the linear regression test result. Temporally, the salinity on wet month tend to has lower level compared to salinity on dry month. This temporal pattern shows that marine waters is more saltiness on dry month than on wet month. It is because wet month has higher rainfall than on dry month so river runoff is also in high intensity. This high intensity of runoff leads high supply of freshwater from mainland into river outlet and marine waters. Spatially, salinity distribution tends to associated and grouped based on its direction from coastline. The closer of coastline makes salinity has lower level in a value of 16 ppt, meanwhile it is get higher on the further to marine water in a value of 31 ppt in study area. This spatial and temporal pattern of salinity distribution has a reverse relationship with the distribution of chlorophyll-a concentration. Temporally, the amount of chlorophyll-a concentration is in high level, meanwhile salinity is in low level. Spatially, both chlorophyll-a concentration and salinity tend to associated or grouped based on its direction from coastline. Chlorophyll-a concentration has high level in the area near with coastal zone, meanwhile salinity has low level in the same zone. Conversely, chlorophyll-a until the direction of approximately 6 km from coastline has low level, meanwhile salinity has high level in the same location.

Spatial appearance of chlorophyll-a concentration and TSS (Figure 6) are also connected to the linear regression test result. Temporally, TSS on wet month tend to has higher level compared to TSS on dry month. This temporal pattern shows that marine waters is more turbid on wet month than on dry month in study area. It is because the high rainfall intensity on wet month period makes the runoff by river get increased. The increasing of runoff on wet month period will cause high amount of suspended solid materials flows from mainland to river outlet then to marine waters at the end. Spatially, TSS distribution in study area tend to associated and grouped based on its distance location from coastline and estuary zone. The closer of coastline makes TSS has high level in a value of 100 mg/L, meanwhile it is get lower on the further to marine water in a value of 100 mg/L in study area. This spatial and temporal pattern of salinity distribution has a direct linear relationship with the distribution of chlorophyll-a concentration. Temporally, the amount of chlorophyll-a concentration is in high level, followed by TSS in high level. Spatially, both chlorophyll-a concentration and TSS tend to associated or grouped based on its direction from coastline. Chlorophyll-a concentration has high level in the area near with coastal zone, followed by TSS with low level in the same zone. Conversely, chlorophyll-a until the direction of approximately 6 km from coastline has low level, meanwhile salinity has high level in the same location.

## 4. Conclusions

This research is done on the year of 2014-2015 based on wet month and dry month in study area of Cirebon marine waters. Wet month in the study area occur on January to April 2014 and December 2014-February 2015 in count of 7 months, meanwhile dry month occur on May-November 2014 and April-November 2015 in count of 13 months. The selection of one of wet and dry season on each year used as the research temporal termination to processing Landsat 8 OLI imagery data. Imagery data processed result combined with in-situ measurement samples shows that Landsat 8 OLI imagery can be used to represent the distribution chlorophyll-a concentration by Wibowo et al. [12] algorithm, salinity by Supriatna et al. [13] algorithm, and TSS by Budhiman [14[ algorithm. Temporally, the distribution of chlorophyll-a concentration shows higher value on wet month than on dry month. Spatially, the distribution of chlorophyll-a concentration has a tendency to grouped based on its direction from coastline, which it has higher level on the areas near with coastline. Salinity and TSS significantly affect the distribution of chlorophyll-a concentration on both wet and dry month in study area. Salinity has reverse relationship pattern with chlorophyll-a, meanwhile TSS has direct relationship. Low level of salinity and high level of TSS can increase the amount of chlorophyll-a concentration.

## Reference

[1] Odum E & Barret G 2004 *Fundamentals of Ecology: 5$^{th}$ Edition*. Cengage Learning, Philadelphia,

Saunders.

[2] Nurdin S, Mustapha M A, Lihan T, & Zainuddin M 2017 Applicability of remote sensing oceanographic data in the detection of potential fishing grounds of Rastrelliger kanagurta in the archipelagic waters of Spermonde, Indonesia. *Journal of Fisheries Research* Vol.196, Scopus.

[3] Hanintyo R & Susilo E 2016 Comparison of Chlorophyll-a Measurement Using Multi Spatial Imagery and Numerical Model in Bali Strait *Journal of Frontiers in Marine Science* Vol.4, Scopus.

[4] Afdal 2014 Chlorophyll-a Variation in Cirebon Waters and Its Relationship With Nutrient Concentration *Jurnal Oseanologi dan Limnologi di Indonesia*, LIPI, Jakarta, Indonesia.

[5] Patty S I 2013 Distribusi Suhu, Salinitas, dan Oksigen Terlarut di Perairan Kema, Sulawesi Utara. *Jurnal Ilmiah Platax*, Vol.1 No.3, Manado, Indonesia.

[6] Parwati E & Purwanto A D 2014 Analisis Algoritma Ekstraksi Informasi TSS Menggunakan Data Landsat 8 di Perairan Berau, In: *Prosiding Sinasinderaja, LAPAN.*

[7] Danoedoro P 2012 *Pengantar Penginderaan Jauh Digital* (Andi Offset: Yogyakarta, Indonesia)

[8] IOCCG 2000 Remote Sensing of Ocean Colour in Coastal, and Other Optically-Complex, Waters. *Report of International Ocean-Colour Coordinating Group No.3*, Dartmouth, Canada.

[9] Snyder J, Boss E, Weatherbee R, Thomas A C, Brady D, & Newell C 2017 Oyster Aquaculture Site Selection Using Landsat 8-derived Sea Surface Temperature, Turbidity, and Chlorophyll-a, *Journal of Frontiers in Marine Science* Vol.4, Scopus.

[10] Oldeman L 1975 *An Agroclimate Map of Java and Madura* (Institut Pertanian Bogor, Indonesia)

[11] Ardiansyah 2015 *Pengolahan Citra Penginderaan Jauh Menggunakan ENVI 5.1 dan ENVI LiDAR (Teori dan Praktek).* Jakarta, Indonesia.

[12] Wibowo A, Sumartono B, & Setyantini W 1993 The Application of Satellite Data For Improvement Site Selection and Monitoring Shrimp Pond Culture, In: *Proceeding National Conference on ERS-1, Landsat, SPOT*, Jakarta, Indonesia.

[13] Supriatna L, Supriatna J, & Koetsoer R 2016 Algorithm Model for The Determination of Cimandiri Estuarine Boundary Using Remote Sensing, In: *AIP Conference Proceedings*.

[14] Budhiman S 2004 *Mapping TSM Concentrations from Multisensor Satellite Images in Turbid Tropical Coastal Waters of Mahakam Delta, Indonesia* (ITC, Netherlands)

# Pattern Analysis of El Nino and La Nina Phenomenon Based on Sea Surface Temperature (SST) and Rainfall Intensity using Oceanic Nino Index (ONI) in West Java Area

**Yudo Prasetyo*** **and Farras Nabilah**

Remote Sensing and Photogrammetry Lab, Geodesy Engineering Dept, Faculty of Engineering, Diponegoro University, Jl. Prof. Soedharto SH, Tembalang-Semarang, Central Java, Indonesia, 50277

yudo.prasetyo@ft.undip.ac.id

**Abstract**. Climate change occurs in 1998-2016 brings significant alteration in the earth surface. It is affects an extremely anomaly temperature such as El Nino and La Nina or mostly known as ENSO (El Nino Southern Oscillation). West Java is one of the regions in Indonesia that encounters the impact of this phenomenon. Climate change due to ENSO also affects food production and other commodities. In this research, processing data method is conducted using programming language to process SST data and rainfall data from 1998 to 2016. The data are sea surface temperature from NOAA satellite, SST Reynolds (Sea Surface Temperature) and daily rainfall temperature from TRMM satellite. Data examination is done using analysis of rainfall spatial pattern and sea surface temperature (SST) where is affected by El Nino and La Nina phenomenon. This research results distribution map of SST and rainfall for each season to find out the impacts of El Nino and La Nina around West Java. El Nino and La Nina in Java Sea are occurring every August to February. During El Nino, sea surface temperature is between 27°C - 28°C with average temperature on 27.71°C. Rainfall intensity is 1.0 mm/day – 2.0 mm/day and the average are 1.63 mm/day. During La Nina, sea surface temperature is between 29°C - 30°C with average temperature on 29.06°C. Rainfall intensity is 9.0 mm/day - 10 mm/day, and the average is 9.74 mm/day. The correlation between rainfall and SST is 0,413 which is expresses a fairly strong correlation between parameters. The conclusion is, during La Nina SST and rainfall increase. While during El Nino SST and rainfall decrease. Hopefully this research could be a guideline to plan disaster mitigation in West Java region that is related extreme climate change.

**Keywords**: Rainfall, El Nino, La Nina, ONI Index and SST Reynold.

## 1. Introduction

### *1.1. Problem Background*

Rainfall is one of the most important climatic elements, but its existence is spatially and temporally still difficult to predict. In addition to its dynamic nature, the physical processes involved are also very complex. The uncertainty of rainfall pattern is complicated by anomalous climatic phenomenon like El Nino and La Nina. According to [3], various disastrous events in Indonesia show that most of the disasters are related to the El Nino Southern Oscilation (ENSO) phenomenon. ENSO is a recurrent pattern of climate variability in the eastern Pacific Ocean that is characterized by sea surface temperature anomalies (sea level warming describes the El Nino incident whereas sea-level cooling illustrates the incidence of La Nina) anomalies [1].

ENSO values can be demonstrated by the Oceanic Nino Index (ONI) as well as changes in sea surface temperature that have an impact on rainfall intensity. So the incidence of El Nino and La Nina causes a decrease and increase in rainfall in Indonesia.Extreme El Nino phenomenon occurred in 2015, causing some areas in Indonesia experiencing drought. According to BMKG Meteorology Deputy, Yunus Subagyo Swarinoto, the results of monitoring the development of El Nino until the beginning of June 2015, show moderate condition of El Nino phenomenon. This condition is expected to last until November 2015 and has higher probability to increase [11].

Extreme La Nina phenomenon occured in 2016 which causes the rainy season becomes longer than the dry season. That conditions resulted in disastrous flood in some areas in Indonesia. According to BNPB Head of Information and Public Relations Data, Sutopo Purwo Nugroho, the rainfall is expected to rise 200 percent. Besides Sumatra and Java, eastern part of Sulawesi, central Papua and Kalimantan are also affected by this phenomenon [3]. Indonesia has lies in the hemisphere with a tropical monsoon climate which is very sensitive to ENSO climate anomaly. In addition to causing the occurrence of El Nino phenomenon, ENSO also increase occurrence of La Nina phenomenon. The phenomenon of El Nino and La Nina also affect the territory of Indonesia, one of which is West Java Province, affected by dry La Nina during the 2016. Further consequences are the occurrence of increasingly long dry season and rainy season. La Nina is characterized by high rainfall intensity in a region that is entering the rainy season, considering the potential for rain is quite larger at that time.

In this research, ONI indicator (Oceanic Nino Index) has been used in observing ENSO phenomena in the form of El Nino and La Nina associated with rainfall in West Java. ONI is one of the most commonly used parameters to state the events of El Nino and La Nina. The Nino Ocean Index represents monthly average value of the SST (Sea Surface Temperature) the month after and before from normal condition, which is then compared with the normal SST of current month. If the value is less than -0.5, then that area is experiencing strong La Nina, -0.5 up to 0.5 experiencing neutral conditions and value of more than 0.5 indicating strong El Nino.

In the end, this research aims to establish SST distribution map and seasonal rainfall. The map will be used to determine the effect of El Nino and La Nina phenomenon in West Java. The final result of this research is the analysis of correlation between the impact of El Nino and La Nina on rainfall and correlation between rainfall data with online BMKG rainfall data.

### *1.2. Research Hypothesis*

a. How to identify the influence of El Nino and La Nina against 1998-2016 rainfall using ONI indicator?
b. What is the analysis of El Nino and La Nina spatial patterns from 1998-2016 based on the ONI indicator?

### *1.3. El Nino and La Nina Phenomenon*

#### *1.3.1. El Nino Phenomenon*

El Nino is defined as a phenomenon of a marked difference between observed sea level temperatures compared to normal conditions in the equatorial Pacific Ocean region. El Nino is a global-scale atmospheric oceanic phenomenon [9]. Such conditions occur repeatedly every 3-8 years period and are usually associated with negative value of the negative oscillation index.

#### *1.3.2. La Nina Phenomenon*

La Nina cannot be seen by the Pacific, even the period is not fixed. In average, La Nina occurs in 3 years to 7 years. It can last 12 months to 36 months; La Nina doesn't have a fixed period so all is expected to happen at 6 months to 9 months beforehand. At the time of La Nina phenomenon, sea surface temperature in the Eastern Equatorial Pacific was lower than its normal condition. While the

IOP Conf. Series: Earth and Environmental Science **98** (2017) 012041 doi:10.1088/1755-1315/98/1/012041

sea surface temperature in Indonesia becomes warmer. So that there are many convection and resulted in air mass gathered in the territory of Indonesia, including air mass from the Eastern Equatorial Pacific. It supports the formation of clouds and rain. So, La Nina phenomenon often result in rainfall far above normal that could cause flooding and landslide, or even followed by strong wind [2].

### *1.4. The Oceanic Nino Index (ONI Index)*
The Oceanic Index is an index showing the division of regions and measures the value of SST (Sea Surface Temperature) in these areas in the Pacific sea [6]. ONI is a new index, which actually is one of El Nino indexes. The ONI is calculated based on the calculation principle for monitoring, assessment and prediction of the ENSO cycle. The ONI also measures the change in the SST value of the regional average of Nino 3.4. The average of three-month is taken and the value of SST change similar to the historical SST analysis is observed [10].

## 2. Materials and Methods
### *2.1. Research Area*
The study area of this study is the territory of Indonesia, especially in the province of West Java lies in the position 104°48'00" E - 108°48'00" E and 5°50'00" S - 7°50'00"S. Research area can be seen in Figure 1.

### *2.2. Data Research*
Data researches needed in this research are:
a. NOAA image of Reynolds SST data from 1998-2016.
b. TRMM image in the form of daily data of rainfall year 1998-2016.
c. BMKG Daily Rainfall Record Years 1998-2016.

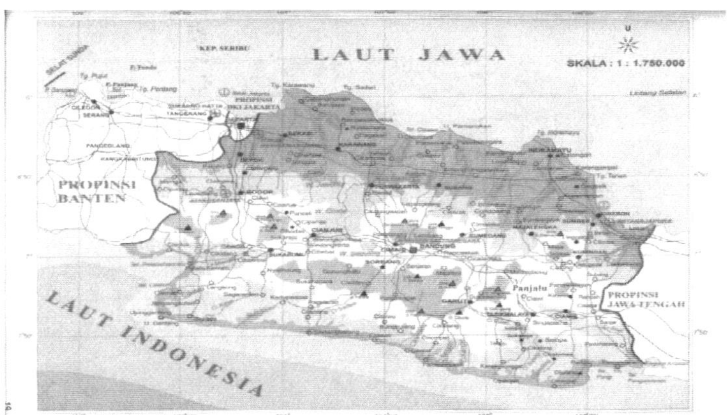


**Figure 1.** Research Area

### *2.3. Research Methodology*
For processing using IDL software divided by several stages can be seen in Figure 5. The final result of this data processing step with IDL is SST and precipitation that has been compiled into monthly climatology in the form (.txt) and (.png), the ONI index of SST is three-month average, and graph of the relationship between SST and rainfall parameters. For data processing flowchart in this research can be seen in Figure 2.

## 3. Results and Discussion
### *3.1. El Nino and La Nina phenomenon identification based on ONI algorithm*
The results of data processing of sea surface temperature and rainfall on a monthly basis are classified into three classes namely El Nino, La Nina and normal which is show in Table 1. The classification is

based on the ONI index that are average of moving obtained from previous month's SST, current SST and SST one month later.

Rainfall and SST data are processed using IDL software resulting monthly data of climatology as the outcome. So, the analysis of the relationship between SST and rainfall on El Nino, normal and La Nina can be established. During El Nino phenomenon, SST tends to be cooler while the intensity of rainfall decreases. Meanwhile, during La Nina phenomenon, SST tends to be warmer and rainfall intensity becomes greater as compared to El Nino. The impact of El Nino and La Nina on Rainfall are shown in Figure 3 and Figure 4.

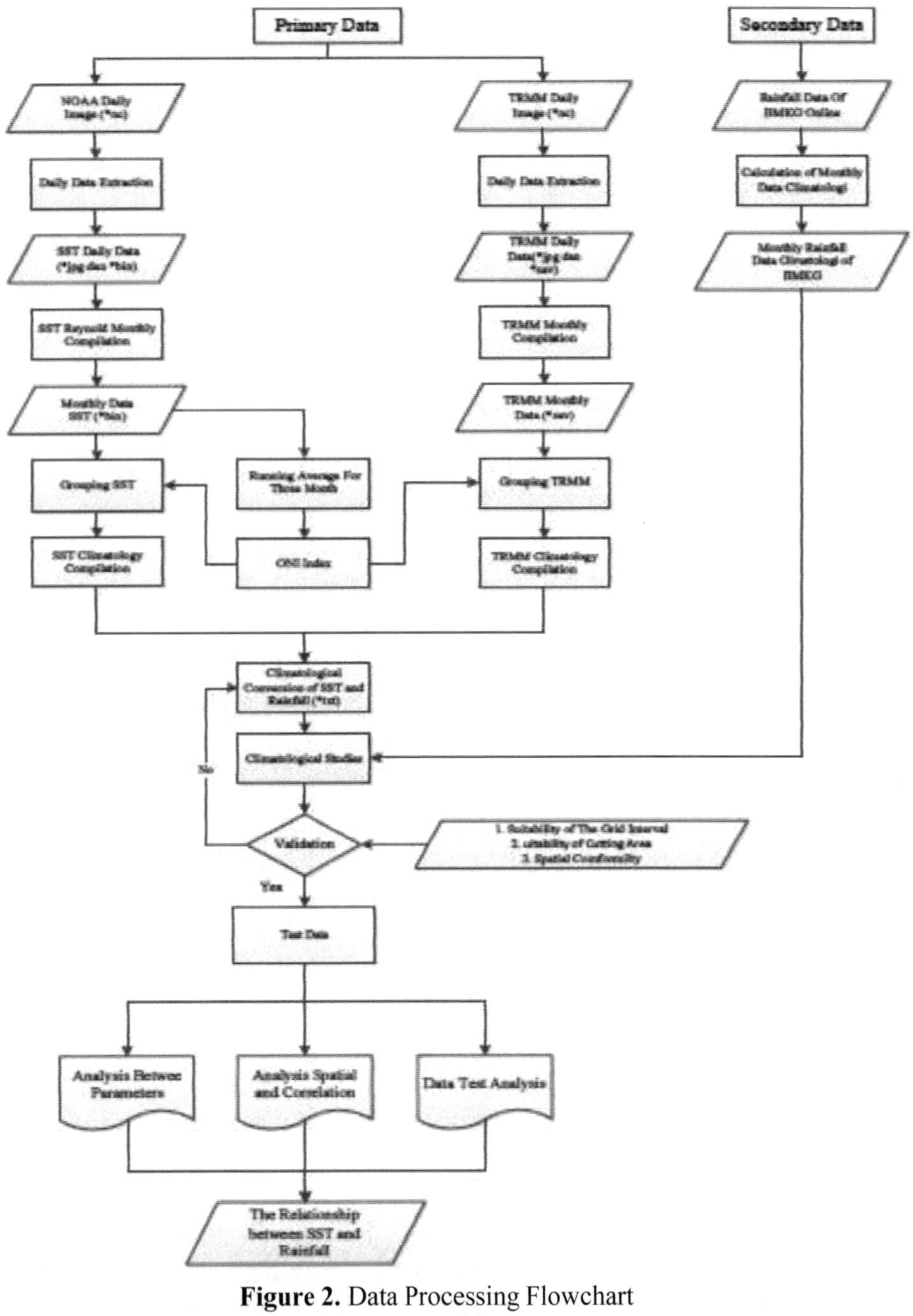


**Figure 2.** Data Processing Flowchart

**Table 1.** Monthly SST Reynolds Classification

| Month | Monthly Temperature | | | | | | | | | | | | | | | | | | |
|---|---|---|---|---|---|---|---|---|---|---|---|---|---|---|---|---|---|---|---|
| | 1998 | 1999 | 2000 | 2001 | 2002 | 2003 | 2004 | 2005 | 2006 | 2007 | 2008 | 2009 | 2010 | 2011 | 2012 | 2013 | 2014 | 2015 | 2016 |
| 1 | 0.00 | -1.87 | -2.03 | -1.16 | -0.58 | 0.71 | -0.07 | 0.16 | -1.17 | 0.32 | -1.85 | -1.13 | 1.19 | -1.68 | -1.09 | -0.86 | -0.89 | 0.26 | 2.01 |
| 2 | 1.54 | -1.37 | -1.65 | -0.73 | -0.22 | 0.64 | -0.03 | 0.25 | -0.90 | 0.10 | -1.55 | -0.84 | 1.20 | -1.33 | -0.70 | -0.56 | -0.75 | 0.40 | 1.95 |
| 3 | 1.43 | -0.75 | -0.93 | -0.24 | 0.27 | 0.69 | 0.31 | 0.53 | -0.23 | 0.24 | -0.86 | -0.18 | 1.30 | -0.64 | -0.24 | 0.07 | -0.09 | 0.93 | 1.90 |
| 4 | 1.38 | -0.29 | -0.35 | 0.19 | 0.71 | 0.62 | 0.67 | 0.92 | 0.38 | 0.44 | -0.17 | 0.48 | 1.22 | 0.03 | 0.27 | 0.52 | 0.75 | 1.35 | 1.66 |
| 5 | 0.86 | -0.20 | -0.04 | 0.42 | 1.12 | 0.48 | 0.90 | 1.07 | 0.81 | 0.59 | 0.10 | 1.00 | 0.79 | 0.38 | 0.67 | 0.63 | 1.19 | 1.72 | 1.28 |
| 6 | 0.11 | -0.39 | -0.16 | 0.40 | 1.15 | 0.37 | 0.89 | 0.86 | 0.72 | 0.30 | 0.12 | 1.07 | 0.07 | 0.39 | 0.76 | 0.33 | 0.93 | 1.69 | 0.50 |
| 7 | -0.81 | -0.82 | -0.34 | 0.19 | 0.99 | 0.25 | 0.74 | 0.45 | 0.52 | -0.10 | -0.01 | 0.89 | -0.65 | 0.03 | 0.72 | -0.06 | 0.43 | 1.64 | -0.13 |
| 8 | -1.21 | -1.12 | -0.60 | -0.15 | 0.76 | 0.10 | 0.62 | 0.00 | 0.33 | -0.70 | -0.21 | 0.65 | -1.24 | -0.45 | 0.46 | -0.26 | 0.05 | 1.58 | -0.65 |
| 9 | -1.44 | -1.36 | -0.78 | -0.38 | 0.73 | 0.05 | 0.48 | -0.23 | 0.36 | -1.17 | -0.39 | 0.58 | -1.57 | -0.82 | 0.30 | -0.40 | 0.03 | 1.72 | -0.87 |
| 10 | -1.50 | -1.52 | -0.95 | -0.54 | 0.90 | 0.07 | 0.43 | -0.46 | 0.51 | -1.49 | -0.51 | 0.82 | -1.68 | -1.04 | 0.21 | -0.30 | 0.27 | 2.06 | -1.03 |
| 11 | -1.77 | -1.77 | -1.17 | -0.69 | 1.02 | 0.09 | 0.38 | -0.75 | 0.66 | -1.63 | -0.76 | 1.12 | -1.67 | -1.17 | -0.14 | -0.41 | 0.34 | 2.23 | -1.05 |
| 12 | -1.92 | -2.04 | -1.26 | -0.73 | 0.92 | -0.03 | 0.30 | -1.11 | 0.61 | -1.81 | -1.03 | 1.29 | -1.75 | -1.21 | -0.50 | -0.60 | 0.29 | 2.24 | 0.00 |

La Nina >-0,5 Normal -0,5<0<0,5 El Nino >0,5

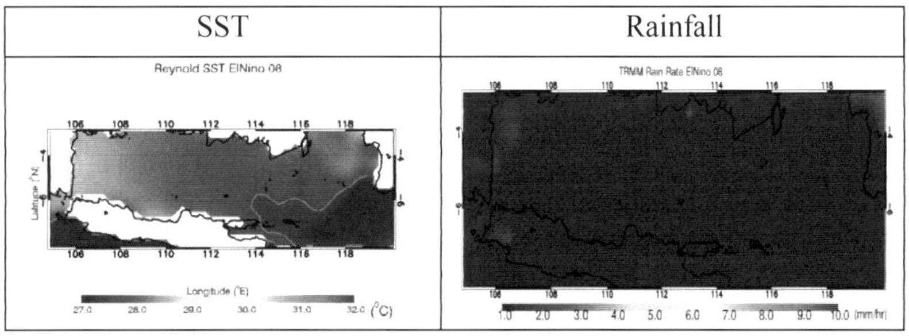


**Figure 3.** El Nino on August

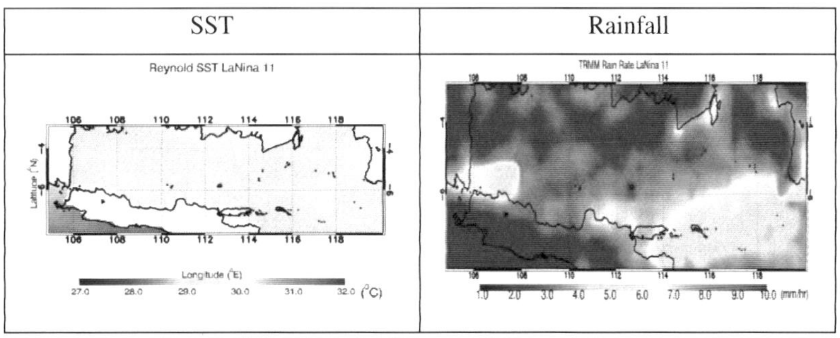


**Figure 4.** La Nina on November

### *3.2. Pattern Analysis of Sea Surface Temperature and Rainfall Variability*

This SST variability assessment aims to determine the spatial pattern of El Nino and La Nina during 1998 until 2016. The spatial distribution value of SST and monthly rainfall of climatology are shown in Table 2. It can be seen that the difference between SST and rainfall from January to December for almost 20 years. It shows a positive change when the sea surface temperature rises and rainfall rises. But in a few months, there is no interrelationship between sea level rise in temperature and rainfall.

Multi temporal graph between SST variable and rainfall in West Java region can be seen in Figure 5. The multi temporal graph shows the relationship between parameters in each month on a seasonal basis.

**Table 2.** SST Value and Rainfall Based on Monthly Climatology

| Months | Sea Surface Temperature (Celsius) | Rainfall (mm/days) |
|---|---|---|
| **January** | 28,63002 | 12,3085 |
| **February** | 28,64475 | 12,18645 |
| **March** | 28,94295 | 9,55454 |
| **April** | 29,42991 | 8,35645 |
| **May** | 29,55567 | 6,07406 |
| **June** | 29,05264 | 4,58059 |
| **July** | 28,29524 | 3,5924 |
| **August** | 27,70947 | 1,63389 |
| **September** | 27,75353 | 1,86311 |
| **October** | 28,42904 | 5,41364 |
| **November** | 29,05697 | 9,74061 |
| **December** | 29,05771 | 11,50177 |

Meanwhile in normal circumstances, the temperature is not remarkably high, but still at the value of 29°C-30°C. In the state of La Nina, SST looks cooler than normal. The temperature range at La Nina is 28°C-29°C. The spread of the temperature began to warm from the northern part of Java Sea so that the temperature spread from North to West. The average value of SST in the vicinity of West Java is 28.63°C.

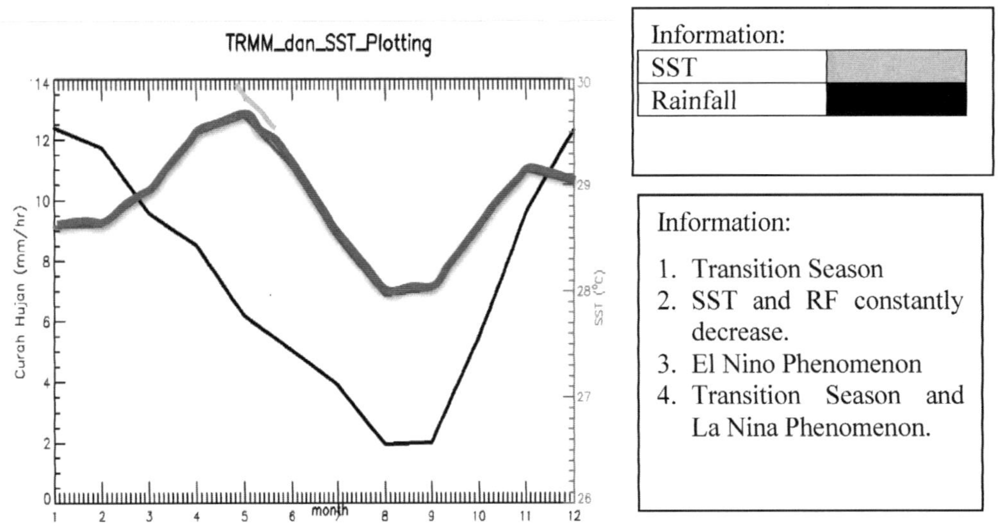


**Figure 5.** Rainfall Intensity in Multi Temporal Season

In January, as can be seen from Figure 6 that rainfall in Java undergoes shifting from La Nina to El Nino. This phenomenon is indicated by the color change from dark red to yellow with a value of 7.0mm/hr-10mm/hr. At time when La Nina tends to occur throughout West Java, these areas experience rainy season with a higher intensity than normal and El Nino. For January, the average value of rainfall in West Java is 12.30 mm/hr.

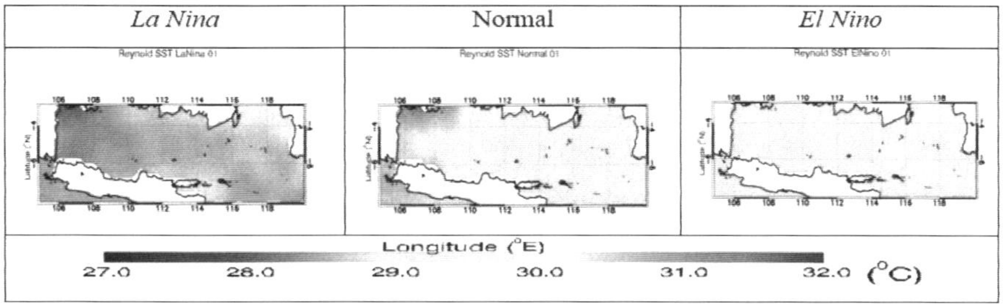


**Figure 6.** SST distribution during January for 20 years Climatology

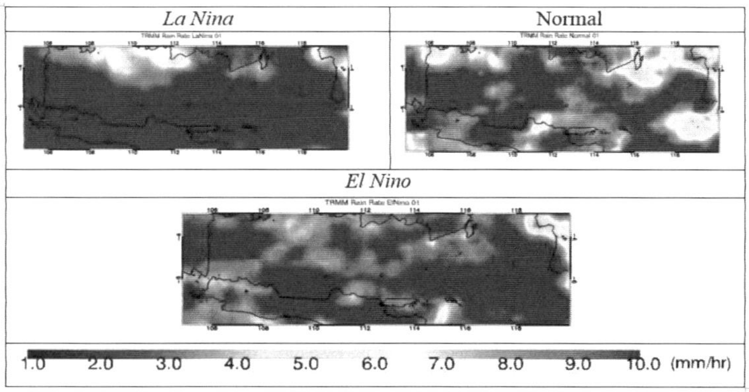


**Figure 7.** Rainfall distribution during January for 20 years Climatology

### *3.3. Pattern Analysis of Multi Temporal Graph of SST and Rainfall Variability*

This analysis aims to see the relationship between rainfall intensity and SST during La Nina and El Nino phenomenon. It was also to see the influence of on the impact of SST changes in Java Sea to the rainfall intensity during El Nino, normal and La Nina from 1998 to 2016 in the three part regions.

A. Northern region

For the northern part of West Java, samples were taken in Bekasi and Karawang region. The objective is to analyze the impact of sea surface temperature change in Java Sea on the intensity of rainfall in that regions. Multi-temporal graph analysis in Bekasi and Karawang region is shown in Figure 8.

As can be seen on Figure 8, rainfall during La Nina from January to June is under climatological rainfall while from July to November is above climatological rainfall. During El Nino, the entire phenomenon is opposite to the state of La Nina. During La Nina from January to June, SST is under the climatological SST while in July to November the value is above the climatological SST. As for December, both values are opposite, as El Nino values rise, whereas at La Nina the value falls.

B. Central region

For the central part of West Java, samples were taken in Bandung. The objective is to analyzed the impact of sea surface temperature change in Java Sea on the intensity of rainfall in that regions. Multi-temporal graph analysis in Bandung region is shown in Figure 9.

As can be seen in Figure 9, the increase of SST does not impact the change of rainfall significantly during January to April. From May to September, rainfall and SST intensity show the same pattern in

IOP Conf. Series: Earth and Environmental Science **98** (2017) 012041 doi:10.1088/1755-1315/98/1/012041

which is the SST value decrease and the rainfall value also decreases. The dotted line indicating the El Nino state has the same pattern but the value differs from the pattern in Bekasi and Karawang area.

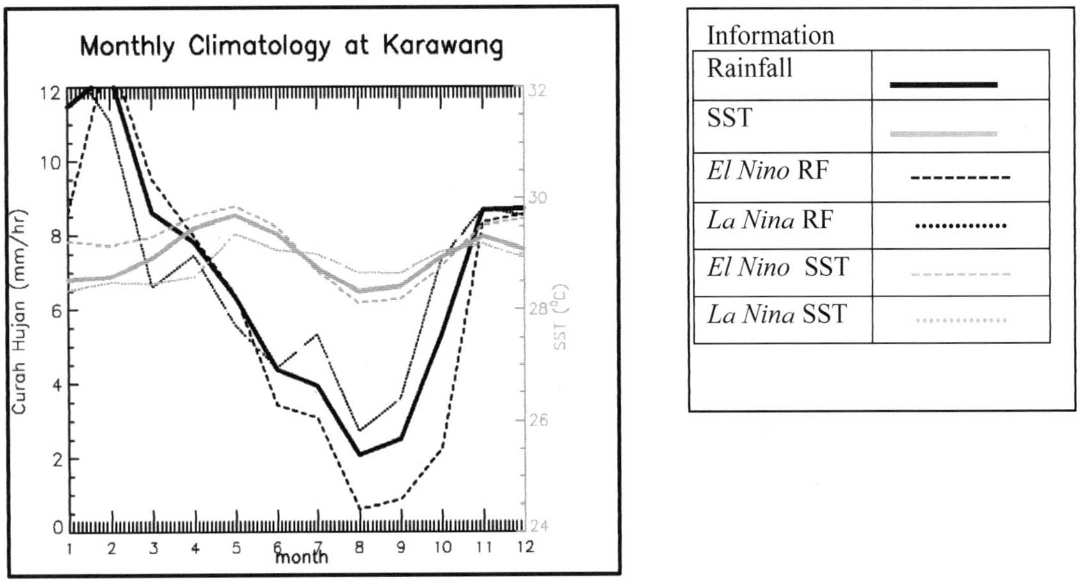


**Figure 8.** Multi Temporal Graph of Karawang Region

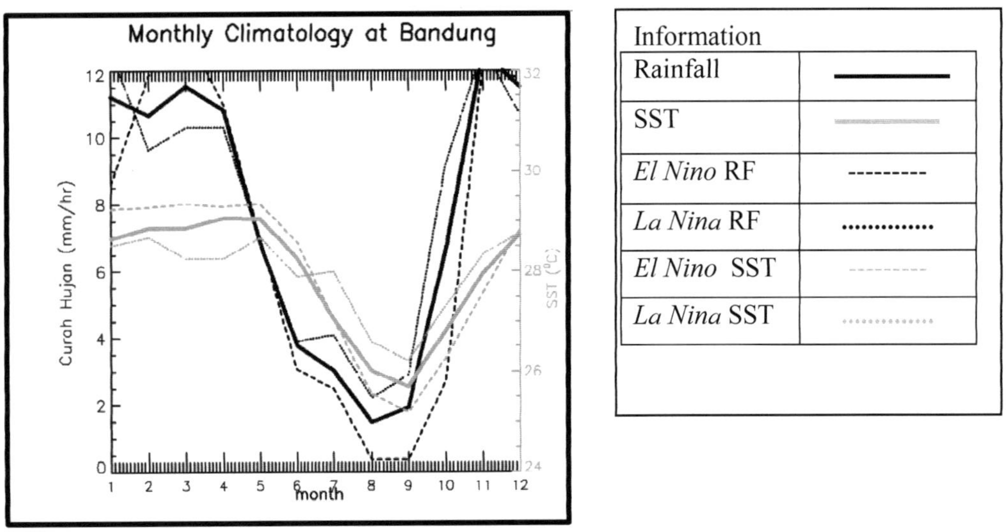


**Figure 9.** Multi Temporal Graph of Bandung Region

C. Southern Region

For the southern part of West Java, samples were taken from Sukabumi area. The objective is to analyze the impact of sea surface temperature change in Java Sea on the intensity of rainfall in that region. Multi-temporal graph analysis in the Sukabumi region is shown in Figure 10.

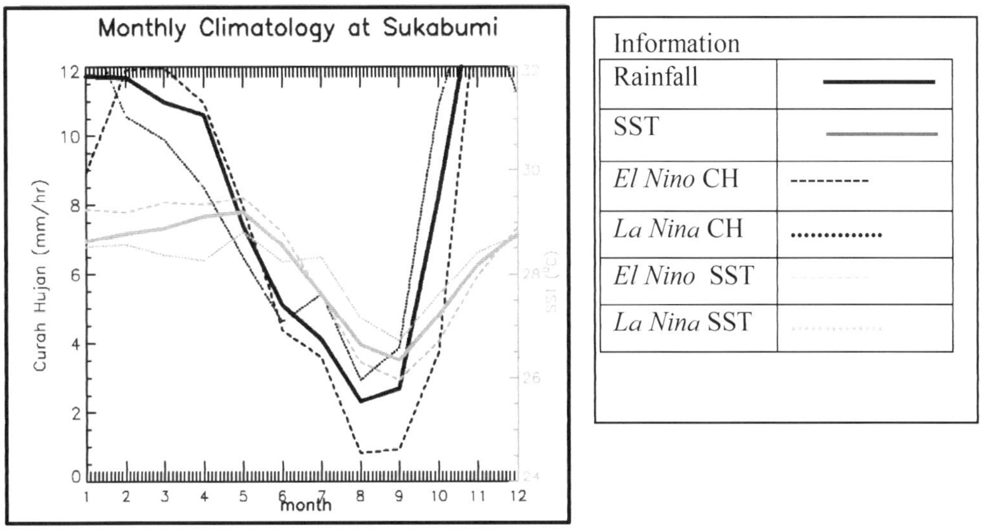


**Figure 10.** Multi Temporal Graph of Sukabumi Region

As can be seen on Figure 10, the increase of SST does not impact significantly on the change of rainfall during January to May. As for June to September, the pattern is similar so that during the decrease of SST, the value of rainfall also decreases. From October to December, the value between SST and rainfall are both increasing. Dashed lines and dots in the picture show the condition when El Nino and La Nina have the same pattern as Bandung area but have different values.

## 4. Conclusion

Based on this research, it can be concluded as follows:

1. This research has been identified the impact of El Nino and La Nina on rainfall during 1998-2016 using ONI indicator by classifying sea surface temperature (SST) data into several categories, namely El Nino, La Nina and Normal condition. Classification of sea surface temperature (SST) were also used to categorized rainfall data. Therefore, the impact of sea surface temperature which is relatively warm on the increase in rainfall intensity and vice versa can be observed.
2. The result of spatial analysis between SST and rainfall on El Nino and La Nina phenomenon in West Java and Java Sea region reveal that in August to September is the highest El Nino condition with average sea surface temperature 27.71°C - 27.75°C followed by decrease in average rainfall intensity of 1.63 mm/hr-1.86 mm/hr. While December to February is the highest La Nina condition with average sea surface temperature 29.06°C - 28.64°C followed by increase in rainfall intensity of 11.50 mm/hr - 12.15 mm/hr.

## Acknowledgment

The authors would like to thank the reviewers for their valuable comments. The authors would also like to thank Dr. Anindya Wirasatriya at faculty of marine and fisheries, Diponegoro for a valuable comment and IDL programming discussion. Also, all staff and researcher in Remote Sensing and Photogrammetry Lab in Geodesy Engineering Department, faculty of Engineering, Diponegoro University for a discussion and data processing support in this research.

## References

[1] As-syakur A R 2010 Pola Spasial Pengaruh Kejadian La Nina Terhadap Curah Hujan di Indonesia Tahun 1998/1999; Observasi Menggunakan Data TRMM Multisatellite Precipitation Analysis (TMPA) 3B43. *Prosiding Pertemuan Ilmiah Tahunan (PIT) XVII dan Kongres V*

*Masyarakat Penginderaan Jauh Indonesia (MAPIN)*. 9 Agustus 2010, Institut Pertanian Bogor (IPB). Bogor-Indonesia. pp. 230-234

[2] Avia L Q & Hidayati R 2001 Dampak Peristiwa Enso Terhadap Anomali CurahHujan di Wilayah Indonesia Selama Periode 1890-1989. *Majalah Lapan*, 3 (2), 62-68.

[3] Boer R, Sutardi, & D Hilman 2007 *Climate Variability and Climate Change and Their Implication on Sectors*. Government of Republic of Indonesia, Jakarta.

[4] Bowman, K.P. 2006. *An Introduction to Programming With Idl. Academic Press.*

[5] Ghozali, Imam, 2006, *Aplikasi Analisis Multivariate dengan Program SPSS*, Cetakan Keempat, Badan Penerbit Universitas Diponegoro, Semarang

[6] Hanafi. 1988. *Klimatologi*. Fakultas Pertanian Universitas Padjadjaran, Bandung.

[7] Reynolds R W 2007 *Daily High-Resolution-Blended Analyses for Sea Surface Temperature*. NOAA/ National Climatic Data Center, 151 Patton Ave., Asheville, NC 28801-5001 (American Meteorological Society)

[8] Salim O 1999 *Distribusi Normal* (Jakarta: Kedokteran Trisakti)

[9] Septicorini E P 2009 *Identifikasi Fenomena ENSO (El Nino Southern Ocillation) dan IOD (Indian Ocean Dipole) Terhadap Dinamika Waktu Tanam Padi di Daerah Jawa Barat (Studi Kasus Kabupaten Indramayu dan Cianjur)* (Bogor: Institut Pertanian Bogor)

[10] Priyanti N S 1999 *Studi Penangkapan Rawai Tuna di Perairan Selatan Jawa Timur – Bali Pada Musim Timur Berdasarkan Pola Distribusi Suhu Permukaan Laut Citra Satelit NOAA-AVHRR dan Data Hasil Tangkapan*. (Bogor: Fakultas Perikanan dan Ilmu Kelautan Institut Pertanian Bogor)

[11] Tongkurut S H J 2011 *El-Nino dan Pengaruhnya Terhadap Curah Hujan di Manado Sulawesi Utara. Manado*. Program Studi Fisika FMIPA Universitas Sam Ratulangi.

# Multi-temporal Land Use Mapping of Coastal Wetlands Area using Machine Learning in Google Earth Engine

**N. M. Farda**

Department of Geographic Information Science, Faculty of Geography, Universitas Gadjah Mada, Sekip Utara Bulaksumur, Yogyakarta, 55281, Indonesia

farda@ugm.ac.id

**Abstract**. Coastal wetlands provide ecosystem services essential to people and the environment. Changes in coastal wetlands, especially on land use, are important to monitor by utilizing multi-temporal imagery. The Google Earth Engine (GEE) provides many machine learning algorithms (10 algorithms) that are very useful for extracting land use from imagery. The research objective is to explore machine learning in Google Earth Engine and its accuracy for multi-temporal land use mapping of coastal wetland area. Landsat 3 MSS (1978), Landsat 5 TM (1991), Landsat 7 ETM+ (2001), and Landsat 8 OLI (2014) images located in Segara Anakan lagoon are selected to represent multi temporal images. The input for machine learning are visible and near infrared bands, PCA band, invers PCA bands, bare soil index, vegetation index, wetness index, elevation from ASTER GDEM, and GLCM (Harralick) texture, and also polygon samples in 140 locations. There are 10 machine learning algorithms applied to extract coastal wetlands land use from Landsat imagery. The algorithms are Fast Naive Bayes, CART (Classification and Regression Tree), Random Forests, GMO Max Entropy, Perceptron (Multi Class Perceptron), Winnow, Voting SVM, Margin SVM, Pegasos (Primal Estimated sub-GrAdient SOlver for Svm), IKPamir (Intersection Kernel Passive Aggressive Method for Information Retrieval, SVM). Machine learning in Google Earth Engine are very helpful in multi-temporal land use mapping, the highest accuracy for land use mapping of coastal wetland is CART with 96.98 % Overall Accuracy using K-Fold Cross Validation (K = 10). GEE is particularly useful for multi-temporal land use mapping with ready used image and classification algorithms, and also very challenging for other applications.

**Keywords**: land use, machine learning, google earth engine, wetland

## 1. Introduction

Land use change plays a major role in the study of global change. Land use is a form of human effort to modify the environment into an established environment such as agricultural land, roads, plantations and settlements. Land use can also be defined as "a number of human arrangements, activities, and inputs on a particular land" [1] [2]. Land use / land cover and human / natural modification have largely resulted in deforestation, loss of biodiversity, global warming and increased flood natural disasters [3–5]. Environmental problems are often related to land use change. Therefore, the data available on land use change can provide important inputs for environmental management decision making and future planning [6] [7].

Land use change is not only happened in urban and suburban areas where the changes are dynamic but also in coastal areas such as in Segara Anakan Lagoon (SAL). Segara Anakan Lagoon is located in the coastal area of Cilacap Regency; this lagoon is surrounded by mangrove forest and muddy land. The lagoon is also surrounded by drains and moats flowing from the mangrove forests, ponds and surrounding rice fields. The existence of this lagoon is strongly influenced by water and sedimentation coming from the Citanduy River and tidal waters of the Indian Ocean through the western slit of Nusakambangan Island and the eastern groove of the Kembangkuning River (Nusakambangan Strait). The river that empties into this lagoon include the river of Citanduy, Kayumati river, Cibereum river, Ujunggalang river and Dangal river, from the rivers the mud is carried away and then will settle in the lagoon. The accumulated sediment then becomes a new land that then gives various impacts, one of which is land use change. The depreciation of the lagoon/waters of Segara Anakan from 185,071.51 Ha in 1978 to 140,693.43 Ha in 1998 and to 123,552.95 Ha in 2003 [8], represents a fairly rapid shrinkage within 25 years. The high sedimentation rate and the formation of new land, accompanied by an increase in the number of people and the level of population demand for food facilities, so that the extent of existing paddy fields to increase, the increase of pond land, and settlements. The expansion of the land is closely related to the shrinking of the existing mangrove and bush land forests. Therefore a technology is needed to measure the rate of land use change.

The application of remote sensing technology to environmental conditions monitoring provides optimum benefits and results, since remote sensing provides ease in spatial analysis, spectral analysis, and can cover a relatively wide area with relatively low cost and fast when compared with terrestrial surveys. Briefly, remote sensing data can provide information quickly, objectively, reliably and economically in measurement, mapping, monitoring and use for modelling in land-use change. Remote sensing data to be used in this research activity are Landsat MSS, TM, ETM +, and OLI data.

The technique of remote sensing image classification that developed today is quite diverse, ranging from pixels-oriented to object-oriented. In addition there is a method of classification supervised and unsupervised. The development of satellite imagery is fast enough, triggering the abundance of enough data, while the information generated from the image of the techniques or methods of classification mentioned above less answer the abundance of data. Supervised machine learning algorithm is useful to explore abundant image data, find the hidden relationship between a number of candidate input variables (parameters) with a target or output variable (classification results), so it is very important and one of the solutions in abundance of satellite imagery data and helps in multi-temporal land use mapping of coastal wetlands area. The Google Earth Engine (GEE) provides a cloud based platform for processing large amount of freely available multi-temporal satellite imagery. The GEE also provides a set of the state-of-the-art machine learning classifiers for pixel-based classification that can be used for multi-temporal land use mapping.

The research objective is to explore machine learning in Google Earth Engine and its accuracy for multi-temporal land use mapping of coastal wetland area.

## 2. Multitemporal Landuse Mapping of Coastal Wetlands Area

Land use can easily be understood as a modification made by humans to the environment into an established environment such as fields, farms and settlements. Land use is defined as "a number of human arrangements, activities, and inputs on a particular land" [1, 2]. The need for land use and the increased of changes have side effects on the emergence of forest reduction, erosion, land degradation, and desert formation.

Detection of land-alterations on the surface of the earth, through remote sensing techniques, is a process of detecting changes that apply a number of multi-temporal systems for quantitative analysis, of an event that changes with time function [9]. From the definition it can be concluded that the detection of land use change is the process of identifying different forms of a land use object through observation at different times.

Lu [10], mentioned that in the application of land use change analysis with remote sensing data there are three main stages that influence: (1) image pre-processing including geometric rectification/image registration, radiometric and atmospheric correction, and topographic correction if area study is in a mountainous region, (2) selection of appropriate change analysis techniques and methods, and (3) the accuracy assessment of the change results.

## 3. Machine Learning

Machine learning is about programming computers to optimize performance criteria by using data samples or past experiences [11]. Machine learning domains focus on how machines learn the rules of the example. The goal of machine learning is to learn the pattern of the example and then to generalize it with a new example. Kotsiantis [12] divides supervised machine learning into 5 groups: logic based algorithms, perceptron based techniques, statistical learning algorithms, instance-based learning, and support vector machines, and detailed examples are shown in Table 1.

**Table 1.** Supervised machine learning and example.

| Supervised Machine Learning | Example |
|---|---|
| Logic based algorithms | • Decision tree: **CART** (Classification and Regression Tree) [13], Decision tree C4.5 [14], **Random Forests** [15], **GMO Max Entropy** [13, 16]<br>• Learning set of rules [17] |
| Perceptron based technique [18] | • Single layered perceptrons: **Winnow** [19]<br>• Multilayered **perceptrons** (Multi class perceptron) [16]<br>• Radial Basis Function (RBF) networks |
| Statistical learning algorithms | • **Fast Naive Bayes** classifiers [16]<br>• Bayesian Networks [20] |
| Instance-based learning [21] | • Nearest neighbour algorithm [22]<br>• k-Nearest Neighbour (kNN) [23] |
| Support vector machines [24] | • **Voting SVM** (Support Vector Machines) [13]<br>• **Margin SVM** [13]<br>• **Pegasos** (Primal Estimated sub-GrAdient SOlver for Svm) [25]<br>• **IKPamir** (Intersection Kernel Passive Aggressive Method for Information Retrieval, SVM) [26] |

Note: Example with text **bold** is machine learning in GEE.

## 4. Materials and methods

The materials and methods described (Figure 1) in this part cover: (a) study area, (b) selection of materials and data, (c) design of field sampling and data collection, (d) transformation and feature (information) extraction from Landsat imagery, (e) machine learning, and (h) accuracy assessment.

Location of the study area is the Segara Anakan Lagoon (SAL), which is included in the administrative area of Kampung Laut district, Cilacap regency, Central Java province, Indonesia (Figure 2). Kampung Laut subdistrict consists of 4 villages, which are Ujung Alang, Ujung Gagak, Panikel and Klaces.

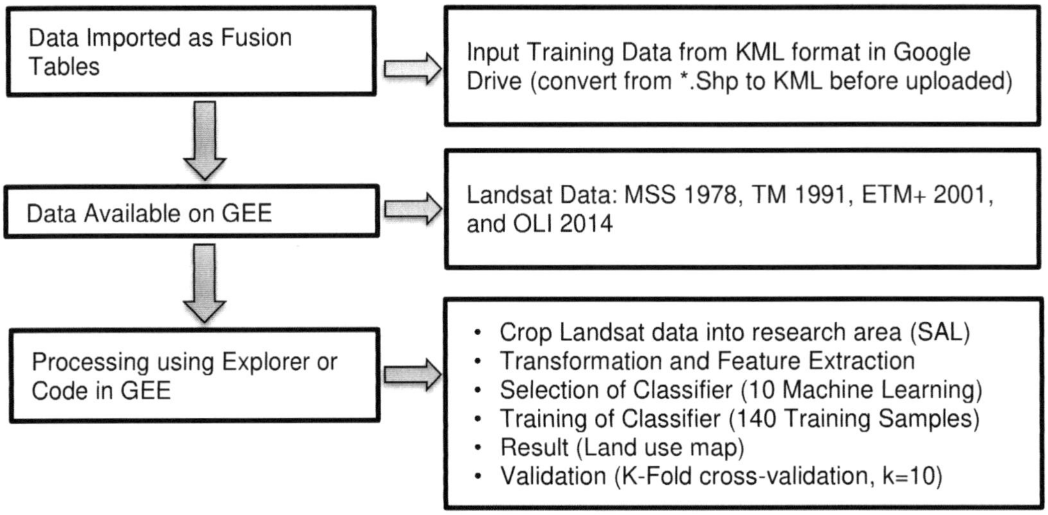


**Figure 1.** Workflow of research methods.

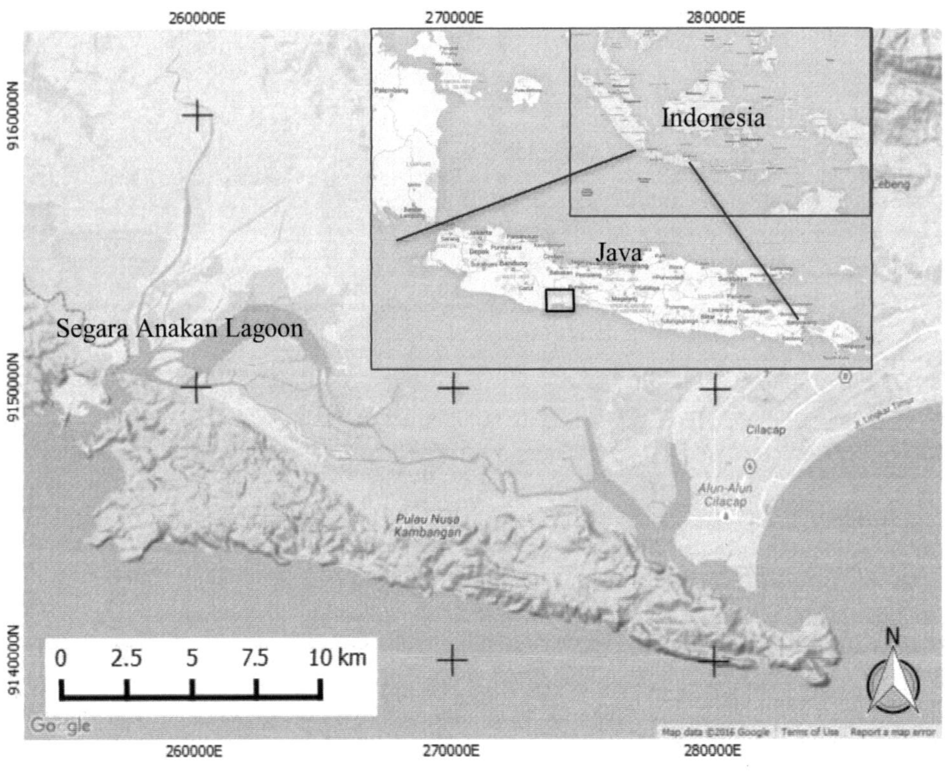


**Figure 2.** Segara Anakan lagoon (SAL) and surroundings, Cilacap District, Central Java.

This study used medium spatial resolution satellite imagery of Landsat 3 MSS (25 April 1978), Landsat 5 TM (5 July 1991), Landsat 7 ETM (22 June 2001), and Landsat 8 OLI (1 May 2014) as primary images and DEM data from ASTER GDEM as additional image for elevation data. Topographic maps (Peta Rupabumi Digital Indonesia) at scale 1:25.000 with 12.5 m contour interval produced by Bakosurtanal (now is Badan Informasi Geospasial) in 2000, covering research area in the

Segara Anakan lagoon, Central Java. These included map sheets Kalipucang (1308-241) and Pengolahan (1308-242).

A stratified random sampling pattern has been applied to collect the land use data for ground truth in the field (Figure 3). Stratified based on land use heterogeneity, the tentative land use map is needed in order to stratify the area and create from unsupervised classification. The unsupervised classification using k-mean clustering is performed in order to obtain preliminary spectral-related land use map. From above classification strata, location selected randomly.

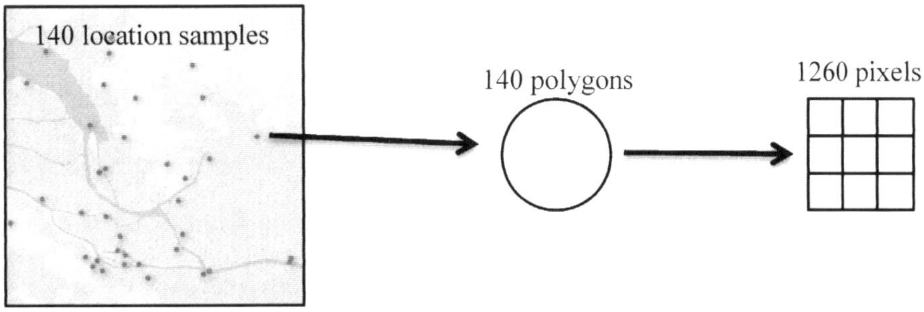


**Figure 3.** Stratified random sampling patterns in 140 polygon samples.

Transformation and feature extraction required for machine learning input are divided into four sections, namely: (1) the spectral factor, (2) textural factors, (3) factors of vegetation, and (4) physical factors (Table 2). Spectral factors include spectral channels that exist on Landsat imagery of the MSS, TM, ETM+, and OLI. Texture factors using Haralick (GLCM) texture. Vegetation factors utilizing NDVI as vegetation index derived from the red and infrared channel. Physical factors are using elevation values from ASTER GDEM, and several indices such as wetness index and bare soil index, and also principle component analysis (PCA).

**Table 2.** Transformation and feature extraction required for machine learning input.

| Land use (coastal wetlands) component | Transformation and Features Extraction |
|---|---|
| Water/Hydro<br>• Salinity (fresh water, brackish water, salt water)<br>• Water Depth < 6 meter | • Spectral bands: Blue, Green, and Principle Component Analysis (PCA)<br>• Normalize Difference Water Index (NDWI), Modified NDWI<br>• Tasseled Cap Wetness (TCW) |
| Landuse/Vegetation<br>• Tree (Mangrove, Swamp)<br>• Shrublands (Muddy Swamp)<br>• Grass (Brackish Swamp) | • Spectral bands: Red, Near Infra-Red, and PCA<br>• Normalize Difference Vegetation Index (NDVI)<br>• Bare Soil Index (BI)<br>• Normalize Difference Bareness Index (NDBaI)<br>• Tasseled Cap Brightness (TCB)<br>• Gray Level Co-occurrence Matrix (GLCM)/Haralick texture (HT) |
| Geomorphology<br>• Basin (close system)<br>• Flow (open system) | • Gray Level Co-occurrence Matrix (GLCM)/Haralick texture (HT)<br>• Digital Elevation Model (DEM) from SRTM and ASTER GDEM |

Machine learning in GEE applied 10 algorithms as follows Fast Naive Bayes, CART (Classification and Regression Tree), Random Forests, GMO Max Entropy, Perceptron (Multi Class

Perceptron), Winnow, Voting SVM (Support Vector Machines), Margin SVM, Pegasos (Primal Estimated sub-GrAdient SOlver for Svm), and IKPamir (Intersection Kernel Passive Aggressive Method for Information Retrieval, SVM).

Accuracy assessment of the results of machine learning is done by creating k-fold cross-validation. The original sample is randomly partitioned into k equal sized subsamples (k=10). A single subsample (1 subsample) is retained as the validation data for testing the model, and the remaining k-1 subsamples (9 subsamples) are used as training data; the cross-validation repeated k times (10 times, the folds). The k results from the folds then calculate the averaged to produce a single estimation.

## 5. Results and discussion

The three set (i1, i2, and i3) of experiments was carried out to select the best combination input and evaluating different machine learning classifiers available in GEE. The first set (i1) of experiment is spectral bands as input for machine learning classifiers. Table 3 show the accuracy assessment of land use classification using 10 machine learning algorithms in Google Earth Engine (GEE). The first set simulation or experiment is to use input from the spectral channel of each image. Landsat 3 MSS uses channels 1, 2, 3, and 4 as inputs. Landsat 5 TM uses channels 1, 2, 3, 4, 5, and 7 as inputs. Landsat 7 ETM+ uses channels 1, 2, 3, 4, and 5 as inputs. Landsat 8 OLI uses channels 1, 2, 3, 4, 5, 6, and 7 as inputs. There are 10 machine learning algorithms on GEE that available to used, and can be grouped into 4 groups. The first group is statistical learning algorithms consisting of Fast Naive Bayes. The second group is a perceptron based technique consisting of Winnow and Perceptron. The third group is logic based algorithms consisting of CART, Random Forests, and GMO Max Entropy. The fourth group is a support vector machine consisting of Voting SVM, Margin SVM, Pegasos, and IKPamir. Highest accuracy is obtained by CART and then Random Forests, both algorithms are logic based algorithms and can also be grouped as decision tree based algorithm. While the lowest accuracy is Pegasos and IKPamir, but both of these algorithms fail to run on the input satellite image.

The second set (i2) of experiment is spectral bands and feature bands as inputs for machine learning classifiers. The feature bands consist of Harralick Texture, NDVI (Normalized Difference Vegetation Index), NDWI ((Normalized Difference Water Index), TCW (Tasseled Cap – Wetness), and ASTER GDEM (Digital Elevation Model). While the third set (i3) of experiment is spectral bands, feature bands, and PCA bands. PCA bands consist of PCA (Principal Component Analysis) and inverse PCA. The highest accuracy is still controlled by CART and Random Forests, just like the first set (i1). But in i2 and i3 there is a SVM group that also has a high accuracy, it's Voting SVM. The lowest accuracy is still the same that Pegasos and IKPamir both failed to run.

The experimental results of input 1 (i1), input 2 (i2) and input 3 (i3) show that there is an increase of accuracy from i1 to i2 and from i2 to i3, so that the highest accuracy is dominated in the last experiment (i3). It is understandable that in machine learning, more and more features of the identifier will facilitate machine learning algorithms in recognizing objects. Figure 4 shows more detailed graph of accuracy on i3, high accuracy (over 80%) is dominated by CART, Random Forests, and Voting SVM. From several studies, the decision-based machine learning algorithm or generally referred to as logic based algorithms in supervised machine learning, also shows the dominance of high yield accuracy and is quite popularly used [27–29]. In the perceptron algorithm (neural network), the accuracy result is always below 70%. Classification methods using neural networks require considerable amount of sample training to improve the accuracy of the results, the training sample at least 10 times the number of weights in neural networks [30]. Pegasos and IKPamir cannot run well on almost all satellite image data from 1978, 1991, 2001, and 2014, as well as experiments with various combinations of inputs (i1, i2, and i3). When Pegasos and IKPamir are executed there is a warning "Server returned HTTP code: 500" which means something has gone wrong on the web site's server. The Google Earth Engine (GEE) is currently in the testing phase for users, so future expectations will be even better.

**Table 3.** The accuracy assessment of land use classification using 10 machine learning algorithms in Google Earth Engine (GEE).

| Supervised Classification (Machine Learning) | K-fold Cross Validation in % (k=10) | | | | | | | | | | | |
|---|---|---|---|---|---|---|---|---|---|---|---|---|
| | 1978 i1 | 1978 i2 | 1978 i3 | 1991 i1 | 1991 i2 | 1991 i3 | 2001 i1 | 2001 i2 | 2001 i3 | 2014 i1 | 2014 i2 | 2014 i3 |
| Fast Naive Bayes | 27.45 | 35.67 | 59.12 | 38.20 | 42.60 | 48.90 | 49.32 | 43.63 | 50.47 | 63.44 | 63.85 | 58.31 |
| GMO Max Entropy | 59.12 | 66.23 | 67.74 | 58.55 | 63.06 | 63.59 | 69.17 | 74.50 | 76.28 | 74.52 | 72.21 | 73.11 |
| Winnow | 25.85 | 26.35 | 26.85 | 39.45 | 43.13 | 36.83 | 41.59 | 6.69 | 10.24 | 7.25 | 10.78 | 25.18 |
| Perceptron | 31.56 | 47.70 | 50.80 | 43.34 | 52,68 | 53.62 | 52.46 | 8.15 | 54.86 | 55.99 | 53.17 | 60.42 |
| Pegasos | 0.00 | 0.00 | 0.00 | 0.00 | 0.00 | 0.00 | 0.00 | 0.00 | 0.00 | 0.00 | 0.00 | 0.00 |
| CART | 95.59 | 94.09 | 95.39 | 81.74 | 87.93 | 96.85 | 84.12 | 82.34 | 85.89 | 91.34 | 93.76 | 96.98 |
| Random Forests | 75.05 | 86.07 | 88.48 | 77.33 | 82.69 | 89.19 | 84.12 | 85.27 | 86.00 | 89.33 | 89.63 | 89.43 |
| IKPamir | 0.00 | 0.00 | 0.00 | 0.00 | 0.00 | 0.00 | 0.00 | 0.00 | 7.42 | 7.25 | 7.25 | 7.25 |
| Voting SVM | 64.03 | 72.44 | 75.45 | 64.53 | 73.66 | 83.95 | 75.24 | 81.19 | 84.95 | 84.39 | 85.70 | 92.45 |
| Margin SVM | 50.30 | 48.80 | 49.59 | 45.96 | 50.37 | 44.60 | 22.15 | 38.56 | 36.89 | 33.23 | 40.48 | 27.90 |

**Notes:**

i1 = spectral bands

i2 = spectral bands + feature bands

i3 = spectral bands + feature bands + pca bands

K-Fold (10) Cross-validation in % :

< 80

80 - 90

> 90

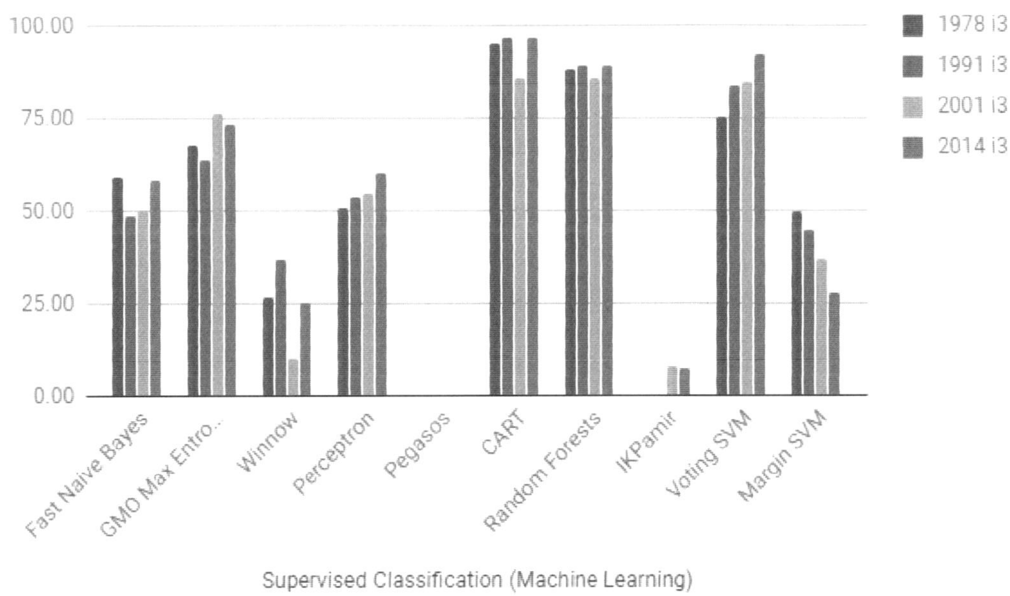


**Figure 4.** K-Fold Cross Validation (k=10) in % for spectral, feature, and PCA bands input (i3).

Multi-temporal analysis on land use maps is strongly influenced by the results of land use classification accuracy from satellite imagery using machine learning. Land use maps from image classification using CART could be selected because the results are consistently in high accuracy (> 90%). The magnitude change for each land use class is calculated by subtracting the area of coverage from class x in 2nd year with 1st year (Magnitude = magnitude of 2nd year - magnitude of 1st year). In Figure 5, the bottom figure shows the statistical class of land use change from 1978 to 2014, either in Ha or %.

Land use change from 1978 to 2014 is very significant in land use class water bodies, intertidal marshes, and irrigated land. Water bodies and intertidal marshes were reduced by -12.92% and -11.05%, respectively, while the change in irrigated land use was 17.26%. Increased irrigated land shows that human need for land becomes a very important issue in the study of land use change. When observed land use change from 1978 to 1991, 1991 to 2001, and 2001 to 2014 a sequence of land use changes occurred in Segara Anakan lagoon from water bodies to intertidal mud flats, subsequently to intertidal marshes, intertidal forested wetlands and turned into aquaculture ponds, seasonally flooded agricultural land, and irrigated land. Although sometimes not in such sequence, it is understandable that there has been a shift from natural wetlands to artificial wetlands.

High accuracy in land use classification results from satellite imagery sometime is not necessarily suitable for analysis of land use change. The accuracy test uses only samples, high accuracy only in the sample locations representing the population. For example in the 1978, the area of settlements land use class are 6648.21 Ha and reduced (-) by 5134.5 Ha to become 1513.71 Ha in 2014. It is illogical, in the case of such a widespread reduction in settlement land use. Analysis of land use change needs to take into account that there are several classes of land use that have low accuracy and are quite difficult to distinguish from other classes. The conditions at the time of satellite image recording should also be noted, whether the climate conditions are the same or not in any multi-temporal image to be analyzed. If the image recorded in the rainy season would be different classification results when compared with the image recorded in the dry season. The 1978 Landsat image was recorded in April and Landsat's image of 2014 was recorded in May, although the moon is close together, checking on climate conditions is essential.

When returning to the goal of machine learning that is learning the pattern of the example and then able to generalize it with a new example, then the accuracy of the above is enough to see the difference in the accuracy of each machine learning performance. If it continues to the process of analysis of land use change, it should be noted that some things as mentioned above are climatic conditions when recording multi-temporal images and more samples to help machine learning to recognize objects better. In addition, the number of land use classes should not be too much and focus on a study, e.g. land use change from non-urban to urban, agriculture to non-agriculture, natural wetland changes to artificial wetlands. The simplification of land use into two classes did to make it easier to see the common threads of land use changes. The simplification of the changed object entity refers to "finding simplicity in complexity" [31] and the land use change conceptual model [32, 33]. The method of updating land use map manually using visual interpretations of satellite imagery or semi-automatic updating land use map using image segmentation, GIS analysis, and visual interpretation could be selected if the area of study is not extensive [34]. GEE is particularly useful for multi-temporal land use mapping with ready used image and classification algorithms (machine learning) and also very challenging for other applications.

IOP Conf. Series: Earth and Environmental Science **98** (2017) 012042 doi:10.1088/1755-1315/98/1/012042

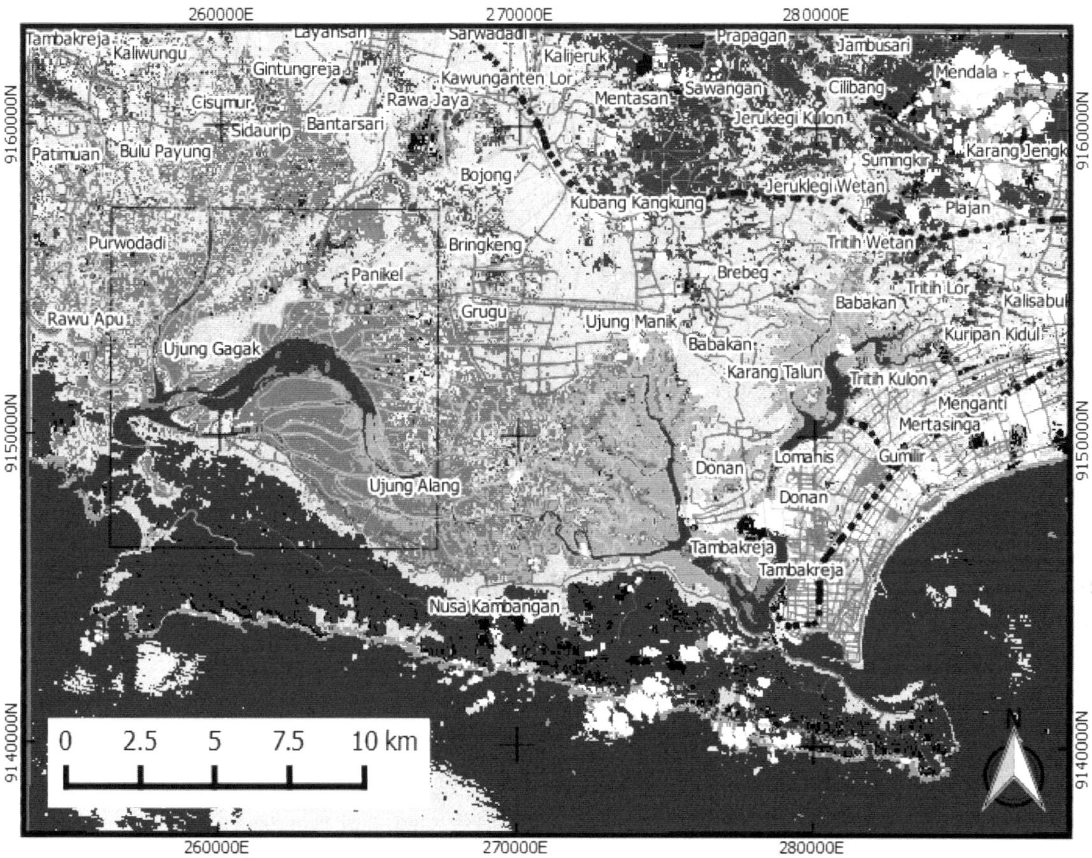


**Legend:**

— Road
- - Railroad
— River

| Class Statistics 1978 - 2014 | Class color | 1978 ha | 2014 ha | Δ ha | 1978 % | 2014 % | Δ % |
|---|---|---|---|---|---|---|---|
| Aquaculture ponds (1) | | 2469.51 | 2375.28 | -94.23 | 2.71 | 2.58 | -0.13 |
| Water body (J) | | 28113.47 | 16507.89 | -11605.58 | 30.86 | 17.93 | -12.92 |
| Intertidal forested wetlands (I) | | 6172.83 | 5781.33 | -391.5 | 6.78 | 6.28 | -0.49 |
| Intertidal marshes (H) | | 13719.23 | 3686.94 | -10032.29 | 15.06 | 4.01 | -11.05 |
| Intertidal mud, sand or salt flats (G) | | 2010.69 | 1058.85 | -951.84 | 2.21 | 1.15 | -1.06 |
| Irrigated land (3) | | 5960.7 | 21907.53 | 15946.83 | 6.54 | 23.80 | 17.26 |
| Ponds (2) | | 6060.33 | 1298.16 | -4762.17 | 6.65 | 1.41 | -5.24 |
| Sand, shingle, or pebble shores (E) | | 767.88 | 1157.58 | 389.7 | 0.84 | 1.26 | 0.41 |
| Seasonally flooded agricultural land (4) | | 4864.95 | 7348.41 | 2483.46 | 5.34 | 7.98 | 2.64 |
| Forest (NW-F) | | 11061.63 | 13808.34 | 2746.71 | 12.14 | 15.00 | 2.86 |
| Industry (NW-I) | | 1434.78 | 4173.3 | 2738.52 | 1.57 | 4.53 | 2.96 |
| Settlements (NW-S) | | 6648.21 | 1513.71 | -5134.5 | 7.30 | 1.64 | -5.65 |
| Cloud cover (NW-Cc) | | 1279.53 | 1124.91 | -154.62 | 1.40 | 1.22 | -0.18 |
| Cloud shadow (NW-Cs | | 548.1 | 1533.96 | 985.86 | 0.60 | 1.67 | 1.06 |

**Figure 5.** Land use map in 2014 extract from Landsat 8 OLI using CART machine learning. The legend has shown the statistics of each land use class and the changes from 1978 to 2014. Detail maps in a square box shown in Figure 6 (next figure).

IOP Conf. Series: Earth and Environmental Science **98** (2017) 012042 doi:10.1088/1755-1315/98/1/012042

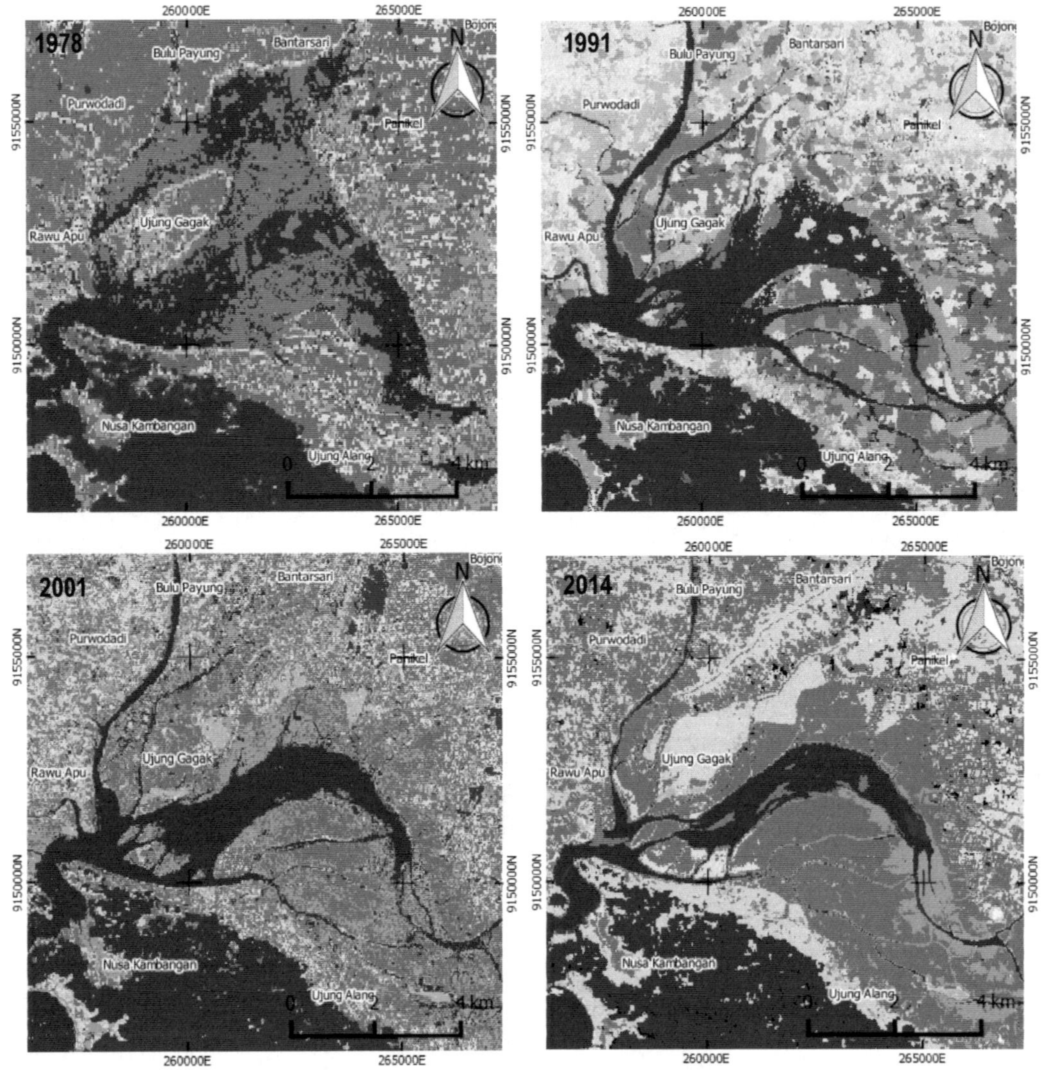


**Legend:**

| Land use class | Class color |
|---|---|
| Aquaculture ponds (1) | |
| Water body (J) | |
| Intertidal forested wetlands (I) | |
| Intertidal marshes (H) | |
| Intertidal mud, sand or salt flats (G) | |
| Irrigated land (3) | |
| Ponds (2) | |

| Land use class | Class color |
|---|---|
| Sand, shingle, or pebble shores (E) | |
| Seasonally flooded agricultural land (4) | |
| Forest (NW-F) | |
| Industry (NW-I) | |
| Settlements (NW-S) | |
| Cloud cover (NW-Cc) | |
| Cloud shadow (NW-Cs | |

**Figure 6.** Land use map in 1978 to 2014 extract from Landsat imagery using CART machine learning.

## 6. Conclusions

Machine learning in Google Earth Engine are very helpful in multi-temporal land use mapping, the highest accuracy for land use mapping of coastal wetland is CART (96,98 % Overall Accuracy). Some of machine learning is not stable (GMO Max Entropy and Pegasos); it's still in testing status. If it is to be used for the analysis of land use change, it is necessary to check the climatic conditions at the time of recording of each multi-temporal image, and need to focus on what land use changes will be assessed. The simplification of land use into two classes did to make it easier to see the common threads of land use changes. GEE is particularly useful for multi-temporal land use mapping with ready used image and classification algorithms (machine learning) and also very challenging for other applications.

## Acknowledgements

The authors would like to convey a gratitude to the Faculty of Geography, Universitas Gadjah Mada (UGM) for funding and supporting this research. This paper is part from Lecturer Research Grant of the first author entitled "Aplikasi Algoritma C4.5 untuk Pemetaan Perubahan Penggunaan Lahan di Laguna Segara Anakan Kabupaten Cilacap".

## References

[1] FAO. *State of the World's Forests*. Rome, Italy, 1997.
[2] FAO, UNEP. *Terminology for Integrated Resources Planning and Management*. Rome, Italy; Nairobi, Kenya, 1999.
[3] Mas JF, Velazquez A, Gallegos JRD, et al. Assessing land use/cover changes: a nationwide multidate spatial database for Mexico. *Int J Appl Earth Obs Geoinf* 2004; 5: 249–261.
[4] Zhao GX, Lin G, Warner T. Using Thematic Mapper Data for Change Detection and Sustainable Use of Cultivated Land: A Case Study in The Yellow River Delta, China. *Int J Remote Sens* 2004; 25: 2509–2522.
[5] Dwivedi RS, Sreenivas K, Ramana K V. Cover: Land-use/land-cover change analysis in part of Ethiopia using Landsat Thematic Mapper data. *Int J Remote Sens* 2005; 26: 1285–1287.
[6] Fan F, Weng Q, Wang Y. Land Use and Land Cover Change in Guangzhou, China, from 1998 to 2003, Based on Landsat TM /ETM+ Imagery. *Sensors* 2007; 7: 1323–1342.
[7] Prenzel B. Remote sensing-based quantification of land-cover and land-use change for planning. *Prog Plann* 2004; 61: 281–299.
[8] Carolita I, Parwaty E, Trisakti B, et al. Model Prediksi Perubahan Lingkungan di Kawasan Perairan Segara Anakan. In: *Pertemuan Ilmiah Tahunan MAPIN XIV*. ITS, Surabaya, 2005.
[9] Jensen JR. *Introductory Digital Image Processing: A Remote Sensing Perspective. Third edition*. 3rd ed. Prentice Hall, 2004.
[10] Lu D, Mausel P, Brondízio E, et al. Change detection techniques. *Int J Remote Sens* 2004; 25: 2365–2407.
[11] Alpaydin E. *Introduction to machine learning*. MIT Press, 2010.
[12] Kotsiantis SB. Supervised machine learning: A review of classification techniques. ... *Appl Comput Eng Real Word ...* 2007; 31: 249–268.
[13] Bishop CM. *Pattern Recognition and Machine Learning*. 2006. Epub ahead of print 2006. DOI: 10.1117/1.2819119.
[14] Quinlan JR. *C4.5: Program for Machine Learning*. San Mateo, California: Morgan Kaufmann Publishers, Inc., 1993.
[15] Breiman L. Random Forests. *Mach Learn* 2001; 45: 5–32.
[16] Haykin SS. *Neural networks and learning machines*. 3rd ed. Prentice Hall/Pearson, 2009.
[17] Fürnkranz J. Separate-and-Conquer Rule Learning. *Artif Intell Rev* 1999; 13: 3–54.
[18] Rosenblatt F. Principles of neurodynamics; perceptrons and the theory of brain mechanisms. 1962; 616 p.
[19] Littlestone N, Warmuth MK. The Weighted Majority Algorithm. *Inf Comput* 1994; 108: 212–

261.
[20] Jensen F V., V. F. *An introduction to Bayesian networks*. Springer, 1996.
[21] Mitchell TM. *Machine Learning*. 1997. Epub ahead of print 1997. DOI: 10.1145/242224.242229.
[22] Aha D. *Lazy Learning*. Dordrecht: Kluwer Academic Publishers, 1997.
[23] Cover T, Hart P. Nearest neighbor pattern classification. *IEEE Trans Inf Theory* 1967; 13: 21–27.
[24] Vapnik VN. *The nature of statistical learning theory*. 1995. Epub ahead of print 1995. DOI: 10.1007/978-1-4757-3264-1.
[25] Shalev-Shwartz S, Singer Y, Srebro N, et al. Pegasos: primal estimated sub-gradient solver for SVM. *Math Program* 2011; 127: 3–30.
[26] Maji S, Berg AC, Malik J. Classification using intersection kernel support vector machines is efficient. In: *2008 IEEE Conference on Computer Vision and Pattern Recognition*. IEEE, 2008, pp. 1–8.
[27] Farda NM, Danoedoro P, Hartono, et al. Remote Sensing Image Mining for Mapping Coastal Wetlands Changes in Segara Anakan Lagoon, Central Java, Indonesia. In: *International Geographical Congress: Down to Earth*. Cologne, DE, 2012.
[28] Hsu W, Lee MLM, Zhang J. Image Mining: Trends and Developments. *J Intell Inf Syst* 2002; 19: 7–23.
[29] Wu X, Kumar V, Quinlan JR, et al. Top 10 algorithms in data mining. *Knowl Inf Syst* 2008; 14: 1–37.
[30] Zhang SL, Chang TC. A Study of Image Classification of Remote Sensing Based on Back-Propagation Neural Network with Extended Delta Bar Delta. *Math Probl Eng* 2015; 2015: 1–10.
[31] Wainwright J, Mulligan M. *Environmental Modelling*. Second Edi. West Sussex, UK: John Wiley & Sons, Ltd., 2013.
[32] Pontius Jr. RG, Huffaker D, Denman K. Useful Techniques of Validation for Spatially Explicit Land-change Models. *Ecol Modell* 2004; 179: 445–461.
[33] Susilo B. *Pemodelan Spasial Dinamika Penggunaan Lahan di Daerah Perkotaan Yogyakarta*. Universitas Gadjah Mada, 2016.
[34] Mas J-F, González R. Change Detection and Land Use / Land Cover Database Updating Using Image Segmentation, GIS Analysis and Visual Interpretation. *ISPRS - Int Arch Photogramm Remote Sens Spat Inf Sci* 2015; XL-3/W3: 61–65.

# Soil Fertility Evaluation and Land Management of Dryland Farming at Tegallalang Sub-District, Gianyar Regency, Bali, Indonesia

**I K Sardiana, D Susila, A A Supadma and M Saifulloh***

Center for Spatial Data Infrastructure Development (PPIDS) Universitas Udayana, Bali, Indonesia

moh9saifulloh@gmail.com

**Abstract.** The landuse of Tegallalang Subdistrict is dominated by dryland farming. The practice of cultivation on agricultural dryland that ignores the carrying capacity of the environment can lead to land degradation that makes the land vulnerable to the deterioration of soil fertility. Soil fertility evaluation and land management of dryland farming in Tegallalang Sub-district, Gianyar Regency were aimed at (1) identifying the soil fertility and it's respective limiting factors, (2) mapping the soil fertility using Geographic Information Systems (GIS) and (3) developing land management for dryland farming in Tegallalang Sub-district. This research implementing explora-tory method which followed by laboratory analysis. Soil samples were taken on each homogene-ous land units which developed by overlay of slope, soil type, and land use maps. The following soil fertility were measured, such as CEC, base saturation, $P_2O_5$, K- Total and C-Organic. The values of soil fertility were mapping using QGIS 2.18.7 and refer to land management evaluation. The results showed that the soil fertility in the research area considered high, and low level. The High soil fertility presents on land units at the flat to undulating slope with different land management systems (fertilizer, without fertilizer, soil tillage and without soil tillage). The low soil fertility includes land units that present on steep slope, and without land managements. The limiting factors of soil fertility were texture, C-Organic, CEC, P2O5, and K- total. It was recommended to applying organic fertilizer, Phonska, and dolomite on the farming area.

**Keywords:** Soil fertility, Land Units, Management Directives, Geography Information System

## 1. Introduction

Tegallalang Sub-District has most of its territory in the form of dry land. Commodities that of-ten planted by farmers in dry land, among others, are oranges, coconut, banana, cocoa, coffee, ginger, and cassava. The main obstacles in the utilization of dry land for agriculture is low levels of soil fertility caused by chemical barriers that limit plant growth, such as, nutrient availability problems. Dry land has a low-level of soil fertility, and low levels of organic material. This condition is exacerbating by the limited use of organic fertilizers, especially during the annual food crop. In addition, naturally the levels of soil oganic matter in tropical areas are fastly declining, reaching 30-60% within 10 years [1]. On the other hand, an intensive tillage is the cause of the decline in productivity of the dry land. The results showed that tillage overload can damage the soil structure [2]. Then good land management efforts (based parameter of chemical soil fertility) are needed in accordance with the requirements for the types of plants that are cultivated. There are five soil parameters used in this study to assess the status of the soil fertility.

Evaluation of soil fertility was done with the analysis of the soil and the plant in total or partial, as fertility recommendations and increase the fertility of a land [3]. Assessment of the status of soil fertility based on the content of Nitrogen, phosphorus and Potassium. This macro nutrients needed in large quantities [4]. Some common ways that is done to determine the status of soil fertility according to Tisdale et al. (1990) is to (1) see symptoms of nutrient deficiency, (2) analysis of plant tissue, (3) analysis of soil biology and (4) a chemical analysis of the soil. The approach was co ducted to find out the status of soil fertility in this study i.e. using chemical analysis of soils. In order to establish the status of soil fertility is required parameters such as soil chemical properties; Cation exchange capacity, Base Saturation, organic-C, P and K levels of total land. To formulate appropriate action, then I need to know what the status of soil fertility. So, known chemical properties of soils which be-come the limiting factor in each land unit, this can be done by evaluating the fertility of the soil [5].

One of the ways that are commonly used in assessing the fertility of the soil is a soil analysis or test with soil samples. Soil test is a chemical in the laboratory analysis activities that are simple, fast, cheap, right, and may be repeated for a presumes the availability of nutrient elements. Soil test in the broad sense i.e. concerning aspects of the interpretation, evaluation and preparation of the fertilizer recommendations from soil test results as well as the taking of soil samples [6]. The spread of soil fertility status can be determined by means of a survey for mapping soils. In addition to this survey aims to determine the land unit also evaluating the potential of soil nutrient elements in providing for the plant through the soil analysis in the laboratory [7]. Given the importance of soil fertility, the necessary evaluation of the status of soil fertility in dry land at Tegallalang Sub-district to establish land management that can be applied to farmers in order to support the production and food security in Tegallalang.

## 2. Methodology

### *2.1. Area of Study*

Tegallalang sub-district is geographically located between the coordinates 8°19'40" to 8°29'38" South Latitude and 115°15'18" - 115°49'8" East Longitude. Tegallalang Sub-district administratively bordering with Bangli Regency in the north, Tampaksiring Sub-district in the East, Ubud sub-district in the south, and Payangan sub-district in the west. Tegallang sub-district has 7 villages namely Keliki Village, Kenderan village, Tegallalang Village, Sebatu Village, Kedisan Village, Pupuan Village and Taro Village. Tegallalang Sub-district area is 61.80 $km^2$ with an altitude ranging between 225-975 meters above sea level. Rainfall in Tegallalang is 1,558.5 mm per year with temperature of 21°C – 24°C [8]. Soil type at Kecamatan Tegallalang have Sub groups of Typic hapludands, with the slope of 3-8% (flat), 8-15% (undulating), 15-25% (rather steep), 25-40% (steep) and >40% ( very steep). Soil analysis was done in the laboratory of soil science of Udayana University. Geographic Information Systems (GIS) analysis done on Evaluation of Land Resources Lab of the Faculty of Agriculture of Udayana University.

### *2.2. Tools and Material*

The materials used in this research includes:

- Digital Topographical Map, scale 1:25,000
- Soil semi detailed map of Gianyar Regency 1:50,000 scale
- Soil samples
- Aquades 0.5 N HCl,
- Solution of peroxide (H2O2) 10% and 30%
- The chemicals for soil analysis in the laboratory.

The tools used in this research include:

- Equipment for the survey include daggers, hoes, meter, drill bhelgy, camera, label, plastic, ring samples
- GPS (Global Positioning System)
- Equipment for the analysis of soil samples includes: sieve diameter 2 mm and 0.5 mm, weigh the bottle, pH meters, measuring cup, test tubes, scales, as well as Erlemeyer.

*2.3. Methodology*

This research using field survey and soil test in the laboratory. Field survey begins with the creation of a map unit of the research area with intersecting thematic maps, namely soil type map, slope map and landuse map. Field survey conducted to re-check set the tentative map units to become permanent land units. The tentative land units is presented in Figure 1 and location sample can be seen on Figure 2. Retrieved five land unit, that unit I, II, III, IV and V. Each unit land soil in composite samples taken at a depth of 0-20 cm. Soil samples have been taken, further analysed in the laboratory. The chemical properties of the soils analysed in laboratory among other things the C-organic (Walkley and Black method); CEC (method 1 N $NH_4OAC$ pH 7); Saturation of the base (Base Saturation/CEC *100%); $P_2O_5$ HCl Extraction method of total (25%); $K_2O$ HCl extraction methods (25%). Analysis of the soil properties with certain specified criteria based on Technical evaluation of soil fertility [5] are presented in Table 1. Primary data obtained from soil analysis results in the laboratory, then conducted the analysis based on the criteria of the status of soil fertility [5]. Combination of soil fertility of low, medium and high can be seen in Table 2. Meanwhile existing condition of the vegetation, fertilizer and land management practice are obtained from interviews with local farmers at the sample location.

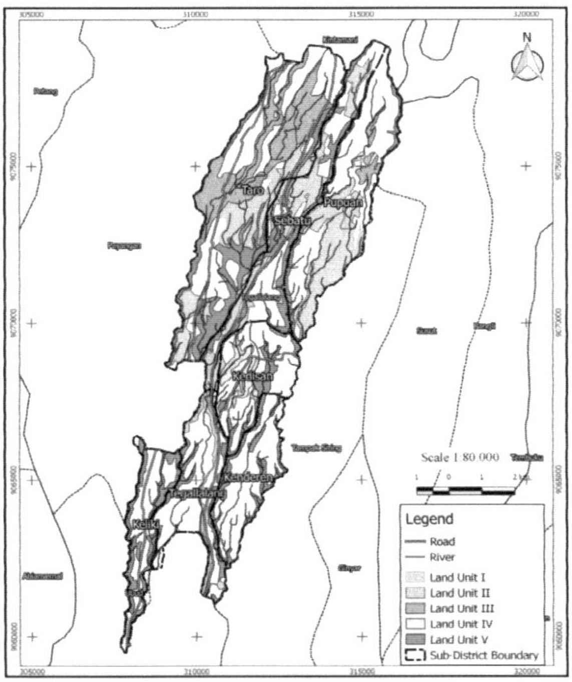


**Figure 1.** Tentative Land Unit Dry Land Farming at Tegallalang Sub-District

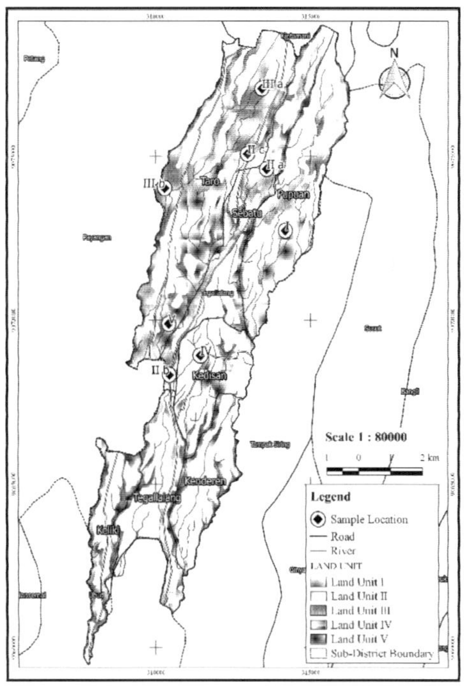


**Figure 2.** Sample Location

**Table 1**. Criteria for Assessment of Chemical Properties of Soil

| Soil Parameter | VL | L | M | H | VH |
|---|---|---|---|---|---|
| C organik (%) | < 1,00 | 1,00 -2,00 | 2,01-3,00 | 3,01 - 5,00 | > 5,00 |
| Base Saturation (%) | < 20 | 20-35 | 36-50 | 51 - 70 | > 70 |
| $P_2O_5$ HCl 25 % | < 10 | 10-20 | 21-40 | 41 - 60 | > 60 |
| $K_2O$ HCl 25 % | < 10 | 10-20 | 21-40 | 41 - 60 | > 60 |
| CEC (me/100 g) | < 5 | 5-15 | 17-24 | 25 - 40 | > 40 |

Description: VL/L/M/H/VH; Very Low/Low/Medium/High/Very High

**Table 2.** Combination of Soil Fertility

| CEC | Base Saturation | $P_2O_5$, $K_2O$, C-organic | Fertility Status |
|---|---|---|---|
| H | H | ≥2 H without R | High |
| H | H | ≥2 H without R | Medium |
| H | H | ≥2 M Without R | High |
| H | H | ≥2 M dengan R | Medium |
| H | H | H > S >R | Medium |
| H | H | ≥2 L dengan H | Medium |
| H | H | ≥2 L dengan S | Low |
| H | M | ≥2 H Without R | High |
| H | M | ≥2 H dengan R | Medium |
| H | M | ≥2 S | Medium |
| H | M | Other Combination | Low |
| H | L | ≥2 H Without R | Medium |
| H | L | ≥2 H dengan R | Low |
| H | L | Other Combination | Low |
| M | H | ≥2 H Without R | Medium |
| M | H | ≥2 S Without R | Medium |
| M | H | Other Combination | Low |
| M | M | ≥2 H Without R | Medium |
| M | M | ≥2 S Without R | Medium |
| M | M | Other Combination | Low |
| M | L | 3 H | Medium |
| M | L | Other Combination | Low |
| L | H | ≥2 H Without R | Medium |
| L | H | ≥2 H dengan R | Low |
| L | H | ≥2 S Without R | Medium |
| L | H | Other Combination | Low |
| L | M | ≥2H Without R | Medium |
| L | M | Other Combination | Low |
| L | L | All Combination | Low |
| VL | H, M, L | All Combination | Very Low |

Description: H/M/L/VL: High/ Medium/ Low/ Vey Low

## 3. Result and Discussion

Intersection processes of soil map, landuse map and slope map resulting land unit map. There are five land unit, namely, land unit I, II, III, IV and V. Each land unit in composite samples taken at a depth of

0-20 cm. Soil samples have been taken, further analyzed in the laboratory. Each soil sample taken at different land unit which indicates different slopes, landuse and vegetation. The slopes, namely, 3-8% slope, 8-15 slope, 15-25% slope, 25-40% slope and more than 40% slope. The soil subgroup at research area indicates the same type which is Typic hapludands and the landuse is dryland farming. The vegetation, fertilizer and land management are obtained from interviews with local farmers. Recent vegetations in dryland farm in Tegalalang are Orange, Papaya, Peanut, Coffee, Snake fruits, Chili, Cassava, Ginger, Chocolate, Coconut, Big Grass, Banana, Sweet potato and Mangosteen. Fertilizer in each land units are differents. No fertilizer occupied in land unit I, IV and V. Meanwhile Organic fertilizer, Phonska, dolomite are the most common fertilizer for other land unit. Land management system is still traditional with Soil tillage with hoe or without soil tillage at all. Table 3 shows the description of each soil samples.

**Table 3.** Land Unit Description

| Land Unit | Sample | Slope | Sub Group | Land Use | Vegetation | Kind of Fertilizer | Land Management |
|---|---|---|---|---|---|---|---|
| I | I | 3-8 % | Typic hapludands | Dry Land Farming | Orange, Papaya, Peanut | Without fertilizer | Without Soil Tillage |
| II | IIa | 8-15 % | Typic hapludands | Dry Land Farming | Coffee, Orange, Snake fruits | Organic Fertilizer, | Soil tillage with hoe |
| | IIb | 8-15 % | Typic hapludands | Dry Land Farming | Orange, Chili, cassava | Organic fertilizer, Phonska, dolomite | Soil tillage with hoe |
| | IIc | 8-15% | Typic hapludands | Dry Land Farming | Banana, sweet potato, mangosteen | Without fertilizer | Soil tillage with hoe |
| III | III a | 15-25 % | Typic hapludands | Dry Land Farming | Orange, Ginger, Chocolate | Organic fertilizer, Phonska | Soil tillage with hoe |
| | III b | 15-25 % | Typic hapludands | Dry Land Farming | Ginger, Orange, Turmeric | Organic fertilizer | Soil tillage with hoe |
| IV | IV | 25-40 % | Typic hapludands | Dry Land Farming | Chocolate, Coffee, Banana | Without fertilizer | Soil tillage with hoe |
| V | V | 40-60 % | Typic hapludands | Dry Land Farming | Coconut, Big Grass, Mango | Without fertilizer | Soil tillage with hoe |

### *3.1. The Measurements of Parameters of Soil Fertility*

*3.1.1 Cation Exchange Capacity.* Cation Exchange Capacity (CEC) is an indicator of soil fertility. The research at Tegallalang Sub-district has indicates a value of cation exchange capacity varies from low to high with the highest average value of 27.92 me/100 g until 15.08 me/100 g. The high value of the CEC allegedly is influenced by dominant of clay fraction content. High value of CEC also influenced by levels of clay, because the land dominated by the fraction of clay that has capacity of ion exchange and holds water, therefore the high aggregate stability is due to soil particles in the bond. The value and criteria of CEC at dry land mixed farming presented in Table 4.

**Table 4.** Values and criteria of Cation Exchange Capacity

| No | Land Unit | Sample | CEC | Criteria |
|---|---|---|---|---|
| I | I | I | 15,93 | Low |
| IIa | | IIa | 21,14 | Medium |
| IIb | II | IIb | 27,92 | High |
| IIc | | IIc | 17,87 | Medium |
| IIIa | III | IIIa | 22,45 | Medium |
| IIIb | | IIIb | 17,47 | Medium |
| IV | IV | IV | 15,08 | Low |
| V | V | V | 19,67 | Medium |

*3.1.2. Base Saturation.* Base saturation is a comparison between the numbers of base cation exchanged with the cation exchange capacity (CEC) of soils that are expressed in percent. Tegallalang Sub-district has value from a high base saturation to very low with the highest median value 95.24% to 29.23%. The value soil CEC is usually directly proportional to the ground, because the saturation KB base is the description of the high number of cation in complex soil colloids [9]. This statement is contrary to the results of this research. This is due to the soil at research area experience a lot of leaching of base cations in the soil. The value and criteria of Base Saturation at dry land mixed farming presented in Table 5.

**Table 5.** Value and criteria of Base Saturation

| No | Land Unit | Sample | Base Saturaion | Criteria |
|---|---|---|---|---|
| 1 | I | I | 95,24 | Very High |
| 2 | | IIa | 38,00 | Medium |
| 3 | II | IIb | 59,37 | High |
| 4 | | IIc | 68,84 | High |
| 5 | III | IIIa | 56,08 | High |
| 6 | | IIIb | 54,76 | High |
| 7 | IV | IV | 66,67 | High |
| 8 | V | V | 29,23 | Low |

*3.1.3. Content of C-Organic.* The results of the determination of the levels of C-Organic in dry land research at the Tegallalang Sub-district varies from very low to high with a percentage of the average value of 4.20% to the lowest is 0.93%. If the C-Organic content on any type of soil in dry land location at Tegallalang Sub-district shows an average 2% still considered low. This is allegedly due to differences of the climatic conditions, rainfall and vegetation of the area so that it can affect the availability of organic matter in the soil. Low content of C-Organic indirectly shows the low production of organic matter al in soil, because soil organic matter is one of the parameters that determine the fertility of the soil. The increase of organic matter also means action against conservation of carbon in the soil, so when the carbon trapped in the ground, then the $CO_2$ emissions into the air [11]. The value and criteria of C-Organic at dry land mixed farming presented in Table 6.

**Table 6.** The value and criteria of C-Organic

| No | Land Unit | Sample | C-Organic | Criteria |
|---|---|---|---|---|
| 1 | I | I | 0,98 | Very Low |
| 2 | | IIa | 6,39 | Very High |
| 3 | II | IIb | 5,56 | Very High |
| 4 | | IIc | 3,17 | Medium |
| 5 | III | IIIa | 8,46 | Very High |
| 6 | | IIIb | 6,44 | Very High |
| 7 | IV | IV | 0,93 | Very Low |
| 8 | V | V | 4,20 | High |

*3.1.4. Content of Phosphate ($P_2O_5$ total).* The results of the measurement of the content of phosphor soil from each land unit, that the content of $P_2O_5$ varies from moderate to high with the highest median value range from 40.43 mg/100 g to the lowest i.e., 12, 21 mg/100 g. The value of P in the research indicates the existence of a low content of Phosphate especially in the IIa, IIc and V that causes a limiting factor or constraint of soil fertility in dryland locations so that it necessary to add reserves of phosphorus into the soil. Land units that containing the element P are high derived from organic matter (manure and the remains of plants), chemical fertilizers (TSP) and the minerals in the soil (apatite). The value and criteria of Phosphate content farming presented in Table 7.

**Table 7.** The Value and criteria Content of Phosphate

| No | Land Unit | Sample | $P_2O_5$ total | Criteria |
|---|---|---|---|---|
| I | I | I | 25,49 | Medium |
| IIa | | IIa | 12,21 | Low |
| IIb | II | IIb | 40,43 | High |
| IIc | | IIc | 11,01 | Low |
| IIIa | III | IIIa | 28,49 | Medium |
| IIIb | | IIIb | 33,74 | Medium |
| IV | IV | IV | 41,03 | High |
| V | V | V | 10,52 | Low |

*3.1.5. Content of Kalium ($K_2O$).* The results of measurements of soil potassium content location research showed a total $K_2O$ content of soil at each land unit is low to very high. The average value of 100 mg/80.71 up 15.91 mg/100 g. Low potassium relates to launder that occur due to the high rainfall [11]. The high Potassium value due to the high value of the CEC on a land unit IV and II b. Cation Exchange Capacity of the soil capability increases K holing at the soil, thus the soil solution was slow to take off lowering the potential leaching, thereby Potassium buildup occurs [12]. The value and criteria of Kalium content at dry land mixed farming presented in Table 8.

**Table 8**. Value and Criteria of Kalium ($K_2O$)

| No | Land Unit | Sample | $K_2O$ | Criteria |
|---|---|---|---|---|
| I | I | I | 50,75 | High |
| IIa | | IIa | 21,72 | Medium |
| IIb | II | IIb | 80,71 | Very High |
| IIc | | IIc | 27, 24 | Medium |
| IIIa | III | IIIa | 15,91 | Low |
| IIIb | | IIIb | 16,22 | Medium |
| IV | IV | IV | 81,41 | High |
| V | V | V | 21,89 | Low |

The graph of parameter of soil fertility in dry land farming at Tegallalang Sub-district presented by Figure 3.

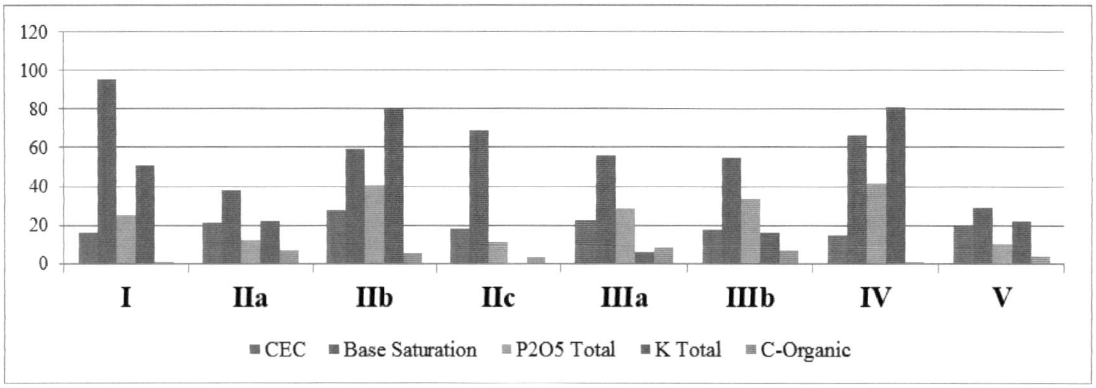


**Figure 3.** Value of Parameter of Soil fertility in Tegallalang Sub-District.

*3.2. Evaluation of Soil Fertility*

Soil fertility is the ability of a soil to provide nutrient elements, in a certain measure and balance on an ongoing basis, to support the growth of a type of plant on the environment with other growth factors in the profitable circumstances. Availability of excellent fertile soil nutrients available to plant will not limiting for plant to growth. The status of soil fertility is the condition of soil fertility in a certain place and time that are judged by the criteria of raw soil fertility parameters according to the instructions of the technical evaluation of the fertility of the soil [5]. The Result of the soil fertility status presented by Table 9. The map of soil fertility in dry land farming at Tegallalang sub-district can be seen in Figure 4.

**Tabel 9**. Evaluation of Soil Fertility

| Land Unit | Sample | CEC (me/100) | Base Saturation (%) | $P_2O_5$ total (mg/100g) | K Total (mg/100g) | C-Organic (%) | Soil Fertility Status |
|---|---|---|---|---|---|---|---|
| I | I | 15,93 (L) | 95,24 (VH) | 25,49 (M) | 50,75 (H) | 0,98 (VL) | L |
| | IIa | 21,14 (M) | 38,00 (M) | 12,21 (L) | 21,72 (M) | 6,39 (VH) | L |
| II | IIb | 27,92 (H) | 59,37 (H) | 40,43 (H) | 80,71 (VH) | 5,56 (VH) | H |
| | IIc | 17,87 (M) | 68,84 (H) | 11,01 (L) | 27, 24 (M) | 3,17 (M) | L |
| III | IIIa | 22,45 (M) | 56,08 (H) | 28,49 (M) | 15,91 (L) | 8,46 (VH) | L |
| | IIIb | 17,47 (M) | 54,76 (H) | 33,74 (M) | 16,22 (L) | 6,44 (VH) | L |
| IV | IV | 15,08 (L) | 66,67 (H) | 41,03 (H) | 81,41 (VH) | 0,93 (VL) | L |
| V | V | 19,67 (M) | 29,23 (L) | 10,52 (L) | 21,89 (M) | 4,20 (H) | L |

Description: H/M/L/VL: High/ Medium/ Low/ Vey Low

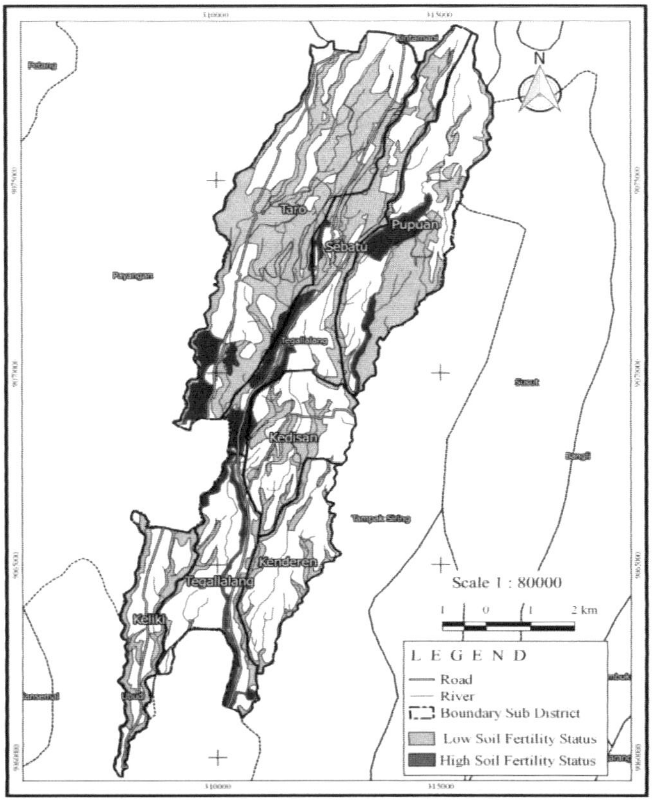


**Figure 4.** Soil Fertility Map at Tegallalang Sub-District

The status of the assessment of the overall fertility of dry land at Tegallalang Sub-district is low except at land unit II which shows high soil fertility status. Low fertility status due to a limiting factor that is the low C content of organic soil and soil CEC. C- organic content (organic material) very influential towards the ability of soils in sustaining soil fertility and productivity through the activity of soil organisms, many soil properties at the land units are good affecting the physical, chemical and biological conditions directly but not affected the organic material.

Absolute organic additions should be given because soil organic matter is very instrumental in creating soil fertility. Base saturation is always connected as a clue about the fertility of the soil. Capability in releasing ions to plants depending on the degree of base saturation. The soil is very fertile when having more than 80% base saturation, and if base saturation is less than 50% then the soil is infertile. This is based on the nature of the ground with 80% base saturation will free alkaline cations that can be exchanged more easily from the ground with 50% base saturation.

### *3.3 Land Management Directives*

Dryland farm management in Tegallalang Sub-district is based on evaluation results of the status of soil fertility and the limiting factor. Land management needed to improve soil fertility in order to increase agricultural production, as well as to achieve a sustainable agricultural system. An important component to note is the use of fertilizers, as well as tillage. The direction of the recommended fertilizer use is manure and fertilizer with phosphate content. Especially on a land unit I, II a, II, c III a, III b, IV and V. The addition of manure and fertilizer must comply with the conditions of the soil and plant needs.

Soil conservation is part of the preparation of the land to make land productivity remains high [13][14]. Conservation tillage is characterized by reduced demolition or reversal of land, intensifying the use of the remaining plants or other materials as mulch. On a land unit with a slope of more than 40% is recommended for making the patio bench. This technique has been developed extensively since 1975 through presidential intrusion [15].

## 4. Conclusions

The status of soil fertility in dry land in the High to low i.e. Tegallang. High soil fertility status shown on the land unit II (II b soil samples). The status of soil fertility low in the show on land units I, II (a sample of IIa, IIb, IIc), III, IV and V. The parameters of the soil fertility which becomes the obstacle of the status of soil fertility in the location of the research on each unit land is CEC soil, C-organic. P and K are low. The recommended management to improve the status of soil fertility is by the addition of organic fertilizer and Phonska Fertilizer in accordance with the needs of the plant. On the other hand need to be done as well as the manufacture of tillage terrace to the efficiency of fertilizer.

## Acknowledgments

Writers would like to thank the Rector of Universitas Udayana, Dean of Agriculture Faculty of Universitas Udayana, Dean of Agriculture Faculty of Universitas Udayana, Head of Center for Spatial Data Infrastructure Development (PPIDS) Universitas Udayana, fellow staf, researchers and professors at Universitas Udayana for all the support, comments and suggestions that makes this paper finished.

## References

[1] Brown S and A E Lugo 1990 Tropical Secondary *Forests Journal of Tropical Ecology* **6**(1) pp 1-32.
[2] Larson W E and G J Osborne 1982 *Tillage Accomplishments and Potential. In Predicting Tillage Effects on Soil Physical Properties and Processes* ASA Special Publ No 44.
[3] FAO 1988 *Soil and Plant Analys* (Rome:FAO Soil Bulletin 38/1) P 241.
[4] Gilman G P 1983 Nutrient Availibility Inn Acid Clearing and Cultivation. Proceedings of The International Workshop on Soils *Research to Resolve Selected Problems of Soils in The Tropics* Townsville, Queensland Australia 12-16 September p 189.
[5] PPT 1995 Technical Evaluation of Soil Fertility *Technical Reports* No 14 Versi 1.0.1. REP II Project CSAR Bogor.
[6] Melsted S W and T R Peck 1972 The Principles of Soil Testing In L M Walsh and J D Beaton (Eds.). *Soil Testing and Plant Analysis*. Soil Science Society of America. Inc. Madison, Wisconsin.
[7] Buol S W, P A Sanchez, R B Cate Jr and M A Granger 1974 Soil Fertility Capability Classification. *Journal Series of The North Carolina Agricultural Experiment Station*. Paper No. 4324. NC State University pp 126-141.
[8] Central Bureau of Statistics of Gianyar Regency 2016 *Gianyar in Figures Year 2015* (Gianyar: Gianyar Regency Government).
[9] Bohn Bernd 2009 *Efficient Parsing of Syntactic and Semantic Dependency Structures*. In Proceedings of Conll-09.
[10] Jastrow J D, J E Amonette and V L Bailey 2007 Mechanism Controlling Soil Carbon Turnover and Their Potential Application for Enhancing Carbon Swquestration *Climate Change* **80** pp 5-23.
[11] Mangel K and E A Kirkby 1982 *Principles of Plant Nutrition* (Bern: International Potash Institute).
[12] Foth H D and L N Turk 1999 *Fundamentals of Soil Science* Fifth Ed (New York : John Waley & Sons).
[13] Brown R E, J L Havlin, D J Lyons, C R Fenster and G A Peterson 1991 Long-Term Tillage and Nitrogen Effects on Wheat Production in a Wheat Fallow Rotation. In Agronomy Abstracts. Annual Meetings ASA, CSSA, And SSSA, Denver Colorado, 27 October–1 November 1991 p 326.
[14] Wagger M G and H P Denton 1991 Consequences of Continuous and Alternating Tillage Regimes on Residue Cover and Grain Yield in a Corn-Soybean Rotation. In Agronomy AbsTracts.

Annual Meetings ASA, CSSA, and SSSA, Denver Colorado, 27 October–1 November 1991 p 344.

[15] Siswomartono D, A N Gintings, K Sebayong and S Sukmana 1990 Development of Conservation Farming Systems, Indonesia Country Review. Regional Action Learning Programme on The Development of Conservation Farming Systems. *Report of The Inaugural Workshop*. Chiang Mai, 23 February-1 March 1990. ASOCON Report No. 2.A reference

IOP Conf. Series: Earth and Environmental Science **98** (2017) 012044 doi:10.1088/1755-1315/98/1/012044

# Development Planning of Tourist Village Using Participatory Mapping (Case study: Mambal Village, Badung Regency, Indonesia)

**I N S Arida[1*], P P K Wiguna[2], I W Narka[2] and N K O Febrianti[2]**

[1] Faculty of Tourism, Universitas Udayana, Bali, Indonesia
[2] Center for Spatial Data Infrastructure Development (PPIDS) Universitas Udayana, Bali, Indonesia

sukma.arida@gmail.com

**Abstract.** Tourism sector is the highest source of income in Badung Regency so it is interesting to see the development of tourist village as one of the alternative tourist destinations in Badung Regency. Most of the village areas in Badung Regency do not have policies, vision and mission as an effort to develop the village into a tourist village. As a result the role of tourist village does not grow in terms of economic and social community. The purpose of this research is to determine and to map the tourism development plan using participatory mapping. The methodology used in this research is field surveys and interviews for data collection and participatory mapping to map the development plan to support tourism. Mambal village is located in Sub-district of Abiansemal, Badung Regency, Indonesia. Mambal village has the potential to become a tourism village because it is supported by the uniqueness of nature and tradition. Mambal village passed by Ayung river, where along the river there are beautiful cliffs which potential to develop as adventure tourism. There is also Senaung Pengibul Cave with a length of more than 15 meters and is wide enough to pass. Mambal village also has a spiritual tour of Pura Demung and Pancoran Pitu, which has a magical story. Currently farmers in Mambal Village are focusing on developing organic farming, of which 38% of the rice fields present in Mambal are pure organic that produces organic rice. Around the rice field area is also created a jogging track for visitors while enjoying the natural beauty of rice fields. Farmers also cultivate oyster mushrooms. In addition, Mambal Village Community also produces handicraft products that are woven in the form of symmetrical Endek (traditional fabrics) and processed products from used goods such as bags, wallets, pencil boxes and others.

**Keywords:** Tourist village, Development plan, Participatory mapping

## 1. Introduction

Tourism as the temporary movement of people to destinations outside their normal places of work and residence, the activities undertaken during their stay in those destinations and the facilities created to cater to their need [1]. Badung Regency is a well-known tourism area in the international arena. South Badung area previously been famous for its coastal charm, but North Badung its existence not yet as famous as South Badung. Characteristic of North Badung is different from South Badung, besides its quieter and cooler environment, its population density is not as dense as South Badung, so Northern part of Badung is very suitable to be a tourist place for tourists who want rural condition.

One approach to developing alternative tourism development is tourist village for sustainable rural development in tourism. Tourism village is a rural area that has some special characteristics to become a tourist destination. In this area, the inhabitants still have tradition and culture that is still original. In addition, several supporting factors such as typical food, agricultural systems and social systems also become potential to a tourist village area. Beyond these factors, nature and environment are still original and awake is one of the most important factor of a tourist destination.

Aspects of natural resources, culture, customs, religion, and character of local community contribute to this holistic [2]. Participation of citizens in management of tourist village not only facilitate their understanding of local tourism [3], but also improve the quality of planning and decision to include the views of locals [4][5]. One of the methods in tourism development planning can be done by geographic information system (GIS). Geographic information system (GIS) has had a positive impact on social, economic and civic progress in order to penetrate the global market [6]. Participatory mapping using GIS applications can be utilized in many ways, such as community and neighborhood planning, environmental and natural resource management, and mapping traditional ecological knowledge of indigenous people [7]. The development of tourism using GIS applications done by the government of Republic Indonesia.

The Government of Republic of Indonesia also showed a strong desire to develop rural tourism by providing funding and guidance to tourist village. In 2012, government of Republic of Indonesia to develop as many as 978 tourist villages through the National Program for Community Empowerment (PNPM) [8].

Through the issuance of Regulation of Badung Regent Number 47 Year 2010 on Stipulation of Tourism Village Area in Badung Regency [10], Regent of Badung is currently focused to develop tourist villages. One of the villages that developed the concept of a tourist village is Mambal Village. Mambal village is very rich in potential that can be developed. Mambal village has a very wide expanse of rice fields that are still very natural with organic farming systems. Mambal village is also passed by Tukad Ayung with along the river there are beautiful cliffs.

The positive effects of tourism are production of income, occupation and new economic resource for local people, assist to reduce on social unbalanced powers, providing job opportunities for maintaining regional population and coefficient of high multiplicity, reviving local culture, enhancing of national dignity and self-confidence, preserving natural and cultural resources [10]. Another positive economic benefit of tourism is that it can provide stability in an economy.

The problem that happens is the potential of natural resources with human resources are not optimized. It is seen that there is no arrangement and management of the existing potential to support the activities of village tourism. People who usually pass through the village of Mambal just look at the natural scenery without any interaction with the local community to know about what is in the village of Mambal whether it's typical food or local products. Although Mambal village has a lot of potential, but there is no clear arrangement from the community, besides the participation to encourage the development of tourism village activity is still low.

## 2. Methodology

### *2.1. Area of Study*

Area of study is located in Mambal Village, Abiansemal Sub-district, Badung Regency, Bali Province, Indonesia. Geographically, Mambal Village located 8.5410° S, 115.2215° E. Mambal Village bordering with Bongkasa Village, Abiansemal Village in the north, Sedang Village, Mekar Bhuana Village in the east, Sibang Kaja Village in the south and Penarungan Village of Badung Regency in the west.

The area of Mambal Village is 3.06 $km^2$ with 184 ha of paddy field (90%), 48.07 ha of moor (16%), 38.03 ha (13%) of yard, 28.4 ha (9%) of plantation, and other landuse for 7 ha (2%) [11]. Mambal Village is 30 kilometers from Ngurah Rai International Airport with travel time about 1 hour 10 minutes with normal traffic.

IOP Conf. Series: Earth and Environmental Science **98** (2017) 012044 doi:10.1088/1755-1315/98/1/012044

### 2.2. *Tools and Material*
The material used in this research are spatial data as follows:
- Questionnaire to conduct interviews related to tourism potential of Mambal Village
- Google Satelite Image year 2017 from obtained from Openlayers Plugin in QGIS to map the location of potential tourism destination
- Openstreet Map data for street and utility data

While the tool used are:
- Computer to process the spatial data
- GPS (Global Positioning System) to determine the coordinate point of observation.
- Camera to record the potential tourism destination in Mambal Village
- Stationary

### *2.3. Data collection*
Data collection method used in this research is primary data and secondary data collection
- Secondary data collection consists of collecting satellite image and maps, including Google Satellite Image and Openstreet Map data. Seconary data will be used as preliminary data for conducting field surveys, interviews and participatory mapping.
- Primary data collection obtained from (1) observing and recording the potential tourism destination in Mambal Village using GPS, (2) conduct interviews with key person who are incharge and understand about the tourism potential in Mambal Village, (3) conduct participatory mapping involving village leaders, communities, and stakeholders to discuss the mapping and planning for tourism destination as well as best management practice for developing tourism in Mambal Village.

### *2.4. Data Analysis*
Data analysis in this research consist of two phase, which are:
- Analysis of potential tourism destination in Mambal Village
  Potential tourism destination as well as challenges to develop tourism in Mambal Village were analysed by interviews with Head of Village, tourism enthusiast or stakeholders at Mambal Village. The result then mapped using QGIS software.
- Analysis of tourism destination planning, supporting facilities and best management practice.
  Tourism destination planning as well as best management practice for developing the tourism in Mambal Village will be conducted using field check and participatory mapping. Field check conducted to map the location of potential tourism destination meanwhile participatory mapping were conducted for establishing tourism development planning and potential tourist destination. Participatory mapping will makes the people of Mambal Village become leading actor to develop their own Village. Besides, the possibility to accommodating suggestions from stakeholders will make the planning become more effectively and become community-based planning. The result of participatory mapping then mapped using QGIS software.

The complete process of methodology used in this research is featured in Figure 1.

## 3. Result and Discussion
### *3.1. Existing Condition and Potential Tourism Destination*
Mambal village has very rich in potential to developed. Mambal village has a very wide expanse of rice fields that are still very natural. Mambal village is also passed by Tukad Ayung (Ayung River), where along the river there are beautiful cliffs and natural panorama. There is also Senaung Pengibul Cave with a length of more than 15 meters and its wide enough to pass with unique characteristics with the height and texture that are formed naturally. Mambal village also has a spiritual tour of Demung Temple,

which surrounding there are pancoran and various unique stone that have a magical story called Pancoran Pitu. Pancoran Pitu and Cave Senaung Pengibull is a symbol that can be developed into a new tourism potential in Mambal Village. Figure 2 shows the atmosphere of Pancoran Pitu, while Figure 3 shows the atmosphere of the Senaung Pengibull Cave.

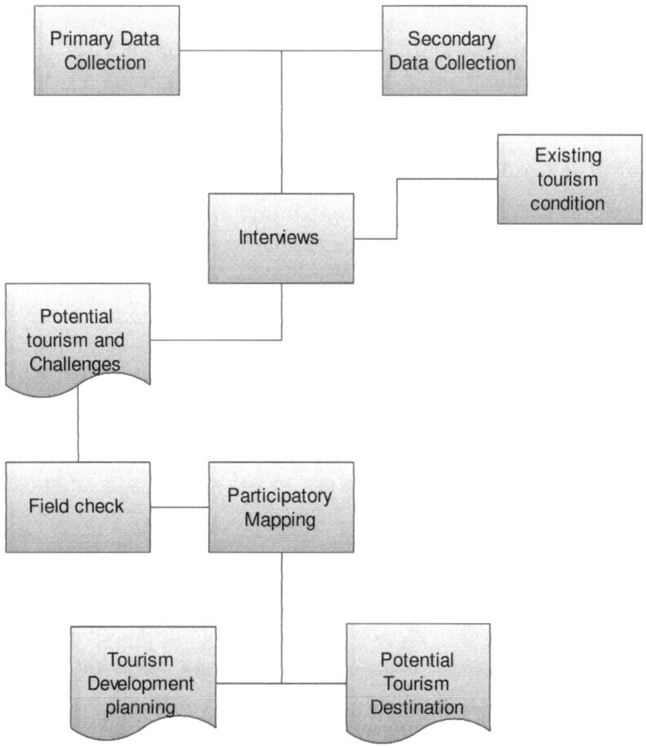


**Figure 1.** Research diagram

**Figure 2.** Demung Temple and Pancoran Pitu

People in Mambal Village are mostly farmers, where all the agricultural activities that occur under the Subak organization named Subak Mambal. Currently farmers in the Mambal Village are focusing on developing organic farming, of which 38% of the rice fields in Mambal are pure organic that produces organic rice. Not only rice cultivation, Mambal villagers also continuously grow sweet corn. Around the rice fields there are also a path / farm road that is intended to jogging tracks for visitors while enjoying

IOP Conf. Series: Earth and Environmental Science **98** (2017) 012044 doi:10.1088/1755-1315/98/1/012044

the natural beauty of rice fields in Mambal. Figure 4 shows the condition of rice fields and jogging tracks in Mambal Village.

In addition to agricultural crops, the condition of Mambal Village tempetatur is cool enough to encourage the cultivation of oyster mushrooms. Oyster mushroom cultivation has been successful in supporting the economy of Mambal villagers. Oganization of oyster mushroom cultivators in Mambal Village is named Jamur Kumbung. In addition to farming, Mambal Village Community also produces handicraft products that are woven in the form of symmetrical endek fabrics and processed products of handicraft from used goods such as bags, wallets, pencil boxes and others. Villagers involved in this handicraft already have an organization called Undagi Lestari. These products have also been able to follow the events both in local and national scale.

This organic farm and handicraft products makes potential for Mambal Village to manage its tourism. In general, the potential of natural resources and culture of Mambal Village able to supports the development of Mambal Village as Tourist Village. Figure 5 shows the results of oyster mushroom cultivation, while Figure 6 shows the handicrafts of Mambal villagers.

**Figure 3.** Senaung Pengibull Cave

**Figure 4.** Ricefields View and Jogging Track

**Figure 5**. Oyster Mushroom Cultivation

All facilities and infrastructure supporting tourism activities are still in small quantities / small scale. For further development it is necessary to re-add facilities and infrastructure, such as restaurant, ATM, and others. All tourism support activities that exist in Mambal still maintained the local culture and are traditional, it can be seen from the arrangement of tourist attractions that carries the concept of Tri

Mandala. In addition, people in Mambal village still adhere to the concept of Tri Hita Karana to maintain a harmonious relationship with God, fellow society and with the natural surroundings.

Mambal village is very open with tourists both local and foreign. To support the development of village tourism, there have been homestays or inns run by individuals. For further development also needs to be made a place to eat / restaurant that serves local food from the local community. Due to the existence of jogging track, it is necessary to make a bicycle rental bike, so that tourists who may have a cycling pleasure can try and enjoy the beauty of Mambal Village. In addition, it has been conceptualized the development for the location of self-cleaning (melukat) Pancoran Pitu by adding bathing location. Figure 7 shows the distribution of tourist attractions and products of Mambal Village.

**Figure 6.** Handicrafts from Used goods

*3.2. Development Planning*

*A. Physical Infrastructure Planning*

Development planning of Mambal Village involving the people of the Village through participatory mapping. The participatory mapping makes the development plan to become community based. the people of Mambal Village are represented by Head of the Village, youth communities, Hindu Priest and tourism communities in Mambal Village. The participatory mapping process shows in Figure 8.

Facilities development plan and infrastructure plan for Mambal Village is conducted to complete the existing facilities and tourism attractions. The planning resulted from the participatory mapping are isted in Table 1. Figure 9 shows the location of tourism facility planning.

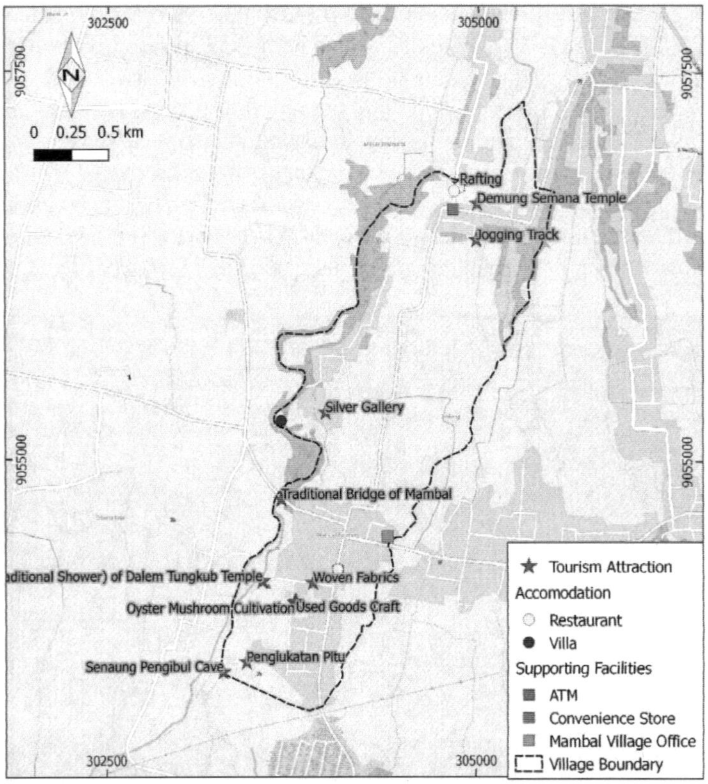


**Figure 7.** Tourism Attractions in Mambal Village

**Figure 8.** Participatory Mapping Process.

**Table 1**. Development for Tourism Facility and Accommodation in Mambal Village

| Plan | Coordinate x | Coordinate y | Type |
|---|---|---|---|
| Restaurant Plan | 303593.78 | 9054232.69 | Accommodation |
| Melukad (Cleansing) Place | 303536.94 | 9054225.47 | Facility |
| Mambal Market Plan | 304117.50 | 9054530.41 | Accommodation |
| Restaurant Plan | 303706.10 | 9054800.16 | Accommodation |
| Dam Modification Plan | 303775.57 | 9054822.72 | Accommodation |
| Center Stage | 303595.73 | 9053978.32 | Facility |
| Parking Lot | 303531.44 | 9053740.37 | Facility |
| Jogging Track | 304510.65 | 9055411.41 | Facility |
| Jogging Track | 303622.61 | 9053772.41 | Facility |

Beside accommodation i.e restaurant plan, Dam modification plan, and market plan, there's also facility planning including melukad (cleansing) place for Hindus, center stage, parking lot and jogging track.

Center stage and parking lot will be facilities for art performing includes traditional dance, painting exhibition and other art gallery. Meanwhile jogging track will be complementary facility for Subak tourism in Mamba Village. Other facilities such as homestays will be using houses of people in Mambal Village to maximize the village experiences of the tourists. Homestays will be decorated in traditional village architecture.

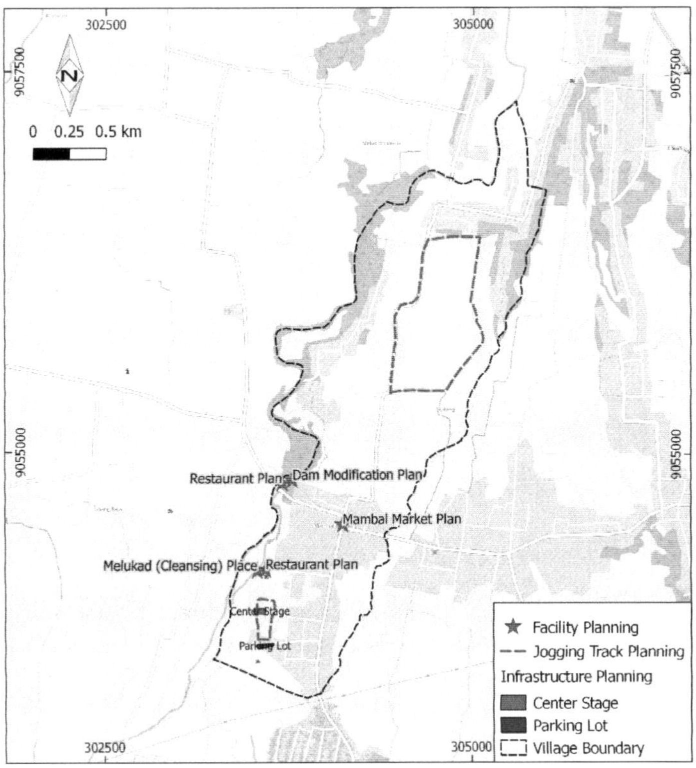


**Figure 9.** Planning Map of Mambal Village

The people of Mambal Village fully support the existence of a tourism village. This will provide a very good impact for the community in addition to developing the potential and preserve the nature, the community can also develop themselves from an economic point of view. In addition, many people will be engaged in tourism.

The managerial capacity of the village has been very good and the whole village apparatus can perform its functions well. But for tourism managerial in the future, mamba Village needs to form a group of tourism awareness (POKDARWIS) so that tourism management can be better organized. In addition, training on human resources who will improve the tourism potential of Mambal village. The ability of local community interaction has been very good, the community is open in receiving visits. In addition, there are also many local people who are engaged in tourism and foreign languages.

*B. Program and Best Practice Planning*
To further increase the selling value and competitiveness Mambal Tourist Village, the Management must make intelligent and creative breakthroughs to conduct structuring and comprehensive improvements to better provide the best and satisfactory services to tourists who come, so that the Mambal Tourist Village growing and become the mainstay attraction in Badung Regency which managed by village community for the benefit of Mambal community itself.

Various programs should be developed divided into 3 (three) Program Phases: Short Term Program, Medium Term Program and Long-Term Program, as follows: (1) Short Term Program, preparing accommodation, such as the provision of local accommodation that is part of the development of Mambal community-based tourism development, preparing local culinary from local with international flavor. This will also provide passion for local communities to create and preserve their local menus, prepare local human resources by providing intensive training and conducting comparative studies to enhance experience in management and exchange of information relating to the management of tourist attraction. In addition, preparing tourist facilities, tour, and products to provide the best service to tourists and also provide the convenience to to make tourists stay longer and enjoy the Mambal Village. Convenient resting places to enjoy the natural beauty of Mambal Village, playground for children, places to hold meetings, decorative interiors, arrange the surrounding parks, arrange the paths for travellers, so that all the traders in Mambal get the same opportunity for tourists to shop at their kiosks.

Medium Term Program, to hold cooperation or MoU with tourism stakeholders, such as Travel Agent, Government, and other tourism industry to create better cooperation and marketing. To support the sustainability of such cooperation, it is necessary to create a kind of MoU to be able to facilitate the preparation of event activities or cultural attractions with assistance from Travel Agent or other tourism industry, to create a calendar of events and tourist events at Mambal Village to be delivered to tourists so that various activities can carried out can be published and known by tourists and to be able to provide appropriate services, making local souvenirs to be known and always remembered the uniqueness and beauty of Mambal Village. These souvenirs are important to make tourist who come to Mambal Village to bring evidence that they have visited and can be a direct promotional to friends or families.

Long Term Program, which is: held Mambal Festival, where various tourism potentials owned by Mambal Village can be explored and arranged a diversity of attraction that provides various tourism options.

## 4. Conclusions
Mambal village has a very wide expanse of rice fields that are still very natural. There is also Senaung Pengibul Cave with unique characteristics with the height and texture that are formed naturally. Mambal village also has a spiritual tour of Demung Temple, which surrounding there are Pancoran Pitu. Pancoran Pitu and Cave Senaung Pengibull is a symbol that can be developed into a new tourism potential in Mambal Village.

To further increase the selling value and competitiveness of Mambal Village, various programs should be developed divided into 3 (three) Program Phases: Short Term Program, Medium Term Program and Long-Term Program. Short term program is to preparing accommodation, medium term program, to hold cooperation or MoU with tourism stakeholders, such as Travel Agent, Government, and other tourism industry to create better cooperation and marketing. Long term program is to held Mambal Festival where various tourism potentials owned by Mambal Village can be explored and arranged a diversity of attraction that provides various tourism options.

**Acknowledgments**
Writers would like to thank the Rector of Universitas Udayana, Dean of Tourism Faculty of Universitas Udayana, Head of Center for Spatial Data Infrastructure Development (PPIDS) Universitas Udayana, fellow staf, researchers and professors at Universitas Udayana for all the support, comments and suggestions that makes this paper finished.

**References**

[1] Mathieson A and G Wall 1982 *Tourism Economics, Physical and Social Impacts* Longman Group Limited.
[2] Pitana I G and P G Gayatri 2005 *Sosiologi Pariwisata*. (Yogjyakarta: Andi Press).
[3] Byrd E T 2007 Stakeholders in Sustainable Tourism Development and Their Roles: Applying Stakeholder Theory to Sustainable Tourism Development *Tourism Review* **62** (2).
[4] Beierle T C and D M Konisky 2000 Values, conflict, and trust in Participatory Environmental Planning *Journal of Policy Analysis and Management* **19**(4).
[5] Carmin J, N Darnall and J Mil-Homens 2003 Stakeholder involvement in design of U.S. Voluntary Environmental Programs: Does Sponsorship Matter? *Policy Studies Journal* **31** (4).
[6] Thomson M M and K D Owens 2012 Introduction to public participation geographic information systems special issue. *Journal of the Urban and Regional Information Systems Association*, **24**(2), pp. 5-6.
[7] Dunn C E 2007 Participatory GIS – A People's GIS? *Progress in Human Geography* **31**(5) pp 616–637.
[8] Suarthana I K, N Madiun, S O Moeljadi and Yuniarsa 2015 Exploring The Community Participation, Tourism Village, and Social-Economic to Environment Impact (Case Study: Pentingsari Village, Yogyakarta) *International Journal of Business and Management Invention* **4** (9) pp85-90.
[9] Regulation of Badung Regent Number 47 Year 2010 on Stipulation of Tourism Village Area in Badung Regency
[10] Smitha S 2014 Village Tourism in Kerala *International Journal of Management and Commerce Innovations* **2** (2) pp 15-20.
[11] Central Bureau of Statistics of Badung Regency 2016 *Badung In Figures 2015* (Badung: Central Bureau of Statistics).

IOP Conf. Series: Earth and Environmental Science **98** (2017) 012045 doi:10.1088/1755-1315/98/1/012045

# Determining Banjar Boundary and Tourism Planning using Participatory Mapping (Case Study in Pesinggahan Village, Klungkung Regency, Bali, Indonesia)

**I N Sunarta[1], T Kusmawati[2], J Hutauruk[2] and N K D A Lestari[2]**

[1]Faculty of Tourism, Universitas Udayana, Bali, Indonesia
[2]Center for Spatial Data Infrastructure Development (PPIDS) Universitas Udayana, Bali, Indonesia

cairns54@yahoo.com

**Abstract.** Banjar is a legal community unity that has territorial boundaries based on origin and customs in Bali. Banjar is the smallest administrative area in Bali Province, Indonesia. Pesinggahan village located in Dawan Sub-district of Klungkung Regency, Bali, Indonesia which consists of five banjar and each has the potential of different areas. Pesinggahan village also established into a ritual village and has tourist attractions such as Goa Lawah and unique landscape oh Subak's ricefields. The tourism in Pesinggahan village has not yet develop due to lack of spatial planning that accommodate the tourism development.The purpose of this research is to determine the banjar boundary in Pesinggahan village and to analyse the tourism development plan using participatory mapping. The Pesinggahan Village is consists of five banjar consist of Banjar Kanginan, Banjar Sukahati, Banjar Suwitrayasa, Banjar Belatung, Banjar Pundukdawa. Tourism area in Pesinggahan Village is Goa Lawah in Banjar Kanginan and Agricultural Land in Subak Pesinggahan in Banjar Suwitrayasa. Boundary Analysis of Villages and Banjar with spatial plan regulation of Klungkung Regency results that the Pesinggahan village are divide into Food Crop Agriculture Zone, Dryland Plantation Zone, Green Open Space and Residential Zone. Participatory mapping process produce several locations of tourism development planning such as homestay area, restaurant, road development around Goa Lawah tourism object, and planning of jogging track, cycling track, farm road, fishing pond, and restaurant around Subak Pesinggahan. Tourism development are located in accordance with zoning allotment of Pesinggahan village that is consists of zone of settlement, agriculture and green open space.

**Keywords:** Banjar Boundary, Participatory Mapping, Tourism Planning

## 1. Introduction

Pesinggahan Village is one of the villages located in District of Dawan, Klungkung Regency, Bali, which is traversed by tourist route in Bali especially for East Bali tourist destination. Pesinggahan Village is a tourist destination that is not as famous as the other tourist areas in Klungkung regency, such as Nusa Penida. Pesinggahan Village better known by local and international community because in this village there is a temple that has become one of the religious destination and also become a tourist destination, namely Pura Goa Lawah. Goa Lawah Temple is one of the sacred temples on the island of

Bali, the cave temple known as Pura Goa Lawah is home to thousands of bats, local legend is to be believed, it also hides a river of healing waters and a titanic snake wearing a crown. The development of Goa Lawah Temple making Pesinggahan Village in February 2017 established as one of the Tourist Village in Klungkung Regency. Tourist village is a rural area that has some special characteristics to become a tourist destination. In this area, the inhabitants still have tradition and culture that is still original. Through Regent's Regulation no.2 Year 2017 on the determination of Pesinggahan Village area as a religious tourist village with an insightful agriculture, making the government, especially in the Pesinggahan Village government to focus more on developing Pesinggahan Village [2].

Pesinggahan Village is rich of potential that can be developed. Recently the Pesinggahan Village government also built a temple that has potential in the development of tourist village,namely Pura Penataran Agung Pasek. Pesinggahan Village also wellknown for its delicious local food named Sate Lilit. In addition, Pesinggahan Village has unique characteristics with topography of the area that located in a hilly and coastal areas that are formed naturally. The problems that occur is the potential of natural resources with human resources in each Banjar (Hamlet) is not optimized considering the Pesinggahan Village is still newly established as a tourist village so that the arrangement and management of existing potential to support the activities of tourist is still lacking.

## 2. Literature Review

### *2.1 Tourism Planning*

Tourism planning was restricted primarily to the measurement of the economic impacts for destination areas, due to the ease with which economic impacts may be measured, compared to environmental and social impacts and the attempt of local governments to optimize economic benefits [2].

### *2.2 Participatory Mapping*

Participatory mapping is the “bottom-up approach” that allows the masses to create maps for all, in contrast to the traditional “top-down approach”, relying on those with the power and resources to create maps that will benefit the masses either directly or indirectly. Mapping is useful for planners, policy-makers, and residents for the collection of knowledge, establishment of boundaries, administration of municipality services, and empowerment of those involved with the land [3]. Participatory mapping is an investigative process and tool, utilizing visual and discursive techniques to better understand local (community) natural resources, together with their management, dynamics and related challenges, with potential solutions to those challenges [4].

## 3. Methodology

### *3.1. Area of Study*

Area of study is located in Candikuning Village, Baturiti Sub-district, Tabanan Regency, Bali Province, Indonesia. Geographically, Candikuning Village located in UTM Zone 5o S, with easting 297258.12 m E and northing 9084770.92 m S. Candikuning Village bordering with Sukasada Village, Sawan Village and Kubutambahan Village of Buleleng Regency in the north, Belok Village of Badung Regency in the east, Batunya Village of Baturiti Sub-District in the south and Pancasari Village of Buleleng Regency in the west. The area of Pesinggahan Village is 365,6 Ha in which 44 Ha is for paddy field (12%), 171,95 Ha is for moor (16%), 126,62 Ha is for plantation (35%), 17,79 Ha is for yard (4,87%), and other land use for 5,24 Ha (1%). Pesinggahan Village divided into 5 (five) hamlet or Banjar Dinas, namely Banjar Dinas Kanginan, Banjar Dinas Suwitrayasa, Banjar Dinas Sukahati, Banjar Dinas Belatung, and Banjar Dinas Pundukdawa. Knowing the boundary of Banjar Dinas or hamlets in the Pesinggahan Village, will facilitate in planning the development to be implemented in the village.

### *3.2. Tools and Material*

The material used in this research are spatial datas as follows:

- Administration Map Scale 1: 25,000

IOP Conf. Series: Earth and Environmental Science **98** (2017) 012045 doi:10.1088/1755-1315/98/1/012045

While the tool used are:

- Computer or laptop used to process the spatial data
- GPS (Global Positioning System) to determine the coordinate point of observation.
- Camera
- Stationary

*3.3. Data collection*

Data collection method used in this research is primary data and secondary data method.

- Primary data collection
  - Field survey. Field surveys were made using scoring table for tourism planning in the Pesinggahan Village.
  - Interview to explore more information related to the development planning of tourism village
  - Participatory Mapping is an advanced process of interviewing to gather stakeholders to explore more information and to mapping the development plan for the development Pesinggahan Village.
- Secondary data collection for tourism planning mainly base on literature review

*2.4. Data Analysis*

- Interview Result Analysis
- Survey Result Analysis
- Focus Group Discussion Analysis

## 4. Result and Discussion

*4.1. Banjar Boundary Mapping*

In accordance with the name of this village, 'Pesinggahan' is a place for a stop or a short break after a long journey. Pesinggahan Village divided into 5 (five) Hamlets or Banjar (Dusun), namely Banjar Kanginan, Banjar Suwitrayasa, Banjar Sukahati, Banjar Belatung, and Banjar Pundukdawa. Knowing the boundary of Banjar in Pesinggahan Village, will facilitating the planning and development to be implemented in the village and to make Pesinggahan Village not only become a stopover but also the main tourist destination.

Determination of Banjar boundary in the Pesinggahan Village implemented using participatory mapping, and at the same time, participatory planning was held to establish development planning in each hamlet to support Pesinggahan Village's tourism. Participatory mapping is done in conjunction with all Banjar Leader and Head of Pesinggahan Village also with delegations from neighboring village in order to avoid overlapping boundaries between Banjar. Determination of Banjar boundaries then approved by all parties, both from the Banjar leaders and Head of Pesinggahan Village and also the neighboring villages. The establishment of Banjar boundary will make it easier for villagers to know the potentials that exist in each Banjar area. Figure 1 shows the participatory mapping process in Pesinggahan Village. Figure 2 shows the map of each Banjar.

**Figure 1.** Determining Banjar Boundary using Participatory Mapping

*4.2. Potential Development*

Banjar Kanginan is categorized as the most developed area. This is because of the existence of Pura Goa Lawah. Pura Goa Lawah (Goa Lawah Temple) is one of the well-known tourist destination which highly contribute to the income of Pesinggahan Village and Banjar Kanginan. Pura Goa Lawah is located on Jalan Raya Goa Lawah, with an easy accessibility because it is located in arterial road. Pura Goa Lawah becomes a tour package of tourism and routinely visited by tourists who are dominated by foreign tourists. Tourism activities especially in Pura Goa Lawah has been arranged and handled by Pokdarwis (Tourism Community Group) of Pesinggahan Village, so the management can run well. Activities in this temple will be quite busy while being carried out rituals such as Piodalan and Ngusaba. The uniqueness of this temple is the presence of thousands of bats nesting in the cave. Tourists who visit Pura Goa Lawah will be amazed by the culture heritage in rituals also the uniqueness of the thousands bats. Tourists who visited Pura Goa Lawah will be accompanied by a tour guide that comes from Pesinggahan Village. Guide will accompany the tourists to the Temple, but tourists are required to obey the rules of the Temple, which one of them is to politely dress. Therefore, before entering the Temple, tourists will be rented cloth and scarves directly on site. Ticket rates to visit Pura Goa Lawah ranges from Rp 6.000, - for adults and Rp 4,000, - for children. In addition to traveling around the temple location, guide that leads tourists will also explain the history of Pura Goa Lawah.

Goa lawah Temple is located on the edge of Goa Lawah Beach. Goa Lawah Beach area also became one of the tourist destination of Banjar Kanginan in Pesinggahan Village, because this beach is used as a location for rituals or traditional ceremony which become a series of events in Pura Goa Lawah itself. Tourists will pack the beach area when the ceremony is in progress. Goa Lawah Beach is a black sandy beach and has beautiful sea views to Nusa Penida Island. Figure 3 shows the conditions of Pura Goa Lawah and Figure 4 shows the Goa Lawah Beach.

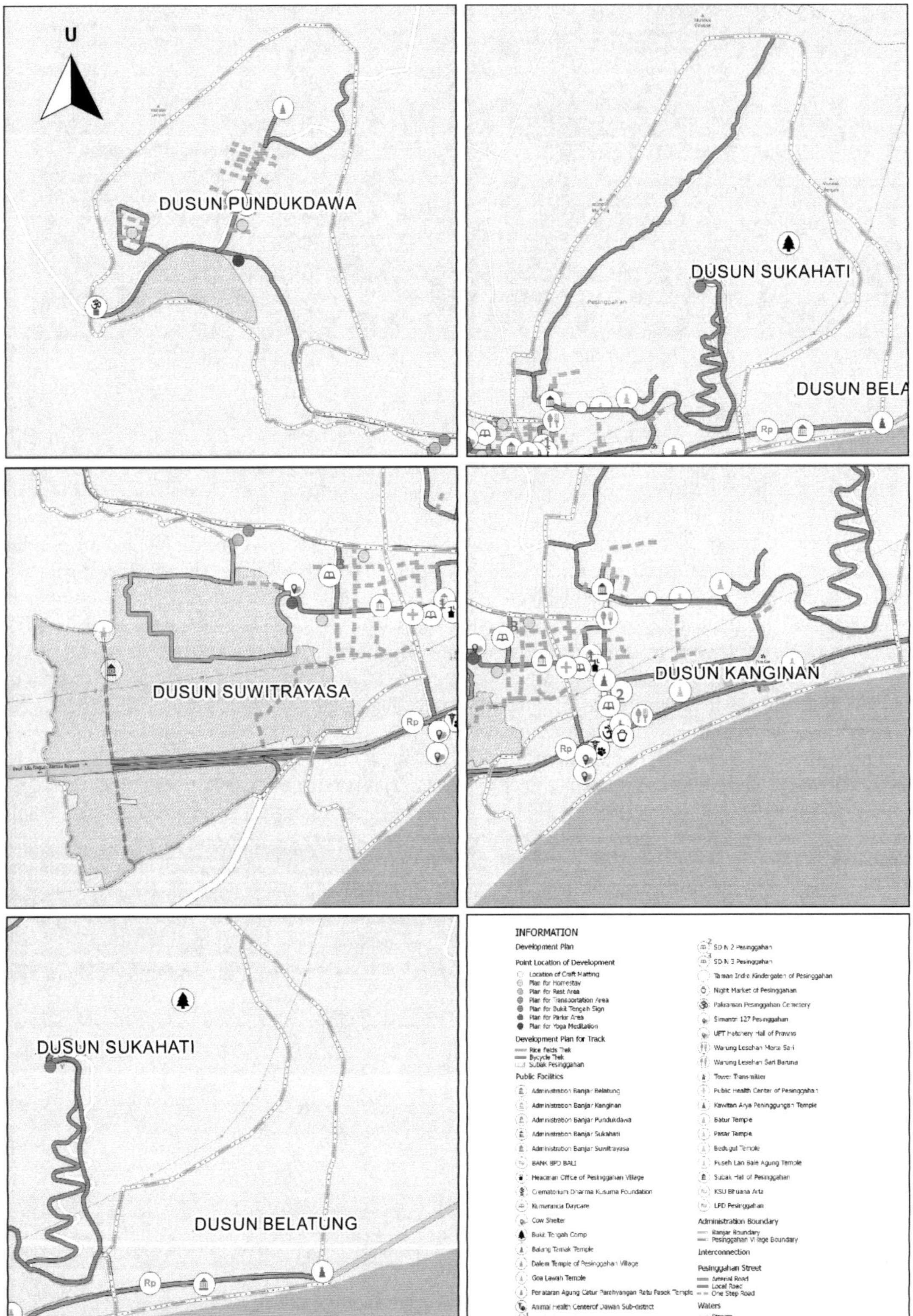


**Figure 2**. Maps of Banjar in Pesinggahan Village

**Figure 3.** Pura Goa Lawah

**Figure 4.** Goa Lawah Beach

Pura Goa Lawah is not the only famous Temple in Pesinggahan Village. Recently there's a new Temple that is also a potential of tourism in Banjar Kanginan, Pesinggahan Village, namely Pura Penataran Agung Catur Parahyangan Ratu Pasek which located in Banjar Pundukdawa. Since this temple was built, Pesinggahan Village becomes more frequently visited by people who want to carry out worship at the Temple. This condition can become a potential ncome and tourism for Banjar Pundukdawa and Pesinggahan Village. Figure 5 shows the conditions of Pura Penataran Agung Catur Parahyangan Ratu Pasek.

**Figure 5.** Penataran Agung Catur Parahyangan Ratu Pasek Temple

IOP Conf. Series: Earth and Environmental Science **98** (2017) 012045 doi:10.1088/1755-1315/98/1/012045

Pesinggahan Village not also blessed with unique culture and beach panorama, it also has agricultural areas in the form of rice fields that beautifully lies between slopes, with views of the hills located in the Banjar Suwitrayasa. Banjar Suwitrayasa also has a potential as a tourist village with agricultural potential, with agriculture activities become one of supporting the development of Banjar Suwitrayasa. Agricultural commodities in this village are rice, maize, and flowers that are usually used for the means of praying for Hindus. Figure 6 shows the panorama of rice fields in Banjar Suwitrayasa.

**Figure 6.** Ricefields in Banjar Suwitrayasa

Agricultural community that also exist in Banjar Suwitrayasa is Simantri 127. Simantri 127 established in 2014 and aims to support the development of agricultural diversification in an integrated and agribusiness-oriented. Simantri 127 is also expected to increase farmers' income. Simultaneous activities of Simantri 127 are more focused for the breeding of Bali cattle. Cow manure also used as fertilizer and used in the cultivation of plants. In addition to saving the cost of fertilizer, farmers in Banjar Suwitrayasa have skills in processing livestock manure and can be utilized. Figure 7 shows the Bali Cattle farm of Simantri 127.

**Figure 7.** Bali Cattle farm of Simantri 127

Pesinggahan Village also famous for its special food, which is a sate seafood stalls and become one of famous culinary in Pesinggahan Village. Pesinggahan village became one of the main culinary destination by the people of Bali. Restaurant scattered along the arterial road to Goa Lawah Temple in Banjar Suwitrayasa and Banjar Kanginan, and some also located in the middle of the settlement. Residents who enjoy this satay culinary usually come at the time of the big religious feast day in Pura Goa Lawah, or just stopover in their travelling. Figure 8 shows the famous culinary in Pesinggahan Village.

**Figure 8.** Satay Food Stall in Pesinggahan Village.

Pesinggahan Village not only potential for its nature beauty, religious and traditional culture but also agricultural potential and culinary. Banjar Kanginan and Banjar Suwitrayasa is two main Banjar in Pesinggahan Village has its own uniqueness that can make Pesinggahan Village not only as stop over for tourists but also become main tourist attractions in Bali. Table 1 shows the uniqueness of Pesinggahan Village.

**Table 1.** Uniqueness of Pesinggahan Village.

| No | Banjar | Uniqueness |
|---|---|---|
| 1 | Kanginan | Pura Goa Lawah, Goa lawah Beach, Culinary, makes it potential for religious tourism. |
| 2 | Suwitrayasa | Ricefield view, agricultural products, Bali cattle, culinary, makes it potential for agrotourism |
| 3 | Pundukdawa | Pura Penataran Agung Catur Parahyangan Ratu Pasek, |
| 4 | Dusun Belatung | Culinary |

*4.3. Development Planning*

The development of tourist village especially in Pesinggahan Village must pay attention to the level of acceptance of local community. It is useful to know the human resource condition of the Pesinggahan Village community that can be acted as subject in the development of the tourist village. The development of a good tourism industry will have an impact on the region's economy, such as opening up employment opportunities, increasing people's incomes and increasing local revenues also to equally distribute the opportunity to involve in tourism business for all Banjar in Pesinggahan Village.

Pesinggahan Village development plan can be done after we know the value of Pesinggahan Village. The value will be calculated using score based on the state of nature, physical environment, culture, infrastructure, village institutions, human resources, community life, accessibility, attraction, village tourism support facilities, and investment opportunities in Pesinggahan Village. Pesinggahan Village has a total score of 199, which means Pesinggahan Village is in the stage of developing tourist village. Appendix 1 shows the complete scoring result of Pesinggahan Village.

The strength of Pesinggahan Village is on the quality of its human resources. This is supported by the number of villagers with productive age, so that there are opportunities for young people to move to the tourism business. Then, lot of people of Pesinggahan Village have worked in the tourism sector especially tourist guide at Pura Goa Lawah. The weakness of Pesinggahan Village is the lack of support facilities for, such as homestays. These weaknesses will be a development opportunity to invest for the private sector and the government itself to improve the quality and quantity of tourist facilities in Pesinggahan Village.

The participatory mapping activities that has undertaken with all Banjar (Dusun) leader head of Pesinggahan Village produced a development plan for Pesinggahan Village. A proper development plan has been established in accordance with the wishes of the village community based on the potential of each banjar.

Through the activities of participatory mapping, obtained the location points that want to be developed by each representatives of Banjar in Pesinggahan Village. More focused development is planned in Banjar Sukahati. The first is the development of Bukit Tengah area in Dusun Sukahati. Bukit Tengah is a hilly area in center of Banjar Sukahati where visitors can see the whole village with sea panorama in the south. The Head of the Pesinggahan Village with the leader of Banjar Sukahati wants the Bukit Tengah area to be developed as one of the attractions of Pesinggahan Village, by carrying the concept of a rest area with the view of the whole village with rice fields and the sea. There's also a plan to make a sign of Bukit Tengah mounted on the side of the hill so it can be seen from a distance.

Another plan to be developed is facilities such as homestay. Currently there is only one homestay located in Dusun Pundukdawa, and the plan is to build more homestay in Banjar Pundukdawa because Banjar Pundukdawa is suitable for homestay construction site and planned for a centralized homestay. Banjar Pundukdawa itself is located close to the expanse of rice fields with hilly topography that adds to the uniqueness of homestay location for tourists. In addition to homestay, another facility to be built is a place of yoga meditation. The place of meditation is very suitable to be developed in the Pesinggahan Village, because in this village is known for its religious tourism. Tourists who come to visit can feel the spiritual experience through yoga meditation. The location of yoga meditation is planned to be built near the paddy fields in the Banjar Pundukdawa. Figure 9 shows the development map of Pesinggahan Village.

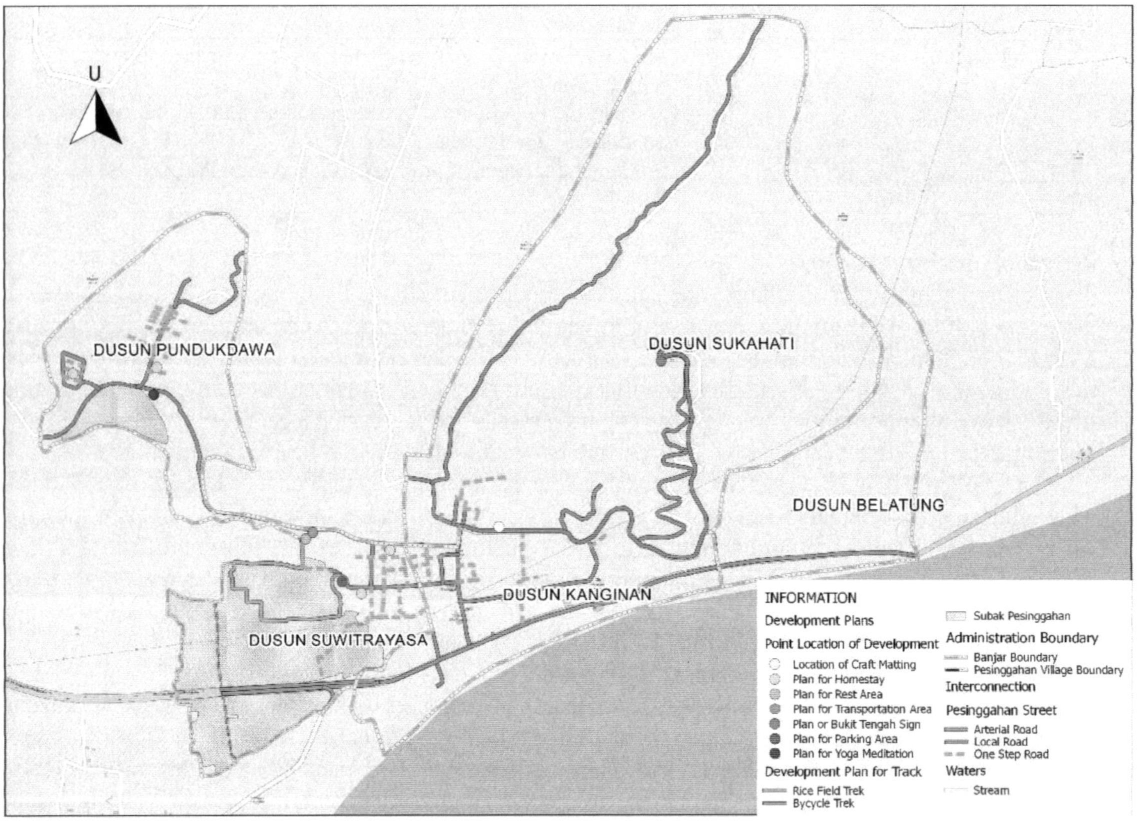


**Figure 9.** Vilage Development Plan

To support agrotourism there's also plan to create ricefield jogging track and bicycle line in Banjar Pundukdawa, and Banjar Sukahati. These two Banjars beside Banjar Suwitrayasa will be planned as agrotourism area with ricefield view and agricultural products. And finals to complete the accommodations for tourists in Pesinggahan Village, there's also plan for rest area and parking place to accommodate the vehicles and also as a income from parking retributions. The culinary also established in each Banjar but with modern taste, different menus and modern management to add the facilities for tourists. Table 2 shows the development plan in Pesinggahan Village.

**Table 2.** Development Plan in Pesinggahan Village

| No | Banjar | Development Plan |
|---|---|---|
| 1 | Suwitrayasa | Agricultural products, Bali cattle, |
| 2 | Kanginan | Handicrafts, Pura Goa Lawah, Goa lawah Beach, Culinary |
| 3 | Belatung | Culinary |
| 4 | Sukahati | Bukit Tengah area, culinary, ricefield jogging track and bicycle line |
| 5 | Pundukdawa | Yoga area, ricefield jogging track and bicycle line, homestays, culinary, Pura Penataran Agung Catur Parahyangan Ratu Pasek |

## 5. Conclusions

Pesinggahan Village has lots of potential to become tourist village. Pesinggahan Village divided into 5 (five) Hamlets or Banjar (Dusun), namely Banjar Kanginan, Banjar Suwitrayasa, Banjar Sukahati, Banjar Belatung, and Banjar Pundukdawa Determining the Banjar boundary in the Pesinggahan Village implemented using participatory mapping, and at the same time, participatory planning was held to establish development planning in each hamlet to support Pesinggahan Village's tourism.

Banjar Kanginan is categorized as the most developed area. This is because of the existence of Pura Goa Lawah (Goa Lawah Temple), and Goa Lawah Beach. There's also ricefield view, agricultural products, Bali cattle and traditional culinary that makes Pesinggahan Village not only potential for its nature beauty, religious and traditional culture but also agricultural potential and culinary.

The strength of Pesinggahan Village is on the quality of its human resources, in which, lot of people in Pesinggahan Village have worked in the tourism sector especially tourist guide at Pura Goa Lawah. The weakness of Pesinggahan Village is the lack of support facilities such as homestay. These weaknesses will be a development opportunity to invest for the private sector and the government itself to improve the quality and quantity of tourist facilities in Pesinggahan Village

Development plan for Pesinggahan Village involving development of new tourism attractions, products and facilities for tourists. New tourism attractions and products such as rest area in Bukit Tengah with Village view, ricefield view and sea view, yoga/ meditation area, handicraft products, agricultural products, ricefield jogging track and bicycle line. Facilities plan such as homestays, parking lot, and traditional culinary.

## Acknowledgments

Writers would like to thank the Rector of Universitas Udayana, Dean of Agriculture Faculty of Universitas Udayana, Dean of Tourism Faculty of Universitas Udayana, Head of Center for Spatial Data Infrastructure Development (PPIDS) Universitas Udayana, fellow staf, researchers and professors at Universitas Udayana for all the support, comments and suggestions that makes this paper finished.

## References

[1] Matthieson and Wall 1982 *Tourism Change, impact and opportunities* (England: Pearson Education Limited).
[2] Alexander L M 1953 *Holiday in the Seychelles* (New York: Purnel).
[3] Warner Cali 2015 *Participatory Mapping: a literature review of community-based research and participatory planning*.
[4] Flintan F 2009 *Participatory Natural Resource Mapping*.

# Spatial Numeric Classification Model Suitability with Landuse Change in Sustainable Food Agriculture Zone in Kediri Sub-district, Tabanan Regency, Indonesia

**N M Trigunasih[1*], I Lanya [2], J Hutauruk [1] and I D M Arthagama [2]**

[1]Center for Spatial Data Infrastructure Development (PPIDS) Udayana University, Indonesia
[2]Study Program Agroecotechnology, Faculty of Agiculture, Udayana University, Indonesia

tri5963@yahoo.com

**Abstract.** The development of rapid population will make the availability and utilization of land resources is increasingly shrinking in number, especially occurs in rice field. Since the last 5 years the numbers of farmland is decrasing by industry, infrastructure development, tourism development and other services. The agricultural problems facing at the moment is the occurrence of a change of use of agricultural land into farming now is not more popular is called over the function of agricultural land into non-farming. According to the Central Bureau of statistics (BPS) of the province of Bali (2013) within a period of 14 years (1999-2013), there has been a change of use of agricultural land be not agriculture/ wetland functions over the 4,906 hectares. When averaged over the function flatten paddy fields per year occurred in Bali approximately 350 ha (0.41%). The highest paddy fields over the function during a period of fourteen years there is in Tabanan area of 1,230 ha. To maintain the existence of the rice fields or subak in Bali in particular, need to be done protection against agricultural lands sustainable. Ninth District/Town in Bali today, haven't had a Perda on protection of agricultural land sustainable food that is mandated by law 41 Year 2009. This will have an impact on food security of the region, and the world's cultural heritage as the water will lose its existence as a system of irrigation organization in Bali. The purpose of this research was done to (1) determine the numerical classification of spatial parameters of sustainable food farm in Tabanan Regency Kediri Subdistrict, (2) determine the model of the zoning of agricultural land area of sustainable food that fits on Years 2020, 2030, 2040, and in district of Kediri, Tabanan Regency. The method used is the kuantitaif method includes the focus group discussion, the development of spatial data, analysis geoprosessing (spatial analysis and analysis of proximity), and statistical analysis, interpolation of digital elevation model raster data, and visualization (cartography) and qualitative methods include the study of the literature (introduction). The research results obtained by as much as 23 rice fields mapped in spatial control system based on its geographical location. The parameters in the classification of sustainable food farming in district of Kediri consists of (1) the suitability of the location of a rice field with spatial Plan area of Tabanan Regency years 2012-2032, (2) land use, (3) Watershed morphology, (4) the type of irrigation, (5) rainfall, (6) the form region, (7) the high place, (8) the suitability of the agroecosystem paddy fields, (9) productivity, (10) the distance from the center of town, (11) minimum area. Spatial numerical classification produces a wide variety of modeling (5 models) and is associated with the projected changes in rice fields by the year 2020, 2030, and 2040. In the year 2020 using model 4 due to sustainable subak in model 4 of 2682.71 ha, approached the farm field area by the year in 2020 of 2684 ha. In the year 2030 using model 3 due to sustainable subak on the model is 1651.37 ha 3 plus ¾ buffer subak of 773.51 ha be 2424.88 ha approached the farm field by the year in 2030 of 2364 ha. In the year 2040 using model 2 due to sustainable subak on the model of 307,99 ha 3 plus ¾ buffer subak of 1781,04 ha be 2089,33 ha approached the farm field by the year in 2040 of 2033 ha.

**Keywords:** Modelling, GIS, Subak, Agriculture, Sustainable

## 1. Introduction

The development of rapid population will affect the availability and utilization of land resources is increasingly shrinking in number especially occur in farmland. Since the last 5 years the number declining farmland due to infiltration by industry trade, infrastructure development, tourism development and other services. A problem that often occurs is the conversion of/over the function of farmland to non-agriculture. This much going on caused the increasing activity of the economy, trade, industry of city centres, as well as the tourism sector which is the underpinning of the economy of Bali. The existence of the attractions as well as the construction of facilities and infrastructure resulted in a very dynamic land conversion while the availability of land for agriculture is very limited, in line with the problems encountered in national food like the degradation of soil fertility, water utilization, competition of food deficits.

According to the BPS province of Bali (2013) within a period of 14 years (1999-2013), there has been a change of use of agricultural land into agricultural/ over the function instead of the paddy fields in Bali province covering an area of 4,906 ha. When averaged over the function flatten paddy fields per year occurred in Bali approximately 350 ha (0.41%). The highest paddy fields over the function during a period of fourteen years there is in Tabanan area of 1,230 ha, later followed by Jembrana Regency of Buleleng Regency 1,078 ha, ha 677 acres, Badung Regency area of 672 ha, town Denpasar area of 659 ha, Gianyar Regency area of 497 ha and ha 173 acres of Klungkung Regency. But the addition of extensive paddy fields occurs in the Karangasem Regency area of 58 hectares and Bangli (village of the subdistrict Mengani Kintamani area of 22 ha). Ninth District/Town in Bali today, haven't had a Perda on protection of agricultural land sustainable food mandated by law 41 Year 2009. This will have an impact on food security of the region, and the world's cultural heritage as the water will lose its existence as a system of irrigation organization in Bali. One of the technologies used is the spatial geographic information systems (GIS) and remote sensing (RS). RS with GIS can provide an excellent framework for obtaining precise and accurate data, measurements and synthesis, all of which are crucial for environmental analysis of eco-environmental comprehensive socio-associated with the changes to the data area of the rice fields [3][5]. Therefore conducted research of spatial model of numerical classifications with the projection of land use change on sustainable food agricultural areas in district of Kediri Regency of Tabanan, Indonesia.

The purpose of this research was done to (1) Determine the Boundaries of The Rice Field subak In Kediri Subdistrict (2) determine the numerical classification of spatial parameters of sustainable food farm in Tabanan Regency Kediri Subdistrict, (3) determine the model of the zoning of agricultural land area of sustainable food that fits on Years 2020, 2030, 2040, and in district of Kediri, Tabanan Regency.

## 2. Literature Review

Protection of agricultural land of sustainable food (PLP2B) is a system and process in planning for establishing, developing, utilizing and fostering, controlling and supervising food and farmland region on an ongoing basis. Under Law Number 41 Year 2009 about the PLP2B, is one of the Government's policy in controlling the pace over the function of agricultural land, in particular rice fields in Indonesia (7). PLP2B has the aim to (1) protect food and farmland area on an ongoing basis, (2) ensure the availability of agricultural land for sustainable food, (3) realize self-reliance, resilience and food sovereignty, (4) protect the ownership of agricultural land that belong to farmers, food (5) enhancing the prosperity and welfare of farmers and society, (6) improving the protection and empowerment of farmers, (7) improving the provision of employment for a decent life, (8) maintain the ecological balance and (9) realize the revitalization of agriculture.

In terms of soil and water conservation, water control system been established since the 7th century with the irrigation system of half technical and simple. Farm water control system also serves as a flood control, capture precipitation, mainly located in the upper reaches of the river. Rice plant as a contributor to oxygen ($O_2$) at the micro as well as macro-scale [4]. Wide range of requirements set out in Act No. 41 of the year 2009, then should the whole Subak in Bali land need special attention about its existence. The existence of the need for the development of agriculture, such as housing, tourism, retail, industry and means of developing the infrastructure, then the necessary of land conversion to a limited extent, especially in the area around the urban centers of Government and tourism.

Law No. 41 of the year 2009 on the protection of agricultural land of sustainable food in article 9 contains a land suitability, availability of infrastructure, land use, land and technical potential of the extents of the unity of the expanse of land. The physical condition of the territory: position of the location of the form against watersheds, a means of irrigation, rainfall, the shape of the area/relief/slope slopes, altitude, rainfall, place land suitability of agroecosystem, compliance against Spatial Plan area and land use, associated with the region that needs to be protected in agriculture sustainable food [4][6].

## 3. Methodology

This research was carried out in district of Kediri, Tabanan Regency. Kediri Subdistrict covering 53.50 km$^2$, with altitude 0-123 m above sea level, and there are 23 Subak irrigation rice field. Extensive Subak in Kecamatan Kediri is 2844 hectares [1].

### *3.1. Materials and Tools*

The tools used in this research is a set of computers and computer software (Qgis 2.10.1).
The materials used in this research are:

- Quickbird satellite imagery of Tabanan Regency Year 2013 from Geospatial Information Agency. Raster DEM, sourced from Aster satellite image,
- Map the Spatial Plan of Tabanan Regency Years 2012-2032,
- Map the watershed morphology Agency of Watershed Management of Northern Bali,
- Indonesia Topographical Map scale 1:25,000,
- Secondary data of rice production sourced from Kediri subdistrict office of Food and Horticulture Farming of Tabanan Regency
- Data of potential ricefield suitability of Kediri Sub-district.

### *3.2. Research Procedures*

Research procedures include preliminary phase, Interpretation of satellite images and the arrangement of the references coordinate system, Focus Group Discussion, updating of rice field data and Spatial Analysis, Classification and Mapping Criteria.

*3.2.1. Preliminary Phase.* Research activities originated from a literature study to obtain preliminary information about the condition of the area from the results of previous research studies, both in the form of data reports, scientific journals, and maps. Preparation of materials research, and and planning of research that will be done.

*3.2.2. Interpretation of satellite images and the arrangement of the references coordinate system.* Interpretation of satellite imagery for identification, description and classification of land use and landform analysis was conducted using three methods of analysis namely: (1) the elements of the analysis, (2) pattern analysis, and (3) fisiografi analysis. Element analysis is carried out with 9 elements: shapes, and sizes, colors and contrasts, textures and patterns, shadows, location, and associations. Nine elements are used for the classification of land use. Preparation of the coordinate system references using the UTM projection system (universal transverse Mercator), World Geodetic System 1984 Datum (WGS84), and 50 southern zones. Then digitation is conducted on screen to make spatial data of rice fields in Kediri Subdistrict.

*3.2.3. Focus Group Discussion (FGD).* Focus groups conducted to determine Subak boundary in district of Kediri which will be followed by 23 Pekaseh (leader) of Subak. Spatial data of rice fields then arranged based on the subak boundary that has been specified in the Focus Group Discussion.

*3.2.4. Updating ricefield data.* Update data is focused on maping the zonation of Subak, digitation of Regional planning (RTRW) map of Tabanan Regency, morphological data obtained from on screen digitation of watershed map, an altitude map obtained from raster DEm data analysis, distance from the town center obtained from buffer analysis, the rainfall data obtained from the analysis of isohyet map and landuse data obtained from landuse map.

*3.2.5. Spatial Analysis, Classification and Mapping Criteria.* Criteria and mapping the area of sustainable food farming by modifying the weights and score corresponds to the physical condition of the area. As the parameters are presented in Table 1 [4] [6].

**Tabel 1.** Numeric Classification Parameter in Kediri

| No. | Parameter | Valuation | | |
|---|---|---|---|---|
| | | **Weight** | **Score** | **Value** |
| 1 | Suitability of the location paddy fields with spatial plan area: | 11 | | |
| | a. Protected areas and Protected strategic | | 3 | 33 |
| | b. Agricultural Cultivation Area | | 2 | 22 |
| | c. Nonagricultural area | | 1 | 11 |
| | **Value** | | | **66** |
| 2 | Landuse: | 10 | | |
| | a. Irrigation Paddy Field | | 3 | 30 |
| | b. Moor | | 2 | 20 |
| | c. Mix Garden | | 1 | 10 |
| | **Value** | | | **60** |

| | | | | |
|---|---|---|---|---|
| 3 | Position and or subak location in watershed: | 9 | | |
| | a. upstream | | | |
| | b. middle | | 3 | 27 |
| | c. downstream | | 2 | 18 |
| | **Value** | | 1 | 9 |
| | | | | **54** |
| 4 | Irrigation System: | 8 | | |
| | a. Technical Irrigation – semi technical | | 3 | 24 |
| | b. Simple irrigation | | 2 | 16 |
| | c. Rainfed | | 1 | 8 |
| | **Value** | | | **48** |
| 5 | Rainfall: | 7 | | |
| | a. > 2500 mm/th | | 3 | 21 |
| | b. 2000 – 2500 mm/th | | 2 | 14 |
| | c. < 2000 mm/th | | 1 | 7 |
| | **Value** | | | **42** |
| 6 | Landform: | 6 | | |
| | a. Hilly up to the mountain, slope > 40 % | | 3 | 18 |
| | b. Wavy up to choppy, slope 25-40% | | 2 | 12 |
| | c. Flat up to ramps, slope < 25 % | | | |
| | **(Table 3.1.)** | | 1 | 6 |
| | **Value** | | | **36** |
| 7 | Altitude: | 5 | | |
| | a. >500 MASL | | 3 | 15 |
| | b. 100 – 500 MASL | | 2 | 10 |
| | c. <100 MASL | | 1 | 5 |
| | **Value** | | | **30** |
| 8 | Suitability of Agroecosystem Land for Paddy Field: | 4 | | |
| | a. Very Suitable (S1) | | 3 | 12 |
| | b. Suitable (S2) | | 2 | 8 |
| | c. Marginal Suitable (S3) | | 1 | 4 |
| | **Value** | | | **24** |
| 9 | Productivity: | 3 | | |
| | a. >5 ton/ha/planting period | | 3 | 9 |
| | b. 2,5 - 5 ton/ha/ planting period | | 2 | 6 |
| | c. < 2,5 ton/ha/ planting period | | 1 | 3 |
| | **Value** | | | **18** |
| 10 | Distance from Center of City: | 2 | | |
| | a. > 5km | | 3 | 6 |
| | b. 2,5 – 5 km | | 2 | 4 |
| | c. < 2,5 km | | 1 | 2 |
| | **Value** | | | **12** |
| 11 | Minimum area: | 1 | | |
| | a. > 10 ha | | 3 | 3 |
| | b. 5 - 10 ha | | 2 | 2 |
| | c. < 5 ha | | 1 | 1 |
| | **Value** | | | **6** |

IOP Conf. Series: Earth and Environmental Science **98** (2017) 012046 doi:10.1088/1755-1315/98/1/012046

All parameters then intersected to combine spatial data and attribute data, then to create total score. The result of the total score then statistically analysed using stastical summary [7] by the formula in equation (1).

$$s = \sqrt{\frac{n\sum_{i=1}^{n} x_i^2 - (\sum_{i=1}^{n} x_1)^2}{n(n-1)}} \quad (1)$$

where :

- $s$ = standard deviation
- $xi$ = value $x$ num-$i$
- $\bar{x}$ = average
- $n$ = data total

Based on the distribution of data, then maximum value, minimum, average, and standard deviation are generated. Data then can be grouped or classified based on average data and standard deviation with various modeling as follows:

a. Model 1 consisting of sustainable subak, buffer and converted, is obtained from the following formula:
   Sustainable subak: x>average +1standarddeviation
   Buffer subak: (x<average-1 standarddeviation) up to (x>average+1 standarddeviation)
   Converted subak: x<rata-rata-1 standarddeviation
b. Model 2 consisting of sustainable Subak, buffer and converted, is obtained from the following formula:
   Sustainable subak: x>average + ½ standarddeviation
   Buffer subak: (x< average - ½ standarddeviation) up to (x> average + ½ standarddeviation)
   Converted subak: x< average - ½ standarddeviation
c. Model 3 consisting of sustainable Subak, buffer and converted, is obtained from the following formula:
   Sustainable Subak: x> average
   Buffer Subak: (x< average – ½ standarddeviation) up to (x> average)
   Converted subak: x< average – ½ standarddeviation
d. Model 4 consisting of sustainable subak, buffer and converted, is obtained from the following formula:
   Sutainable Subak: x> average – ½ standarddeviation
   Buffer Subak: (x< average – 1 standarddeviation) up to (x> average – ½ standarddeviation)
   Converted Subak: x< average - 1 standarddeviation
e. Model 5 consisting of sustainable Subak, buffer and converted, is obtained from the following formula:
   Sustainable Subak: x> average – 1 standarddeviation
   Buffer Subak: (x< average – 1 ½ standarddeviation) up to (x> average - 1 standarddeviation)
   Converted Subak: x< average – 1 ½ standarddeviation

*3.2.6. Analysis of Suitability of Numerical Classification Model with Change Projection of Subak Area.* Changes in rice fields over the last 5 years (2010 to 2015) in Kecamatan Kediri are calculated using the formula on Equation (2)

$$\text{Change of paddy fields per year} = \frac{\text{paddy field area of 2015} - \text{paddy field area of 2010}}{5 \text{ years}} \quad (2)$$

From the data of rice fields changes per year can be determined from the projection of changes in Subak area by the year 2020, 2030 and 2040. The suitability of the modeling is determined from the area of the sustainable Subak and the buffer of Subak approached by the Subak area in each specified year. Parameter data was performed by visualization (cartography) for map making of 11 parameters and also making Subak zonation map in various modeling in Kediri sub-district of Tabanan regency.

## 4. Result and Discussion

The result of Subak mapping in Kediri Sub-district which obtained from Focus Group Discussion presented in Figure 1. Subak Map of Kediri Sub-district which consists of 23 Subak is presented in Appendix 1.

The map of parameters in the classification of sustainable food farming in Kediri Sub-district consists of (1) the suitability of the location of a rice field with Regional Spatial Plan of Tabanan Regency of 2012-2032, (2) exsisting land use, (3) Watershed morphology, (4) the type of irrigation, (5) rainfall, (6) the landform, (7) altitude, (8) the suitability of the agroecosystem of rice fields, (9) productivity, (10) the distance from the city center, (11) minimum area, are presented in Figure 2 to Figure 12.

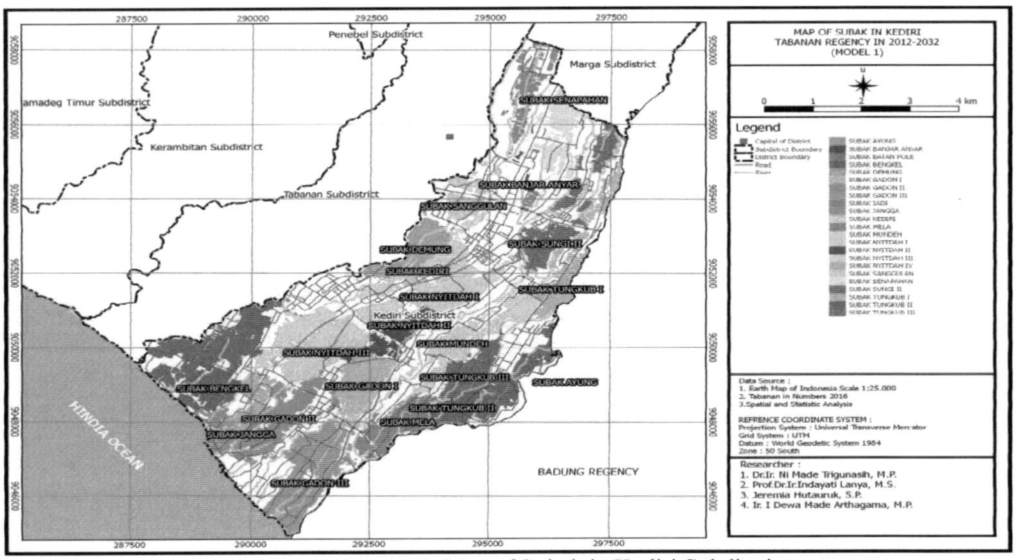


**Figure 1.** Map of Subak in Kediri Subdistrict

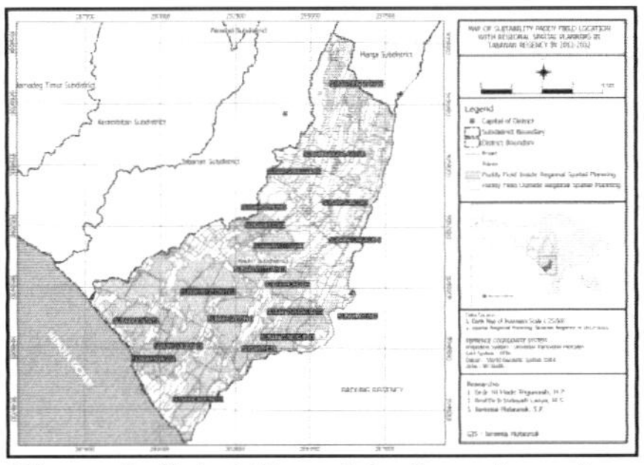

**Figure 2.** Suitability of the Location of Rice Field with Regional Spatial Plan of Tabanan Regency of 2012-2032

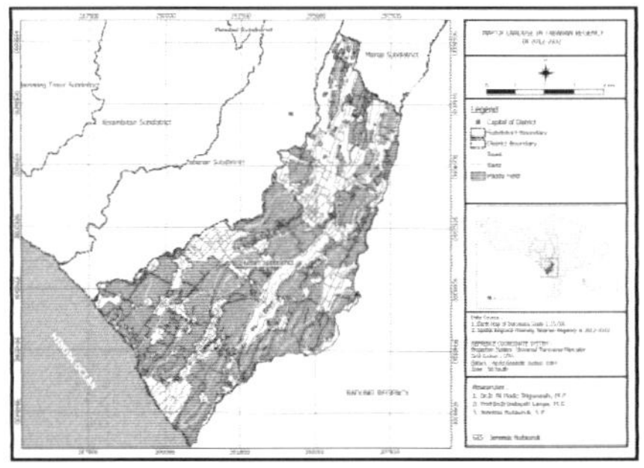

**Figure 3.** Exsisting Landuse of Kediri Sub-district

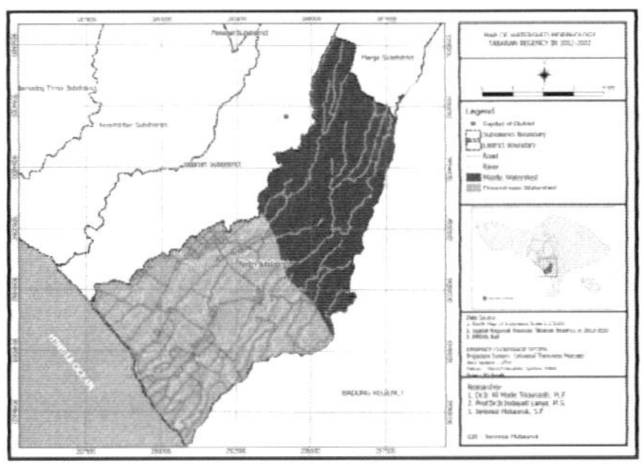

**Figure 4.** Watershed morphology

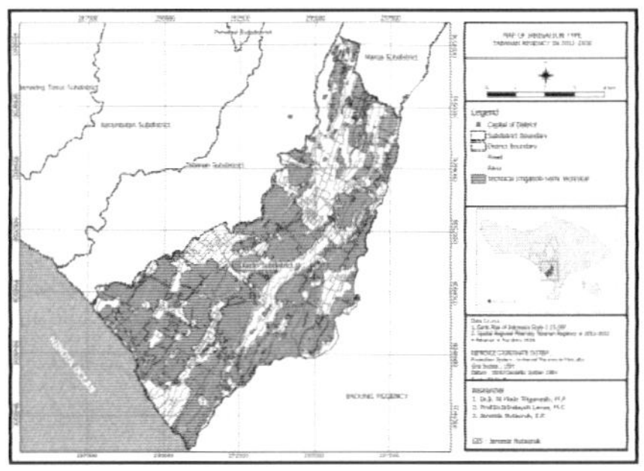

**Figure 5.** Irrigation Type

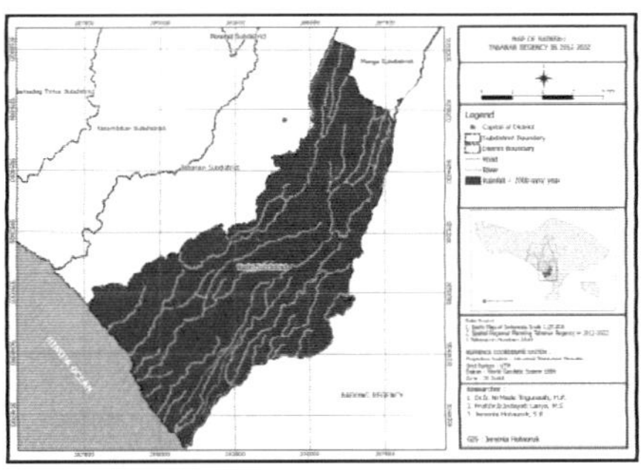

**Figure 6.** Rainfall Intensity

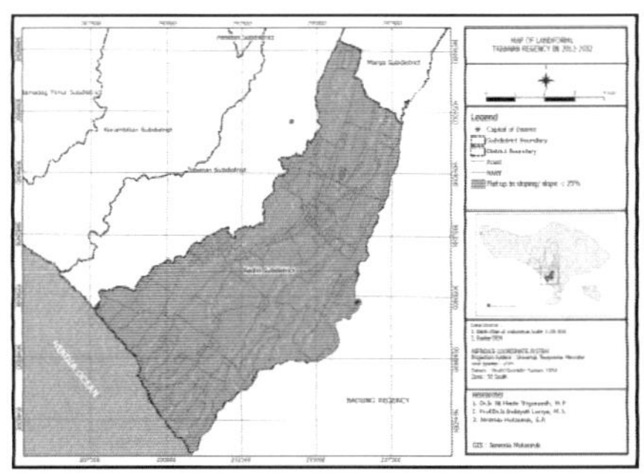

**Figure 7.** Landform

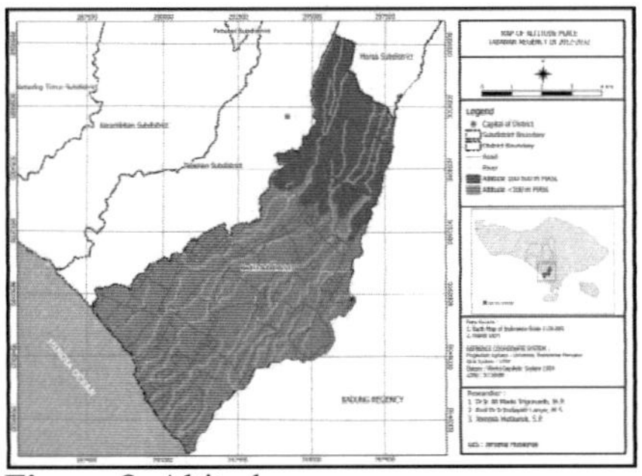

**Figure 8.** Altitude

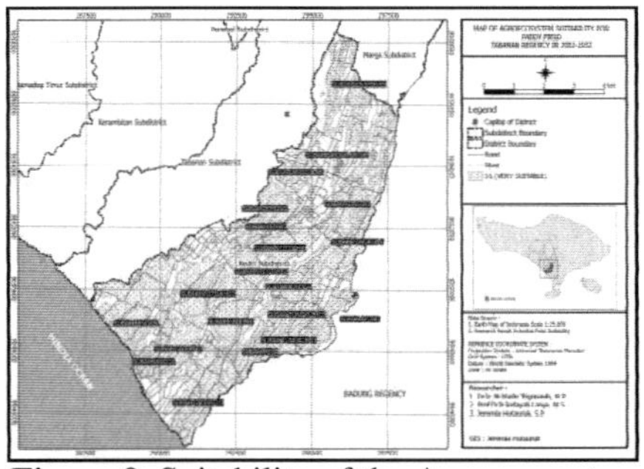

**Figure 9.** Suitability of the Agroecosystem of Subak rice Fields

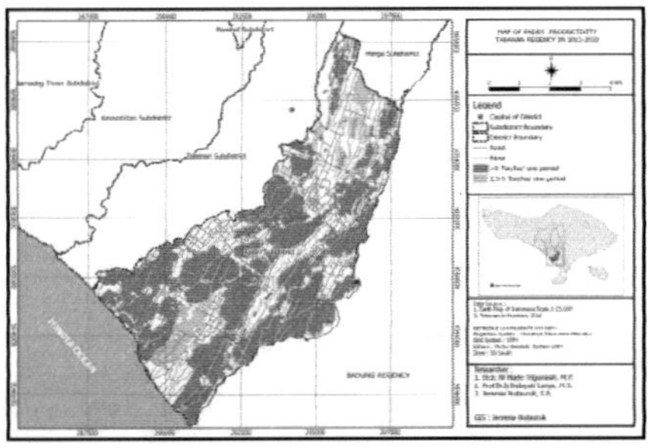

**Figure 10.** Ricefield productivity

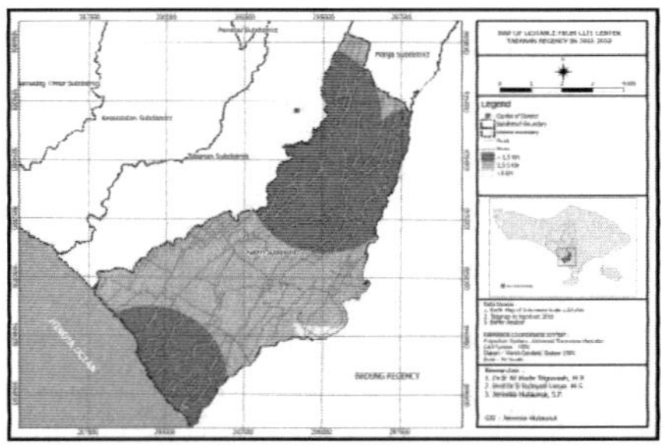

Figure 11. Distance from The City Center

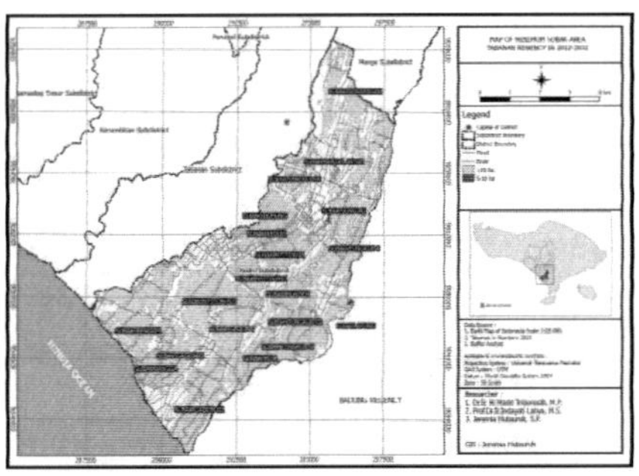

**Figure 12.** Minimum Area of Subak

The results of the intersection of eleven-parameter in Kecamatan Kediri resulted a modeling based on the standard deviation of the population data which divided into three regions: Subak sustainable, subak buffer, and subak converted, shown in Table 2.

**Table 2.** Value of Subak Sustainable, Subak Buffer, and Subak Converted in The Modeling

| Modeling | Formula | Value |
|---|---|---|
| **I. Sustainable Subak** | | |
| model 1 | x>average +1standarddeviation | x>136,4 |
| model 2 | x> average + ½ standarddeviation | x>132,7 |
| model 3 | x> average | x>129 |
| model 4 | x> average – ½ standarddeviation | x>125,3 |
| model 5 | x> average – 1 standarddeviation | x>121,6 |
| **II. Buffer Subak** | | |
| model 1 | (x< average -1standardeviasi) up to (x> average +1 standarddeviation) | 121,5>x<136,4 |
| model 2 | (x< average - ½ standardeviasi) up to (x> average + ½ standarddeviation) | 125,3>x<132,7 |
| model 3 | (x< average – ½ standarddeviation) up to (x> average) | 125,3>x<129 |
| model 4 | (x< average – 1 standarddeviation) up to (x> average – ½ standardeviasi) | 121,6>x<125,3 |
| model 5 | (x< average – 1 ½ standarddeviation) up to (x> average - 1 standarddeviation | x>117,9<121,6 |
| **III. Converted Subak** | | |
| model 1 | x< average -1 standarddeviation | x<121,6 |
| model 2 | x< average - ½ standarddeviation | x<125,3 |
| model 3 | x< average – ½ standarddeviation | x<125,3 |
| model 4 | x< average - 1 standarddeviation | x<121,6 |
| model 5 | x< average – 1 ½ standarddeviation | x<117,9 |

The results of the area of each model are presented in Figure 13 and maps of each region in each model in Kediri Subdistrict are presented in Figure 14.

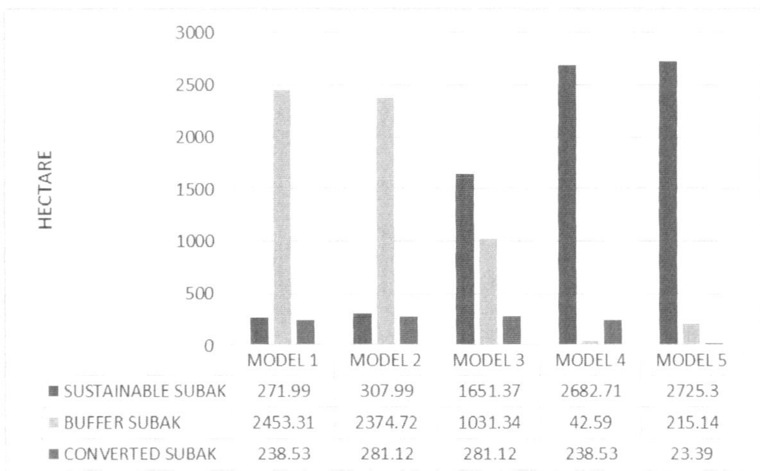


**Figure 13.** Graph of Subak Zoning Area in Various Models

IOP Conf. Series: Earth and Environmental Science **98** (2017) 012046 doi:10.1088/1755-1315/98/1/012046

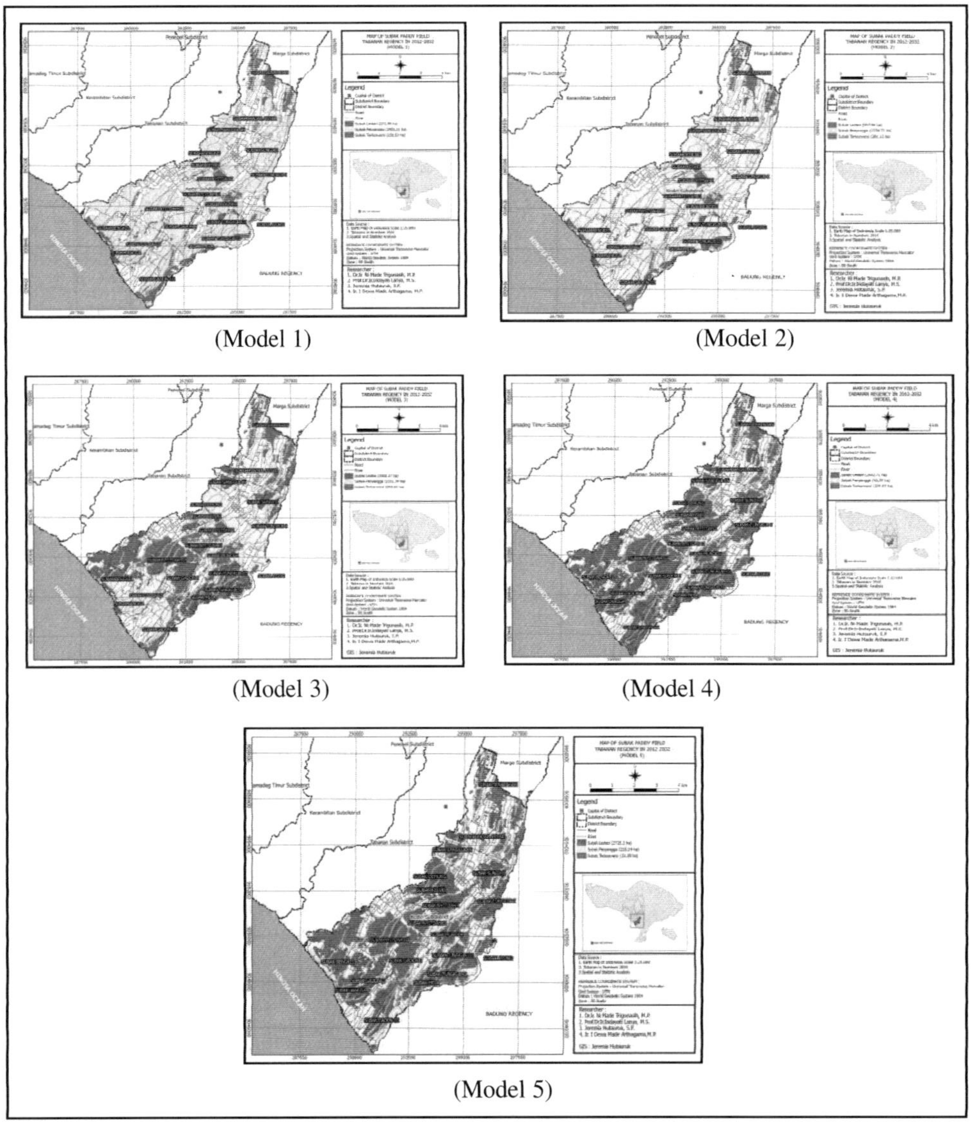


**Figure 14.** Map of Subak Zonation in Various Models

In 2016 irrigated rice fields approximately reached 2844 ha [1] and in Year 2010 irrigated rice fields approximately reached 3036 ha [2]. Area of rice fields which converted into non-farming each year can be calculated as Equation (3).

Extensive rice converted per year = year 2010 extensive rice – extensive rice year 2016/ 6 years (3)

The extensive of rice fields which converted per year is 32 hectares. To calculate the projected changes in rice fields for the years 2020, 2030 and 2040 calculated by subtracting the total change of the year and multiplied by the change of rice fields annually as shown in Table 3.

**Table 3.** Projection Data of Rice Field Change Year 2020, 2030 and 2040 in Kediri Sub-district

| Year | Area (hectare) |
|---|---|
| 2015 | 2844 |
| 2020 | 2684 |
| 2030 | 2364 |
| 2040 | 2044 |

Based on the data of subak zonation modeling and projection of rice field change in 2020, 2030, and 2040 in Kecamatan Kediri, adjustment of the model to see the area of Subak Sustainable and Subak Buffer, shown in Tabel 4.

**Table 4.** Conformity of Subak Zoning Modeling with Projection of Paddy Field Change Year 2020, 2030 and 2040

| Modeling in the year | | |
|---|---|---|
| 2020 | 2030 | 2040 |
| Model 4 | Model 3 | Model 2 |

Modeling in year 2020 using model 4 because of the area of Subak sustainable in model 4 of 2682,71 ha approaching rice field area in the year 2020 which amounted to 2684 ha. In 2030 using model 3 due to the subak sustainable area in model 3 of 1651,37 ha plus ¾ width of subak buffer of 773,51 ha to 2424,88 ha nears rice field area in year 2030 which equal to 2364 ha. In 2030, the area of land modeling and projection is still under the zoning area of crop agriculture in the spatial plan of Tabanan Regency of the year 2012-2033 at 2382,71 ha. In 2040 using model 2 because of the Subak Sustainable area in model 2 of 307.99 ha plus ¾ width of the buffer subak of 1781.04 ha to 2089.33 ha is close to 2040 ha of rice field area in 2040 ha.

From the suitability modeling with the projection of Subak area can be determined sub-zonation of Subak that must be preserved, Subak Buffer zonation that can be preserved and converted, also Subak zonation to be converted. The width of the area and location based on the geographical location of the Subak in Kediri Subdistrict can be explained and presented in Appendix 1.

## 5. Conclusions and Recomendations

### *5.1. Conclusions*

Based on the results and discussion can be concluded:

1. Kediri sub-district consists of 23 Subak of rice fields and mapped s based on geographical location.
2. Parameters in the classification of sustainable food farming in Kediri sub-district consists of (1) the suitability of the location of a rice field with Regional Spatial Plan of Tabanan Regency years 2012-2032, (2) land use, (3) Watershed morphology, (4) the type of irrigation, (5) rainfall, (6) the regional landform, (7) altitude, (8) the suitability of the agroecosystem paddy fields, (9) productivity, (10) the distance from the city center, (11) minimum area.
3. Spatial numerical classification produces a wide variety of modeling (5 models) and is associated with the projected changes in rice fields by the year 2020, 2030, and 2040. In the year 2020 using model 4

due to sustainable subak in model 4 of 2682.71 ha approached the farm field area by the year in 2020 of 2684 ha. In the year 2030 using model 3 due sustainable subak on the model of 1651.37 ha 3 plus ¾ buffer subak of 773.51 ha be 2424.88 ha approached the farm field by the year in 2030 of 2364 ha. In the year 2040 using model 2 due sustainable subak on the model of 307,99 ha 3 plus ¾ buffer subak of 1781,04 ha be 2089,33 ha approached the farm field by the year in 2040 of 2033 ha.

*5.2. Recomendations*

1. This modeling can be used for other areas according to the characteristics of the given areas.
2. To meet the needs of food, food supply can be increased through agricultural intensification, so that productivity obtained at least 8 tons / ha.

**Acknowledgments:**
Our thanks go to the Rector of Udayana University and Center for Spatial Data Infrastructure Development (PPIDS) Universitas Udayana, Bali who has provided funding and facilities during the course of this research.

**References**

[1] Central Bureau of Statistics of Tabanan Regency 2017 *Tabanan in Number 2016* (Tabanan: Central Bureau of Statistics of Tabanan Regency).
[2] Central Bureau of Statistics of Tabanan Regency 2011 *Tabanan in number 2010* (Tabanan: Central Bureau of Statistics of Tabanan Regency).
[3] Herold M, J Scepan and K C Clarke 2002 The Use of Remote Sensing and Landscape Metrics to Describe Structures and Changes in Urban Land Uses Environment and Planning A **34** (8) 1443-1458.
[4] Lanya I, I N N Subadiyasa, K Sardiana, I G P Ratna Adi 2015 Numerical Classification, Subak Zoning and Land Tranfer Function rice field in The Province of Bali Based on Remote Sensing and GIS *Journal of Procedia Enviromental Science* **24** pp 47-55.
[5] Mallinis G, D Emmandoloudis, V Giannakopoulos, F Maris and N Koutsias 2011 Mapping and interpreting historical land cover/ land use changes in a Natura 2000 site using earth ovservational data: The case of Nestos Delta, Greece *Applied Geography* **31**(1) pp 312-320.
[6] Laws of Republic Indonesia Number 41 in 2009 about sustainable agriculture food protection Ministry of Agriculture, Directorate General of Land and Water Management.
[7] Steel R G D and J H Torrie 1991 *Principles and Procedures of Statistics A Biometric Approach* (Jakarta: Publisher PT Gramedia Pustaka Utama).

## Appendix 1 Data of Subak Zonation in some Modeling in Kediri Subdistrict

| Subak Name | Wide_Hectare | Village | Model 2_hectare | | | Model3_hectare | | | Model4_hectare | | |
|---|---|---|---|---|---|---|---|---|---|---|---|
| | | | Sustainable Subak | Buffer Subak | Converted Subak | Sustainable Subak | Buffer Subak | Converted Subak | Sustainable Subak | Buffer Subak | Converted Subak |
| Subak ayung | 113.51 | Cepaka | 13.11 | 69.27 | 25.70 | 16.07 | 66.31 | 25.70 | 82.38 | 25.70 | 0 |
| Subak banjar anyar | 10.02 | Banjar Anyar | 0 | 50.11 | 0 | 0 | 50.11 | 0 | 50.11 | 0 | 0 |
| Subak batan pole | 7.05 | Kediri | 0 | 53.04 | 0 | 0 | 53.04 | 0 | 53.04 | 0 | 0 |
| Subak bengkel | 329.66 | Bengkel | 0 | 294.95 | 26.31 | 245.57 | 49.38 | 26.31 | 294.95 | 0 | 26.31 |
| Subak demung | 112.14 | Kediri | 2.27 | 108.82 | 1.03 | 21.68 | 89.41 | 1.03 | 111.09 | 0 | 1.03 |
| Subak gadon i | 141.54 | Pandak Gede | 0 | 33.45 | 9.20 | 33.45 | 0 | 9.20 | 33.45 | 0 | 9.20 |
| Subak gadon ii | 169.26 | Beraban | 0 | 145.87 | 23.39 | 0 | 145.87 | 23.39 | 145.87 | 0 | 23.39 |
| Subak gadon iii | 255.80 | Beraban | 0 | 315.75 | 72.4 | 152.92 | 162.83 | 72.40 | 378.75 | 0 | 72.40 |
| Subak jadi | 78.97 | Banjar Anyar | 70.53 | 8.40 | 0 | 78.93 | 0 | 0 | 78.93 | 0 | 0 |
| Subak jangga | 60.22 | Belalang | 0 | 43.55 | 24.8 | 0 | 43.53 | 24.80 | 43.53 | 0 | 24.80 |
| Subak kediri | 113.60 | Kediri | 10.81 | 99.16 | 3.89 | 30.26 | 79.55 | 3.89 | 109.71 | 0 | 3.89 |
| Subak mela | 115.18 | Buwit | 0 | 157.75 | 18.57 | 157.75 | 0 | 18.57 | 157.75 | 0 | 18.57 |
| Subak mundeh | 161.39 | Nyambu | 34.11 | 229.24 | 1.91 | 157.69 | 105.66 | 1.91 | 263.35 | 0 | 1.91 |
| Subak nyitdah i | 113.27 | Pandang Mandung | 34.93 | 18.73 | 0 | 34.93 | 18.73 | 0 | 53.66 | 0 | 0 |
| Subak nyitdah ii | 80.16 | Pandak Gede | 0 | 53.6 | 6.73 | 53.60 | 0 | 6.73 | 53.6 | 0 | 6.73 |
| Subak nyitdah iii | 275.84 | Nyitdah | 270.52 | 0 | 5.33 | 270.52 | 0 | 5.33 | 0 | 0 | 5.33 |
| Subak nyitdah iv | 68.88 | Belalang | 289.92 | 0 | 3.89 | 289.92 | 0 | 3.84 | 289.92 | 0 | 3.84 |
| Subak sanggulan | 76.32 | Banjar Anyar | 0 | 76.31 | 0 | 67.32 | 8.99 | 0 | 76.31 | 0 | 0 |
| Subak senapahan | 117.14 | Banjar Anyar | 58.71 | 4.40 | 0 | 63.11 | 0 | 0 | 63.11 | 0 | 0 |
| Subak sungi ii | 158.31 | Abian Tuwung | 40.46 | 115.83 | 0 | 125.72 | 31.57 | 0 | 156.29 | 0 | 0 |
| Subak tungkub i | 129.66 | Nyambu | 11.63 | 14.06 | 0 | 11.63 | 14.06 | 0 | 25.69 | 0 | 0 |
| Subak tungkub ii | 197.69 | Kaba-Kaba | 31.43 | 125.32 | 40.88 | 111.92 | 44.83 | 40.88 | 156.75 | 16.89 | 23.99 |
| Subak tungkub iii | 100.13 | Kaba-Kaba | 61.3 | 21,86 | 17.14 | 0 | 21.86 | 17.14 | 21.86 | 0 | 17.14 |

# Evaluation of Base Station CORS UDIP and CSEM for monitoring Ground Deformation Sayung Demak Indonesia

**B. D. Yuwono*, M. Awaluddin, Kun F H and Lutfi E R**

Geodesy Department, Faculty of Engineering, Diponegoro University, Tembalang-Semarang, 50277, Indonesia

bdyuwono92@gmail.com

**Abstract**. Sayung is a subdistrict in Demak Regency which is located on the north coast is very vulnerable to natural disasters such as rob flood, abrasion and deformation of land subsidence. The condition is suspected, among others, by several factors, among others, geological structure as a large area dominated by young alluvium layers are still experiencing compression, loading and retrieval of ground water. It is necessary to do research related to ground deformation. The geodetic method used for monitoring ground deformation by satellite surveys with GNSS. The research was conducted to observe GPS survey in 2015 and 2016. GNSS data would be processed with scientific processing GAMIT 10.6. Strategic of GPS data proccesing is the important to reach a better accuracy. The purpose of this research is to evaluate the result of calculation of coordinate value and spatial deformation obtained by both base station that is CORS UDIP and CORS CSEM for monitoring ground deformation.

**Keywords:** Vulnerable, GNSS, CORS, Deformation

## 1. Introduction

Ground deformation is a change in ground level in the vertical direction to the bottom of a height reference plane which is geometrically represented by height collection point on the ground surface [1]. The speed of ground deformation expecially land subsidence is slow (mm) in a relatively long period of time. The phenomena are often found not only in big cities in Indonesia such as Jakarta, Semarang and Bandung [2] but also in several major cities in the world including in the United States [3], Venezuela [4], Choshuichi - Taiwan [5] and Iran [6].

Demak is a district located in the north of Java Island. Historically, most of the areas of Demak regency, especially on the northern coast, have unstable soil structures, since they were originally swamps. Geographically Demak is highly vulnerable due to enviromental stress, such as abration, tidal, land subsidence, and coastal flooding in the rainy season the northern coast of Demak is often flooded and during the dry season the characteristics of the soil are cracked because it comes from the muddy ground structure. The condition is suspected, among others, by several factors, among others, geological structure as a large area dominated by young alluvium layers are still experiencing compression, loading and retrieval of ground water. So it is necessary to do research related to ground deformation. There are several reasearch for Preliminaery study in monitoring landsubsidence. using geodetic method [7] [9]

IOP Conf. Series: Earth and Environmental Science **98** (2017) 012047 doi:10.1088/1755-1315/98/1/012047

[11]. It important to know about strategic data processing for having a good acuracy and precition, Comparing method would give description which it suitable processing method due to enhace acuracy.

Administratively, the area of Demak Regency is 89,743 ha divided into 14 districts, 243 villages and 6 kelurahan. Most of Demak Regency is wetland area reaching 51,799 ha (57,72%) and the rest is dry land. 13.77%. used for tegalan / garden, 0.05% while not used and 11.16% used for ponds [14]

Demak is one of the districts in Central Java geographically located at coordinates 6 degrees 43 "26" - 7 degrees 09 "43" south latitude and 110 degrees 27 "58" - 110 degrees 48 "47" east longitude. The farthest distance from west to east 49 km and from north to south along 41 km, with an area of 89.743 Ha. lies in 14 villages of Sriwulan, Bedono, Timbulsloko and Surodadi (Sayung sub district), then Tambakbulusan Village, Karangtengah Subdistrict, Morodemak Village, Purworejo and Betahwalang Village (Bonang Subdistrict), Wedung, Berahankulon, Berahanwetan, Wedung and Babalan (Wedung District) .. The farthest distance from west to east 49 km and from north to south along 41 km, with an area of 89,743 Ha [13] shown in figure 1.

## 2. Data and Method

### *2.1. Data Aquisition*

Data Observation of deformation of coastal land in Demak Regency has been done by 2015 and 2016. Location Bench Mark separate at 6 Villages / kelurahan all of which are in the District Sayung. 2 Observation marks are in Bedono Village in 2015. GPS measurements made at 7 geodetic monuments. The GPS observation was conducted at May-June 2015 and April-May 2016. Bench Mark for monitoring deforamtion were located in Sidodadi, Gemah, Tugu, Purwosari, Bedono and Timbulsloko Distribution of GPS Survey shown at figure 2. GPS survey were conducted in May – June 15, and the second observation in April - May 2016. GPS survey done using dual frequency geodetic-type GPS receivers Hiper II and Hiper Gb.

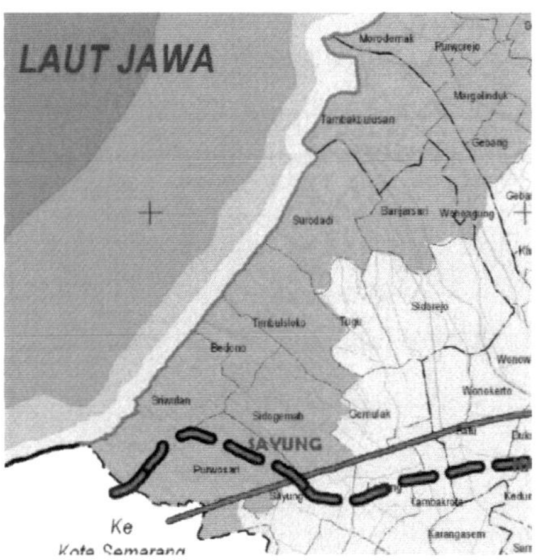


**Figure 1.** Studi Area of Sayung Demak Regency [13]

**Figure 2.** Distribution on GPS Survey 2016

*2.2. Method*

This research conducted with two method of GPS data processing as base station. First using GNSS CORS UDIP and station IGS namely, BAKO, CNMR, COCO, DARW, IISC, KAT1, PIMO, SOLO, XMIS and CORS Udip second using CSEM as Base Station. CORS CSEM as the reference station was located on the roof of Telkom Building Kalibanteng, Semarang. GNSS CORS UDIP was located in the top of building B Geodetic Departement, Engineering Faculty Diponegoro University. In the GPS measurements, each epoch has 5 until 6 hour obsevation. Flowchart can be seen in figure 3. The GPS Data Processing would done using GAMIT 10.6 scientific GPS [ Herring , 2010] Data Proccessing.

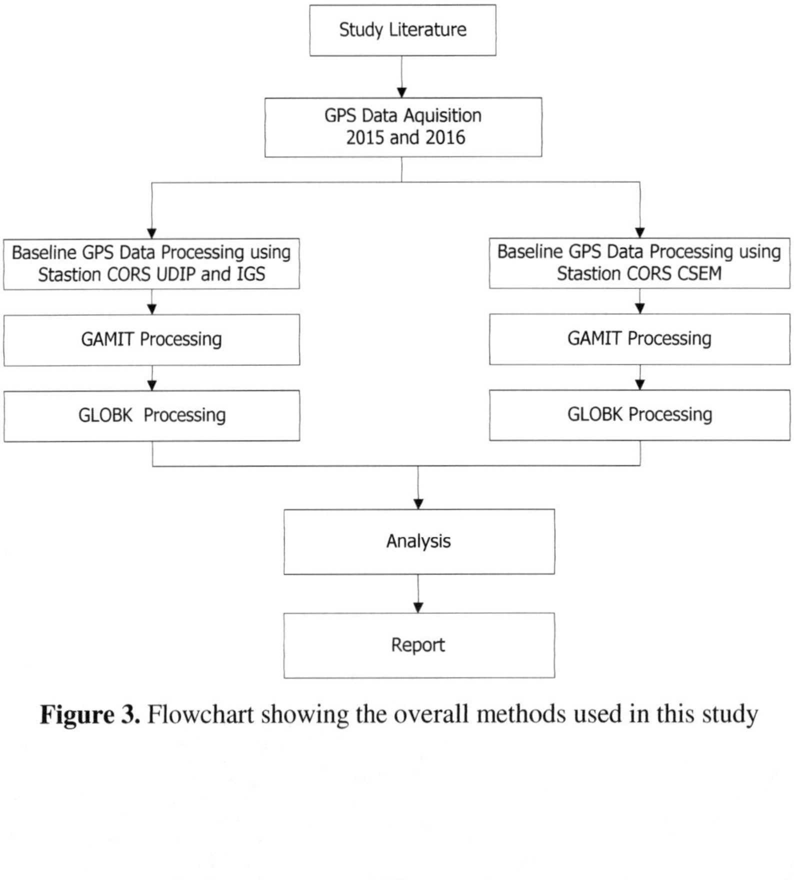


**Figure 3.** Flowchart showing the overall methods used in this study

IOP Conf. Series: Earth and Environmental Science **98** (2017) 012047 doi:10.1088/1755-1315/98/1/012047

To test whether there is any deformation in each bench mark then t test is done. The t test (also called Student's T Test) compares two averages and will describe they are different from each other. The t test shown differences could have happened by chance. Reference [9] Height change defined as *Δdhij* and its rate *υΔdhij* at each station are derived using the following relation

$$\Delta dhij = dh(tj) - dh(ti) \quad . \tag{1}$$

$$\upsilon\Delta dhij = \Delta dhij \,/\, (tj - ti) \tag{2}$$

Null hypothesis Ho : dhij = 0
Alternative hyphotesis Δdhij ≠ 0

Test statistic is

$$t = \frac{\Delta dh_{ij}}{\sigma(dh_{ij})} \tag{3}$$

Which has the customary Student t-distribution if Ho is true. The null hypothesis is rejected if.

$$|t| > t_{df,\alpha/2}$$

At a confidence level of 99% (i.e. α=1%), the critical value is $t_{\infty,005=2.576}$

F- test is one of statistical test in which the test statistic has an F-distribution. It is most often used to compare two variances, in order to identify the model that best fits the population from which the data were sampled. The test compares the ratio of two variances. If found that the variance is equal, the ratio of the variances will equal 1. The population variances are equal when running an F Test, and assumed that the variances area equal to 4.

The null hypothesis will always be that the variance is equal.

$$F = \frac{s_1^2}{s_2^2} \text{ or } F = \frac{s_2^2}{s_1^2} \; F = \frac{large\ sampel\ variance}{smaller\ sampel\ variance} \tag{4}$$

Table F Distributions

$$F_{\propto/2,vi,v2} = \frac{1}{F1_{-\propto,v2,v1}} \tag{5}$$

The null hypothesis used in this statistical test is the method of processing there is no significant difference so that:

$$H_0 \quad H_o: dV = 0 \tag{6}$$

$$H_1 \quad H_o: dV \neq 0$$

IOP Conf. Series: Earth and Environmental Science **98** (2017) 012047 doi:10.1088/1755-1315/98/1/012047

## 3. Results and Discussion

### *3.1. Data Quality Control*

GPS observation data in advance is checked to know the quality of data from each point of observation by using TEQC program. GPS observation data have good quality whether or not seen from the value of MP1 and MP2. MP1 and MP2 are moving averaging values, i.e. RMS values from a combination of multipath data recorded

**Table 1.** Quality Control GPS Data

| Year | Station | MP1 | MP2 | GPS Type | DOY |
|---|---|---|---|---|---|
| 2015 | WNSR | 0,570400 | 0,623026 | Topcon Hiper II | 122 |
| 2015 | PDSR | 0,078153 | 0,126307 | Astech Promark800 | 122 |
| 2015 | TMBL | 0,111019 | 0,217499 | Astech Promark800 | 122 |
| 2015 | SURO | 0,480724 | 0,494162 | Topcon Hiper II | 124 |
| 2015 | DADI | 0,089762 | 0,160813 | Astech Promark800 | 124 |
| 2015 | PRWS | 0,071471 | 0,163779 | Astech Promark800 | 124 |
| 2015 | SIDO | 0,669758 | 0,736448 | Topcon Hiper II | 126 |
| 2015 | GEMA | 0,070251 | 0,124984 | Astech Promark800 | 126 |
| 2015 | TUGU | 0,142555 | 0,201344 | Topcon Hiper GB | 126 |
| 2016 | PRWS | 0.300293 | 0.314016 | Topcon Hiper GB | 114 |
| 2016 | WNSR | 0.518471 | 0.544309 | Topcon Hiper GB | 114 |
| 2016 | SIDO | 0.481011 | 0.48126 | Topcon Hiper II | 114 |
| 2016 | GEMA | 0.311381 | 0.305413 | Topcon Hiper II | 115 |
| 2016 | TUGU | 0.343395 | 0.357062 | Topcon Hiper GB | 115 |
| 2016 | TMBL | 0.681992 | 0.571695 | Topcon Hiper GB | 115 |
| 2016 | SURO | 0.353386 | 0.338689 | Topcon Hiper GB | 116 |
| 2016 | DADI | 0.621246 | 0.583523 | Topcon Hiper GB | 116 |

In 2015 shows that the moving average value of MP1 and MP2 has value les than 0.5, It means that there are no signicant multipath as long as gps observation. A considerable in 2016 multipath effect occurs at three points WNSR, TMBL and DADI. The poor quality of data is influenced by several factors that occur in the field, such as rain, the environment around the observation point that many obstruction and trees as well as the factors of equipment that is quite influential

### *3.2. Result of Data GPS Proccesing using CSEM*

The results of processing using Base Station CSEM can be seen in table 2. The result shows that the standard value of deviation in the direction X 0.0039 – 0.0045, Y 0.0047 – 0.0049 and Z 0.0035 – 0.0047 Z

**Table 2.** Coordinate 3D Cartesian Network CSEM Method

| Station | Year | X | Y | Z | std X | std Y | std Z | h |
|---|---|---|---|---|---|---|---|---|
| PRWS | 2015 | -2217365,724 | 5930844,540 | -765128,9478 | 0,0039 | 0,0047 | 0,0035 | 28,52408 |
| WNSR | 2015 | -2215939,663 | 5931507,728 | -764119,404 | 0,0040 | 0,0048 | 0,0037 | 27,78359 |
| SIDO | 2015 | -2219010,166 | 5930248,619 | -764969,655 | 0,0040 | 0,0047 | 0,0037 | 27,09083 |
| GEMA | 2015 | -2219058,490 | 5930468,196 | -763140,325 | 0,0039 | 0,0047 | 0,0037 | 27,44666 |
| TUGU | 2015 | -2219848,755 | 5930322,289 | -761984,428 | 0,0039 | 0,0047 | 0,0047 | 27,64855 |
| TMBL | 2015 | -2218703,448 | 5930946,940 | -760463,879 | 0,0042 | 0,0048 | 0,0038 | 27,26408 |
| SURO | 2015 | -2219836,606 | 5930737,920 | -758794,972 | 0,0040 | 0,0048 | 0,0037 | 27,07961 |
| DADI | 2015 | -2220521,107 | 5930610,887 | -757795,196 | 0,0043 | 0,0049 | 0,0040 | 27,56375 |
| PRWS | 2016 | -2217365,728 | 5930844,529 | -765128,940 | 0,0041 | 0,0048 | 0,0037 | 28,51381 |
| WNSR | 2016 | -2215939,630 | 5931507,682 | -764119,408 | 0,0043 | 0,0048 | 0,0040 | 27,72983 |
| SIDO | 2016 | -2219010,166 | 5930248,619 | -764969,655 | 0,0040 | 0,0047 | 0,0037 | 27,09083 |
| GEMA | 2016 | -2219058,486 | 5930468,138 | -763140,329 | 0,0043 | 0,0048 | 0,0039 | 27,39212 |
| TUGU | 2016 | -2219848,742 | 5930322,234 | -761984,421 | 0,0044 | 0,0049 | 0,0041 | 27,5924 |
| TMBL | 2016 | -2218703,413 | 5930946,903 | -760463,894 | 0,0045 | 0,0049 | 0,0043 | 27,21941 |
| SURO | 2016 | -2219836,564 | 5930737,933 | -758794,970 | 0,0044 | 0,0049 | 0,0041 | 27,07694 |
| DADI | 2016 | -2220521,158 | 5930610,882 | -757795,195 | 0,0045 | 0,0049 | 0,0042 | 27,57672 |

Almost all GPS-derived ellipsoidal height changes indicated that they passed the stastical testing, and it could be concluded that with 99% confidence level, there were significant ellipsoidal changes at all the stations during the period between May-June 2015 and April-May 2016 . In table 3. shows that most stations experience a high change marked by the value of T> count of T table and the SURO point does not change high, because the result of T count is smaller than 1.96. The high change in question is the decline that occurs in 7 points, although the SURO point has decreased, but the decline at that point is not significant enough, so the statistical test shows no high change

**Table 3.** T test

| Station | $Dh_{12}$ cm | s($\sigma$ $dh_{12}$) (mm) | $t_{tes}$ | Deformation |
|---|---|---|---|---|
| PRWS | 1.027 | 4.90 | 2.095918 | Yes |
| WNSR | 5.376 | 4.94 | 10.88259 | Yes |
| SIDO | 2.487 | 4.92 | 5.054878 | Yes |
| GEMA | 5.454 | 4.91 | 11.10794 | Yes |
| TUGU | 5.615 | 4.94 | 11.3664 | Yes |
| TMBL | 4.467 | 4.96 | 9.006048 | Yes |
| SURO | 0.267 | 4.93 | 0.541582 | No |
| DADI | 1.297 | 4.95 | 2.620202 | Yes |

### *3.3. Result of Data GPS Proccesing using UDIP and IGS*

The results of processing using Base Station UDIP and IGS can be seen in table 4. The result shows that the standard value of deviation in the direction X 0.0042 – 0.0047, Y 0.0047 – 0.0049 and Z 0.0033 – 0.0045 Z.

IOP Conf. Series: Earth and Environmental Science **98** (2017) 012047 doi:10.1088/1755-1315/98/1/012047

**Table 4.** Coordinate 3D Cartesian Network UDIP and IGS

| Station | Year | X | Y | Z | std X | std Y | std Z | h |
|---|---|---|---|---|---|---|---|---|
| PRWS | 2015 | -2217365,806 | 5930844,518 | -765128,990 | 0,0042 | 0,0048 | 0,0033 | 28,5374 |
| WNSR | 2015 | -2215939,747 | 5931507,711 | -764119,445 | 0,0042 | 0,0048 | 0,0036 | 27,8019 |
| SIDO | 2015 | -2219010,242 | 5930248,600 | -764969,695 | 0,0040 | 0,0047 | 0,0035 | 27,1043 |
| GEMA | 2015 | -2219058,563 | 5930468,172 | -763140,365 | 0,0042 | 0,0048 | 0,0034 | 27,4548 |
| TUGU | 2015 | -2219848,835 | 5930322,227 | -761984,456 | 0,0038 | 0,0047 | 0,0032 | 27,6221 |
| TMBL | 2015 | -2218703,520 | 5930946,922 | -760463,917 | 0,0042 | 0,0048 | 0,0037 | 27,2767 |
| SURO | 2015 | -2219836,657 | 5930737,849 | -758795,002 | 0,0038 | 0,0047 | 0,0032 | 27,0352 |
| DADI | 2015 | -2220521,173 | 5930610,792 | -757795,221 | 0,0044 | 0,0049 | 0,0039 | 27,5014 |
| PRWS | 2016 | -2217365,830 | 5930844,464 | -765128,993 | 0,0041 | 0,0048 | 0,0036 | 28,496 |
| WNSR | 2016 | -2215939,745 | 5931507,623 | -764119,461 | 0,0044 | 0,0049 | 0,0040 | 27,721 |
| SIDO | 2016 | -2219010,226 | 5930248,477 | -764969,709 | 0,0045 | 0,0048 | 0,0040 | 26,986 |
| GEMA | 2016 | -2219058,579 | 5930468,035 | -763140,357 | 0,0043 | 0,0048 | 0,0039 | 27,332 |
| TUGU | 2016 | -2219848,861 | 5930322,148 | -761984,449 | 0,0047 | 0,0049 | 0,0045 | 27,557 |
| TMBL | 2016 | -2218703,509 | 5930946,799 | -760463,925 | 0,0046 | 0,0049 | 0,0045 | 27,159 |
| SURO | 2016 | -2219836,666 | 5930737,847 | -758795,020 | 0,0043 | 0,0048 | 0,0039 | 27,038 |
| DADI | 2016 | -2220521,270 | 5930610,781 | -757795,241 | 0,0046 | 0,0049 | 0,0043 | 27,528 |

Result of T test indicates that qlmost all GPS-derived ellipsoidal height changes passed the stastical testing.It could be stated that with 99% confidence level, there were significant ellipsoidal changes at all the stations during the period between May-June 2015 and April-May 2016 exept Suro the station. In table 5. The high subsidence level that occurs in 7 points, although the SURO point has decreased, but the subsidence level at that point is not significant enough, so the statistical test shows no high change

**Table 5.** T test Network GNSS UDIP and IGS as Reference Base Station

| Station | h 2015 (m) | h2016 (m) | $Dh_{12}$ cm | s($\sigma dh_{12}$) (mm) | t-test | Deformation |
|---|---|---|---|---|---|---|
| WNSR | 27.8 | 27.72 | -8.1 | 5 | -16.3 | Yes |
| SIDO | 27.1 | 26.99 | -11.9 | 4.9 | -24.1 | Yes |
| GEMA | 27.45 | 27.33 | -12.3 | 4.9 | -24.9 | Yes |
| TUGU | 27.62 | 27.56 | -6.5 | 5 | -13.1 | Yes |
| TMBL | 27.28 | 27.16 | -11.7 | 5 | -23.6 | Yes |
| SURO | 27.04 | 27.04 | 0.3 | 4.9 | 0.7 | No |
| WNSR | 27.8 | 27.72 | -8.1 | 5 | -16.3 | Yes |

### *3.4. Spatial and geometry analysis*

Differences Base Station using to station CSEM and UDIP can be seen from the coordinate value and strandart defiatio, can be seen in table 6. The highest difference value on the x-axis is located at WNSR of 11.5 cm while for the Y component of 14.2cm is located, for the Z direction of 5.4 cm ini SIDO. The maximum diffrence for ellipsoid height reach 10.48 cm at SIDO .

**Table 5.** Coordinates Difference between UDIP and CSEM as Base Station

| Station | Year | Δ X(m) | ΔY(m) | Δ Z(m) | Δ Std X (m) | Δ Std Y (m) | Δ Std Z (m) | Δ h (cm) |
|---|---|---|---|---|---|---|---|---|
| PRWS | 2015 | 0,082 | 0,022 | 0,042 | 0,0004 | 0,0001 | -0,0002 | -1,33 |
| WNSR | 2015 | 0,084 | 0,017 | 0,041 | 0,0001 | 0,0000 | -0,0001 | -1,83 |
| SIDO | 2015 | 0,076 | 0,019 | 0,040 | 0,0000 | 0,0000 | -0,0002 | -1,35 |
| GEMA | 2015 | 0,073 | 0,024 | 0,040 | 0,0003 | 0,0001 | -0,0002 | -0,81 |
| TUGU | 2015 | 0,080 | 0,062 | 0,028 | -0,0001 | 0,0000 | -0,0015 | 2,65 |
| TMBL | 2015 | 0,072 | 0,018 | 0,038 | 0,0000 | 0,0000 | -0,0002 | -1,26 |
| SURO | 2015 | 0,051 | 0,071 | 0,030 | -0,0002 | 0,0000 | -0,0005 | 4,44 |
| DADI | 2015 | 0,066 | 0,095 | 0,025 | 0,0000 | 0,0000 | 0,0000 | 6,23 |
| PRWS | 2016 | 0,102 | 0,065 | 0,053 | 0,0000 | 0,0000 | -0,0001 | 1,78 |
| WNSR | 2016 | **0,115** | 0,059 | 0,053 | 0,0001 | 0,0000 | 0,0000 | 0,88 |
| SIDO | 2016 | 0,060 | **0,142** | **0,054** | 0,0006 | 0,0001 | 0,0003 | 10,48 |
| GEMA | 2016 | 0,093 | 0,103 | 0,028 | 0,0000 | 0,0000 | 0,0000 | 6,01 |
| TUGU | 2016 | 0,119 | 0,086 | 0,028 | 0,0003 | 0,0001 | 0,0004 | 3,54 |
| TMBL | 2016 | 0,096 | 0,104 | 0,031 | 0,0001 | 0,0000 | 0,0002 | 6,04 |
| SURO | 2016 | 0,102 | 0,086 | 0,050 | -0,0001 | 0,0000 | -0,0002 | 3,89 |
| DADI | 2016 | 0,112 | 0,101 | 0,046 | 0,0001 | 0,0000 | 0,0000 | 4,87 |

F statistical test is used to determine whether there is difference in the measurement results by processing 2 methods from two years observation. This test is done by comparing the variance of the two processing gps data methods. The result of f test using gps data in 2015 and 2016 gets smaller values than the F table, ($F_{tes}$ < F table), so the null hypothesis is accepted all. The conclusions were the GPS data processing using the GAMIT scientific software with the CSEM or UDIP as base station is not significantly different. Shown on table 5.

**Tabel 6.** F test

| Method | Year | Variance | $F_{tes}$ | $F_{table}$ |
|---|---|---|---|---|
| CSEM | 2015 | 0,000134941 | 0,483 | 6.61 |
| UDIP + IGS | 2105 | 0,000062255 | | 6.61 |
| CSEM | 2016 | 0,000042055 | 0,461 | 6.61 |
| UDIP + IGS | 2106 | 0,000087028 | | 6.61 |

*3.5. Deformation Map*

Due limitation GPS data observation not covering of whole of area study cause the contour map doest not good enough. Shown at the figure 4 dan 5, pattern of typical subsidence have not look similar because of the type of interpolation. figure 4 using IDW interpolation form result of CSEM Base Station and figure 5 using Kriggig interpolation UDIP IGS Network.

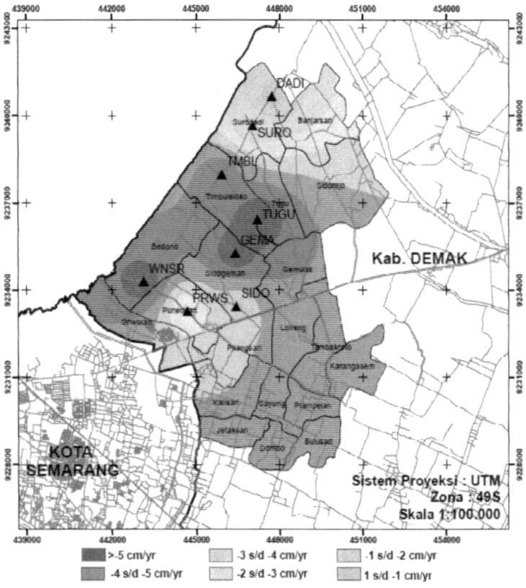


**Figure 4.** Deformation Map Base CSEM as Base Station ( Lutfi, 2016)

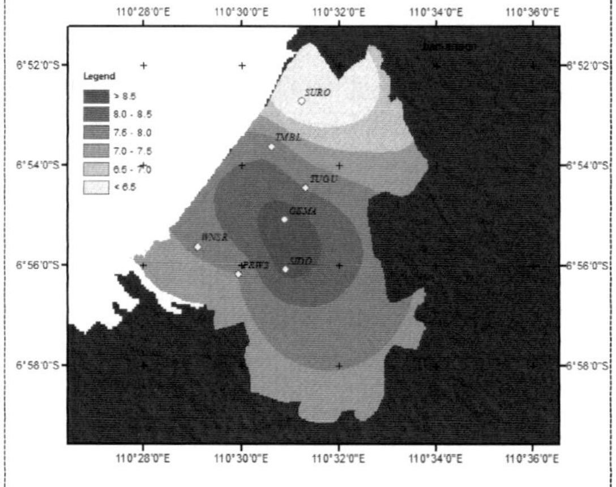


**Figure 5.** Deformation Map using CORS UDIP as Base Station .

## 4. Conclusions

The statistical t test shown that each method has similarity result that station SURO has a test smaller than 1.96, it means there was not have deformation in vertical direction. Root Mean Square Error (RMSE) CSEM as base station shows a relatively small value compared to the RMSE UDIP IGS Network. because the distance baseline CSEM is shorter than UDIP IGS Network. With limitation of time occupation abourt five – six hours, CSEM base station is more better than UDIP IGS Network..For increasing accuracy. UDIP IGS network can be done with extend time of occupation to have better accuracy.

## References

[1] Yuwono B D, Abidin H Z, Gumilar I, Andreas H, M Awaluddin , Haqqi K F, Khoirunisa R 2015 " Preliminary Survey and Performance of Landsubsidence in North Semarang Demak ", The5 International Symposium on Earth-hazard and Disaster Mitigation ITB Bandung, Date : 19 Oktober 2015; *5th International Symposium on Earth-hazard and Disaster Mitigation* ITB Bandung, Date : 19 Oktober 2015.

[2] Abidin H Z, H Andreas, R Djaja, M. Gamal, T Deguchi, Y Maruyama 2008 Penurunan Tanah di Wilayah Jakarta : Karakteristik, Penyebab dan Dampaknya, *Jurnal DKP*,2008

[3] Galloway D, Jones R.D.Ingebritsen S.E 1999 *Landsubsidence in The United State US Geological Survey*.

[4] Chrzanowski A & Y Q Chen 1991 *Use of the Global Positioning System (GPS) for Ground Subsidence Measurements in Western Venezuela Oil Fields*

[5] Rong-Kang Shang, Yi-Shiang Shiu, Kuo-Chen Ma 2011 Using geographical weighted regression to explore the spatially varying relationship between land subsidence and groundwater level variations: A case study in the Choshuichi alluvial fan, Taiwan. 978-1- 4244-8351-8 /11/$26.00 ©2011 *IEEE*.

[6] Motagh M 2007 Land Subsidence in Mashhad Valey, Northeast Iran; Result from InSar, Levelling and GPS. *Geophysic J. Int* (2007) 168, 518-526

[7] Haqqi, M.K. Fariqul 2015 Survei Pendahuluan Deformasi Muka Tanah dengan Pengamatan GPS

IOP Conf. Series: Earth and Environmental Science **98** (2017) 012047 doi:10.1088/1755-1315/98/1/012047

di Kabupaten Demak. *Skripsi* Teknik Geodesi Universitas Diponegoro.

[8] Herring T A dkk 2010 *Introduction to GAMIT/GLOBK*, Department of Earth,Atmospheric, and Planetary Science, Massachusetts Institute of Technology

[9] Lutfi E R, 2016, Survei Pemantauan Deformasi Muka Tanah Kawasan Pesisir Menggunakan Metode Pengukuran Gps Di Kabupaten Demak Tahun 2016 (Studi Kasus : Pesisir Kecamatan Sayung, Demak), Skripsi Teknik Geodesi Universitas Diponegoro.

[10] Leick A 2004 *GPS Satellite Surveying*. Third edition. John Wiley & Sons. New York, USA.

[11] Yuwono B D, H Z Abidin, H Andreas, I.Gumilar, M Awaluddin, Najib 2017 ".Development of Static Differential Method GNSS CORS UDIP For Monitoring Land Subsidence In Semarang Demak,

[12] Leick, A 2004 *GPS Satellite Surveying*. Third edition. John Wiley & Sons. New York, USA.

[13] http://demakkab.go.id/profil/geografi-dan-kependudukan/ Diakses 3 April 2017

# Modeling of Soil Water Availability for Agricultural Planning at Pelaga Village, Badung Regency, Bali, Indonesia

**R Suyarto*, I N Sunarta, Wiyanti and N K H Padmayani**

Center for Spatial Data Infrastructure Development (PPIDS) Universitas Udayana, Bali, Indonesia

rsuyarto@yahoo.co.id

**Abstract.** Pelaga Village is located in Badung regency which has the advantage in agriculture with the cultivation of coffee plants, oranges, carrots, cabbage, and chili. The physical condition of Pelaga Village which has high rainfall, bumpy areas, and sandy-sandy ground texture causes air to air to be available for plants. Based on these questions then conducted a study to determine the comparison between the available water and water requirement for agriculture. Available water was difference field capacity and permanent wilting point method and crop water requirement was using Blaney-Criddle method. The results from this research was deficit between available air and crop water requirements. Available water was 12,12% and crop water requirement in initial stage, dev. Stage, mid-season stage, and late season stage respectively, coffee 11.28%, 24.19%, 35.49%, 29.04%; cabbage 19.58%, 19.58%, 33.10%, 27.74%: carrot 14.82%, 28.61%, 28.61%, 27.95%: Orange 14.82%, 28.61%, 28.61%, 27.23%; chili, 17.37%,17.37%, 34.80%, 30.46%. Soil management that must be done is by short-term land management by sprinkling long-term soil management by means of organic material valuation, irrigation making, and terracing making.

**Key words:** Soil water hydrological, Water requirement, Agriculture Planning.

## 1. Introduction

Water is very important for plants grow because it is the main component in the cells to making up the plant tissue. The amount of water in the plant ranges from 80-90 percent of the dry weight of the plant. This percentage will become even greater in the parts of the plant that are actively growing [1]. Water requirement for every plant is deferent depends on the shape, type, age, planting medium, environmental conditions around the plant and season so that each plant has a certain water content limit for its growth. In soil the presence of water is indispensable to the plants that must be available to meet the need for evapotranspiration and as a solvent, together with dissolved nutrients to form a soil solution to be absorbed by plant roots. The crop water requirement is highly dependent on the water available in the soil for the growing process of soil's ability to retain water. The crop water requirements are based on crops evapotranspiration [8].

The ability to hold water on soil differs according to the texture and the pore count. In general, dominant clay textured soil has the ability to hold and store high water, because the water molecule is strongly absorbed on the surface of the soil. While on the sand-textured soil has the ability to hold and store water is very low. Good water supply is 60-80% of field capacity [1]. Pelaga Village is an administrative area of Petang District, Badung Regency. The topography of the area is bumpy to hilly with temperatures of 20-30°C (Pelaga website). Based on the regional spatial plan, Pelaga Village is planned to plant horticultural crops, garden plants and food crops. Planning serves to improve the economy of the community. Plants planned for horticulture plants and gardens adapted to the plants that have been planted by the Pelaga Village community. The plants are cabbage, carrots, chili, tomatoes, and oranges and coffee. The types of plants to be planted will be greatly influenced by the hydrological nature of Pelaga Village. So, it will be very important to know the needs of water plants and water available in the soil. Based on the above description, the authors are interested to conduct research with modeling of available water for and planning of horticultural plants and mixed plantations. The usual plants that planted by the farmers in Pelaga Village are coffee, cabbage, carrots, orange and pepper. The results of this study aim to find out the needs of water for horticultural plants associated with water available in the soil and planning of agricultural planning as input to the community in Pelaga Village, Petang District, Badung Regency, Bali, Indonesia.

## 2. Methodology

### *2.1. Area of Study*

The research was conducted in Pelaga Village, Petang District, Badung Regency. Total area of pelaga village is 3545,204 Ha with height 1017 meter above sea level. Average temperature 20° - 30° C with population of 5903 people [10]. Pelaga Village is located in 8.30561 Latitude South and 115.21399 Longitude east.

### *2.2 Tools and Material*

The material used in this research are spatial datas as follows:

- *Pelaga Village map* 1: 25,000
- *planning map* 1: 25,000
- *soil sample*

While the tools are:

- Computer or laptop used to process the spatial data
- GPS (Global Positioning System) to determine the coordinate point of observation.
- Camera
- Stationary
- Abney level
- Ring sample
- Double ring infiltrometer
- Soil munshel color chart
- Soil drill

### *2.3 Methods and Implementation of Research*

The method used in this research is observation method with direct observation in field and laboratory analysis. Research done in accordance with several stages namely; (1) preliminary planning and preparation of land surveys aimed at finding out the location points of soil sampling, (2) soil sampling, (3) analysis of some soil hydrological characteristics such as permeability, infiltration, moisture capacity, permanent wilting water point, water available, and water requirement, (4) data analysis with descriptive method.

### *2.4 Field Survey and Soil Sampling*

The location of the study is a located in a land that is not cultivated. In the research area some sample points were taken. Soil samples obtained in the field were mixed composites and analyzed in the laboratory. Soil sampling was done at the degradation layer with a depth of 0-30 cm. Soil samples are taken in disturbed and undisturbed form. Soil sampling in the form is disturbed for texture observation, particle density, moisture content of field capacity, moisture content of permanent wilting point and organic matter. Direct observation in the field to observe: infiltration, soil structure, soil color, effective soil depth, slope, slope direction, soil moisture depth, and growing vegetation.

Field observation was done on the soil of each sample point sampling in Pelaga Village, Petang District, Badung Regency. Parameters observed include Infiltration, using double ring infiltrometer. The term infiltration refers specifically to entry of water into the soil surface. The movement of water into the soil by infiltration may be limited by any restriction to the flow of water through the soil profile [5]. The area of study and the soil sample location can be seen in Figure 1.

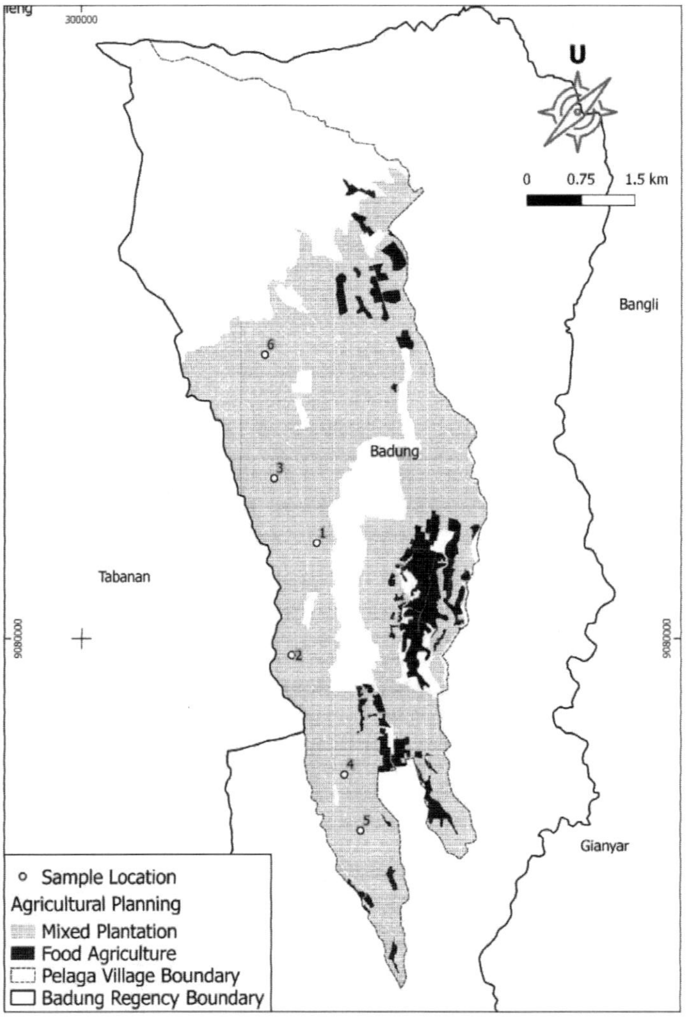


**Figure 1.** Area of Study and Sample Location

*2.5 Soil Analysis Method*

Soil analysis is done in Soil Laboratory, of Udayana University. Parameters analyzed include:

1. Soil text with pipe method
2. Water content of field capacity with gravimetric method.
3. The water content of the permanent wilting point by looking at the change of color change limit (Atterberg's determination) of the soil paste becomes dry, then searched the water content by gravimetric method.

4. Permeability of soil by constant head permeameter method
5. Soil organic matter with Walkley and Black method.

*2.6. Water Requirements*
Water requirements for crop calculated based on the water needed for crop to replace the water loss due to evapotranspiration process. The water requirements include the calculations of available water content, evapotranspiration rate and consumptive water requirement for crops based on plant coefficient.

*2.6.1. Available Water content.* Available Water content calculation is available by the formula of equation (1).

Available Water Content = Field Capacity Water Level - Permanent Wilting Point (1)

*2.6.2. Evapotranspiration.* Evapotranspiration is calculated using Blaney-Criddle method with the formula on equation (2)

$$Eto = p. (0.46t + 8.13) \tag{2}$$

p = comparison of the average length of day time for a particular month with the amount of time spent in the day of the year
t = daily average temperature ($^oC$)

Mean Daily Percentage (p) of Annual Daytime Hours for different latitudes, listed in Table 1 [8].

**Table 1.** P Factor of Blaney – Criddle

| **Lat** | **N** | **Jan** | **Feb** | **Mar** | **Apr** | **May** | **June** | **July** | **Aug** | **Sept** | **Oct** | **Nov** | **Des** |
|---|---|---|---|---|---|---|---|---|---|---|---|---|---|
| | S | July | Aug | Sept | Oct | Nov | Dec | Jan | Feb | Mar | Apr | May | June |
| **60** | | 0.15 | 0.20 | 0.26 | 0.32 | 0.38 | 0.41 | 0.40 | 0.34 | 0.28 | 0.22 | 0.17 | 0.13 |
| **55** | | 0.17 | 0.21 | 0.26 | 0.32 | 0.36 | 0.39 | 0.38 | 0.33 | 0.28 | 0.23 | 0.18 | 0.16 |
| **50** | | 0.19 | 0.23 | 0.27 | 0.31 | 0.34 | 0.36 | 0.35 | 0.32 | 0.28 | 0.24 | 0.20 | 0.18 |
| **45** | | 0.20 | 0.23 | 0.27 | 0.30 | 0.34 | 0.35 | 0.34 | 0.32 | 0.28 | 0.24 | 0.21 | 0.20 |
| **40** | | 0.22 | 0.24 | 0.27 | 0.30 | 0.32 | 0.34 | 0.33 | 0.31 | 0.28 | 0.25 | 0.22 | 0.21 |
| **35** | | 0.23 | 0.25 | 0.27 | 0.29 | 0.31 | 0.32 | 0.32 | 0.30 | 0.28 | 0.25 | 0.23 | 0.22 |
| **30** | | 0.24 | 0.25 | 0.27 | 0.29 | 0.31 | 0.32 | 0.31 | 0.30 | 0.28 | 0.26 | 0.24 | 0.23 |
| **25** | | 0.24 | 0.26 | 0.27 | 0.29 | 0.30 | 0.31 | 0.31 | 0.29 | 0.28 | 0.26 | 0.25 | 0.24 |
| **20** | | 0.25 | 0.26 | 0.27 | 0.28 | 0.29 | 0.30 | 0.30 | 0.29 | 0.28 | 0.26 | 0.25 | 0.25 |
| **15** | | 0.26 | 0.26 | 0.27 | 0.28 | 0.29 | 0.29 | 0.29 | 0.28 | 0.28 | 0.27 | 0.26 | 0.25 |
| **10** | | 0.26 | 0.27 | 0.27 | 0.28 | 0.28 | 0.29 | 0.29 | 0.28 | 0.28 | 0.27 | 0.26 | 0.26 |
| **5** | | 0.27 | 0.27 | 0.27 | 0.28 | 0.28 | 0.28 | 0.28 | 0.28 | 0.28 | 0.27 | 0.27 | 0.27 |
| **0** | | 0.27 | 0.27 | 0.27 | 0.27 | 0.27 | 0.27 | 0.27 | 0.27 | 0.27 | 0.27 | 0.27 | 0.27 |

Southern Latitude: Apply 6 month difference as shown

*2.6.3. Consumptive Water Requirements.* Consumptive Water Requrements is calculated using Blaney-Criddle method [5] with the formula on Equation (3)

$$CWR = Kc. Eto \tag{3}$$

CWR = consumptive water requirement (mm / 0.5 bln)

Kc = coefficient of plant
Eto = evapotranspiration (mm / 0.5 bln)

Values of The Crop Factor (Kc) For Crops and Growth Stage are listed in Table 2 [8].

**Table 2.** Values of The Crop Factor (Kc) For Crops dan Growth Stage

| Crop | Initial Stage | Crop dev. Stage | Min-season Stage | Late season stage |
|---|---|---|---|---|
| Coffee (bean) | 0.35 | 0.70 | 1.10 | 0.90 |
| Cabbage/Carrot | 0.45 | 0.75 | 1.05 | 0.90 |
| Orange | 0.45 | 0.70 | 0.90 | 0.75 |
| Chili | 0.35 | 0.79 | 1.05 | 0.90 |

## 3. Results and Discussion

### *3.1 Results*

Results of direct observation in the field and laboratory analysis from each parameter in the assessment of available water and crop water requirements, physical and chemical parameters include texture, permeability, organic matter, water characteristic parameters include infiltration, field capacity, permanent wilting point, and available water. Soil texture analysis result is clay soil with soil type of Andisol vitrik, permeability 93.40 cm/hr with very fast category, soil organic material is 5.04%. While the result of infiltration water analysis is 39.60 cm/ hour, field capacity 24,80%, permanent wilting point 12.58%, and available water content 12.12%.

Crop water requirements deferent by growth phase. The highest average water requirement is required by plants on the crop definitive stage and at mid-season stage. In coffee plants water demand increases in the crop definitive stage and continues to increase until mid-season stage, but decreases at the late season stage. This also applies to cabbage, carrot, squash, and pepper. Water requirement for each crop shows in Table 3. Water requirements for each crop of coffee, cabbage, squash, carrot, and pepper shows in Figure 2.

**Table 3.** Water Requirement Plant coffee, cabbage, carrots, orange and pepper.

| CWR | Coffee | Cabbage | Carrot | Squash | Pepper |
|---|---|---|---|---|---|
| Initial Stage | 11.28 | 19.58 | 14.82 | 14.82 | 17.37 |
| Crop dev. Stage | 24.19 | 19.58 | 28.61 | 28.61 | 17.37 |
| Mid-season Stage | 35.49 | 33.10 | 28.61 | 28.61 | 34.80 |
| Late season stage | 29.04 | 27.74 | 27.95 | 27.23 | 30.46 |

### *3.2. Discussions*

Water available in the soil tends to be low with 12.12% result. The results depend on the parameters of infiltration, organic material, soil texture and soil type in the study area. The soil type in the research area is Andisol. The soil properties of Andisol are thick, dark, brown to black, very porous, very loose, non-tough, and unsteady, crumb or granular structures, oily as they contain 8-30% organic matter [9]. This nature causes the water content available in Pelaga Village tends to be low, because the soil cannot store water for too long. Texture of clay soil with a more dominant percentage of dust causes water not to be infiltrated properly, because the pores of the soil are covered with dust, so that the falling rain water will become more of a surface runoff. Low organic matter with a value of 5.04%, causing soil aggregates not formed and water entering into the soil through rainwater cannot be stored.

Texture of clay soil has more dominant percentage of dust. Dust is a fraction of the soil that is very sensitive to erosion / easily eroded because of the rainfall, so it can close the pores of the soil so the falling rain will be converted into runoff. The level of organic matter that is relatively ideal for the soil can lead to the formation of a solid soil aggregate, but in the Andisol soil the role of organic matter in absorbing water is not optimal. Andisol soil has the ability to absorb and store water that cannot revert as previous when experiencing drought (irreversible drying). This irreversible nature causes a change in particle size, because the allophane mineral contained in Andisol soil will tend to form a pseudosand fraction of allophane aggregation with other particles including organic matter [11]. Allophane will increase the humus retention against microbiological decomposition so that humus does not bind water.

The water needs of coffee crops, oranges, cabbage, chilies, and carrots differed according to the growing period. The most water is needed during the crop definitive stage and mid-season stage. Water demand during vegetative and generative growth is very important. Water shortages in the vegetative growth phase cause cell development and division to become late which affects plant growth and production [3]. Water demand increases during this phase because the plant has begun to multiply the growth of its leaves and the plants begin to bloom and bear fruit. It causes more water are needed for evapotranspiration process. The availability of groundwater tends to deficit against the need for water for coffee, citrus, cabbage, carrots, and chilli in Pelaga Village when there is no addition of water from outside such as rainwater and irrigation. This water shortage can decrease the productivity of plants, but it is strongly influenced by the respective response of each type of plant to the water shortage [4].

To meet the needs of crop water should be done some long-term and short-term land management. Short-term management that must be done is by watering. To accelerate growth and improve crop yields need watering according to water requirements so that water can be utilized efficiently [2]. While long-term management that must be done is the provision of organic materials and making irrigation networks. Irrigation is limited largely by the available water supply. Relatively large quantity of water required to satisfy the needs of the crop and to supply conveyance, evaporation, and seepage lessees [5]. The role of organic matter to the physical soil is as one of the aggregate soil formers, which has a role as adhesive agent between soil particles to unite into soil aggregates, so that organic matter is important in the formation of soil structure [7]. Increased soil organic matter causes the water content to increase as well, adding organic matter will increase the soil's ability to retain water so that the ability to provide soil-water for plant growth increases [10].

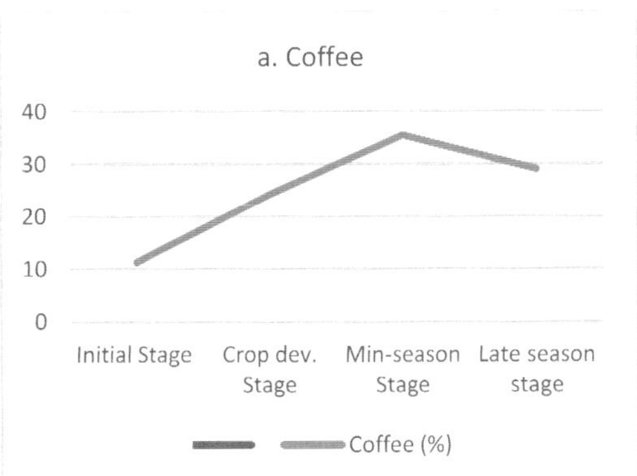


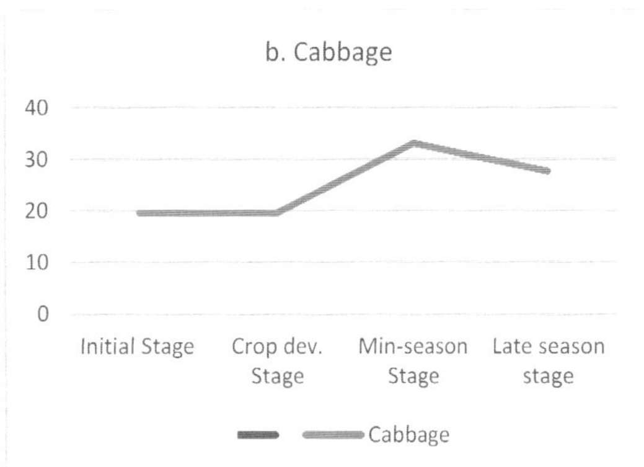

IOP Conf. Series: Earth and Environmental Science **98** (2017) 012048 doi:10.1088/1755-1315/98/1/012048

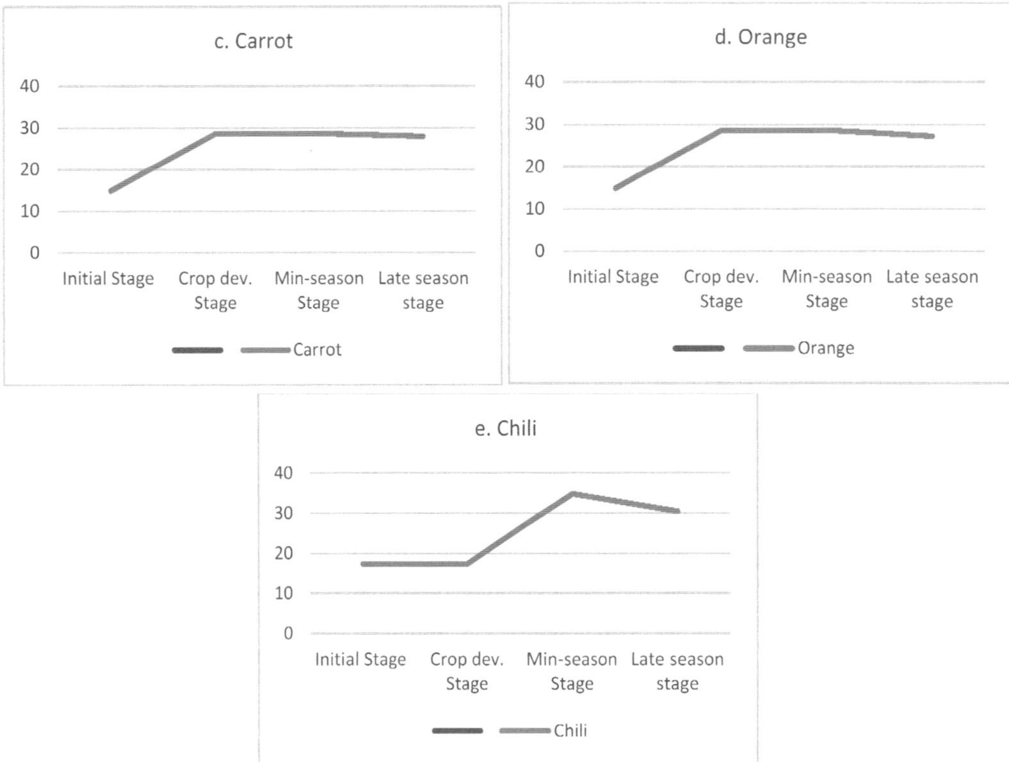


**Figure 2.** Water Requirement for (a) coffee, (b) cabbage, (c) carrot, (d) squash, (e) pepper

## 4. Conclusions

Based on the results of statistical analysis and discussion, it can be concluded as follows:

1. Water available Pelaga Village is 12,12%. Soil texture analysis result is clay soil with soil type of Andisol vitrik, permeability 93,40 cm/ hr with very fast category, soil organic material is 5.04% with low category. While the result of infiltration water analysis is 39,60 cm/ hour, field capacity 24,80%, permanent wilting point 12,58%, and available water content 12,12%.
2. Crop water requirement for each planting period respectively was, coffee 11.28%, 24.19%, 35.49%, 29.04%; cabbage 19.58%, 19.58%, 33.10%, 27.74%: carrot 14.82%, 28.61%, 28.61%, 27.95%: Orange 14.82%, 28.61%, 28.61%, 27.23%; chili, 17.37%,17.37%, 34.80%, 30.46%.
3. Comparison between water requirement and water available in the soil was deficit. Planning and management of the soil that can be done by watering, irrigation, giving organic materials and making terracing to reduce the water deficiency for the crops.

## Acknowledgments

Writers would like to thank the Rector of Universitas Udayana, Dean of Agriculture Faculty of Universitas Udayana, Head of Center for Spatial Data Infrastructure Development (PPIDS) Universitas Udayana, fellow staff, researchers and professors at Universitas Udayana for all the support, comments and suggestions that makes this paper finished.

## References

[1] Williams C N and K T Joseph 1973 *Climate, Soil and Crop Production in the Humid Tropics* (Kuala Lumpur: Oxford University Press).
[2] Doorenbos J and A H Kassam 1979 *Yield Response to Water*. (Rome: FAO Irrigation and Drainage Paper 33)

[3] Tjitrosomo S S, S Harran, A Sudiarto and Hadisunaryo 1985 *Botani Umum* (Bandung: Angkasa Bandung).
[4] Solichatun et al 2005 Pengaruh ketersediaan air terhadap pertumbuhan dan kandungan bahan aktif saponin tanaman ginseng jawa (Talinum paniculatum Gaertn.) *Biofarmasi* **3** (2) p 4751.
[5] Schwab Glenn O, F Richard K, W Talcott, Edminster and B Kennet K 1981 *Soil and Water Conservation Engenering* Third Edition (Canada: Wiley).
[6] Soil Research Center 1983 *Kriteria Penilaian Sifat Kimia Tanah.* (Bogor: Soil research Center).
[7] Utomo M S, B Rusman, T Subrina, J Lumbanraja and Wawan 2016 *Ilmu Tanah Dasar-Dasar dan Pengelolaan* (Jakarta: Kencana).
[8] Doorenbos J 1977 *Guidelines for Predicting Crop Water Requirements* (Rome: Food and Agriculture Organization of The United Nation).
[9] Darmawijaya I 1990 *Klasifikasi Tanah, Dasar-dasar Teori Bagi Penelitian Tanah dan Pelaksanaan Penelitian* (Yogyaakrta: UGM Press).
[10] Atmojo S W 2003 *Peranan Bahan Organik Terhadap Kesuburan Tanah Dan Upaya Pengelolaannya. Ilmu Kesuburan Tanah* (Surakarta: Fakultas Pertanian. Universitas Sebelas Maret)
[11] Tan K H 1998 *Dasar-Dasar Kimia Tanah. Cetakan Kelima.* Terjemahan D H Geonadi. (Yogyakarta: Gadjah Mada University Press).

IOP Conf. Series: Earth and Environmental Science **98** (2017) 012049 doi:10.1088/1755-1315/98/1/012049

# Land Boundary Delineation to Supporting of Program *Systematic Complete Land Registration* (PTSL) Using Multicopter-RTF Data (Case study: Wotan Village, Panceng Sub District, Gresik district)

**A B Cahyono and U W Deviantari***

Geomatics Engineering Departement – Sepuluh Nopember Institute of Technology. Surabaya 60111

dianada87@gmail.com

**Abstract.** According to statutory regulation issued by Ministry of Land and Spatial Planning/ Head of National Land Agency (BPN) number 35/2016, Comprehensive Systematic land registration is a sequential activity of which continuously and systematically carried out by the government ranging from collecting, processing, recording and presenting, as well as maintaining the physical and juridical data in the form of map and list of land-plots and flats, including the transfer of legal title for land plots and flats with their inherent rights. Delineation is one method to identify land plots by utilizing map image or high resolution photo and defining the boundaries by drawing lines to determine the valid and recognizable boundaries. A guideline to delineate the unregistered land plots may be determined from this two methods' accuracy result, using general boundary applied to aerial photo taken by multicopter RTF. Data taken from a height of 70 meter on an area obtained a number of 156 photos with 5 GCP resulting in an photo map with GSD 2.14 cm. The 11 samples parcels are selected in the sites of ± 7 ha. There are 11 samples of land parcels are tested. The area difference test for every parcel using a average standard deviation of 17,043 indicates that there are three land parcels which have significant area difference and 8 others do not have significant area difference. Based on the tolerance of National Land Agency, among 11 parcels studied, there are 8 parcels that fullfill the tolerances and three others do not fullfill tolerances. The percentage of area difference average between land registration map and orthophoto is 4,72%. The result shows that the differences in boundaries and areas that may be caused by a systematic error of method in describing the boundaries of the ground.

**Keywords:** Delineation, Land Registration, Multi Copter

## 1. Introduction

To carry out the tasks and functions of BPN in the region formed Regional Office BPN in the province and Land Office in the district/city. Land Office can be established more than one Land Office in each district / city (Presidential Regulation of The Republic of Indonesia Number 20 of 2015 article 7). Currently, the Gresik District Land Office establishes a land-based database per village called a complete cadastre (KAKAP) where data from each plot of land is obtained from measurements and delineations for uncharted fields. The stages in the construction of complete cadastre (KAKAP), among others, are 1) Provision of large-scale basic map for viewing maps of air with a scale of 1: 1000; 2) Graphical Index Mapping (GIM) of all parcels registered to the corrected map; 3) Delineation of all parcels of land not yet registered; 4) Validation of land books, textual letters, spatial measurements, and registration maps; 5) Zonation making based on KAKAP field boundary (map of land value zone and spatial zoning map); and 6) Updating of land value in new zones. Land delineation is one of the components in complete cadastral construction where delineation is performed on high resolution imagery for unregistered terrain. The photo map is a map depicting the detail of the field from aerial photography / satellite imagery to a certain scale. The photo map has been through the process of photogrammetry mapping, therefore the measurements on the photo map are correct, so the detail on the photo map and can be identified in the field has the correct position on the map [1]. The delineation of the parcels is done by identifying the plots of land using the photo map and drawing the measuring line for clear and eligible plot boundaries. There are two types of boundary categories that can be used as a reference in the delineation method, including fixed boundary and general boundary. Fixed boundary is a line defined by authorized through a formal survey and expressed mathematically by relationship and distance, or by coordinates. While the general boundary is a line that appears to exist in reality but has not been established by the authorities, usually the limit is physical, natural or man-made appearance such as fences, ditches, or roads (Enemark, et al. 2010). The boundaries of parcels identified on the map should be measured in the field [2]. The existence of technological advances in the field of photogrammetry, especially the UAV RTF; the more cheap, easy and ready to use needs to be a study, especially to know the accuracy by comparing with the results of size in the field. With the research on the analysis of the delineation method of the area of land in the image image with this type of UAV is expected to be known geometric and wide position of the resulting field and the accuracy of the use of the delineation results that have been done. This is mainly in finding the extent of the land area in accordance with the standard of wide accuracy that has been issued by BPN. So, it is expected that this wide delineation of land can be one of the factors to support the acceleration in the program of Systematic Complete Land Registration (PTSL).

## 2. Photogrammetry UAV TRF System

Currently drones have a variety of shapes, sizes, configurations, and functions. The presence of a camera device with a lightweight and compact size but has good quality. DJI Phantom Aerial UAV Drone Quadcopter one that weighs not more than 2 kilograms where there is already a camera installed safely and neatly. The drone can be controlled by remote control within 1000 meters. Ready To Fly: Complete fuselage ready to be directly flown consists of Aircraft Kit, servo ESCMotor, BrushlessTx/Transmitter/RemotRx Receiver, Propane Rover, Adapter, Propeler, Battery Charger and AutoPilot. Phantom version 3 is a drone that have models waypoint navigation and can now be used for photogrammetry. This quadcopter uses the latest IMU, flight control stabilization technology to fly super still along with a 4k stabilized integrated gimbal and camera. This will create accurate point clouds and perfect 3D maps stitched together using software. Waypoint navigation is very important for creating accurate 3D photogrammetry images. The Dji Phantom 3 Advance used the "Draw Waypoints" for it's autonomous programmed flight.

This type of drone is known as Rotary-wing or Vertical Take-Off and Landing type which is currently commonly used by the public because it does not require special skill in controlling it as well as a lot of market with affordable price. But now it has been equipped with various high-tech features that make it possible to do photogrammetric mapping with ease. Among them is the ability to autopilot the waypoint tracking that will make it easier when making the flight path. As a complement in this system also has many types of data processing from the Open Source (VSFM, MeshLab, etc.) to the license software (AgiSoft, Pix4D, Menci, etc.) to get the results of orthophoto [3].

IOP Conf. Series: Earth and Environmental Science **98** (2017) 012049 doi:10.1088/1755-1315/98/1/012049

**Figure 1.** Spesification Drone Dji Phantom 3 Advanced. Weight (Battery & Propellers Included) : 1280 g. Diagonal Size (Propellers Excluded) : 350 mm. Max Flight Time : Approx. 23 minutes. Satellite Positioning Systems : GPS/GLONASS. Sensor : 1/2.3" CMOS. Effective pixels: 12.4 M (total pixels: 12.76 M). Camera Sony EXMOR FC300S, Focal length : 3,64 mm. Lens : FOV 94° 20 mm (35 mm format equivalent), f/2.8. Electronic Shutter Speed : 8 - 1/8000 s. Image Size : 4000 × 3000. Range distances of up to 3.1 miles (5 kilometers).

## 3. Results and Discussion

The location used in this research is in Wotan Village, Paceng District, Gresik Regency Geographically the Gresik District is located at geographic coordinates within latitude -6 ° 49 '- 7 ° 24' South and its longitude 112 ° 22 '- 112 ° 40' East. The first preparation of the image acquisition and GCP measurement, a reconnaissance activity to get an overview of the location was conducted. The important element which might affect the image acquisition procedure and GCP measurement such as land cover.

**Figure 2.** Base Station at photograph

**Figure 3.** Premarking & measurement at Base Point with HyperPro Topcon GPS

Premarking activities begin by installing a quadrilateral premarking made of plastic in contrasting colours (orange) with its surroundings. The markers consist of 4 pieces of rectangular sized use 20 cm x 40 cm rectangular. To connect to be united then sewn. While the GPS stop and go positioning survey have a centimetre to decimetre accuracy. Based on the post-processing of data using software Topsurv software, the GCP point consisting of 5 points has an accuracy of 1.00 cm – 6.00 cm.

**Table 1.** GCP Measurement Results

| Name | Grid Northing (m) | Grid Easting (m) | Elevation (m) | Std Dev n (m) | Std Dev e (m) | Std Dev u (m) | Std Dev Hz (m) |
|---|---|---|---|---|---|---|---|
| Base | 9230574,429 | 664175,155 | 66,607 | 0 | 0 | 0 | 0 |
| GCP 1 | 9230718,416 | 664209,795 | 66,682 | 0,001 | 0,002 | 0,006 | 0,002 |
| GCP 2 | 9230639,46 | 664313,905 | 63,999 | 0,001 | 0,001 | 0,003 | 0,002 |
| GCP 3 | 9230545,727 | 664333,164 | 62,072 | 0,001 | 0,002 | 0,004 | 0,002 |
| GCP 4 | 9230555,282 | 664259,779 | 64,819 | 0,001 | 0,001 | 0,003 | 0,001 |

**Figure 4**. Multicopter RTF on base station

**Figure 5**. Flight route mission

The next step UAV and the remote control as the ground station was set up, the flight parameter was established, such as 70 m height; percentage of overlap (80% overlap and 70% sidelap); and size of the area to be captured. The area of the field was divided into one block. The whole area of fieldwork ± 7 ha. But to make a better result in the residential area then 1 more blocks were added, therefore in total 2 flights with 156 images were executed. The UAV flew for around 12 minutes for each flight, but to avoid the chance of the falling due to low battery, every time the UAV was flown, it was taken back to the base to change the battery. The processing aerial photo processing by structure from motion algorithm and bundle adjustment with self-calibration has been done with 5 GCP and 4 check points.

**Figure 6.** Paddy field boundary

**Figure 7.** Urban boundary

The analysis is done by comparing the area and the field distance that has been registered, between the results of the delineation method fixed boundary and general boundary methods. Delineation method fixed boundary is performed on the Land Registration Map of the data plots of land obtained from the results of registration measurements land in the field. While delineation method of general boundary performed on orthophoto UAV image.

**Table 2.** Area comparison

| No | Parcel (NIB) | *Area ($m^2$)* | |
|---|---|---|---|
| | | **Existing Maps** | **Orthophoto** |
| **1** | 470 | 103,977 | 105,118 |
| **2** | 472 | 191,202 | 203,853 |
| **3** | 473 | 89,571 | 71,663 |
| **4** | 10015 | 86,233 | 104,581 |
| **5** | 478 | 153,645 | 160,895 |
| **6** | 479 | 269,446 | 265,910 |
| **7** | 538 | 152,858 | 156,226 |
| **8** | 532 | 181,624 | 186,118 |
| **9** | 468 | 163,407 | 185,118 |
| **10** | 1014 | 211,021 | 207,896 |
| **11** | 470 | 103,977 | 135,118 |

Data which is used as a reference is the data area and distance of results delineation of the fixed boundary method. Comparative analysis of area to know the difference of area is done by using paired t-test (paired sample t-test). The large value of the residuals illustrated the disparities of the boundaries and areas of the parcel which might have caused by the systematic error of the method to delineate the boundaries. From the visual comparison of the digitized output and the referenced parcels observed with the existing method, some of the points were missed to be marked or marked in the different location with the referenced one. Table 2 shows the comparison of area calculation the methods. The comparison between the existing maps and orthophoto shows that there are some significant differences exist in the parcel boundary with the NIB 10015, 468 and 470 with a different area of 18.334 $m^2$, 22.021 $m^2$ and 32.132 $m^2$. According to [4], some of the causes significant area difference, among others; (1) the one side of the corner boundary grow large trees that pre-mark is not visible, (2) some of the boundary monument has been move, and (3) the physical parcel of paddies rice in the field is not clear. In this research the statistical test show that there are significant differences between the results of UAV data processing and a land registration map. The area difference test for every parcel using an average standard deviation of 17,043 indicates that there are three land parcels which have significant area difference and 8 others do not have significant area difference. Based on the tolerance of National Land Agency, among 11 parcels studied, there are 8 parcels that fulfil the tolerances and three others do not fulfil tolerances. The percentage of area difference average between land registration map and orthophoto is 4,72%.

## 4. Conclusions

By using UAV Multicopter RTF type rotary wing or multicopter model can be generated an orthophoto map which can be used as data for delineation of land boundaries. The important thing is about the regulation of technical regulations, this is because more and more public that use this tool for mapping. For the problem of accuracy achieved, many studies have shown that with orthophoto maps shooting with this technology can achieve the accuracy of GSD to 5-10 cm [4] or 3 cm [5] where the results have sufficient standards to be used as a base map. The residuals between the observed and predicted value from the result represent the model fitness towards the data, the smaller the value means the closer the result is to the data. It is remarked that the result generated from both boundaries and areas of the parcel has shown a significance discrepancy between the parcels. For future research will be applied the research about automatically boundary delineation calculations with UAV data, that is currently being developed in some countries about boundary delineation algorithm [6].

## Acknowledgement

Authors would like thank to Institute of Research and Public Service (LPPM) Sepuluh November Institute of Technology for supporting the research funding, and also for ATR/BPN Kanta Gresik, Jawa Timur, who has permitted field data collection at research area.

## References

[1] Sudarsono B and A Laila 2008 Cadastral Measurement and Mapping Using Aerial Photo Identification Methods *Engineering Journal Diponegoro University* **29**(1).

[2] Regulation of The State Minister of Agrarian Affairs / Head of National Land Agency on Provision of Land Registration, Pub. L. No. 3, 1997.

[3] Cahyono A B, H H Handayani, H Hidayat, Y Budisusanto 2016 Procurement of Village Geospatial Data Using Drone-Quadcopter (Case Study: Sugeng Village, Kec.Trawas, Mojokerto Regency) *Geoid Journal* **12**(1).

[4] Rokhmana C A and S Utomo 2016 The Low-Cost UAV-Based Remote Sensing System Capabilities for Large Scale Cadaster Mapping *IOP Conference Series: Earth and Environmental Science* **47**(1).

[5] Ramadhani S A 2016 Using Unmanned Aircraft System Images To Support Cadastral Boundary Data Acquitition in Indonesia *Thesis* Faculty of Geoinformation and Earth Observation – Universites of Twente.

[6] Crommelinck et al 2017 Contour Detection for UAV-Based Cadastral Mapping *Remote Sens* **9**(2) p 171; doi:10.3390/rs9020171

[7] Hidayat H and A B Cahyono 2016 Combined Aerial And Terrestrial Images For Complete 3D Documentation Of Singosari Temple Based On Structure From Motion Algorithm. *IOP Conference Series: Earth and Environmental Science* **47**(1).

# Model of Peatland Vegetation Species using HyMap Image and Machine Learning

**Muhammad Dayuf Jusuf[1*], Projo Danoedoro[2], Bangun Muljo Sukojo[3], Hartono[4]**

[1] PhD student remote sensing, Fakultas Geografi – Universitas Gadjah Mada, Yogyakarta
[2, 4] Fakultas Geografi – Universitas Gadjah Mada, Yogyakarta
[3] Fakultas Teknik Geomatika, Institut Teknologi Sepuluh Nopember – ITS, Surabaya

dayuf.jusuf@mail.ugm.ac.id

**Abstract**. Species Tumih / Parepat (*Combretocarpus-rotundatus* Mig. Dancer) family *Anisophylleaceae* and Meranti (*Shorea Belangerang, Shorea Teysmanniana* Dyer ex Brandis) family *Dipterocarpaceae* is a group of vegetation species distribution model. Species pioneer is predicted as an indicator of the succession of ecosystem restoration of tropical peatland characteristics and extremely fragile (unique) in the endemic hot spot of Sundaland. Climate change projections and conservation planning are hot topics of current discussion, analysis of alternative approaches and the development of combinations of species projection modelling algorithms through geospatial information systems technology. Approach model to find out the research problem of vegetation level based on the machine learning hybrid method, wavelet and artificial neural networks. Field data are used as a reference collection of natural resource field sample objects and biodiversity assessment. The testing and training ANN data set iterations times 28, achieve a performance value of 0.0867 MSE value is smaller than the ANN training data, above 50%, and spectral accuracy 82.1 %. Identify the location of the sample point position of the Tumih / Parepat vegetation species using HyMap Image is good enough, at least the modelling, design of the species distribution can reach the target in this study. The computation validation rate above 90% proves the calculation can be considered.

**Keywords**: Peatland species, HyMap image, WANN hybrid algorithm model.

## 1. Introduction

Some methods of uncertainty around the effectiveness of biodiversity make the researchers and resource managers ambiguous, as well as conservation planners. Summarized the model distribution method of species distribution models (SDMs), niche concepts of environmental habitat suitability and climate prediction distribution, spatial species attribute data of scale, space, and time. In the last five years there have been many published papers on various aspects of modelling methods of species distribution that enable the achievement of various advances. Theoretical modelling of species, ecological data provides comprehensive guidance for the production of statistical models, machine learning and environments for the classification of spatial distribution species [3].

The intention of modelling is usually done based on the interpolation and spatial distribution of geospatial species of non-linear geographically broad. Such modelling practices are widespread and supported by particular modelling contexts in various comparative results. The species distribution model is an empirical model related to field observations for predictors of world-level environmental response variables, statistically or theoretically based. Vegetation species data may be present or not and appear to be abundant or not, depending on the results of field observation, done at random or stratified sampling for ease of observation of the object in nature, the layout can be shown in Figure 1.

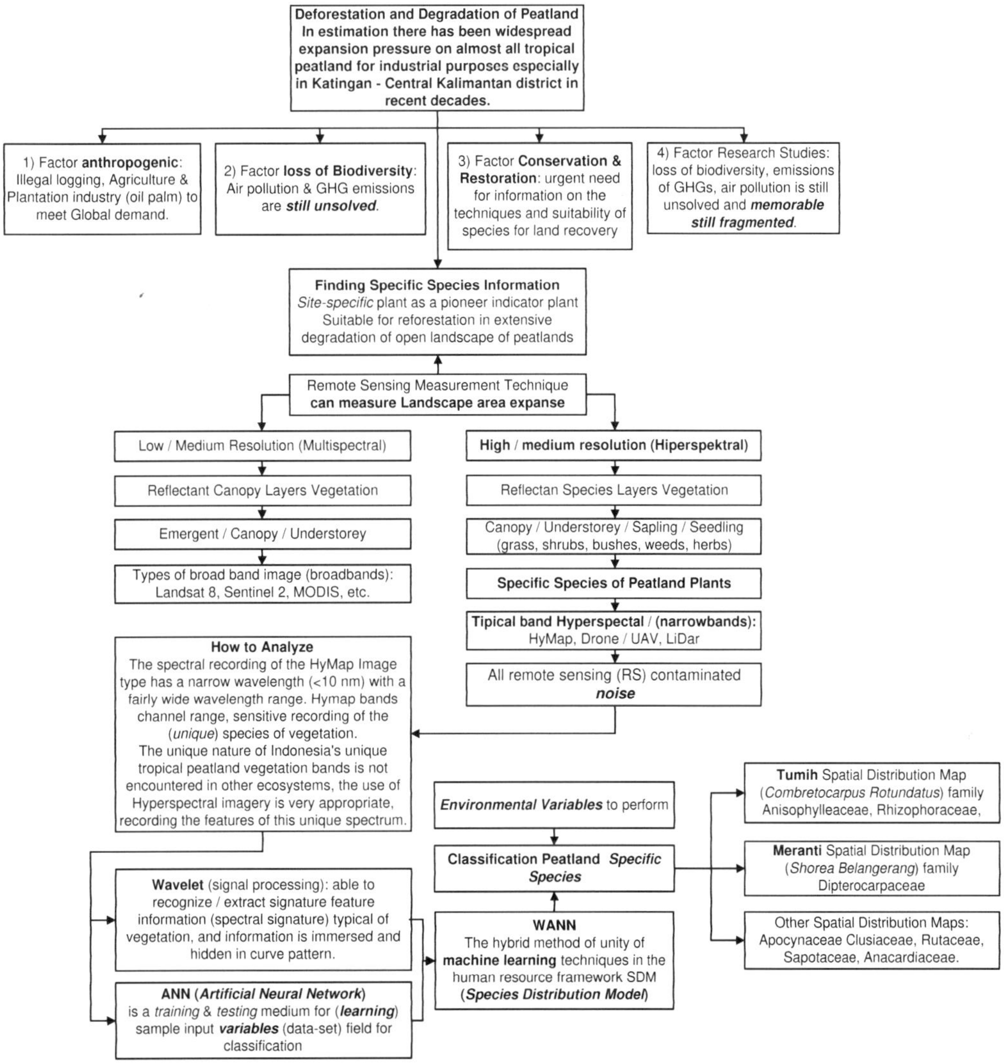


**Figure 1.** Research Flowchart

The actual species distribution maps or habitat potentials are required in many facets of environmental inquiry, such as: resource management, and preservation planners. It includes biodiversity assessment, biological availability design, habitat management and restoration, species preservation, habitat design, population survival analysis, environmental hazard assessment, invasive species management, community ecosystem modelling, and predicted impacts of global environmental ecosystem changes in species [1].

The dispersion model of species results and benefits are relatively inadequate in the appropriate resource analysis. Integration of Geographic Information System (GIS) model map modelling is not even complete enough to plump for the standardization of the organization model of the environment variable and produce prediction map of species, predictive evaluation method of errors and doubt, can inhibit effective modelling of collecting key components of problem approach apply all over the world [7] [3].

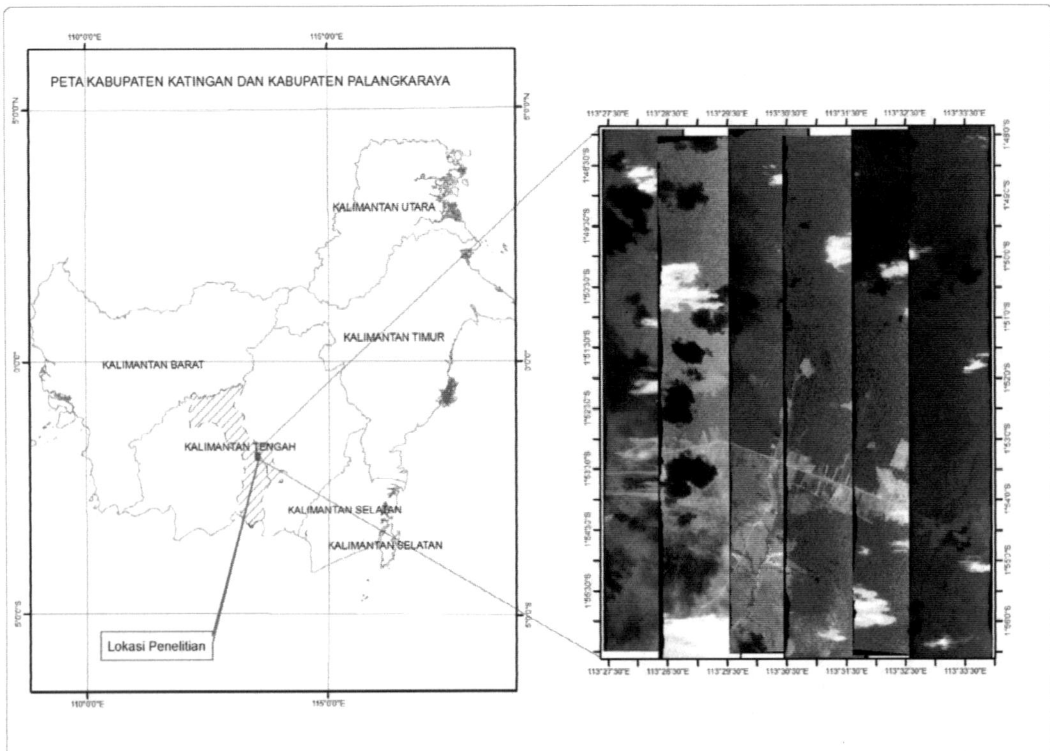


**Figure 2.** Research study area

The subject area of this research was covered area surrounding HEF (hampengan ecology, forest) disturb forest in Katingan district - Central Kalimantan, Indonesia is shown in figure 2. This subject area selected due to various objects types and vegetation density. The use of Hyperspectral (HyMap) imagery with high spatial resolution expected to be capable to distinguish the various objects types and vegetation in the study area accurately.

The Hyperspectral imagery used was (HyMap) imagery obtained from an aerial scan survey. The imagery was recorded in 15-16 July$^{th}$ 2011 and it was already corrected automatically since proceed it on air. Thus, the Hyperspectral (HyMap) imagery had already 4.2 m spatial resolution in a multi spectral band in 4 groups [VINIR, NIR (VISNIR & NIR), SWIR (NIR) and SWIR 1_2]. Detailed information about Hyperspectral (HyMap) imagery shown in Table 1.

IOP Conf. Series: Earth and Environmental Science **98** (2017) 012050 doi:10.1088/1755-1315/98/1/012050

**Table 1.** Detail information about the Hyperspectral (HyMap) imagery used in this research.

| Data | Processing 4 level data | Time Acquire | Spectral Resolution | Spatial Resolution |
|---|---|---|---|---|
| HyMap (*Hyperspectral Mapper*) | Geocorrection, radiance, raw_dn, reflectance | 15 – 16 July 2011 | Interval Wavelength | 4, 2 |

| Channel Type | Bands | | Channel Length |
|---|---|---|---|
| VNIR | VISNIR | 1 – 20 | 425.50000 - 714.50000 |
| NIR | VISNIR | 21 – 32 | 729.70000 - 895.60000 |
| | NIR | 1 – 17 | 868.30000 – 1119.40000 |
| SWIR | NIR | 18 – 32 | 1134.20000 - 1333.19995 |
| SWIR | SWIR-1 | 1 – 32 | 1418.09998 - 1817.90002 |
| | SWIR-2 | 1 – 32 | 1959.00000 - 2490.00000 |

## 2. Methods

Machine learning, is a discipline of science that develops algorithms or models that can memorize from the data from the knowledge, information that is on [5]. Briefly, learning machine can be interpreted as the procedure of changing data into information. Learning machine is divided into Supervised Learning and Unsupervised Learning.

Simulation of Wavelet transformation and Artificial Neural Network (WANN) is performed by using MatLab software. These hybrid transformation programming processes, takes four main steps to receive the test solutions and simulation practice. The Sequence programming includes: 1) design features files dataset species, vegetation, 2) extraction sample feature dataset in DWT decomposition + power average calculation, 3) constructs programming species pioneer identification Tumih and several pioneer species peatlands, 4) Training and testing dataset in ANN. The results indicate an increase in accuracy and speed in recognition of high level error and opposition.

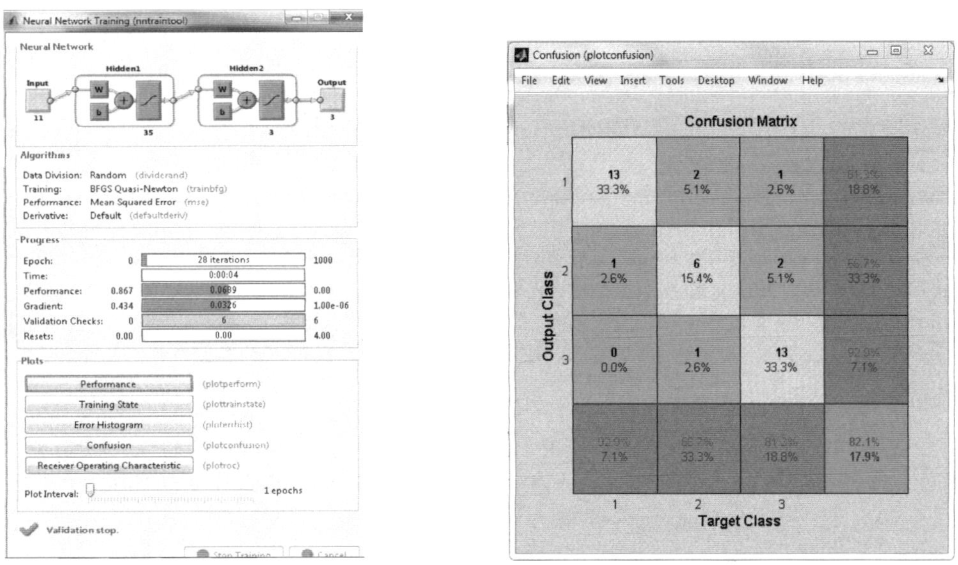


**Figure 3.** Result execution of ANN training dataset in these experiments, achievement spectral accuracy (82.1%) for 3 class species pioneers (Tumih, Meranti and others).

The basic simulation design as input parameters 11 dataset and input 15 works on the *Artificial Neural Network* (ANN) layer of *Levenberg–Marquardt* algorithm (LMA or just LM), used to solve nonlinear least squares problems in mathematics and computing. These minimization problems arise, especially in least squares curve fitting [4]. Experimental results obtained vary widely in pigment concentration (per unit leaf area and unit area) and leaf area index, based on a plant canopy scale, and independent variation between chlorophyll and carotenoid leaf area index. Survey data were used to evaluate the nature and strength of the relationship between the amount of spectral transformation and pigment concentration (per unit area land) [8].

The hyperspectral narrowband reflectance in the region's electromagnetic spectrum appears to be related to chlorophyll concentrations, while the wavelength ratio index works near-infrared (NIR), especially the more successful green wavelength. While the red edge position (REP) of the red edge is connected to the chlorophyll concentration, the foremost and second amplitude characteristics of the reflectance and pseudo absorption are stronger correlated with chlorophyll. The research approaches the information, analysis method of Hymap image integration and study data using algorithm design:

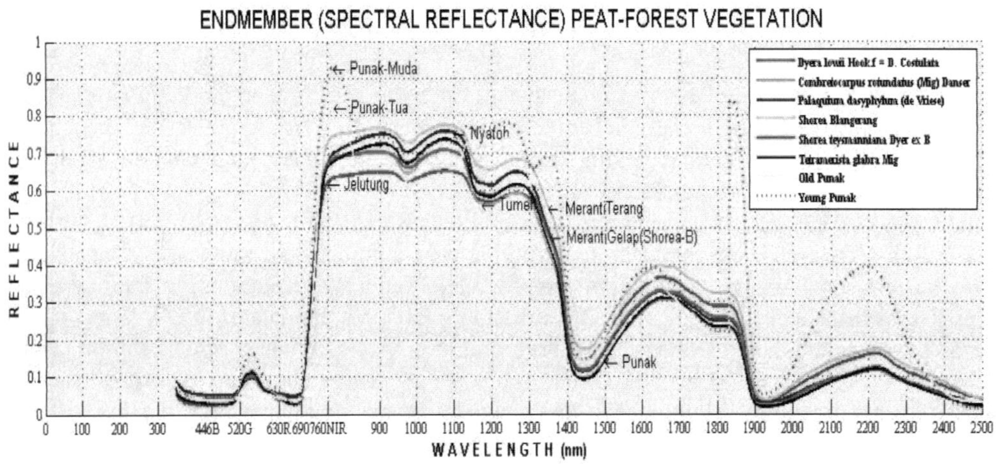


**Figure 4.** Development methods WANN for peatland species pioneer.

No spectral reflectance transformation was tested for Carotenoid concentrations, but the two index ratios were strongly associated with carotenoid ratios for chlorophyll a. Leaf area index (LAI) is not related to all spectral ratio indices, including broadband normalization difference and simple vegetation ratio index. This indicates that the spectral index may take in a limited application to estimate the LAI of vegetation canopy when the leaves comprising chlorophyll concentrations (per unit of leaf area) of the canopy have been different [5].

The multi sensor data application of the remote sensing field develops combine satellite data sources, opening up a new field research. Basic investigation of satellite imagery, related to the land cover classification factor. Optical data in the collection of high-resolution radiometer sensors for thematic mapping, in the specific classification capabilities of specific spectral signatures, some cover varies by scale and type of information. Traditional image processing resampling leads to significant differences in the quality of information extracts. The role of the spectral signature scale helps to secure the correct interpretation of the classification results [8].

Hyperspectral data analysis provides detailed spectral information of electromagnetic spectrum band subjects. The content of hyperspectral image information is extremely correlated to the neighbouring bands beyond and riding the size of its large dimension. Hyperspectral endmember validation analysis data are simulated in ENVI programming. The information of spectral bands of overlapping hyperspectral data should be extracted by the characteristics of the estimate of the exact properties of vegetation and soil [2].

Whereas a strong relationship exists between the image's spectral values and the object, the image can be used for further modelling. The undisturbed environment is not homogeneous, in contrast to highly heterogeneous-related topography. The slope of the slope and the direction of the slope affect the magnitude of the energy of the reflected object, and recording the result. The model relationship between spectral values and vegetation species should be corrected for topographic effects [5].

## 3. Results and Discussion

To obtain the testing and training ANN data set, shown are: several components: epoch iterations times 28, performance 0.067 means that a good enough because above 50 %, and spectral accuracy 82.1 %. However, the calculations show that the MSE (Minimum Square Error) value is smaller than the ANN training data that can achieve a performance value of 0.0837 and MSE = 0.0547. Identify the location of the sample point position of the Tumih / Parepat vegetation species using HyMap Image is good enough, at least the modelling, design of the species distribution can reach the target in this study. The computation validation rate above 90% proves the calculation can be considered. The aim of the development programming algorithm of peatland vegetation species can be utilized for other optical images such as: Multispectral, Hyperspectral, and Drone image.

## 4. Conclusions

Hyperspectral remote sensing image, the recorded spectrum has a narrow wavelength (<10 NM), but the overall wavelength range is broad enough. In this example, a sensor that receives only 20 bands, only includes a wavelength of 500 ~ 700 NM, can be categorized included in hyperspectral. The narrow reach of each channel of the hyperspectral channel is really sensitive to recording for unique objects. This is called for a spectral library field of peatland vegetation species.

The channels in the hyperspectral image data in this survey should consist of 10 image strips, but used only 6 (six) image strips where the availability of successful field samples in the collection: HU1_01, HU1_02, HU1_03, HU1_04, HU1_05 and HU1_06.

## Acknowledgement

I would like to thank are conveyed on behalf of deceased our brother Mr. Muhammad Evri (Agency for the Assessment and Application of Technology, BPPT) who delivered previous talks and advised me to this research and Mr. Hendrik Segah from the Faculty of Forestry, University of Palangkaraya (UNPAR) - Central Kalimantan, which has been willing to provide a means of handheld spectrometer measurements in the field. Which supply me of hyperspectral data (HyMap) images from cooperation projects between BPPT- UNPAR, Indonesia and JST-JICA Japan.

## References

[1] Burhanuddin. 2011. "The Association of Arbuscular Mycorrhizal Mushrooms with Perepat (Combretocarpus Rotundatus Miq) and Jelutung (Dyera Lowii Hook) In Peatlands." *Post-Graduate Program* - Gadjah Mada University Yogyakarta, Indonesia.

[2] Harris, A., R. Charnock, and R.M. Lucas. 2015. "Hyperspectral Remote Sensing of Peatland Floristic Gradients." *Remote Sensing of Environment* 162. Elsevier Inc.: 99–111. Doi:10.1016/j.rse.2015.01.029.

[3] Janet Franklin, Jennifer A. Miller. 2010. *Mapping Species Distributions: Spatial Inference and Prediction*. Edited by, and Jeremy Wilson, Paul Adam, H. J. B. Birks, Lena Gustafsson, Jeff McNeely, R. T. Paine, David Richardson, Birds. Cambridge CB2 8RU, UK: Cambridge University Press. www.cambridge.org/9780521876353.

[4] Kalacska, M., G. a. Sanchez-Azofeifa, B. Rivard, T. Caelli, H. Peter White, and J. C. Calvo-Alvarado. 2007. "Ecological Fingerprinting of Ecosystem Succession: Estimating Secondary Tropical Dry Forest Structure and Diversity Using Imaging Spectroscopy." *Remote Sensing*

*of Environment* 108 (1): 82–96. Doi:10.1016/j.rse.2006.11.007.

[5] Kalacska, M., M. Lalonde, and T.R. Moore. 2015. "Estimation of Foliar Chlorophyll and Nitrogen Content in an Ombrotrophic Bog from Hyperspectral Data: Scaling from Leaf to Image." *Remote Sensing of Environment* 169. Elsevier Inc.: 270–79. doi:10.1016/j.rse.2015.08.012.

[6] Murfi, H. 2017. *Model Combination*. Available at: http://ocw.ui.ac.id/course/view.php?id=21 [Access date: 14 Agustus. 2017].

[7] Raes, L.G. Saw, P.C. Van Welzen, T. Yahara. 2013. "Legume Diversity as Indicator for Botanical Diversity on Sundaland, South East Asia." *South African Journal of Botany* 89: 265–72.

[8] Heikkinen, J, and R Makipaa. 2010. "Testing Hypotheses on Shape and Distribution of Ecological Response Curves." *Ecological Modelling* 221 (3): 388–99. Doi: 10.1016/j. ecolmodel. 2009.10.030.

# Comparison of Spectral Characteristic between LAPAN-A3 and Sentinel-2A

**Z Zylshal*, N M Sari, J T Nugroho and D Kushardono**

Remote Sensing Applications Center, Indonesian National Institute of Aeronautics and Space (LAPAN), Jl. Kalisari No. 8, Pekayon, Pasar Rebo, Jakarta 17310, Indonesia.

zylshal@lapan.go.id

**Abstract**. Indonesian National Institute of Aeronautics and Space (LAPAN) started building its experimental microsatellite back in 2007 and finally able to launch its first microsatellite dubbed as LAPAN-A1/LAPAN-Tubsat. With the launch of LAPAN-A3/LAPAN-IPB, Indonesian experimental satellite programme hit its third generation. LAPAN-A3 is carrying multiple payloads including multispectral push-broom imager, digital matrix camera, as well as video camera. This paper aims to highlight the spectral differences between LAPAN-A3 and the well-established Sentinel-2A multispectral to investigate the potential of using LAPAN-A3 data to complement the other well-established medium resolution satellite data. Comparisons between corresponding bands and band transformations were performed over a dataset. Three areas of interest were chosen as the test sites. Linear regression and Pearson correlation coefficient were then calculated between the corresponding bands. The preliminary results showed a moderate correlation between the two sensors with Pearson correlation coefficient ranging from 0.39 to 0.65. Some issues were found regarding the radiometric quality over the whole scene of LAPAN-A3.

**Keywords:** LAPAN-A3, Sentinel-2A, Microsatellite, Spectral Comparison

## 1. Introduction

LAPAN-A3/LAPAN-IPB was launched on October 2016 as part of the Indonesian National Institute of Aeronautics and Space program on Space Technology. LAPAN-A3 is carrying multiple payloads including multispectral push-broom imager, digital matrix camera, as well as video camera. The multispectral payload, dubbed as Line Imager Space Application (LISA), has four multispectral bands ranging from visible to near infrared spectrum. With 123 km swath width, it falls between the 185 km of Landsat-8 OLI and the 100km tiled Sentinel-2A/2B MSI data. With this successful launch, there have not been many comparative analyses between these platform in terms of image quality, classification, and accuracy assessment. The spectral comparison between multiple different sensors [1][2][3][4][5][6], especially for the newly launched platforms often performed to identify the sensors performance on board the satellite after its launch. Tabel 1 shows the multispectral characteristics of LAPAN-A3 compared to two aforementioned operational medium resolution EO satellites.

The nature of high cloud cover in tropical region remains a challenge on the availability of usable optical remote sensing data. The joint use of different satellite sensor boasting similar spatial and spectral characteristic offers the opportunity to build usable time series data for long time monitoring and disaster mitigation [1][7]. Thus, the application of combining multiple Earth Observation (EO) satellite sensors for a combined use is attracting researchers and users in order to fill the gaps in observations [5].

**Table 1.** Multispectral Sensor characteristics of LAPAN-A3 LISA and Sentinel-2A MSI. Bands with bold font marked the corresponding bands of each sensor.

| Satellite | Sensor | Band | Spectral Range | Spatial Resolution | Swath Width | Revisit time | Radiometric Quantification |
|---|---|---|---|---|---|---|---|
| LAPAN-A3 | LISA (*Line Imager Space Application*) | **B1 - Blue** | **0.41 – 0.49 µm** | **18 m** | 122.4 km | 21 days | 16 bit |
| | | **B2 - Green** | **0.51 – 0.58 µm** | **18 m** | | | |
| | | **B3 - Red** | **0.63 – 0.70 µm** | **18 m** | | | |
| | | **B4 - NIR (Near Infrared)** | **0.77 – 0.99 µm** | **18 m** | | | |
| Sentinel-2A | MSI (*Multi-Spectral Instrument*) | B1 - Coastal | 0.433 – 0.453 µm | 60 m | 290 km | 10 days (5 days with Sentine l-2B) | 12 bit |
| | | Aerosol | **0.458 – 0.523 µm** | **10 m** | | | |
| | | **B2 - Blue** | **0.543 – 0.578 µm** | **10 m** | | | |
| | | **B3 - Green** | **0.650 – 0.680 µm** | **10 m** | | | |
| | | **B4 - Red** | 0.698 – 0.713 µm | 20 m | | | |
| | | B5 - Red Edge 1 | 0.733 – 0.748 µm | 20 m | | | |
| | | B6 - Red Edge 2 | 0.765 – 0.785 µm | 20 m | | | |
| | | B7 - Red Edge 3 | **0.785 – 0.900 µm** | **10 m** | | | |
| | | **B8 - NIR** | 0.855 – 0.875 µm | 20 m | | | |
| | | B8A - NIR Narrow | 0.930 – 0.950 µm | 60 m | | | |
| | | B9 - Water Vapor | 1.365 – 1.385 µm | 60 m | | | |
| | | B10 - Cirrus | 1.565 – 1.655 µm | 20 m | | | |
| | | B11 - SWIR 1 | 2.100 – 2.280 µm | 20 m | | | |
| | | B12 - SWIR 2 | | | | | |

Source: [8][9][10]

As the satellite itself is still an experimental satellite and yet reached its full operational capacity, we were not yet able to exploit the necessary information to perform the necessary radiometric and atmospheric correction. Therefore, the spectral comparison performed in this study was meant to be a first step towards a more comprehensive radiometric correction of LAPAN-A3 using relative radiometric normalization technique [11]. The image analysed in this study was provided by LAPAN's Satellite Technology Center (PUSTEKSAT) for first assessment and feedback. This paper aims to point out the spectral differences between LAPAN-A3 LISA and Sentinel-2A MSI.

## 2. Material and Methods

LAPAN-A3 and Sentinel-2A were acquired on October 2016 in some part of Jember Regency, in East Java Province, Indonesia (Figure 1). This particular data was the one with the least cloud coverage However, these datasets were not perfectly covered each other, and only portions of them were overlapped as shown in figure 1A.

In this preliminary assessment, we only used a single date, early stage LAPAN-A3 multispectral data. Considering the availability of LAPAN-A3 images was still limited due to its early stages and the cloud cover in Indonesian region, we were not able to obtain the same day image between the two sensors. What we have, instead, was images recorded in the same month, which is in October 2016, and within 1 day of each other. LAPAN-A3 was acquired on October 19th, and Sentinel-2A was acquired on October 18th.

The data preparation involved the geometric co-registration between LAPAN-A3 and Sentinel-2A. Sentinel-2A was used as the reference/base image; hence the final product would match Sentinel-2A coordinates (Universal Transverse Mercator, WGS 1984 Datum). Geometric co-registration involves image resampling and is often reduced the image quality [12]. To minimize the influence of resampling procedure on spectral information, we selected the nearest neighbour interpolation method for our study, as it found to be able to maintain the original gray level compared to bilinear and cubic convolution interpolation [2]. We then resampled Sentinel-2A data to match LAPAN-A3 data to 18 meter using the aforementioned interpolation method.

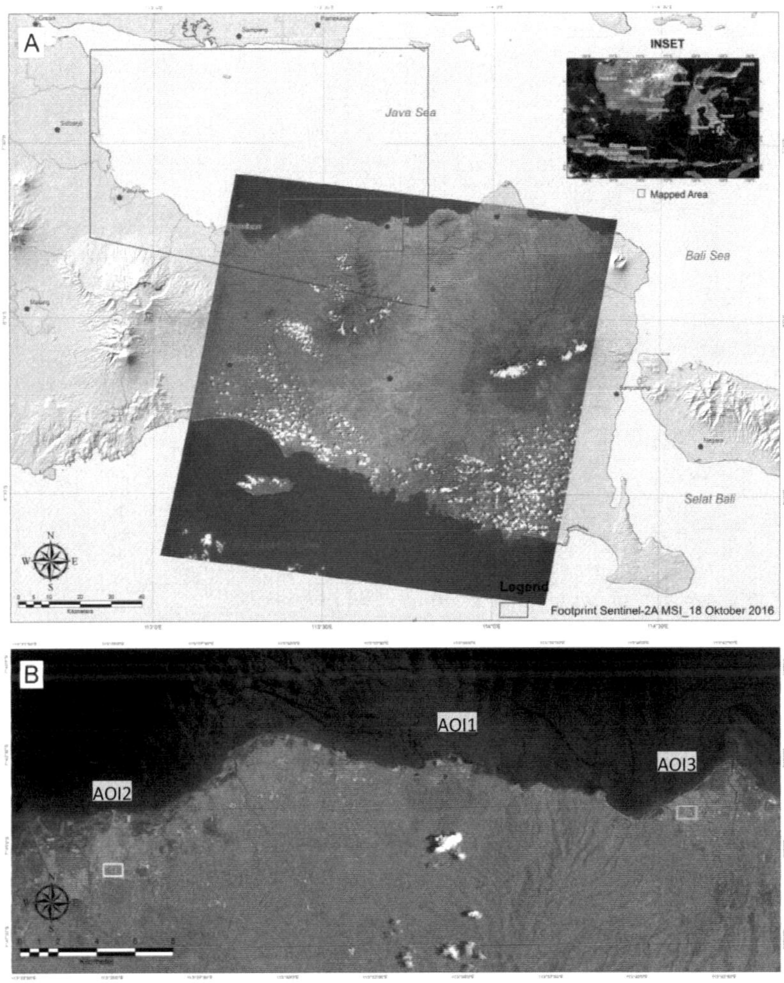


**Figure 1.** Overview of study area. (**A**). LAPAN-A3 full scene with RGB 432. Blue rectangle is the Sentinel-2A footprint. (**B**) Detailed look at the overlapped area. Coloured rectangles on the top figure represent three areas of interest in this study. Blue, yellow, and green rectangles are AOI 1, AOI 2, and AOI 3, respectively.

On this initial approach, three rectangular shape areas of interest (AOI) were created. These AOIs were chosen to represent different landuse/landcover characteristic. The AOIs distribution is shown in figure 2. AOI 1 is a steam power plant, known as PLTU Paiton Unit 9. As a power plant, it is mostly covered with concrete and asphalt. Large building blocks are also dominant with little vegetation. The vegetation is minimum compared to other impervious surface, which consist of well-maintained grass and sparse broadleaf trees. Being a steam generated power plant, a 50 meter wide canal is also existed within its parameter. AOI 2, Located in Kraksaan, Probolinggo, represents rural area that consists of a mixed of settlements and sparse vegetation. Impervious surface, however, considered more dominant than vegetation. AOI 3 represents a denser impervious surface than AOI 2, located in Besuki, Situbondo.

IOP Conf. Series: Earth and Environmental Science **98** (2017) 012051 doi:10.1088/1755-1315/98/1/012051

The spectral comparison were conducted on these areas are considered less affected by the seasonal and biological cycles [13]. These data then clipped based on previous AOIs.

Four original bands (B, G, R, NIR) and two spectral indices, NDVI and NDWI [14] were computed and compared using original digital number value without any prior radiometric correction. Considering the similar bandwidth between LAPAN-A3's band 4 and Sentinel-2A band 8, we decided to use it, instead of the narrower Sentinel-2A's band 8A for infrared spectrum. The use of spectral indices was to explore common land cover type, as well as to have a more comparable data in terms of radiometric quantifications, and computed as follows:

$$NDVI = \frac{\rho_{NIR} - \rho_R}{\rho_{NIR} + \rho_R} \quad (1)$$

$$NDWI = \frac{\rho_G - \rho_{NIR}}{\rho_G + \rho_{NIR}} \quad (2)$$

Where $\rho_{NIR}$, $\rho_R$, $\rho_G$ are the digital number (DN) of near infrared, red, and green, respectively.

The comparisons were evaluated to obtain the correlation, by computing the coefficients of linear regression and the Pearson Correlation coefficient. Linear regression was computed with the assumption that the atmospheric differences between images on the same geographic location acquired at different times are linearly correlated [11][13][15]. The workflow used in this study is shown in figure 3.

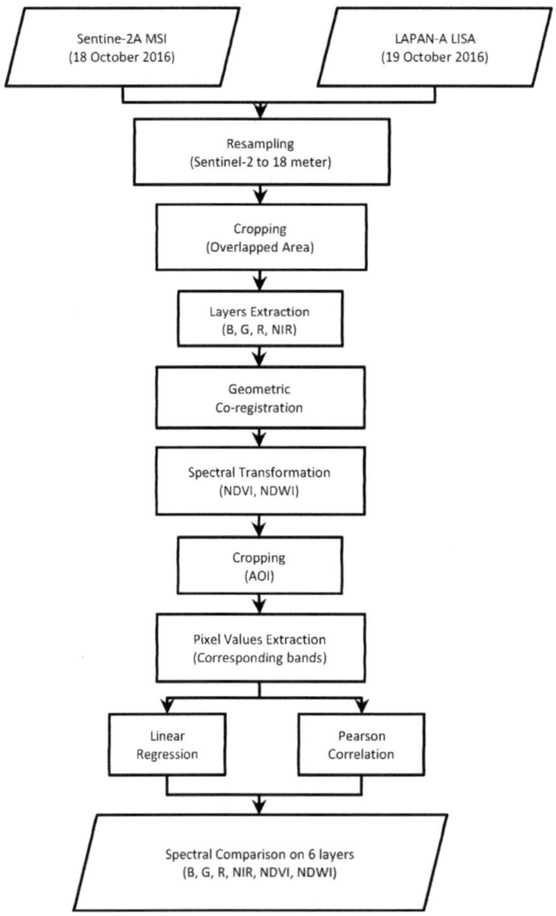


**Figure 2**. Workflow of the study

## 3. Results and Discussion

The co-registration between the corresponding bands was performed using 25-30 tie-points. The total RMSE on the tie-points coordinates was ranging from 5 to 9 meters (less than LAPAN-A3 pixel size). These values are sufficient in order to do pixel-by-pixel comparison for the corresponding data [5][6][16][17]. Each RMSE for the corresponding bands are shown in Table 2. Finally, the Sentinel-2A data were resampled to LAPAN-A3 spatial resolution (18 m) with nearest neighbour algorithm.

**Table 2.** RMSE for each corresponding band

| Bands | | RMSE |
|---|---|---|
| LAPAN-A3 | Sentinel-2A | (meter) |
| 1 | 2 | 6.5 |
| 2 | 3 | 8.8 |
| 3 | 4 | 4.9 |
| 4 | 8 | 9.2 |

Six pairs of corresponding band were generated for each AOI. Linear regression and the correlation coefficient of the corresponding bands over these three AOIs were calculated. The results are as follows.

**Table 3.** Spectral comparison between LAPAN-A3 and Sentinel-2A data on three different AOIs

| Band | | AOI | Linear Regression | | Pearson Correlation Coefficient |
|---|---|---|---|---|---|
| LAPAN-A3 | Sentinel-2A | | Intercept | Slope | |
| 1 | 2 | 1 | -309.09 | 0.1446 | 0.582 |
| | | 2 | - 704.96 | 0.1698 | 0.514 |
| | | 3 | 51.534 | 0.1266 | 0.522 |
| 2 | 3 | 1 | -15.634 | 0.0454 | 0.544 |
| | | 2 | - 84.356 | 0.0457 | 0.398 |
| | | 3 | 59.672 | 0.0546 | 0.479 |
| 3 | 4 | 1 | -122.54 | 0.054 | 0.528 |
| | | 2 | - 149.75 | 0.0506 | 0.530 |
| | | 3 | 104.62 | 0.0514 | 0.420 |
| 4 | 8 | 1 | -189.83 | 0.1082 | 0.557 |
| | | 2 | - 962 | 0.1259 | 0.548 |
| | | 3 | 588.26 | 0.1256 | 0.534 |
| NDVI (3,4) | NDVI (4,8) | 1 | 0.298 | 0.9255 | 0.599 |
| | | 2 | 0.3228 | 1.1387 | 0.655 |
| | | 3 | 0.2763 | 1.0253 | 0.527 |
| NDWI (2,4) | NDWI (3,8) | 1 | 0.2906 | 0.2625 | 0.449 |
| | | 2 | - 0.3546 | 1.1964 | 0.655 |
| | | 3 | - 0.3135 | 1.0156 | 0.578 |

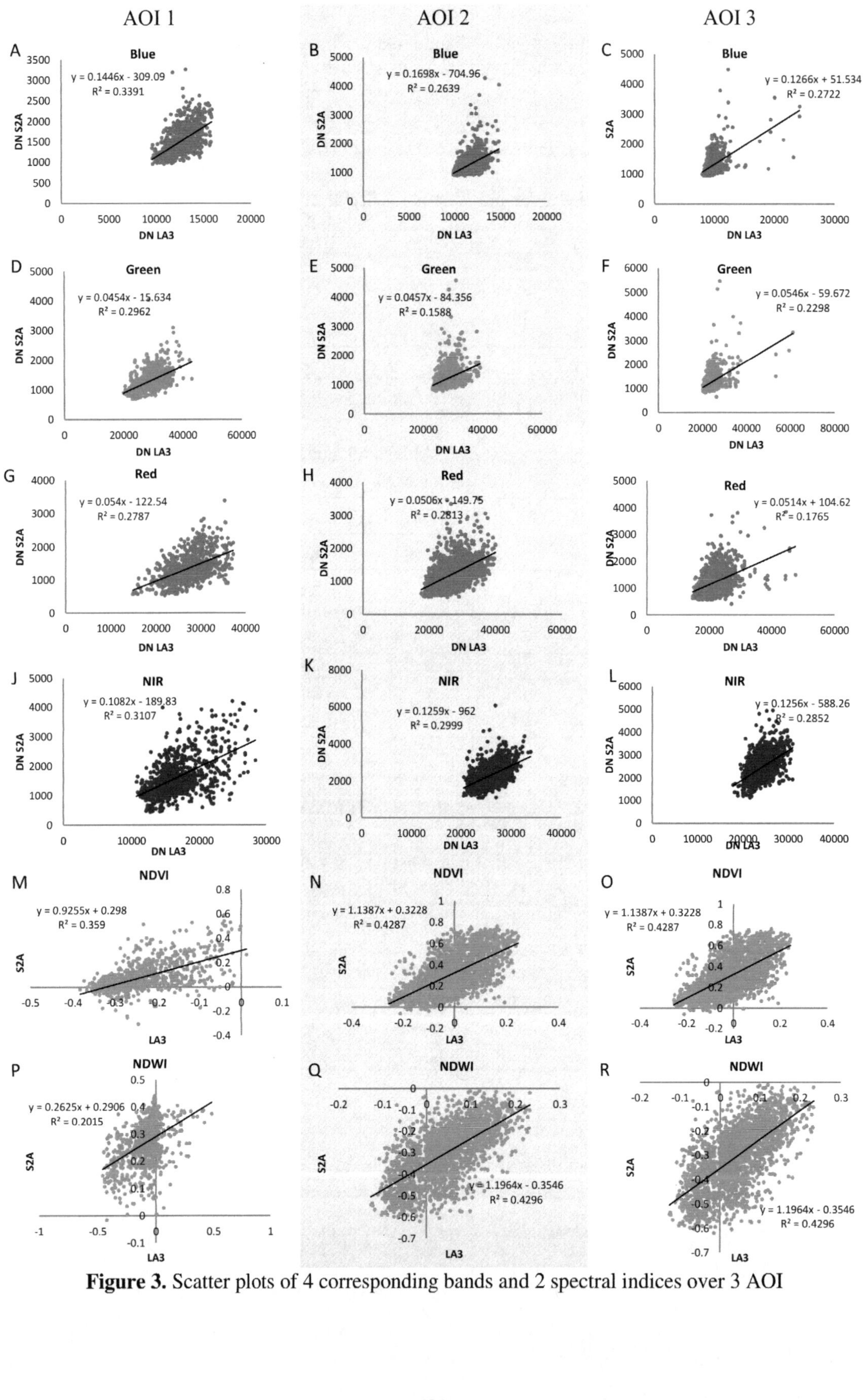


**Figure 3.** Scatter plots of 4 corresponding bands and 2 spectral indices over 3 AOI

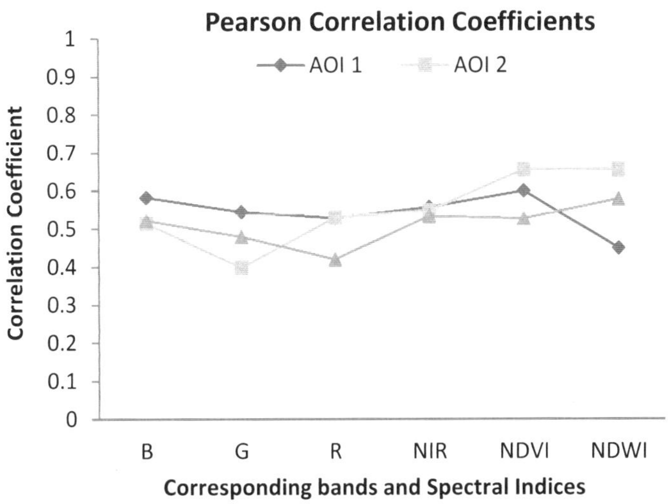


**Figure 4.** Pearson Correlation Coefficients for 4 corresponding bands and 2 spectral indices for three AOIS

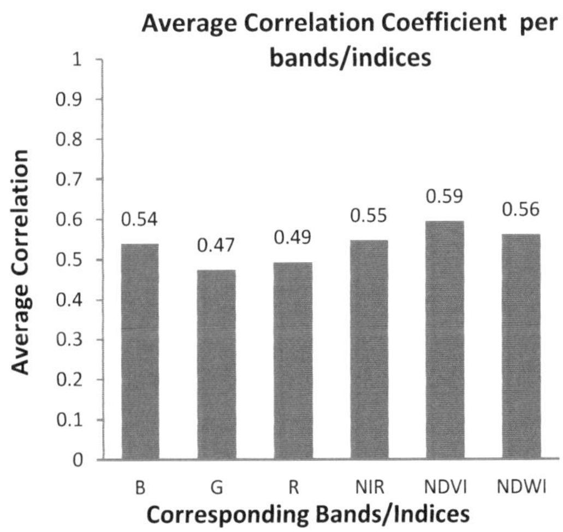


**Figure 5.** Average Correlation Coefficient within 3 AOI per corresponding bands and indices.

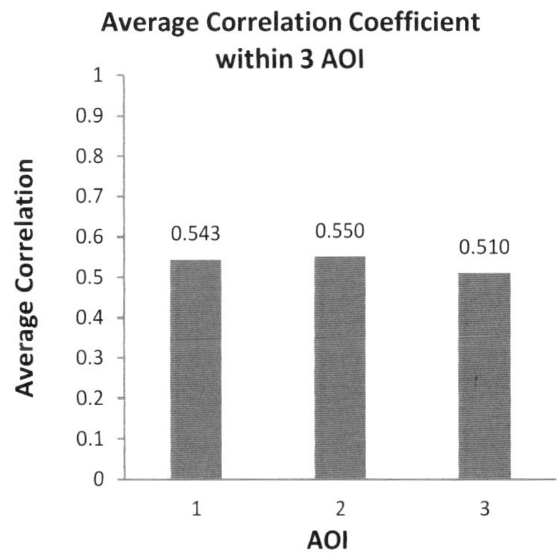


**Figure 6.** Average Correlation Coefficient for three AOI.

Linear models for three AOI are shown in figure 4. Visually, linear correlation can be seen on all of the corresponding spectral indices, except for NDWI at AOI 1 (Figure 3.P). Blue and green band on AOI 3 yield less linear correlation compared to the other (Figure 3B, 3C, 3E, 3F, 3I). The highest correlation coefficient was found on AOI 2 with NDWI and NDVI both scored 0.655 (table 2). Table 2 shows a complete correlation value. The lowest correlation coefficient was also found on AOI 2 with green band correlation at 0.398. Average correlation coefficient for three AOI gave moderate value (> 0.5) (figure 6). Highest average correlation for all three AOIs was found on NDVI at 0.59. Over three AOIs, red and green band gave less than 0.5 correlation, with the latter was at the lowest with 0.47.

Correlations on NIR band for all three AOI are relatively constant. This was expected as the near infrared band with longer wavelength was less affected by atmospheric condition compared to visible bands, as explained by [18][19]. NDWI correlation on ROI 1 was poorer due to the existence of canal within the AOI. Upon visual evaluation on very high resolution Geo-eye data, this canal is a highly dynamic in terms of its current. Coupled with the difference in acquisition time, this would introduce an error within NDWI value. As reported by [20], dynamic water would cause high variability in the reflectance value of green and infrared band. A calm, clear and deep open waterbody might fix this problem as it will reduce the aforementioned variability.

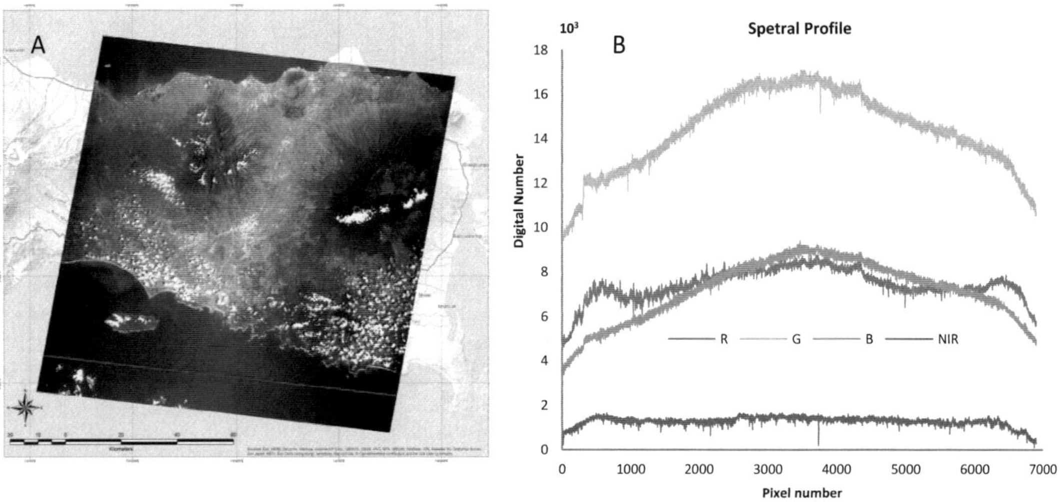


**Figure 7.** Uneven illumination across-track of LAPAN-A3 LISA Scene. (A) Visual evaluation for uneven illumination over LAPAN-A3 LISA's whole scene. Red line indicates the transect line for spectral profile drawn from west to east. (B) Spectral profile over the transect line.

It is also worth to note that the spectral indices gave higher average value compared to the original bands as shown by figure 5. It is slightly different with [1], where they found how correlation coefficient may decrease when the number of bands involved increase, due to noises. To explain this phenomenon, it is better to look at the bands involved in the spectral indices that being used. Our study uses only two bands, for each index. Both of them used NIR band, which was less affected by atmospheric condition. In the other hand, [1] compared the FII (ferrous iron index) [21] which involved two more bands than NDVI or NDWI.

The 1-day difference on the acquisition time should also be carefully considered as it may cause some differences in radiometric value because of different atmosphere and illumination conditions. This should be minimized on our future work by using the same day image pair.

For our future works, we will try to makes the same day comparison between two datasets, provided that we managed to get a relatively cloud free data over different data scenes covered different area in Indonesia. The above explanation also emphasizes how the selection of test area should carefully be evaluated. A relatively homogenous area, located in flat topography, or calm, clear water body are great test site to perform spectral comparison.

This study found an issue regarding LAPAN-A3 LISA's radiometric quality that needs to be addressed in order to fully utilize multispectral data acquired by the satellite. As shown in Figure 8, LAPAN-A3 suffers an uneven cross-track illumination (CTI). This type of error is a systematic error, which means that it can be resolved further along the way, provided that all the necessary parameters were acquired. Furthermore, several approachs have also been developed in order to correct this error[22][23][24]. While this approach works well with Hyperion data, its performance on LAPAN-A3 are still needs to be further evaluated. Alternatively, a relative radiometric normalization (RRN) could be applied to the image prior to spectral comparison process. [15] described these several RRN method, specifically on the methods that only used reference-subject image pairs. As mentioned in the introduction, this will be included on our future research.

The initial findings can be used as a reference point on how LAPAN-A3 LISA imagery correlates to other well-established medium resolution EO satellite data. A better radiometric correction, such as converting digital number to at-satellite reflectance would reduce the variation caused by sunlight and atmosphere [2]. [25] further explained that this process would reduce the discrepancies between multi-platform satellite data acquired at different imaging times in terms of radiant quantities of incidence and reflections. Thus, would improve the correlation between the two sensors. This information also important as it can be used as a reference on choosing the appropriate method on utilizing LAPAN-A3 data. A landuse/landcover extraction where the training data extracted from the image itself is less affected by the difference, rather than imported from another image [26]. Another potential use is a change detection analysis where the images are analysed independently using post-classification change detection method [19][27][28].

## 4. Conclusions

The study provides an initial evaluation of LAPAN-A3 spectral characteristics compared to Sentinel-2A MSI data. This paper found moderate correlation, ranging from 0.39 to 0.65, between LAPAN-A3 and Sentinel-2A spectral information on visible-near infrared spectrum with one-day time gap. Near infrared and blue band performed better compared to two other bands. The use of spectral indices, particularly NDVI and NDWI can increase the correlation coefficient. A careful examination on the test area should be considered as it can affect the correlation coefficient. This initial finding will act as a reference point on how to better utilize LAPAN-A3 spectral information. However, several problems were found within LAPAN-A3 regarding the striping effect as well as uneven illumination across whole scene that needs to be addressed before LAPAN-A3 multispectral data can be fully utilized.

## Acknowledgments

Authors would like to say thank you to LAPAN's Satellite Technology Center for delivering LAPAN-A3 data. Comments received from Prof. Lilik Budi Prasetyo and Dr. Yudi Setiawan of Institut Pertanian Bogor (IPB) are also sincerely appreciated. This research is funded by Remote Sensing Applications Center, Indonesian National Institute of Aeronautics and Space (LAPAN).

## References

[1] Mandanici E and G Bitelli 2016 Preliminary Comparison of Sentinel-2 and Landsat 8 Imagery for a Combined Use. *Remote Sens* [Internet] **8**(12) p 1014. Available from: http://www.mdpi.com/2072-4292/8/12/1014
[2] Li G, Li X, Li G, Wen W, Wang H, Chen L, et al 2013 Comparison of spectral characteristics between China HJ1-CCD and landsat 5 TM imagery. *IEEE J Sel Top Appl Earth Obs Remote Sens* **6**(1) pp 139–48.
[3] Wang L, Yang R, Tian Q, Yang Y, Zhou Y, Sun Y, et al 2015 Comparative analysis of GF-1 WFV, ZY-3 MUX, and HJ-1 CCD sensor data for grassland monitoring applications *Remote Sens* **7**(2) pp 2089–108.
[4] Quintanilha JA, Filho LE and Beltrame AMK 2007 Spatial and spectral comparison among IKONOS, CBERS and ASTER images to identify and detect land occupation changes around urban railway in Sao Paulo - Brazil *Int Geosci Remote Sens Symp* pp 659–62.
[5] Phongaksorn N, Tripathi NK, Kumar S, Soni P 2012 Inter-sensor comparison between THEOS and landsat 5 TM data in a study of two crops related to biofuel in Thailand. *Remote Sens* **4**(2) pp 354–76.
[6] Hill J and Aifadopoulou D 1990 Comparative analysis of landsat-5 TM and SPOT HRV-1 data for use in multiple sensor approaches *Remote Sens Environ* **34**(1 pp :55–70.
[7] van der Werff H and van der Meer F 2016 Sentinel-2A MSI and Landsat 8 OLI provide data continuity for geological remote sensing *Remote Sens* **8**(11).
[8] ESA 2015 *SENTINEL-2 User Handbook. European Space Agency* 64 p.
[9] Hakim PR and Permala R 2017 Analysis of LAPAN-IPB image lossless compression using differential pulse code modulation and huffman coding. In: *IOP Conference Series: Earth and Environmental Science* [Internet]. IOP Publishing; [cited 2017 Mar 6]. p. 12096. Available

from: http://stacks.iop.org/1755-1315/54/i=1/a=012096?key=crossref.1a61d1a04f592578ef866b15f20559b9
[10] Hasbi W and Suhermanto 2013 Development of LAPAN-A3 / IPB Satellite an Experimental Remote Sensing Microsatellite. In: *34th Asian Conference on Remote Sensing 2013*, ACRS 2013 [Internet]. Bali p. 8. Available from: http://a-a-r-s.org/acrs/index.php/acrs/acrs-overview/proceedings-1?view=publication&task=show&id=1365
[11] Hall FG, Strebel DE, Nickeson JE and Goetz SJ 1991 Radiometric rectification: Toward a common radiometric response among multidate, multisensor images. *Remote Sens Environ* [Internet]. Jan [cited 2017 Jul 17];35(1):11–27. Available from: http://linkinghub.elsevier.com/retrieve/pii/003442579190062B
[12] Parker J a, Kenyon R V and Troxel DE 1983 Comparison of interpolation methods for image resampling *IEEE Trans Med Imaging* [Internet]. **2**(1) pp 31–9. Available from: http://www.ncbi.nlm.nih.gov/pubmed/18230471
[13] Schott JR, Salvaggio C and Volchok WJ 1988 Radiometric scene normalization using pseudoinvariant features *Remote Sens Environ* **26**(1) pp 1–16.
[14] McFeeters SK 1996 The use of the Normalized Difference Water Index (NDWI) in the delineation of open water features *Int J Remote Sens* **17**(7) pp 1425–32.
[15] Yang XX and Lo CP 2000 Relative radiometric normalization performance for change detection from multi-date satellite images *Photogramm Eng Remote Sensing* **66**(August) pp 967–80.
[16] Yuan D and Elvidge CD 1996 Comparison of relative radiometric normalization techniques. *ISPRS J Photogramm Remote Sens* **51**(3) pp 117–26.
[17] Chauhan HB and Dwivedi RM 2008 Inter sensor comparison between RESOURCESAT LISS III, LISS IV and AWiFS with reference to coastal landuse/landcover studies. *Int J Appl Earth Obs Geoinf* **10**(2) pp 181–5.
[18] Lillesand TM, Kiefer RW and Chipman JW 2008 *Remote sensing and image interpretation*. 6th ed. (Minnesota: John Wiley & Sons) 756 p.
[19] Jensen JR 2005 *Introductory Digital Image Processing: A Remote Sensing Perspective*. 3rd Ed. Pearson Education
[20] Phan MH, Reniers A, Ye Q and Stivei M 2016 Response of the Mekong deltaic coast to its changing sediment sources and sinks. In: *Coastal Dynamis* 2017 [Internet]. Helsingor, Denmark: TU Delft; 2016. Available from: https://data.4tu.nl/repository/uuid:2d1a8002-53a8-4b60-93cc-0b2dac94af05
[21] Aleks K and Oliver S 2004 *ASTER Mineral Index Processing Manual. Remote Sensing Applications*, Geoscience Australia. Canberra, ACT, Australia.
[22] B. Taner San and M. Lütf I Süzen 2011 Evaluation of cross-track illumination in EO-1 Hyperion imagery for lithological mapping *Int J Remote Sens* **32**(22) pp 7873–89.
[23] Pearlman JS, Barry PS, Segal CC, Shepanski J, Beiso D and Carman SL 2003 Hyperion, a space-based imaging spectrometer *IEEE Trans Geosci Remote Sens*41(6 PART I) pp 1160–73.
[24] Goodenough DG, Dyk A, Niemann KO, Pearlman JS, Chen H, Han T, et al 2003 Processing Hyperion and ALI for forest classification *IEEE Trans Geosci Remote Sens* **41**(6 PART I) pp 1321–31.
[25] Teillet PM, Barker JL, Markham BL, Irish RR, Fedosejevs G and Storey JC 2001 Radiometric cross-calibration of the Landsat-7 ETM+ and Landsat-5 TM sensors based on tandem data sets *Remote Sens Environ* **78**(1–2) pp 39–54.
[26] Lu D, Mausel P, Brondízios E e and Moran E 2003 Change detection techniques *Int J Remote Sens* **25**(12) pp 2365–407.
[27] Singh A 1989 Review Article: Digital change detection techniques using remotely-sensed data *Int J Remote Sens* **10**(6) pp 989–1003.
[28] Foody GMG 1996 Approaches for the production and evaluation of fuzzy land cover classifications from remotely sensed data *Int J Remote Sens* [Internet] **17**(1996) pp 1317–40. Available from: file:///C:/Documents and Settings/nadia.rochdi/My

Documents/Dropbox/NWT_BiBlio/DOC/ePrints-2014-ijrs96.pdf

# Analysis of Horizontal Accuracy for Large Scale Rural Mapping Using Rotary Wing UAV Image

**Husnul Hidayat*, Bangun Muljo Sukojo**

Department of Geomatics Engineering, ITS, Surabaya

hidayat@geodesy.its.ac.id

**Abstract**. In order to fulfil the need of large scale map for rural mapping, Indonesian government is now looking for alternatives of geospatial data sources. With newly developed technology, nowadays rotary wing UAV can be used to acquire very high resolution aerial imagery quickly with low cost. This research assesses the horizontal accuracy of rural mapping in 1:2000 scale using orthophoto derived from rotary wing UAV image data. The test site of this research is Kebonwaris village, Pandaan, East Java which has an area of approximately 167 hectares. Image data was taken with approximately 80% overlap each other and processed using Structure from Motion approach. Twelve GCPs coordinates were measured using differential GPS observations for georeferencing purpose. For accuracy assessment, 22 test points were established and their coordinates were measured using static differential GPS observations. The results show that the mean absolute horizontal errors are 0.071 m and 0.142 m for easting and northing respectively and the Root Mean Square Errors are 0.088 m and 0.169 m for easting and northing respectively. These RMSE values represents horizontal RMSE 0.190 m. According to Peraturan Kepala BIG No. 15 Tahun 2014, this RMSE value represents the 0.289 m horizontal accuracy based on CE90 criterion. Therefore, with this level of accuracy the UAV image data can be used to make a class 1 base map in 1:2500 scale or class 2 base map in 1:1000 scale.

**Keywords**: UAV, rural, accuracy, mapping

## 1. Introduction

### *1.1. Background*

Indonesian government is now seeking for alternatives of geospatial data acquisition for rural mapping. On the Undang-undang Nomor 14 Tahun 2014 Pasal 8 Ayat 3 states that the location and administrative boundary of a village must be shown in the form of rural or village map. Currently the most detailed official topographic map released by Badan Informasi Geospasial (BIG, Geospatial Information Agency) are available only in 1:10000th scale. On the other hand, Peraturan Kepala BIG Nomor 3 Tahun 2016 [1] states that the rural map must be presented in 1:2500th, 1:5000th, or 1:10000th scale according to the total area and boundary extent of the village. Since the size of many villages are varying, for small villages there are no data for producing any map with 1:2500th or 1:5000th scale.

An alternative for this problem is high resolution optical satellite imagery. But in reality this image is not cloud-free at all, mainly in tropical region which is prone to rain. Besides that, the generated image is not truly orthographic imagery mainly in hilly and mountainous region. This effect is known as relief displacement and can only be corrected by using Digital Elevation Model (DEM) data. Nowadays the worldwide available DEM data is only in 30-meter resolution which is not accurate for correcting optical imagery with submeter resolution.

Recent development in robotic and automation technology have widener the possibility of large scale photogrammetric mapping using Unmanned Aerial Vehicle (UAV). The UAV can be defined as any aerial platform which can be operated without a pilot sitting in it. This platform can be remotely controlled or fly autonomously with preprogrammed flight trajectory [2]. One advantage of UAV image data is the ability to extract topographic information along with land cover information. This platform is effective for mapping an area less than 1000 ha [3]. This platform has been used in many ways, such as disaster monitoring [4][5], archaeological documentation [6][7][8], and high accuracy topographic mapping [9][3].

### *1.2. Problems*

One type of popular UAV is rotary wing UAV. This platform is more preferred due to ease of use, great manoeuvrability, and its ability to take off without special runway. But main the drawbacks of this platform are limited operational range which requires more flight and limited payload. Most of consumer-grade UAV are equipped with small format nonmetric camera which have lower image quality compared to metric camera image. Therefore, the final product of orthophoto derived from UAV images must be assessed mainly in horizontal accuracy. This accuracy test is fundamental to determine the scale of the most detailed map which can be derived from this type of imagery.

The main problems of this research is how to create orthophoto mosaic from rotary-wing UAV image data and assessing its horizontal accuracy for large scale map production. The study location is limited only in Kebonwaris Village, Pandaan, East Java. Image acquisition is only using rotary-wing UAV and ground measurements for georeferencing and accuracy testing were conducted using Global Positioning System. This data acquisition method will be described more in next section.

## 2. Methodology

### *2.1. Research location*

The test site of this research is Kebonwaris Village, District of Pandaan, Pasuruan Regency, East Java. It is located about 60 km from Surabaya to the south and can be reached within 1 hour driving via Surabaya-Malang arterial street or Gempol-Pandaan toll road. The village has mean elevation about + 150 m above mean sea level with an area ± 167 hectares and ± 10 km perimeter and mainly dominated by agricultural land, rural settlements, and vegetation.

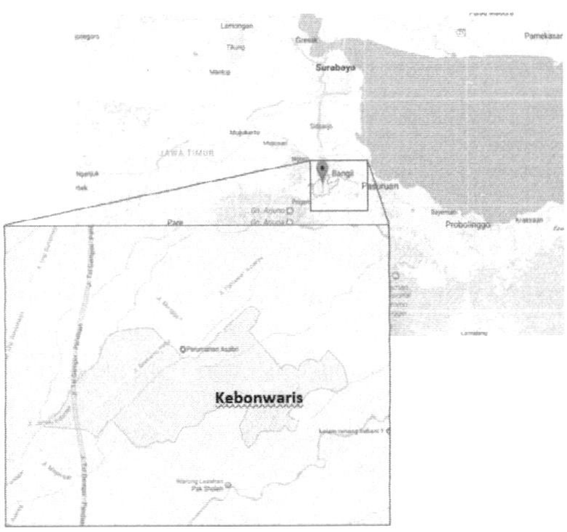


**Figure 1**. Research location (http://maps.google.com)

IOP Conf. Series: Earth and Environmental Science **98** (2017) 012052 doi:10.1088/1755-1315/98/1/012052

### *2.2. Aerial image acquisition*

The main instrument for image acquisition is a consumer grade UAV namely DJI Phantom 3. This platform was chosen because of its ease of use in operation. In addition, it has a built-in camera, GPS antenna, and compass. Therefore, the operator can monitor the position, orientation, and image stream from the platform anytime during the flight. The main specifications of this platform is as follows:

**Figure 2**. DJI Phantom 3 (www.dji.com)

**Table 1.** DJI Phantom 3 specifications

| Information | Description |
|---|---|
| Weight | 1280 g |
| Maximum speed | 16 m/s |
| Maximum flight time | 23 minutes |
| Hovering accuracy | $\pm$ 0.5 m |
| Maximum cruising range | 3.5 km |
| Sensor size | 1/2.3" CMOS |
| Image size | 4000x3000 pixels |
| Focal length/35 mm focal length | 4 mm/20 mm |
| File format | JPEG, DNG (raw) |
| Storage | Micro SD card up to 64 GB |

The platform was operated by a pilot and the assistant by manually controlling the movement of the aircraft. Cruising altitude was set to 150 m above ground level and camera shutter was released every 5 seconds. The pilot attempted to achieve at least 80% overlap between images. To ensure sufficient overlap between photos and checking area coverage the pilot must carefully watch the monitor. The assistant helped the pilot to spot the aircraft in the sky and guiding the pilot for landing, since in remote area there is no perfect surface for landing. Due to limited flight time, image acquisition was performed in several missions. Each mission has its own take off and landing spot to optimize image coverage. In this research the total number of acquired images is 2528.

**Figure 3**. UAV operator: pilot and assistant

*2.3. Ground control measurement*

To derive final georeferenced raster image, the presence of Ground Control Point (GCP) is required. The GCPs were established across the study area to ensure same horizontal accuracy of final product in all study area. There were 12 GCPs on the study area. All points were measured using differential Global Positioning System with 5 seconds data interval during at least 30 minutes. This will ensure that the baselines have sufficient data for differential GPS processing. All GCPs final coordinates are presented in UTM Zone 49S with WGS 1984 reference ellipsoid.

**Figure 4**. GCP distribution at research area

For GCP identification on the images, a large premark with 1-m length was installed on each GCP. This ensures that the GCPs will be easily identified especially at the places where natural target such as road corners or signs do not exist.

*2.4. Image orientation and Georeferencing*

The main data processing strategy in this research is Structure from Motion (SfM) approach. This approach is selected because of its ability to reconstruct camera pose and 3D object simultaneously and automatically. The SfM approach can handle many unordered sets of images and doesn't require regularities in image blocks. Therefore, this method is very suitable for this research since the manual control of UAV unit can result irregularities in image blocks.

Since the UAV has a GPS, the acquired images also have geographic coordinate information. This position data recorded from the UAV can be used to orient the stereo model into approximated absolute coordinate system. But since the accuracy of GPS data from the UAV hasn't reached centimetre level, the final georeferencing process must be done based on the GCPs coordinates. In this research, the accuracy of final georeferencing process is 0.030 m. This accuracy is sufficient since the ground sample distance of the raw image is 0.064 m.

*2.5. Photogrammetric restitution*

After the images have been oriented properly in absolute coordinate system, the 3D reconstruction process can be started. In this step the object geometry is reconstructed with higher detail. This process mainly relies on space intersection and digital image matching technique. By knowing the exterior orientation parameters of the images and corresponding image point in other images, the 3D position many points can be determined. To speed up the processing, digital image matching technique only applied along the epipolar lines on other images. The epipolar lines were generated from exterior orientation parameters.

**Figure 5.** Dense point cloud

The result of this step is 3D model of research area with fine details. This 3D model represents topographic features such as flat land, inclined land, riverbanks, etc. The high vegetation and manmade features such as building, houses, etc are also presented well. The elevation information of these objects is very important in eliminating relief displacement to produce accurate orthophoto.

*2.6. DSM and orthophoto generation*

The resulting dense point cloud from photogrammetric restitution has represented the 3D appearance of research area. But it is still presented in discrete data that must be interpolated to be 3D surface model. This 3D surface model is essential for Digital Surface Model and orthophoto generation. The 3D surface model is very essential for removing relief displacement.

**Figure 6.** Digital Surface Model

Due to various lighting condition and sun elevation, there are several inconsistencies of image brightness in some regions. Higher sun elevation combined with reflective surface such as water bodies can cause strong reflection which makes some flares at final orthophoto. To avoid this, the sun's reflections on water bodies were excluded in texture blending process.

## 3. Result and Discussion

### *3.1. Image acquisition and flight trajectory*

This research has acquired totally 2528 images of research area. The redundant images ensure that all area have been covered well. But in further processing some images were filtered out due to blurring and excessive overlap which can slow down the processing and decreasing the image sharpness. Due to manual control of platform's movement, the resulting image block and flight trajectory becomes irregular. But in terms of image coverage and overlap the resulting images are sufficient enough. All area is covered by at least 4 images.

**Figure 7.** Image coverage, overlap, and flight trajectory

Flying the platform by relying on pilot's skill is very difficult in order to get regular image blocks and flight trajectory. Beside that it needs more effort to ensure that all study area has good image coverage and overlap. Notice the area which has less overlap on the north-eastern part of the area. Therefore, in the future the presence of automated flight planning will become more important.

### *3.2. Orthophoto of research area*

Although the ground sample distance is up to 6 cm resolution, the final orthorectified image was rendered at 10 cm resolution for storage and memory efficiency. This orthorectified image covers all area of Kebonwaris Village and has good visual details. There are no significant variations in image brightness. Many appereances of manmade structures have been corrected from relief displacement. Thus, this image is ready for further usage such as map production.

**Figure 8**. Orthophoto of research area

### *3.3. Horizontal accuracy analysis*

In order to analyse the horizontal accuracy of orthophoto, the coordinate measurement of some Independent Check Points is required. The check points are distributed across the research area. In this research there are 22 check points whose coordinates are measured by using differential GPS technique with at least 15 minutes observation with 5 seconds data interval. Then, the resulting coordinate from GPS measurement is compared against coordinate measurement from the orthophoto. The distribution of these check points is shown in Figure 9.

**Figure 9**. Distribution of Independent Check Points

The results show that the mean absolute horizontal errors are 0.071 m and 0.142 m for easting and northing respectively. These horizontal errors are ranging from -0.198 m to 0.089 m and from -0.256 to 0.335 for easting and northing respectively. These absolute horizontal errors are corresponding to resultant error 0.170 m. The Root Mean Square Error (RMSE) of easting and northing component are 0.088 m and 0.169 m respectively. These values correspond to horizontal RMSE 0.190 m. Compared to the ground sample distance of raw image, i.e. 0.064 m, the reached RMSE value is 3 times of GSD. This value can be lowered by using pre-calibrated camera sensor for more optimal lens distortion removal.

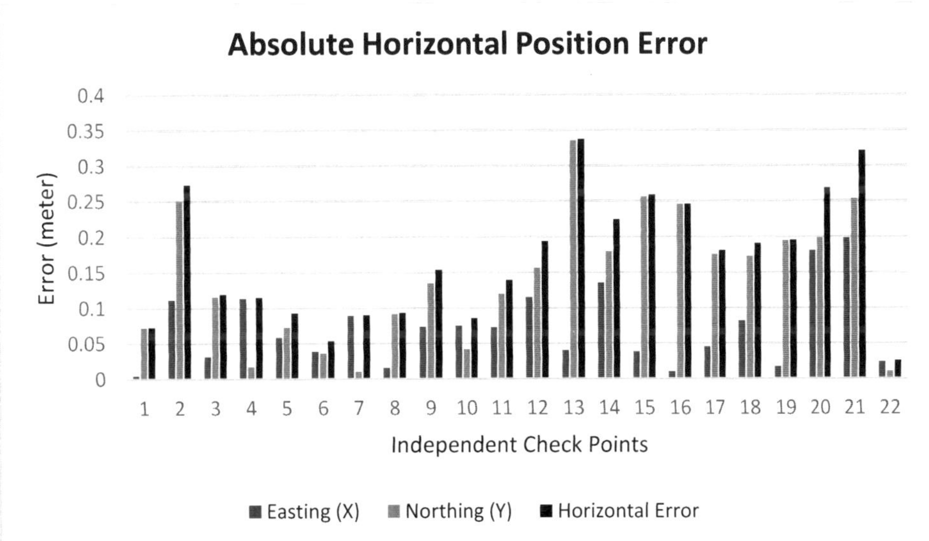


**Figure 10.** Horizontal position error

According to Peraturan Kepala BIG No. 15 Tahun 2014 [10], one of criteria for assessing the horizontal accuracy of mapping product is CE90 criterion. In this document the CE90 criterion is simply defined as:

$$CE\ 90 = 1.5175 \times RMSE_r$$

where $RMSE_r$ is the overall horizontal Root Mean Square Error. Based on this formula the CE 90 value in this research is 0.289 m. This value is acceptable for making a base map in 1:2500th scale with the best quality (class 1 quality) or a base map in 1:1000th scale with good quality (class 2 quality).

## 4. Conclusions

In this research the method of large scale mapping for rural map production has been demonstrated. The use of rotary wing UAV could produce acceptable result according to current standard by Badan Informasi Geospatial. By using this method, the horizontal accuracy could reach 0.289 m and it is acceptable for making base map in 1:2500th scale or class 2 base map in 1:1000th scale. Some improvements should be made in order to achieve more efficient process and better results. The use of flight planning becomes very important in order to ensure overall coverage of mapped area. Pre-calibrated camera sensor should be used in next research to achieve an accuracy nearly with desired Ground Sample Distance.

## Acknowledgement

Authors would like thank to Lembaga Penelitian dan Pengabdian Masyarakat (LPPM) Institut Teknologi Sepuluh November for supporting the research funding, and also for Geodesy and Surveying Laboratory, and Laboratory of Cadastre and Land Policy for field data acquisition support.

## References

[1] Badan Informasi Geospasial. 2016. Peraturan Kepala Badan Informasi Geospasial Nomor 3 Tahun 2016 tentang Spesifikasi Teknis Penyajian Peta Desa. Bogor: Badan Informasi Geospasial.

[2] Eisenbeiss, H. 2009. UAV Photogrammetry. Disertasi. Zürich: Swiss Federal Institute of Technology Zürich

[3] Rokhmana, C. A. 2014. Mapping the Outermost Small Islands Utilizing UAV–Based Aerial Photography. FIG Congress 2014, 16-21 Juni 2014, Kuala Lumpur, Malaysia.

[4] Li, C., dkk. 2011. Quick Image Processing Method of UAV Without Control Points Data in Earthquake Disaster Area. Transactions of Nonferrous Metals Society of China 21 (2011) s523-s528.

[5] Niethammer, U., dkk. 2012. UAV-based Remote Sensing of The Super-Sauze Landslide: Evaluation and Results. Engineering Geology 128 (2012) 2-11.

[6] Seitz, C., dan Altenbach, H. 2011. Project Archeye-The Quadrocopter as The Archaeologist's Eye. International Archives of Photogrammetry, Remote Sensing, and Spatial Information Sciences, Vol. XXXVIII-1/C22, UAV-g 2011, Conference on Unmanned Aerial Vehicle in Geomatics, Zurich, Switzerland.

[7] Mozas-Calvache, A. T., dkk. 2012. Method for Photogrammetric Surveying of Archaeological Sites with Light Aerial Platforms. Journal of Archaeological Science 39 (2012) 521-530.

[8] Lo Brutto, M., dkk. 2014. UAV Platforms for Cultural Heritage: First Results. ISPRS Annals of Photogrammetry, Remote Sensing, and Spatial Information Sciences, Volume II-5, 2014, ISPRS Technical Commission V Symposium, 23-25 Juni 2014, Riva del Garda, Italia.

[9] Mancini, F., dkk. Using Unmanned Aerial Vehicles (UAV) for High-Resolution Reconstruction of Topography: The Structure from Motion Approach on Coastal Environments. Remote Sensing 2013, 5, 6880-6898; doi: 10.3390/rs5126880.

[10] Badan Informasi Geospasial. 2014. Peraturan Kepala Badan Informasi Geospasial Nomor 15 Tahun 2014 tentang Pedoman Teknis Ketelitian Peta Desa. Bogor: Badan Informasi Geospasial.

# Monitoring of Drought Events in Gorontalo Regency

**S Koem* and Rusiyah**

Department of Earth Science and Technology, Universitas Negeri Gorontalo, Gorontalo, 96128, Indonesia

s.koem@ung.ac.id

**Abstract.** Gorontalo Regency is a region vulnerable to drought. Drought is one of meteorological disaster because it tends to bring negative impact on various sectors. This study used rainfall data from 1981 to 2016 (35 years). The research employed Standardized Precipitation Index (SPI) to monitor and calculate the level of drought from the duration, intensity, and frequency in different time scales. The SPI value was calculated using the DrinC and ArcGIS software is used to create drought spatial distribution maps. The mean intensity of drought simultaneously followed the drought magnitude in Bilato station. The peak of drought in SPI-3 occurs in 1982, 2009 and 2016. In 1982, about 76.5% of the stations showed that the peak of drought events for SPI-3 in October to December. Moreover, 94% of the stations reveals that the peak of drought events for SPI-6 occur in July to December 1982. This shows that drought in 1982 was more severe than other years in the last three decades. Linear trends of drought for the period of 1981 to 2016 in most stations show an increasing trend, hence, it can be concluded that Gorontalo Regency experienced an increase in the wet period. Changes in time-scale caused the tendency for a high number of dry period frequencies. Drought spatial distribution could be used to determine the priority plans in finding the solutions due to droughts that occur in drought-vulnerable areas. Drought analysis using SPI could contribute to the decision-making in the future as an effort to minimize the impact of drought.

**Keywords:** drought severity, spatial analysis, standardized precipitation index

## 1. Introduction

Gorontalo Regency is one of the regencies in Gorontalo Province with highest rice production with total rice production in 2015 is about 153.515 tons and maize production in the same year is about 142.863 tons [3]. Therefore, Gorontalo Regency is the largest food producer in the province. Furthermore, Gorontalo Regency is also vulnerable to drought. Drought is one of meteorological disaster because it tends to bring negative impact on various sectors, such as in social and economic sectors as well as in other areas. The potential drought area in Gorontalo Regency is about 172.894 hectares, whereas the possible environmental damage due to drought is about 59.311 hectares [2]. The impact of the drought is economically and socially felt by the people in this regency. Therefore, efforts to assess the level of drought hazard are needed. This effort can be made by utilizing the climate data to monitor the drought events. Hence, adaptation and mitigation of drought hazard can be done.

Monitoring of drought events needs an exact calculation to assess the severity classification of the drought. Assessing the severity classification and providing information for early drought warning can support adaptation and mitigation actions of the drought hazard. Therefore, tools to be used in evaluating the level of drought severity are needed. In the recent years, meteorological drought index has become a widely used index for assessing the severity of the drought. The often-used meteorological drought index are Standardized Precipitation Index (SPI), Standardized Precipitation Evapotranspiration Index (SPEI), Palmer Drought Severity Index (PDSI), Crop Moisture Index (CMI), Surface Water Supply Index (SWAZEE), and Reconnaissance Drought Index (RDI). SPI is highly valuable meteorological drought index used to estimate the hazard level of drought [10]. Also, SPI can monitor and measure the level of drought duration, intensity, frequency, and drought probability in various time scales [12].

SPI application in monitoring and measuring drought level is straightforward because SPI only needs monthly rainfall data. Hence, in detail, SPI detects rainfall deficit in giving period. The flexibility of the time scale becomes one of the SPI advantages in monitoring short-term and long-term drought. This study aims at (1) estimating the duration, magnitude, and intensity of the drought; (2) analyzing the frequency of wet, normal, and dry periods; (3) to consider spatial distribution of drought in Gorontalo Regency.

## 2. Materials and Method

### *2.1. Climate data*

This study used rainfall data from 1981 to 2016 (35 years). Assessing the severity level of the long-term rainfall data was ($\geq$ 30 years) [8,14]. Usage of rainfall data to produce drought characteristic is trusted, because the longer the data entry period, the more accurate the produced value. The SPI value was calculated using the DrinC (Drought Indices Calculator) software developed by the Laboratory of Reclamation Works and Water Resources Management of the National Technical University of Athens, Greece [20]. Monthly rainfall data were obtained from 17 stations located all around Gorontalo Regency and 2 stations located outside of Gorontalo Regency (Figure 1). The data were collected from the Balai Wilayah Sungai Sulawesi II (Watershed Agency of Sulawesi II, henceforth called as BWS), Meteorological Station of Meteorology, Climatology, Geophysics Agency (henceforth called as BMKG) and 0.25-degree data grid of BMKG, were obtained from the sacad.database.bmkg.go.id/grid/web/ website. The additional data were obtained from the iridl.ldeo.columbia.edu/SOURCES/NOAA/ website.

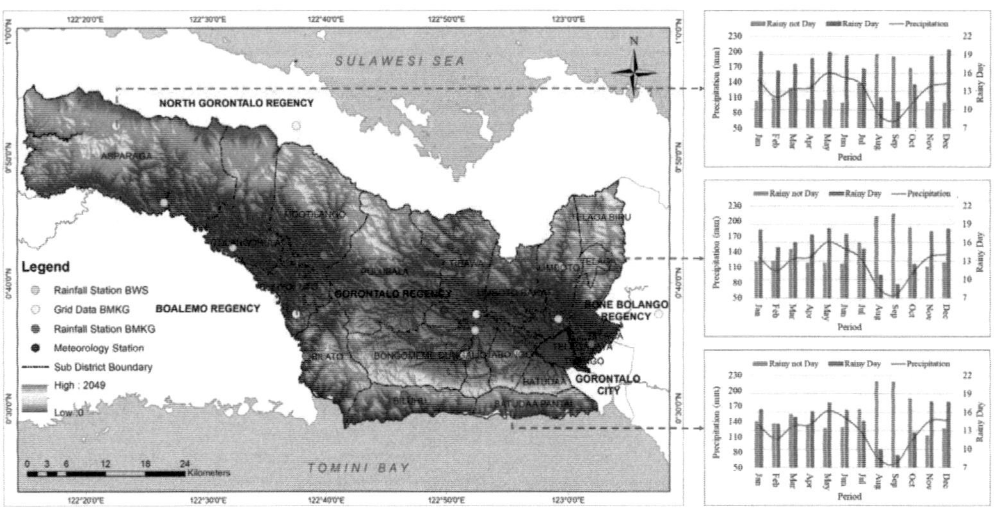


**Figure 1.** The administrative map of Gorontalo Regency and the station location (left), rainfall averages for the period of 1981 to 2016 at the Pangahu station (above), Motoduto station (middle), Isimu station (below).

*2.2. Estimated monthly rainfall*

Rainfall data obtained from some stations still contain missing data and data not available. The estimated rainfall data using Inverse Distance Weighting method (IDW), this method can be estimated as follows:

$$X_p = \frac{\sum_{i=1}^{n}\left(\frac{1}{d_i^2}.X_i\right)}{\sum_{i=1}^{n}\left(\frac{1}{d_i^2}\right)} \quad (1)$$

where Xp: the estimated value of rainfall, $X_i$: the value of rainfall at the nearest station in the same period, $d_i$: distance between the estimated station and the nearest station [13].

*2.3. Calculating the SPI*

SPI is calculated using DrinC, providing drought index calculation based on the method of the gamma distribution. The gamma distribution to fit well to the rainfall time series data [18]. Gamma distribution is a method to describe rainfall variability. Gamma distribution is a probability density function. The equations used are:

$$g(x) = \frac{1}{\beta^{\alpha}\Gamma(\alpha)} x^{\alpha-1} e^{-x/\beta}, \text{ for } x > 0 \quad (2)$$

where,

$$\Gamma(\alpha) = \int_0^{\infty} y^{\alpha-1} e^{-y} dy \quad (3)$$

where $\alpha$ and $\beta$ are the shape and scale parameters, x is the amount of rainfall and $\Gamma(\alpha)$ is a gamma function. The parameters of $\alpha$ and $\beta$ gamma probability density function of each station to estimate rainfall in the period a scale of 1, 3, 6, and 12 months.

The method used to estimate the value of the gamma distribution parameter is the maximum likelihood method is:

$$\alpha = \frac{1}{4A}\left(1 + \sqrt{1 + \frac{4A}{3}}\right), \beta = \frac{\bar{x}}{\alpha}, \text{ where } A = \ln(\bar{x}) - \frac{\sum \ln(x)}{n} \quad (4)$$

n= number of rainfall observations

The resulting parameters are then used to determine the cumulative probability value of the observed rainfall events in a month and the time period recorded on the observed station [6]. The gamma distribution is undefined (x=0) and rainfall distribution likely contains a zero value, the cumulative probability becomes:

$$H(x) = q + (1-q)G(x) \quad (5)$$

where q is the probability of zero rainfall dan G(x) is the cumulative probability of the incomplete gamma function. If *m* is the number of zeros of the rainfall time series, then *q* can be estimated by *m/n*. The cumulative probability of H (x) is then transformed to a normal standard random variable Z with a mean value of zero and a variance of one [1]. Moreover, according to Edwards [6] the standard normal random variable Z which is also the SPI value is more easily obtained by using the approach proposed by [1] which converts the cumulative probability to the normal standard random variable Z (SPI):

$$Z = SPI = -\left( t - \frac{c_0 + c_1 t + c_2 t^2}{1 + d_1 t + d_2 t^2 + d_3 t^3} \right), \; t = \sqrt{\ln\left( \frac{1}{(H(x))^2} \right)} \qquad \text{for } 0 < H(x) \leq 0.5 \qquad (6)$$

$$Z = SPI = \left( t - \frac{c_0 + c_1 t + c_2 t^2}{1 + d_1 t + d_2 t^2 + d_3 t^3} \right), \; t = \sqrt{\ln\left( \frac{1}{(1.0 - H(x))^2} \right)} \qquad \text{for } 0.5 < H(x) < 1.0 \qquad (7)$$

with, c0 = 2.515517, c1 = 0.802853, c2 = 0.010328, d1 = 1.432788, d2 = 0.189269, d3 = 0.001308. Rainfall data recorded on the observation station is not normally distributed so that the transformation will follow basically normal distributed SPI values [19]. The calculation of the SPI values at different time scales are classified based on wet and dry as shown in Table 1.

**Table 1.** Drought classification according to SPI value.

| SPI range | Drought category |
|---|---|
| 2.0 or more | Extremely wet |
| 1.5 to 1.99 | Very wet |
| 1.0 to 1.49 | Moderately wet |
| -0.99 to 0.99 | Near normal |
| -1.0 to -1.49 | Moderately dry |
| -1.5 to -1.99 | Severely dry |
| -2.0 or less | Extremely dry |

*2.4. Duration, magnitude, and intensity of drought*

Drought is identified by looking at the number of months since the first month, where drought starts if SPI values show negative values up to -1, drought finished if the SPI values showed a positive value [12][16]. Based on these indicators can be determined longest drought period is referred to as the drought duration (DD) at any rainfall station. Drought duration since the beginning of the case until the finished of the drought can be calculated the amount drought magnitude (DM). DM is defined as the accumulation of SPI values in the events of drought duration [12], DM is formulated with:

$$DM = -\left( \sum_{i=1}^{n} SPI_{ij} \right) \qquad (8)$$

where $n$ is the number of months with the occurrence of drought on the time scale $j$. Based on the value of DM and DD can be calculated the mean intensity value (MI). The mean intensity is the average accumulation of water deficit during the period of drought [15].

*2.5. Frequency and spatial distribution of drought*

Calculate the frequency of drought events using the equation $f = n/N$ x 100, where: $n$ is a number of months with drought events (negative SPI), $N$ is a number of total months [15]. Drought spatial distribution using Inverse Distance Weighted (IDW) method. For this purpose, ArcGIS software is used to create drought spatial distribution maps.

## 3. Results and Discussion

### *3.1. Description of rainfall in the research site*

Gorontalo Regency is one of the areas in the equator, where convective clouds are developed. This phenomenon has caused Gorontalo Regency to receive abundant rainfall all year round. The average annual rainfalls in Pangahu, Motoduto, and Isimu stations for the period of 1981 to 2016 were 1449, 1397 and 1436 mm with two peak patterns of rainfall. The peak of rainfall occurs during April to June and November to January, whereas the lowest rainfall occurs in August to October (Figure 1). The same pattern also found in other weather observation stations in the Gorontalo Regency era. The number of the highest rainy day was at the Pangahu station with 17 days per month, whereas, in Isimu and Motoduto stations, the number of rainy days was 15 days per month.

### *3.2. Rainfall estimation result*

The result of rainfall estimation data shows that IDW method had similarity with the data from observed rainfall, even though the estimation data showed higher or lower data than observed data on certain periods. In addition, stations density is one of the main factors in selecting the spatial interpolation data in estimating the rainfall [4]. The analysis reveals good correlation values were obtained from stations that are close to each other. The optimum distance between stations used in interpolation of rainfall data is ten to thirty kilometers [5]. In this study, the maximum distance between stations used was 20.5 km. Table 2 shows that observed monthly rainfall and rainfall estimated in Datahu, Talumelito and Pilolalenga stations show good results with correlations values of 0.84, 0.80, and 0.77 (high correlation). These correlation values ensure that estimated rainfall data using the IDW method could be used to monitor drought events in Gorontalo Regency.

**Table 2.** Correlation between the monthly rainfall observations and estimation.

| Station | $r_{value}$ | Category |
|---|---|---|
| Datahu | 0.84 | H |
| Talumelito | 0.80 | H |
| Pilolalenga | 0.77 | H |
| Batudaa | 0.77 | H |
| Tabongo | 0.71 | H |
| Mootilango | 0.68 | M |
| Biyonga | 0.67 | M |
| Hepuhulawa | 0.62 | M |
| Bilato | 0.54 | M |

H = High correlation, M = Moderate correlation

### *3.3. Drought events from 1981 to 2016*

Rainfall variability in Gorontalo Regency during 1981 to 2016 was influenced by El Nino Southern Oscillation (ENSO) phenomenon. This phenomenon has caused low rainfall in Gorontalo Regency. Hence, it triggered drought. In addition, drought had caused water deficit when the recorded data in one rainfall station had been below the standard for several months [7]. Drought analysis for the period of 1981 to 2016 had different duration, magnitude and drought intensity in various time scales. Drought duration, magnitude, and intensity identification were conducted in one, three, and six months' timescale. As shown in Table 3, the duration of SPI-1 was six to seven months, with the drought magnitude was found in Bilato station with the value of -22.2 and mean intensity of -2.3 (extremely dry). The mean intensity of drought simultaneously followed the drought magnitude in Bilato station. About 59% of the stations showed that the peak of drought in SPI-1 occurs in November 1984, regardless that in that year no El-Nino phenomenon that occurs. This indicates that SPI was able to identify below normal rainfall in short time-scale.

The SPI-3 analysis shows that the longest drought duration occurs in Tabongo and Talumelito with the value of drought magnitude in each station was -5.5 and -6.6, with a mean intensity of -1.4 (moderately dry) for Tabongo and -1.6 (severely dry) for Talumelito (Table 3). The duration of drought in Tabongo and Talumelito stations were not simultaneously followed by the magnitude and the intensity of the drought. This was because, in other stations, it was revealed that the drought

**Table 3.** Duration, magnitude, intensity and peak of drought during the period 1981 to 2016 in Gorontalo Regency

| Station | DD | | | DM | | | MI | | | P | | |
|---|---|---|---|---|---|---|---|---|---|---|---|---|
| | SPI-1 | SPI-3 | SPI-6 | SPI-1 | SPI-3 | SPI-6 | SPI-1 | SPI-3 | SPI-6 | SPI-1 | SPI-3 | SPI-6 |
| Hepuhulawa | 6 | 3 | 2 | -17.2 | -7.6 | -2.2 | -1.8 | -2.5 | -1.1 | Feb-98 | Jan-Mar 2016 | Jul-Dec 1982 |
| Biyonga | 7 | 3 | 2 | -16.9 | -6.9 | -2.7 | -1.7 | -2.3 | -1.4 | Des-15 | Oct- Dec 1982 | Jul-Dec 1982 |
| Tabongo | 7 | 4 | 2 | -17.8 | -5.5 | -3.4 | -1.7 | -1.4 | -1.7 | Nov-84 | Oct- Dec 1982 | Jul-Dec 1982 |
| Datahu | 7 | 3 | 2 | -17.5 | -6.6 | -5.5 | -1.6 | -2.2 | -2.7 | Feb-98 | Oct- Dec 2009 | Jul-Dec 1982 |
| Pilolalenga | 7 | 3 | 3 | -18.4 | -6.4 | -4.8 | -1.9 | -2.1 | -1.6 | Nov-84 | Oct- Dec 1982 | Jul-Dec 1982 |
| Lakeya | 7 | 3 | 2 | -20.6 | -9.1 | -2.1 | -2.1 | -3.0 | -1.1 | Nov-84 | Oct- Dec 1982 | Jul-Dec 1982 |
| Mohiolo | 7 | 3 | 2 | -17.4 | -6.6 | -4.8 | -1.7 | -2.2 | -2.4 | Feb-98 | Apr-Jun 2009 | Jul-Dec 1982 |
| Biluhu | 6 | 3 | 3 | -14.6 | -6.3 | -4.3 | -1.6 | -2.1 | -1.4 | Feb-98 | Jul-Sep 1982 | Jul-Dec 1982 |
| Bilato | 7 | 3 | 2 | -22.2 | -9.3 | -4.8 | -2.3 | -3.1 | -2.4 | Jul-82 | Jul-Sep 1982 | Jul-Dec 1982 |
| Talumelito | 6 | 4 | 1 | -16.0 | -6.6 | -3.8 | -1.7 | -1.6 | -3.8 | Nov-84 | Oct- Dec 1982 | Jan-Jun 2012 |
| Mootilango | 7 | 3 | 2 | -20.7 | -8.9 | -4.9 | -2.2 | -3.0 | -2.5 | Nov-84 | Oct- Dec 1982 | Jul-Dec 1982 |
| Bongomeme | 7 | 3 | 2 | -19.3 | -8.3 | -2.5 | -1.9 | -2.8 | -1.2 | Nov-84 | Oct- Dec 1982 | Jul-Dec 1982 |
| Batudaa | 6 | 3 | 5 | -16.1 | -5.9 | -8.5 | -1.7 | -2.0 | -1.7 | Feb-98 | Oct- Dec 1982 | Jul-Dec 1982 |
| Pangahu | 6 | 3 | 2 | -17.0 | -7.9 | -4.7 | -1.8 | -2.6 | -2.3 | Nov-84 | Oct- Dec 1982 | Jul-Dec 1982 |
| Motoduto | 7 | 3 | 2 | -19.5 | -7.6 | -5.0 | -1.8 | -2.5 | -2.5 | Nov-84 | Oct- Dec 1982 | Jul-Dec 1982 |
| Isimu | 7 | 3 | 2 | -21.3 | -8.9 | -2.8 | -2.1 | -3.0 | -1.4 | Nov-84 | Oct- Dec 1982 | Jul-Dec 1982 |
| Djalaluddin | 7 | 3 | 2 | -16.7 | -7.0 | -2.4 | -1.6 | -2.3 | -1.2 | Nov-84 | Oct- Dec 1982 | Jul-Dec 1982 |

DD = Drought duration, DM = Drought Magnitude, MI = Mean Intensity, P = Peak of Drought

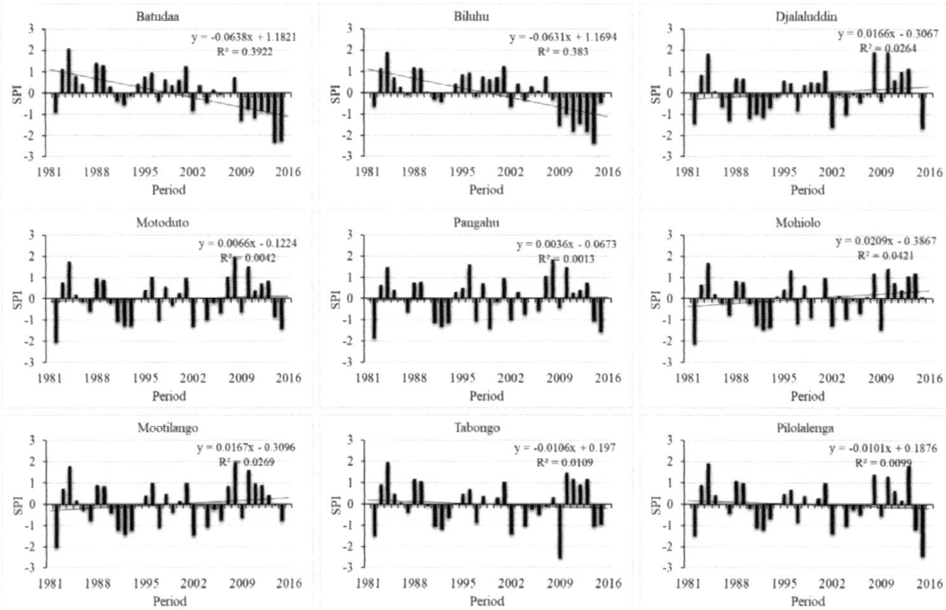


**Figure 2.** The linear trend of SPI-12 from 1981 to 2016

duration values were lower than those of Tabongo and Talumelito stations but the drought magnitude and drought intensity was higher. This indicates that the area with longer drought duration does not necessarily mean that the area is drier than other areas. As shown in Table 3, The peak of drought in SPI-3 occurs in 1982, 2009 and 2016. In 1982, about 76.5% of the stations showed that the peak of drought events for SPI-3 in October to December. Moreover, 94% of the stations reveals that the peak of drought events for SPI-6 occur in July to December 1982. This shows that drought in 1982 was more severe than other years in the last three decades. Also, based on the analysis, during 1982, all stations showed the value of SPI < -2 or within the extremely dry category.

The longest drought duration in SPI-6 occurs in Batudaa station during January 2014 to June 2016, whereas the drought magnitude of -8.5 and a mean intensity of -1.7 (severely dry). The shortest drought duration occurs in Talumelito station with the drought magnitude of -3.8 and mean intensity value of -3.8 (extremely dry), has the most severe impact on the area (Table 3). Description of duration, magnitude, and mean intensity of drought in SPI-1 and SPI-3 were different with SPI-6, where, the drought duration in SPI-6 tend to be longer whereas the drought magnitude SPI-6 tend to be lower than those in SPI-1 and SPI-3 (Table 3). These conditions indicate that rainfall accumulation would be bigger if the SPI time-scale is increased. Hence, it will influence the duration, magnitude, and intensity of the drought.

The drought magnitude and the length of drought duration were heavily influenced by the rainfall recorded in observation stations, the low rainfall intensity for an extended period of time will trigger a drought with a long duration. Figure 2 shows a number of events and level of drought severity in nine observation stations from 1981 to 2016. Analysis of drought dynamic in SPI-12 showed four drought events in Batudaa station, where two of the events was categorized as extremely dry. In Biluhu station there were six drought events recorded, one of them has been registered as extremely dry. Moreover, in Mootilango and Pilolalenga stations, there were seven drought events recorded for each station, and one of them was categorized as extremely dry, in 1982 for Mootilango station and extremely dry in 2015 for Pilolalenga station. In addition, Motoduto, Mohiolo, and Tabongo stations each recorded eight drought events, with only one drought events recorded as extremely dry for each station, namely in 1982 for Motoduto and Mohiolo stations and in 2009 for Tabongo station. Further, in Pangahu and Djalaluddin stations there were many drought events recorded, eight drought events for Djalaluddin station and nine events for Pangahu station. However, there was no drought categorized as extreme drought in these two stations from 1981 to 2016.

In general, the drought events varied in each observation station in Gorontalo Regency. Based on the SPI-12 analysis, most of the stations show that drought occurs in 1982, 1991, 1992, 2002 and 2015, with most of the station reported extremely dry events in 1982. From 1981 to 2016, the number of the dry years was more than the number of wet years. Therefore, the drought events potential for Gorontalo Regency in average was two to three times in each decade, with reference to a number of drought events in SPI-12. Nevertheless, the number of drought events depended on global warming that caused changes in rainfall and long-term climate change [21]. Linear trends of drought for the period of 1981 to 2016 in most stations show an increasing trend (Figure 2), hence, it can be concluded that Gorontalo Regency tends to increase in the wet period.

### *3.4. Frequency of drought*

Drought frequency was identified in SPI-3, SPI-6 and SPI-12 time-scale. SPI-3 analysis showed that the wet period frequency in Djalaluddin, Motoduto, and Pangahu stations were 13, 14 and 16% whereas the dry periods were 12, 11 and 15%. The normal period for SPI-3 in Djalaluddin and Motoduto stations was 75%, whereas for Pangahu station was 69%. Overall, the wet period frequency was higher than the dry period. SPI-6 time-scale showed that the dry period frequency in Motoduto and Pangahu station was higher with the percentage of 17% (Figure 3). Also, the SPI-12 analysis indicates that frequency for all wet period in Djalaluddin and Motoduto stations were 18% whereas for Pangahu station was 15%. In reverse, the incidence for the dry period in Djalaluddin and Motoduto stations were 23%, while for Pangahu station was 26%. The condition for a normal period in Djalaluddin, Motoduto and Pangahu stations show the similar frequency of 59% (Figure 3). Drought events changes as timescale changes [11].

The results are shown by SPI-3, SPI-6, and SPI-12 indicate that the longer time-scale used the higher the dry period compared to the wet period. However, the short time-scales usage (SPI-3 and SPI-6) were able to identify the level of drought more clearly and able to indicate the influence of seasonal rainfall influence toward the drought distribution. Therefore, short time-scale SPI can be used to monitor short-term drought for the agricultural sector. One of the drought characteristics was the change in drought frequency along with the change in time scale [21]. Changes in time-scale, for the short-time scale to

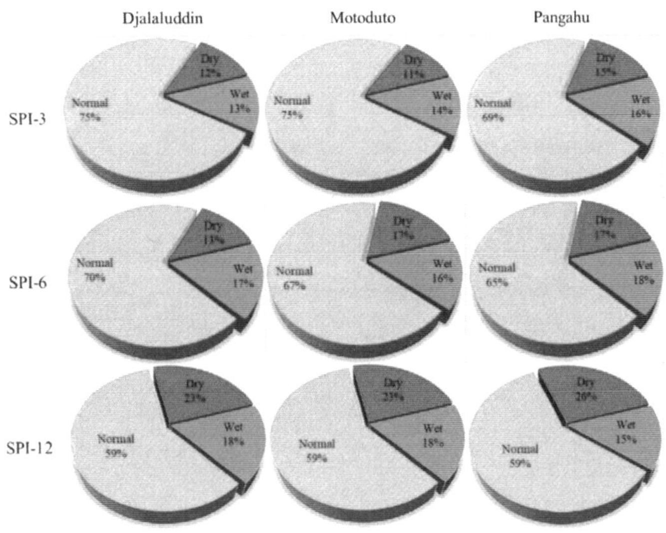


**Figure 3.** Frequency of wet, normal and dry period of SPI-3, SPI-6 and SPI-12.

longer time-scale, will be followed by the change in frequency. Changes in time-scale caused the tendency for a high number of dry period frequencies. However, the magnitude of the drought will wear off due to the increasing accumulation of rainfall in longer time-scale. Drought monitoring through SPI in different time-scale can assist in planning efficient water resource usage for farming and hydropower [21].

### *3.5. Spatial distribution of drought*

The value of SPI in all stations in Gorontalo Regency indicated severe drought in 1982 and 1992. July to September 1982 period which showed that drought occurs in all areas of Gorontalo Regency with extremely dry category (Figure 4). The 1982 drought is one of the most severe droughts that occur not only in Gorontalo Regency but also all over Indonesia. From April to June 1997 all area in Gorontalo Regency experienced the dry condition, the same condition also occurs in 1982. However, the 1997 drought magnitude was lower than the one in 1982. In April to June 1997, 67% of the area was in the extremely dry category, and 33% of the area was in several dry and moderately dry condition (Figure 4). The dry condition for July to December 2009 and January to June 2014 was more varied, from moderately wet to extremely dry. The normal condition in 2009 covered 53% of the area, whereas 47% of the area was in dry condition. The normal condition in 2014 covered 58% of the area, whereas 40% of the area was in dry condition and 2% in wet conditions. Spatial analysis approach provides additional information to assess the drought characteristics [9].

In general, based on the spatial drought analysis in 1982, 1997, 2009 and 2014, it shows that the drought events varied for all area in Gorontalo Regency. Seasonal drought events due to the seasonal rainfall variability need to be anticipated. Hence, it can be concluded that SPI for July to December indicates the high severity of drought due to the variability of seasonal rainfall. Other evidence to back this up is the drought peaks in the last three decades mostly occur in July to December. One of the main factors in the severity of the drought is the El Nino phenomenon that occurs periodically. The impact of this El Nino phenomenon is different for every area depending on the local climate characteristics. For Gorontalo, for instance, the high drought risk due to El Nino occurs in March to April and in July to October [17]. On those months, the observed drought events for the period of 1981 to 2016 was in line with the El Nino events. Thus, the SPI in the month could help assess the drought impact, especially in the agricultural sector in Gorontalo Regency. Also, it could increase the capacity in risk management due to the El Nino events [17]. Spatial drought analysis assists in identifying and describing the areas that are vulnerable to drought [14].

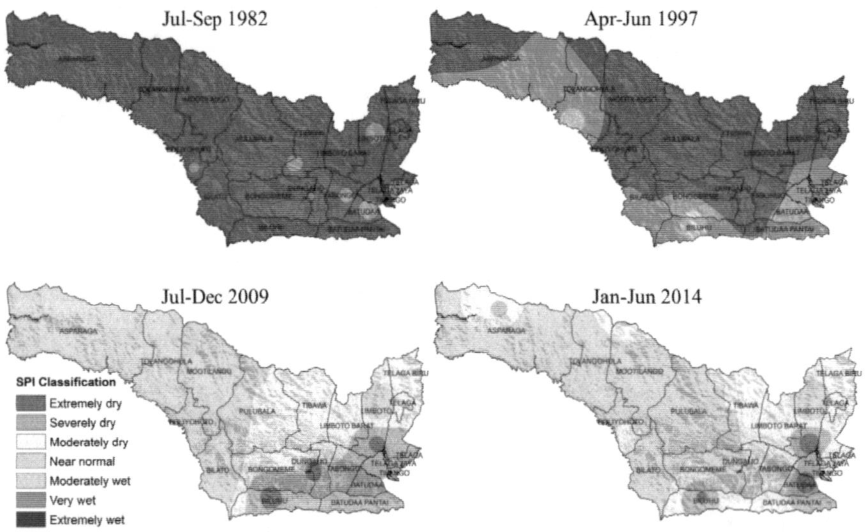


**Figure 4.** Spatial distribution of SPI-3 in 1982 and 1997, SPI-6 in 2009 and 2014

## 4. Conclusions

Monitoring of drought events in Gorontalo Regency used the monthly rainfall data during the period of 1981-2016. SPI was recommended to measure the severity level of drought because it is simple and it is flexible in observing drought in different timescale [8]. The result showed that rainfall lower is the main factors for the drought events. In addition, the duration and magnitude of drought were heavily influenced by the accumulation of rainfall in each time scale. The droughts that occur in Gorontalo Regency was also heavily determined by seasonal climate variability. For instance, droughts with the longest durations which occurred in 1982, 1986, 1997 and 2015 and the wet conditions that took place in 1984, 2001, 2008, 2010 and 2013, was closely related to El Nino Southern Oscillation (ENSO) phenomenon that periodically occur. Moreover, drought events frequency also related to the frequency and the strength of ENSO.

Drought monitoring using SPI in various time-scales was critical to be done to formulate mitigation plans. In addition, it also provided effective early warning information based on the severity of the drought [21]. Drought spatial distribution could be used to determine the priority plans in finding the solutions due to droughts that occur in drought-vulnerable areas. Therefore, drought analysis using SPI could contribute to the decision-making in the future as an effort to minimize the impact of drought in Gorontalo Regency. This study is an initial step in assessing the characteristics of drought in Gorontalo Regency. Hence, advance drought analysis needs to be done to determine the extent of ENSO influence toward the drought severity level in Gorontalo Regency.

## Acknowledgements

We would like to thank LP2M Universitas Negeri Gorontalo for the support of this research. Thanks also to BMKG and BWS for the support of climate data. Thanks to the head of the Geography Laboratory and the assistants for the support of the facilities provided during the completion of this research.

## References

[1] Abramovitz M and Stegun I A 1964 Handbook of mathematical functions *National Bureau of Standards. Washington D C*

[2] [BNPB] Badan Nasional Penanggulangan Bencana 2015 Kajian Risiko Bencana Gorontalo 2016-2020 *Deputi Bidang Pencegahan Dan Kesiapsiagaan* 10-15

[3] [BPS] Badan Pusat Statistik 2016 Prov. Gorontalo Dalam Angka *BPS Prov. Gorontalo* 161-163

[4] Camera C, Bruggeman A, Hadjinicolaou P, Pashiardis S and Lange M A 2014 *J. Geophys. Res. Atmos.* 119 693

[5] Chen F W and Liu C H 2012. *Paddy Water Environ.* **10** 209

[6] Edward D C 1997 Characteristics of 20th century drought in the united states at multiple time scales *Thesis Graduate School, Colarado State University. Colorado.*

[7] Gocic M and Trajkovic S 2014 *J. Hydrol.* **510** 110

[8] Hayes M J, Svoboda M D, Wilhite D A and Vanyarkho O V 1998 *Bull. Am. Meteorol. Soc.* **80** 429

[9] Hughes B L 2012 *Int. J. Climatol.* **32** 406

[10] Keyantash J and Dracup J A 2002 *Bull. Am. Meteorol. Soc.* **83** 1167

[11] Labedzki L 2007 *Irrig. And Drain.* **56** 67

[12] McKee T B, Doesken N J and Kleist J 1993 *Proceedings of the 8th Conference on Applied Climatology* **17-12** 179

[13] Mello C R, Lima J M, Silva A M, Mello J M and Oliveira M S 2003 *R. Bras. Ci. Solo.* **27** 925

[14] Rahman M R and Lateh H 2016 *Environ. Earth Sci.* **75** 1

[15] Saravi MM, Safdari AA and Malekian A 2009 *Hydrol. Earth Syst. Sci.* **6** 1347

[16] Spinoni J, Antofie T, Barbosa P, Bihari Z, Lakatos M, Szalai S, Szentimrey T and Vogt J 2013 *Adv. Sci. Res.* **10** 21

[17] Supari, Muharsyah R and Sopaheluwakan A 2016 *AIP Conf. Proc.* **1730** 07000-1

[18] Thom H C S 1958 *Mon. Weather Rev.* **86** 117

[19] Tigkas D, Vangelis H and Tsakiris G 2014 *Earth Sci. Infor.* **8** 697

[20] Tsakiris G, Tigkas D, Vangelis H and Pangalou D 2013 DrinC-Drought Indices Calculator. Getting Started Guide, Version 1.5 1-7

[21] Wagan B, Zhang Z, Baopeing F, Wagan H, Han S, Ahmad I and Kabo-Bah A T 2015 *Outlook Agr.* **44** 235

# Spatial Analysis for Potential Water Catchment Areas using GIS: Weighted Overlay Technique

**Disyacitta Awanda*, Anugrah Nurul H, Zahrotul Musfiroh, Dinda Dwi N P**

Undergraduate Student of Universitas Gadjah Mada, Bulaksumur, Yogyakarta, Indonesia 55281

disyaawanda@hotmail.com

**Abstract**. The development of applied GIS is growing rapidly and has been widely applied in various fields. Preparation of a model to obtain information is one of the benefits of GIS. Obtaining information for water resources such as water catchment areas is one part of GIS modelling. Water catchment model can be utilized to see the distribution of potential and ability of a region in water absorbing. The use of overlay techniques with the weighting obtained from the literature from previous research is used to build the model. Model builder parameters are obtained through remote sensing interpretation techniques such as land use, landforms, and soil texture. Secondary data such as rock type maps are also used as water catchment model parameters. The location of this research is in the upstream part of the Opak river basin. The purpose of this research is to get information about potential distribution of water catchment area with overlay technique. The results of this study indicate the potential of water catchment areas with excellent category, good, medium, poor and very poor. These results may indicate that the Upper river basin is either good or in bad condition, so it can be used for better water resources management policy determination.

**Keywords**: water catchment area, GIS, modelling, recharge zone

## 1. Introduction

The existence of groundwater is strongly influenced by the vegetation contained on it. In the scope of water ecosystem, known as Watershed which will further influence the existence of ground water. The development of applied GIS is growing rapidly and has been widely applied in various fields. Preparation of a model to obtain information is one of the benefits of GIS. Obtaining information for water resources such as water catchment areas is one part of GIS modeling. Water catchment model can be utilized to see the distribution of potential and ability of a region in water absorbing. The existence of water in the earth is very important for the sustainability of living beings.

The Opak River flows through the Sleman and Bantul districts where these two districts are fairly populated areas, thus causing relatively high levels of development in the region. Increased development in urban areas causes the amount of runoff [1]. Meanwhile, exploitation of water resources excessively causing a decrease in soil surface air, salt water intrusion, soil degradation, and drought [2]. The potential constriction of the water catchment area will directly affect the groundwater recharge zone. The groundwater recharge zone is the main determinant of the presence of water in the soil. If the land that serves as a catchment area is experiencing a continuous decline, then the groundwater recharge zone cannot work optimally [3].

The catchment area is an area that serves as a water catchment area that has important benefits to maintain the sustainability of the function of water sources in a. catchment area is closely related to watercourses. Watershed is an area that limits the mountain ridges where rainfall that falls on the area will be accommodated by the ridge and will be channeled through small rivers to the main river [4].

In general, the process of groundwater entry occurs through two sequential processes, namely infiltration (movement of water from the top into the soil surface) and percolation of water movement down from the unsaturated zone into the water saturated zone. The infiltration power is the maximum possible infiltration rate, determined by the ground surface conditions. The percolation power is the maximum possible rate of percolation, which is determined by the soil conditions in the unsaturated zone [5]. Infiltration rate will be equal to rain intensity if infiltration rate is still less than infiltration power. Percolation will not occur if the porosity in the zone is not saturated yet contains maximum water. The infiltration process plays an important role in replenishing the soil moisture and ground water. Replenishment of soil moisture is equal to the difference between infiltration and percolation [6]. So, the process of water entry into the soil is greatly influenced by water catchment areas and groundwater recharging zones. If the potential distribution of water catchment areas for the recharge zone is well distributed then the availability of water in the soil will be balanced.

GIS is an organized support of computer hardware, software, geographic data, and personnel designed to efficiently store, store, update, manipulate, analyze and display all geographic information [7]. The application of GIS is an effective way to map water catchment zones where modeling can use data from several parameters: rainfall, land use, geological data, slope, and water table height [8]. Modeling is done by using the overlay technique with tiered method by giving weight to each parameter in use. The purpose of this study is to obtain information on the potential distribution of water catchment areas for the recharge zone with overlay techniques using weighted methods, so that the results of the research can be used for better determination of water resources management policy. The purpose of this research is to get information about potential distribution of water catchment area with overlay technique. The results of this study indicate the potential of water catchment areas with excellent category, good, medium, poor and very poor.

## 2. Methods

The study sites are located in the Upper Opak Sub-watershed until the Opak-Oyo river meeting. In general, the research area covers the southern slopes of Mount Merapi which is still active today. Determination of the air absorption area absorbed some physical parameters that contribute to the infiltration and percolation process. Each parameter has a different role with different constituent parameters. The layers of each parameter are overlapped so that a new result will be classified as a potential for the airflow region. Simply put, the method used to construct a water catchment model is a weighted, tiered overlay that weighs on the guidance and guidance of the existing literature.

Parameters used in this study include rock types, soil texture, land use, rainfall, slope and ground air. Rock type parameters increase through geological map scale 1: 100.000 then known geological information that is equal to the level of permeability. Soil texture is known based on morphological approach and form of land through image and DEM data. Land use is obtained by digitally performing with a pitcher with maximum likelihood. The land use class is then reviewed based on the permeability level of each land use. Rainfall results of interpolation of rainfall data from several observation stations from related institutions. Interpolation method used by inverse Distance Weighted. Slope information is extracted through DEM processing data and its classification on Cook. Ground water is obtained through interpolation by Kriging method based on direct field observation data on the population wells. All parameters are done based on each weight (Table 1).

IOP Conf. Series: Earth and Environmental Science **98** (2017) 012054 doi:10.1088/1755-1315/98/1/012054

**Table 1.** Scoring of Recharge Zone (Wibowo, 2006 with modified)

| Parameter | Score |
|---|---|
| Geology | 6 |
| Rainfall Intensity | 5 |
| Soil Texture | 4 |
| Land use/cover | 3 |
| Slope | 2 |
| Water table | 1 |

## 3. Result and Disscusion

Water catchment areas are affected by several physical parameters such as geology, soil texture, slope, rainfall, water table fluctuation and land use. Each of these parameters has its own contribution in the ability to absorb water to become a groundwater reserve. Some extraction of physical parameters is done by using remote sensing technique with certain approach. Several other parameters were obtained through direct measurement in the field and also the acquisition of secondary data. The technique used to get the water catchment model is to utilize GIS application with weighted overlay technique.

### *3.1. Land Use*

Land use Opak River Basin Map made by image data processing, supervised classification method using maximum likelihood. Land use maps are classified into: settlements, grasses, gardens / moorlands, pine forests, forests, rivers, paddy fields, wet fields, open land. Images used for land use information extraction are Landsat 8 OLI images (Fig.1).

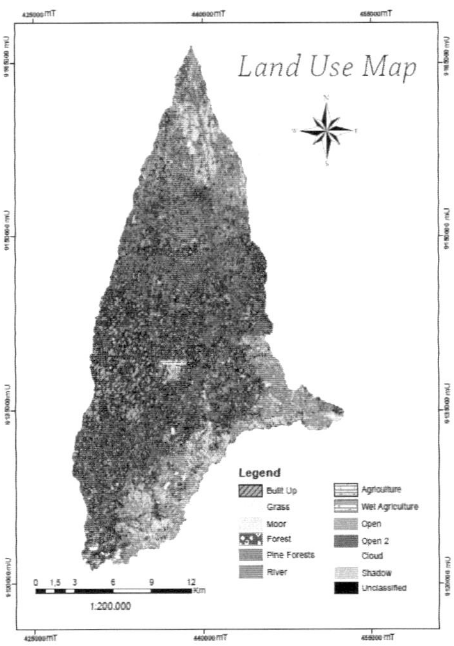


**Figure 1.** Land Use Map in Opak sub-watershed

**Table 2.** Scoring of Land Use

| Infiltration | Land Use | Score |
|---|---|---|
| Very High | Forest | 5 |
| High | High Production Forest, Garden, Open Land | 4 |
| Moderate | Shrubs, Grasses | 3 |
| Low | Field | 2 |
| Very Low | Urban, Rice Field | 1 |

Land use information is also used to estimate the difference in the amount of groundwater due to changes in land use and vegetation by measuring changes in groundwater levels. Areas overgrown with trees have the ability to absorb water runoff is very good so that the amount of ground water in the area more. Areas overgrown by trees will be better able to absorb larger runoff water, so the amount of groundwater contained in the area quite a lot, but runoff water flows slowly because of the roots and trunks of the tree so that the run time of water runs longer to get to the river. Land use of the Opak watershed is classified into; forests, gardens / fields, rice fields, wet rice fields, grasses, pine forests, urban, open land and rivers. Land use in the OPAK basin is largely a settlement in the transport zone or in the middle and downstream areas, while in the north or upstream areas there are fairly dense forests, especially the area around Merapi Volcano. Dense urban in the central and downstream areas has resulted in flood vulnerability in the middle to downstream areas quite higher than the level of flood vulnerability in the upstream or northern parts.

*3.2. Geology*
Map of geology made with reference to the geological map of Yogyakarta Province. Classification is done by giving weight to each geology of rocks. The map is created with scale 1: 450.000. The result consists five class; low, moderate, sufficient, high and very high. One of the parameters to be considered for the determination of water absorption areas is, the constituent rocks. Constituent rocks have a close relationship with permeability levels. The greater the permeability level of a rock, the higher the rate of air passage. Constituent rocks have an influence on water absorption into the soil distinguished by weight values.

The result of the classification of the constituent rocks shows that upstream of the permeability level of the Opak watershed is sufficient because it is dominated by the rock of the quarter era (Fig.2). In the middle of Opak watershed, permeability levels are moderate and dominated by tertiary age rocks. While the downstream area leading directly to the Indian Ocean the permeability level is sufficient. In general, the influence of the constituent rocks against the water absorption zone in the Opak watershed is in the middle region often floods. In addition to the level of permeability that is leading to low, high sedimentation in the middle region also affect the shallow watershed. So the possibility of flooding in the area is quite high.

**Table 3.** Scoring of Geology

| Permeability | Rock | Score |
|---|---|---|
| $>10^3$ | Aluvial Deposits | 5 |
| $10^1$ - $10^3$ | Young Quarter Deposits | 4 |
| $10^{-2} – 10^1$ | Old Quarter Depostis | 3 |
| $10^{-4} – 10^{-2}$ | Tertiary Depostis | 2 |
| $<10^{-4}$ | Intrusion rock | 1 |

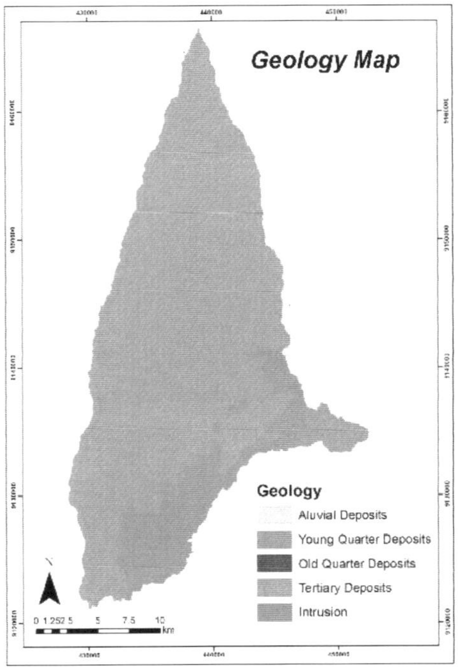


**Figure 2.** Geology map of Opak sub-watershed

### *3.3. Rainfall*

Rainwater data is obtained through secondary data through related institutions. The distribution of rain stations to be used as interpolation is spread evenly throughout the study area. The rain stations used are not only within the study area but also outside the study area so that interpolation results can cover all areas of study. The interpolation method used is Inverse Distance Weighted (IDW). Interpolation results with the method is good enough to give an overview of the intensity of rain that occurred in the last 10 years. There are 3 different classes in the study area.

The last 10 years of precipitation in Opak sub-watershed ranges below 1500 to 2500 mm / year (Fig.3). In the upstream part, rainfall is higher than the area below it. It is influenced by the topographical condition or the phenomenon of orographic rain. The high rainfall in the upper reaches the water absorption potential, because the higher the rainfall, the higher the infiltration process that occurs in the soil surface to the subsurface. The pressure of water in the soil due to the high infiltration process will also enlarge the ground water supply. In the middle and downstream part of the rainfall it tends not to be as high as the upstream, it is very natural to happen. Based on these results rainfall has a big role in determining the potential of water catchment areas in the upstream, so that the parameter of rainfall for the water catchment area is quite potential.

**Table 4.** Scoring of Rainfall

| Rainfall (mm/yr) | Rainfall Infiltration Factor *) | Score |
|---|---|---|
| < 1.500 | < 2.775 | 1 |
| 1.500 – 2.000 | 2.775 – 3.700 | 2 |
| 2.000 – 2.500 | 3.700 – 4.625 | 3 |
| 2.500 – 3.000 | 4.625 – 5.550 | 4 |
| > 5.550 | > 5.550 | 5 |

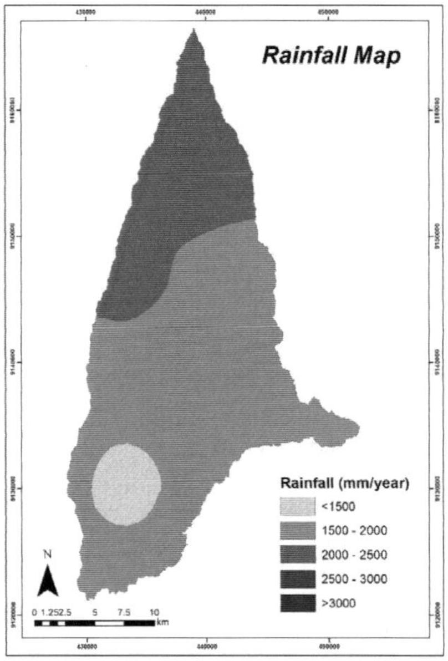


**Figure 3.** Inverse Distance Weighted for Rainfall Intensity

*3.4. Soil Texture*

Soil texture is one of the physical parameters that can be detected deductively through remote sensing image. Gradually by using the approach of landform and topography aspects can be known texture of soil in a region. Opak sub-watershed is located around a volcano that is quite young so that most of the area is still strongly affected by the activity of the mountain. Based on the interpretation results, most of the land has sand texture mixed with clay texture and dust. At the top is dominated by the texture of sand and dust, while the bottom is dominated by the texture of sandy clay and clay. Soil texture conditions allow water to be impregnated fairly quickly, because the influence of sand texture is quite dominating. At the upstream the permeability level is the fastest because the soil texture of the upstream area is mostly sand texture (Fig.4). At the center and downstream are already mixed with the texture of the clay so that the permeability level is lower. It shows physically reviewed based on sub-watershed Opak soil texture has good upflow potential at upstream.

**Table 5.** Scoring of Soil Texture

| Permeability ($10^{-5}$m/dt) | Rock | Score |
|---|---|---|
| Very slow (<2) | Clay | 1 |
| Slow (2-7) | Silty Clay | 2 |
| moderate (7-15) | Sandy Clay | 3 |
| Fast (15-30) | Silty Sand | 4 |
| Very Fast (>30) | Sand | 5 |

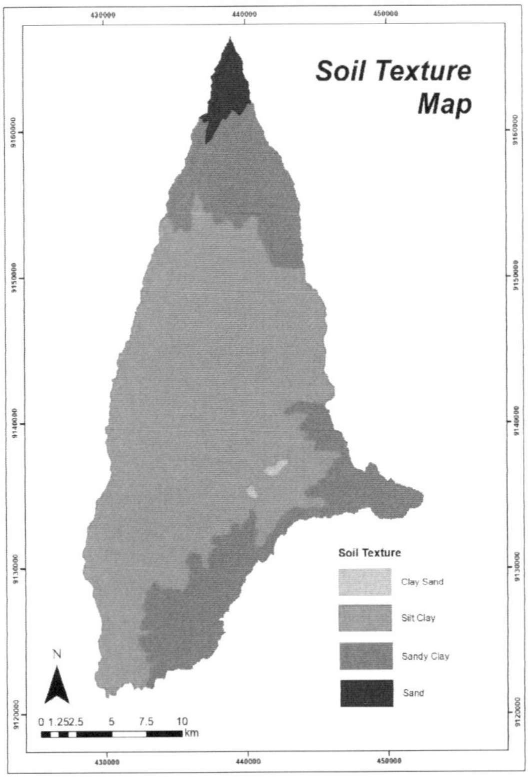


**Figure 4.** Soil Texture based on Landform approach

*3.5. Slope*

Slope can be obtained from the processing of elevation data or radar image data or commonly called with DEM data. The slope slope of this research is obtained through SRTM data which is processed and recycled according to the classification according to Cook classification. The slope at the upper sub-watershed Opak ranges from 10% to more than 25%, while the lower part tends to be relatively data with slope slope ranging from 5% to 10% (Fig.5). In the East, there are hills that have a slope similar to the upstream. It shows the variation of the gradual gradient slope from the upstream to the downstream and the eastern to the downstream. Judging from the spread of most of the slope of the slope on Opak sub-watershed relatively flat to choppy.

**Table 6.** Scoring of Slope

| Slope (%) | coefficient of infiltration | Score |
|---|---|---|
| < 5 | > 0,95 | 4 |
| 5 – 10 | 0,8 | 3 |
| 10 – 25 | 0,7 | 2 |
| >25 | 0,5 | 1 |

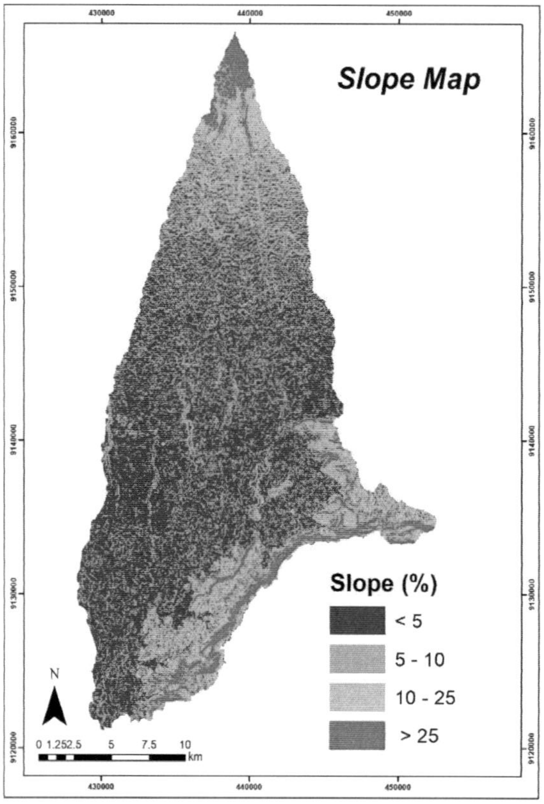


**Figure 5.** Land Use Map in DAS Opak

*3.6. Water Table Fluctuation*

Water table fluctuation is obtained through direct measurement of the field in the wells of the settlement. Measurements are made at specified sample points and scattered throughout the Opak sub-watershed region. Groundwater depths vary greatly from 2 meters to a depth of more than 12 meters. The depth of the groundwater level can indicate the level of use or consumption of ground water. In addition, groundwater can show the direction of groundwater flow. Groundwater contours or interpolation results of groundwater using Kriging method indicate that in areas with settled land use the depth of water level is deep enough. This may indicate the linkage of groundwater use to a fairly massive region of acceptance (Fig.6).

**Table 7.** Scoring of Water Tabel

| Fluctuation | Score |
|---|---|
| > 12 | 5 |
| 8 – 12 | 4 |
| 6 – 8 | 3 |
| 4 – 6 | 2 |
| < 4 | 1 |

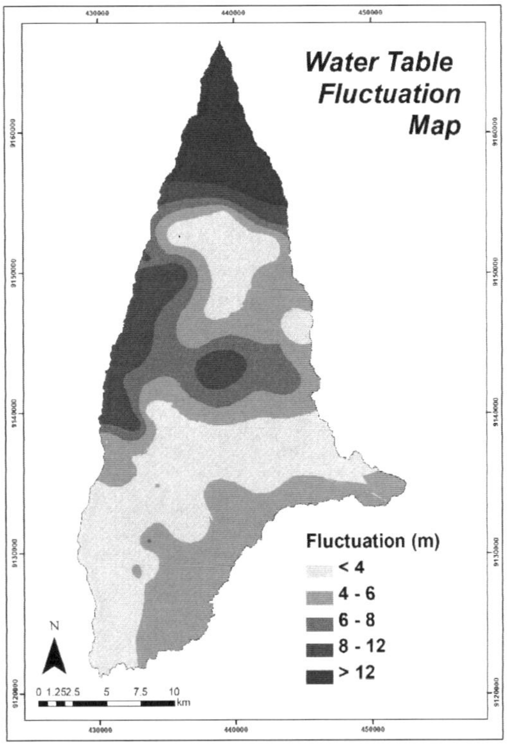


**Figure 6.** Water Table fluctuation

*3.7. Recharge Zone Potential*

GIS processing of the parameters used to construct a potential water catchment potential model shows good results. Weighted tiered method based on weighting or weighted literature guidance (Wibowo, 2006) still needs further study, since the weighting still has drawbacks especially on the less detailed geological class on the medium scale. The result of overlay and weighting resulted in 5 potential classes of water catchment area.

The upstream section has the greatest potential for recharge areas, since their supporting parameters are highly suitable as water catchment areas (Fig.7). Texture tends to sand, geologically has a good permeability, high rainfall and adequate land use despite the start of many settlements. This makes the upstream area has the potential for a good water catchment area. At the center of potential water catchment area sebgian large into the category of moderate and on the downstream belong to the category poor. In general, the water catchment maps made based on weighted tiered overlay techniques show good results, but this still needs to be further developed in order to obtain more optimal and accurate results. The weight assessment on each parameter obtained from the literature study still has limitations related to the level of class detail as well as its effect on the potential of water catchment areas.

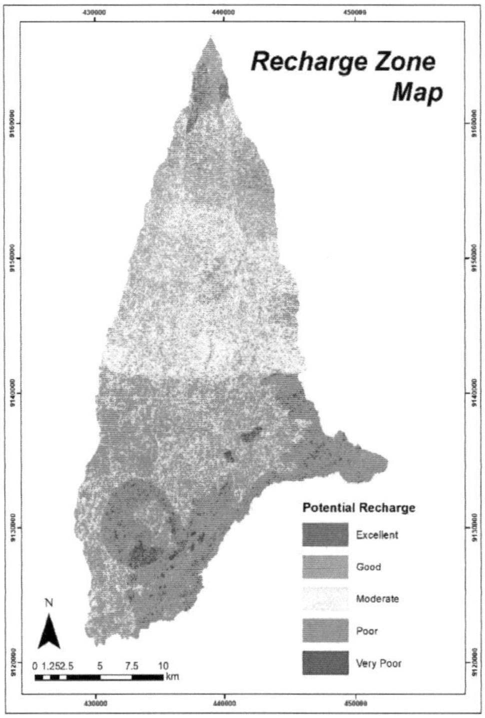


**Figure 7.** Recharge zone potential map

## 4. Conclusions

GIS is very helpful in doing the processing of spatial data that has information parameters used to build a model. In addition, remote sensing is also very helpful to perform the primary data extraction quite effectively. Spatial data processing to build a water catchment model needs an in-depth study of the relationship of biophysical characteristics with hydrological properties, since hydrological properties are highly dynamic to biophysical characteristics. Applied GIS in deepening hydrological phenomena can provide a more complete picture because it can be analyzed spatially and the relationships that occur between different locations. A distinct advantage in hydrological research involving applied GIS. In this case, the study of water catchment areas with weighted tiered overalay method can give a good enough result to give an idea of potential or potential model of recharge zone.

In general, Opak sub-watershed has a good potential water absorption potential, especially at the very top. Each influential physical factor shows that the Opak sub-watershed still has good potential, although most Opak sub-watershed should have excellent potential because it is a mofologi upstream area whose function as a place to absorb water. Limitations of this study include the accuracy of each paramater that has not been tested, so the accuracy of the model results built is still not fully used for decision-making. The limited methods in this study are cited in a weighted literature study that refers to a single source. In future work, this research can be developed more optimally to get more accurate results.

## References

[1] Raymond, A.Duraiswami. 2009. Geospatial Mapping of Potential Recharge Zones in Parts of Pune City. *JOURNAL GEOLOGICAL SOCIETY OF INDIA Vol.73, May 2009, pp.621-638*

[2] Faisal, K.Zaidi., et al. 2016. Identification of artificial groundwater recharging zone using a GIS-based fuzzy logic approach: a case study in a coal mine area of the Damodar Valley, India. *Appl Water Sci DOI 10.1007/s13201-017-0603-8*

[3] Kaliraj, et al. 2012. Identification of potential groundwater recharge zones in Vaigai upper basin, Tamil Nadu, using GIS-based analytical hierarchical process (AHP) technique. *Arab J Geosci (2014) 7:1385–1401 DOI 10.1007/s12517-013-0849-x*

[4] Asdak, Chay. 1995. *Hidrologi dan Pengelolaan Daerah Aliran Sungai.* Yogyakarta : Gadjah Mada University Press.

[5] Sudjarwadi.1983.Drainase dan Aerase Dalam Tanah.

[6] Wibowo, Mardi. 2006. Model Penentuan Kawasan Resapan Air Untuk Perencanaan Tata Ruang Berwawasan Lingkungan. Jakarta: *Jurnal Hidrosfer Vol 1, Hal 1-7.*

[7] Prahasta, Edi. 2005. *Konsep-Konsep Dasar Sistem Informasi Geografis.* Bandung: CV. Informatika.

[8] W.Chingombe et al. 2014. A Participatory Approach in GIS Data Collection for Flood Risk Management, Muzarabani District, Zimbabwe. *Arabian Journal of Geosciences Volume 8, Issue 2, pp 1029–1040*

[9] Undang-Undang Nomor 23 *Tahun 2006* tentang Administrasi

# Remote Sensing and Water Quality Indicators in the West Flood Canal Semarang City: Spatio-temporal Structures of Lansat-8 Derived Chlorophyll-a and Total Suspended Solids

**Sawitri Subiyanto**

Department of Geodesy Engineering, Faculty of Engineering, Diponegoro University, Semarang – Indonesia

sawitrisubiyanto66@gmail.com

**Abstract**. One of the waters that has been contaminated by industrial waste and domestic waste is the waters of West Flood Canal in Semarang City which is the estuary of the river system, which passes through the Western City of Semarang which is dense with residential and industrial. So, it is necessary to have information about the assessment of water quality in the estuary of the West Flood Canal. Remote sensing technology can analyze the results of recording the spectral characteristics of water with water quality parameters. One of the parameters for assessing water quality is Chlorophyll-a and Total Suspended Solid, can be estimated through remote sensing technology using multispectral Lansat-8 Satellite images data from April, June, and August, 2017 and there are three selected algorithms. Based on the results of TSS and Chlorophyll-A processing, the TSS shows values greater than or equal to 100 which can be said that West Flood Canal is damaged (hypertrophic). While the chlorophyll-a shows a value less than 100 indicating Eutrophic status (threatened). This is caused by the number of suspended materials in the water surface and also because of the disturbance of water vegetation in the form of weeds that destroy the function of the actual West Canal Flood.

**Keywords:** Chlorophyll-a, Suspended solids, Remote sensing, West Flood Canal

## 1. Introduction

Coastal environments throughout the world are now experiencing unprecedented anthropogenic eutrophication, hypoxia and harmful algal blooms due to increased human activities and subsequently increased nutrient loadings [1]. Because of their significant role in human well-being relying upon them, deterioration of water quality indicators (e.g., water temperature, dissolved oxygen, chlorophyll, inorganic nutrients, heavy metals, total suspended solid) can be potentially catastrophic for marine ecosystems as species are threatened by conditions which are no longer suitable for their survival [2]. As a result, monitoring changes in spatio-temporal patterns of water quality (WQ) in both terrestrial and marine ecosystems, and interpretation of the implications have been important across disciplines these days, such as environmental/marine sciences and socio-economics. However, it is inherently difficult to properly address varying spatial and temporal scales of WQ for application to coastal marine waters monitoring and assessment [2].

West Flood Canal (BKB) is a union of three major rivers namely Kreo, Kripik, and Garang River. In Gunungpati District, the three rivers merge and become upstream of BKB. The three rivers have upstream in Semarang regency. Nevertheless, about 90 percent of the river is located in the city of Semarang. One of the waters that has been contaminated by industrial waste and domestic waste is the waters. The West Flood Canal River crossing the Western city of Semarang is dense residential and industrial. Many industrial activities around this watershed (DAS). Among other things are the textile industry, foodstuff, plastic, karoseri, printing, pharmaceutical and herbal medicine, paint, furniture, lubricating oil, workshop, even there is a fish auction. These waters become domestic and urban waste disposal sites and industrial wastes generated by activities around the river basin (Riyanto, 2004 in [3]).

In coastal waters, total suspended solid (TSS) can be an important factor controlling biological processes [4] and its distribution and variations can be strongly influenced by tides and river discharge [5]. TSS is also important in light penetration and water movements in the coastal waters (Sonetal, 2014). Thus, accurate estimation of TSS is critical to understand the thermal structure of upper water columns and physical processes [6] as well as biological processes (e.g., phytoplankton photosynthesis) in the ocean euphotic zone [7]. With the amount of waste discharged along the river of West Canal Flood, Kreo, Kripik, and Garang River is so that there is need for information on the assessment of water quality in the estuary of the West Flood Canal. One of the parameters for assessing water quality is Chlorophyll-a and Total Suspended Solid. Chlorophyll-a is a parameter used to determine the level of water fertility while Total Suspended Solid (TSS) is one of the parameters used to determine the level of water pollution. By knowing the content of Chlorophyll-a and Total Suspended Solid (TSS) in the waters of this West Canal Flood will be obtained information about water quality in estuary waters of West Canal Flood.

Distribution of Chlorophyll-a and Total Suspended Solid (TSS) can be estimated through remote sensing technology, such as from Landsat satellite imagery. Remote sensing technology can analyze the results of recording the spectral characteristics of water with water quality parameters. Remote sensing technology using satellite imagery offers a variety of facilities for obtaining data on fertility and water pollution compared to using more complex and costly laboratory tests. One of the benefits of using remote sensing technology is its wide coverage area, having an uncomplicated processing and low cost. In Landsat satellite image processing to obtain information on fertility distribution data and water pollution, an algorithm is required to obtain suitable distribution value between image data and laboratory test data so that the selection of this algorithm is expected to get the value of distribution of fertility and water contamination matching in the study area.

## 2. Materials and Literature Study

### *2.1. Materials*

Location of study in this research is Semarang area which densely populated. West Canal Floods pass 5 sub-districts of West Semarang, Semarang Utara, Semarang Tengah, South Semarang and Gajah Mungkur. The West Canal Flood has limited land because its left right is flanked by 4 (four) highways between Sugijopranoto bridge to the sea flanked by Madukoro highway Kokrosono road and between Sugijopranoto bridge to Simongan dam is flanked by Bodjong Salaman road - Basudewo road. The West Flood Canal River is one of the means important urban infrastructure in Semarang City. The main function of the West Flood Canal River is as the city's flood and drainage controller that is actively used since it was first built in 1880 until today, in addition to the West Flood Canal River into a waterfront icon in Semarang.

IOP Conf. Series: Earth and Environmental Science **98** (2017) 012055 doi:10.1088/1755-1315/98/1/012055

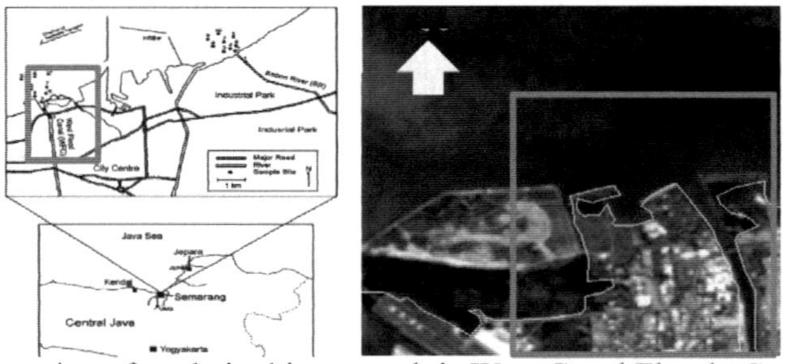


**Figure 1.** Location of study in this research is West Canal Floods, Semarang City.

### *2.2. Literature Study*

*2.2.1. Remote Sensing.* The application of satellite and airborne remote sensing imagery to collect WQ information has become more frequent [8]. Remotely sensed datasets are generally more comprehensive than those directly measured *insitu* in that they provide greater spatial coverage with finer resolution and often increased temporal frequency and resolution. This makes remote-sensing approach a rich source of data. However, the large amount of data also embeds challenges for the extraction of meaningful information of WQ parameters. Remotely-sensed sea surface temperature (SST) is one of the most important oceanic and atmospheric variables which has been widely used in a variety of researches to develop an understanding of ocean dynamics, as well as physical and biogeochemical processes in the upper ocean [9].

*2.2.2. Water Pollution.* Water pollution caused by humans can arise from a variety of activities, whether intentionally or not, and in general greatly affect the environment resulting from pollution by living things. Pollution if not prevented or reduced is essentially harmful and detrimental to human beings in terms of health as well as in terms of social life or the survival of living things [10]. Water pollution affects organisms and plants in water bodies. In many cases this effect destroys not only individual populations and species but also natural biological communities.
According to [11], the indicator or sign that water has been polluted is that there are changes or observable signs:

1. Water temperature changes.
   The existence of this temperature change is generally due to the presence of industrial waste in terms of water cooling process to remove heat from the machines used. The hot water is then discharged into the environment. If the hot water is dumped into the river, then the river water will become hot, river water that rises temperature will disrupt the life of aquatic animals and other aquatic organisms due to dissolved oxygen levels in water will decrease along with rising temperatures.

2. Changes in pH or Hydrogen Ion concentration.
   The normal water that qualifies for a life has a pH ranging from 6.5 to 7.5. Water can be acidic or alkaline depending on the size of the water pH or the amount of hydrogen ion concentration in the water. Water having a pH smaller than normal pH will be acidic, while water having a pH greater than normal pH will be alkaline.

3. Color Changes, Odor and Flavor.
   Waste or waste materials in the form of organic and inorganic materials are often soluble in water. If the waste material and waste water dissolves in water then there will be a change of water color. Water in normal circumstances and clean will not be colored, so it looks clear and clear. Water pollution does not necessarily depend on the color of the water.

4. Occurrence of Deposition, Colloidal, and Substances Dissolved.
   The precipitate and colloidal as well as the solute material are derived from the presence of solid waste materials. Solid waste materials that are not fully soluble form will settle at the bottom of the river and which may dissolve in part will become colloidal. The precipitate before it reaches the bottom of the river will float in water together with colloidal.

5. Microorganisms
   Microorganisms are instrumental in the process of degradation of waste materials from industrial or other activities that are discharged into the environment water, both rivers, lakes and seas. If the waste material that must be degraded quite a lot, means that microorganisms will participate multiply. In the proliferation of these microorganisms did not rule out the possibility that microbial pathogens also develop. Microbial pathogen is the cause of various diseases.

*2.2.3. Chlorophyll-a.* Chlorophyll comes from the Greek, which consists of two syllables, chloros which means green and phylum which means leaf. Chlorophyll captures life force or solar energy and is used to split $H_2O$ molecules into elements of H and $O_2$, then combine them between elements H with $CO_2$ gas and produce sugar or carbohydrates. From the process of photosynthesis is also produced byproducts of gas $O_2$. Chlorophyll is better known as green leaf substance which is a pigment found in a producer organism that functions as a converter of carbon dioxide into carbohydrates, through photosynthesis. Chlorophyll has a chemical formula $C_{55}H_{72}O_5N_4Mg$ with Mg atom as its center. Chlorophyll-a is one of the most decisive parameters of primary productivity in the ocean. Chlorophyll-a is an important component that is supported by phytoplankton and water plants which are both a source of natural food for fish. Chlorophyll-a is an active pigment in plant cells that play an important role in the ongoing process of photosynthesis (Prezelin, 1981 in [12]).

*2.2.4. Trophic Water Status.* Eutrophication is a water enrichment with nutrients in the form of inorganic materials needed by plants and leads to an increase in primary productivity of waters. Nutrients in question are nitrogen and phosphorus. Eutrophication is classified into two, namely artificial or cultural eutrophication and natural eutrophication. Eutrophication is classified as artificial (cultural) eutrophication if nutrient uptake in waters is caused by human activity; And are classified as natural eutrophication if nutrient uptake in waters is not caused by human activity but natural activity [13].
Eutrophication is classified into four categories of trophic status [14], namely:

*Oligotrophic*; Is trophic status of lake water and / or reservoir containing low nutrient content, this status indicates that water quality is still unspoiled nature from nutrient sources N and P.

*Mesotrophic*; Is the trophic status of lake water and / or reservoir containing nutrients with moderate levels, this status indicates an increase in levels of N and P, but still within the tolerance limit because it has not shown any indication of water pollution.

*Eutrophic*; Is the trophic status of lake water and / or reservoir containing high levels of nutrients, this status indicates that water has been contaminated by increased levels of N and P.

*Hypereutrophic*; Is the trophic status of lake water and / or reservoir containing very high levels of nutrients, this status indicates that water has been heavily polluted by increasing levels of N and P.

Next on the criteria of trophic water status according to [15] are listed in the following table:

**Table 1.** Water status trophic criteria

| **Trophic Status** | **Average Chlorophyll a (μg/l)** |
|---|---|
| Oligotroph | <2.6 |
| Mesotrophic | 2.6-7.3 |
| Eutrophic | 7.3-56 |
| Hypereutrophic | >56 |

*2.2.5. Total Suspended Solid.* TSS are suspended materials (diameter> 1 μm) retained on a sieve of Millipore with a pore diameter of 0.45 μm. TSS consists of mud and fine sand and microorganisms. TSS concentrations in waters generally consist of phytoplankton, zooplankton, human waste, animal waste, sludge, crop residues and animals, and industrial waste. Suspended materials in natural waters are not toxic, but if excessive amounts can increase the turbidity value which further inhibits the penetration of sunlight into the water column [16]. TSS is closely related to soil erosion and erosion of river channels. TSS is highly variable, ranging from less than 5 mg.L-1 is the most extreme 30.000 mg.L-1 in some rivers. TSS is not only an important measure of erosion in the river channel, it is also closely related to transport through the river system of nutrients (especially phosphorus), and various industrial and agricultural chemicals [17].

*2.2.6. Pollution Index Determination*. [18], University of Texas, A.S., proposed an index relating to a significant pollutant compound for a designation. This index is expressed as the Pollution Index used to determine the level of pollution relative to permitted water quality parameters [19]. This index has a different concept with the Water Quality Index. The pollution index (PI) is determined for a designation, then it can be developed for some designation for all parts of the water body or part of a river. Water quality management on the basis of pollution index (PI) can provide input to decision makers in order to assess the quality of water bodies for a designation and take action to improve quality in case of quality deterioration due to the presence of pollutant compounds. PI encompasses a variety of independent and meaningful quality parameter groups (KMNLH No. 115 Year 2003). Results of image data processing on Total Suspended Solid (TSS) concentration and subsequent laboratory test results are calculated by pollution index in accordance with Decree of State Minister of Environment No.115 Year 2003 about Water Quality Status as comparison with quality standard to know pollution level at point location The sampling. The formula for calculating the pollution index is as follows:

$$PI=Ci/Lij \tag{1}$$

Where:
Ci: concentration of water quality parameters obtained from laboratory test results (TSS).
Lij: concentration of water quality parameters specified in the Water Quality Standard.

Evaluation of PI values are: $0 \leq PI \leq 1.0$: meet the quality standard, $1.0 < PI \leq 5.0$: lightly contaminated, $5.0 < PI \leq 10$: moderately polluted, $PI > 10$: heavily polluted

*2.2.7. Chlorophyll-a Calculation Algorithm.* There are 1 algorithms that will be used in this research with the aim to get the best algorithm that can be used as formula of chlorophyll-a content processing in research area and can be used by multitemporal data of multilayered area study.

*Nuriya Algorithm et a.* The algorithm used for the determination of chlorophyll-a content using the Nuriya algorithm. So in the processing of image data the algorithm is used as follows:

$$C = 0{,}2818 \times \left(\frac{B5+B6}{B4}\right)^{3{,}497} \tag{2}$$

Information:
C: Chlorophyll-a concentration (mg / l)
B4: The reflectance value of channel 4
B5: Reflectance value of channel 5
B6: The channel reflectance value of 6

IOP Conf. Series: Earth and Environmental Science **98** (2017) 012055 doi:10.1088/1755-1315/98/1/012055

Because the algorithm uses the reflectance value, then the image data must be extracted first from the value of DN (digital number) to the radiance value. Then from the value of radiance to the value of Reflectance.

*2.2.8. TSS Concentration Determination Algorithm.* There are 1 algorithms that will be used in this research with the aim to get the best algorithm that can be used as a formula for processing Total Suspended Solid (TSS) concentration in research area and can be used by multitemporal data of Landsat study area.

a. Doxaran algorithm (*Red Band*): $5.1271 * \exp(0.0027 * i1)$ (3)

b. Laili algorithm (*Band* 2 and *Band* 4): $TSS\ ({}^{mg}/{}_{l}) = 31.42 * \left(\frac{\log(RRS2)}{\log(RRS4)}\right) - 12.719$ (4)

c. Trisakti algorithm (*Red Band*): $TSS\ \left({}^{mg}/{}_{l}\right) = 4.9453e^{0.0028x}$ (5)

## 3. Results and Discussion

### *3.1. Total Suspended Solid Algorithm Results*

Distribution of water quality results of TSS parameters by using the doxaran algorithm on April 4, 2017, June 3, 2017, and August 22, 2017 1. Distribution of water quality results of TSS parameters by using the doxaran algorithm on April 4, 2017, June 3, 2017, and August 22, 2017.

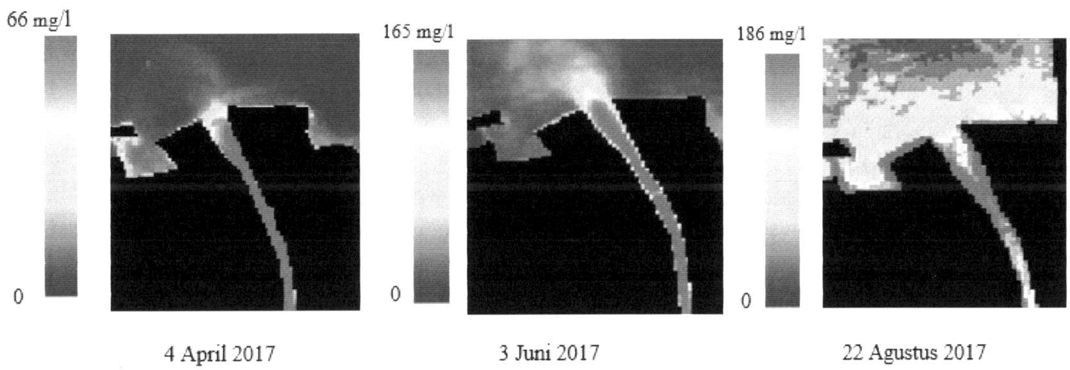


**Figure 2.** Water quality results with Doxaran Algorithm

Distribution of water quality results of TSS parameters by using the Laili algorithm on April 4, 2017, June 3, 2017, and August 22, 2017.

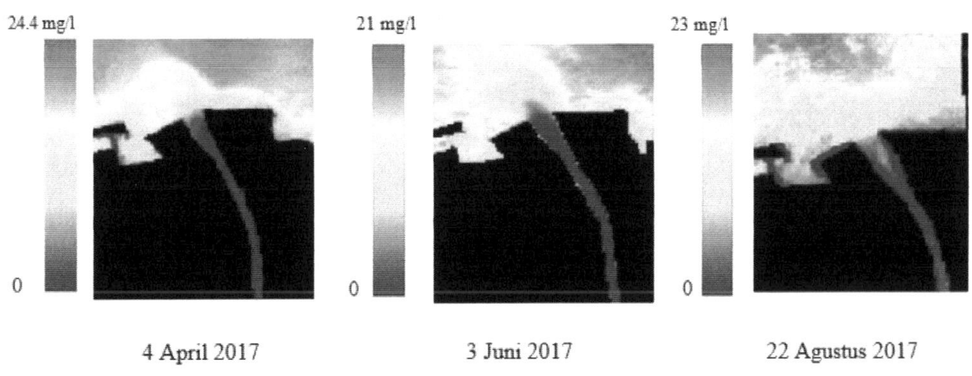


**Figure 3.** Water quality results with Doxaran Algorithm Laili

Distribution of water quality results of TSS parameters by using the Trisakti algorithm on April 4, 2017, June 3, 2017, and August 22, 2017.

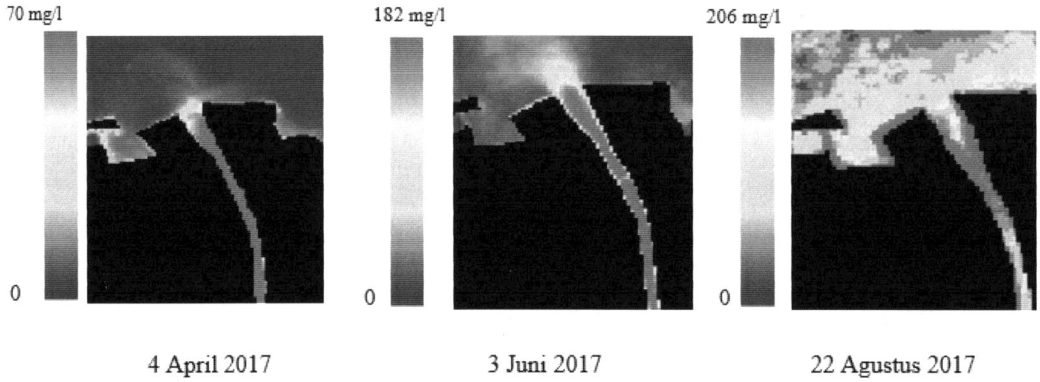


**Figure 4.** Water quality results with Doxaran Algorithm Trisakti

*3.2 Results of Distribution Value of Chlorophyll-a*

Results of Distribution Value of Chlorophyll-A April, June, and August 2017 with Algorithm Nuriya.

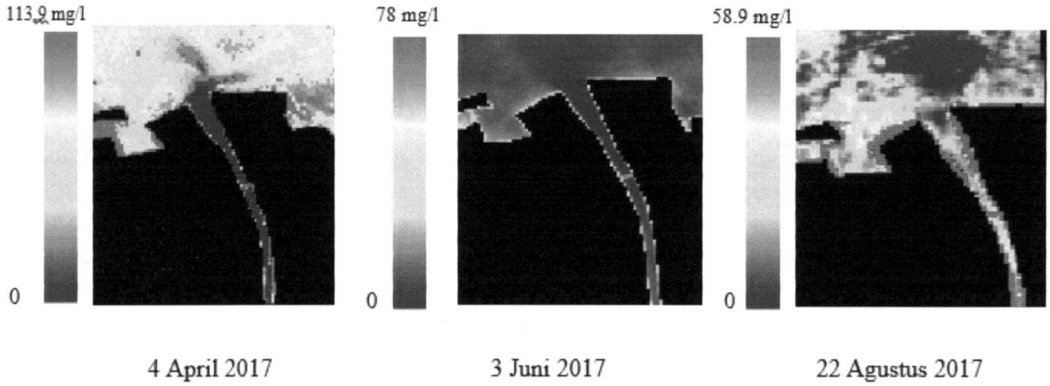


**Figure 5.** Water quality results with Doxaran Algorithm Nuriya

*3.3 Analysis of Tropical Flood Status of Western Canal*

Monitoring of water quality level in West Canal Flood can be done by qualitative method which use TSS and Chlorophyll-a parameter where the hypothesis is the bigger the TSS concentration value, the smaller the water volume. Based on the results of TSS and chlorophyll-A we get the most significant result is to use the algorithm Doxaran and Trisakti where the two algorithms are not much different results. But to test the tropical status of West Canal Flood using the Doxaran algorithm. Here are the results of trophic status data of West Canal Flood with TSS and Brightness parameters using Doxaran algorithm.

**Table 2**. Tropical data status data of west canal floods

| Month | TSS Value Algorithm Doxaran (mg / l) | Value of Chlorophyll-a (mg/l) | Tropic Status |
|---|---|---|---|
| 4 April 2017 | 66 | 113.9 | Hypereutrophic |
| 3 June 2017 | 165 | 78 | Hypereutrophic |
| 22 August 2017 | 186 | 58.9 | Hypereutrophic |

## 4. Conclusions

Based on the results of TSS and Chlorophyll-A processing, the TSS shows values of less than or equal to 00 on June 3, 2017 and August 22, 2017 and Chlorophyll-A shows values of more than or equal to 200 on April 4, 2017 which can be said that the West Canal Flood status damaged (Hypertrophic). While on June 3, 2017 and August 22, 2017 Chlorophyll shows a value less than 100 indicating Eutrophic status (threatened). This is caused by the number of suspended materials in the water surface and also because of the disturbance of water vegetation in the form of weeds that destroy the function of the actual West Canal Flood.

## References

[1] Anderson D M, P M Gilbert and J M Burkholder 2002 Harmful Algal Blooms and Eutrophication: Nutrient Sources, Composition, and Consequences *Estuaries* **25** pp 704-726

[2] Bierman P, M Lewis, B Ostendorf and J Tanner 2011 A Review of Methods for Analysing Spatial and Temporal Patterns in Coastal Water Quality *Ecol. Indic* **11** pp 103-114

[3] Wulandari S Y 2012 Status Perairan Banjir Kanal Timur Semarang Ditinjau dari Kadar Logam Berat Chromium dalam Air, Sedimen dan Jaringan Lunak Kerang Darah (Anadara granossa) *Buletin Oseanografi Marina* **1** pp 1-7

[4] Doxaran D, J M. Froidefond and P Castaing 2003 Remote-sensing Reflectance of Turbid Sediment-dominated Waters. Reduction of Sediment Type Variations and Changing Illumination Conditions Effects by Use of Reflectance Ratios *Appl. Opt* **42** pp 2623-2634

[5] Kim Y H, J Im, H K Ha, J K Choi and S. Ha 2014 Machine Learning Approaches to Coastal Water Quality Monitoring Using GOCI Satellite Data *Gisci Remote Sen* **51** pp 158-174

[6] Sathyendranath S, A D Gouveia, S R Shetye, P Ravindran dan T. Platt 1991 Biological Control of Surface Temperature in the Arabian Sea *Nature* **349** pp 54-56

[7] Sathyendranath S, T Platt, C M Caverhill, R E Warnock and M R Lewis 1989 Remote Sensing of Oceanic Primary Production: Computations Using a Spectral Model *Deep Sea Res* **36** pp 431-453

[8] Goetz S, N Gardiner and J Viers 2008 Monitoring Freshwater, Estuarine and Near-shore Benthic Ecosystems with Multi-sensor Remote Sensing: An Introduction to the Special Issue *Remote Sensing of Environment* **112** pp 3993-3995

[9] Park K A, E Y Lee, E Chang and S Hong 2015 Spatial and Temporal Variability of Sea Surface Temperature and Warming Trends in the Yellow Sea *J. Mar. Syst* **143** pp 24-38

[10] Dix H M 1981 *Environmental Pollution* (New York: John Wiley & Sons)

[11] Wardhana W 2004 *Dampak Pencemaran Lingkungan* (Yogyakarta: Andy Offset)

[12] Krismono 2010 Hubungan antara Kualitas Air dengan Klorofil-A dan Pengaruhnya terhadap Populasi Ikan di Perairan Danau Limboto *Limnotek* **17**(2), pp 171-180

[13] Effendi H 2003 *Telaah Kualitas Air bagi Pengelolaan Sumber Daya dan Lingkungan* (Yogyakarta: Kanisius)

[14] MENLH 2004 *Keputusan Menteri Lingkungan Hidup Nomor: 51/MENLH/2004 Tahun 2004* (Jakarta: Kementerian Lingkungan Hidup)

[15] Carlson R E and J Simpson 1996 A Coordinator's Guide to Volunteer Lake Monitoring Methods," *North American Lake Management Society* **96**

[16] Effendi H 2000 Telaahan Kualitas Air bagi Pengelolaan Sumberdaya dan Lingkungan Perairan," *Manajemen Sumberdaya Perairan* pp 12-18

[17] Bird T 1987 *Kimia Fisik untuk Universitas* (Jakarta: Gramedia)

[18] Nemerow N L and H Sumitomo 1970 Benefits of Water Quality Enhancement (New York: U.S. Environmental Protection Agency)

[19] Nemerow N L 1974 *Scientific Stream Pollution Analysis* (McGraw-Hill Book)

# The Application of Forest Cover Density (FCD) Model for Structural Composition of Vegetation Changes in Part of Lore Lindu National Park, Central Sulawesi Province

**Muhammad Ismail[1*], Hartono[2], Projo Danoedoro[3]**

[1]Alumnus of Remote Sensing Programme, Faculty of Geography, Gadjah Mada University
[2]Geography Information Science Department, Faculty of Geography, Gadjah Mada University
[3]Geography Information Science Department, Faculty of Geography, Gadjah Mada University

muhammadismael89@gmail.com

**Abstract.** One method that is considered to be more accurate for the identification of forest density related to the structural composition of vegetation is using a FCD model. However, the FCD model needs to use terrain correction in the identification of structural compositions. The aims of this study are to analyse the ability of Landsat image in FCD model to map structural composition of vegetation and to study the rate of changes. Remote sensing imagery used in this study is the Landsat 7 ETM + in September 2002 and Landsat 8 OLI in September 2015. Based on the results of this study concluded that the FCD model and multitemporal Landsat imagery can be utilized for the identification of structural composition without and with terrain correction. Result of identification structural composition of vegetation in 2002 and 2015 have accuracy respectively equal to 67,2% and 78,3%. Changes in the structural composition of vegetation without terrain correction consisted of decrease is 66,255.5 ha. In addition, the result of the structural composition of vegetation with terrain correction in 2002 and 2015 has an accuracy of 75.3% and 79.4%. Changes in the structural composition of vegetation with terrain correction consisted of a decrease is 45,456.8 ha.

**Keywords:** FCD, vegetation, Landsat 7 ETM+

## 1. Introduction

Rapid technological developments produce a science and method that is considered capable of identifying various problems. Remote sensing is the result of these technological developments. Remote sensing is the science and art of obtaining information about an object, region, or phenomenon through analysis of data obtained by means without a direct contact [1]. Remote sensing applications have an important role in helping the process of monitoring forest changes, especially the structural composition of vegetation. Periodic measurements of forest changes can be obtained with remote sensing applications and are capable of providing the best accuracy [2]. With remote sensing, information related to geosphere phenomena can be quickly obtained and analyzed systematically.

In digital analysis related to spectral transformation there is Forest Cover Density (FCD) model. According to [3] explains that FCD is the development of spectral transformation in density estimate and tropical forest vegetation structure. The FCD model was developed by [4] which became the model of vegetation mapping used by the International Timber Trade Organization (ITTO). The FCD model has advantages over other generic vegetation indices. The superiority of the FCD model is the vegetation structure density information in the form of vegetative structural composition obtained from four FCD modeling index comprising AVI (Advanced Vegetation Index), BI (Baresoil Index), SI (Shadow Index), and TI (Thermal Index). The FCD model utilizes various remote sensing imagery bands from the visible band to the thermal infrared band of Landsat satellite imagery.

The result of FCD model is forest density information both vertically and horizontally (structural composition). For example, there is case of changes in forest structural composition in Lore Lindu National Park area. This study aims to analyze the ability of Landsat's image capability in the Forest Cover Density (FCD) model to map structural composition of vegetation and to study the rate of change in the class of vegetation structural composition.

## 2. Study Area

The research was conducted in part of Lore Lindu National Park, which is administratively located in Sigi Regency, Central Sulawesi Province. The study area has an area of ± 97,204,7 ha that stretches from north to south in the western part of Lore Lindu National Park area and consisting of eight districts. Climatic characteristics of the study area that has a rainfall between 2000 - 4000 mm/year with climate type A to climate type D. Elevation in the study area ranges from ± 100 - 1900 mdpl which is dominated by mountain morphology.

## 3. Data

The data used in the extraction of structural compositions is Landsat 7 ETM+ recording of September 28, 2002 and Landsat 8 OLI recording image dated September 8, 2015. Landsat 7 ETM+ has 8 spectral bands meanwhile Landsat 8 OLI contains 11 spectral bands. In addition, SRTM image data is also used for slope extraction and aspect in terrain correction. The RBI map is also used as a reference for the pre-processing stage, as well as the Zoning Map of the Lore Lindu National Park Area as the reference for the study area.

## 4. Methods

### *4.1. Transformation of FCD (Forest Cover Density) Model*

The transformation of the FCD model is the method used to identify the structural composition of vegetation. This model uses four kind of index calculation: vegetation index using Advance Vegetation Index (AVI), soil index using Bare Soil Index (BI), Shadow Index (SI) and temperature index using Thermal Index (TI). Here are the formulas of the four indices [5].

$$\text{AVI} = (\text{NIR Band} \times (256 - \text{Red Band}) \times (\text{NIR Band} - \text{Red Band}) + 1)^{1/3} \quad (1)$$

$$\text{BI} = \frac{((\text{SWIR Band} + \text{Red Band}) - (\text{Blue Band} + \text{NIR Band}))}{((\text{SWIR Band} + \text{Red Band}) + (\text{Blue Band} + \text{NIR Band}))} \quad (2)$$

$$\text{SI} = ((256 - \text{Blue Band}) \times (256 - \text{Green Band}) \times (256 - \text{Red Band}))^{1/3} \quad (3)$$

The determination of the surface temperature values obtained from the thermal infrared bands is changed from the following radians values.

$$L_\lambda = (\text{Gain} * \text{DN}) + \text{Bias} \quad (4)$$

$$T = K2 / \text{In} ((K1 / L_\lambda) + 1) \quad (5)$$

Information:

- $L_\lambda$ = Radians value watt/(m²* m*ster*µm)
- Gain = Lmax – Lmin / DNmax – DNmin
- DN = Digital number
- Bias = Lmin
- T = Surface temperature value (C)
- K1 = 666.09 watt/(m²* m*ster*µm)
- K2 = 1282.71 watt/(m²* m*ster*µm)

A high FCD value that is a forest appears if the vegetation index and the shadow index are high. A low FCD value that is a shrubs object appears if the ground index and thermal index are moderate. Grassland objects have a high vegetation index and a low shadow index. If the vegetation index and the shadow index are low then the visible object is open ground. Based on the appearance so that it becomes the basis of the relationship between the indexes that build the FCD model The following combinations of four index characteristics in the FCD are presented in Table 1.

**Table 1.** Combination of four index characteristics

| | Forest | Shrubs | Grass | Soil |
|---|---|---|---|---|
| AVI | High | Middle | High | Low |
| BSI | Low | Low | Low | High |
| SI | High | Middle | Low | Low |
| TI | Low | Middle | Middle | High |

Source from [4]

The results of the index are then downgraded to the values of vegetation density (VD) and Scaled Shadow Index (SSI). VD is a derivative of a combination of Advance Vegetation Index (AVI) and Bare Soil Index (BI). While SSI is derived from a combination of Shadow Index (SI) and Thermal Index (TI). The next stage to get FCD is by integration between VD and SSI using the following formula.

$$FCD = (VD \times SSI + 1)^{1/2} - 1 \quad (6)$$

*4.2. Changes Detection*

According to [6] explained that the study of change is capable of producing data and information that is (1) changing zones / regions as well as the rate of change, (2) the spatial distribution of changes, (3) the stages of change, (4) the accuracy of the results of the change study. The change study is a step to see changes in vegetative structural composition resulting from FCD models in 2002 (early) and 2015 (end) monitoring. To identify such changes through algorithms available in digital image processing software and geographic information systems. Generally, the method used is the overlay method is a method that operates by overlapping the map to produce a map of the object changes studied. Through integration of FCD maps of early and final year yields result an information on changes in the vegetation structural composition as well as study changes in the study area.

The field check step is done by observing / checking the existing condition in the field. Checking was done on crown density, tree height and vegetation composition using 30 sample points. The accuracy test process in this research uses the accuracy test method with Standard error of estimation and Confidence level analysis to test the result of the vegetative structural composition of the FCD model. Selection of this method because the result of image processing that is map of structural composition of vegetation is ordinal data with certain range.

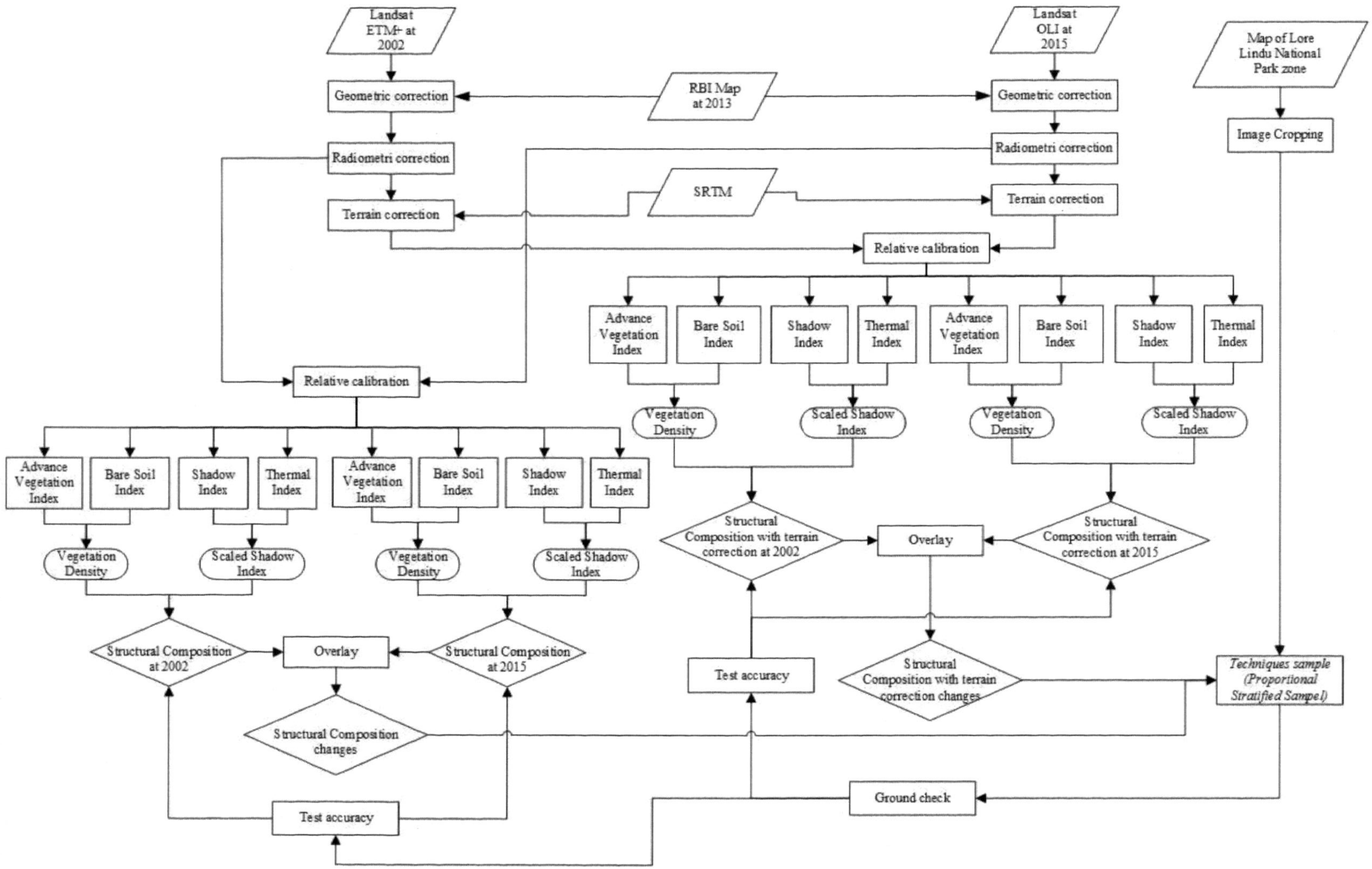


**Figure 1.** Flowchart of research

## 5. Result and Discussion

### *5.1. Structural Composition of Vegetation*

The structural composition is a description of the horizontal density and the vertical density of vegetation in relation to the layer structure of vegetation. The structural composition is the result of the extraction of the Forest Cover Density (FCD) model consisting of several indices. Vegetation index, bare soil index, shadow index and thermal index as the parameters that make up the FCD model. The relationship between vegetation index and bare soil index shows negative correlation. So also in the shadow index and temperature index also showed negative correlation. The correlation is derived from a principle component analysis (PCA) which in principle aims to reduce the dimensions of information that can provide more complete new information. The principle component analysis (PCA) on the vegetation index with open soil index formed the percentage of vegetation density (VD), meanwhile principle component analysis (PCA) on shadow index and thermal index formed a scaled shadow index (SSI). Next step is integrating VD and SSI to produce FCD that shows structural composition of vegetation. The vegetative structural composition of the FCD model results is illustrated in the vegetation layer structure associated with the percentage of vegetation density from low to high that can be interpreted from grass, shrubs, bush, bushes and forest.

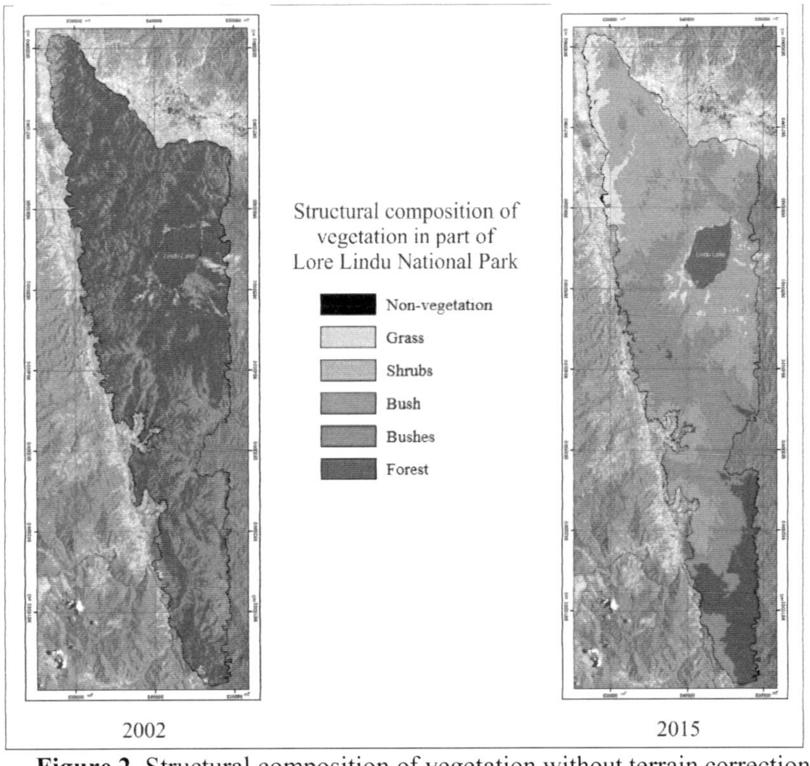


**Figure 2.** Structural composition of vegetation without terrain correction

The structural composition of vegetation in 2002 presented 5 classes dominated by forest classes with a range of values between 214 - 255 which has an area of 59,838.1 ha. Based on Figure 2. the appearance of the forest class is spread over the topography of the surface from flat (plains) to very steep (mountain), but the majority of the appearance lies on the topography of the steep surface.

**Table 2.** Structural composition of vegetation classes in 2002

| Class Structural Composition | Area (ha) | Percentage |
|---|---|---|
| Floor/Ground Layer (Grass) | 190,8 | 0,2 % |
| Shrub Layer (Shrubs) | 421,9 | 0,4 % |
| Understorey (Bush) | 2.270,9 | 2,3 % |
| Midstorey (Bushes) | 31.001,4 | 31,9 % |
| Upperstorey (Forest) | 59.838,1 | 61,6 % |

Source: Data Analysis, 2016

The result of the vegetative structural composition of the FCD model transformation on Landsat image 2015 also shows 5 classes. The shrub class with a range of values 132 - 173 dominates the area of study which has an area of 41,686.3 ha. In general, the class is scattered throughout the northern and central parts of the study area that the majority lies on the topography of the surface from flat to very steep.

**Table 3.** Structural composition of vegetation classes in 2015

| Class Structural Composition | Area (ha) | Percentage |
|---|---|---|
| Floor/Ground Layer (Grass) | 1.137,6 | 1,2 % |
| Shrub Layer (Shrubs) | 7.124 | 7,3 % |
| Understorey (Bush) | 41.687,3 | 42,9 % |
| Midstorey (Bushes) | 37.198,4 | 38,3 % |
| Upperstorey (Forest) | 9.902,9 | 10,2 % |

Source : Data Analysis , 2016

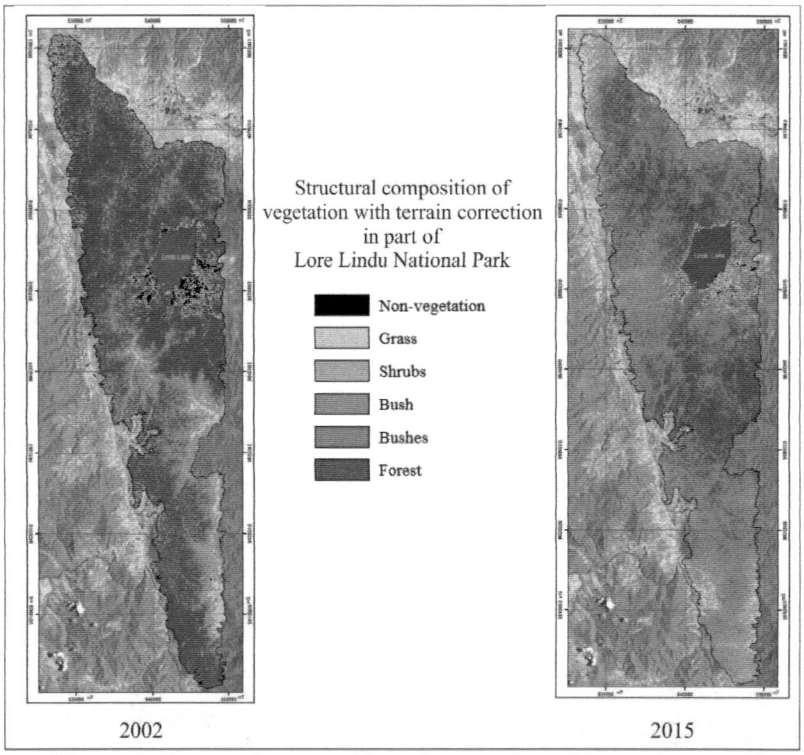


**Figure 3.** Structural composition of vegetation with terrain correction

The structural composition of vegetation with terrain correction in 2002 also featured 5 classes dominated by forest classes with a range of values between 214 to 255 which has an area of 50,949.7 ha. Based on Figure 3. the appearance of forest class is the most dominant class located on the topography of the surface from flat to very steep.

**Table 4.** Structural composition of vegetation classes with terrain correction in 2002

| Class Structural Composition | Area (ha) | Percentage |
|---|---|---|
| Floor/Ground Layer (Grass) | 2.281,7 | 2,3 % |
| Shrub Layer (Shrubs) | 5.248,1 | 5,4 % |
| Understorey (Bush) | 11.152,4 | 11,5 % |
| Midstorey (Bushes) | 24.456,6 | 25,2 % |
| Upperstorey (Forest) | 50.949,7 | 52,4 % |

Source: Data Analysis, 2017

The result of the vegetation structural composition of the FCD model transformation with terrain correction in Landsat image 2015 also shows 5 classes. The result of the structural composition with terrain correction is dominated by shrub class with a value range between 173 to 214 and has an area of 54,923 ha. The majority of the vegetation structural composition classes that dominate the study area lie on the steep surface topography.

**Table 5.** Structural composition of vegetation classes with terrain correction in 2015

| Class Structural Composition | Area (ha) | Percentage |
|---|---|---|
| Floor/Ground Layer (Grass) | 1.596,2 | 1,6 % |
| Shrub Layer (Shrubs) | 3.991 | 4,1 % |
| Understorey (Bush) | 11.328,1 | 11,7 % |
| Midstorey (Bushes) | 54.923 | 56,5 % |
| Upperstorey (Forest) | 21.441,6 | 22,1 % |

Source: Data Analysis, 2017

The next step after obtaining the map of the vegetation structural composition is the accuracy test. The accuracy test used standard analysis of estimation and Confidence level analysis on the vegetation structure composition map of 2002 and 2015. The accuracy test result is the accuracy value of map of vegetation structural composition both in 2002 and 2015 with field observation data. Accuracy on the map of structural composition obtained by 67.2% in 2002 and 78.3% in 2015. In addition, the accuracy test conducted on the structural composition of vegetation FCD model with terrain correction in 2002 and 2015 each has a value Amounted to 75.3% and 79.4%.

*5.2. Structural Composition of Vegetation Changes*

The multitemporal study is one of the hallmarks of remote sensing making it possible to see changes in the object of study. The study of changes in this research resulted in a map of changes in the vegetation structural composition of each pixel of the FCD model transformed map. The map of the change is a map compiled based on overlay techniques on the structural composition of the vegetation map of 2002 and 2015. All pixels on both maps are arranged to see how the changes and extent of the changes are. The assumptions constructed in the consideration of changes in vegetation structural composition are changes in the composition layer and vegetation structure in relation to the density and height of vegetation structure stands.

The change in the structural composition of vegetation is evident with the visualization of a 30-class change map that is 30 possible from changes in the structural composition of vegetation. The change is an assumption of changes in every class such as grass, bush, bushes, shrub and forest as well as non-vegetation objects that indicate changes in the structural composition of vegetation is reduced and increased, for example from forest class to shrub class. In addition there are 5 classes of vegetation structural composition that is not changed.

Based on data analysis and map of change of structural composition of vegetation, it can be explained that the total area of class of vegetation structural composition change is decreased 66.255,5 ha, while the class of change of vegetation structural composition is increased 10,699,9 ha and vegetation structural composition class still have area of 20,249, 3 ha. The majority of the class changes in the structural composition of vegetation encountered in the class is reduced especially in the forest class into a shrub with an area of 31,712 ha. The class of vegetation structural compositional changes is increasing which has the largest area of the shrub class into a forest with an area of 5,673.6 ha.

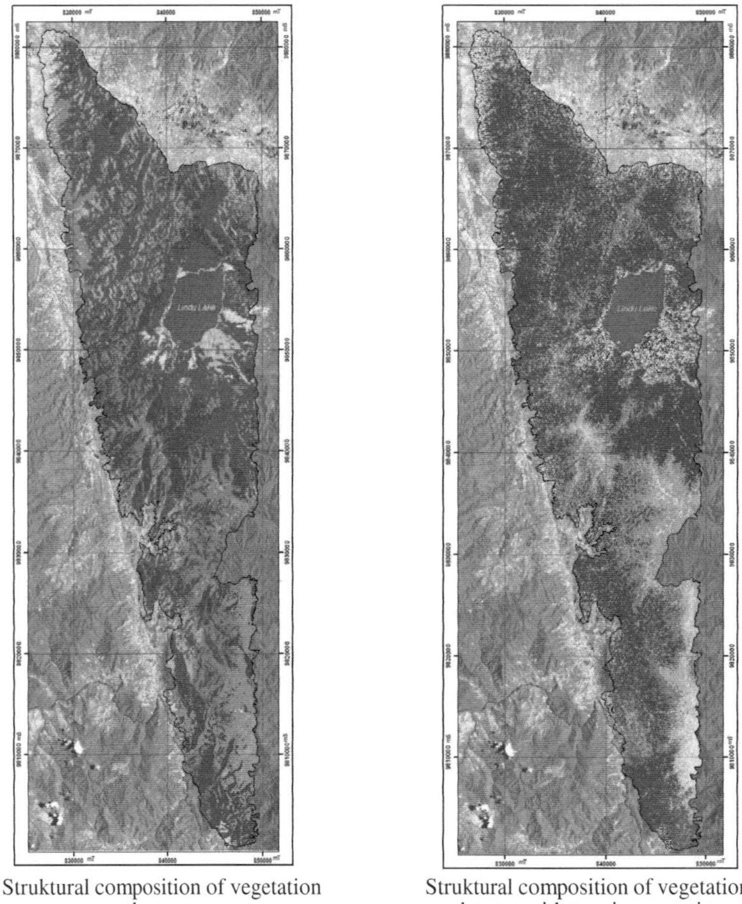

Struktural composition of vegetation changes | Struktural composition of vegetatior changes with terrain correction

| Structural Composition of Vegetation Changes | | | | | | |
|---|---|---|---|---|---|---|
| | Structural Composition of Vegetation in 2002 | | | | | |
| | | Non-vegetation | Grass | Shrubs | Bush | Bushes | Forest |
| Structural Composition of Vegetation in 2015 | Non-vegetation | | | | | | |
| | Grass | | | | | | |
| | Shrubs | | | | | | |
| | Bush | | | | | | |
| | Bushes | | | | | | |
| | Forest | | | | | | |

**Figure 4.** Structural composition of vegetation changes in part of Lore Lindu National Park

## 6. Conclusions

Transformation of FCD model can produce data and information structural composition of vegetation. Class structural compositions consist of 5 classes based on layers or layers of vegetation structures. The accuracy test on the result structural composition of vegetation using FCD model in 2002 was 67,2%, while the accuracy test of FCD model 2015 was 78,3%. The accuracy test of vegetation structural composition with terrain correction using FCD model in 2002 was 75.3%, while the accuracy test of the FCD model in 2015 was 79.4%. Accuracy to the structural composition of vegetation includes the intermediate level due to the difference in radiometric resolution of the image used. Decreasing in the structural composition of vegetation is 66,255.5 ha. On the other hand, the increasingly changing class has is 5,673.6 ha, while stable structural composition of vegetation is 20,249.3 ha. Decreasing in the structural composition of vegetation with terrain correction is 45,456.8 ha. In addition, the increasingly changing class is 20,866.7 ha, while stable structural composition of vegetation with terrain correction is 30,881.2 ha.

## References


[1] Lillesand, T. M., & Kiefer, R. W. 1979, *Remote Sensing and Image Interpretation*. John Wiley and Sons, New York.

[2] Miller, A. B., Bryant, E. S., and Birnie, R. W. 1998. *An Analysis of Land Cover Changes in The Northern Forest of New England Using Multi-temporal LANDSAT MSS Data*. International Journal Remote Sensing 19(2) : 215-265.

[3] Danoedoro, P. 2012, *Pengantar Penginderaan Jauh Digital*. Penerbit Andi, Yogyakarta

[4] Rikimaru, A., Roy, P. S., and Miyatake, S. 2002, *Tropical Forest Density Mapping*. Tropical Ecology 43(1) : 39-47.

[5] ITTO/JOFCA. 2003, *Optimum Utilization of RADARSAT-SAR Data in Conjuction with Enchanced FCD Model to Monitor Change in Status of Forest Resources*, Project Report. JOFCA

[6] Lu, D., Mausel, P., Brondizio, E., Moran. E. 2004, *Change Detection Technique*. International Journal Remote Sensing 25(12) : 2365-2407.

# The Combination of Spherical Photogrammetry and UAV for 3D Modeling

**T Ihsanudin[1*] and A R Affriani[2]**

[1]Postgraduate student at Geomatic Engineering, Engineering Faculty, UGM
[2]Lecturer staff at Mapping Survey and Geographic Information (SPIG), Faculty of Social Science Education, UPI

taufiq.ihsanudin9@gmail.com

**Abstract**. The complete of 3D models required the object that was recorded from both side and top. If the object recorded from above, then the object from the side can not be covered, and if the objects recorded from the side, it can not be covered from the top. Recording of objects from the side using spherical photogrammetry method and from the top using UAV method. The merge of both models using a conform transformation, by bringing the spherical photogrammetry coordinates system to the UAV model. The object of this research is Ratu Boko temple, Sleman, Yogyakarta. The spherical photogrammetry recording was performed by rotating the camera in 360° angle on the entire area of the temple. The area consists of 12 stations. The UAV method uses a drone with flight attitude of 20 meters. The merge of the both models produced the completeness of the temple model from the top and side.

**Keywords:** Spherical Photogrammetry, UAV, 3D model

## 1. Introduction

The technology for the acquisition of earth surface objects using aerial photographs develops in close range photogrammetry. Close range photogrammetry is a spatial data acquisition technique on the surface of the earth using photogrammetric methods with the distance of the camera to the object is 10 meters up to 100 meters. The application of close range photogrammetry is spherical photogrammetry or panoramic imaging. Spherical photogrammetry is a photogrammetry method that used panoramic images as the basic information [1]. The excess of this photo has a 360 ° perspective in a horizontal direction, 180 ° in a vertical direction and can form a projection field of a spherical surface.

The complete three-dimensional (3D) model cannot be done using a single survey tool. It needs some tools which can fill the void of the measurement object that cannot be reached from one survey tools. This research aims to combine spherical photogrammetry photo with Unmanned Aerial Vehicle (UAV) model. Spherical photogrammetry model was recorded the object from the side. As a complement of data from the top view, it used the UAV method.

IOP Conf. Series: Earth and Environmental Science **98** (2017) 012057 doi:10.1088/1755-1315/98/1/012057

## 2. Method

### *2.1. Location*

The object of this research is Ratu Boko temple, which located at Dawung, Bokoharjo village and Sumberwatu, Sambirejo village, Prambanan, Sleman, Yogyakarta.

### *2.2. Data*

The temple cannot be captured by single view photo for detail 3D modeling purpose; it needs to be captured from terrestrial photo with spherical photogrammetry. This data acquisition method are panoramic and UAV.

**Table 1.** Acquisition of 3D modelling

| Type | Equipment | Camera | Average photo distance | Number of Image |
|---|---|---|---|---|
| **Spherical Photogrammetry** | 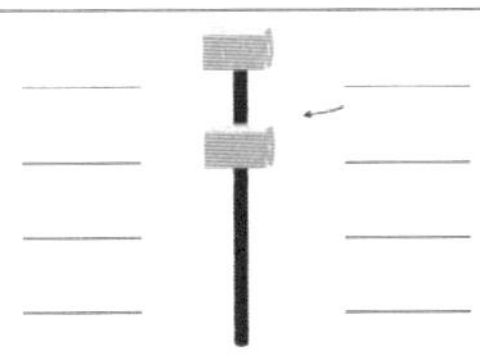 | Action camera Xiaomi YI | 7 m | 982 |
| **UAV** | 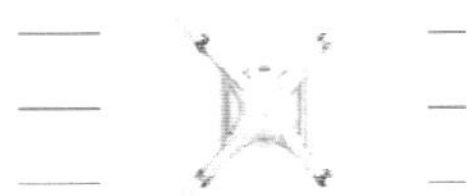 | Action camera Xiaomi YI | 20 m | 90 |

### *2.3. Spherical Photogrammetric Method*

Photo data was taken by positioning the camera at one exposure point. This point was used as the camera's rotational axis when shooting at 360 ° viewing angle. Overlap panoramic camera was made be vertical to get stereo photo of the temple facade which will be made 3D model. Shooting configuration of spherical photogrammetry method can be seen in figure 1.

### *2.4. UAV Recording*

UAV data collection using quadcopter with a flight attitude of 20 meters above the object. Yi's Xiaomi camera was mounted on the quadcopter body and set the exposure time every 1 second. The numbers of Ground Control Point (GCP) are 4 points and it measured using GPS rapid static method for 30 minutes observation. The GCP was used for reference binding during 3D UAV model processing.

## 3. Build 3D Model

### *3.1. Camera Calibration*

The process of 3D modeling must have accurate position. The accurate position needs the internal parameters of a camera. So, it must be known. The internal parameters need to reconstruct the beam and to know the size of camera systematic error. Calibration was performed to obtain the focal length, the main position of the photo (Xp, Yp), radial distortion (K1, K2, K3), and tangential distortion (P1, P2).

### *3.2. Object Reconstruction*

This research uses image-based modelling method. 3D modelling with IBM used to reconstructs the geometry object by identifying the same set of points from two or more photographs [2]. These points were connected with the same object or feature in each photo to produce the camera position and estimate the shape of the 3D model. The points from the same feature identification on several photographs resulted a very tight bounding point, it become a point cloud. The result of point clouds using spherical photogrammetry can be seen in Figure 1. The result of point cloud using the UAV can be seen in Figure 2.

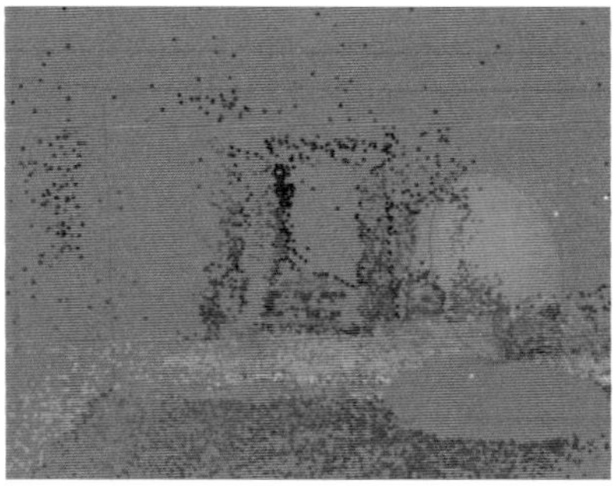

**Figure 1.** Point cloud of Spherical photogrammetry

**Figure 2.** Point cloud of UAV

### *3.3. Meshing and Texturing*

Meshing and texturing becomes the final part of 3D modelling (Kwiatek, 2016). The surface of the object can be represented in mesh form. Mesh (in the computer graph community) was derived from integration of the point which connected with the same features into a triangular net, it known as Triangular Irregular Network (TIN) in photogrammetry. The TIN has not been able to display the objects in real and accurate position. The meshing process can be visualized by texturing. It was done by overlaid the photo on 3D surface. The internal and external orientation parameters of each photo were adjusted by photo coordinates for each vertex mesh, so the 3D surface will have a texture with realistic photographs [3].

The result of texturing process using spherical photogrammetry model can be seen in Figure 3. The result of texturing process using UAV model can be seen in Figure 4.

**Figure 3.** Spherical photogrammetry 3D model

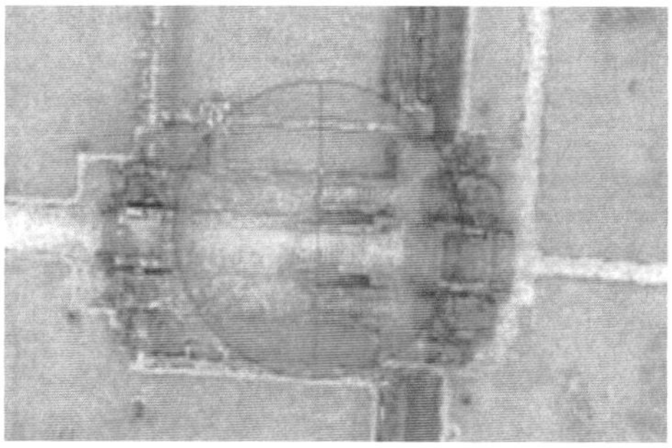

**Figure 4.** UAV 3D model

### *3.4. Combination of 3D Model*

The merge of both 3D model done by transformation conform between two 3D models. Merging is done by searching for a minimum of three Allied point between these two models. In the process of transformation, 3D roof model used as a reference, while the 3D facade models transformed to the roof model. This is because the 3D roof model was already on the ground coordinate system (georeferenced), while the 3D facade model was in the local coordinate system. The selection of ally point in both models generated a matrix transformation. It used to transform the 3D facade models to 3D roof models.

IOP Conf. Series: Earth and Environmental Science **98** (2017) 012057 doi:10.1088/1755-1315/98/1/012057

**Table 2.** Combination of spherical photogrammetry and UAV

| No. | Method | Top View | Side View |
|---|---|---|---|
| 1 | 3D model of Spherical Photogrammetry |  |  |
| 2 | 3D model of UAV |  |  |
| 3 | 3D Combine of Spherical Photogrammetry + UAV |  |  |

Each data processing produces its 3D model. UAV data produces a top-view 3D model, while spherical photogrammetry data produces a side 3D model (facade). Both models have their respective deficiencies the combination was done to get a complete three-dimensional modelling.

## 4. Conclusions

This research explains the combination of spherical photogrammetry and UAV. The advantage of combination 3D model result was fulfilled completeness side view (facade) of the temple. The previous UAV data only can model the visible part of a temple, then after the combination it can model the whole temple object. The disadvantage of the 3D model combination was can not unifying colour between the 3D roof model and 3D facade model. It happens because the difference time when shooting both object models. The difference time affects the brightness of the colours in both models.

## References


[1] Fangi G 2015 Towards an Easier Orientation for Spherical Photogrammetry The International Archives of the Photogrammetry *Remote Sensing and Spatial Information Sciences.*

[2] Cohen A et al 2012 *Discovering and Exploiting 3D Symmetries in Structure from Motion.*

[3] Remondino F 2006 *Image-Based 3D Modeling: A Review* 21(September) pp 269–291.

[4] Huang F, R Klette, K Scheibe 2008 *Panoramic Imaging Sensor-line Camera and laser Range-Finders* (United Kingdom: John Wiley & Sons Ltd).

[5] Michail Bethel and McGlone 2001 *Introduction to Modern Photogrammetry* (New York: John Willey and Sons. INC).

[6] Lowe M B a D G 2003 *Recognising Panoramas* Nice, France, ICCV2003 pp 1218-1225.

# Remote Sensing Studies of Suspended Sediment Concentration Variation in Barito Delta

**Deasy Arisanty[1*], Aswin Nur Saputra[1]**

[1]Geography Education, Faculty of Teaching and Education Science, Universitas Lambung Mangkurat, Brigjend H. Hasan Basry Banjarmasin, Indonesia

deasyarisanty@unlam.ac.id

**Abstract**. The dynamic of suspended sediment concentration in Barito Delta depend on the process in upstream. Agriculture, mining, and illegal logging in Barito River upstream has an effect for suspended sediment concentration in Barito Delta. The objective of research is to estimate the variation of suspended sediment concentration in Barito Delta. The data of research consist of Landsat 7 in year 2011 and measurement result data of suspended sediment concentration both in wet season and dry season in year 2011. Data analysis is regression analysis to estimates the variation of suspended sediment concentration in Barito Delta. The method of research compares three types of spectral transformation for suspended sediment that is Normalized Suspended Material Index (NSMI), Normalized Differences Suspended Sediment Index (NDSSI), and band ratio (green/blue). The result of the transformation is compared with the value of the field measurement. Based on the result of the comparison can be known the suitable type of transformation for the suspended sediment estimation in Barito Delta. The result of research explains that NSMI has the highest value to estimate the variation of suspended sediment concentration in Barito Delta.

**Keywords:** remote sensing, suspended sediment concentration

## 1. Introduction

Barito River is main river in Barito Delta. Barito River flows from Muller Mountain into Java Sea [1]. Barito River has the length about 900 km, width average about 800 m, and depth average about 8 m [2] [1].

Land use changes in upper part of watershed have increased the sediment in Barito Delta. Degradation has been occurred in upper part of watershed due to illegal logging, mining, forest encroachment, and forest fire [3] [1]. Land use may affect the sediment yield from the watershed [4].

Suspended sediment is transported by the river discharge. Barito River has the river discharge about 1731 $m^3/s$ in wet season and 1008.75 $m^3/s$ in dry season. River discharge in mouth of Barito River can transport about 17.67 mg/l in dry season. River discharge in mouth of Barito River can transport the suspended sediment about 25 mg/l in wet season [1] [5].

Variation of Suspended Sediment is the important for studying about coastal protection, management, engineering, restoration, and development. To estimate variation of suspended sediment in simultaneously and short time periods needs the satellite remote sensing [6]. Remote sensing can be used to obtain spasio-temporal information of some water quality parameter such as suspended sediment concentration, especially in large river [7] [8].

Suspended material can be identified using remote sensing data. The interaction of light with a water-sediment mixture is complex because both materials scatter and absorb radiation. the apparent upwelling radiance measured remotely is not only a function of water and sediment properties, but also is a function of initial solar input, atmospheric transmission of the radiation, the upwelling radiance at the water surface, and specular reflection off the water surface due to both sunlight and skylight. These variables in turn are depend on the illumination geometry, atmospheric conditions, and water surface parameters such as water surface roughness [9]. If all of these variables can be accounted for, it may be possible to determine accurately the relationship between remotely sensed reflectance and sediment concentration.

The equation to identify the suspended material is Normalized Suspended Material Index (NSMI). The equation has two principles, i.e. a peak reflectance in the blues range of the visible spectrum is contained in water, and reflectance through the visible spectrum can be increased by the presence of SM. Suspended sediment can be identified using Normalized Difference Suspended Sediment Index (NDSSSI) and a Band Ratio (green/blue) [10].

## 2. Methods

The research objective is to analyze the best index for suspended material information in Barito Delta. The index to generate information about suspended material content through satellite imagery consists of Normalized Suspended Material Index (NSMI), Normalized Differences Suspended Sediment Index (NDSSI), and band ratio (green/blue) [10]. The research uses Landsat 7 image with recording time is on 12, May 2011 and 3, October 2011. In this image has been done geometric correction and atmospheric correction to the level of reflectance at surface. Recording time related to the sampling time season during wet season and dry season.

Suspended solid sampling point in Barito Delta area is 18 points of observation. Suspended sediment is measured in mouth of river and near shore. Transect method is used to measure and to collect data. Suspended load is measured in dry season and wet season. Suspended load is collected with sediment sampler [1].

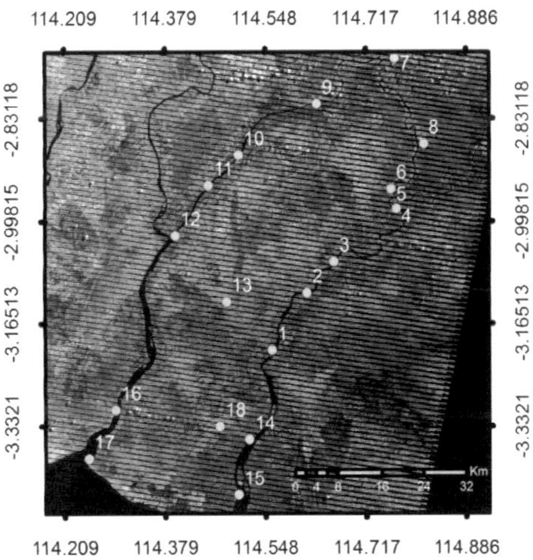


**Figure 1.** Sample area of Barito Delta

The estimation method of suspended material content in Barito delta uses spectral transformation of satellite imagery. Spectral transformations use several variations of band deletion. The types of band intercepts are NSMI, NDSSI, and Band Ratio. The result of spectral transformation based on index is done by linear regression analysis with field measurement value. The highest regression value is used as the most appropriate index for Barito Delta.

NSMI is a transformation index that exploits band delimitation between red, green, and blue bands. The result of channeling produces a range of values from -1 to +1. The higher value of transformation will be the brighter of hue. NDSSI is almost similar to the NSMI index. NDSSI is limited to the near infrared and blue bands, but NSMI is also utilizes the wavelength of green and red. The resulting range on the NDSSI index is between -1 to +1.

$$NSMI = \frac{(\rho_{red})+(\rho_{green})-(\rho_{blue})}{(\rho_{red})+(\rho_{green})+(\rho_{blue})} \quad (1)$$

$$NDSSI = \frac{(\rho_{blue})-(\rho_{nir})}{(\rho_{blue})+(\rho_{nir})} \quad (2)$$

The Band Ratio method is a type of band incision applied to 2 image bands. In this study, the band ratio uses green and blue bands to estimate the content of suspended materials. The result of this index is a range of values from 0 to infinity.

$$\text{Band Ratio} = \frac{(\rho_{green})}{(\rho_{blue})} \quad (3)$$

## 3. Result and Discussion

Based on the result of regression test it is found that the highest value is found in the method using NSMI transformation with $R^2$ of 0.514 in wet season and $R^2$ equal to 0.700 in dry season. The result is made possible using of visible bands compared to the NDSSI method ($R^2 = 0.010$) using the infrared band. The equation uses to understand the suspended solid content in water. Visible bands more penetrating in the water surface than infrared wavelength, due to infrared wavelength tends to be absorbed, when it comes to the water surface.

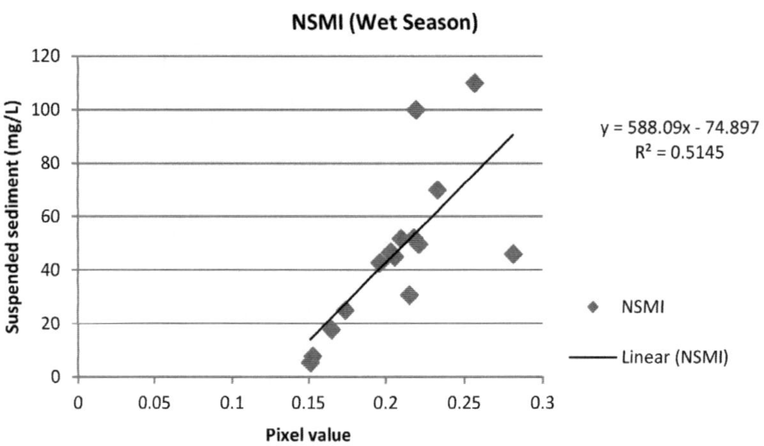


**Figure 2.** NSMI in Wet Season

**Table 1.** Suspended sediment during wet season

| No. | Location | Suspended Sediment (mg/l) |
|---|---|---|
| 1 | Barito River 1 | 45 |
| 2 | Barito River 2 | 7.67 |
| 3 | Barito River 3 | 17.67 |
| 4 | Barito River 4 | 30.67 |
| 5 | Barito River 5 | 49.67 |
| 6 | Barito River 6 | 60 |
| 7 | Barito River 7 | 46 |
| 8 | Barito River 8 | 99.97 |
| 9 | Barito River 9 | 25 |
| 10 | Pulau Petak River 1 | 46.67 |
| 11 | Pulau Petak River 2 | 51.67 |
| 12 | Pulau Petak River 3 | 30.67 |
| 13 | Pulau Petak River 4 | 25.33 |
| 14 | Kapuas Murung River 1 | 5.33 |
| 15 | Kapuas Murung River 2 | 70 |
| 16 | Kapuas Murung River 3 | 42.67 |
| 17 | Talaran Channel | 110 |
| 18 | Serapat Channel | 52 |

Source: Arisanty (2013)

**Table 2.** Suspended sediment during dry season

| No. | Location | Suspended sediment (mg/l) |
|---|---|---|
| 1 | Barito River 1 | 25 |
| 2 | Barito River 2 | 14 |
| 3 | Barito River 3 | 15.33 |
| 4 | Barito River 4 | 12 |
| 5 | Barito River 5 | 55 |
| 6 | Barito River 6 | 45.67 |
| 7 | Barito River 7 | 53.67 |
| 8 | Barito River 8 | 66.33 |
| 9 | Barito River 9 | 36 |
| 10 | Pulau Petak River 1 | 56.33 |
| 11 | Pulau Petak River 2 | 130 |
| 12 | Pulau Petak River 3 | 60 |
| 13 | Pulau Petak River 4 | 82 |
| 14 | Kapuas Murung River 1 | 71 |
| 15 | Kapuas Murung River 2 | 13.67 |
| 16 | Kapuas Murung River 3 | 17.67 |
| 17 | Talaran Channel | 95 |
| 18 | Serapat Channel | 66.67 |

Source: Arisanty (2013)

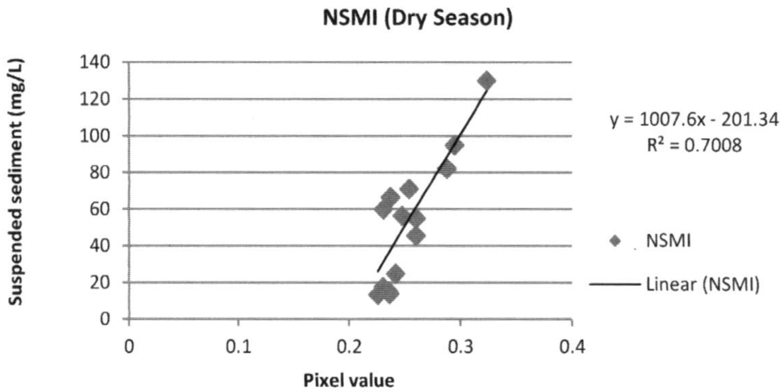


**Figure 3.** NSMI in Dry Season

Band Ratio between green and blue has a lower value ($R^2 = 0.05$) when compared with the other two methods, due to the absence of red band. The red band is a band with a fairly sensitive wavelength. Red band has a higher reflection to the ground when compared to a blue or green band.

Plotting the location of sampling in the field on the right images is in the striping line at Barito River 6, Kapuas Murung River 1 and 2 on the image of wet season. While on dry season the sample points overlap with striping line at Barito River 6 and 9. Striping line is still present in Landsat 7 ETM + image and has not done striping line correction so that in some locations the pixel value can't be used. Another problem is the existence of measurements made in small channel areas such as in Talaran and Serapat are relatively not too large. It becomes unrepresentative to compare the value in the field with the index value through the utilization of Landsat-7 ETM + image which has a spatial resolution of 30m on the visible band. Based on the results obtained by applying the algorithm obtained through Band Math gets the description of the distribution of suspended material in Delta Barito. In the resulting image looks suspended material content is higher in the dry season with a content of more than 70 mg/l. This is possible because of the low flow of river water, so that the suspended material content is more clearly visible from the surface [5].

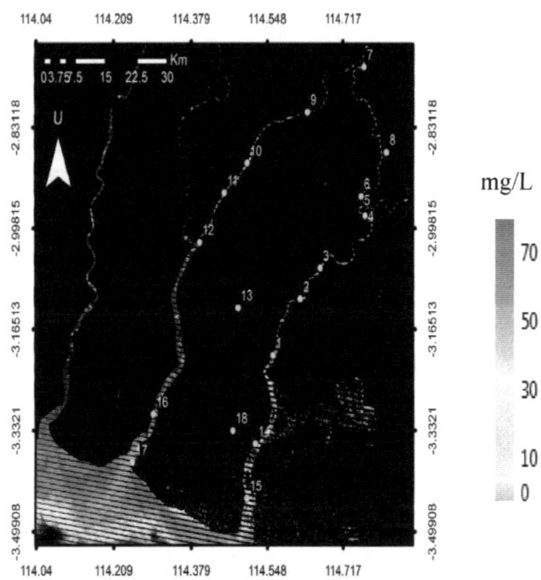


**Figure 4.** Distribution of dry season Suspended Material

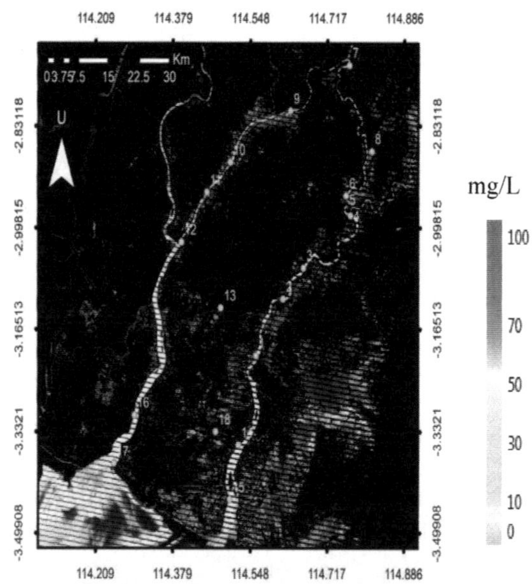


**Figure 5.** Distribution of wet season Suspended Material

## 4. Conclusions

The result of research shows that NSMI has the highest value to estimate the variation of suspended sediment concentration in Barito Delta. NSMI value in wet season is 0.514 and 0.700 in dry season. NDSSI and Band Ratio has the lower value of regression than NSMI. NSMI has the higest value due to visible bands wavelenght more penetrating in the water surface than infrared wavelength.

## References

[1] Arisanty D 2013 *Morphodynamic of Barito Delta, Southern Kalimantan (*Yogyakarta: Faculty of Geography Universitas Gadjah Mada)

[2] Mulyahadi 2010 Accessibility Assessment of Road Infrastructure for Support of River Transportation in Kalimantan *Jurnal Transportasi Darat* **12**(4) pp 155-175

[3] Ministry of Environment *Study and Implementation of Barito Watershed Integrated Management* 25 September 2010. [Online]. Available: http://www.menlh.go.id/kajian-dan-implementasi-penanganan-das-barito-terpadu-national-side-event/. [Diakses 20 October 2010]

[4] Rovira A and C Ibanez 2007 Sediment Management Options for the Lower Ebro River and Its Delta *Journal of Soil Sediment* **7**(5) pp 285-295

[5] Arisanty D, J Sartohadi, M A Marfai and D S Hadmoko 2014 Sediment Dynamic of Barito Delta, Southern Kalimantan, Indonesia *Journal of Environment* **1**(1) pp 30-37

[6] Li Y, C Li and X Li 2017 Remote Sensing Studies of Suspended Sediment Concentration Variations in a Coastal Bay during the Passages of Atmospheric Cold Fronts *IEEE Journal of Selected Tropics in Applied Earth Observations and Remote Sensing* **10**(6)

[7] Cui L, Y Q T Fei, Y Liu and G Wu 2013 Using Remotely Sensed Suspended Sediment Concentration Variation to Improve Management of Poyang Lake, China *Lake and Reservoir Management Journal* **29**(1)

[8] Wang J J, X X Lu, S C Liew and Y Zhou 2010 Remote Sensing of Suspended Sediment Concentration of Large Rivers Using Multi-temporal MODIS Image: an Example in the Middle and Lower Yangtze River, China *International Journal of Remote Sensing* **31**(4)

[9] Mertens L A, M O Smith and J B Adams 1993 Estimating Suspended Sediment Concentration in Surface Waters of the Amazon River Wetlands from Landsat Images *Remote Sensing of Environment* **43**(3) pp 281-301

[10] Montalvo 2010 *Spectral Analysis of Suspended Material in Coastal Waters: A Comparison between Band Math Equations* (Mayaguez: Departement of Geology University of Puerto Rico)

# The Effect of Field Spectral Reflectance Measurement Distance to the Spectral Reflectance of *Rhizophora stylosa*

**M Kamal[1,2*], M U L Ningam[1], F Alqorina[1]**

[1]Department of Geographic Information Science, Faculty of Geography, Universitas Gadjah Mada, Yogyakarta, Indonesia
[2]PUSPICS, Faculty of Geography, Universitas Gadjah Mada, Yogyakarta, Indonesia

m.kamal@ugm.ac.id

**Abstract.** Mapping mangrove species from remote sensing data through its spectral reflectance pattern collected in the field is challenging. There are high variations in light condition, leaf orientation, canopy structure, background objects and measurement distance when measuring mangrove spectral reflectance in the field. Spectral measurement distance to the object is one of the most important aspects controlling the result of spectral reflectance pattern. This research is aimed to assess the effect of spectral reflectance pattern of *Rhizophora stylosa* collected at various distances. Specific objectives of this research are to collect samples of mangrove spectral reflectance pattern in the field, to assess the effect of the observation scale to the result of the spectral reflectance pattern, and to characterize the mangrove spectral reflectance pattern resulted from different observation scales. Spectral reflectance data collection in the field was conducted using JAZ EL-350 field spectrometer at 2cm, 50cm, 1m, 2m, and 5m distance and was conducted in Karimunjawa Island, Jepara, Central Java, Indonesia. A visual comparison of the spectral reflectance curve was conducted to understand the effect of measurement distance. The results of this study indicate that the difference in the measurement distance of *Rhizophora stylosa* species was highly influential to the resulting spectral reflectance curve. The spectral reflectance curve recorded at close range to the leaf (i.e. 2 cm) has the lowest curve variation, as well as the furthest distance (i.e. 5 m). This study is a basic study that supports the development of the use of remote sensing imagery for mangrove species mapping.

**Keywords:** Spectral reflectance, *Rhyzophora stylosa*, JAZ EL-350

## 1. Introduction

Mangrove is one of the most important objects in the wetland ecosystem, which forms the relationship between terrestrial and oceanic ecosystems in the tropics and sub-tropics. Mangroves have many ecological and economical functions, such as keeping coastal areas from ocean currents and winds, providing spawning grounds for marine organisms, maintaining food cycles of marine organisms, maintaining coastal water quality, wildlife fauna and as tourist attractions [1][2][3][4][5]. Nevertheless, the health and existence of mangroves are under considerable threat from human disturbance and natural disasters. Major threats to mangroves include logging for fuels and timber [6], conversion of mangrove habitats to other land uses such as agriculture, ponds, industry and urban development [7][8][9], and relative sea level rise [10]. These threats corroborate the importance of efforts to provide accurate and up-to-date data on mangrove status in Indonesia.

Over the past 20 years, remote sensing data have been used for mapping and monitoring of mangrove ecosystems [11][12]. Mangroves have distinct appearance in the optical remote sensing images due to their specific spectral reflectance pattern, and make them recognisable from their spectral reflectance pattern. The mangrove distinctive spectral patterns are mainly found in near infrared wavelengths (NIRs, 700-1200 nm) and shortwave infrared (SWIR 1200-2500 nm) [12][13]. This spectral pattern able to distinguish them from nearby objects, such as forested forests, brackish marshes, and mud flats. Several studies have been conducted using a spectral reflectance pattern to map the mangrove ecosystem [14][15][16] and identification of mangrove species [17][18][19][20][21], with varying degrees of success. The largest spectral-based species separability can be found at the NIR wavelength due to differences in the internal structure of the leaves of each species and the difference in the geometry of the leaves and canopy [12]. SWIR wavelength is also good for this study because there are no visible variations of pigment conditions among mangrove species at this wavelength, but the leaf components (salinity, sugar, water, protein, oil, lignin and cellulose) can be distinguished [20].

According to Hirano et al. [19] and Vaiphasa et al. [20], each mangrove species provides specific spectral reflectance characteristics that make it possible to be used as a key species identification of remote sensing imagery. However, when we examined more deeply, the mechanism of interaction between light (i.e. electromagnetic waves) and mangrove objects holds two major problems. First, the mangrove spectral reflectance pattern is strongly influenced by the tidal effect of sea level on the substrate of the mud soil at its base, thus causing mixed-pixel pixels. Secondly, many environmental factors influence the mangrove spectral reflectance so that the species identification based on spectral reflectance can be inaccurate [22][23]. According to Diaz and Blackburn [23], the spectral reflectance variation of mangrove canopy is a function of some optical characteristics of objects, such as leaf area index (LAI), substrate base reflectance, and leaf orientation direction. For a single species, the pattern of spectral reflectance is affected by the age, health, and phenological characteristics and plant physiology [22]. So, the measurement of the spectral reflect pattern for the identification of mangrove species should be done carefully to avoid reading errors due to the influence of reflectance other than mangrove objects.

It is therefore necessary to conduct a systematic study to assess the effect of distance measurement of spectral reflectance patterns of mangrove species in the field to the characteristics of the resulting spectral reflectance curves. This research is aimed to assess the effect of spectral reflectance pattern of *Rhizophora stylosa* collected at various distances. Specific objectives of this research are to collect samples of mangrove spectral reflectance pattern in the field, to assess the effect of the observation scale to the result of the spectral reflectance pattern, and to characterize the mangrove spectral reflectance pattern resulted from different observation scales. In Indonesia, research on these effects has not been conducted and it is important to establish a field spectral reflectance measurement procedure for the identification and mapping of mangroves. This research is a basic research that can be developed for mapping, inventory and monitoring of mangrove species using remote sensing imagery.

## 2. Methods

### *2.1. Study site*

The location of this research is mangrove forest in part of Karimunjawa Islands National Park, Jepara, Central Java, Indonesia (between 110° 24' 10" - 110° 30' 10" E and 4° 47' 48" - 5° 50' 12" S). Karimunjawa islands consist of 22 islands with total area of 111,625 ha, where only five of these islands are inhabited [24]. This area has a wet tropical maritime climate with an average temperature between 26-30 °C and an average humidity between 70-85%. The mean annual rainfall is 2632 mm; while the average rainfall of dry months (April to September) is 60 mm and the wet month (October to

March) is 400 mm. Karimunjawa Islands represent several habitats including lowland rainforests, seagrass and algae, coastal forests, mangrove forests and coral reefs [25]. In general, along the coast of the island is composed of mangrove formations that protect the coast from the waves of the sea and wind. The largest mangrove area in Karimunjawa National Park is located in the west of the two main islands; Karimunjawa and Kemujan. Based on reports from Karimunjawa National Park Authority, there are 45 mangrove species in this area (27 true mangroves and 18 mangroves), where *Rhizophora stylosa* is the most dominant species.

### *2.2. Equipment*

The equipment used to collect mangrove spectral reflectance in this study was JAZ EL-350 portable spectrometer. JAZ is a portable spectrometer product from Ocean Optics (https://oceanoptics.com/) that is designed to support fast, precise, non-destructive optical measurements and without direct contact with object. JAZ spectrometer system is designed in the form of modular structure consisting of several autonomous modules that form a whole spectrometer system. The JAZ module used in this study consists of a Data Processing Unit (DPU), battery, and VIS-NIR spectrometer modules. Other than that, the JAZ system is supported by several components, including 2 m fiber optic cable, collimating lens, and white panel (spectralon). The JAZ detector wavelength ranges from 350 to 1000 nm, which covers visible to near infrared spectrum of the light. A collimating lens is attached at the front of the fibre optic cable. This component used to align the scattering light coming into the JAZ sensor to optimize the light signal. It covers the spectrum range from 200 to 2000 nm, and the field of view is set up 14.25°. This system can be used for various spectral-based measurements, including absorption, reflectance, emission or transmittance of the object or irradiance. This system is designed to be used for spectra sample measurements both in the laboratory and in the field.

### *2.3. Field spectral reflectance measurement*

Spectrometers and all supporting components need to be prepared and setup prior to the field measurements. The preparation includes what environment will we work in, what components are needed and whether they work well or need calibration. The field work was conducted on 5 April 2016 between 09.00 to 11.00 a.m. During the field campaign, we followed the procedure of spectral reflectance measurement developed by our Remote Sensing Laboratory [26] (Figure 1). To normalize the light condition, a white and dark calibration was performed prior to each measurement set. The angle of spectrometer sensor (collimating lens) was setup at 45° against the direction of the sun light to avoid the shadow on the target object. The selection of sample location was guided by WorldView-2 pan-sharpened image (0.5m pixel size), that is by identifying accessible *Rhizophora stylosa* at the edge of the mangrove forest where measurement of spectral reflectance at different distance can be performed. From previous research [27] *Avicennia marina*, *Rhizophora stylosa*, *Bruguiera gymnorrhiza*, and *Lumnitzera racemosa* can be distinguished by visual interpretation of WV-2 images because they have specific canopy patterns and clustered at specific locations. From this site, a single sampling point was taken purposively based on the availability of target species and measured at several distances.

Spectral reflectance measurements of object samples were performed at several distance variations to the spectrometer sensor, i.e. 2 cm, 50 cm, 1 m, 2 m, and 5 m. The 2 cm distance represents the measurement of individual leaf objects, 50 cm and 1 m represent the variation of leaf orientation in the canopy, and a distance of 2 and 5 m represent the object's canopy condition as seen by remote sensing image sensors. Each sample measurement was repeated at least 10 times to evaluate the variation of readings that can be used for further analysis.

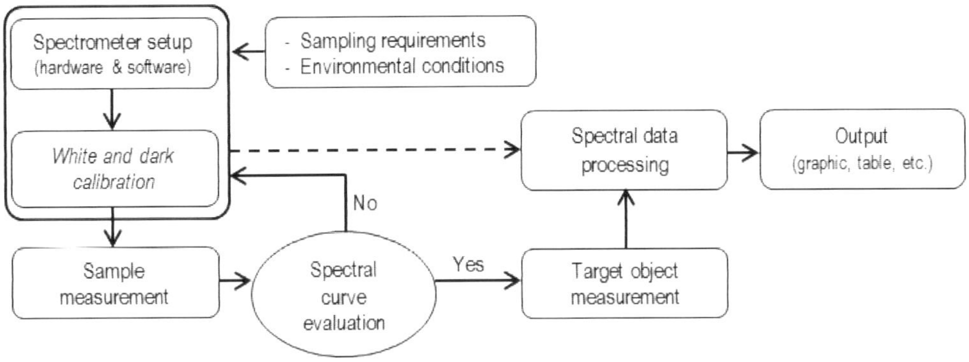


**Figure 1.** The procedure of spectral reflectance measurement.

## 3. Results and Discussion

### *3.1. Field spectral reflectance measurement*

Measurement of the spectral reflectance sample of *Rhizophora stylosa* was performed on the part of mangrove forest adjacent to sea water where this species is commonly encountered. The optical measurement module performed by JAZ EL-350 for this research was reflectance, which measures the amount of reflected energy from the sun by an object at a wavelength of 350-1000 nm. Each measurement of spectral reflectance begins with the light calibration at the measuring location. The calibration requires the sampling of the 'white reference' obtained from spectralon to represent almost perfect reflection of incident light or energy, and the 'dark reference' spectrum to represent fully absorbed energy. Based on the comparison between the white and dark reference, the spectral reflectance sample of the object can be calculated. The magnitude of the reflectance of the object sample relative to the white reference and the dark reference is as follows [28]:

$$\%R_\lambda = \frac{S_\lambda - D_\lambda}{R_\lambda - D_\lambda}\ x\ 100\% \qquad (1)$$

where $\%R_\lambda$ is the percentage of object spectral reflectance, $S_\lambda$ is the sample of the object's measurement, $D_\lambda$ is the dark reference, and $R_\lambda$ is the white reference.

Spectral reflectance measurements of objects using JAZ EL-350 in default mode produce a spectral reflectance curve that is full of noise in the near-infrared wave spectrum section (750-950 nm). In fact, this part of the light spectrum is theoretically very important to distinguish the spectral response between mangrove species [12][20]. Therefore, adjustments were made to the spectrometer setup; those were the 'boxcar width' and 'scan to average' options. Boxcar width averages the spectral values of neighbouring pixels along the wavelength axis. High value of boxcar width involved more neighbouring pixels involved in the averaging. This needs to be done primarily to minimize noise, but the implication of the increased value of boxcar is decreased its optical resolution. Another setup option, scan to average, allows the spectral reading of the object several times in one sample and then the readings are averaged. This step is generally not necessary, but is useful for reducing noise or increasing the reflectance of objects with low-intensity reflectance. Based on our experiments, the value of boxcar width and scan to average is 10. By giving this value, the resulting spectral reflectance curve is smoother and free of noise (Figure 2).

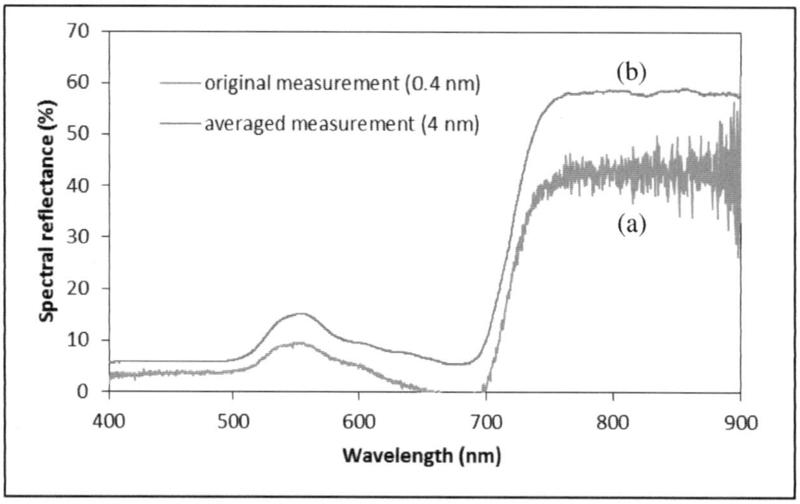


**Figure 2.** Spectral reflectance reading, (a) original reading at 0.4 nm spectral resolution, and (b) averaged reading at 4 nm.

*3.2. Spectrometer measurement distance comparison*

Spectral reflectance curve readings were performed at 2 cm, 50 cm, 1 m, 2 m, and 5 m range of sensors to simulate different recording levels of vegetation objects. Each measurement was repeated 10 times of consecutive recording to assess the consistency of reading results. From the plot of spectral reflectance results (Figure 3), it can be seen that the spectral reflectance pattern follows the spectral reflectance pattern of healthy and green vegetation. From the measurement results it is noticeable that there are variations of spectral pattern of objects caused by differences in the distance measurement of the spectrometer sensor. The reading of the spectral reflectance response at the leaf level with a distance of 2 cm provides a narrow variation of the curve, meaning that there is not much change in the readings between repeatability measurements. This indicates that at a measurement distance of 2 cm the consistency of measurement is high.

At 50 cm, 1 m, and 2 m variations in the pattern of spectral curves between repeated measurements were very large. Even at a distance of 50 cm and 2 m saturation effect occurred where the upper limit curve value exceeds 100% of reflectance. This high curve pattern variation shows inconsistencies of spectral reflectance readings. This is probably due to a combination of three things, (1) the measurement is done manually by hand, so that the possibility of the sensor's coverage was fluctuating, (2) a slight movement of leaves and canopy was noticeable at these distances, and (3) the sensor reading was influenced by other objects such as branch and soil or water background. At a distance measurement of 5 m, we found variation of spectral reflectance pattern but not as large as 50 cm, 1 m, and 2 m. The distance of the sensor away from the object causes the sensor field of view recorder the whole canopy and any movement occurred on the object was undetected by the spectrometer sensor, thus decreased the readings variations. From the resulted pattern, we found that variation of spectral reflectance is a function of distance measurement. It is also showed that the consistency of measurement results was obtained at close range distance at leaf level, or long-range distance at canopy level.

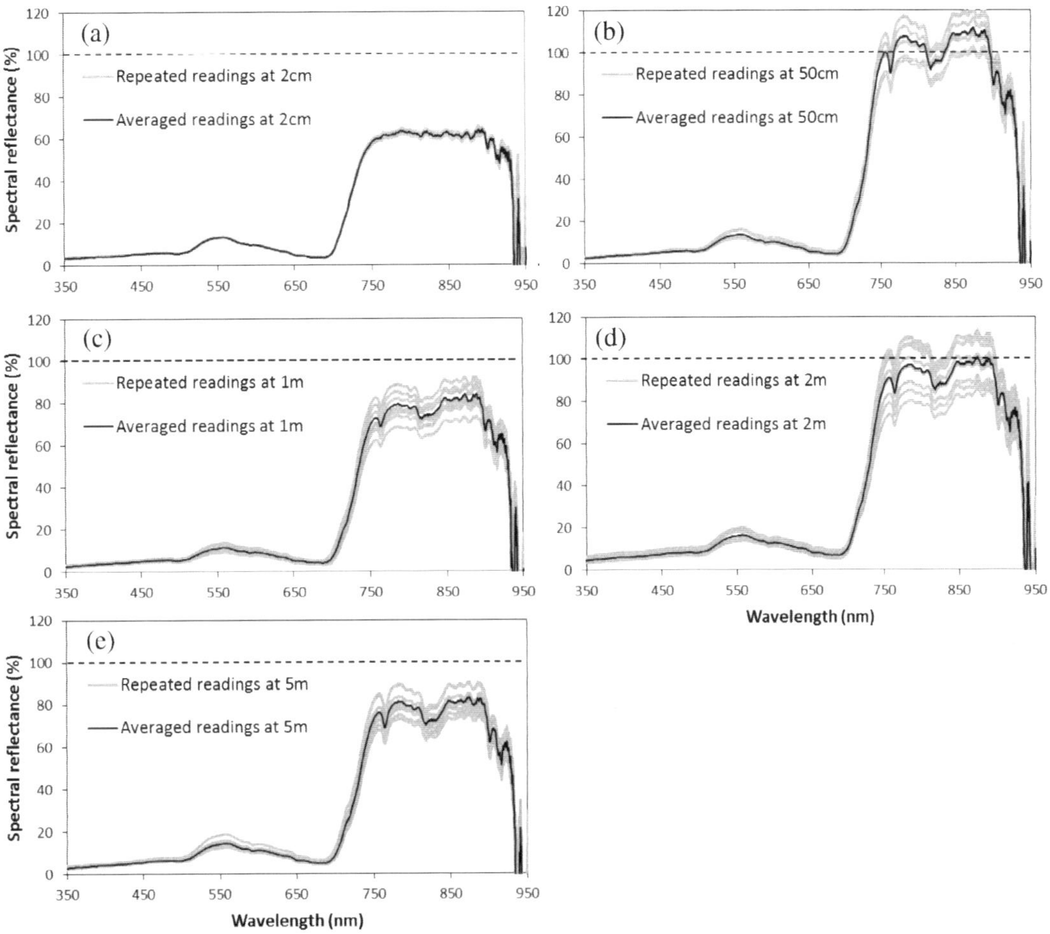


**Figure 3.** Results of the spectral reflectance measurement at different distances showing the repeated and averaged graph at (a) 2 cm, (b) 50 cm, (c) 1 m, (d) 2 m, and (e) 5 m.

### *3.3. Object spectral characteristics*

The vegetation spectral reflectance pattern has a specific characteristic relative to the wavelength of the recording. In the visible spectrum (± 300-600 nm) the interaction between sunlight and leaf energy is strongly influenced by leaf pigment, whereas in the near infrared spectrum is affected by the internal cell structure of leaves [12]. In blue bands (± 300-400 nm) and red (± 500-600 nm) the energy from the sun is absorbed by chlorophyll a, b, and leaf carotenoids for the process of photosynthesis of vegetation. While the green band (± 400-500 nm) does not occur energy absorption so reflected by the surface of leaves and leaves appear green by our eyes [29].

In the near infrared spectrum, in theory there will be spectral reflectance spikes in the region between 700 and 1200 nm. The mesophyll sponge layer on the green leaves controls the amount of near infrared energy reflected by the leaf, ranging from 40-60% [29]. The main information that can be drawn on this spectrum are leaf and canopy structures, and water content in vegetation [12]. In the near infrared spectrum, the differences in the spectral reflect curve between the species of mangroves are strikingly different, so they can be used to distinguish mangrove objects from the surrounding environment [13], even among mangrove species [12, 21].

The spectral reflectance pattern of *Rhyzophora stylosa* recorded by the field spectrometer follows the healthy and green vegetation reflectance pattern in general (Figure 3a-e). The difference in the pattern of spectral reflectance was due to variation in distance between the spectrometer and object sensor. In general, as described above, the near infrared spectrum is very sensitive to changes in spectral reflectance especially at the range 740-890 nm. So, the reading difference as a function of distance measurement can be seen clearly at that range.

When we look into a more detailed level, in both the visible and near infrared spectra (Figure 4b and c), there is no consistent correlation or sequence pattern of spectral reflectance curves. In the visible spectrum (Figure 4b) the highest spectral reflectance curve is at a distance of 2 m, the lowest being 1 m, and the others are somewhere in between. While in the near infrared spectrum (Figure 4c) the highest is the average of 50 cm, the lowest is 2 cm, and the others are in between. The results of this measurement indicate that: (1) the measurement distance strongly determines the shape of the spectral reflectance curve of the object, (2) there is no specific pattern showing the relationship between the measurement distance and the wavelength, and (3) the near infrared spectrum is very sensitive to the measurement distance indicated by high curve variations compared to visible spectrum.

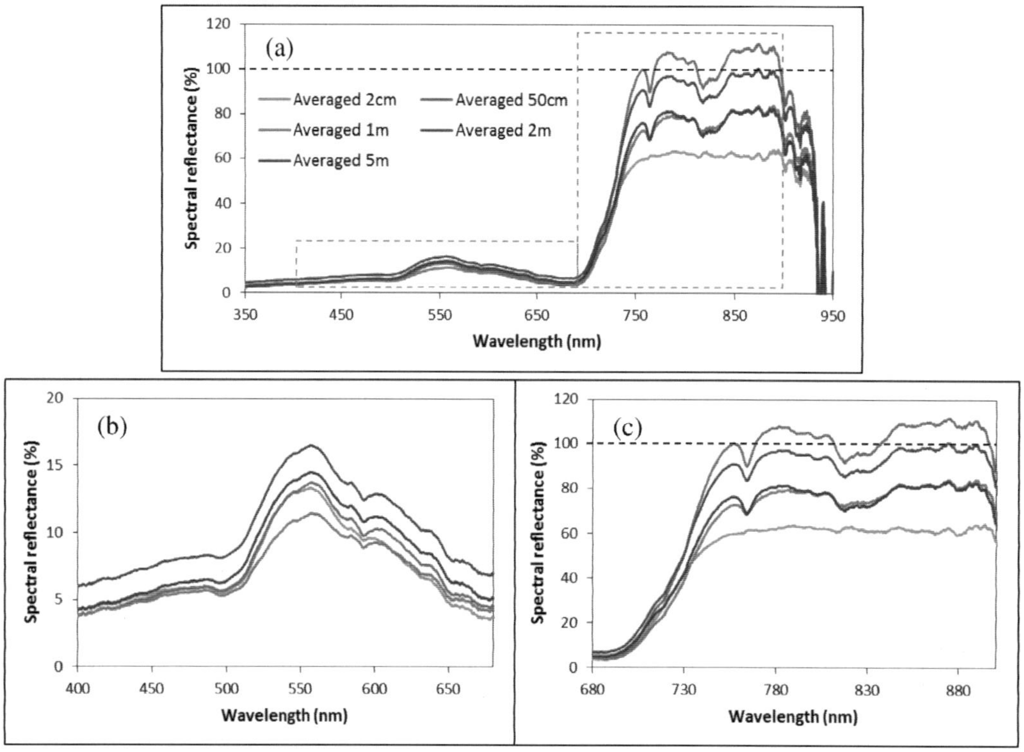


**Figure 4.** Averaged of *Rhyzophora stylosa* spectral reflectance (a) at different measurement distances, (b) in visible, and (c) in near infrared spectrum.

## 4. Conclusions and Future Work

This study was aimed to assess the effect of spectral reflectance measurement at various distances. The measurement distance between the spectrometer sensor and the target object greatly influence the pattern of spectral reflection curves of *Rhizophora stylosa*. At measurement distances of 2 cm and 5 m, the repeated readings of spectral reflectance have low variation, indicating the high stability of the readings. On the other hand, the measurement distance of 50 cm, 1 m, and 2 m have a large curve variation. In terms of the spectral differentiation, as expected, the near infrared spectrum region (680-900 nm) has the most significant curve spacing to distinguish between the measurement distances. Based on the current findings, some research agenda are proposed in the future, including (1) measurement of object spectral reflectance using several scenarios such as different sunlight, measurement angle and tidal conditions, (2) field measurements using fixed and controlled measurement, and (3) measurement for various mangrove species.

**Acknowledgement**
The authors would like to thank the following institution/person: the Faculty of Geography (FGE) Universitas Gadjah Mada (UGM) for providing the research funding (No. UGM/GE/1002.49/M/03/16), PUSPICS FGE UGM for the JAZ-EL350 field spectrometer, Pramaditya Wicaksono for the analysis discussion and validation data, Muhammad Hafizt, and Yaya Ihya Ulumuddin for assisting the field work.

**References**

[1] Lugo, AE and Snedaker, SC 1974, 'The ecology of mangroves'. *Annual Review of Ecology and Systematics,* vol. 5, pp. 39-64.
[2] Hutchings, P and Saenger, P 1987, *Ecology of Mangroves,* University of Queensland Press,Brisbane.
[3] Robertson, AI and Duke, NC 1987, 'Mangroves as nursery sites: comparisons of the abundance and species composition of fish and crustaceans in mangroves and other nearshore habitats in tropical Australia'. *Marine Biology,* vol. 96, pp. 193-205.
[4] Green, EP, Clark, CD, Mumby, PJ, Edwards, AJ and Ellis, AC 1998, 'Remote sensing techniques for mangrove mapping'. *International Journal of Remote Sensing,* vol. 19, pp. 935-956.
[5] Giri, C, Ochieng, E, Tieszen, LL, Zhu, Z, Singh, A, Loveland, T, Masek, J and Duke, N 2011, 'Status and distribution of mangrove forests of the world using earth observation satellite data'. *Global Ecology and Biogeography,* vol. 20, pp. 154-159.
[6] Gopal, B and Chauhan, M 2006, 'Biodiversity and its conservation in the Sundarban mangrove ecosystem'. *Aquatic Sciences - Research Across Boundaries,* vol. 68, pp. 338-354.
[7] Alongi, DM 2002, 'Present state and future of the world's mangrove forests'. *Environmental Conservation,* vol. 29, pp. 331-349.
[8] Manson, FJ, Loneragan, NR and Phinn, SR 2003, 'Spatial and temporal variation in distribution of mangroves in Moreton Bay, subtropical Australia: a comparison of pattern metrics and change detection analyses based on aerial photographs'. *Estuarine, Coastal and Shelf Science,* vol. 57, pp. 653-666.
[9] Giri, C, Zhu, Z, Tieszen, LL, Singh, A, Gillette, S and Kelmelis, JA 2008, 'Mangrove forest distributions and dynamics (1975–2005) of the tsunami-affected region of Asia'. *Journal of Biogeography,* vol. 35, pp. 519-528.
[10] Gilman, EL, Ellison, J, Duke, NC and Field, C 2008, 'Threats to mangroves from climate change and adaptation options: a review'. *Aquatic Botany,* vol. 89, pp. 237-250.
[11] Heumann, BW 2011, 'Satellite remote sensing of mangrove forests: recent advances and future opportunities'. *Progress in Physical Geography,* vol. 35, pp. 87-108.
[12] Kuenzer, C, Bluemel, A, Gebhardt, S, Quoc, TV and Dech, S 2011, 'Remote sensing of mangrove ecosystems: a review'. *Remote Sensing,* vol. 3, pp. 878-928.
[13] Spalding, M, Kainuma, M and Collins, L 2010, *World Atlas of Mangroves,* Earthscan,London.
[14] Clark, CD, Ripley, HT, Green, EP, Edwards, AJ and Mumby, PJ 1997, 'Mapping and measurement of tropical coastal environments with hyperspectral and high spatial resolution data'. *International Journal of Remote Sensing,* vol. 18, pp. 237-242.
[15] Green, EP and Mumby, PJ 2000, Mapping mangroves. *In:* Edwards, AJ (ed.) *Remote Sensing Handbook for Tropical Coastal Management*, UNESCO Publishing, Paris.
[16] Jensen, R, Mausel, P, Dias, N, Gonser, R, Yang, C, Everitt, J and Fletcher, R 2007, 'Spectral analysis of coastal vegetation and land cover using AISA+ hyperspectral data'. *Geocarto International,* vol. 22, pp. 17-28.
[17] Demuro, M and Chisholm, L. 2003, 'Assessment of Hyperion for characterizing mangrove communities'. Available:

ftp://popo.jpl.nasa.gov/pub/docs/workshops/03_docs/Demuro_AVIRIS_2003_web.pdf [Accessed 28 October 2007].

[18] Held, A, Ticehurst, C, Lymburner, L and Williams, N 2003, 'High resolution mapping of tropical mangrove ecosystems using hyperspectral and radar remote sensing'. *International Journal of Remote Sensing,* vol. 24, pp. 2739-2759.

[19] Hirano, A, Madden, M and Welch, R 2003, 'Hyperspectral image data for mapping wetland vegetation'. *Wetlands,* vol. 23, pp. 436-448.

[20] Vaiphasa, C, Ongsomwang, S, Vaiphasa, T & Skidmore, AK 2005, 'Tropical mangrove species discrimination using hyperspectral data: a laboratory study'. *Estuarine, Coastal and Shelf Science,* vol. 65, pp. 371-379.

[21] Kamal, M and Phinn, SR 2011, 'Hyperspectral data for mangrove species mapping: a comparison of pixel-based and object-based approach'. *Remote Sensing,* vol. 3, pp. 2222-2242.

[22] Blasco, F, Gauquelin, T, Rasolofoharinoro, M, Denis, J, Aizpuru, M & Caldairou, V 1998, 'Recent advances in mangrove studies using remote sensing data'. *Marine and Freshwater Research,* vol. 49, pp. 287-296.

[23] Díaz, BM and Blackburn, GA 2003, 'Remote sensing of mangrove biophysical properties: evidence from a laboratory simulation of the possible effects of background variation on spectral vegetation indices'. *International Journal of Remote Sensing,* vol. 24, pp. 53-73.

[24] BTNK 2008, 'Statistik Balai Taman Nasional Karimunjawa (BTNK) 2008', BTNK, Dirjen Perlindungan Hutan dan Konservasi Alam, Departemen Kehutanan, Semarang.

[25] BTNK 2001, 'Rencana Pengelolaan 25 Tahun Taman Nasional Karimunjawa Periode 2002-2027', Balai Taman Nasional Karimunjawa, Semarang.

[26] Kamal, M, Adi, NS, and Arjasakusuma, S 2012, *Jaz EL-350 Portable Spectrometer: panduan operasional pengukurandan pengelolaan data pantulan spektral objek*, Laboratorium Penginderaan Jauh Dasar, Fak. Geografi UGM, Yogyakarta.

[27] Kamal, M 2015, '*Remote sensing for multi-scale mangrove mapping*', PhD Thesis, The University of Queensland, Australia.

[28] Ocean Optics 2010, *Jaz Installation and Operation Manual*, Ocean Optics Inc., Dunedin, Florida.

[29] Jensen, JR 2005, *Introductory Digital Image Processing: a remote sensing perspective*, 3rd edn,Pearson Prentice Hall, Sydney.

**Institute of Physics**
Dirac House, Temple Back
Bristol BS1 6BE UK

ISSN: 1755-1307
ISBN 978-1-5108-5490-1

CURRAN ASSOCIATES INC.
proceedings.com